ESSENTIALS OF GENETICS

ESSENTIALS OF GENETICS

William S. Klug
Trenton State College at Hillwood Lakes

Michael R. Cummings
University of Illinois at Chicago

With Essays contributed by
Elizabeth Savage
University of Alberta at Edmonton

Macmillan Publishing Company
New York

Maxwell Macmillan Canada
Toronto

Editor: Sheri S. Walvoord
Development Editor: Richard E. Morel
Production Supervision and Art Direction: Hudson River Studio
Production Manager: Aliza Greenblatt
Text and Cover Designer: Hudson River Studio
Cover Photograph: Grant Heilman Photography, Inc.
Photo Researcher: Diane Austin

E-01 Chris Mihulka/The Stock Market
E-02 CNRI/Science Photo Library/Photo Researchers, Inc.
E-04 Dr. Robin Williams/Science Photo Library/Photo Researchers, Inc.
E-07 Carlos Goldin/Science Photo Library/Photo Researchers, Inc.
E-11 Courtesy of Cold Spring Harbor Laboratory
E-14 Ted Horowitz/The Stock Market
E-15 Dr. Thomas Broker/Phototake
E-17 Dr. A.R. Lawton/Science Photo Library/Photo Researchers, Inc.
E-19 Moredun Animal Health Ltd./Science Photo Library/Photo Researchers, Inc.
E-21 Uniphoto/Pictor

This book was set in Garamond Light by York Graphic Services,
and was printed and bound by R.R. Donnelley & Sons Company.
The cover was printed by Phoenix Color Corp.

Printed in the United States of America

Macmillan Publishing Company
866 Third Avenue, New York, New York 10022

Macmillan Publishing Company is part of the Maxwell Communication Group of Companies.

Maxwell Macmillan Canada, Inc.
1200 Eglinton Avenue East
Suite 200
Don Mills, Ontario M3C 3N1

Library of Congress Cataloging-in-Publication Data

Klug, William S.
Essentials of genetics / William S. Klug, Michael R. Cummings.
p. cm.
Includes bibliographical references and index.
ISBN 0-02-364797-3
1. Genetics. I. Cummings, Michael R. II. Title.
QH430.K578 1993 92-38098
575.1—dc20 CIP

Printing: 1 2 3 4 5 6 7 Year: 3 4 5 6 7 8 9 0 1 2

DEDICATION

To those who, by virtue of heredity, have places in our hearts, and to those who have had to find their way on their own. Thanks for being there.

ABOUT THE AUTHORS

WILLIAM S. KLUG is currently a Professor of Biology at Trenton State College at Hillwood Lakes in Ewing, New Jersey. He is Chairman of the Biology Department, a position to which he was first elected in 1974. He received his B.A. at Wabash College in Crawfordsville, Indiana and his Ph.D. at Northwestern University in Evanston, Illinois. Before coming to Trenton State, he returned to Wabash College as an Assistant Professor, where he first taught Genetics as well as Electron Microscopy. His research interests prior to succumbing to educational administration and textbook authorship centered around the genetic control of development in *Drosophila*. He has taught the General Genetics course to undergraduates for the last 25 years.

MICHAEL R. CUMMINGS is currently Associate Professor of Biological Sciences and Associate Professor of Genetics at the University of Illinois at Chicago. He is also Research Associate Professor in the Institute for the Study of Developmental Disabilities at the University. He received his B.A. at St. Mary's College in Winona, Minnesota, and his M.S. and Ph.D. degrees from Northwestern University in Evanston, Illinois. His research interests include the molecular organization and physical mapping of the region surrounding the centromere of human chromosome 21. At the undergraduate level, he teaches courses in Mendelian/Molecular Genetics and in Human Genetics.

PREFACE

ESSENTIALS OF GENETICS is designed for courses requiring a text that is shorter and more basic than its more comprehensive companion text, *Concepts of Genetics.* While coverage is still thorough and of high quality, *Essentials* is written to be more accessible to students with minimal backgrounds in biology and chemistry. While it will serve well courses offered to biology majors early in their undergraduate careers, it will be particularly useful in courses populated by a mixture of majors who are pursuing careers in agriculture, chemistry, psychology, etc. Because the text is substantially shorter than most other available books, *Essentials of Genetics* will be much more manageable in courses of both one-quarter and one-semester length.

GOALS

While *Essentials of Genetics* is about 200 pages shorter than its companion volume, most of our goals have not changed, namely to:

- Emphasize concepts rather than excess detail.
- Write clearly and directly to the students of genetics in order to provide straightforward explanations of complex, analytical topics.
- Provide careful organization within and between chapters.
- Maintain constant emphasis on scientific analysis as the means by which our knowledge in genetics has been acquired.
- Propagate the rich history of genetics that so beautifully illustrates how information is extended within a discipline as it develops and grows.
- Design inviting, beautiful, and pedagogically sound full-color figures and select equally useful photographs to support the textual material.

These are worthy goals, and ones that serve as the cornerstones of this text. Providing such a pedagogic foundation will allow *Essentials of Genetics* to accommodate courses with many different organizational approaches and lecture formats. Chapters are written to be as independent of one another as possible, allowing instructors to utilize them in various sequences. We believe that by combining these various approaches, we have created a text that provides in an optimal way the type of support students require as they study genetics.

FEATURES

The major features that distinguish *Essentials of Genetics* from *Concepts of Genetics* include:

- Length—a more manageable, streamlined text in the 500 page range compared to the 700–800 page ranges characteristic of most Genetics texts.
- Number of Chapters—21 compared to 25.
- A Revised Organization—a chapter sequence designed to provide a smooth flow of topics from start to finish.

- Two New Chapters—representing the most significant, emerging topics in Genetics: Cancer and Immunology.
- Modernization of Topics—to remain current in areas that are at the forefront of genetic studies, including:

  — DNA Replication
  — Genetic Expression
  — Gene Regulation
  — Recombinant DNA Technology
  — Human Genetics
  — Gene Therapy
  — Transposable Elements

- Essays in Genetics—included in most chapters, providing brief introductions to a variety of current topics of particular interest during the study of this discipline.
- Showcase Art—over a dozen carefully designed and beautifully developed figures provided to enhance learning involving a variety of complex topics, e.g. cell structure, mitosis, meiosis, transcription, and translation.
- Revision of Many Figures—to further improve the premier art program in Genetics texts.

EMPHASIS ON CONCEPTS

As in its companion volume, *Essentials of Genetics* continues to emphasize the conceptual framework of this discipline. The concepts derived from each major topic represent those ideas and information that students "take with them" once the course is completed. To aid the student in identifying the conceptual aspects of a major topic, each chapter begins with a section called "Chapter Concepts," which in a few sentences captures the essence of the material about to be presented. Then, each chapter ends with a section called "Chapter Summary," which enumerates the 5–10 key points that have been established.

Between these two sections, the text is written with an emphasis on concepts rather than detail. Reinforcing examples and carefully designed figures have been selected to support this approach.

PROBLEM SOLVING AND ANALYSIS

Genetics, more than any other discipline within Biology, lends itself to problem solving and analytical thinking. At the end of each chapter, we have maintained the popular and successful section "Insights and Solutions," first developed in *Concepts of Genetics.* In this section, we have stressed:

- Problem Solving
- Quantitative Analysis
- Analytical Thinking
- Experimental Rationale

Problems or questions are proposed and detailed solutions or answers are provided. This section is designed to "prime" the students as they move on to "Problems and Discussion Questions," which allows students to confront problem solving and analysis on their own. Brief answers to about half of the problems are available in Appendix A. For those faculty wishing to expose their students to detailed answers to all problems and questions, the "Student Handbook" is available (see Supplements section).

Taken together, we have attempted to optimize the opportunities for student growth in these important areas of Biology by stimulating their analytical thinking skills and facilitating their success at problem solving.

SUPPLEMENT PACKAGE

Students Handbook: A Guide to Concepts and Problem Solving by Harry Nickla of Creighton University reviews vocabulary, concepts, and problem solving chapter by chapter. In addition, it provides a very detailed solution or lengthy discussion of every problem and question in the text, and it supplies additional problems to be used for practice by students.

The *Instructor's Sourcebook,* also by Harry Nickla, contains questions and problems an instructor can use to prepare exams.

Overhead transparencies of approximately 150 illustrations from the text will be available to adopters of the text.

ACKNOWLEDGEMENTS

All comprehensive texts are dependent on the valuable input provided by reviewers and colleagues. While we take full responsibility for any errors herein, we gratefully acknowledge the help provided by those individuals involved in this and previous editions. With regard to *Essentials of Genetics* we wish to specifically recognize:

James Blahnik — Lorraine County Community College

James Bricker	Trenton State College
Jeremy Bruenn	SUNY-Buffalo
Thomas C. Dent	West Virginia University
Paul Doerder	Cleveland State University
Marianne Eleuterio	West Chester University
Dwayne Englert	South Illinois University at Carbondale
Michael Gates	Cleveland State University
Dr. David Nash	University of Alberta at Edmonton
Malcolm Vidrine	Louisiana State University
Oscar H. Will, III	Augustana College
Michael C. Wooten	Auburn University

Authors can never underestimate the importance of, nor forget to acknowledge the secretarial staff who has brightened the arduous process of producing and transmitting the many versions of each chapter. Many thanks to Bette Baier and Monica Zrada at Trenton State College for their help and patience.

At Macmillan Publishing, we wish to acknowledge Mr. Bob Rogers, who conceived and helped plan this project. Thanks are also due to Kristin Watts Peri and Bob Pirtle, who guided the preparation of this manuscript. We are grateful to Chris Migdol and Diane Austin for procuring the many photos that enhance the text. Their efficiency and help are very much appreciated.

Skillful developmental editing of both the manuscript and art was provided by Dick Morel of Strong House Associates, whose steadying and pleasant manner was critical to and greatly enhanced the entire effort.

The high quality production of this text and its art program was dependent on the professional efforts of Lorraine and Edward Burke and their staff at Hudson River Studio, all of whom place a premium on human interactions as well as quality workmanship. Thanks are due to them all, and particularly Inez Sovjani, who coordinated the art program.

Last, but not least, we wish to recognize and thank Elizabeth Savage at the University of Alberta for contributing the "Essays" found in most chapters. We believe that these add immeasurably to the pedagogic value of the text, and we are grateful for her efforts.

It has been very pleasurable working with all of the above individuals, who deserve to share in any success enjoyed by this text. Our gratitude is equalled only by the extreme dedication evident during their many efforts.

BRIEF CONTENTS

CONTENTS

ESSENTIALS OF GENETICS

1 An Introduction to Genetics

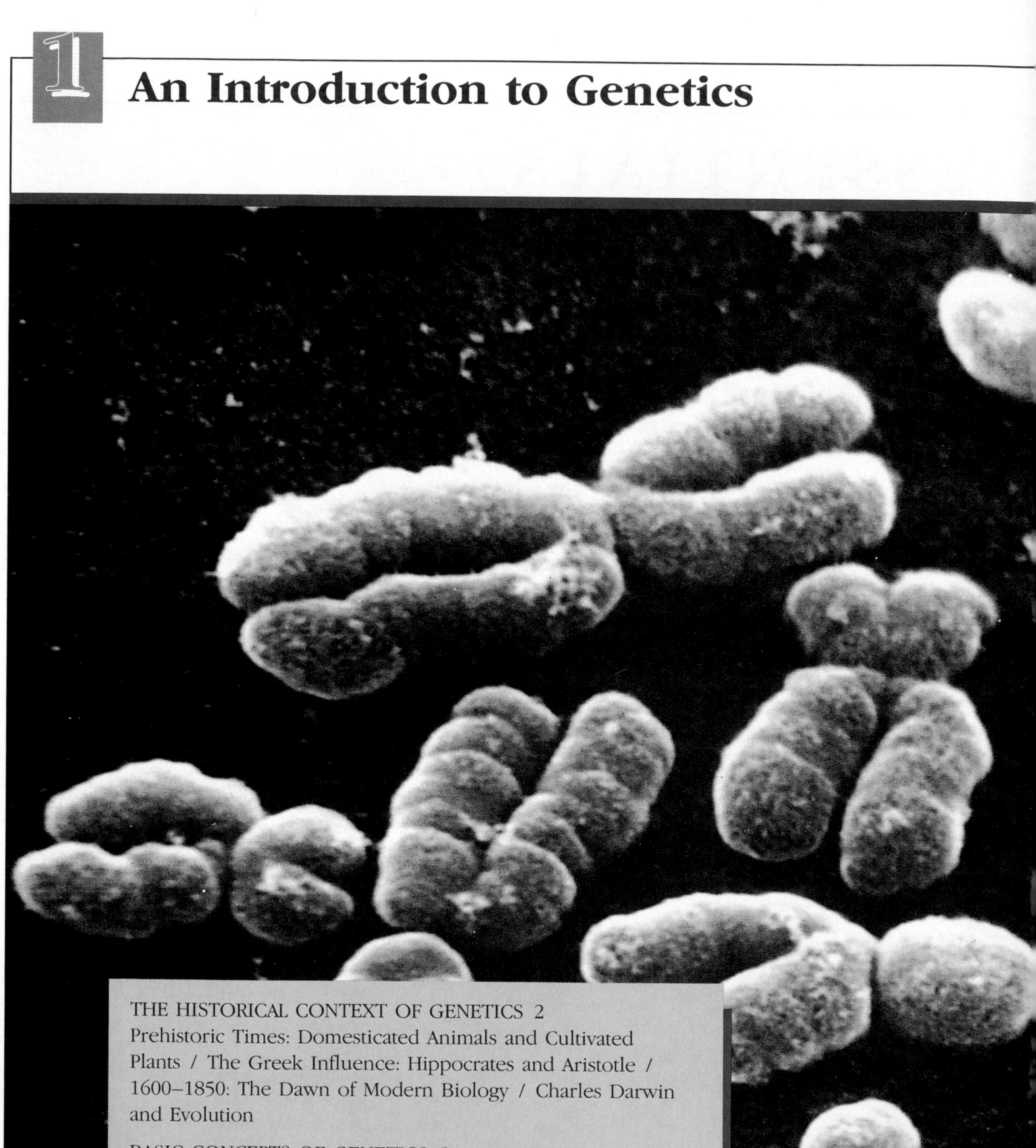

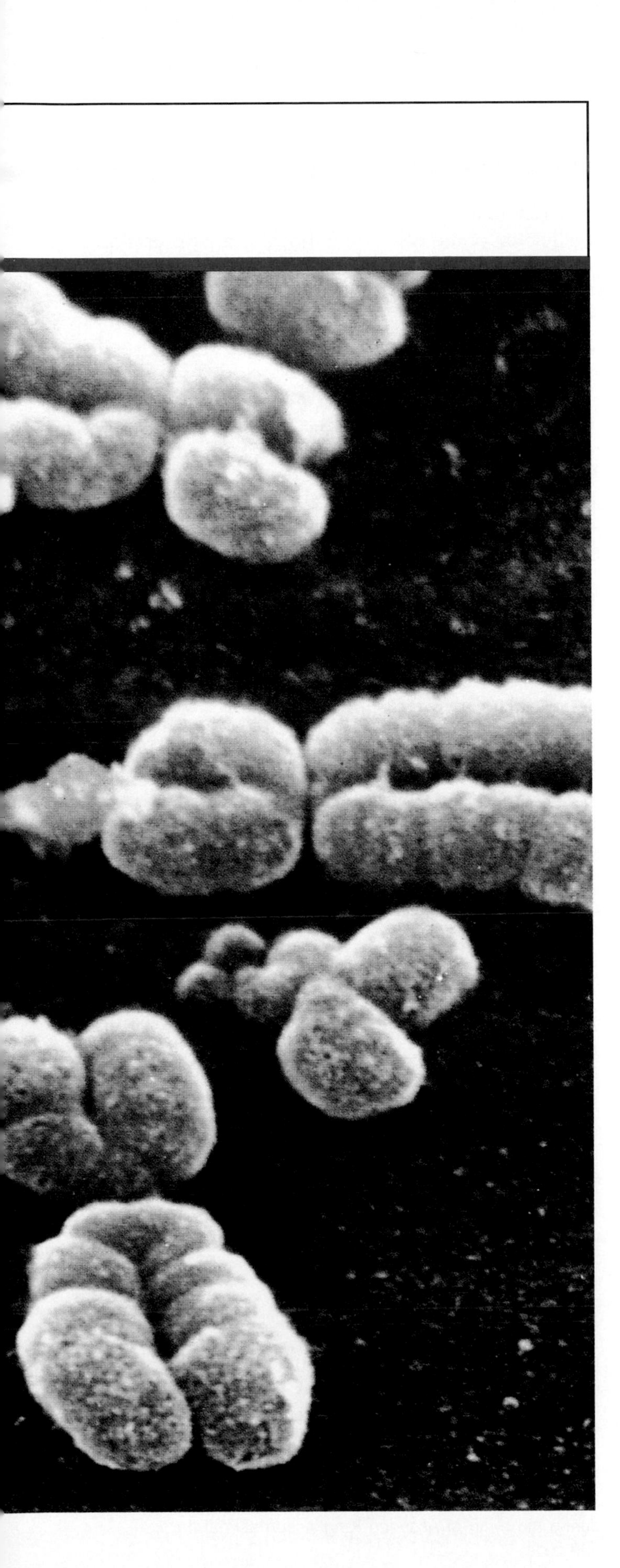

Chapter Concepts

Genetics is the science of heredity. The discipline has a rich history and involves investigations of molecules, cells, organisms, and populations, using many different experimental approaches. Not only does genetic information play a significant role during evolution, but its expression influences the function of individuals at all levels. Thus, genetics unifies the study of biology and has a profound impact on human affairs.

Welcome to the study of genetics. You are about to explore a subject that many students before you have found to be the most interesting and fascinating in the field of biology. This is not surprising because an understanding of genetic processes is fundamental to the comprehension of life itself. Genetic information directs cellular function, determines an organism's external appearance, and serves as the link between generations in every species. Knowing how these processes occur is important in understanding the living world. Knowledge of genetic concepts also helps us to understand the other disciplines of biology. The topics studied in genetics overlap directly with molecular biology, cell biology, physiology, evolution, ecology, systematics, and behavior. The study of each of these disciplines is incomplete without the knowledge of the genetic component underlying each of them. Genetics thus unifies biology and serves as its "core."

Interest in and fascination with this discipline further stem from the fact that, in genetics, many initially vague and abstract concepts have been so thoroughly investigated that subsequently, they have become clearly and definitively understood. As a result, genetics has a rich history that exemplifies the nature of scientific inquiry and the analytical approach used to acquire information. Scientific analysis, moving from the unknown to the known, is one of the major forces that attracts students to biology.

There is still another reason why the study of genetics is so appealing. Every year large numbers of new findings are made. Although it has been said that scientific knowledge doubles every ten years, one estimate holds that the doubling time in genetics is less than five years. Certainly, over the past four decades, no five-year period has passed without new discoveries in genetics causing us to revise our thinking or to extend our knowledge beyond a major frontier. Each advance becomes part of an ever-expanding cornerstone upon which further progress is based. It is exciting to be in the midst of these developments, whether you are studying or teaching genetics.

THE HISTORICAL CONTEXT OF GENETICS

In the chapters that follow, we will discuss the behavior of chromosomes, the way in which genes are transmitted from one generation to the next, and the way in which genetic information is stored, altered, expressed, and regulated. The most significant scientific findings, which serve as the foundation for this information, were obtained in the nineteenth century. As the twentieth century dawned, a period of integration of this knowledge occurred, clarifying the physical basis of living organisms and their relationship to one another. The ideas that: 1) matter is composed of atoms; 2) cells are the fundamental units of living organisms; 3) cells contain nuclei that house threadlike structures called chromosomes; and 4) chromosomes are constant in number in a species and thus might be important in heredity together provided the basis for an important synthesis of ideas. When these ideas were combined with the newly rediscovered findings of Gregor Mendel and integrated with Darwin's theory of evolution and the origin of species, a more complete picture of life at the level of the individual and of the population emerged. The era of modern-day biology was initiated on this foundation.

But what of the many important ideas and hypotheses that, valid or not, served as forerunners of nineteenth-century thought? In the following short section, we consider some of these important ideas. Two of these ideas persisted incorrectly for over 2000 years, yet they served to propel the quest for a better understanding of the basis for the existence of all living things.

Prehistoric Times: Domesticated Animals and Cultivated Plants

We may never know when people first recognized the existence of heredity. However, a variety of archeological evidence (primitive art, preserved bones and skulls, dried seeds, and soforth) have provided many insights. Such evidence documents the successful domestication of animals and cultivation of plants thousands of years ago. Such efforts represent artificial selection of genetic variants within populations.

Between 8000 and 1000 B.C., horses, camels, oxen, and even various breeds of dogs (derived from the wolf family) were domesticated to serve various roles. Cultivation of plants paralleled these developments. Several, including maize, wheat, rice, and the date palm are thought to have been developed around 5000 B.C. Maize remains dating to this period have been recovered in caves in the Tehucan Valley of Mexico. Assyrian art depicts artificial pollination of the date palm, thought to have originated in Babylonia (Figure 1.1). Such cultivation very likely influenced the types of modern-day palms found in the region. Today, there are over 400 kinds of date palm in just four oases in the Sahara, differing from one another in various traits such as fruit taste.

The presence of cultivated plants and domestic animals during prehistoric times documents our ancestors' successful attempts to manipulate the genetic composition of useful species. There is little doubt that it was soon learned that desirable and undesirable traits were passed to successive generations and that selection could be performed to produce more desirable varieties of animals and plants. Such efforts document human awareness of heredity during prehistoric times.

The Greek Influence: Hippocrates and Aristotle

While few, if any, ideas attempted to seriously explain heredity during prehistoric times, much more is known about such attempts during the Golden Age of Greek Culture. Considerable attention was devoted to the subjects of reproduction and heredity, particularly as they related to the origin of humans.

This is most evident in the writings of the Hippocratic school of medicine (fifth–fourth century B.C.) and subse-

FIGURE 1.1
Relief carving depicting artificial pollination of date palms during the reign of Assyrian King Assurnasirpal II (883–859 B.C.).

quently of the philosopher and naturalist Aristotle (384–322 B.C.).

Central to an explanation of the hereditary basis of reproduction of animals was: (1) the source of the physical substance of the offspring, and (2) the generative force behind the direction taken by that substance as it materializes into a living organism.

The Hippocratic treatise *On the Seed* argues that male semen is formed in numerous parts of the body and is transported through blood vessels to the testicles. Active "humors," as the bearer of hereditary traits, are drawn from various parts of the body. These humors could be healthy or diseased, the latter condition accounting for the appearance of newborns exhibiting congenital disorders or deformities. Furthermore, these humors could be altered in individuals, and in this new form, be passed on to offspring. Thus, newborns could "inherit" traits that their parents had "acquired in their environment."

Aristotle, who had studied under Plato for some twenty years, was more critical and more expansive than Hippocrates in his analysis of the origin of humans and their heritage. Aristotle proposed that male semen was formed from blood rather than from each organ and that its generative power resided in a vital heat that it contained. This vital heat had the capacity to produce an offspring of the same "form" (i.e., basic structure and capacities) as the parent. It generated the offspring by cooking and shaping the menstrual blood produced by the female, which was the "matter" for the offspring. The properties of the offspring were determined primarily by the heat supplied by the male, but partly by the material supplied by the female. The embryo would develop from the initial "setting" of the menstrual blood by the semen into a mature offspring, not because it already contained the parts in miniature (as some Hippocratics had thought), but because of the shaping power of that vital heat. These ideas constituted only one part of Aristotelian philosophy of order in the living world.

Although the ideas of Hippocrates and Aristotle may sound primitive and naive today, we should recall that neither sperm nor eggs had yet been observed in any mammal, let alone humans. In fact, eggs were not discovered until 1827! Thus, in their own right, these explanations were worthy ones in their time and, in fact, for centuries to come. As we will see, their thinking was not so different from that of Charles Darwin in his formal proposal of the **theory of pangenesis** during the nineteenth century, which will soon be defined and discussed.

1600–1850: The Dawn of Modern Biology

During the ensuing 1900 years (300 B.C.–A.D. 1600), the theoretical understanding of genetics was not extended by significant, new ideas, but interest in applied genetics remained strong. As early as Roman times, plant grafting and animal breeding were greatly emphasized. By the Middle Ages, naturalists, well aware of the impact of heredity on organisms they studied, were faced with reconciling their findings with current religious beliefs. The theories of Hippocrates and Aristotle still prevailed at that time and when applied to humans, they no doubt conflicted with the religious doctrine of special creation.

Beginning in the seventeenth century and continuing through the nineteenth century, major strides were made in experimental biology that provided much greater insights into the basis of life. The many findings set the scene for the revolutionary work and principles presented by Charles Darwin and Gregor Mendel. A brief review of some of the most significant observations and interpretations illustrates this point.

In the 1600s the English anatomist William Harvey (1578–1657) studied early development. Better known for his experiments demonstrating that the blood is pumped by the heart through a circulatory system made up of the arteries and veins, he also wrote a treatise on reproduction and development, patterned after Aristotle's work. In it he is credited with the earliest statement of the **theory of epigenesis**—that an organism is derived from substances present in the egg, which are assembled and differentiate during embryonic development. Epigenesis holds that new structures, such as the many body organs, are not present initially, but instead arise *de novo* during development. This theory is consistent with Aristotle's ideas describing procreation. Indeed, Harvey had studied Aristotle and was well aware of his ideas.

The theory of epigenesis conflicted directly with the **theory of preformationism,** first put forward in the seventeenth century. Stating that sex cells contain a complete miniature adult called the homunculus, perfect in every form, preformationism was popular well into the eighteenth century. Some people, called ovists, believed it was the ovum, while others (spermists) believed it was the sperm that housed the homunculus. However, work by the embryologist Casper Wolff (1733–1794) and others clearly disproved this theory and strongly favored epigenesis. Wolff was quite convinced that several structures, such as the alimentary canal, were not present in the earliest embryos he studied, but instead, were formed later during development.

During this same period, there were other significant chemical and biological findings. In 1808, John Dalton expounded his atomic theory, which stated that all matter is composed of small invisible units called atoms. Improved microscopes became available, and around 1830, Matthias Schleiden and Theodor Schwann proposed that all organisms are composed of basic visible units called cells. By this time, the idea of **spontaneous generation,** the creation of living organisms from nonliving components, had clearly been disproved by the experiments of Francesco Redi (1621–1697), Lazzaro Spallanzani (1729–1799), and Louis Pasteur (1822–1895), among others. Thus, all living organisms were considered to be derived from preexisting organisms and to consist of cells made up of atoms.

Another prevailing notion had a major influence on nineteenth-century thinking: **the fixity of species.** According to this doctrine, animal and plant groups remain unchanged from that moment of their appearance on earth. Embraced particularly by those also adhering to a belief in special creation, this doctrine was attributed to several people, including the Swedish physician and plant taxonomist, Carolus Linnaeus (1707–1778), who is better known for devising the binomial system of classification.

The influence of this tenet is illustrated by considering the work of the German plant breeder Joseph Gottlieb Kolreuter (1733–1806), who was considered to be the first to successfully hybridize a variety of plants. Several of his findings were potentially quite far-reaching. In work with tobacco, he crossbred two groups and derived a new hybrid form, which he then converted back to one of the parental types by repeated backcrosses. In other breeding experiments, using carnations, he clearly observed segregation of traits, which was to become one of Mendel's principles of genetics. These results are in contradiction to the idea of species not changing with time. Because of Kolreuter's belief in both special creation and the fixity of species, he was puzzled about these outcomes, and he failed to recognize the real significance of his findings.

Neither was the prevailing scientific climate kind to plant hybridizers such as Kolreuter. The German school of thought, known as *Naturphilosophie,* believed that nature was unity and that only the study of the whole organism was significant. Like Kolreuter, Karl Friedrich Gartner (1772–1850), experimenting with peas, obtained results similar to Mendel's monohybrid cross, upon which the principles of dominance/recessiveness and segregation are based. However, Gartner did not concentrate on the analysis of individual traits and failed to grasp the significance of his own work.

Unfortunately, proponents of the *Naturphilosophie* actively discouraged hybridization studies, creating an atmosphere that made it difficult for the work of Kolreuter, Gartner, and others to find acceptance. Undoubtedly, this influence also adversely affected the acceptance of Mendel's work half a century later.

Charles Darwin and Evolution

With the above information as background, we conclude our coverage of the historical context of genetics with a brief discussion of the work of Charles Darwin, who in 1859 published the book-length statement of his evolutionary theory, *The Origin of Species.* His many geological, geographical, and biological observations convinced Darwin that existing species arose by descent with modification from other ancestral species. Greatly influenced by his now famous voyage on the Beagle (1831–1836), Darwin's thinking culminated in his formulation of the **theory of natural selection,** a theory of the causes of evolutionary change. This theory, formulated and proposed at the same time but independently by Alfred Russell Wallace, hypothesized that populations tend to leave more offspring than the environment can support, leading to a struggle for existence among them. In such a struggle, those organisms with heritable traits that better adapt them to their environment are better able to survive and reproduce than those with less adaptive traits. Over a long period of time, numerous slight, advantageous variations will accumulate. Once reproductive isolation results, a new species is formed as evolution continues.

The primary gap in the theory was a lack of understanding of the genetic basis of variation and inheritance, a gap that left Darwin's theory open to reasonable criticism well into the twentieth century. Aware of this weakness in his theory of evolution, in 1868 Darwin published a second book, *Variations in Animals and Plants under Domestication,* in which he attempted to provide a more definitive explanation of how heritable variation arises gradually over time. Two of his major ideas have their roots in the theories involving "humors," as put forward by Hippocrates and Aristotle.

In his **provisional hypothesis of pangenesis,** Darwin coined the term **gemmules** (rather than humors) to describe the physical units representing each body part that were gathered by the blood into the semen. Darwin felt that these gemmules determined the nature or form of each body part. He further believed that gemmules could respond in an adaptive way to an individual's external environment. Once altered, such changes would be passed on to offspring, allowing for the inheritance of acquired characteristics. Lamarck formalized this latter theory in his 1809 treatise, *Philosophie Zoologique.* Lamarck's theory, which became known as the **doctrine of use and disuse,** proposed that organisms acquire or lose characteristics that then become heritable.

As Darwin's work ensued, the experiments of Gregor Johann Mendel (Figure 1.2) were performed between 1856 and 1863, forming the basis for his classic 1865 paper. In it, Mendel demonstrated a number of statistical patterns underlying inheritance and developed a theory involving hereditary factors in the germ cells to explain these patterns. His research was virtually ignored until it was partially duplicated and then cited by Carl Correns, Hugo de Vries, and Eric Von Tschermak in 1900, and championed by William Bateson. During this interval, chromosomes were discovered and support for the epigenetic interpretation of development grew considerably. It gradually became clear that heredity and development were dependent on "information" contained in

FIGURE 1.2
Gregor Johann Mendel, who in 1865 put forward the major postulates of transmission genetics as a result of experiments with the garden pea.

chromosomes, which were contributed by gametes to each individual.

As we have seen, a rich history of scientific endeavor and thinking preceded and surrounded Mendel's work. In Chapter 3, we will return to a thorough analysis of his findings, which have served to this day as the foundation of genetics. And his work was but one important part of the body of knowledge that would initiate the era of modern biological thought in the twentieth century.

BASIC CONCEPTS OF GENETICS

In this introductory chapter we review some of the simple but basic concepts in genetics, which you have undoubtedly already studied. By reviewing them at the outset, we can establish an initial vocabulary and proceed through the text with a common foundation. We shall approach these basic concepts by asking and answering a series of questions. You may wish to write or think through an answer before reading the explanation of each question. Throughout the text, the answers to these questions will be expanded as more detailed information is presented.

What does "genetics" mean?

Genetics is the branch of biology concerned with heredity and variation. This discipline involves the study of cells, individuals, their offspring, and the populations within which organisms live. Geneticists investigate all forms of inherited variation and the nature of the underlying genetic basis of such characteristics.

What is the center of heredity in a cell?

In eukaryotic organisms, the **nucleus** contains the genetic material. In prokaryotes, such as bacteria, the genetic material exists in an unenclosed but recognizable area of the cell called the **nucleoid region.** In viruses, which are not true cells, the genetic material is ensheathed in the protein coat referred to as the viral head.

What is the genetic material?

In eukaryotes and prokaryotes, **DNA** serves as the molecule storing genetic information. In viruses, either DNA or **RNA** serves this function.

What do DNA and RNA stand for?

DNA and RNA are abbreviations for **deoxyribonucleic acid** and **ribonucleic acid,** respectively. These are the two types of nucleic acids found in organisms. Nucleic acids, along with carbohydrates, lipids, and proteins, compose the four major classes of organic biomolecules found in living things.

How is DNA organized to serve as the genetic material?

DNA, although single stranded in a few viruses, is usually a double-stranded molecule organized as a double helix. Contained within each DNA molecule are hereditary units called **genes,** which are part of a larger element, the **chromosome.**

What is a gene?

In simplest terms, the gene is the functional unit of heredity. In chemical terms, it is a linear array of nucleotides—the chemical building blocks of DNA and RNA. A more sophisticated approach is to consider it as an informational storage unit capable of undergoing replication, mutation, and expression. As investigations have progressed, the gene has been found to be a very complex genetic element.

What is a chromosome?

In viruses and bacteria, the chromosome (Figure 1.3) is most simply thought of as a long, usually circular DNA molecule organized into genes. In eukaryotes, the chromosome is more complex. It is composed of a linear DNA molecule associated with proteins. In addition to

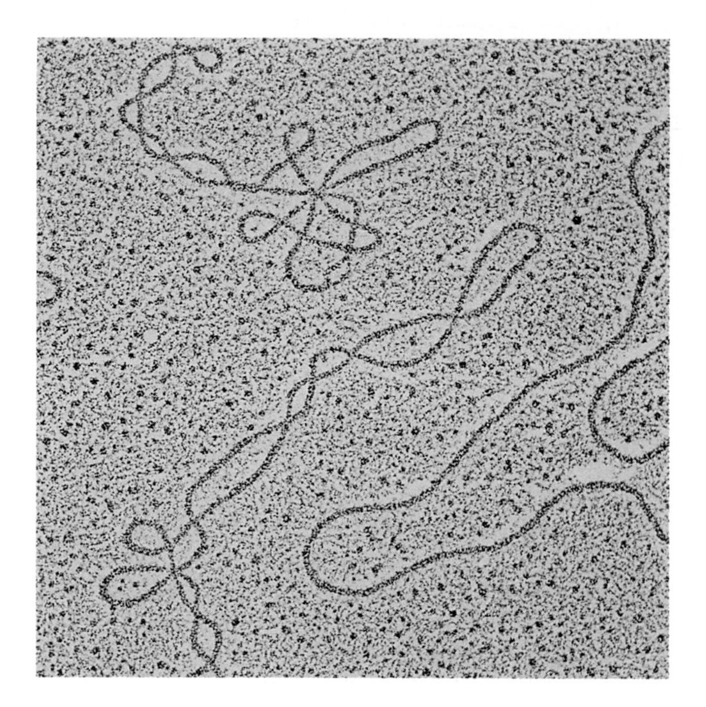

FIGURE 1.3
The DNA of the chromosome of a bacterial virus, also called a bacteriophage, viewed under the electron microscope.

genes, the chromosome contains many regions that do not store genetic information. It is not yet clear what role, if any, is played by some of these regions of chromosomes. Our knowledge of the chromosome, like that of the gene, is continually expanding.

When and how can chromosomes be visualized?

If the chromosomes are released from the viral head or the bacterial cell, they can be visualized under the electron microscope. In eukaryotes, chromosomes are most easily visualized under the light microscope when they are undergoing **mitosis** or **meiosis.** In these division processes, the chromosomes are tightly coiled and condensed. Following division, they uncoil and exist as **chromatin fibers** during interphase, when they can be studied under the electron microscope.

How many chromosomes does an organism have?

Although there are many exceptions, members of most species have a specific number of chromosomes present in each somatic cell called the **diploid number (*2n*).** Upon close analysis, these chromosomes are found to occur in pairs, each member of which shares a nearly identical appearance when visible during cell division. Called **homologous chromosomes,** the members of each pair are identical in their length and in the location of the **centromere,** the point of spindle fiber attachment during division. They also contain the same sequence of gene sites, or **loci,** and pair with one another during meiosis. Thus, the number of different *types* of chromosomes in any diploid species is equal to half the diploid number and is called the **haploid (*n*)** number. Some organisms, such as yeast, are haploid and contain only one "set" of chromosomes. Other organisms, notably plants, are sometimes characterized by more than two sets of chromosomes and are said to be **polyploid.**

What is accomplished during the processes of mitosis and meiosis?

Mitosis is the process by which the genetic material of eukaryotic cells is duplicated and distributed during cell division. **Meiosis** is the process whereby cell division produces gametes in animals and spores in most plants. While mitosis occurs in somatic tissue and yields two progeny cells with an amount of genetic material identical to the progenitor cell, meiosis creates cells with precisely one-half of the genetic material. Each gamete receives one member of each homologous pair of chromosomes, and is haploid. This accomplishment is essential if offspring arising from two gametes are to maintain a constant number of chromosomes characteristic of their parents and other members of the species.

What are the sources of genetic variation?

Classically, there are two sources of genetic variation: **chromosomal mutations** and **gene mutations.** The former, also called **chromosomal aberrations,** includes duplication, deletion, or rearrangement of chromosome segments. Gene mutations result from a change in the stored chemical information in DNA. Such a change may include substitution, duplication, or deletion of nucleotides, which compose this chemical informa-

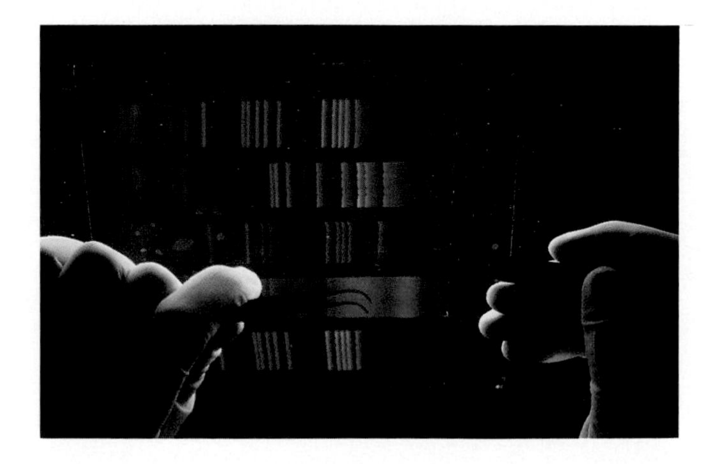

FIGURE 1.4
Visualization of DNA fragments under ultraviolet light. The bands were produced using recombinant DNA technology.

FIGURE 1.5
A bumblebee pollinates a flower, achieving cross-fertilization and enhancing genetic variability within the species.

tion. Alternate forms of the gene, which result from mutation, are called **alleles.** Genetic variation frequently, but not always, results in a change in some characteristic of an organism.

How does DNA store genetic information?
The sequence of nucleotides in a segment of DNA constituting a gene is present in the form of a **genetic code.** This code specifies the chemical nature (the amino acid composition) of proteins, which are the end product of genetic expression. Mutations are produced when the genetic code is altered.

How is the genetic code organized?
There are four different nucleotides in DNA, each varying in one of its components, the **nitrogenous base.** The genetic code is a triplet; therefore, each combination of three nucleotides constitutes a code word. Almost all possible codes specify one of twenty **amino acids,** the chemical building blocks of proteins.

How is the genetic code expressed?
The coded information in DNA is first transferred during a process called **transcription** into a **messenger RNA (mRNA)** molecule. The mRNA subsequently associates with the cellular organelle, the **ribosome,** where it is **translated** into a protein molecule.

Are there exceptions where proteins are not the end product of a gene?
Yes. For example, genes coding for **ribosomal RNA (rRNA),** which is part of the ribosome, and for **transfer RNA (tRNA),** which is involved in the translation process, are transcribed but not translated. Therefore, RNA is sometimes the end product of stored genetic information.

Why are proteins so important to living organisms that they serve as the end product of the vast majority of genes?
Many proteins serve as highly specific biological catalysts, or **enzymes.** In this role, these proteins control cellular metabolism, determining which carbohydrates, lipids, nucleic acids, and other proteins are present in the cell. Many other proteins perform nonenzymatic roles. For example, hemoglobin, collagen, immunoglobulins, and some hormones are proteins.

Why are enzymes necessary in living organisms?
As biological catalysts, enzymes lower the activation energy required for most biochemical reactions and speed the attainment of equilibrium. Otherwise, these reactions would proceed so slowly as to be ineffectual in organisms living under the conditions on earth. Thus, some genes control the variety of enzymes present in any cell type, which in turn dictates its overall biochemical composition.

INVESTIGATIVE APPROACHES IN GENETICS

The scope of topics encompassed in the field of genetics is enormous. Studies have involved viruses, bacteria, and a wide variety of plants and animals and have spanned all levels of biological organization, from molecules to populations. It is helpful, before we embark on a detailed study of genetics, to know the types of investigations that have been used most often in this field. Although some overlap exists, most have used one of four basic approaches.

The most classical investigative approach is the study of **transmission genetics,** in which the patterns of inheri-

FIGURE 1.6
A spider protects her egg case, ensuring genetic continuity between generations.

tance of traits are examined. Experiments are designed so that the transmission of traits from parents to offspring can be analyzed through several generations. Patterns of inheritance are sought that will provide insights into more general genetic principles. The first significant experimentation of this kind to have a major impact on understanding heredity was performed by Gregor Mendel in the middle of the nineteenth century. Information derived from his work serves today as the foundation of transmission genetics. In human studies, where designed matings are neither possible nor desirable, **pedigree analysis** is used. In pedigree analysis, patterns of inheritance are traced through as many generations as possible, leading to inferences concerning the mode of inheritance of the trait under investigation.

The second approach involves **cytological investigations** of the genetic material. The earliest such studies used the light microscope. The initial discovery in the early twentieth century of chromosome behavior during mitosis and meiosis was a critical event in the history of genetics. In addition to playing an important role in the rediscovery and acceptance of Mendelian principles, these observations served as the basis of the **chromosomal theory of inheritance.** This theory, which viewed the chromosome as the carrier of genes and the functional unit of transmission of genetic information, was the cornerstone for further studies in genetics throughout the first half of this century.

The light microscope continues to be an important research tool. It is useful in the investigation of chromosome structure and abnormalities and is instrumental in preparing **karyotypes,** which illustrate the chromosomes characteristic of any species arranged in a standard sequence.

With the advent of electron microscopy, the repertoire of investigative approaches in genetics has grown. In high-resolution microscopy genetic molecules and their behavior during gene expression can be directly visualized.

The third general approach, **molecular and biochemical analysis,** has had the greatest impact on the recent growth of genetic knowledge. Molecular studies beginning in the early 1940s have consistently expanded our knowledge of the role of genetics in life processes. Although experimental sources were initially bacteria and the viruses that invade them, extensive information is now available concerning the nature, expression, replication, and regulation of the genetic information in eukaryotes as well. The precise nucleotide sequence has been determined for genes cloned in the laboratory. **Recombinant DNA studies** (Figure 1.4), where genes from any organism are literally spliced into bacterial or viral DNA and cloned *en masse,* is the most significant and far-reaching research technology used in molecular genetic investigations. As a result, it is now possible to probe gene structure and function with a resolution heretofore impossible. Such molecular and biochemical analysis has had profound implications in medicine and agriculture.

The final approach involves the study of the **genetic structure of populations.** In these investigations scientists attempt to define how and why certain genetic variation is preserved in populations, while other variation diminishes or is lost with time. Such information is critical to the understanding of the evolutionary process. Population genetics also allows us to predict gene frequencies in future generations.

Together these approaches have transformed a subject poorly understood in 1900 into one of the most advanced disciplines in biology today. As such, the impact of genetics on society has been immense. We shall discuss many examples of the applications of genetics in the following section and throughout the text.

THE SIGNIFICANCE OF GENETICS IN SOCIETY

In addition to acquiring information for the sake of extending knowledge in any discipline of science—an experimental approach called **basic research**—scientists conduct investigations to solve problems facing society

FIGURE 1.7
A coral reef environment, demonstrating extensive genetic variation among organisms.

or simply to improve the well-being of members of that society—an approach called **applied research.** The history of applied research shows that it is usually possible only as an extension of prior basic research.

In the case of genetics, both types of research have combined to enhance the quality of our existence on this planet, and to provide a more thorough understanding of the life process. As we shall see throughout this text, there is very little in our lives that genetics fails to touch.

Genetic Advances in Agriculture and Medicine

The major benefits to society as a result of genetic study have been in the areas of agriculture and medicine. Although cultivation of plants and domestication of animals had begun long before, the rediscovery of Mendel's work in the early twentieth century spurred scientists to apply genetic principles to these human endeavors. The use of selective breeding and hybridization techniques has had the most significant impact in agriculture (Figure 1.8).

Plants have been improved in four major ways: (1) more efficient energy utilization during photosynthesis, resulting in more vigorous growth and increased yields; (2) increased resistance to natural predators and pests, including insects and disease-causing microorganisms; (3) production of hybrids exhibiting a combination of superior traits derived from two different strains or even two different species; and (4) selection of genetic variants with increased protein value or an increased content of limiting amino acids, which are essential in the human diet.

These improvements have resulted in a tremendous increase in yield and nutrient value in such crops as barley, beans, corn, oats, rice, rye, and wheat. It is estimated that in the United States the use of improved genetic strains has led to a threefold increase in crop yield per acre. In Mexico, where corn is the staple crop, a significant increase in protein content and yield has occurred. A substantial effort has also been made to improve the growth of Mexican wheat. Led by Norman Borlaug, a team of researchers was able to develop a strain of wheat that incorporated favorable genes from other strains found in various parts of the world, creating a superior variety that is now grown in many underdeveloped countries besides Mexico. Because of this effort, which led to the well-publicized "Green Revolution," Borlaug received the Nobel Peace Prize in 1970. There is little question that this application of genetics has contributed to the well-being of our own species by improving the quality of nutrition worldwide.

Applied research in genetics has also developed superior breeds of animals. Enormous increases in usable meat supplies produced per unit of food intake have occurred. For example, selective breeding has produced chickens that grow faster, produce more high-quality meat per chicken, and lay greater numbers of larger eggs. In larger animals, including pigs and cows, the use of artificial insemination has been particularly important.

TRANSGENIC PLANTS

Domesticated plants and animals have been genetically manipulated for thousands of years by selection and selective breeding. During our life time a new approach, which shows great promise in the production of superior plants and animals, has been developed. This approach is called genetic engineering; potentially useful genes can be identified, extracted, and transferred from one organism to another. Altering the genotypes of plants by the introduction of new genes has the potential to overcome some of the agricultural problems that cannot be solved using the genes normally found in a particular plant species. With genetic engineering it is possible to transfer genes affecting yield, nutritional value, hardiness, and disease resistance. Plants whose genotypes have been altered by the introduction of new genes are called transgenic plants. The first transgenic plants were created less than ten years ago and since then more than 50 plant species have been "genetically engineered."

Resistance to insect predation is one of the problems which has been addressed by this technique. An insecticidal protein is naturally produced by the bacterium *Bacillus thuringiensis.* This protein binds to the gut membrane of certain caterpillar larvae and disrupts the insects' feeding. This natural insecticide has no toxic effect on mammals and has been used by farmers and gardeners for the last 30 years. The genes that produce this insecticidal protein have been isolated from the bacteria and transferred directly into several plant species. Cotton plants carrying these genes have undergone 2 years of field trials and have proven to be effectively resistant to certain major caterpillar pests. It is suggested that the use of these transgenic plants would reduce the use of insecticides on cotton by about 40 percent to 60 percent.

Scientists have also turned their attention to the development of crops that could be grown on land that has been considered either too cold, too hot, too dry, too salty, or too wet to support stable crop production. For example, new efforts are being made to move cold-hardiness genes of winter rye into wheat so that wheat farming could extend to Alaska and to the northern regions of Canada. Drought resistance is also receiving considerable attention. However, recent findings indicate that this trait is controlled, not by one or two genes, but by numerous hereditary factors. This problem points out one of the limitations of the current technology; at this time, genetic engineers can only modify traits determined by a small number of genes.

Technical limitation is not the only problem encountered in agricultural biotechnology; genetic engineering is a controversial area. It has been widely debated whether or not plant varieties that have been produced by genetic technology are potentially hazardous, either to human health or to the environment. The result of this debate is that, although dozens of potential crop varieties produced through genetic engineering have been tested, no transgenic plant is yet available for human consumption.

New guidelines are forthcoming. A recent National Academy of Science report has concluded that "crops modified by molecular and cellular methods should pose risks no different than those modified by classical genetic methods for similar traits." The U.S. Food and Drug Administration (FDA), which is the primary government agency with the responsibility to insure the safety of the United States food supply, has decided that the government should regulate the products of genetic engineering, and not the process by which they are created. Under certain conditions, however, special labeling may be required for plant food that is produced by genetic engineering. Recently, a transgenic tomato with a foreign gene that retards ripening has received government approval and should be commercially available in 1993.

Sperm samples derived from a single male with superior genetic traits may now be used to fertilize thousands of females located in all parts of the world.

Equivalent strides have been made in medicine as a result of advances in genetics, particularly since 1950. Numerous disorders in humans have been discovered to result from either a single mutation or a specific chromosomal abnormality. For example, the genetic basis of sickle-cell anemia, erythroblastosis fetalis, cystic fibrosis, hemophilia, muscular dystrophy, Tay-Sachs disease, Down syndrome, and countless metabolic disorders is now well documented and understood at the molecular level. It is estimated that more than 10 million children or adults in the United States suffer from some form of genetic affliction.

Recognition of the genetic basis of these disorders has provided direction for the development of detection and treatment. **Genetic counseling** provides parents with objective information upon which they can base rational decisions about parenting. It is estimated that every childbearing couple stands an approximately three percent risk of having a child with some form of genetic anomaly. There are currently thousands of documented genetic disorders.

Applied research in genetics has provided other medical benefits. Increased knowledge in **immunogenetics** has made possible compatible blood transfusions as well as organ transplants. The discovery of genetically determined, tissuebound antigens has led to the important concepts of **histocompatibility** and tissue typing. In conjunction with immunosuppressive drugs, transplant operations involving human organs, including the heart, liver, pancreas, and kidney, are increasing annually and are now considered routine surgery.

Recombinant DNA technology is also an important part of applied genetics. Cloned human genes that code for such medically important molecules as insulin and interferon serve as the source of mass production of many essential molecules. Recombinant DNA techniques will also play an increasing and essential role in human genetic engineering, which involves the direct manipulation of the genetic material. In the not-so-distant future, human genetic engineering will undoubtedly be used to alter the genetic constitution of individuals harboring genetic defects and to correct such defects in the developing fetus. Although such processes present ethical questions, they may correct serious genetic errors in members of our species.

FIGURE 1.8
Triticale, a hybrid grain derived from wheat and rye, produced as a result of genetic research.

In later chapters, these and other examples in agriculture and medicine are discussed in greater detail. Although other scientific disciplines are also expanding in knowledge, none have paralleled the growth of information that is occurring annually in genetics. By the end of this course, we are confident you will agree that the present truly represents the "Age of Genetics."

CHAPTER SUMMARY

1 The history of genetics, which emerged as a fundamental discipline of biology early in the twentieth century, dates back to prehistoric times.

2 Numerous concepts and a basic vocabulary essential to the study of genetics have been presented.

3 Four investigative approaches are most often used in the study of genetics, including transmission genetic studies, cytogenetic analyses, molecular-biochemical experimentation, and inquiries into the genetic structure of populations.

4 Genetic research can be either basic or applied. Basic genetic research extends our knowledge of the discipline; the objective of applied genetics research is to solve specific problems affecting the quality of our lives.

5 Recombinant DNA technology has greatly expanded our research capability. It also has had a profound impact in elucidating the basis of inherited diseases and has made possible the mass production of medically important gene products.

KEY TERMS

allele
amino acid
chromatin
chromosomal mutation (aberration)
chromosome
diploid number ($2n$)
DNA (deoxyribonucleic acid)
enzyme
epigenesis
fixity of species
gemmules
gene
gene mutation
genetics
genetic code
haploid (n)
homologous chromosomes
karyotypes
locus
meiosis
mitosis
mRNA (messenger RNA)
natural selection
nitrogenous base
nucleoid
pangenesis
pedigree analysis
polyploid
RNA (ribonucleic acid)
rRNA (ribosomal RNA)
spontaneous generation
transmission genetics
tRNA (transfer RNA)

PROBLEMS AND DISCUSSION QUESTIONS

1 Describe and contrast the ideas of Hippocrates and Aristotle related to the genetic basis of life.

2 Define and contrast pangenesis, epigenesis and preformationism.

3 Which ideas and doctrines that preceded Darwin were central to his thinking?

4 Describe Darwin's and Wallace's theory of natural selection. What information was lacking from it, i.e., what gap remained in it?

5 Contrast chromosomes and genes and describe their role in heredity.

6 Describe the four major investigative approaches used in studying genetics.

7 Contrast basic vs. applied research.

8 Norman Borlaug received the Nobel Peace Prize for his work in genetics. Why do you think he was awarded this prize?

9 How has genetic research been applied to the field of medicine?

SELECTED READINGS

ANDERSON, W. F., and DIRCUMAKOS, E. G. 1981. Genetic engineering in mammalian cells. *Scient. Amer.* (July) 245:106–21.

BORLAUG, N. E. 1983. Contributions of conventional plant breeding to food production. *Science* 219:689–93.

BOWLER, P. J. 1989. *The Mendelian revolution: The emergence of hereditarian concepts in modern science and society.* London: Athione.

COCKING, E. C., DAVEY, M. R., PENTAL, D., and POWER, J. B. 1981. Aspects of plant genetic manipulation. *Nature* 293:265–70.

DAY, P. R. 1977. Plant genetics: Increasing crop yield. *Science* 197:1334–39.

DUNN, L. C. 1965. *A short history of genetics.* New York: McGraw-Hill.

GARDNER, E. J. 1972. *History of biology.* 3rd ed. New York: Macmillan.

GARFIELD, E. 1981. Medical genetics: The new preventive medicine. *Current Contents—Life Sciences,* vol. 24, no. 36, pp. 5–20.

HOPWOOD, D. A. 1981. The genetic programming of industrial microorganisms. *Scient. Amer.* (Sept.) 245:91–102.

HUTTON, R. 1978. *Bio-revolution: DNA and the ethics of man-made life.* New York: New American Library.

KING, R. C., and STANSFIELD, W. D. 1990. *A dictionary of genetics.* 4th ed. New York: Oxford University Press.

MOORE, J. A. 1985. Science as a way of knowing—Genetics. *Amer. Zool.* 25:1–165.

OLBY, R. C. 1966. *Origins of Mendelism.* London: Constable.

STUBBE, H. 1972. *History of genetics: From prehistoric times to the rediscovery of Mendel.* (Translated by T. R. W. Waters.) Cambridge, MA: MIT Press.

TORREY, J. G. 1985. The development of plant biotechnology. *Amer. Scient.* 73:354–63.

WEINBERG, R. A. 1985. The molecules of life. *Scient. Amer.* (Oct.) 253:48–57.

2 Cell Division and Chromosomes

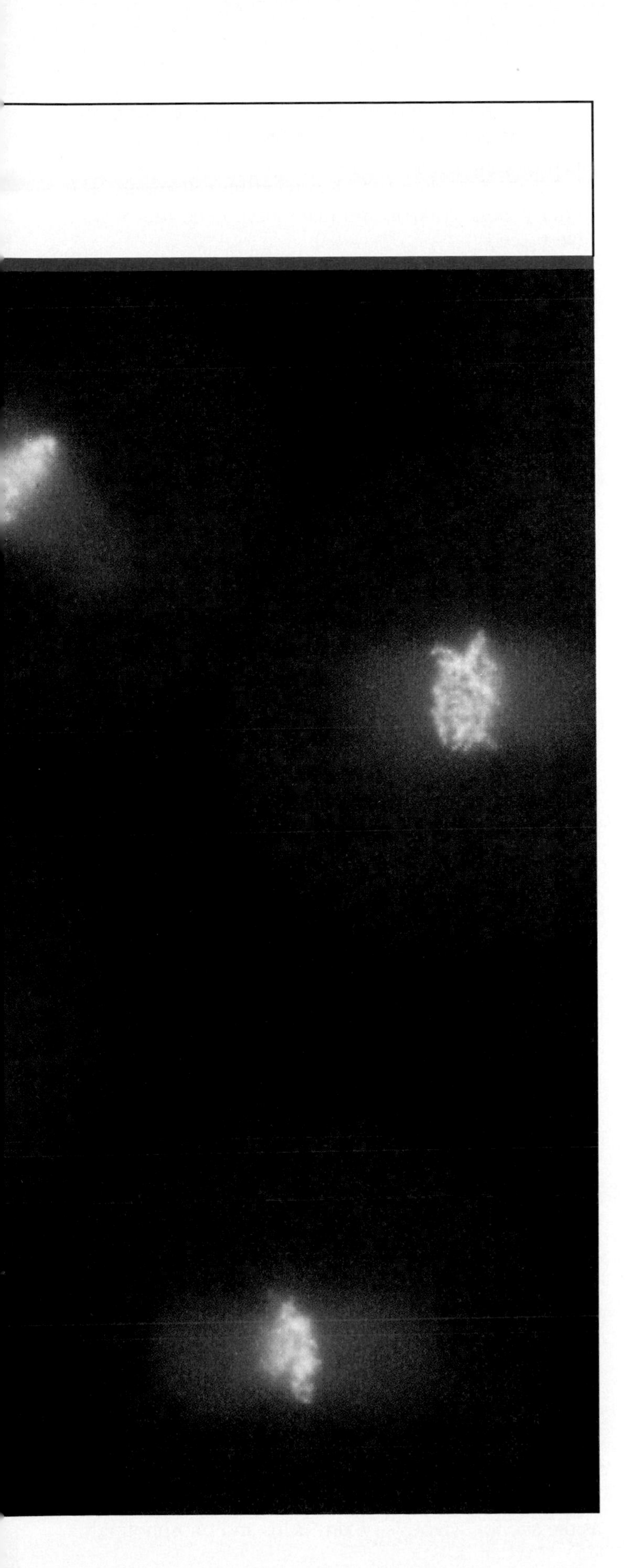

Chapter Concepts

Genetic continuity between cells and organisms of any sexually reproducing species is maintained by the processes of mitosis and meiosis. The processes are orderly and efficient, serving to produce diploid somatic cells and haploid gametes, respectively.

In every living thing there exists a substance referred to as the **genetic material.** Except in certain viruses, this material is composed of the nucleic acid, DNA. A molecule of DNA is organized into units called **genes,** which direct all metabolic activities of cells. Genes are organized into **chromosomes,** structures that serve as the vehicle for transmission of genetic information. In this chapter, we consider the topic of genetic continuity between cells and organisms. The manner in which chromosomes are transmitted from one generation of cells to the next, and from organisms to their descendants, is exceedingly precise.

Two major processes are involved in eukaryotes: **mitosis** and **meiosis.** Although the mechanisms of the two processes are similar in many ways, the outcomes are quite different. Mitosis leads to the production of two cells with an identical number of chromosomes. Meiosis, on the other hand, reduces the genetic content and the number of chromosomes by precisely half. This reduction is essential if sexual reproduction is to occur without doubling the amount of genetic material at each generation. Strictly speaking, mitosis is that portion of the cell cycle during which the hereditary components are precisely and equally divided into daughter cells. Meiosis is part of a special type of cell division leading to the production of sex cells: gametes and spores. This process is an essential step in the transmission of genetic information from an organism to its offspring.

Normally, chromosomes can be seen only during mitosis and meiosis. But, in this chapter, we will also consider two specialized cases where chromosomes can be seen in nondividing cells: polytene and lampbrush chromosomes. These specialized structures extend our knowledge of genetic organization.

CELL STRUCTURE

Before describing mitosis and meiosis, we will briefly review the structure of cells. As we shall see, many components, such as the nucleolus, ribosome, and centriole, are involved directly or indirectly with genetic processes. Other components, the mitochondria and chloroplasts, contain their own unique genetic information. It is also useful for us to compare the structural differences of the prokaryotic bacterial cell with the eukaryotic cell. Variation in cell structure is dependent on the specific genetic expression by each cell type.

Before 1940, knowledge of cell structure was based on information obtained with the light microscope. Around 1940 the transmission electron microscope was in its early stages of development, and by 1950 many details of cell ultrastructure were unveiled. Under the electron microscope cells were seen as highly organized, precise structures. A new world of whorling membranes, miniature organelles, microtubules, granules, and filaments was revealed. These discoveries revolutionized thinking in the entire field of biology. We will be concerned with those aspects of cell structure relating to genetic study. As the parts of the cell are described, refer to Figure 2.1, which depicts a typical animal cell.

The cell is surrounded by a **plasma membrane,** an outer covering that defines the cell boundary and delimits the cell from its immediate external environment. This membrane is not passive, but instead it controls the movement of materials such as gases, nutrients, and waste products into and out of the cell. In addition to this membrane, plant cells have an outer covering called the **cell wall.** One of the major components of this rigid structure is a polysaccharide called **cellulose.** Bacterial cells also have a cell wall, but its chemical composition is quite different from that of the plant cell wall. The major component in bacteria is a complex macromolecule called a **peptidoglycan.** As its name suggests, the molecule consists of peptide and sugar units. Long polysaccharide chains are cross-linked with short peptides, which impart great strength and rigidity to the bacterial cell. Some bacterial cells have still another covering, a **capsule.** It is a mucuslike polysaccharide that protects these bacteria from phagocytic activity by the host during their pathogenic invasion of eukaryotic organisms.

Many, if not most, animal cells have a covering over the plasma membrane called a **cell coat.** Consisting of glycoproteins and polysaccharides, the chemical composition of the cell coat differs from comparable structures in either plants or bacteria. One function served by the cell coat is to provide biochemical identity at the surface of cells. Among other forms of molecular recognition, various antigenic determinants are part of the cell coat. For example, the **AB** and **MN** antigens are found on the surface of red blood cells. In other cells, the **histocompatibility antigens,** which elicit an immune response during tissue and organ transplants, are part of the cell coat. All forms of biochemical identity at the cell surface are under genetic control, and many have been thoroughly investigated.

The presence of the **nucleus** and other membranous organelles characterizes eukaryotic cells. The nucleus houses the genetic material, DNA, which is found in association with large numbers of acidic and basic proteins. During nondivisional phases of the cell cycle this DNA/protein complex exists in an uncoiled, dispersed state called **chromatin.** As we will soon discuss, during mitosis and meiosis this material coils up and condenses into structures called **chromosomes.** Also present in the nucleus is the **nucleolus,** an amorphous component where ribosomal RNA is synthesized and where the initial stages of ribosomal assembly occur. The areas of DNA encoding rRNA are collectively referred to as the **nucleolus organizer region** or the **NOR.**

The lack of a nuclear envelope and membraneous organelles is characteristic of prokaryotes. In bacteria such as *E. coli* the genetic material is present as a long, circular DNA molecule that is compacted into a region referred to as the **nucleoid.** Part of the DNA may be attached to the cell membrane, but in general, the nucleoid constitutes a large area throughout the cell. Although the DNA is compacted, it does not undergo the extensive coiling characteristic of the stages of mitosis where, in eukaryotes, chromosomes become visible. Nor is the DNA in these organisms associated as extensively with proteins as is eukaryotic DNA. Figure 2.2 shows the formation of two bacteria during cell division and illustrates the bacterial chromosomes in the nucleoid regions. Prokaryotic cells do not have a distinct nucleolus, but do contain genes that specify rRNA molecules.

The remainder of the eukaryotic cell enclosed by the plasma membrane, and excluding the nucleus, is composed of **cytoplasm.** Cytoplasm consists of a nonparticulate, colloidal material referred to as the **cytosol,** which surrounds and encompasses the numerous types of cellular organelles. Beyond these components, an extensive system of tubules and filaments comprising the **cytoskeleton** provides a lattice of support structures within the cytoplasm. Consisting primarily of tubulin-derived microtubules and actin-derived microfilaments, this structural framework maintains cell shape, facilitates cell mobility, and anchors the various organelles. Tubulin and actin are both proteins found abundantly in eukaryotic cells.

One such organelle, the membranous **endoplasmic reticulum (ER),** compartmentalizes the cytoplasm, greatly increasing the surface area available for biochemical synthesis. ER may be smooth, in which case it serves as the site for synthesis of fatty acids and phospholipids.

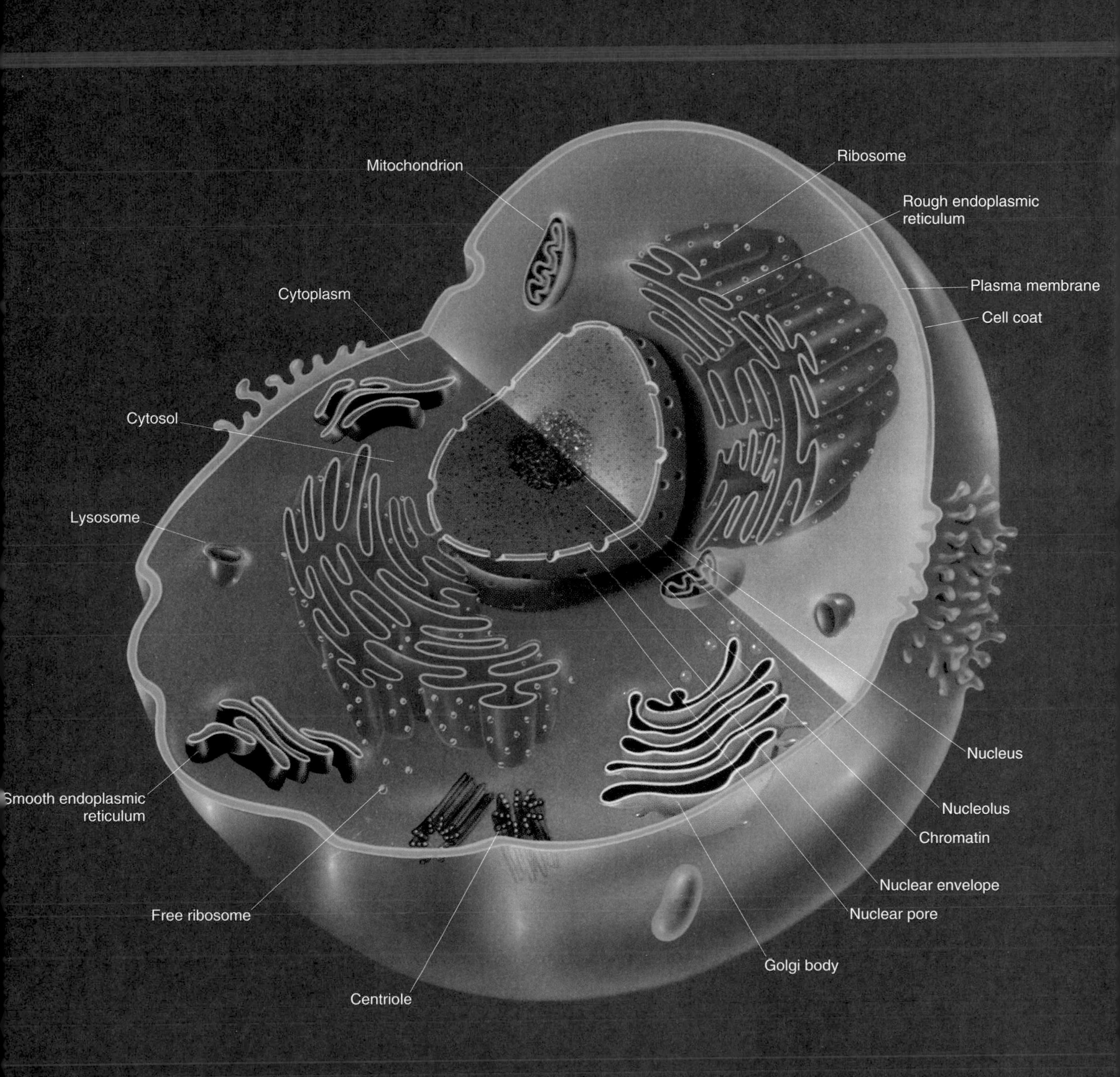

FIGURE 2.1
Drawing of a generalized animal cell. Emphasis has been placed on the cellular components discussed in the text.

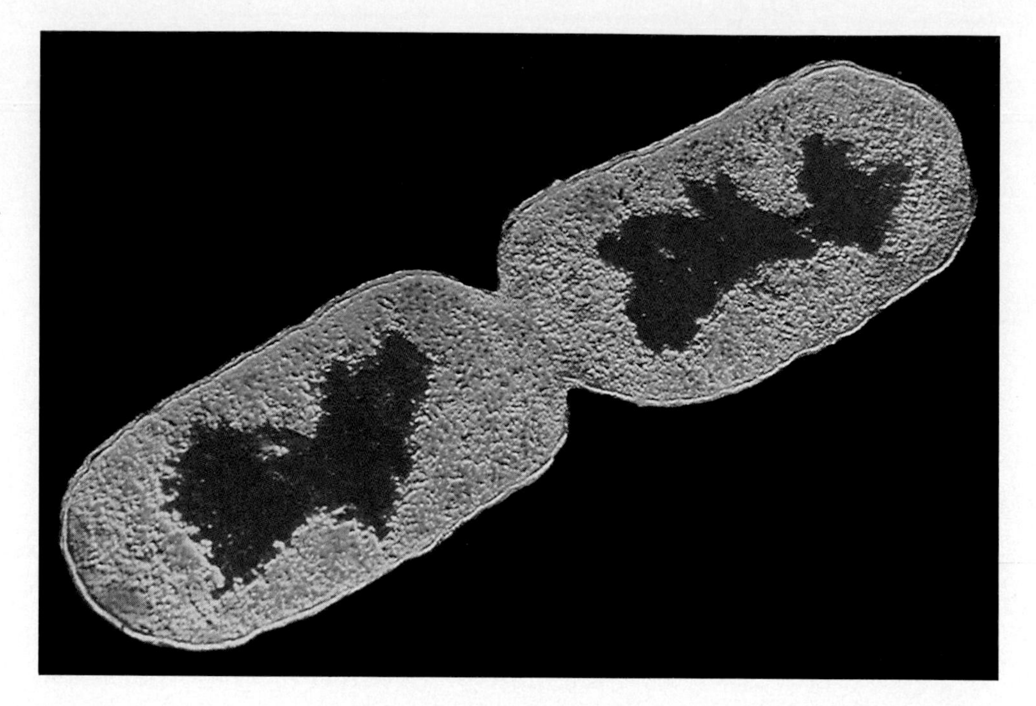

FIGURE 2.2
Color-enhanced electron micrograph of *E. coli* undergoing cell division. Particularly prominent are the two chromosomal areas that have been partitioned into the daughter cells.

Or, the ER may be rough because it is studded with ribosomes. Ribosomes, which will later be discussed in detail, serve as sites for the translation of genetic information contained in messenger RNA (mRNA) into proteins.

Three other cytoplasmic structures are very important in the eukaryotic cell's activities: mitochondria, chloroplasts, and centrioles. Mitochondria are found in both animal and plant cells and are the sites of the **oxidative phases of cell respiration.** These chemical reactions generate large amounts of adenosine triphosphate (ATP), an energy-rich molecule. Chloroplasts are found in plants, algae, and some protozoans. This organelle is associated with **photosynthesis,** the major energy-trapping process on earth. Both mitochondria and chloroplasts contain a type of DNA distinct from that found in the nucleus. Furthermore, these organelles can duplicate themselves and transcribe and translate their genetic information. It is interesting to note that the genetic machinery of mitochondria and chloroplasts closely resembles that of prokaryotic cells. This and other observations have led to the proposal that these organelles were once primitive free-living structures that established a symbiotic relationship with a primitive eukaryotic cell. This theory, which describes the evolutionary origin of these organelles, is called the **endosymbiont theory.**

Animal and some plant cells also contain a pair of complex structures called the **centrioles.** These cytoplasmic bodies, contained within a specialized region called the **centrosome,** are associated with the organization of those spindle fibers that function in mitosis and meiosis. In some organisms, the centriole is derived from another structure, the **basal body,** which is associated with the formation of cilia and flagella. Over the years, there have been many reports that have suggested that centrioles and basal bodies contain DNA, which is involved in the replication of these structures. Recently, very convincing evidence has been presented that this is most certainly the case in the single-celled green alga, *Chlamydomonas.* Several genes are contained within a circular DNA structure associated with the basal body and centriole.

The organization of **spindle fibers** by the centrioles occurs during the early phases of mitosis and meiosis. Composed of arrays of microtubules, these fibers play an important role in the movement of chromosomes as they separate during cell division. The microtubules consist of polymers of alpha and beta subunits of the protein tubulin. The interaction of the chromosomes and spindle fibers will be considered later in this chapter.

HOMOLOGOUS CHROMOSOMES, HAPLOIDY, AND DIPLOIDY

To describe mitosis and meiosis, we must employ the concept of **homologous chromosomes.** An understanding of this concept will also be of critical importance in the following chapters on transmission genetics.

Chromosomes are most easily visualized during mitosis; when they are examined carefully, they are seen to take on distinctive lengths and shapes. Each contains a condensed or constricted region called the **centro-**

mere, which establishes the general appearance of each chromosome. Figure 2.3 illustrates chromosomes with centromere placements at different points along their lengths. Extending from either side of the centromere are the arms of the chromosome. Depending on the position of the centromere, different arm ratios are produced. As Figure 2.3 illustrates, chromosomes are classified as **metacentric, submetacentric, acrocentric,** or **telocentric** on the basis of the centromere location. The shorter arm of the latter three types, by convention, is shown above the centromere and is called the **p arm.** The longer arm is shown below the centromere and is called the **q arm.**

When studying mitosis, we may make several other important observations. First, each somatic cell within members of the same species contains an identical number of chromosomes. This is called the **diploid number ($2n$).** When the lengths and centromere placements of all such chromosomes are examined, a second general feature is apparent. Nearly all of the chromosomes exist in pairs with regard to these two criteria. The members of each pair are called **homologous chromosomes.** For each chromosome exhibiting a specific length and centromere placement, another exists with identical features.

Figure 2.4 illustrates the nearly identical physical appearance of members of homologous chromosome pairs. There, the human mitotic chromosomes have been photographed, cut out of the print, and matched up, creating a **karyotype.** As you can see, humans have a $2n$ value of 46 and exhibit a diversity of sizes and centromere placements. Note also that each of the 46 chromosomes is actually a double structure. Had these chromosomes been allowed to continue dividing, the two parts of each chromosome—called **sister chromatids**—would have separated into the two new cells as division continued.

Collectively, the total set of genes contained on one member of each homologous pair of chromosomes constitutes the **haploid genome** of the species. The **haploid number (n)** of chromosomes is one-half of the diploid number. Table 2.1 demonstrates the wide range of n values found in a variety of plants and animals.

Homologous pairs of chromosomes have important genetic similarity. They contain identical gene sites, or **loci,** along their lengths. Thus, they have identical genetic potential. In sexually reproducing organisms, one

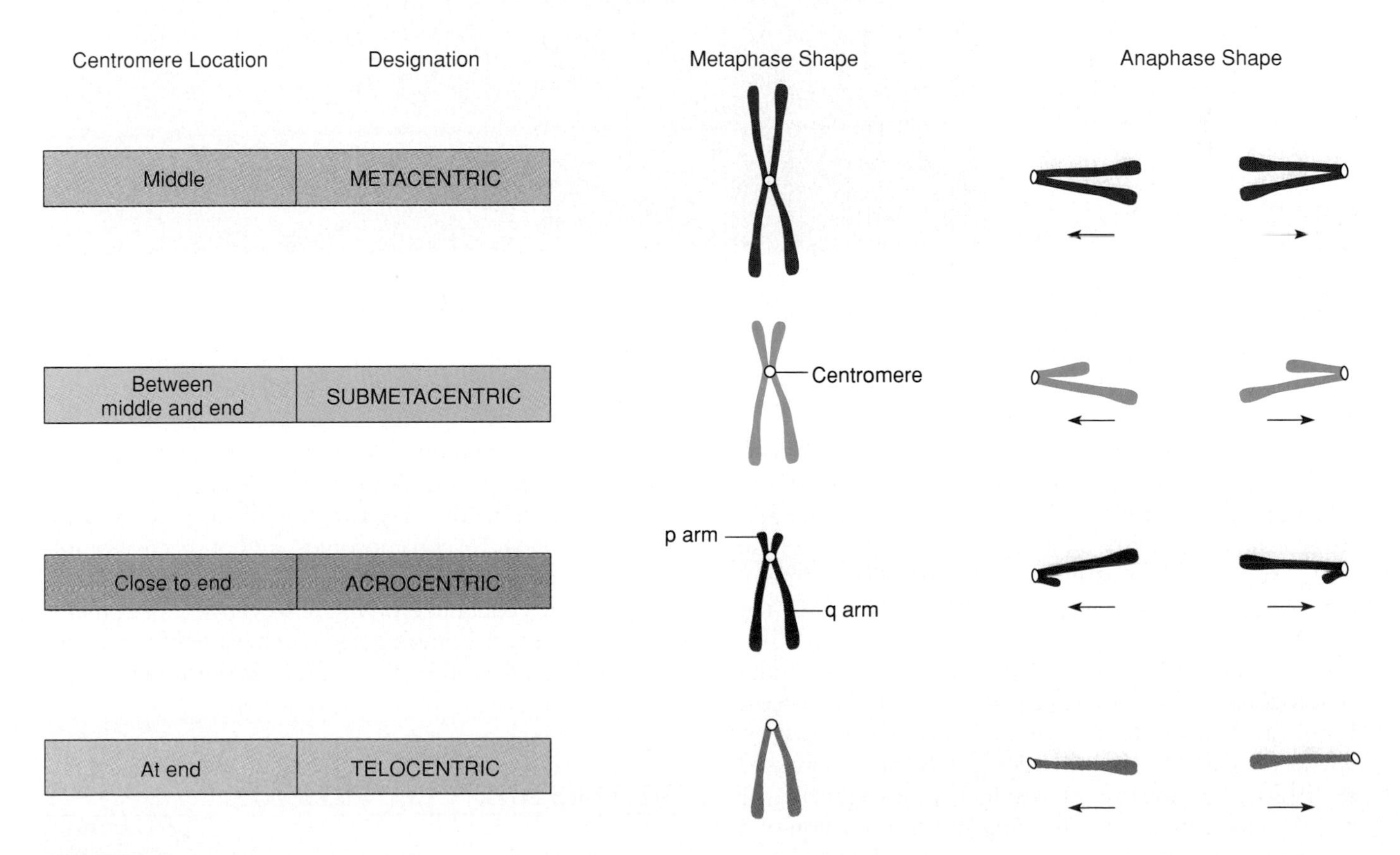

FIGURE 2.3
Centromere locations and designations of chromosomes based on their location. Note that the shape of the chromosome during anaphase is determined by the position of the centromere.

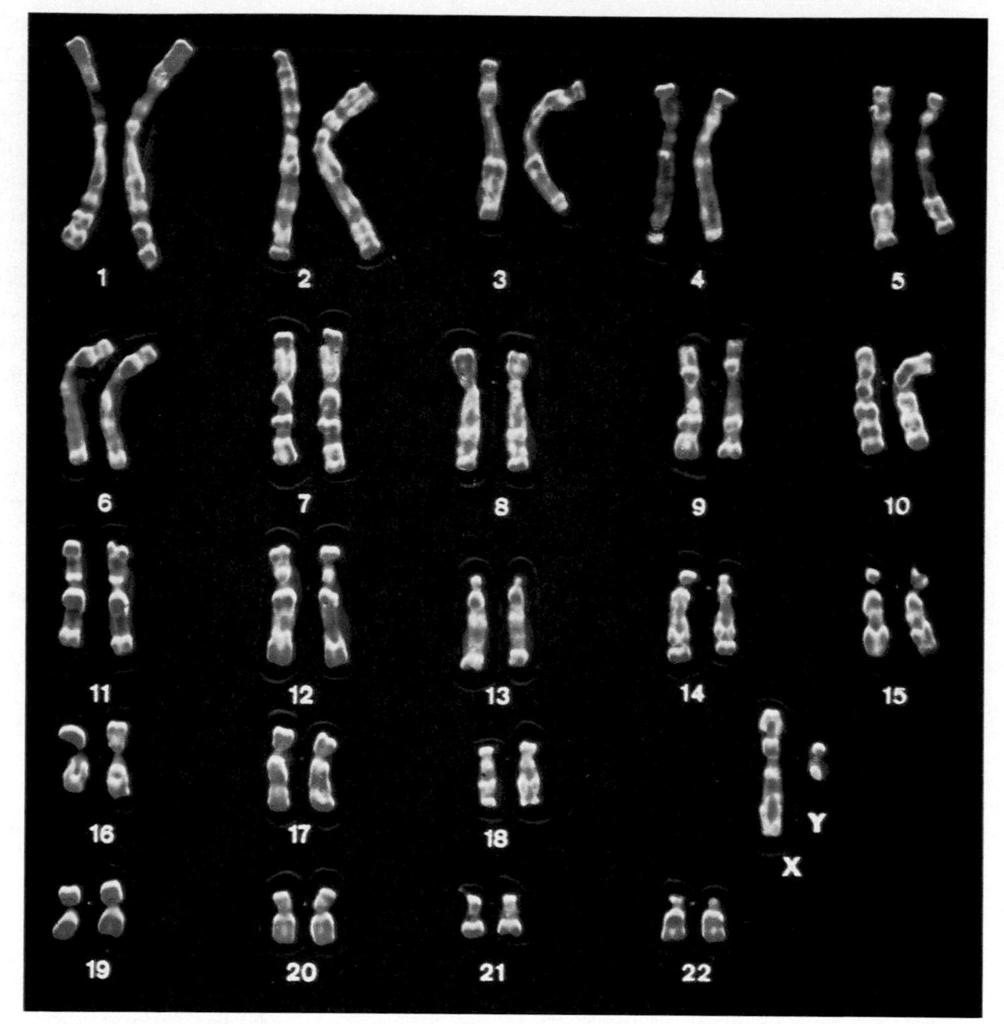

FIGURE 2.4
Mitotic chromosomes constituting the human male karyotype. All but the X and Y chromosomes are present in homologous pairs. Each chromosome is actually a double structure.

member of each pair is derived from the maternal parent (through the ovum) and one from the paternal parent (through the sperm). Thus, each diploid organism contains two copies of each gene as a consequence of **biparental inheritance.** As will be seen in the following chapters on transmission genetics, the members of each pair of genes, while influencing the same characteristic or trait, need not be identical. Alternate forms of the same gene are called **alleles.** In a population of members of the same species, many different alleles of the same gene may exist.

The concepts of haploid number, diploid number, and homologous chromosomes may be related to the process of meiosis. During the formation of gametes or spores, meiosis converts the diploid number of chromosomes to the haploid number. As a result, haploid gametes or spores contain precisely one member of each homologous pair of chromosomes, i.e., one complete set. Following fusion of two gametes in fertilization, the diploid number is reestablished, i.e., two complete sets. The constancy of genetic material from generation to generation is thus maintained.

There is one important exception to the concept of homologous pairs of chromosomes. In many species, one pair, the **sex-determining chromosomes,** may not be homologous in size, centromere placement, arm ratio, or genetic potential. For example, in humans, males contain a Y chromosome in addition to one X chromosome, while females carry two homologous X chromosomes. The X and Y chromosomes are not strictly homologous. The Y is considerably smaller and lacks most of the gene sites contained on the X. Nevertheless, in meiosis they behave as homologues so that gametes produced by males receive either the X or Y chromosome.

MITOSIS AND CELL DIVISION

The process of mitosis is critical to all eukaryotic organisms. In single-celled organisms, such as protozoans, fungi, and algae, mitosis is a part of cell division, which provides the basis of asexual reproduction. Multicellular diploid organisms begin life as single-celled fertilized

TABLE 2.1
The haploid number of chromosomes for representative organisms.

Common Name	Genus-Species	Haploid No.	Common Name	Genus-Species	Haploid No.
Black bread mold	*Aspergillus nidulans*	8	House fly	*Musca domestica*	6
Broad bean	*Vicia faba*	6	House mouse	*Mus musculus*	20
Cat	*Felis domesticus*	19	Human	*Homo sapiens*	23
Cattle	*Bos taurus*	30	Jimson weed	*Datura stramonium*	12
Chicken	*Gallus domesticus*	39	Mosquito	*Culex pipiens*	3
Chimpanzee	*Pan troglodytes*	24	Pink bread mold	*Neurospora crassa*	7
Corn	*Zea mays*	10	Potato	*Solanum tuberosum*	24
Cotton	*Gossypium hirsutum*	26	Rhesus monkey	*Macaca mulatta*	21
Dog	*Canis familiaris*	39	Roundworm	*Caenorhabditis elegans*	6
Evening primrose	*Oenothera biennis*	7	Silkworm	*Bombyx mori*	28
Frog	*Rana pipiens*	13	Slime mold	*Dictyostelium discoidium*	7
Fruit fly	*Drosophila melanogaster*	4	Snapdragon	*Antirrhinum majus*	8
Garden onion	*Allium cepa*	8	Tobacco	*Nicotiana tabacum*	24
Garden pea	*Pisum sativum*	7	Tomato	*Lycopersicon esculentum*	12
Grasshopper	*Melanoplus differentialis*	12	Water fly	*Nymphaea alba*	80
Green alga	*Chlamydomonas reinhardi*	18	Wheat	*Triticum aestivum*	21
Horse	*Equus caballus*	32	Yeast	*Saccharomyces cerevisiae*	17

eggs or **zygotes.** The mitotic activity of the zygote and the subsequent daughter cells is the foundation for development and growth of the organism. In adult organisms, mitotic activity is prominent in wound healing and other forms of cell replacement in certain tissues. For example, the epidermal skin cells of humans are continuously being sloughed off and replaced. Cell division also results in a continuous production of reticulocytes, which eventually shed their nuclei and replenish the supply of red blood cells in vertebrates. In abnormal situations, somatic cells may exhibit uncontrolled cell divisions, resulting in cancer.

It is generally observed that following cell division, the initial size of each new daughter cell is approximately one-half the size of its parent. However, the nucleus of each new cell is not appreciably smaller than the nucleus of the original cell. Quantitative measurements of DNA confirm that there are equivalent amounts of genetic material in the daughter nuclei as in the parent cell.

The process of cytoplasmic division is called **cytokinesis.** The division of cytoplasm requires a mechanism that results in a partitioning of the volume into two parts, followed by the enclosure of both new cells within a distinct plasma membrane. Cytoplasmic organelles either replicate themselves, arise from existing membrane structures, or are synthesized *de novo* (anew) in each cell. The subsequent proliferation of these structures is a reasonable and adequate mechanism for reconstituting the cytoplasm in daughter cells.

The division of the genetic material into daughter cells is more complex than cytokinesis and requires more precision. The chromosomes must first be exactly replicated and then accurately distributed into daughter cells. The end result is the production of two daughter cells, each with a chromosome composition identical to the parent cell.

Interphase and the Cell Cycle

Many cells undergo a continuous alternation between division and nondivision. The interval between each mitotic division is called **interphase.** It was once thought that the biochemical activity during interphase was devoted solely to the cell's growth and its normal function. However, we now know that another biochemical step critical to the next mitosis occurs during interphase: **the replication of the DNA of each chromosome.** Occurring midway through interphase, this period during which DNA is synthesized is called the **S phase.** The initiation and completion of synthesis can be detected by monitoring the incorporation of radioactive DNA precursors.

Investigations of this nature have demonstrated two periods during interphase, before and after S, when no DNA synthesis occurs. These are designated $\mathbf{G_1}$ (gap I) and $\mathbf{G_2}$ (gap II), respectively. During both of these periods, as well as during S, intensive metabolic activity, cell growth, and cell differentiation occur. By the end of G_2, the volume of the cell has roughly doubled, DNA has been replicated, and mitosis (M) is initiated. Continuously dividing cells then repeat this cycle (G_1, S, G_2, M) over and over. This concept of such a **cell cycle** is illustrated in Figure 2.5.

Much is known about the cell cycle based on *in vitro* (test tube) studies. When grown in culture, most cell types traverse the complete cycle in about 20 hours. The actual process of mitosis occupies only a small part of the cycle. The length of the S and G_2 stages of interphase are fairly consistent among different cell types. Most variation is seen in the length of time spent in the G_1 stage. Figure 2.6 illustrates the relative length of these periods in a typical cell.

G_1 is of great interest in the study of cell proliferation and its control. At a point late in G_1, all cells follow one of two paths. They either withdraw from the cycle and become quiescent in the **G_0 stage,** or they become committed to initiate DNA synthesis and complete the cycle. The time when this decision is made has been referred to as

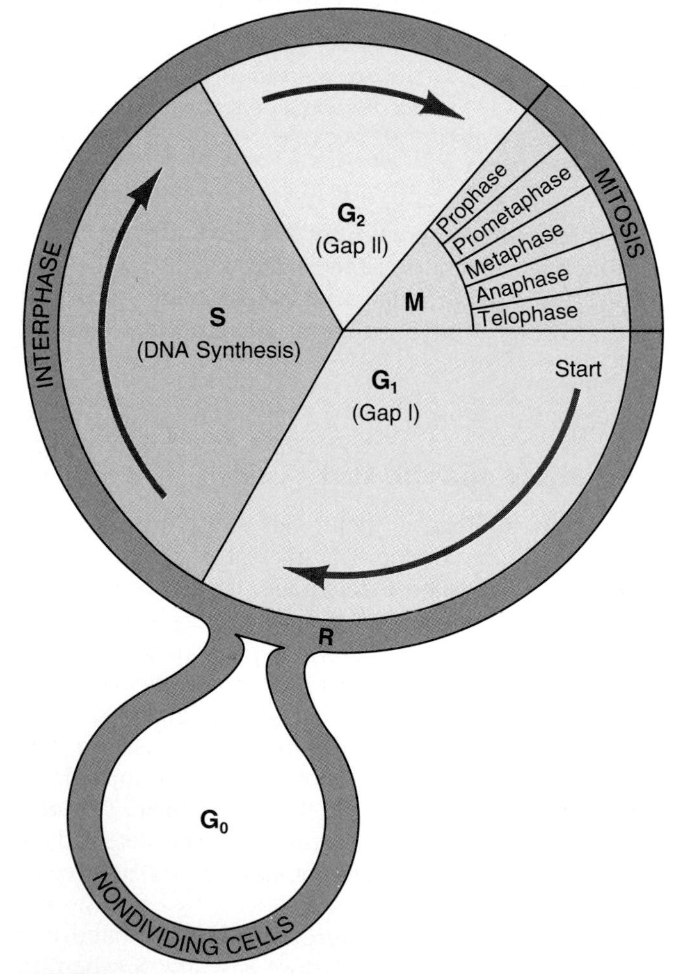

FIGURE 2.5
Diagrammatic representation of the stages that compose an arbitrary cell cycle. Following mitosis (M), cells initiate a new cycle (G_1). Cells may become nondividing (G_0) or continue through the restriction point (R) where they become committed to begin DNA synthesis (S) and complete the cycle (G_2 and M). Following mitosis, two daughter cells are produced.

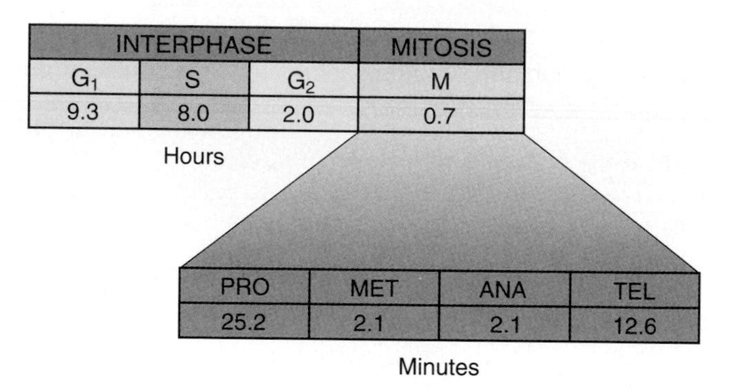

INTERPHASE
MITOSIS
G1 S G2 M
9.3 8.0 2.0 0.7
Hours
PRO MET ANA TEL
25.2 2.1 2.1 12.6
Minutes

FIGURE 2.6
The time spent in each phase of one complete cell cycle of a human cell in culture. Times vary according to cell types and conditions.

the **restriction** or **R point.** Cells that enter G_0 remain viable and metabolically active but are nonproliferative. Cancer cells appear to avoid entering G_0. Other cells enter G_0 and never reenter the cell cycle. Still others can remain quiescent in G_0, but they may be stimulated to return to G_1, reentering the cycle.

Prophase

Once G_1, S, and G_2 are completed, mitosis is initiated. Mitosis is a dynamic period of vigorous and continual activity. For discussion purposes, the entire process is subdivided into discrete phases, and specific events are assigned to each stage. These stages, in order of occurrence, are **prophase, prometaphase, metaphase, anaphase,** and **telophase.** Each of these stages is depicted in a drawing in Figure 2.7A. Many are shown as they actually occur in Figure 2.7B.

Almost one-third of mitosis is spent in prophase; significant activities occur during this stage. One of the early events in animal cell and lower plant prophase involves the migration of two pairs of centrioles to opposite ends of the cell. These structures are found just outside the nuclear envelope in an area of differentiated cytoplasm called the **centrosome.** It is thought that each pair of centrioles consists of one mature unit and a smaller, newly formed centriole.

The direction of migration of the centrioles is such that two poles are established at opposite ends of the cell. This creates an axis along which chromosomal separation occurs. Following their migration, the centrioles are responsible for the organization of cytoplasmic microtubules into a series of **spindle fibers.** Although the cells of plants, fungi, and certain algae seem to lack centrioles, spindle fibers are nevertheless apparent during mitosis. If some other center that organizes microtubules into spindle fibers exists in these cells, it has yet to be discovered.

GENETIC REGULATION OF THE CELL CYCLE

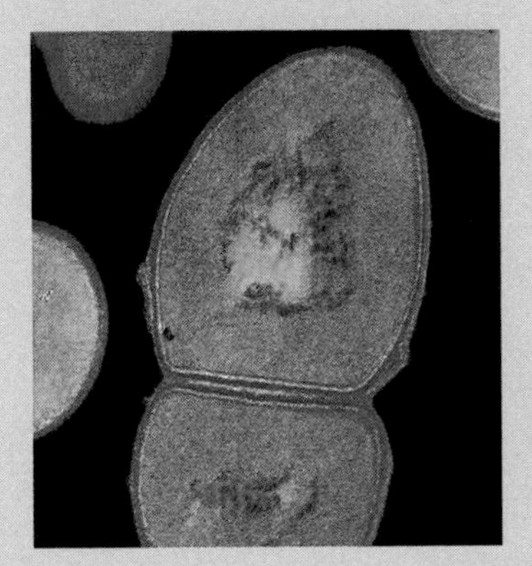

Within the life cycle of an eukaryotic cell, the first phase G1 is the growth phase of the cell. The time a cell spends in G1 is variable: Some cells may pass through this stage in hours, some may spend years in G1. During G1, a normal cell must make a decision whether or not to divide. Once the decision to divide is made the cell is irreversibly programmed to enter the next phase S, where DNA is synthesized, and to complete the cell cycle.

The cell cycle is tightly regulated under the control of various genes. Thus, mutations in certain genes can interrupt its orderly progression. Checks are in place to ensure that one phase is completed properly before entry into the next phase occurs. These checks are also subject to genetic control. Mammalian cells that lose cell cycle regulation can become cancerous cells.

In order to gain insight into the genes that regulate and control cell division, scientists looked for mutants that altered the cell's ability to successfully complete the cell cycle. The rational is that the nonmutated form of the gene is thus necessary for cell cycle regulation. Analysis of the mutants and their interactions with each other have made it possible to determine the normal sequence of some of the genetic events in the cell cycle.

One of the genes isolated proved to be an important gene in the cell cycle of all eukaryotes, from yeast to humans. This gene is called *cdc2* (cell division cycle) and functions at several stages in the cycle. It is necessary for entry into mitosis and has been shown recently in yeast to be necessary for entry into S phase.

The *cdc2* gene encodes "*cdc2* kinase", an enzyme that, like all protein kinases can transfer phosphate groups from ATP to certain amino acids in particular proteins. The addition and removal of phosphates (the removal involves a different kind of enzyme) is a means to change the behavior of proteins. For example, at the beginning of mitosis, *cdc2* kinase adds phosphate groups to certain proteins in the nuclear membrane. This addition causes the proteins to dissociate and leads to nuclear membrane breakdown, an important early step in mitosis. Although *cdc2* kinase is made throughout the cell cycle, it becomes active only when complexed with yet another protein called *cyclin*.

There are several different cyclins known; each type accumulates at specific times during the cell cycle. In yeast it has been shown that *cdc2* kinase is complexed with one kind of cyclin at the end of G1 when the cell becomes committed to enter S and with a different cyclin when the cell is undergoing mitosis. Successive appearance of different cyclins and their interactions with cellular components like *cdc2* kinase and the nuclear membrane appear to propel the cell through the cell cycle.

There are other proteins that can delay that progression under certain circumstances. If the nutrient pool is not sufficient for proper completion of DNA synthesis in S phase, the cell cycle can be delayed in G1. If DNA is damaged or incompletely replicated the cell cycle can be delayed in G2 to allow time for repair before moving into mitosis.

A gene called *p53* has been shown to cause the arrest of cells in G1 under certain conditions. This gene is particularly interesting since it was first discovered as a mutant in many types of human cancer cells. Because impairment of normal *p53* function leads to tumors, the normal *p53* gene became known as tumor suppressor gene. It was predicted that the normal *p53* gene would have an important function in cell cycle regulation.

Experiments have shown that *p53* helps regulate the transition of a cell from G1 to S phase. Under some stress conditions, the entry into S phase is delayed in cells that have a normal *p53* gene; in cells that carry mutated copies of the gene, the cell is not delayed and proceeds into S phase. Those cells that enter S phase and replicate their chromosomes are found to exhibit chromosomal abnormalities.

The mechanisms by which *p53, cdc2,* the cyclins, and other factors regulate the cell cycle are poorly understood at present. It is hoped that the eventual understanding of these normal processes will also define the processes that turn a normal cell into a cancerous cell and lead to new ways to treat or cure this disease.

MITOSIS

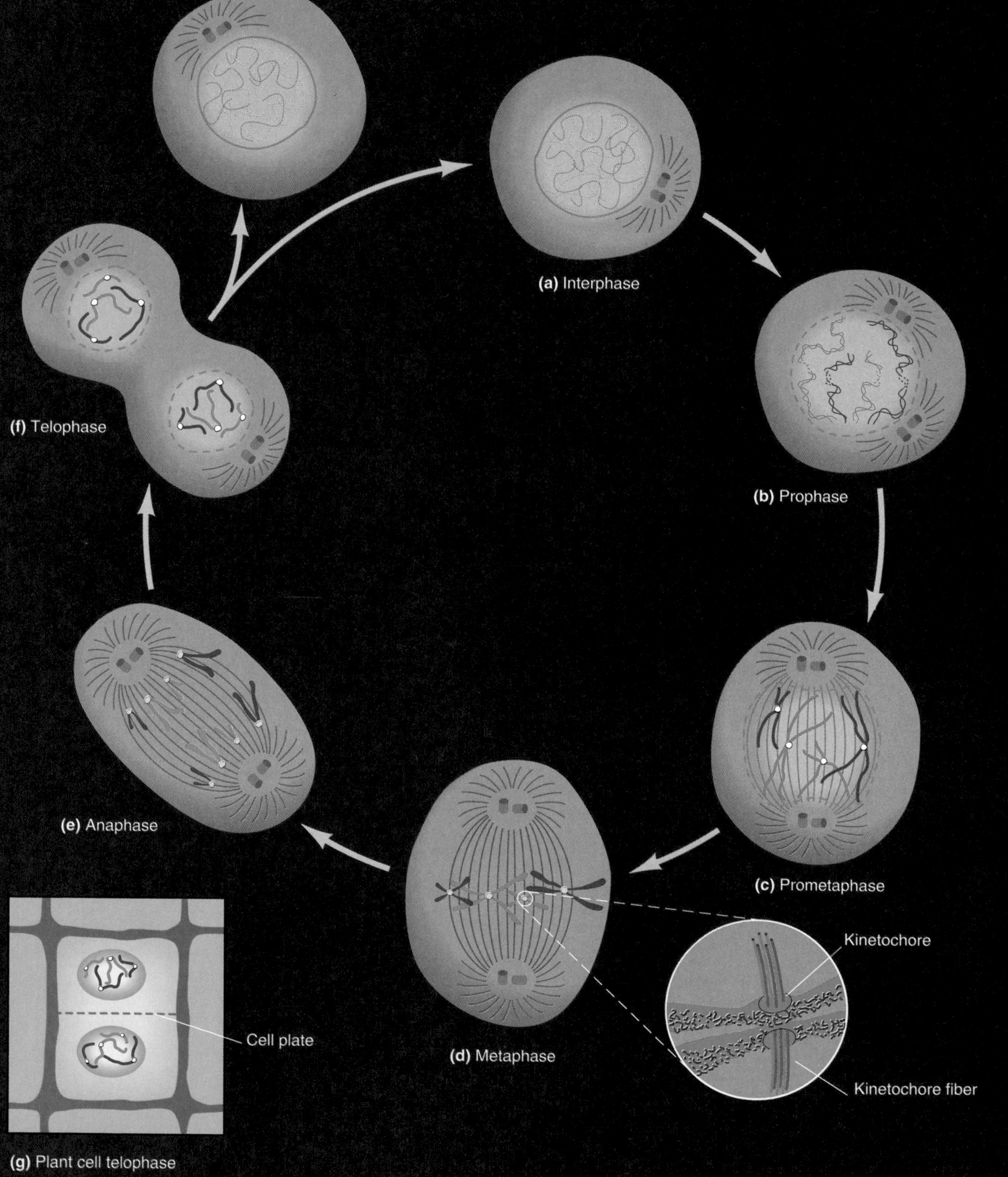

(a) Interphase

(b) Prophase

(c) Prometaphase

(d) Metaphase

Kinetochore

Kinetochore fiber

(e) Anaphase

(f) Telophase

Cell plate

(g) Plant cell telophase

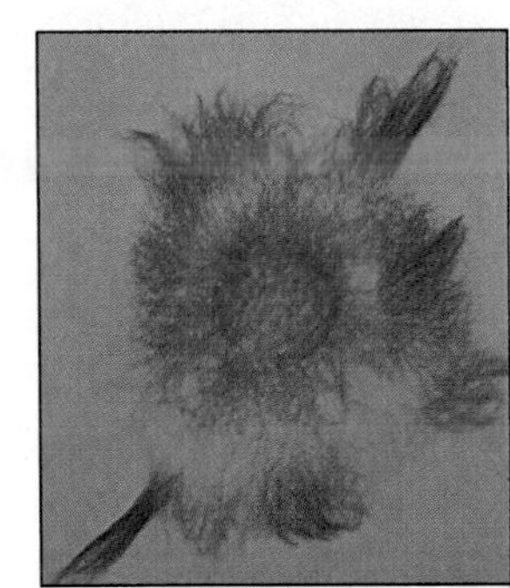

Interphase

Prophase

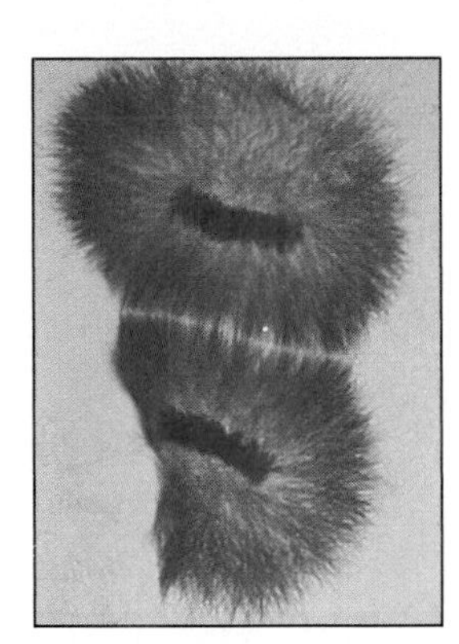

Late telophase

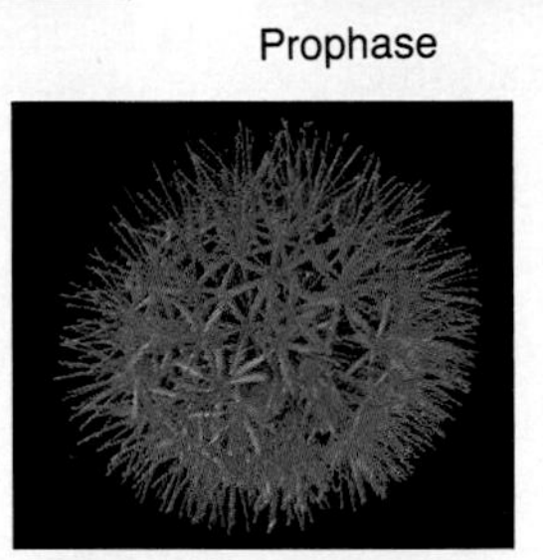

Prometaphase

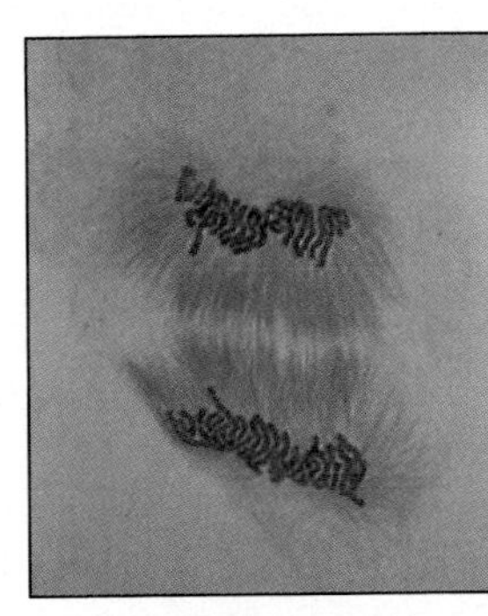

Early telophase

Anaphase

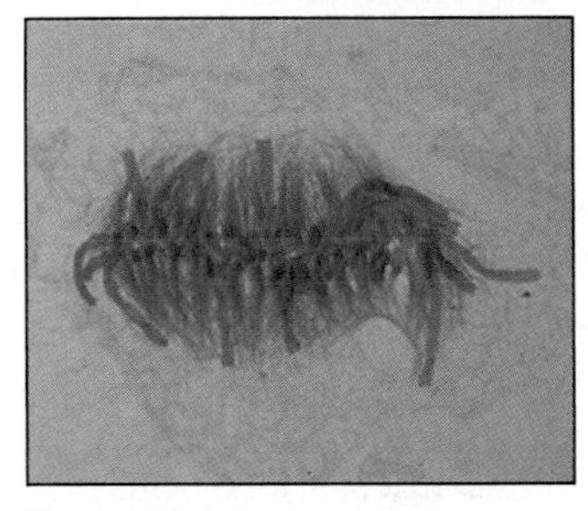

Metaphase

FIGURE 2.7B
Light micrographs illustrating the stages of mitosis depicted in Figure 2.7A. These stages are derived from the flower of *Haemanthus,* shown in the center of the figure.

As the centrioles migrate, the nuclear envelope begins to break down and gradually disappears. In a similar fashion, the nucleolus disintegrates within the nucleus. While these events are taking place, the diffuse chromatin—the characteristic uncoiled form of the genetic material during interphase—condenses until distinct threadlike structures, or chromosomes, are visible. As condensation continues it becomes apparent near the end of prophase that each chromosome is a double structure split longitudinally except at a single point of constriction, the centromere.* The two parts of each chromosome are called **chromatids.** The DNA contained in each pair of chromatids constituting a chromosome was derived from a replication event during the S phase of the previous inter-

*You may sometimes see the term *kinetochore* used synonymously with *centromere.* The kinetochore is actually a granule within the centromere that attaches to spindle fibers during mitosis. Two kinetochores form on opposite faces of the centromere, each attaching one or the other member of a pair of sister chromatids to the spindle fibers.

◄ **FIGURE 2.7A**
Mitosis in an animal cell with a diploid number of 4. The events occurring in each stage are described in the text. Of the two homologous pairs of chromosomes, one contains longer, metacentric members and the other shorter, submetacentric members. The maternal chromosomes and the paternal chromosomes are shown in different colors. The insert (g), showing the telophase stage in a plant cell, illustrates the formation of the cell plate and lack of centrioles.

phase. Thus both members of each chromosome are genetically identical and are called **sister chromatids.** In humans, with a diploid number of 46, a cytological preparation of late prophase will reveal 46 such chromosomal structures (Figure 2.4). In the cell, these are found randomly distributed in the area formerly occupied by the nucleus.

By the completion of prophase, spindle fibers have been laid down between the centrioles, which are now at opposite ends of the cell. The nucleolus and nuclear envelope are no longer visible, and sister chromatids are apparent. Parts (a) and (b) of Figure 2.7A illustrate interphase and prophase.

Prometaphase and Metaphase

The distinguishing event of the ensuing stages is the migration of the centromeric region of each chromosome to the equatorial plane. In some descriptions the term **metaphase** is applied strictly to the chromosome configuration following this movement. In such descriptions, **prometaphase** refers to the period of chromosome movement, as depicted in part (c) of Figure 2.7A. The equatorial plane, also referred to as the **metaphase plate,** is the midline region of the cell, a plane that lies perpendicular to the axis established by the spindle fibers.

Migration is made possible by the binding of one or more spindle fibers to the kinetochore contained within the centromere of each chromosome. At the completion of metaphase, each centromere is aligned at the plate with the chromosome arms extending outward in a random array. This configuration is shown in part (d) of Figure 2.7A.

Anaphase

Events critical to chromosome distribution during mitosis occur during its shortest stage, **anaphase.** It is during this phase that sister chromatids of each double chromosomal structure separate from each other and migrate to opposite ends of the cell. In order for complete separation to occur, each centromeric region must be divided into two. Once this has occurred, each chromatid is now referred to as a **daughter chromosome.** Depending on the location of the centromere along the chromosome, different shapes are assumed during separation (review Figure 2.3).

As the chromosomes begin their migration, the spindle elongates, extending the distance between the two centrosome regions. At the completion of anaphase, the chromosomes have migrated to the opposite poles of the cell. The steps occurring during anaphase are critical in providing each subsequent daughter cell with an identical set of chromosomes. In human cells there would now be 46 chromosomes at each pole, one from each original sister pair. Part (e) of Figure 2.7A illustrates anaphase prior to its completion.

Telophase

Telophase is the final stage of mitosis. At its beginning, there are two complete sets of chromosomes, one at each pole. The most significant event is **cytokinesis,** the division or partitioning of the cytoplasm. Cytokinesis is essential if two new cells are to be produced from one. The mechanism differs greatly in plant and animal cells. In plant cells, a **cell plate** is synthesized and laid down across the region of the metaphase plate. Animal cells, however, undergo a constriction of the cytoplasm in much the same way a loop might be tightened around the middle of a balloon. The end result is the same: Two distinct cells are formed.

It is not surprising that the process of cytokinesis varies among cells of different organisms. Plant cells, which are more regularly shaped and structurally rigid, require a mechanism for the deposition of new cell wall material around the plasma membrane. The cell plate, laid down during telophase, becomes the **middle lamella.** Subsequently, the primary and secondary layers of the cell wall are deposited between the cell membrane and middle lamella on both sides of the boundary between the two daughter cells. In animals, complete constriction of the cell membrane produces the **cell furrow** characteristic of newly divided cells.

Other events necessary for the transition from mitosis to interphase are also initiated during late telophase. They represent a general reversal of those that occurred during prophase. In each new cell, the chromosomes begin to uncoil and become diffuse chromatin once again, while the nuclear envelope reforms around them. The nucleolus gradually reforms and is completely visible in the nucleus during early interphase. The spindle fibers also disappear. Telophase in animal and plant cells is illustrated in part (f) and part (g), respectively, of Figure 2.7A.

MEIOSIS AND SEXUAL REPRODUCTION

The process of meiosis, unlike mitosis, reduces the amount of genetic information. While mitosis produces daughter cells with a full diploid complement, meiosis produces gametes or spores with only one haploid set of chromosomes. During sexual reproduction, gametes then combine in fertilization to reconstitute the diploid complement found in parental cells. The process itself must be very specific, because it is insufficient to produce gametes with a random array of one-half of the total number of chromosomes. Instead, each gamete must receive one member of each homologous pair of chromosomes, ensuring genetic continuity from generation to generation.

The process of sexual reproduction also ensures genetic variety among members of a species. Each offspring receives one complete set of chromosomes from one parent as well as a second complete set from the other parent. For any given gene site on a chromosome, called a **locus,** alternate forms of that gene may exist. These alternate forms are called **alleles** and have arisen from genetic changes or mutations. Sexual reproduction, therefore, reshuffles the combinations of alleles, producing an offspring that is never identical to either parent. This process constitutes one form of genetic recombination within species. A second type of genetic recombination—**crossing over,** or genetic exchange between homologous chromosomes during meiosis—may also occur. Crossing over produces an even greater potential for genetic variability among individuals.

An Overview of Meiosis

We have already established what must be accomplished during meiosis. Before systematically considering the stages of this process, we will briefly describe how diploid cells are converted to haploid gametes or spores. Unlike mitosis, in which each paternally and maternally derived member of any given homologous pair of chromosomes behaves autonomously during division, in meiosis homologous chromosomes pair together, or **synapse.** Each synapsed structure, called a **bivalent,** gives rise to a unit, the **tetrad,** consisting of four chromatids. The presence of four chromatids demonstrates that both chromosomes have duplicated. In order to achieve haploidy, two divisions are necessary. In the first division, described as **reductional,** each tetrad separates and yields two **dyads,** each of which contains two sister chromatids joined at a common centromere. During the second division, described as **equational,** each dyad splits into two **monads** of one chromosome each. Thus, the two divisions may potentially produce four haploid cells.

The First Meiotic Division: Prophase I

As in mitosis, DNA synthesis precedes meiosis, although the products of replication are not visible until the first prophase. The first meiotic prophase is a fairly lengthy one, and it has been further subdivided into five substages: leptonema,* zygonema,* pachynema,* diplonema,* and diakinesis.

Leptonema During the **leptotene stage,** the interphase chromatin material begins to condense, and the chromosomes, although still extended, become visible. Along each chromosome are **chromomeres,** localized condensations that resemble beads on a string.

Zygonema The chromosomes continue to shorten and thicken in the **zygotene stage.** Homologous chromosomes are attracted to each other and begin to undergo zipperlike pairing or synapsis. Pairing is accompanied by the appearance of a unique ultrastructural cell component, the **synaptonemal complex.** As prophase continues, it is found in association with the synapsed homologues. Thus, it is speculated that the function of the synaptonemal complex is associated with chromosome pairing. At the completion of zygonema, the paired structures are sometimes referred to as **bivalents.** Although each member of any pair has already replicated, this is not visually apparent until the following pachytene stage. The number of bivalents is equal to the number of different types (n) of chromosomes.

Pachynema In the **pachytene stage,** coiling and shortening of the synapsed chromosomes continue. While such configurations are not emphasized in Figure 2.8, cytological preparations reveal the chromosomes to overlap and be twisted around one another. During this stage, it first becomes evident that each chromosome is really a double structure, thus providing visual evidence that chromosome replication has already occurred. As in mitosis, the two members of each chromosome are connected by a common centromere and are called **sister chromatids.** Thus, each bivalent now contains four members, (two pairs of sister chromatids) and is referred to as a **tetrad.** The number of tetrads is equivalent to the haploid number of the species.

Diplonema During observation of the ensuing diplotene stage it is readily apparent that each tetrad consists of two pairs of sister chromatids. Within each tetrad, each pair of sister chromatids begins to separate. However, one or more areas remain in contact where chromatids are intertwined. Each such area, called a **chiasma,** is thought to represent a point where nonsister chromatids

---

*These are the noun forms of these stages. The adjective forms (leptotene, zygotene, pachytene, and diplotene) are also used in the text of this chapter.

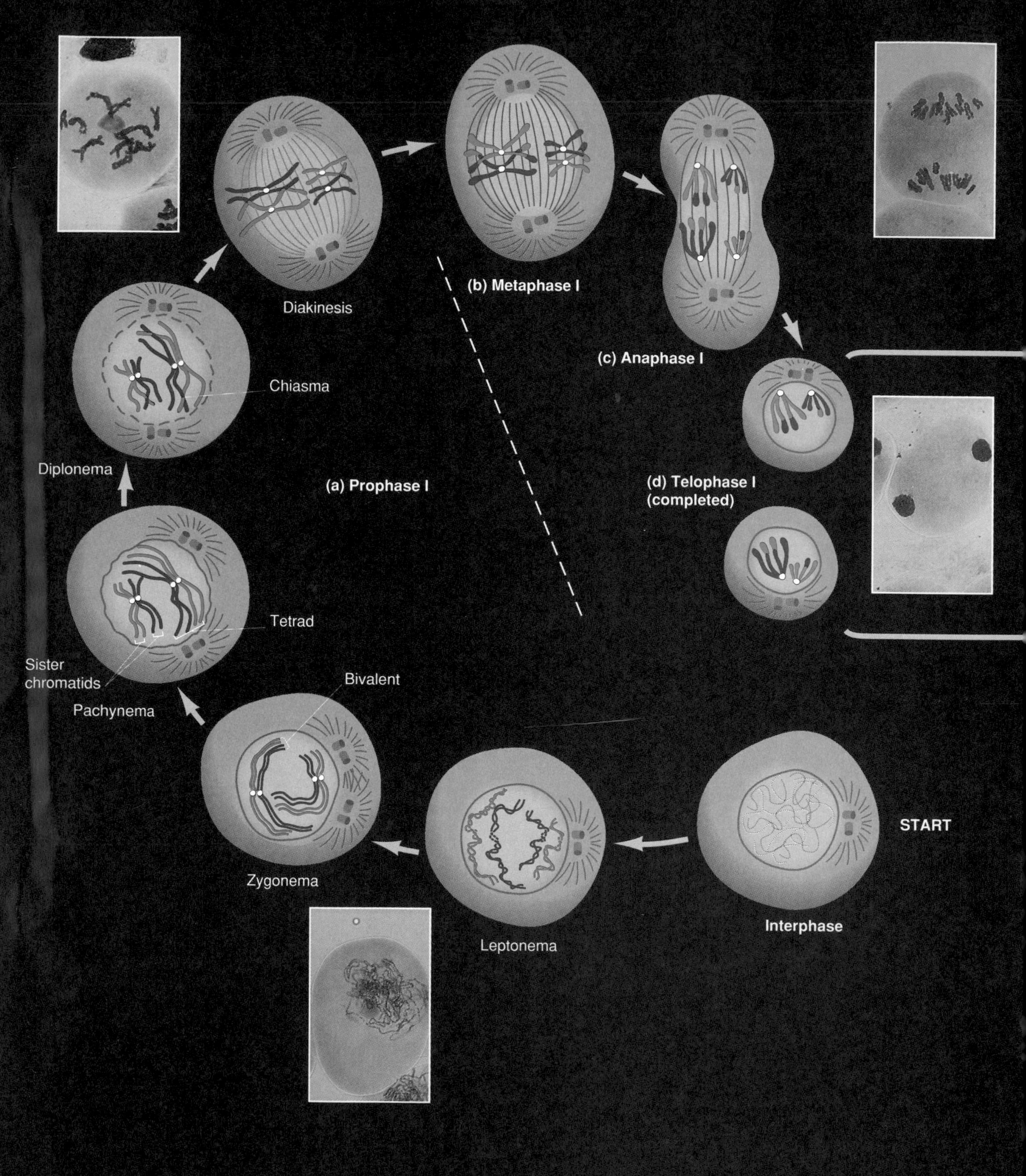
MEIOSIS I
Diakinesis
(b) Metaphase I
(c) Anaphase I
Chiasma
Diplonema
(a) Prophase I
(d) Telophase I
(completed)
Tetrad
Sister
chromatids
Pachynema
Bivalent
START
Zygonema
Interphase
Leptonema

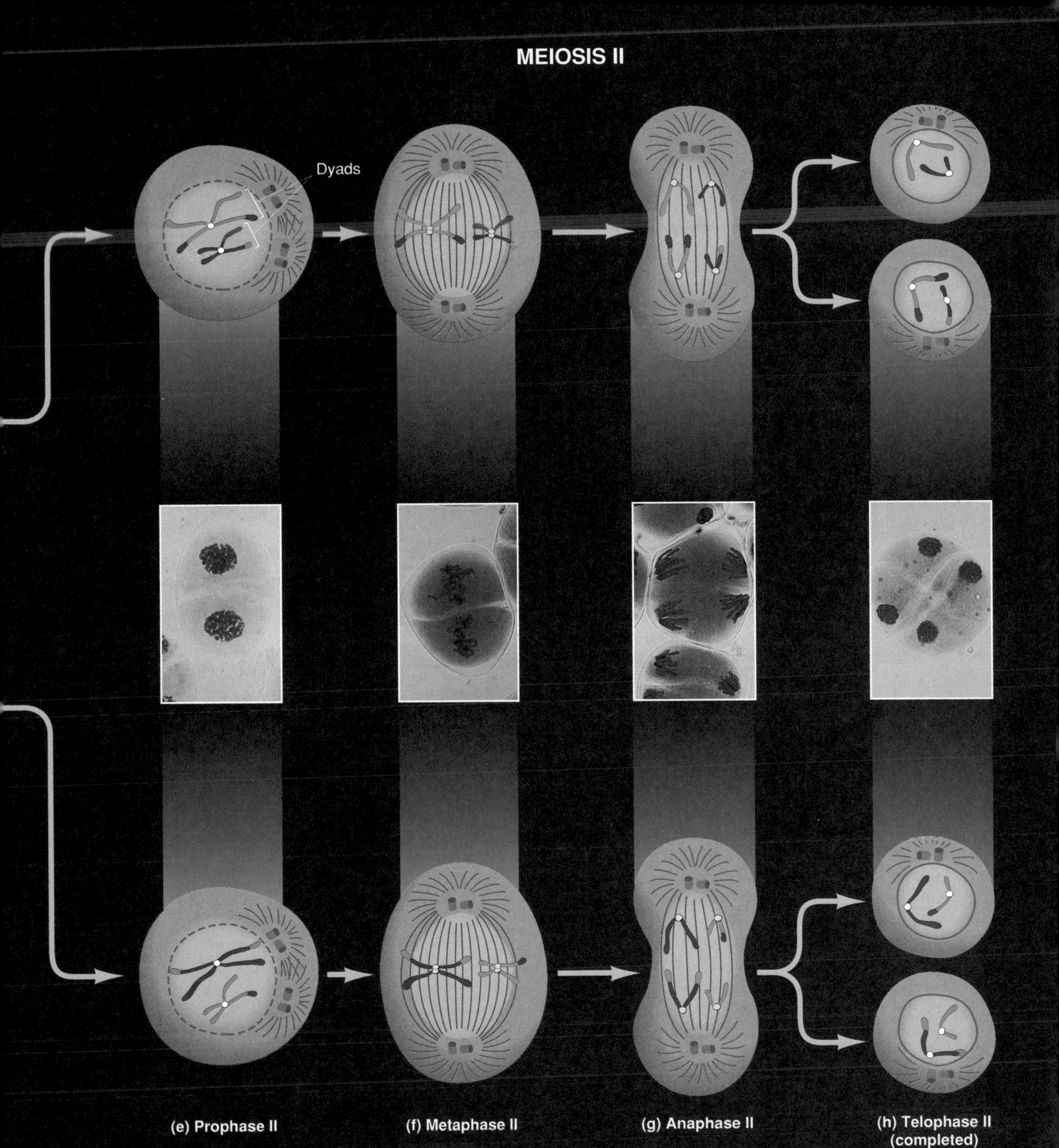

FIGURE 2.8
A diagrammatic representation of the major events occurring during meiosis in a male animal with a diploid number of 4. The same chromosomes described in Figure 2.7 are followed, as described in the legend there. Note that the combination of chromosomes contained in the cells produced following telophase II is dependent on the random alignment of each tetrad and dyad on the equatorial plate during metaphase I and metaphase II. Several other combinations, which are not shown, can be formed. The events depicted in this figure should be correlated with the description in the text. Light micrographs illustrate many of the stages of meiosis depicted in the diagrams.

have undergone **crossing over.** Although the physical exchange between chromosome areas occurs during the previous pachytene stage, the result of crossing over is visible only when the duplicated chromosomes begin to separate.

Diakinesis Further shortening and condensing of the chromatids take place during the final stage of the first meiotic prophase, **diakinesis.** The chromosomes pull farther apart, but sister chromatids on either side of the chiasmata remain loosely associated. As this separation proceeds, the chiasmata move toward the ends of the tetrad. This process, called **terminalization,** begins in late diplonema, and is completed during diakinesis. During this final period of prophase I, the nucleolus and nuclear envelope break down, and the two centromeres of each tetrad become attached to the recently formed spindle fibers.

Metaphase, Anaphase, and Telophase I

Following the first meiotic prophase stage, steps similar to those of mitosis occur. In the **metaphase stage of the first division,** the chromosomes have maximally shortened and thickened. The terminal chiasmata of each tetrad are visible and appear to be the only factor holding the nonsister chromosomes together. Each tetrad interacts with spindle fibers, facilitating movement to the metaphase plate.

During the first division, a single centromere holds each pair of sister chromatids together. It does *not* divide. At the **first anaphase,** one-half of each tetrad (one pair of sister chromatids—called a **dyad**) is pulled toward each pole of the dividing cell. At the completion of anaphase I, there is a series of dyads equal to the haploid number present at each pole.

If no crossing over had occurred in the first meiotic prophase, each dyad at each pole would consist solely of either paternal or maternal chromatids. However, the exchanges produced by crossing over create mosaic chromatids of paternal and maternal origin. The alignment of each tetrad prior to this first anaphase stage is random. One-half of each tetrad will be pulled to one or the other pole at random, and the other half will move to the opposite pole. This random **segregation** of dyads is the basis for the Mendelian principle of **independent assortment,** which we will discuss in Chapter 3. You may wish to return to this discussion when you study this principle.

In many organisms, **telophase of the first meiotic division** reveals a nuclear membrane forming around the dyads. Then, the nucleus enters into a short interphase period. In other cases, the cells go directly from the first anaphase into the second meiotic division. If an interphase period occurs, the chromosomes do not replicate since they already consist of two chromatids. In general, meiotic telophase is much shorter than the corresponding stage in mitosis.

The Second Meiotic Division

A second division of the sister chromatids is essential to achieve haploidy in the meiotic products. During **prophase II,** each dyad is composed of one pair of sister chromatids attached by a common centromere. During **metaphase II,** the centromeres are directed to the equatorial plate. Then, the centromere divides, and during **anaphase II** the sister chromatids of each dyad are pulled to opposite poles. Since the number of dyads is equal to the haploid number, **telophase II** reveals one member of each pair of homologous chromosomes present at each pole. Each chromosome is referred to as a **monad.** Not only has the haploid state been achieved, but, if crossing over has occurred, each monad is a combination of maternal and paternal genetic information. As a result, the offspring produced by any gamete will receive from it a mixture of genetic information originally present in his or her grandparents. Following cytokinesis in telophase II, potentially four haploid gametes may result from a single meiotic event.

SPERMATOGENESIS AND OOGENESIS

Although events that occur during the meiotic divisions are similar in all cells that participate in gametogenesis, there are certain differences between the production of a male gamete (spermatogenesis) and a female gamete (oogenesis) in most animal species.

Spermatogenesis takes place in the testes, the male reproductive organs. The process begins with the expanded growth of an undifferentiated diploid germ cell called a **spermatogonium.** This cell enlarges to become a **primary spermatocyte,** which undergoes the first meiotic division. The products of this division are called **secondary spermatocytes.** Each secondary spermatocyte contains a haploid number of dyads. The secondary spermatocytes then undergo the second meiotic division, and each of these cells produces two haploid **spermatids.** Spermatids go through a series of developmental changes, **spermiogenesis,** and become highly specialized, motile **spermatozoa** or **sperm.** All sperm cells produced during spermatogenesis receive equal amounts of genetic material and cytoplasm. Figure 2.9 summarizes these steps.

Spermatogenesis may be continuous or occur periodically in mature male animals, with its onset determined by the nature of the species' reproductive cycle. Animals that reproduce year-round produce sperm continuously, while those whose breeding period is confined to a particular season produce sperm only during that time.

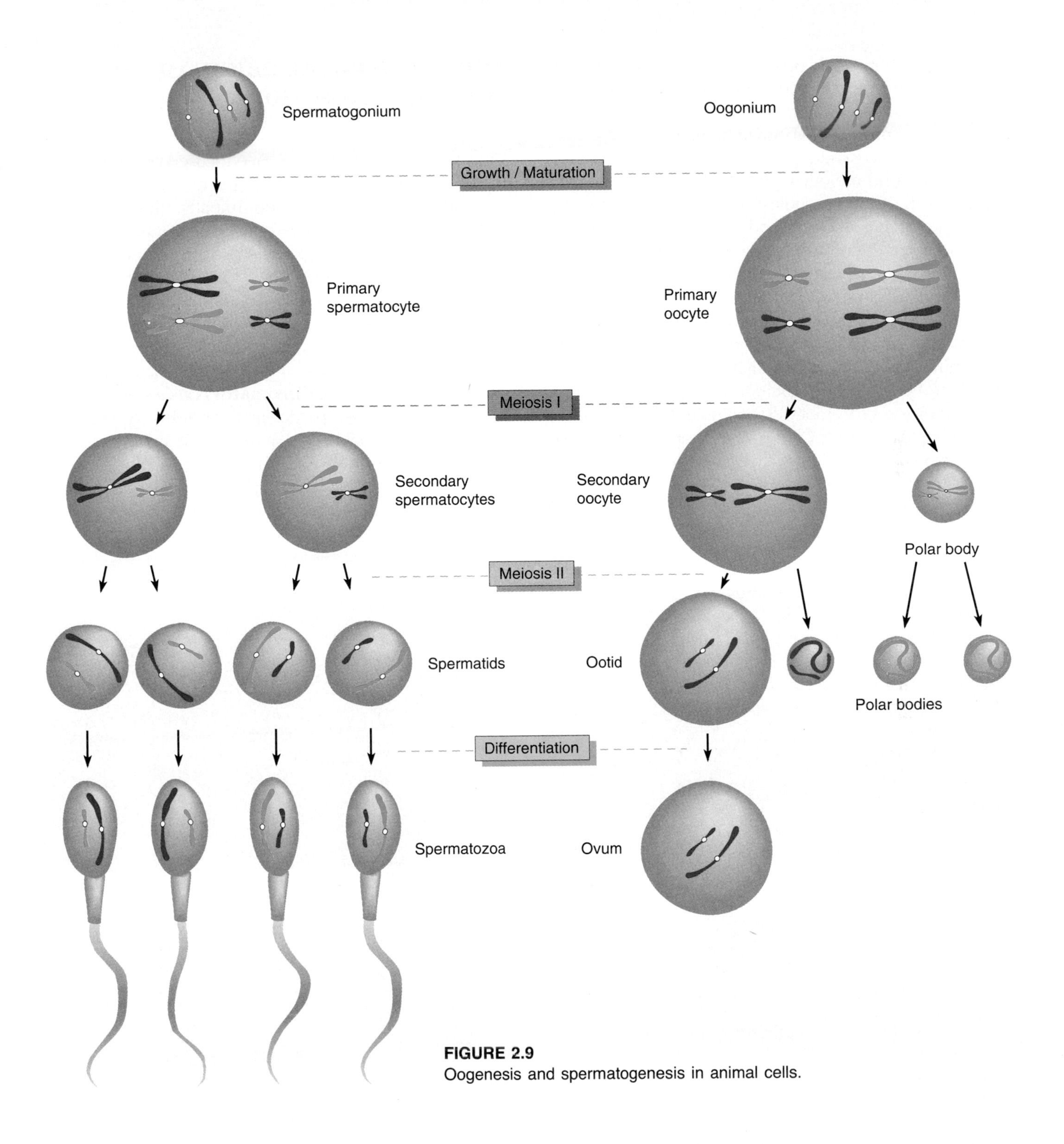

FIGURE 2.9
Oogenesis and spermatogenesis in animal cells.

In animal **oogenesis,** the formation of **ova** (singular: **ovum**), or eggs, occurs in the ovaries, the female reproductive organs. The daughter cells resulting from the two meiotic divisions receive equal amounts of genetic material, but they do *not* receive equal amounts of cytoplasm. Instead, during each division, almost all the cytoplasm of the **primary oocyte,** itself derived from the **oogonium,** is concentrated in one of the two daughter cells. The concentration of cytoplasm is necessary because the function of the mature ovum is to nourish the developing embryo following fertilization.

During the first meiotic anaphase in oogenesis, the tetrads of the primary oocyte separate, and the dyads move toward opposite poles. During the first telophase, the dyads present at one pole are pinched off with very little surrounding cytoplasm to form the **first polar body.**

The other daughter cell produced by this first meiotic division contains most of the cytoplasm and is called the **secondary oocyte.** The first polar body may or may not divide again to produce two small haploid cells. The mature ovum will be produced from the secondary oocyte during the second meiotic division. During this division, the cytoplasm of the secondary oocyte again divides unequally, producing an **ootid** and a **second polar body.** The ootid then differentiates into the mature ovum. Figure 2.9 illustrates the steps leading to formation of the mature ovum and polar bodies.

Unlike the divisions of spermatogenesis, the two meiotic divisions of oogenesis may not be continuous. In some animal species the two divisions may directly follow each other. In others, including the human species, the first division of all oocytes begins in the embryonic ovary, but arrests in prophase I. Many years later, the first division is reinitiated in each oocyte upon its ovulation. The second division is completed only after fertilization, if it occurs.

THE CYTOLOGICAL ORIGIN OF THE MITOTIC CHROMOSOME

Thus far in this chapter, we have focused on mitotic and meiotic chromosomes, emphasizing behavior during cell division and gamete formation. Initially, biologists knew about these chromosomes only from routine observations made with the light microscope. While chromosomes are invisible during interphase, they appear during the prophase stage of mitosis. Geneticists were curious as to how this could happen. We are now quite clear as to why chromosomes are visible during mitosis.

During interphase, only dispersed chromatin fibers are present in the nucleus. **Chromatin** consists of DNA and associated proteins, particularly proteins called **histones.** It is now believed that during interphase, starting in G_1, mitotic chromosomes unwind to form these long chromatin fibers. It is in this physical arrangement that DNA can function during transcription and can be repli-

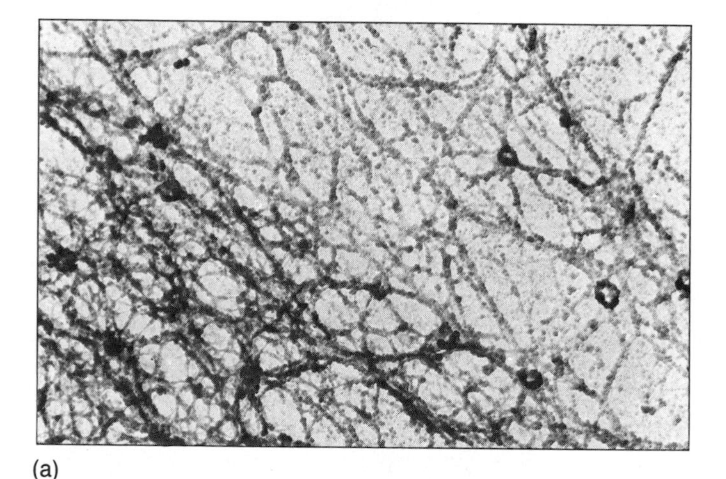

(a)

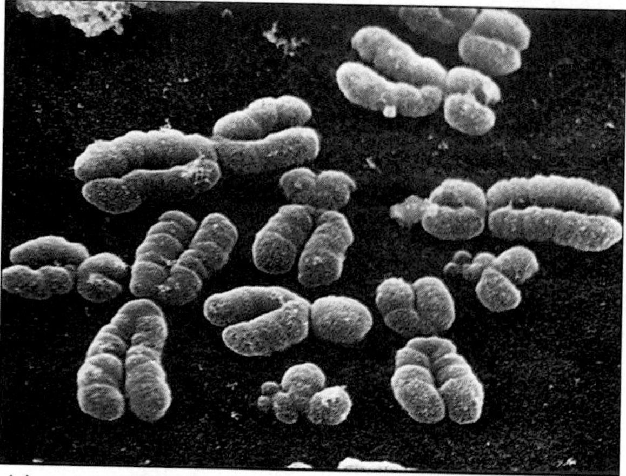

(c)

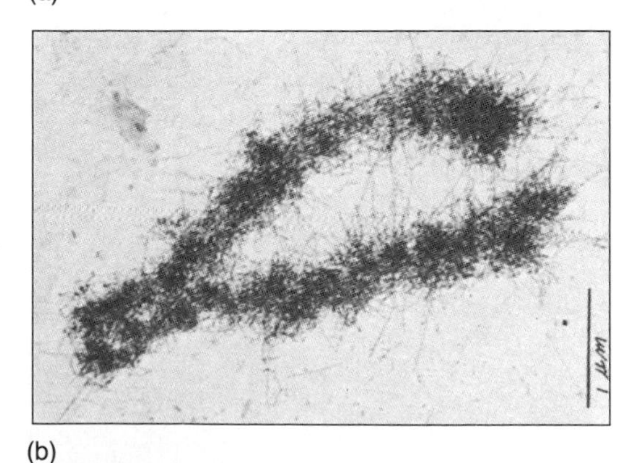

(b)

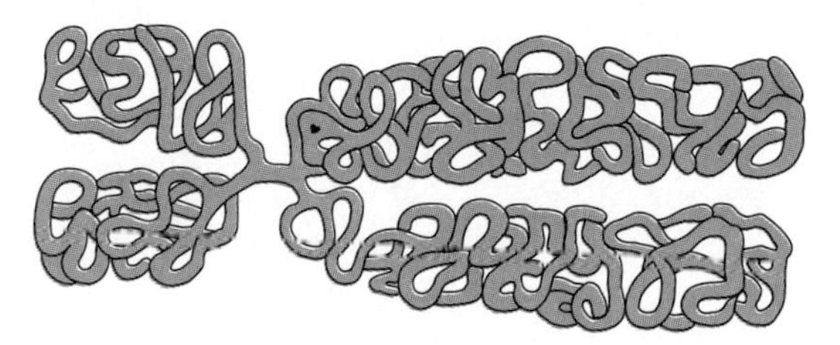

(d)

FIGURE 2.10
A comparison of the chromatin fibers (a) characteristic of the interphase nucleus with metaphase chromosomes (b) and (c) that are derived from chromatin during mitosis. Part (d) depicts the folded fiber model showing how chromatin is condensed into a metaphase chromosome. Parts (a) and (b) are transmission electron micrographs, while Part (c) is a scanning electron micrograph.

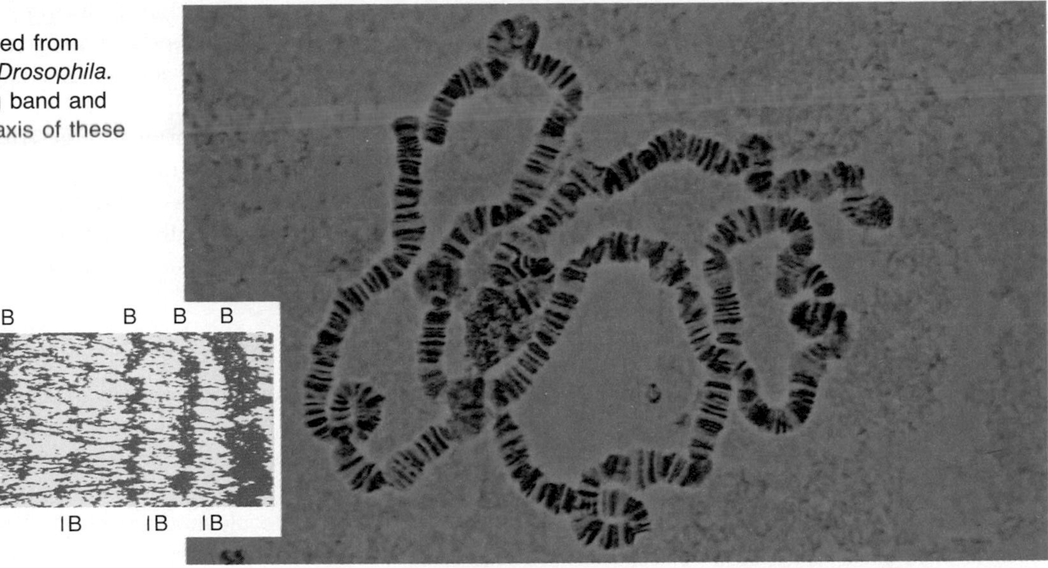

FIGURE 2.11
Polytene chromosomes derived from larval salivary gland cells of *Drosophila*. The insert depicts alternating band and interband regions along the axis of these giant chromosomes.

cated. Once mitosis begins, however, the fibers begin to coil and fold up, condensing into typical mitotic chromosomes.

Electron microscopic observations of such chromosomes support this interpretation, which is described as the **folded fiber model.** In Figure 2.10, several informative electron micrographs are presented, including one of chromatin fibers, as well as both a transmission and scanning electron micrograph of mitotic chromosomes. An interpretive drawing depicts the modern interpretation of the mitotic chromosome; that it is derived from condensed chromatin is widely accepted.

During the transition from interphase to prophase, it is estimated that a 5000-fold contraction occurs in the length of DNA within the chromatin fiber! This process must indeed be extremely precise, given the highly ordered nature and consistent appearance of mitotic chromosomes in all eukaryotes. Note particularly in the micrographs the clear distinction between the sister chromatids constituting each chromosome. They are joined only by the common centromere that they share prior to anaphase. In a later chapter, after we have provided a more thorough description of DNA structure, we will return to this topic and explore the molecular basis of the chromatin fiber.

SPECIALIZED CHROMOSOMES

We conclude this chapter by introducing two other unique types of chromosomes visible under the light microscope. It is useful to do so here in order to demonstrate specialized forms chromosomes may achieve besides their mitotic-like structures. Both types were studied extensively long before we understood the rationale for the appearance of chromosomes during mitosis.

Polytene Chromosomes

Cells from a variety of organisms contain giant **polytene chromosomes.** They are found in various insect dipteran larval cells (salivary, midgut, rectal, and malpighian excretory tubules) and in several species of protozoans and plants. The large amount of information obtained from studies of these genetic structures has provided a model system for more recent investigations of chromosomes. Such structures were first observed by E. G. Balbiani in 1881 and are illustrated in Figure 2.11. What is particularly intriguing about these chromosomes is that they can be seen in the nuclei of interphase cells!

Each polytene chromosome observed under the light microscope is seen to be composed of a linear series of alternating bands and interbands (see insert in Figure 2.11). The banding pattern is distinctive for each chromosome in any given species. Individual bands are sometimes called **chromomeres,** a more generalized term describing lateral condensations of material along the axis of a chromosome. Each polytene chromosome is 200–600 μm long.

Following extensive study, it is now clear how such giant chromosomes are formed. Their large size and distinctiveness result from the many DNA strands that compose them. Many replication cycles of paired homologues occur, but without strand separation or cytoplasmic division. Thus, as replication proceeds, chromosomes are created having 1000–5000 DNA strands that

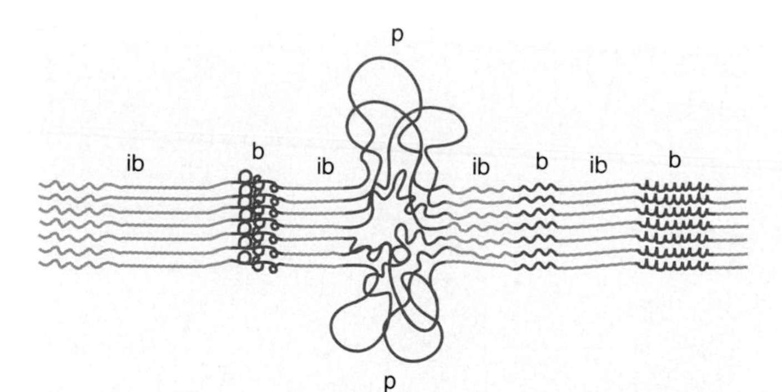

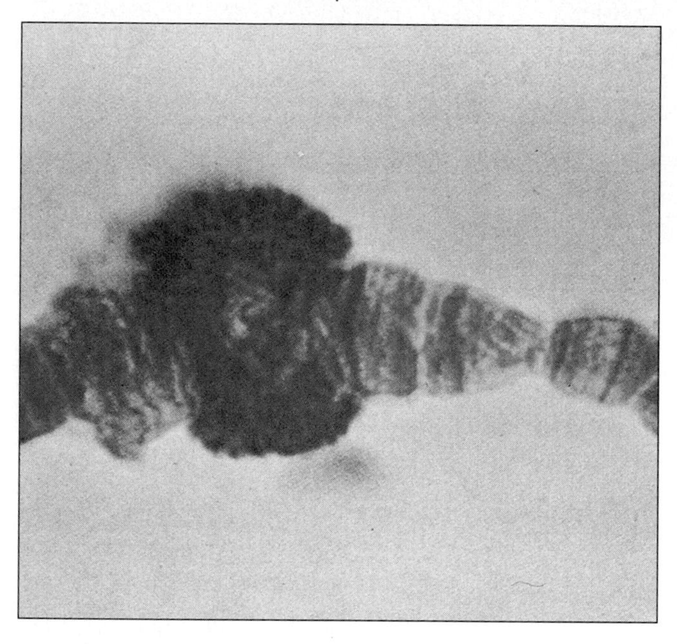

FIGURE 2.12
Photograph of a puff within a polytene chromosome. The schematic representation depicts the uncoiling of strands within a band (b) region to produce a puff (p) in polytene chromosomes. Interband regions (ib) are also labeled.

remain in parallel register with one another. It is apparently the parallel register of so many DNA strands that gives rise to the distinctive band pattern along the axis of the chromosome.

It is obvious that these chromosomes bear genes that store genetic information. The presence of bands was initially interpreted as the visible manifestation of individual genes. When it became clear that the strands present in bands undergo localized uncoiling during genetic activity, this view was further strengthened. Each such uncoiling event results in what is called a **puff** because of its appearance (see Figure 2.12). That puffs are visible manifestations of gene activity (transcription) is evidenced by their high rate of incorporation of radioactivity labeled RNA precursors, as assayed by autoradiography. Bands that are not extended into puffs incorporate much less radioactive precursor or none at all.

The study of bands during development in insects such as *Drosophila* and the midge fly *Chironomus,* reveals differential gene activity. A characteristic pattern of band formation that is equated with gene activation is observed as development proceeds. This topic is pursued in more detail in Chapter 18. In spite of many recent genetic investigations, it is still not clear whether only a single gene is contained within each band. There is much more DNA present per band than is needed to encode a single gene!

Lampbrush Chromosomes

Another type of specialized chromosome that has provided insights into chromosomal structure is the **lampbrush chromosome.** It was given this name because it was imagined to be similar in appearance to the brushes used to clean lamp chimneys in centuries past. Lampbrush chromosomes were first discovered in 1892 in the oocytes of sharks and are now known to be characteristic of most vertebrate oocytes as well as spermatocytes of some insects. Therefore, they are meiotic chromosomes. Most experimental work has been done with material taken from amphibian oocytes.

These unique chromosomes are easily isolated from oocytes in the diplotene stage of the first prophase of meiosis, where they are active in directing the metabolic activities of the developing cell. The homologues are seen as synapsed pairs held together by chiasmata, but instead of condensing, as do most meiotic chromosomes, the lampbrush chromosomes are often extended to lengths of 500–800 μm. Later in meiosis they revert to their normal length of 15–20 μm. Thus, lampbrush chromosomes are interpreted as uncoiled and unfolded versions of the normal meiotic chromosomes.

Figure 2.13 shows several views of these structures. Collectively, they provide significant insights into the morphology of these chromosomes. In Part (a), the meiotic configuration is visualized under the light microscope. The linear axis of each structure contains large numbers of repeating condensations. As in polytene chromosomes, the more general term, chromomere, describes each condensation. Each chromomere appears to support a pair of **lateral loops,** which give the chromosome its distinctive appearance. In Part (b), the scanning electron microscope (SEM) reveals an extended axis and adjacent loops. As with bands in polytene chromosomes, there is much more DNA present in each loop than is needed to encode a single gene.

Part (c) of Figure 2.13 presents another SEM, but at a much higher magnification. This micrograph provides a detailed view of one chromomere and the loop emanating from it. Each loop is thought to be composed of one DNA double helix, while the central axis is made up of two DNA helices. This hypothesis is consistent with the

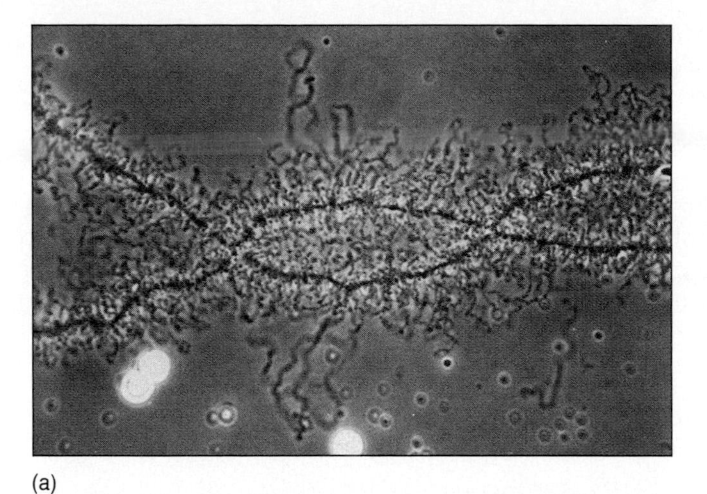
(a)

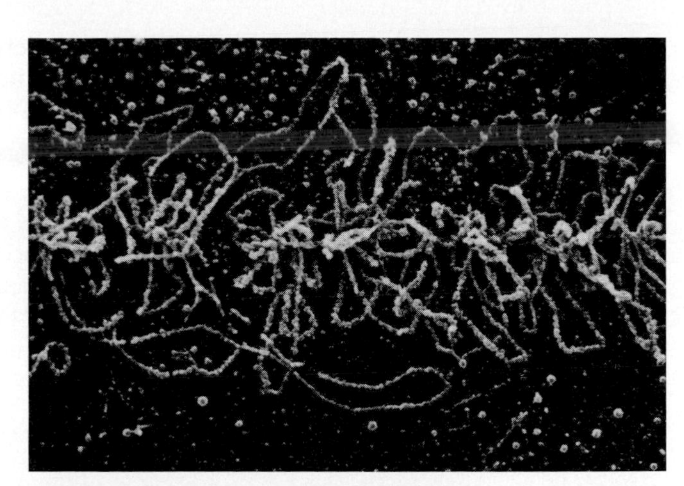
(b)

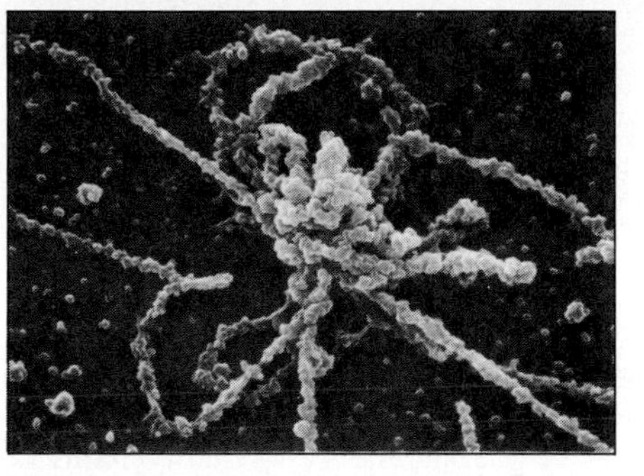
(c)

FIGURE 2.13
Lampbrush chromosomes derived from amphibian oocytes. Part (a) is a photomicrograph, while parts (b) and (c) are scanning electron micrographs.

belief that each chromosome is composed of a pair of sister chromatids. Studies using radioactive RNA precursors reveal that the loops are active in the synthesis of RNA. The lampbrush loops, in a way similar to puffs in polytene chromosomes, represent DNA that has been reeled out from the central chromomere axis during transcription. As with polytene chromosomes, the study of lampbrush chromosomes has provided many insights into the arrangement and function of the genetic information.

CHAPTER SUMMARY

1 Chromosomes, in diploid organisms, exist in homologous pairs. Each pair shares the same size, centromere placement, and gene sites. One member of each pair is derived from the maternal parent and one from the paternal parent.

2 Mitosis and meiosis are mechanisms by which cells distribute genetic information contained in these chromosomes to their descendants in a precise, orderly fashion.

3 Mitosis, or nuclear division, is part of the cell cycle and leads to cellular reproduction. Daughter cells are produced that are genetically identical to their progenitor cell.

4 Meiosis, the underlying basis of sexual reproduction, results in the conversion of a diploid cell to a haploid gamete or spore. Each haploid cell contains one member of each homologous pair of chromosomes.

5 Mitosis is subdivided into discrete phases: prophase, prometaphase, metaphase, anaphase, and telophase. Condensation of chromatin into chromosome structures occurs during prophase. During prometaphase, chromosomes take on the appearance of double structures, each represented by a pair of sister chromatids. In metaphase, chromosomes line up on the equatorial plane of the cell. During anaphase, sister chromatids of each chromosome are pulled apart and directed toward opposite poles. Telophase completes daughter cell formation and is characterized by cytokinesis, the division of the cytoplasm.

6 During cytokinesis in animal cells, the cytoplasm is pinched into two cells. In plants, a cell plate is synthesized and laid down, dividing the original cell into two daughter cells.

7 Meiosis results in extensive genetic variation by virtue of the exchange between homologous chromosomes during crossing over and the random segregation of maternal and paternal chromatids.

8 A major difference exists between meiosis in males and females. Spermatogenesis partitions cytoplasmic volume equally and produces four haploid sperm cells. Oogenesis, on the other hand, accumulates the cytoplasm around one egg cell and reduces the other haploid sets of genetic material to polar bodies. The extra cytoplasm contributes to zygote development following fertilization.

9 Mitotic chromosomes are produced as a result of the coiling and condensation of chromatin fibers characteristic of interphase.

10 Polytene and lampbrush chromosomes are examples of specialized structures that have extended our knowledge of genetic organization and function.

KEY TERMS

acrocentric
anaphase
antigen
autoradiography
bivalent
cdc mutations
cell cycle
centriole
centromere
centrosome
chiasma (chiasmata)
chromatid
chromatin
chromomere
crossing over
cytokinesis
cytoplasmic inheritance
diakinesis
diploid
diplotene
dyad
endoplasmic reticulum (ER)
endosymbiont theory
gamete
genome
haploid
homologous chromosome
independent assortment
interphase
karyotype
kinetochore
lampbrush chromosome
leptotene
metacentric
metaphase
metaphase plate
monad
monosomic
nondisjunction
oncogene
oocyte
oogenesis
pachytene
photosynthesis
polar body
polytene chromosome

prokaryote
proto-oncogenes
R point
segregation
sex chromosome
sister chromatid exchange (SCE)
spermatogenesis
S phase
spindle fibers
submetacentric
synapsis
synaptonemal complex
telocentric
telophase
tetrad
trisomic
zygote
zygotene

With this initial appearance of "Insights and Solutions," it is appropriate to provide a brief description of its value to you as a student. This section will precede the "Problems and Discussion Questions" in each chapter. One or more examples will be provided. Solutions to these problems and answers to these questions will illustrate approaches useful in **genetic analysis.** Our initial emphasis will be on insights that will help you arrive at correct solutions to ensuing problems.

1 In an organism with a haploid number of 3, how many individual chromosomal structures will align on the metaphase plate during (a) mitosis; (b) meiosis I; and (c) meiosis II? Describe each configuration.

ANSWER: (a) In mitosis, where homologous chromosomes do not synapse, there will be 6 double structures, each consisting of a pair of sister chromatids. The number of structures is equivalent to the diploid number.
(b) In meiosis I, the homologues have synapsed, reducing the number of structures to 3. Each is called a tetrad and consists of two pairs of sister chromatids.
(c) In meiosis II, the same number of structures exist (3), but in this case, they are called dyads. Each consists of a pair of sister chromatids. When crossing over has occurred, each chromatid may contain part of one of its nonsister chromatids obtained during exchange in prophase I.

2 For the chromosomes illustrated in Figure 2.8, draw all possible alignment configurations that may occur during metaphase of meiosis I. How many different configurations can occur with three pairs of chromosomes ($n = 3$)?

ANSWER: As shown in the illustration below, there are four configurations possible when $n = 2$. If $n = 3$, then 8 different configurations would be possible. The formula 2^n, where n equals the haploid number, will allow you to calculate the number of potential alignment patterns. As we will see in the next chapter, these patterns serve as the physical basis of the Mendelian postulate of independent assortment.

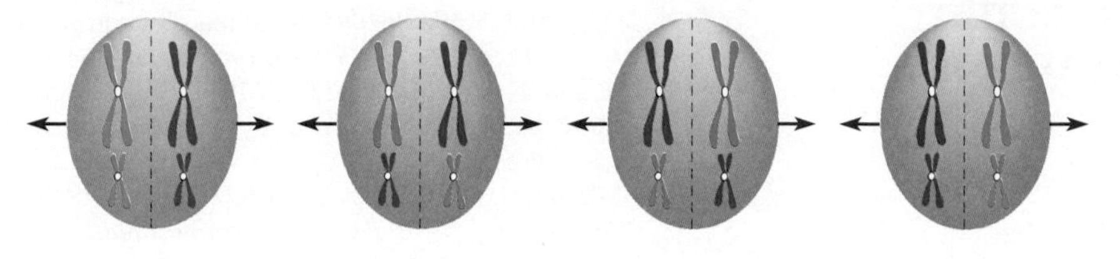

PROBLEMS AND DISCUSSION QUESTIONS

1 What role do the following cellular components play in the storage, expression, or transmission of genetic information: (a) chromatin, (b) nucleolus, (c) ribosome, (d) mitochondrion, (e) centriole, (f) centromere?

2 Discuss the concepts of homologous chromosomes, diploidy, and haploidy. What characteristics are shared between two chromosomes considered to be homologous?

3 If two chromosomes of a species are the same length and have similar centromere placements yet are *not* homologous, what *is* different about them?

4 Describe the events that characterize each stage of mitosis.

5 If an organism has a diploid number of 16, how many chromatids are visible at the end of mitotic prophase? How many chromosomes are moving to each pole during anaphase of mitosis?

6 How are chromosomes generally named on the basis of centromere placement?

7 Contrast telophase in plant and animal mitosis.

8 Outline and discuss the events of the cell cycle.

9 Contrast the end results of meiosis with those of mitosis.

10 Define and discuss the following terms: (a) synapsis, (b) bivalents, (c) chiasmata, (d) crossing over, (e) chromomeres, (f) sister chromatids, (g) tetrads, (h) dyads, (i) monads, and (j) synaptonemal complex.

11 If an organism has a diploid number of 16 in an oocyte,
(a) How many tetrads are present in the first meiotic prophase?
(b) How many dyads are present in the second meiotic prophase?
(c) How many monads migrate to each pole during the second meiotic anaphase?

12 Contrast spermatogenesis and oogenesis. What is the significance of the formation of polar bodies?

13 Assume that a diploid cell contains three pairs of homologous chromosomes designated C^1C^2, M^1M^2, and S^1S^2 and that no crossing over occurs. What possible configurations of chromosomes will be present in: (a) two daughter cells following mitosis?; (b) the first meiotic metaphase?; and (c) haploid cells following meiosis?

14 Contrast the chromatin fiber with the mitotic chromosome. How are the two structures related?

15 Describe and compare polytene and lampbrush chromosomes.

SELECTED READINGS

ALBERTS, B., et al. 1989. *Molecular biology of the cell.* 2nd. ed. New York: Garland.
ANGELIER, N., et al. 1984. Scanning electron microscopy of amphibian lampbrush chromosomes. *Chromosoma* 89:243–53.
BEERMAN, W., and CLEVER, U. 1964. Chromosome puffs. *Scient. Amer.* (April) 210:50–58.
BRACHET, J., and MIRSKY, A. E. 1961. *The cell: Meiosis and mitosis.* Vol. 3 Orlando: Academic Press.
CALLAN, H. G. 1986. *Lampbrush chromosomes.* New York: Springer-Verlag.
DuPRAW, E. J. 1970. *DNA and chromosomes.* New York: Holt, Rinehart and Winston.
GALL, J. G. 1963. Kinetics of deoxyribonuclease on chromosomes. *Nature* 198:36–38.

GOLOMB, H. M., and BAHR, G. F. 1971. Scanning electron microscopic observations of surface structures of isolated human chromosomes. *Science* 171:1024–26.

HILL, R. J., and RUDKIN, G. T. 1987. Polytene chromosomes: The status of the band-interband question. *BioEssays.* 7:35–40.

MAZIA, D. 1961. How cell divide. *Scient. Amer.* (Jan.) 205:101–20.

———. 1974. The cell cycle. *Scient. Amer.* (Jan.) 235:54–64.

McINTOSH, J. R., and McDONALD, K. L. 1989. The mitotic spindle. *Scient. Amer.* (Oct.) 261:48–56.

PRESCOTT, D. M., and FLEXER, A. S. 1986. *Cancer, the misguided cell.* 2nd ed. Sunderland, MA: Sinauer.

SWANSON, C. P., MERZ, T., and YOUNG, W. J. 1981. *Cytogenetics, the chromosome in division, inheritance, and evolution.* 2nd ed. Englewood Cliffs, NJ: Prentice-Hall.

WESTERGAARD, M., and vonWETTSTEIN, D. 1972. The synaptinemal complex. *Ann Rev. Genet.* 6:71–110.

WHEATLEY, D. N. 1982. *The centriole: A central enigma of cell biology.* New York: Elsevier/North-Holland Biomedical.

YUNIS, J. J., and CHANDLER, M. E. 1979. Cytogenetics. In *Clinical diagnosis and management by laboratory methods,* ed. J. B. Henry, Vol. 1. Philadelphia: W. B. Saunders.

ZIMMERMAN, A. M., and FORER, A., eds., 1981. *Mitosis/Cytokinesis.* New York: Academic Press.

3 Mendelian Genetics

Chapter Concepts

Inherited characteristics are the result of particulate factors called genes that are transmitted from generation to generation on vehicles called chromosomes, according to rules first described by Gregor Mendel.

Although inheritance of biological traits has been recognized for thousands of years, the first significant insights into the mechanisms involved occurred only a little over a century ago. In 1866, Gregor Mendel published the results of a series of experiments that would lay the foundation for the formal discipline of genetics. In the ensuing years the concept of the gene as a distinct hereditary unit was established, and the ways in which genes are transmitted to offspring and control traits were clarified. Research in these areas was accelerated in the first half of the twentieth century, generating the interest so important to the acquisition of the knowledge derived in this field since about 1950. It is safe to say that studies in genetics, particularly at the molecular level, have remained continually at the forefront of biological research since that time.

In this chapter we focus on the development of the principles established by Mendel, now referred to as **Mendelian** or **transmission genetics.** These principles describe how genes are transmitted from parents to offspring and were derived directly from his experimentation.

When Mendel began his studies of inheritance using *Pisum sativum,* the garden pea, there was no knowledge of chromosomes nor the role and mechanism of meiosis. Nevertheless, he was able to determine that distinct **units of inheritance** exist and to predict their behavior during the formation of gametes. Subsequent investigators, with access to cytological data, were able to relate their observations of chromosome behavior during meiosis to Mendel's principles of inheritance. Once this correlation had been made, Mendel's postulates were accepted as the basis for the study of transmission genetics.

Even today, they serve as the cornerstone of the study of inheritance.

GREGOR MENDEL

In 1822, Johann Mendel was born of a peasant family in the European village of Heinzendorf, now part of Czechoslovakia. An excellent student in high school, Mendel studied philosophy for several years afterward and was admitted to the Augustinian Monastery of St. Thomas in Brno in 1843. There, he took the name of Gregor and received support for his studies and research throughout the rest of his life. In 1849, he was relieved of pastoral duties and received a teaching appointment that lasted several years. From 1851 to 1853 he attended the University of Vienna, where he studied physics and botany. In 1854 he returned to Brno, where, for the next sixteen years, he taught physics and natural science.

In 1856 Mendel performed the first set of hybridization experiments with the garden pea. The research phase of his career lasted until 1868, when he was elected abbot of the monastery. Although his interest in genetics remained, his new responsibilities demanded most of his time. In 1884 Mendel died of a kidney disorder. The local newspaper paid him the following tribute: "His death deprives the poor of a benefactor, and mankind at large of a man of the noblest character, one who was a warm friend, a promoter of the natural sciences, and an exemplary priest. . . ."

Mendel's Experimental Approach

In 1865, Mendel first reported the results of some simple genetic crosses between certain strains of the garden pea. Although, as we saw in Chapter 1, his was not the first attempt to provide experimental evidence pertaining to inheritance, Mendel's work is an elegant model of experimental design and analysis.

Mendel showed remarkable insight into the methodology necessary for good experimental biology. He chose an organism that was easy to grow and interbreed. The pea plant is self-fertilizing in nature, but is easily crossbred in designed experiments. It reproduces well and grows to maturity in a single season. Mendel worked with seven unit characters, visible features that were each represented by two contrasting forms or traits. For the character stem height, for example, he experimented with the traits *tall* and *dwarf.* He selected six other contrasting pairs of traits involving seed shape and color, pod shape and color, and pod and flower arrangement. True-breeding strains were available from local seed merchants. Each trait appeared generation after generation in self-fertilizing plants; that is, the strains exhibiting them "bred true."

Mendel's success in an area where others had failed may be attributed to several factors in addition to the choice of a suitable organism. He restricted his examination to one or very few pairs of contrasting traits in each experiment. He also kept accurate quantitative records, a necessity in genetic experiments. From the analysis of his data, Mendel derived certain postulates that have become the principles of transmission genetics.

The significance of Mendel's experiments was not realized until the early twentieth century, well after his death. Once Mendel's publications were rediscovered by geneticists investigating the function and behavior of chromosomes, the implications of his postulates were immediately apparent. He had discovered the basis for the transmission of hereditary traits!

THE MONOHYBRID CROSS

The simplest crosses performed by Mendel involved only one pair of contrasting traits. Each such breeding experiment is called a **monohybrid cross.** A monohybrid cross is made by mating individuals from two parent strains, each of which exhibits one of the two contrasting forms of the character under study. Initially we will examine the first generation of offspring of such a cross, and then we will consider the offspring of **selfing** or **self-fertilizing** individuals from this first generation. The original parents are called the $\mathbf{P_1}$ or **parental generation,** their offspring are the $\mathbf{F_1}$ or **first filial generation,** and the individuals resulting from the selfing of the F_1 generation are called the $\mathbf{F_2}$ or **second filial generation.** We can, of course, continue to follow subsequent generations, if desirable.

The cross between peas with tall stems and dwarf stems is representative of Mendel's monohybrid crosses. *Tall* and *dwarf* represent contrasting forms or traits of the character of stem height. Unless tall or dwarf plants are crossed together or with another strain, they will undergo self-fertilization and breed true, producing their respective trait generation after generation. However, when Mendel crossed tall plants with dwarf plants, the resulting F_1 generation consisted only of tall plants. When members of the F_1 generation were selfed, Mendel observed that 787 of 1064 F_2 plants were tall, while 277 of 1064 were dwarf. Note that in this cross (Figure 3.1) the dwarf trait disappears in the F_1, only to reappear in the F_2 generation.

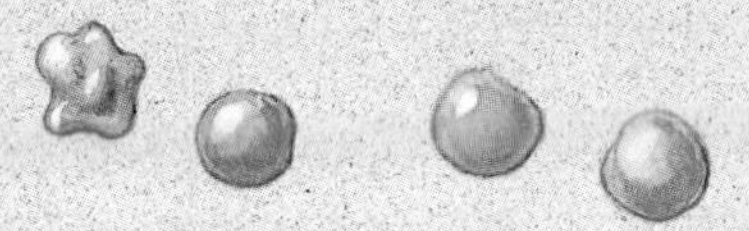

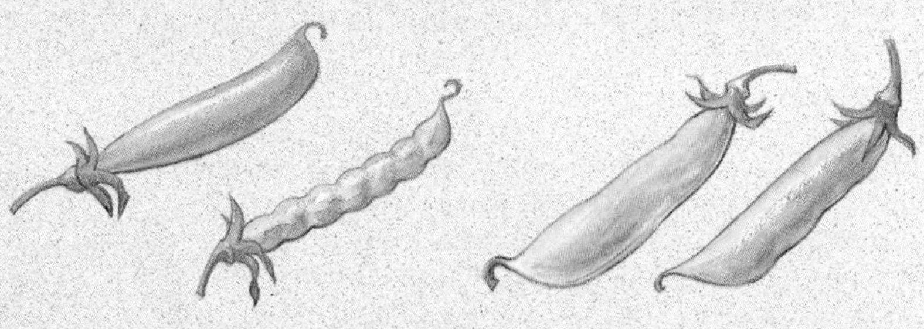

Character	Contrasting traits	F_1 Results	F_2 Results	F_2 Ratio
Seeds	Round/wrinkled	Round	5474 round 1850 wrinkled	2.96:1
	Yellow/green	Yellow	6022 yellow 2001 green	3.01:1
Pods	Axial/terminal	Axial	651 axial 207 terminal	3.14:1
	Full/constricted	Full	882 full 299 constricted	2.95:1
	Yellow/green	Green	428 green 152 yellow	2.82:1
Flowers	Violet/white	Violet	705 violet 224 white	3.15:1
Stem	Tall/dwarf	Tall	787 tall 277 dwarf	2.84:1

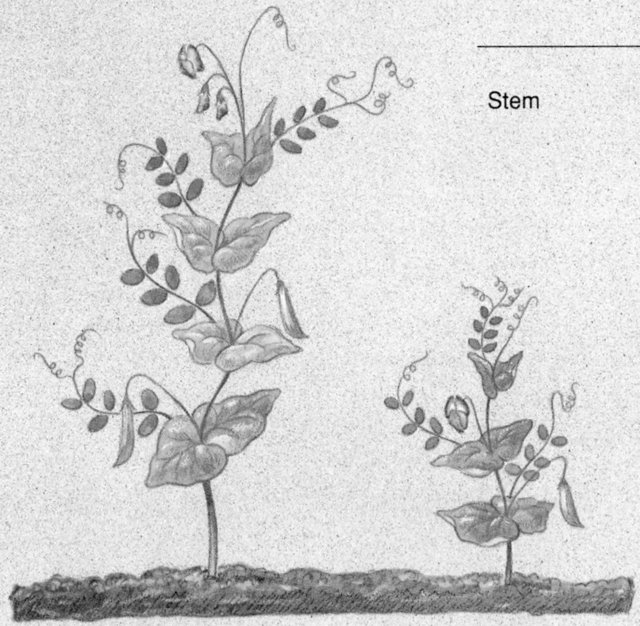

FIGURE 3.1
A summary of the seven pairs of contrasting traits and the results of Mendel's seven monohybrid crosses. In each case, pollen derived from plants exhibiting one contrasting trait was used to fertilize the ova of plants exhibiting the other contrasting trait. In the F_1 generation one of the two traits, referred to as dominant, was exhibited by all plants. The contrasting trait, referred to as recessive, then reappeared in approximately one-fourth of the F_2 plants. The garden pea (*Pisum sativum*) is also shown.

Genetic data are usually expressed and analyzed as ratios. In this particular example, many identical P_1 crosses were made and many F_1 plants—all tall—were produced. Of the 1064 F_2 offspring, 787 were tall and 277 were dwarf—a ratio of approximately 2.8:1.0, or about 3:1.

Mendel made similar crosses between pea plants exhibiting each of the other pairs of contrasting traits. The results of these crosses are also shown in Figure 3.1. In every case, the outcome was similar to the tall/dwarf cross just described. All F_1 offspring were identical to one of the parents. In the F_2, an approximate ratio of 3:1 was obtained. Three-fourths appeared like the F_1 plants, while one-fourth exhibited the contrasting trait, which had disappeared in the F_1 generation.

It is appropriate to point out one further aspect of the monohybrid crosses. In each, the F_1 and F_2 patterns of inheritance were similar regardless of which P_1 plant served as the source of pollen, or sperm, and which served as the source of the ovum, or egg. The crosses could be made either way—that is, pollen from the tall plant pollinating dwarf plants, or vice versa. These are called **reciprocal crosses.** Therefore, the results of Mendel's monohybrid crosses were not sex-dependent.

To explain these results, Mendel proposed the existence of particulate **unit factors** for each trait. He suggested that these factors serve as the basic units of heredity and are passed unchanged from generation to generation, determining various traits expressed by each individual plant. Using these general ideas, Mendel proceeded to hypothesize precisely how such factors could account for the results of the monohybrid crosses.

Mendel's First Three Postulates

Using the consistent pattern of results in the monohybrid crosses, Mendel derived the following three postulates or principles of inheritance.

1 UNIT FACTORS IN PAIRS
Genetic characters are controlled by unit factors that exist in pairs in individual organisms.
In the monohybrid cross involving tall and dwarf stems, a specific unit factor exists for each trait. Because the factors occur in pairs, three combinations are possible: two factors for tallness, two factors for dwarfness, or one of each factor. Every individual contains one of these three combinations, which determines stem height.

2 DOMINANCE/RECESSIVENESS
When two unlike unit factors responsible for a single character are present in a single individual, one unit factor is dominant to the other, which is said to be recessive.
In each monohybrid cross, the trait expressed in the F_1 generation is controlled by the dominant unit factor. The trait that is not expressed is controlled by the recessive unit factor. Note that this dominance/recessiveness relationship only pertains when unlike unit factors are present in pairs. The terms **dominant** and **recessive** are also used to designate the traits. In this case, tall stems are said to be dominant to the recessive dwarf stems.

3 SEGREGATION
During the formation of gametes, the paired unit factors separate or segregate randomly so that each gamete receives one or the other.
If an individual contains a pair of like unit factors (e.g., both are specific for tall), then all gametes receive one tall unit factor. If an individual contains unlike unit factors (e.g., one for tall and one for dwarf), then each gamete has a 50 percent probability of receiving either the tall *or* dwarf unit factor.

These postulates provide a suitable explanation for the results of the monohybrid crosses. The tall/dwarf cross will be used to illustrate this explanation. Mendel reasoned that P_1 tall plants contained identical paired unit factors, as did the P_1 dwarf plants. The gametes of tall plants all received one tall unit factor as a result of segregation. Likewise, the gametes of dwarf plants all received one dwarf unit factor. Following fertilization, all F_1 plants received one unit factor from each parent, a tall factor from one and a dwarf factor from the other, reestablishing the paired relationship. Because tall is dominant to dwarf, all F_1 plants were tall.

When F_1 plants form gametes, the postulate of segregation demands that each gamete randomly receive *either* the tall *or* dwarf unit factor. Following random fertilization events during F_1 selfing, four F_2 combinations will result in equal frequency:

(1) tall/tall
(2) tall/dwarf
(3) dwarf/tall
(4) dwarf/dwarf

Combinations (1) and (4) will result in tall and dwarf plants, respectively. According to the postulate of dominance/recessiveness, combinations (2) and (3) will both yield tall plants. Therefore, the F_2 is predicted to consist of three-fourths tall and one-fourth dwarf, or a ratio of 3:1. This is approximately what Mendel observed in the cross between tall and dwarf plants. A similar pattern was observed in each of the other monohybrid crosses.

Modern Genetic Terminology

In order to illustrate the monohybrid cross and Mendel's first three postulates, we must introduce several new terms as well as a set of symbols for the unit factors. Traits such as tall or dwarf are visible expressions of the information contained in unit factors. The physical appearance of a trait is called the **phenotype** of the individual.

All unit factors represent units of inheritance called **genes** by modern geneticists. For any given character, such as plant height, the phenotype is determined by alternate forms of a single gene called **alleles.** For example, the unit factors representing tall and dwarf are alleles determining the height of the pea plant.

By one convention, the first letter of the recessive trait is chosen to symbolize the character in question. The lowercase letter designates that allele for the recessive trait, and the uppercase letter designates the allele for the dominant trait. Therefore, we let *d* stand for the dwarf allele and *D* represent the tall allele. When alleles are written in pairs to represent the two unit factors present in any individual (*DD*, *Dd*, or *dd*), these symbols are referred to as the **genotype.** This term reflects the genetic makeup of an individual whether it is haploid or diploid. By reading the genotype, it is possible to know the phenotype of the individual: *DD* and *Dd* are tall, and *dd* is dwarf. When both alleles are the same (*DD* or *dd*), the individual is said to be **homozygous** or a **homozygote;** when alleles are different (*Dd*), we use the term **heterozygous** or **heterozygote.** These symbols and terms are used in Figure 3.2 to illustrate the complete monohybrid cross.

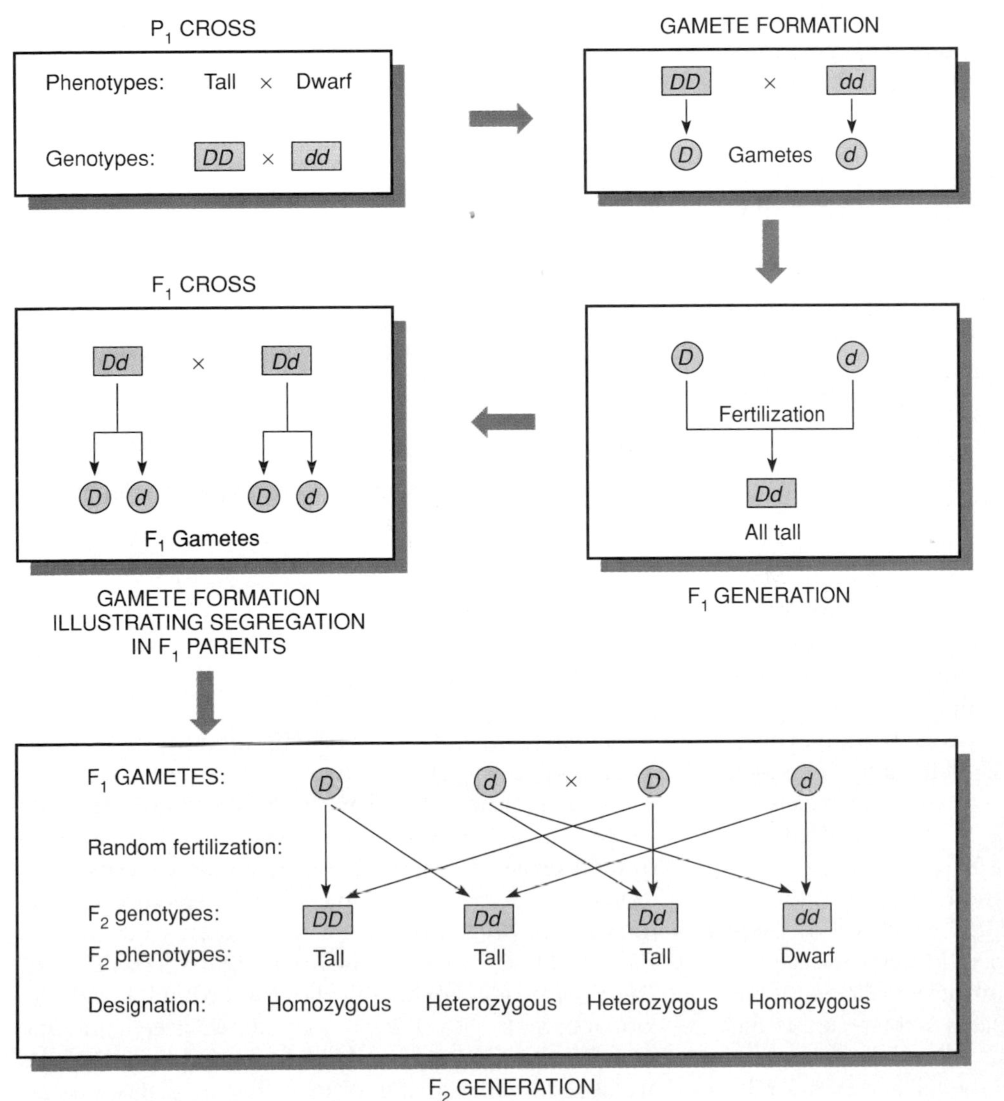

FIGURE 3.2
An explanation of the monohybrid cross between tall and dwarf pea plants. The symbols *D* and *d* are used to designate the tall and dwarf unit factors, respectively, in the genotypes of mature plants and gametes. All individuals are shown in rectangles. All gametes are shown in circles.

Because he operated without the hindsight that modern geneticists enjoy, Mendel's analytical reasoning must be considered a truly outstanding scientific achievement. On the basis of rather simple, but precisely executed breeding experiments, he proposed that discrete particulate units of heredity exist, and he explained how they are transmitted from one generation to the next!

Punnett Squares

The genotypes and phenotypes resulting from the recombination of gametes during fertilization can be easily visualized by constructing a **Punnett square,** so named after the person who first devised this approach, Reginald C. Punnett. Figure 3.3 illustrates this method of analysis for the $F_1 \times F_1$ monohybrid cross. All possible gametes are assigned to a column or a row, with the vertical column representing those of the male parent and the horizontal row those of the female parent. After entering the gametes in rows and columns, the new generation is predicted by combining the male and female gametic information for each combination and entering the resulting genotypes in the boxes. This process represents all possible random fertilization events. The genotypes and phenotypes of all potential offspring are ascertained by reading the entries in the boxes.

The Punnett square method is particularly useful when one is first learning about genetics and how to solve problems. In Figure 3.3, note the ease with which the 3:1 phenotypic ratio and the 1:2:1 genotypic ratio may be derived in the F_2 generation.

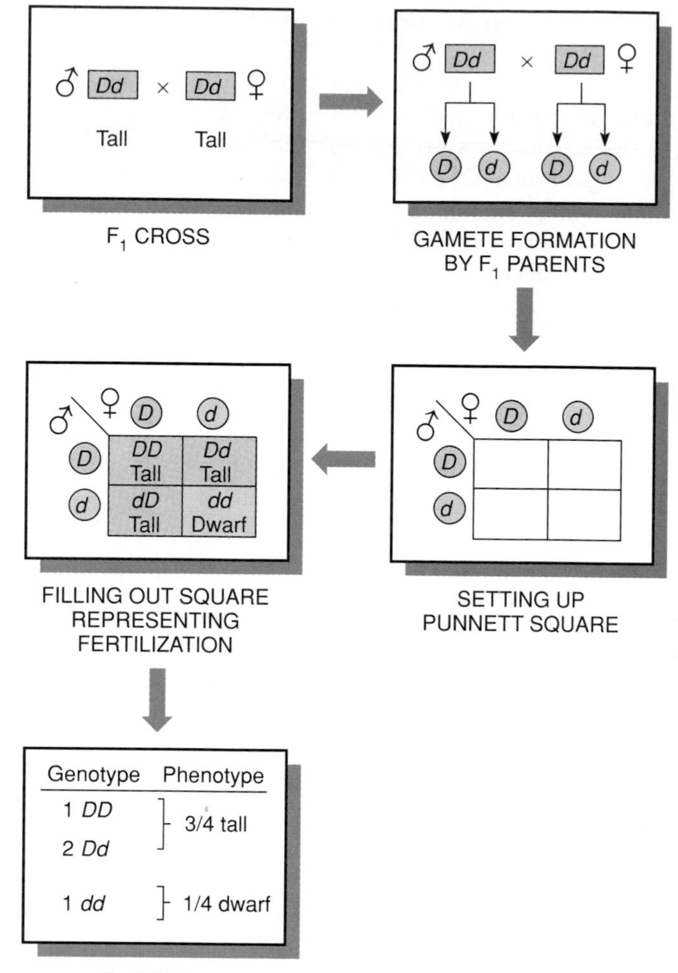

FIGURE 3.3
The use of a Punnett square in generating the F_2 ratio of the $F_1 \times F_1$ cross shown in Figure 3.2.

The Test Cross: One Character

Tall plants produced in the F_2 generation are predicted to be of either the *DD* or *Dd* genotypes. We might ask if there is a way to distinguish the genotype. Mendel devised a rather simple method that is still used today in breeding procedures of plants and animals: the **test cross.** The organism of the dominant phenotype but unknown genotype is crossed to a **homozygous recessive individual.** For example, if a tall plant of genotype *DD* is test-crossed to a dwarf plant, which must have the *dd* genotype, all offspring will be tall phenotypically and *Dd* genotypically. However, if a tall plant is *Dd* and is crossed to a dwarf plant (*dd*), then one-half of the offspring will be tall (*Dd*) and the other half will be dwarf (*dd*). Therefore, a 1:1 tall/dwarf ratio demonstrates the heterozygous nature of the tall plant of unknown genotype. The basis for these conclusions is illustrated in Figure 3.4. The test cross reinforced Mendel's conclusion that separate unit factors control the tall and dwarf traits.

THE DIHYBRID CROSS

A natural extension of performing monohybrid crosses was for Mendel to design experiments where two characters were examined simultaneously. We will refer to such a cross, involving two pairs of contrasting traits, as a **dihybrid cross.** It is also called a **two-factor cross.** For example, if pea plants having yellow seeds that are also round were bred with those having green seeds that are also wrinkled, the results shown in Figure 3.5 will occur. The F_1 offspring will be all yellow and round. It is therefore apparent that yellow is dominant to green, and that round is dominant to wrinkled. When the F_1 individuals are selfed, approximately 9/16 of the F_2 plants express

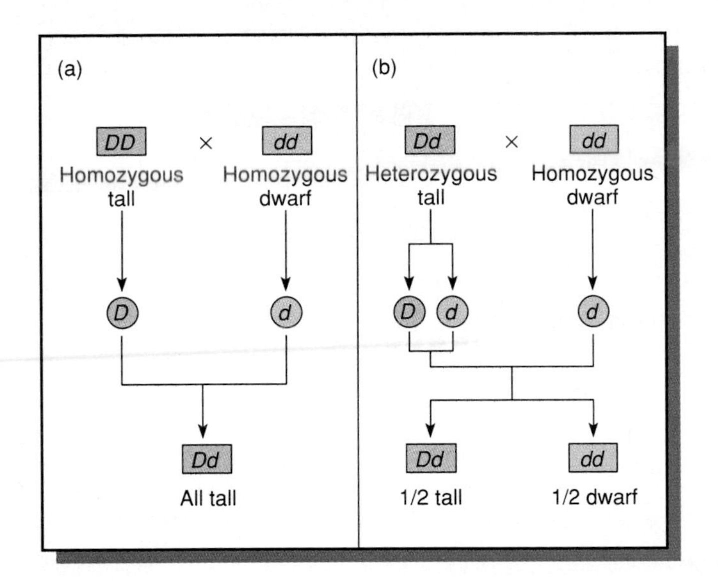

FIGURE 3.4
The test cross illustrated with a single character. In (a), the tall parent is homozygous. In (b), the tall parent is heterozygous. The genotypes of each tall parent may be determined by examining the offspring when each is crossed to the homozygous recessive dwarf plant.

yellow and round, 3/16 express yellow and wrinkled, 3/16 express green and round, and 1/16 express green and wrinkled.

A variation of this cross is also shown in Figure 3.5. If, instead of having one P_1 parent with both dominant traits (yellow, round) and one with both recessive traits (green, wrinkled), plants with yellow wrinkled seeds can be bred with those having green, round seeds in a P_1 cross. In spite of the change in parental phenotypes, both the F_1 and F_2 results remain unchanged. It will become clear in the next section why this is so.

Mendel's Fourth Postulate: Independent Assortment

We can most easily understand the results of a dihybrid cross if we consider it theoretically as consisting of two monohybrid crosses conducted separately. Think of the two sets of traits as being inherited independently of each other; that is, the chance of any plant becoming tall or dwarf is not at all influenced by the chance that this plant will have round or wrinkled seeds. Thus, because yellow is dominant to green, all F_1 plants in the first theoretical cross would have yellow seeds. In the second theoretical cross, all F_1 plants would have round seeds because round is dominant to wrinkled. When Mendel examined the F_1 plants of the dihybrid cross, all were yellow and round, as predicted.

The predicted F_2 results of the first cross are 3/4 yellow and 1/4 green. Similarly, the second cross would yield 3/4 round and 1/4 wrinkled. Figure 3.5 shows that in the dihybrid cross, 12/16 F_2 plants are yellow while 4/16 are green, exhibiting the 3:1 ratio. Similarly, 12/16 F_2 plants have round seeds while 4/16 have wrinkled seeds, again revealing the 3:1 ratio.

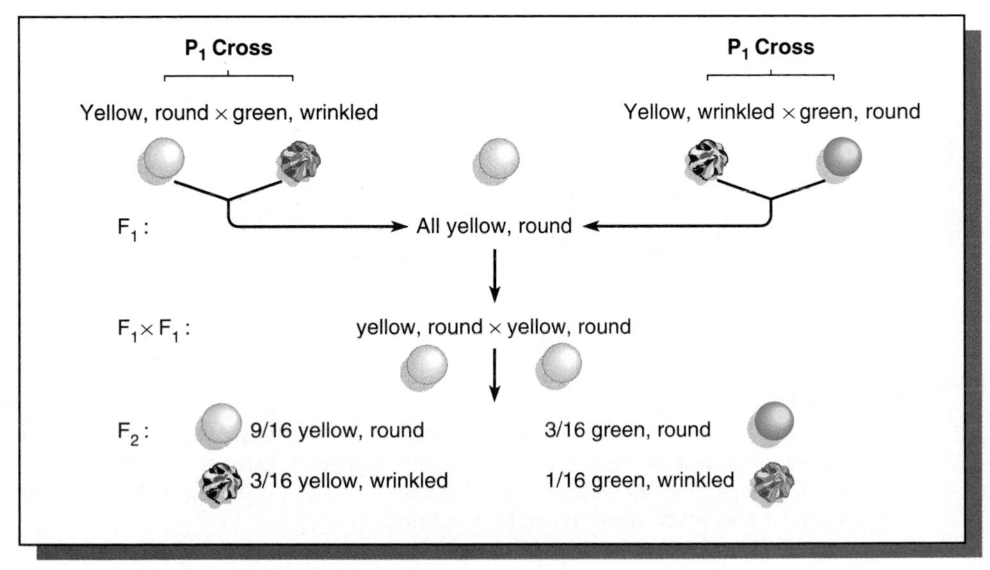

FIGURE 3.5
The F_1 and F_2 results of Mendel's dihybrid crosses between yellow, round and green, wrinkled pea plants, and between yellow, wrinkled and green, round pea plants.

FIGURE 3.6
The determination of the combined probabilities of each F_2 phenotype for two independently inherited characters. The probability of each plant being yellow or green is independent of the probability of it being round or wrinkled.

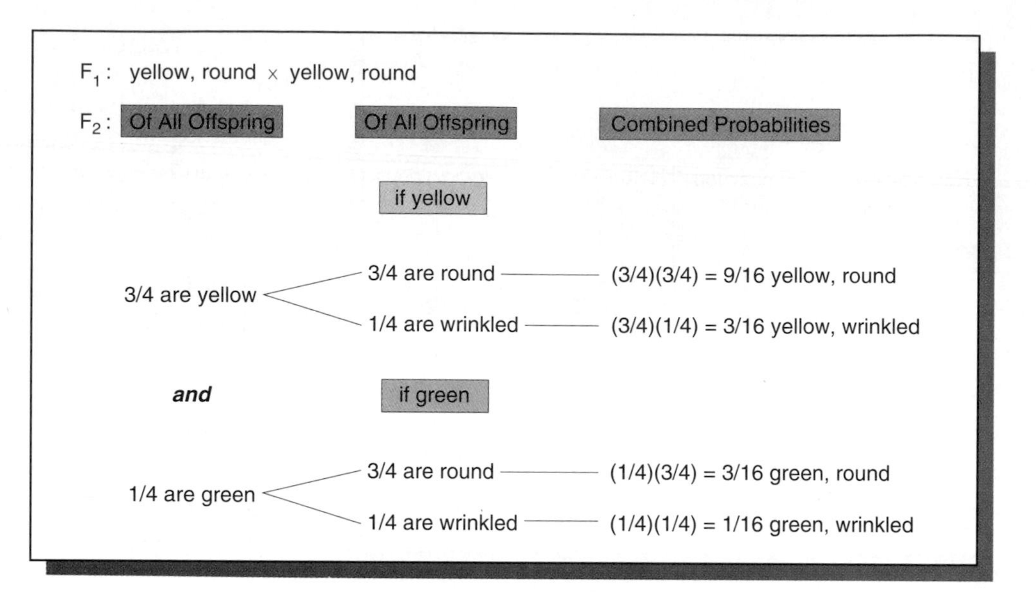

Because it is evident that the two pairs of contrasting traits are inherited independently, we can predict the frequencies of all possible F_2 phenotypes by applying the "product law" of probabilities: **When two independent events occur simultaneously, their combined probability is equal to the product of their individual probabilities of occurrence.** For example, the probability of an F_2 plant having yellow *and* round seeds is (3/4)(3/4), or 9/16, because 3/4 of all F_2 plants should be yellow and 3/4 of all F_2 plants should be round.

In a like way, the probabilities of the other three F_2 phenotypes may be calculated: yellow (3/4) *and* wrinkled (1/4) are predicted to be present together 3/16 of the time; green (1/4) *and* round (3/4) are predicted 3/16 of the time, and green (1/4) *and* wrinkled (1/4) are predicted 1/16 of the time. These calculations are illustrated in Figure 3.6. It is now apparent why the F_1 and F_2 results are identical whether the initial cross is yellow, round bred with green, wrinkled or if yellow, wrinkled are bred with green, round. In both crosses, the F_1 genotype of all plants is identical. Each plant is heterozygous for both gene pairs. As a result, the F_2 generation is also identical in both crosses.

On the basis of similar results in numerous dihybrid crosses, Mendel proposed a fourth postulate called **independent assortment:** During gamete formation, segregating pairs of unit factors assort independently of each other.

This postulate stipulates that segregation of any pair of unit factors occurs independently of all others. As a result of segregation, each gamete receives one member of every pair of unit factors. For one pair, whichever unit factor is received does not influence the outcome of segregation of any other pair. Thus, according to the postulate of independent assortment, all possible combinations of gametes will be formed in equal frequency.

Independent assortment is illustrated in the formation of the F_2 generation, shown in the Punnett square in Figure 3.7. Examine the formation of gametes by the F_1 plants. Segregation prescribes that every gamete receives either a *G* or *g* allele *and* a *W* or *w* allele. Independent assortment stipulates that all four combinations (*GW*, *Gw*, *gW*, and *gw*) will be formed with equal probabilities.

In every $F_1 \times F_1$ fertilization event, each zygote has an equal probability of receiving one of the four combinations from each parent. If a large number of offspring is

FIGURE 3.7 ▶
Diagram of the dihybrid crosses shown in Figure 3.5. The F_1 heterozygous plants are self-fertilized to produce an F_2 generation, which is computed using a Punnett square. Both the phenotypic and genotypic F_2 ratios are shown. The photographs illustrate the round/wrinkled and yellow/green phenotypes of peas.

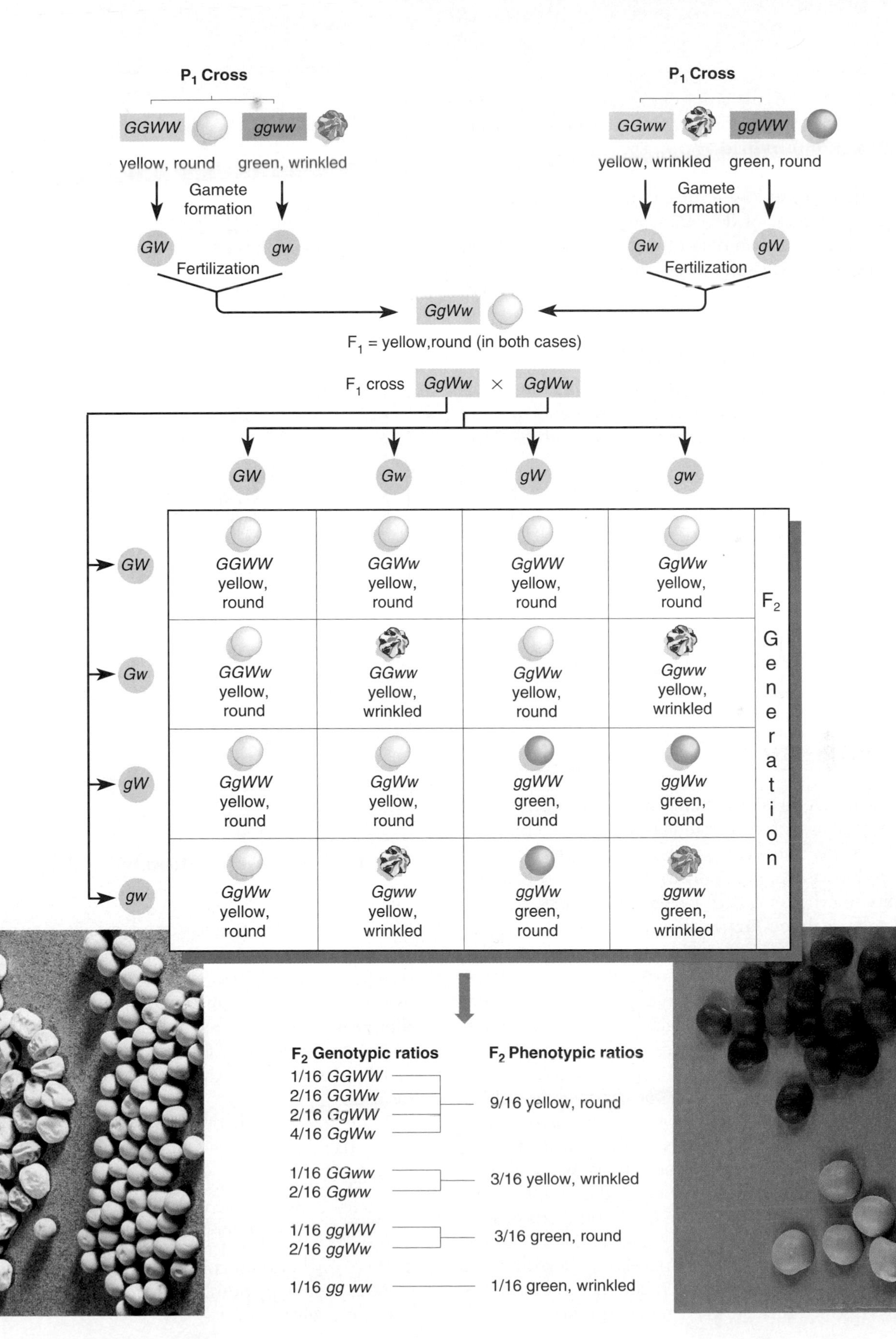
P₁ Cross
GGWW
ggww
yellow, round
green, wrinkled
Gamete formation
GW
gw
Fertilization
P₁ Cross
GGww
ggWW
yellow, wrinkled
green, round
Gamete formation
Gw
gW
Fertilization
GgWw
F₁ = yellow,round (in both cases)
F₁ cross GgWw × GgWw
GW
Gw
gW
gw
GGWW yellow, round
GGWw yellow, round
GgWW yellow, round
GgWw yellow, round
GGWw yellow, round
GGww yellow, wrinkled
GgWw yellow, round
Ggww yellow, wrinkled
GgWW yellow, round
GgWw yellow, round
ggWW green, round
ggWw green, round
GgWw yellow, round
Ggww yellow, wrinkled
ggWw green, round
ggww green, wrinkled
F₂ Generation
F₂ Genotypic ratios
1/16 GGWW
2/16 GGWw
2/16 GgWW
4/16 GgWw
1/16 GGww
2/16 Ggww
1/16 ggWW
2/16 ggWw
1/16 gg ww
F₂ Phenotypic ratios
9/16 yellow, round
3/16 yellow, wrinkled
3/16 green, round
1/16 green, wrinkled

produced, 9/16 are yellow and round, 3/16 are yellow and wrinkled, 3/16 are green and round, and 1/16 are green and wrinkled, yielding what is designated as **Mendel's 9:3:3:1 dihybrid ratio.** This ratio is based on probability events involving segregation, independent assortment, and random fertilization. Therefore, it is an ideal ratio. Because of deviation due strictly to chance, particularly if small numbers of offspring are produced, the perfect ratio will seldom be approached.

The Test Cross: Two Characters

The test cross may also be applied to individuals that express two dominant traits, but whose genotypes are unknown. For example, the expression of the yellow, round phenotype in the F_2 generation just described may result from the *GGWW*, *GGWw*, *GgWW*, and *GgWw* genotypes. If an F_2 yellow, round plant is crossed with the homozygous recessive green, wrinkled plant (*ggww*), analysis of the offspring will indicate the correct genotype of that yellow, round plant. Each of the above genotypes will result in a different set of gametes, and in a test cross, a different set of phenotypes in the resulting offspring.

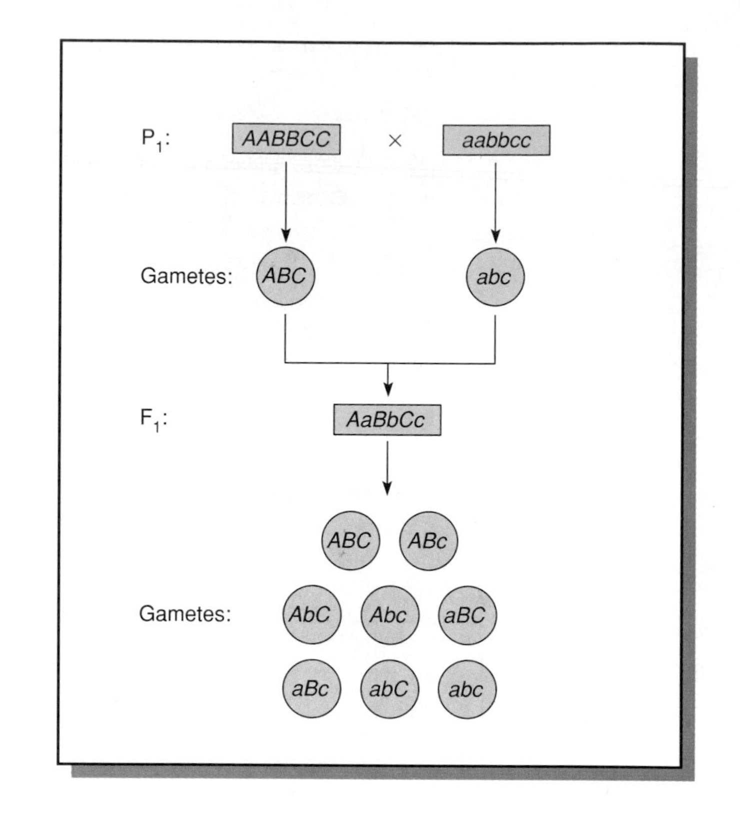

FIGURE 3.8
The formation of P_1 and F_1 gametes in a trihybrid cross.

THE TRIHYBRID CROSS

We have thus far considered inheritance by individuals of up to two pairs of contrasting traits. Mendel demonstrated that the identical processes of segregation and independent assortment apply to three pairs of contrasting traits in what is called a **trihybrid cross,** also referred to as a **three-factor cross.**

Although a trihybrid cross is somewhat more complex than a dihybrid cross, its results are easily calculated if the principles of segregation and independent assortment are followed. For example, consider the cross shown in Figure 3.8 where the gene pairs representing theoretical contrasting traits are symbolized *A/a*, *B/b*, and *C/c*. In the cross between *AABBCC* and *aabbcc* individuals, all F_1 individuals are heterozygous for all three gene pairs. Their genotype, *AaBbCc*, results in the phenotypic expression of the dominant *A*, *B*, and *C* traits. When F_1 individuals are parents, they all produce 8 different gametes in equal frequencies. At this point, we could construct a Punnett square with 64 separate boxes and read out the phenotypes. Because such a method is cumbersome in a cross involving so many factors, another method has been devised to calculate the predicted ratio.

The Forked-Line Method, or Branch Diagram

It is much less exacting to consider each contrasting pair of traits separately and then to combine these results using the **forked-line method,** which was first illustrated in Figure 3.6. This method, also called a **branch diagram,** relies on the simple application of the laws of probability established for the dihybrid cross. Each gene pair is assumed to behave independently during gamete formation.

When the monohybrid cross AA × aa is made, we know that:

1 All F_1 individuals have the genotype Aa and demonstrate the phenotype represented by the *A* allele, which is called the *A* phenotype in the following discussion.

2 The F_2 generation consists of individuals with either the *A* phenotype or the *a* phenotype in the ratio of 3:1, respectively.

F_2 TRIHYBRID PHENOTYPES

A or *a*	*B* or *b*	*C* or *c*	Combined Proportion
3/4 *A*	3/4 *B*	3/4 *C*	(3/4)(3/4)(3/4) *ABC* = 27/64 *ABC*
		1/4 *c*	(3/4)(3/4)(1/4) *ABc* = 9/64 *ABc*
	1/4 *b*	3/4 *C*	(3/4)(1/4)(3/4) *AbC* = 9/64 *AbC*
		1/4 *c*	(3/4)(1/4)(1/4) *Abc* = 3/64 *Abc*
1/4 *a*	3/4 *B*	3/4 *C*	(1/4)(3/4)(3/4) *aBC* = 9/64 *aBC*
		1/4 *c*	(1/4)(3/4)(1/4) *aBc* = 3/64 *aBc*
	1/4 *b*	3/4 *C*	(1/4)(1/4)(3/4) *abC* = 3/64 *abC*
		1/4 *c*	(1/4)(1/4)(1/4) *abc* = 1/64 *abc*

FIGURE 3.9
The generation of the F_2 trihybrid phenotypic ratio using the forked-line or branch diagram, method.

The same generalizations may be made for the *BB* × *bb* and *CC* × *cc* crosses. Thus, in the F_2 generation, 3/4 of all organisms will have phenotype *A*, 3/4 will have *B*, and 3/4 will have *C*. Similarly, 1/4 of all organisms will have phenotype *a*, 1/4 will have *b*, and 1/4 will have *c*. The proportions of organisms that express each phenotypic combination may be predicted by assuming that fertilization, following the independent assortment of these three gene pairs during gamete formation, is a random process. We must simply apply once again the product law of probabilities.

The phenotypes of the F_2 generation calculated in this way using the forked-line method are illustrated in Figure 3.9. They fall into the trihybrid ratio of 27:9:9:9:3:3:3:1. The same method may be applied when solving crosses involving any number of gene pairs, *provided* that all gene pairs assort independently from each other. We will see later that this is not always the case. However, it appeared to be true for all of Mendel's characters.

THE REDISCOVERY OF MENDEL'S WORK

Mendel's work, initiated in 1856, was presented to the Brünn Society of Natural Science in 1865 and published one year later. However, his findings went largely unnoticed for about 35 years! Many reasons have been suggested to explain why the significance of his research was not immediately recognized.

First of all, Mendel's adherence to mathematical analysis of probability events was quite an unusual approach in biological studies. Perhaps his approach seemed foreign to his contemporaries. More importantly, his conclusions drawn from such analyses did not fit well with the existing theories involving the cause of variation between organisms. The source of natural variation intrigued students of evolutionary theory. These individuals, stimulated by the proposal developed by Charles Darwin and Alfred Russell Wallace, believed that variation was of a **continuous nature** and that offspring were a blend of their parents' phenotypes. As we have mentioned earlier, Mendel theorized that variation was due to discrete or particulate units and was therefore of a **discontinuous nature.** For example, Mendel proposed that the F_2 offspring of a dihybrid cross were merely expressing traits produced by new combinations of previously existing unit factors. As a result, Mendel's theories did not fit well with the evolutionists' preconceptions about causes of variation.

Near the end of the nineteenth century, a remarkable observation set the scene for the rebirth of Mendel's work—Walter Flemming's **discovery of chromosomes** in the nuclei of salamander cells. Occurring in 1879, Flemming was able to describe the behavior of these threadlike structures during cell division. As a result of the findings of Flemming and many other cytologists, the presence of a nuclear component soon became an integral part of ideas surrounding inheritance. It was in this setting that scientists were able to reexamine Mendel's findings.

In the early twentieth century, research led to the rebirth of Mendel's work. Hybridization experiments similar to Mendel's were independently performed by three botanists, Hugo DeVries, Karl Correns, and Erich Tschermak. DeVries's work, for example, had focused on unit characters, and he demonstrated the principle of segregation in his experiments with several plant species. Apparently, he had searched the existing literature and found that Mendel's work had anticipated his own conclusions! Correns and Tschermak had independently reached conclusions similar to those of Mendel.

In 1902 two cytologists, Walter Sutton and Theodor Boveri, independently published papers linking their discoveries of the behavior of chromosomes during meiosis to the Mendelian principles of segregation and independent assortment. They pointed out that the separation of chromosomes during meiosis could serve as the cytological basis of these two postulates. Although they thought that Mendel's unit factors were probably chromosomes rather than genes on chromosomes, their findings made Mendel's work the foundation of ensuing genetic investigations.

Unit Factors, Genes, and Homologous Chromosomes

Because the correlation between Sutton's and Boveri's observations and Mendelian principles is the foundation for the modern interpretation of transmission genetics, we will examine this correlation before moving to the topics of the next several chapters.

As pointed out in Chapter 2, each species possesses a specific number of chromosomes in each somatic (body) cell nucleus. For diploid organisms, this number is called the **diploid number (2*n*)** and is characteristic of that species. During the formation of gametes, this number is precisely halved (n), and when two gametes combine during fertilization, the diploid number is reestablished. The chromosome number is not reduced in a random manner, however. It was apparent to early cytologists that the diploid number of chromosomes is composed of homologous pairs identifiable by their morphological appearance and behavior. The gametes contain one member of each pair. The chromosome complement of a gamete is thus quite specific, and the number of chromosomes in each gamete is equal to the haploid number.

With this basic information, we can see the correlation between the behavior of unit factors and chromosomes and genes. Figure 3.10 shows three of Mendel's postulates and the accepted explanation of each. Unit factors are really genes located on homologous pairs of chromosomes [Figure 3.10(a)]. Members of each pair of homologues separate, or segregate, during gamete formation [Figure 3.10(b)].

To illustrate the principle of independent assortment, it is important to distinguish between members of any given homologous pair of chromosomes. One member of each pair is derived from the **maternal parent,** while the other comes from the **paternal parent.** We represent their different origins by different colors. In Figure 3.10(c) the two pairs of segregating homologues behave independently during gamete formation. Each gamete always receives one homologue from each pair. All possible combinations are shown. If we add the symbols used in Mendel's dihybrid cross (*D*, *d* and *W*, *w*) to the diagram, we see why equal numbers of the four types of gametes are formed. The independent behavior of Mendel's pairs of unit factors (*D* and *W* in this example) was due to the fact that they were on separate pairs of homologous chromosomes.

From observations of the phenotypic diversity of living organisms, we see that it is logical to assume that there are many more genes than chromosomes. Therefore, each homologue must carry genetic information for more than one trait. The currently accepted concept is that a chromosome is composed of a large number of linearly ordered, information-containing units called **genes.** Thus, Mendel's unit factors (which determine tall or dwarf stems, for example) actually constitute a pair of genes located on one pair of homologous chromosomes. The location on a given chromosome where any particular gene occurs is called its **locus.** The different forms taken by a given gene, called **alleles** (*D* or *d*), contain slightly different genetic information that determines the same character (stem length). Alleles are alternate forms of the same gene. Although we have only discussed genes with two alternative alleles, most genes have *more* than two allelic forms. We discuss the concept of **multiple alleles** in Chapter 4.

We conclude this section by reviewing the criteria necessary to classify two chromosomes as a homologous pair:

1 During mitosis and meiosis, when chromosomes are visible as distinct figures, both members of a homologous pair are the same size and exhibit identical centromere locations.

2 During early stages of meiosis, homologous chromosomes pair together, or synapse.

3 Although not generally microscopically visible, homologues contain identical, linearly ordered, gene loci.

INDEPENDENT ASSORTMENT AND GENETIC VARIATION

One of the major consequences of independent assortment is the production by individuals of genetically dissimilar gametes. Genetic variation results because the two members of any homologous pair of chromosomes are rarely, if ever, genetically identical. Therefore, because independent assortment leads to the production of all possible chromosome combinations, extensive genetic diversity results.

The number of possible gametes, each with different chromosome compositions, is 2^n, where n equals the

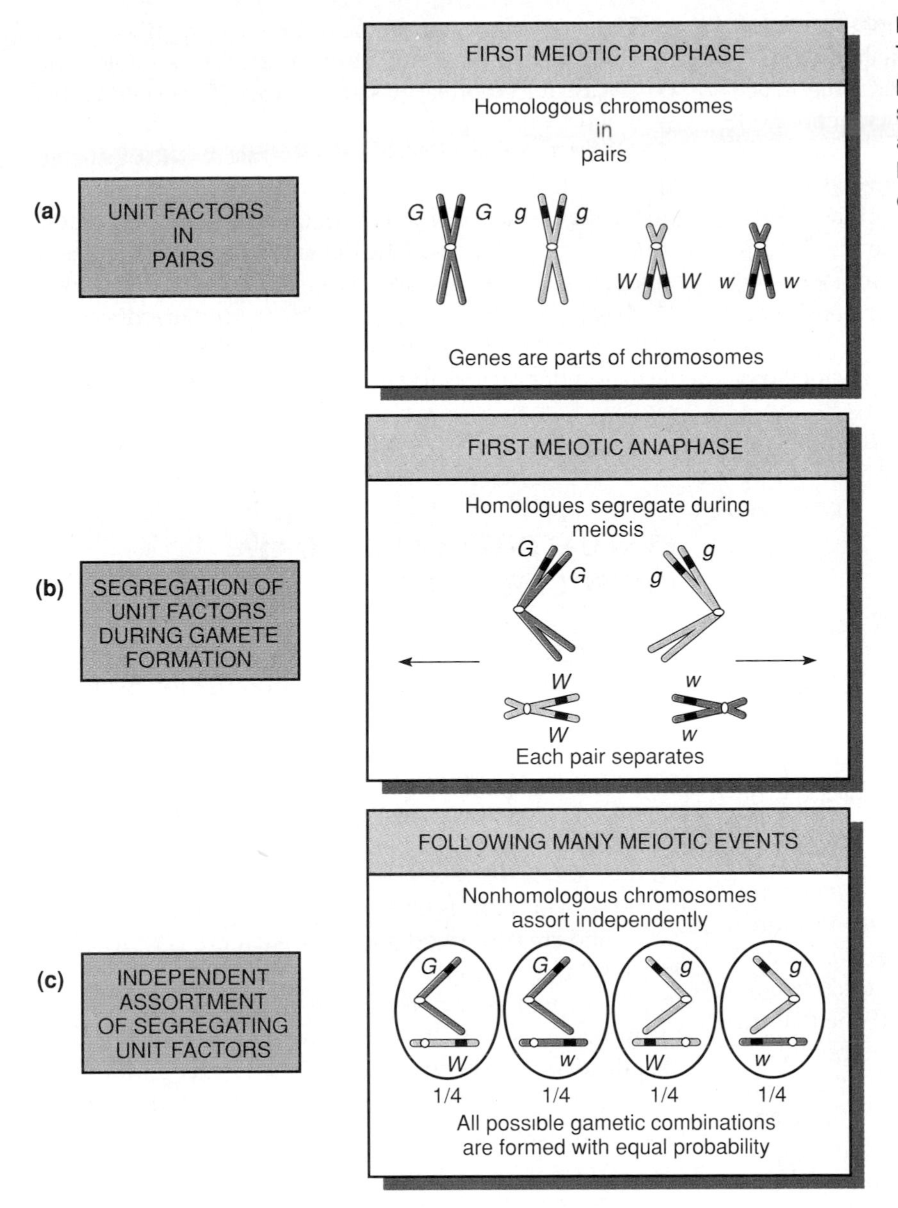

FIGURE 3.10
The correlation between the Mendelian postulates of (a) unit factors in pairs, (b) segregation, and (c) independent assortment, and the presence of genes located on homologous chromosomes and their behavior during meiosis.

haploid number. Thus, if a species has a haploid number of 4, then 2^4 or 16 different gamete combinations can be formed as a result of independent assortment. Although this number is not great, consider the human species, where $n = 23$. If 2^{23} is calculated, we find that in excess of 8×10^6, or over 8 million, different types of gametes are represented. Because fertilization represents an event involving only one of approximately 8×10^6 possible gametes from each of two parents, each offspring represents only one of $(8 \times 10^6)^2$, or 64×10^{12}, potential genetic combinations! It is no wonder that, except for identical twins, each member of the human species demonstrates a distinctive appearance and individuality. This number of combinations is far greater than the number of humans who have ever lived on earth! Genetic variation resulting from independent assortment has been extremely important to the process of organic evolution in all organisms.

PROBABILITY AND GENETIC EVENTS

Genetic ratios are most properly expressed as probabilities, e.g., 3/4 tall : 1/4 dwarf. These values predict the outcome of each fertilization event, such that the probability of each zygote having the genetic potential for becoming tall is 3/4, while the potential for becoming dwarf is 1/4. Probabilities range from 0, where an event is certain *not* to occur, to 1.0, where an event is certain *to* occur.

When two or more events occur independently of one another, but at the same time, we can calculate the probability that they will indeed occur together. This is accomplished by applying the **product law.** As mentioned in our earlier discussion of independent assortment, it states that the probability of two or more events occurring simultaneously is equal to the *product* of their individual probabilities. Two or more events are independent of one another if the outcome of each one does not affect the outcome of any of the others under consideration.

To illustrate the use of the product law, consider the possible results if you were to toss a penny (*P*) and a nickel (*N*) at the same time and examined all combinations of heads (*H*) and tails (*T*) that can occur. There are four possible outcomes:

$$(P_H{:}N_H) = (1/2)(1/2) = 1/4$$
$$(P_T{:}N_H) = (1/2)(1/2) = 1/4$$
$$(P_H{:}N_T) = (1/2)(1/2) = 1/4$$
$$(P_T{:}N_T) = (1/2)(1/2) = 1/4$$

The probability of obtaining a head or a tail in the toss of either coin is 1/2 and unrelated to the outcome of the toss of the other coin. All four possible combinations are predicted to occur with equal probability.

If we were interested in calculating the probability where the possible outcomes of two events are independent of one another, but can be accomplished in more than one way, we would apply the **sum law.** For example, we can ask, what is the probability of tossing our penny and nickel and obtaining one head and one tail? In such a case, we do not care whether it is the penny or nickel that comes up heads, provided that the other coin has the alternate outcome. As we can see above, there are two ways in which the desired outcome can be accomplished, each with a probability of 1/4. Thus, according to the sum law, the overall probability is equal to:

$$(1/4) + (1/4) = 1/2$$

One-half of all such tosses are predicted to yield the desired outcome.

These simple probability laws will be useful throughout our discussions of transmission genetics, and as you solve genetics problems. In fact, we have already applied the product law earlier when the forked-line method was used to calculate the phenotypic results of Mendel's dihybrid and trihybrid crosses. When we wish to know the results of a cross, we need to only calculate the probability of each possible outcome. The results of this calculation then allow us to predict the proportion of offspring expressing each phenotype or each genotype.

There is a very important point to remember when dealing with probability. Predictions of possible outcomes are based on large sample sizes. If we predict that 9/16 of the offspring of a dihybrid cross will express both dominant traits, it is very unlikely that 9 out of 16 offspring will express this phenotype. Instead, our prediction is that, of a large number of offspring, approximately 9/16 of them will do so. The deviation from the predicted ratio in smaller sample sizes is attributed to deviation due to chance, a subject we deal with in our discussion of statistics in the next section. As we will see, the impact of deviation due strictly to chance is diminished as the sample size increases.

EVALUATING GENETIC DATA: CHI-SQUARE ANALYSIS

Mendel's 3:1 monohybrid and 9:3:3:1 dihybrid ratios are hypothetical predictions based on the following assumptions: (1) each allele is dominant or recessive; (2) segregation is operative; (3) independent assortment occurs; and (4) fertilization is random. The last three assumptions are influenced by chance events and therefore are subject to random fluctuation. This concept, called **chance deviation,** is most easily illustrated by tossing a single coin numerous times and recording the number of heads and tails observed. In each toss, there is a probability of 1/2 that a head will occur and a probability of 1/2 that a tail will occur. Therefore, the expected ratio of many tosses is 1:1. If a coin were tossed 1000 times, usually *about* 500 heads and 500 tails would be observed. Any reasonable fluctuation from this hypothetical ratio (e.g., 486 heads and 514 tails) would be attributed to chance.

As the total number of tosses is reduced, the impact of chance deviation increases. For example, if a coin were tossed only 4 times, you wouldn't be too surprised if all 4 tosses resulted in only heads or only tails. But, for 1000 tosses, 1000 heads or 1000 tails would be most unexpected. In fact, you might believe that such a result would be impossible. Actually, all heads or all tails in 1000 tosses would be predicted to occur with a probability of only $(1/2)^{1000}$. Because $(1/2)^{20}$ is equivalent to less than 1 in 1 million times, an event occurring with a probability of only $(1/2)^{1000}$ would be virtually impossible to achieve.

Two major points are significant here:

1. The outcomes of segregation, independent assortment, and fertilization, like coin tossing, are subject to random fluctuations from their predicted occurrences as a result of chance deviation.

TABLE 3.1
Chi-square analysis

(a) Hypothetical monohybrid cross.

Expected Ratio	Observed (o)	Expected (e)	Deviation ($o - e$)	Deviation² (d^2)	Deviation²/Expected (d^2/e)
3/4	740	3/4 (1000) = 750	740 − 750 = −10	$(-10)^2 = 100$	100/750 = 0.13
1/4	260	1/4 (1000) = 250	260 − 250 = +10	$(+10)^2 = 100$	100/250 = 0.40
TOTAL = 1000					$\chi^2 = 0.53$
					$p = 0.48$

(b) Hypothetical dihybrid cross.

Expected Ratio	o	e	$o - e$	d^2	d^2/e
9/16	587	567	+20	400	0.71
3/16	197	189	+8	64	0.34
3/16	168	189	−21	441	2.33
1/16	56	63	−7	49	0.78
TOTAL = 1008				$\chi^2 = 4.16$	
				$p = 0.26$	

2 As the sample size increases, the average deviation from the expected decreases. Therefore, a larger sample size diminishes the impact of chance deviation on the final outcome.

It is important in genetics to be able to evaluate observed deviation. When we assume that data will fit a given ratio such as 1:1, 3:1, or 9:3:3:1, we establish what is called the **null hypothesis.** It is so named because the hypothesis assumes that there is no real difference between the **measured values** (or ratio) and the **predicted values** (or ratio). Evaluation of the null hypothesis is accomplished by statistical analysis. On this basis, the null hypothesis may either: (1) be rejected, or (2) fail to be rejected. If it is rejected, any observed deviation from the expected cannot be attributed to chance alone. The null hypothesis and the underlying assumptions leading to it must be reexamined. If the null hypothesis fails to be rejected, any observed deviations are attributed to chance.

One of the simplest statistical tests devised to assess the null hypothesis is **chi-square analysis (χ^2).** This test takes into account the observed deviation in each component of an expected ratio as well as the sample size and reduces them to a single numerical value. This value (χ^2) is then used to estimate how frequently the observed deviation can be expected to occur strictly as a result of chance. The formula used in chi-square analysis is:

$$\chi^2 = \Sigma \frac{(o - e)^2}{e}$$

In this equation, o is the observed value for a given category and e is the expected value for that category. Σ (sigma) represents the sum of the calculated values for each category of the ratio. Because $(o - e)$ is the deviation (d) in each case, the equation can be reduced to

$$\chi^2 = \Sigma \frac{d^2}{e}$$

Table 3.1(a) illustrates the step-by-step procedure necessary to make the χ^2 calculation for the F_2 results of a hypothetical monohybrid cross. If you were analyzing these data, you would work from left to right, calculating

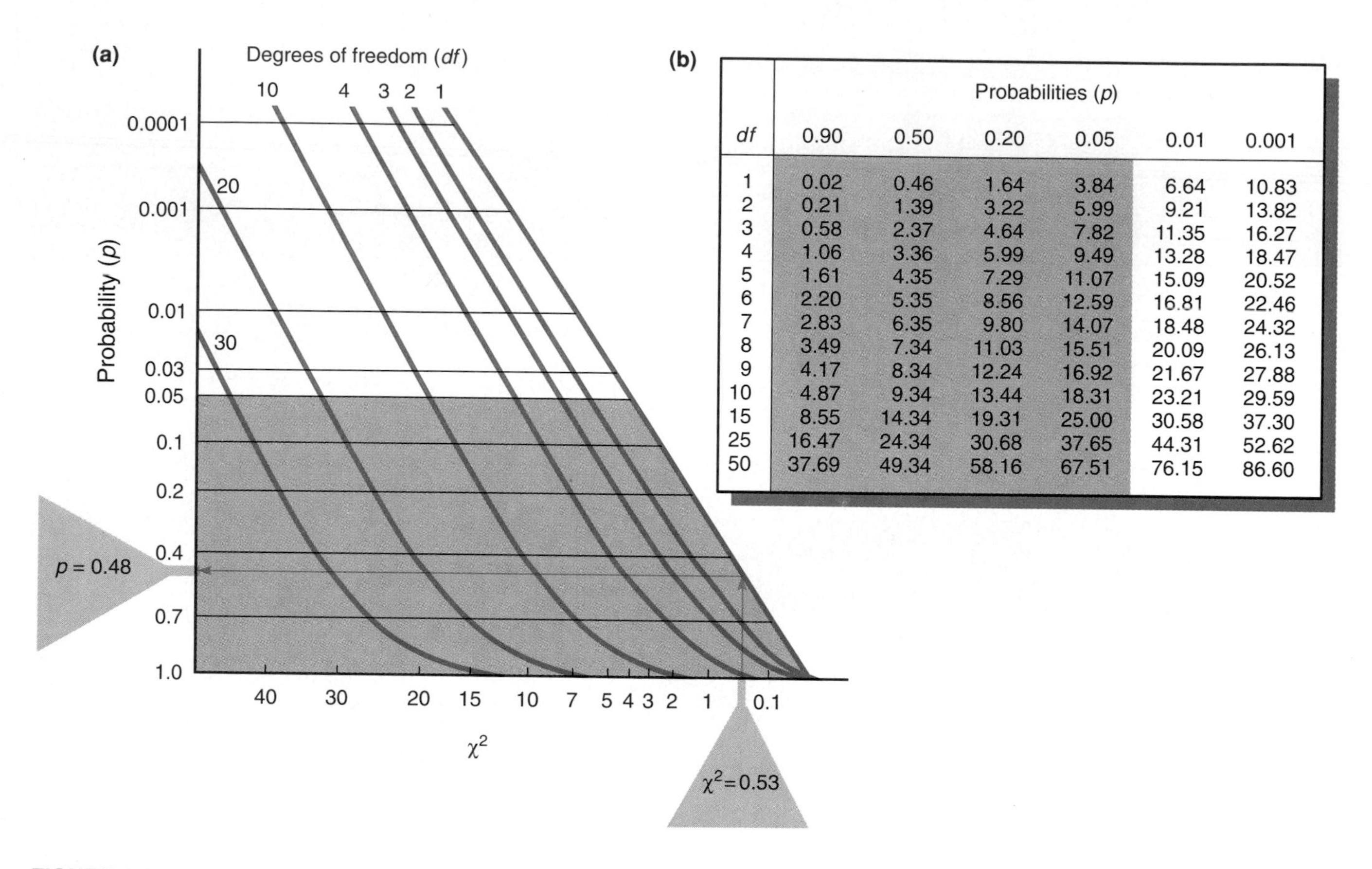

df	Probabilities (p) 0.90	0.50	0.20	0.05	0.01	0.001
1	0.02	0.46	1.64	3.84	6.64	10.83
2	0.21	1.39	3.22	5.99	9.21	13.82
3	0.58	2.37	4.64	7.82	11.35	16.27
4	1.06	3.36	5.99	9.49	13.28	18.47
5	1.61	4.35	7.29	11.07	15.09	20.52
6	2.20	5.35	8.56	12.59	16.81	22.46
7	2.83	6.35	9.80	14.07	18.48	24.32
8	3.49	7.34	11.03	15.51	20.09	26.13
9	4.17	8.34	12.24	16.92	21.67	27.88
10	4.87	9.34	13.44	18.31	23.21	29.59
15	8.55	14.34	19.31	25.00	30.58	37.30
25	16.47	24.34	30.68	37.65	44.31	52.62
50	37.69	49.34	58.16	67.51	76.15	86.60

FIGURE 3.11
(a) A graph used to convert χ^2 values to p values. The conversion of a χ^2 value of 0.53 with 1 degree of freedom to an estimated probability value of 0.48 is illustrated. (b) A table showing χ^2 values for a variety of combinations of *df* and *p*. Any χ^2 value greater than that shown at the $p = 0.05$ level for a particular *df* serves as the basis to reject the null hypothesis in question. In our example, a χ^2 value of 0.53 for a *df* of 1 is converted to a probability value between 0.20 and 0.50. From our graph in (a), the more precise value ($p = 0.48$) was estimated by interpolation. All values that serve to fail to reject the null hypothesis are shaded in both the graph and the chart.

and entering the appropriate numbers in each column. Regardless of whether the calculated deviation $(o - e)$ is initially positive or negative, it becomes positive after the number is squared. Table 3.1(b) illustrates the analysis of the F_2 results of a hypothetical dihybrid cross. Based on your study of the calculations involved in the monohybrid cross, check to make certain that you understand how each number was calculated in the dihybrid example.

The final step in the chi-square analysis is to interpret the χ^2 value. To do so, you must initially determine the value of the **degrees of freedom (*df*)**, which is equal to $n - 1$ where n is the number of different categories into which each datum point may fall. For the 3:1 ratio, $n = 2$, so $df = 2 - 1 = 1$. For the 9:3:3:1 ratio, $df = 3$. Degrees of freedom must be taken into account because the greater the number of categories, the more deviation is expected as a result of chance.

With this accomplished, the χ^2 value must now be interpreted in terms of a corresponding **probability value (*p*)**. Because this calculation is complex, the p value is usually located on a table or graph. Figure 3.11 shows the wide range of χ^2 and p values for numerous degrees of freedom in both forms. We will use the graph to determine the p value. The caption for Figure 3.11(b) explains how to use the table.

These simple steps must be followed to determine p:

1 Locate the χ^2 value on the abscissa (the horizontal axis).

2 Draw a vertical line from this point up to the line on the graph representing the appropriate *df*.

3 Extend a horizontal line from this point to the left until it intersects the ordinate (the vertical axis).

4 Estimate, by interpolation, the corresponding *p* value.

For our first example (the monohybrid cross) in Table 3.1, the *p* value of 0.48 may be estimated in this way and is illustrated in Figure 3.11(a). For the dihybrid cross, use this method to see if you can determine the value. χ^2 is 4.16 and *df* equal 3. A *p* value of 0.26 is the approximate value. Use of the table rather than the graph confirms that both *p* values are between 0.20 and 0.50. Examine the table to confirm this.

So far, we have been concerned only with the determination of *p*. The most important aspect of χ^2 analysis is understanding what the *p* value actually means. We will use the example of the dihybrid cross ($p = 0.26$) to illustrate. In these discussions, it is simplest to think of the *p* value as a percentage (e.g., 0.26 = 26 percent).

The *p* value indicates the probability of obtaining as great or even greater deviation by chance alone. In our dihybrid example, the probability is 26 percent that the observed deviation or more will occur owing to chance. The nearer the *p* value is to 1.0, the closer the data are to the predicted or ideal ratio.

Another description provides more specific statistical information. In our example, the *p* value indicates that, were the same experiment repeated many times, 26 percent of the trials would be expected to exhibit chance deviation as great or greater than that seen in the initial trial. Conversely, 74 percent of the repeats would show less deviation as a result of chance than initially observed.

These interpretations of the *p* value reveal that a hypothesis (a 9:3:3:1 ratio in this case) is never proved or disproved absolutely. Instead, a relative standard must be set to serve as the basis for either supporting or rejecting the hypothesis. This standard is often a probability value of 0.05. In chi-square analysis, a *p* value less than 0.05 makes it unlikely that the deviation in the observed set of results could be obtained by chance alone. Instead, such a *p* value indicates that the difference between the observed and predicted results is substantial and thus serves as the basis for rejecting the null hypothesis.

On the other hand, *p* values of 0.05 or greater (1.0–0.05) fail to reject the null hypothesis. In our example where $p = 0.26$, the hypothesis of independent assortment is supported by the experimental data. That is, the data do not provide any reason to reject the hypothesis. Therefore, the observed deviation can be attributed to chance.

HUMAN PEDIGREES

In all crosses discussed so far, one of the two traits for each character has been dominant to the other. Based on this observation, two significant questions may be asked:

1 Does the expression of all genes occur in this fashion?

2 Is it possible to ascertain the mode of inheritance of genes in organisms where designed crosses and the production of large numbers of offspring are impossible?

The answer to the first question is no. As we will see in Chapter 4, many modes of inheritance exist that modify the monohybrid and dihybrid ratios observed by Mendel.

The answer to the second question is yes. Even in humans the pattern of inheritance of a specific phenotype can be studied.

The simplest way to study this pattern is to construct a family tree indicating the phenotype of the trait in question for each member. Such a family tree is called a **pedigree.** By analyzing the pedigree, we may be able to determine how the gene controlling the trait is inherited.

Figure 3.12 shows the conventions used in constructing a pedigree. Circles represent females, and squares designate males. If the sex is unknown, a diamond may be used. If a pedigree traces only a single trait, as Figure 3.12 does, the circles, squares, and diamonds are shaded if the phenotype being considered is expressed. Heterozygotes who fail to express a recessive trait, when known with certainty, have only the left half of their square or circle shaded.

The parents are connected by a horizontal line, and vertical lines lead to their offspring. All such offspring are called **sibs** and are connected by a horizontal **sibship line.** Sibs are placed from left to right according to birth order and are labeled with Arabic numerals. Each generation is indicated by a Roman numeral.

Twins are indicated by connected diagonal lines. **Monozygotic** or **identical twins** stem from a single line itself connected to the sibship line (see III-5,6 in Figure 3.12). **Dizygotic** or **fraternal twins** are connected directly to the sibship line (see III-8,9). A number within one of the symbols (II-10–13) represents numerous sibs of the same or unknown phenotypes. A male whose phenotype drew the attention of a physician or geneticist is called the **propositus** (a female is a **proposita**) or sometimes the **proband** and is indicated by an arrow (see III-4).

The pedigree shown in Figure 3.12 traces the theoretical pattern of inheritance of the human trait **albinism.** By analyzing the pedigree, we will see that albinism is inherited as a recessive trait.

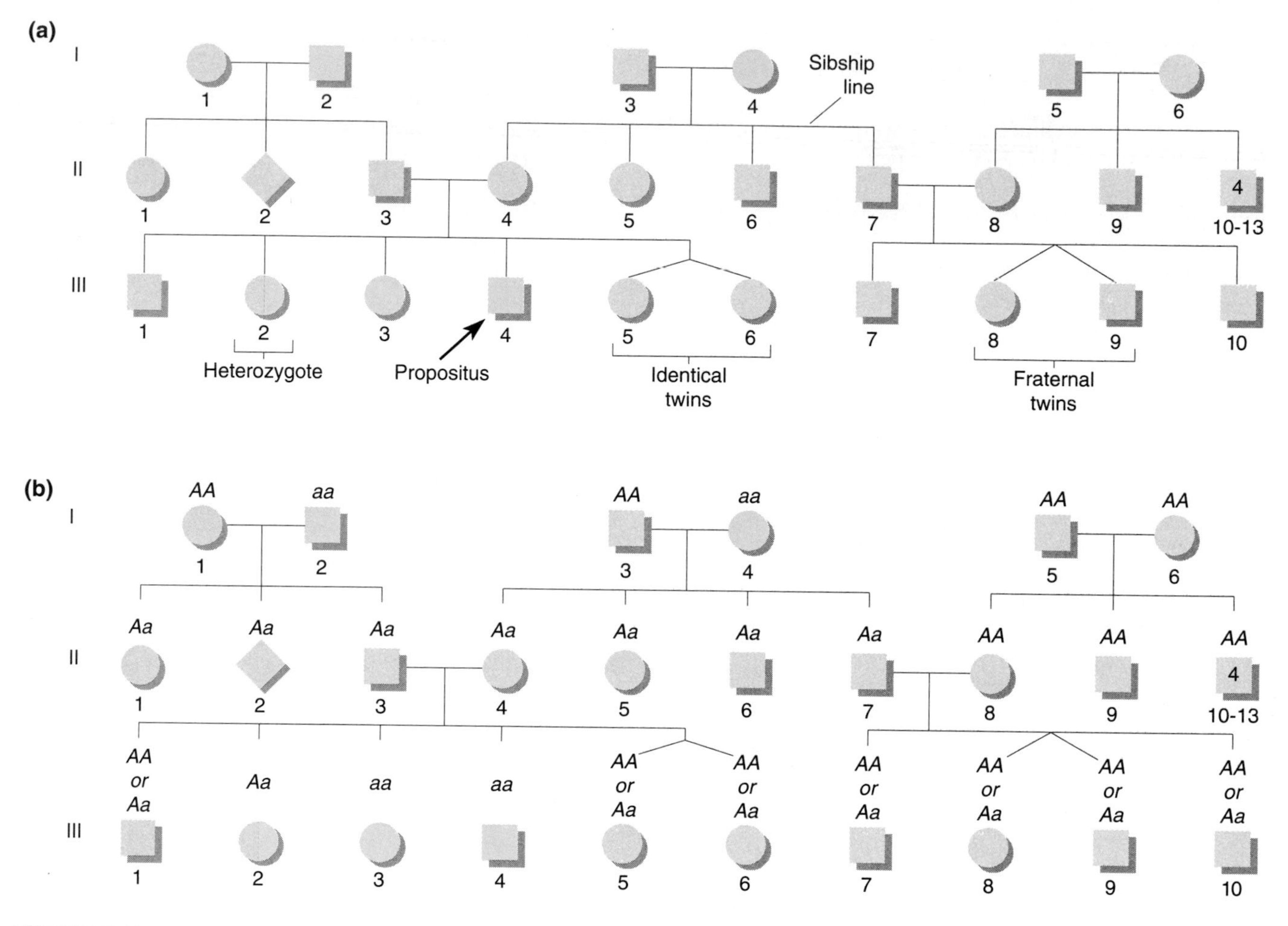

FIGURE 3.12
(a) A representative pedigree for a single character through three generations. (b) The most probable genotypes of each individual in the pedigree.

Two of the parents of the first generation, I-2 and I-4, are affected. Because none of their offspring show the disorder, it is reasonable to conclude that the unaffected parents (I-1 and I-3) were homozygous normal individuals. Had they been heterozygous, one-half of their offspring would be expected to exhibit albinism.

An unaffected second generation is characteristic of a rare recessive trait. If albinism were inherited as a dominant trait, one-half of the second generation would be expected to exhibit the disorder in the crosses involving the I-2 and I-4 parents. Inspection of the offspring constituting the third generation (row III) provides further support for the hypothesis that albinism is a recessive trait. Parents II-3 and II-4 are apparently both heterozygous, and approximately one-fourth of their offspring should be affected. Two of the six offspring do show albinism. This deviation from the expected ratio is characteristic of crosses with few offspring.

Individual II-7 is undoubtedly heterozygous, while II-8 is most likely homozygous normal, given the low frequency of the *a* allele in the population. If so, we would predict that none of their offspring (III-7,8,9, and 10) would be albino, and this is borne out.

Based on this pedigree analysis and the conclusion that albinism is a recessive trait, the genotypes of all individuals can be predicted. For both the first and second generations, this can be done with some certainty. In the third generation, for most normal individuals, we can only guess whether or not they are homozygous or heterozygous.

Pedigree analysis of many traits has been an extremely valuable research technique in human genetic studies.

TABLE 3.2
Representative recessive and dominant human traits

Recessive Traits	Dominant Traits
Albinism	Achondroplasia
Alkaptonuria	Brachydactyly
Ataxia telangiectasia	Congenital stationary night blindness
Color blindness	Ehler-Danlos syndrome
Cystic fibrosis	Fascio-scapulo-humeral muscular dystrophy
Duchenne muscular dystrophy	Huntington disease
Galactosemia	Hypercholesterolemia
Hemophilia	Marfan syndrome
Lesch-Nyhan syndrome	Neurofibromatosis
Phenylketonuria	Phenylthiocarbamide tasting (PTC)
Sickle-cell anemia	Widow's peak
Tay-Sachs disease	

However, this approach does not usually provide the certainty in drawing conclusions that is afforded by designed crosses yielding large numbers of offspring. Nevertheless, when many independent pedigrees of the same trait or disorder are analyzed, consistent conclusions can often be drawn. Table 3.2 lists numerous human traits and classifies them according to their recessive or dominant expression. As we will see in Chapter 4, the genes controlling some of these traits are located on the sex-determining chromosomes.

CHAPTER SUMMARY

1 Over a century ago, Mendel studied inheritance patterns in the garden pea, establishing the principles of transmission genetics.

2 Mendel's postulates help describe the basis for the inheritance of phenotypic expression. He showed that unit factors, later called alleles, exist in pairs and exhibit a dominant/recessive relationship in determining the expression of traits.

3 Mendel postulated that these alleles must segregate during gamete formation, such that each receives only one of the two factors with equal probability.

4 Mendel's final postulate of independent assortment states that each pair of unit factors segregates independently of other such pairs. As a result, all possible combinations of gametes will be formed with equal probability.

5 The discovery of chromosomes in the late 1800s and subsequent studies of their behavior during meiosis led to the rebirth of Mendel's work, linking the behavior of his unit factors with that of chromosomes during meiosis.

6 The Punnett square and the forked-line methods are used to predict the probabilities of phenotypes (and genotypes) from crosses involving two or more gene pairs.

7 Genetic ratios are expressed as probabilities. Thus, deriving outcomes of genetic crosses relies on an understanding of the laws of probability.

8 Statistical analysis is used to test the validity of experimental outcomes. In genetics, variations from the expected ratios are anticipated owing to chance deviation. A chi-square analysis tests the probability of these variations being generated from chance alone. It provides the basis for assessing the null hypothesis.

9 Pedigree analysis provides a method for studying the inheritance pattern of human traits over several generations. This often provides the basis for determining the mode of inheritance of human characteristics and disorders.

KEY TERMS

albinism
chi-square (χ^2) analysis
dihybrid cross
discontinuous variation
dizygotic twins
dominance
F_1 (first filial generation)
F_2 (second filial generation)
forked-line method
genotype
heterozygote
homozygote
monohybrid cross
monozygotic twins
null hypothesis
P_1 (parental generation)
pedigree
phenotype
proband
product law
propositus (proposita)
recessive
reciprocal cross
sum law
test cross
trihybrid cross

Students demonstrate their knowledge of transmission genetics by solving genetics problems. Success at this task represents not only comprehension of the theory but its application to more practical genetic situations. Most students find problem solving in genetics to be challenging but rewarding. This section is designed to provide basic insights into the reasoning essential to this process.

Genetics problems are in many ways similar to algebraic word problems. The approach taken should be identical: (1) analyze the problem carefully; (2) translate words into symbols, defining each one first; and (3) choose and apply a specific technique to solve the problem. The first two steps are the most critical. The third step is largely mechanical.

The simplest problems are those that state all necessary information about the P_1 generation and ask you to find the expected ratios of the F_1 and F_2 genotypes and/or phenotypes. The following steps should always be followed when you encounter this type of problem:

1 Determine the genotypes of the P_1 generation.

2 Determine what gametes may be formed by the P_1 parents.

3 Recombine gametes either by the Punnett square method, the forked-line method, or, if the situation is very simple, by inspection. Read the F_1 phenotypes directly.

4 Repeat the process to obtain information about the F_2 generation.

Determining the genotypes from the given information requires an understanding of the basic theory of transmission genetics. For example, consider the following problem:

A recessive mutant allele, *black*, causes a very dark body in *Drosophila* when homozygous. The wild-type color is described as gray. What F_1 phenotypic ratio is predicted when a black female is crossed to a gray male whose father was black?

To work this problem, you must understand dominance and recessiveness as well as the principle of segregation. Further, you must use the information about the male parent's father. You can work out the problem as follows:

1 Because the female parent is black, she must be homozygous for the mutant allele (*bb*).

2 The male parent is gray; therefore, he must have at least one dominant allele (*B*). Because his father was black (*bb*) and he received one of the chromosomes bearing these alleles, the male parent must be heterozygous (*Bb*).

From here, the problem is simple:

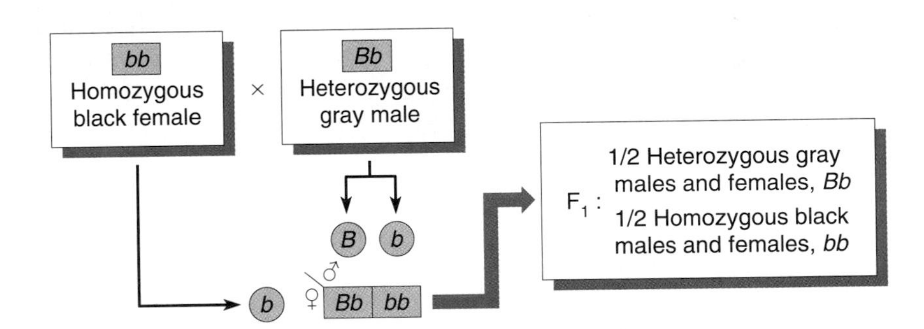

Apply this approach to the following problems.

1 In Mendel's work, he found that full pods are dominant to constricted pods while round seeds are dominant to wrinkled seeds. One of his crosses was between full, round plants and constricted, wrinkled plants. From this cross, he obtained an F_1 that was all full and round. In the F_2, Mendel obtained his classic 9:3:3:1 ratio. Using the above information, determine the expected F_1 and F_2 results of a cross between constricted, round and full, wrinkled plants.

SOLUTION: First of all, define gene symbols for each pair of contrasting traits. Select the lowercase forms of the first letter of the recessive traits to designate those phenotypes, and use the uppercase forms to designate the dominant traits. Thus, use *C* and *c* to indicate full and constricted, and use *W* and *w* to indicate the round and wrinkled phenotypes, respectively.

Now, determine the genotypes of the P_1 generation, form gametes, reconstitute the F_1 generation, and read off the phenotype(s):

P_1:	*ccWW* constricted, round	×	*CCww* full, wrinkled
	↓		↓
Gametes:	*cW*		*Cw*
F_1:	*CcWw* full, round		

You can see immediately that the F_1 generation expresses both dominant phenotypes and is heterozygous for both gene pairs. Thus, we can expect that the F_2 generation will yield the classic Mendelian ratio of 9:3:3:1. Let's work it out anyway, just to confirm this, using the forked-line method. Because both gene pairs are heterozygous and can be expected to assort independently, we can predict the F_2 outcomes from each gene pair separately and then proceed with the forked-line method.

Every F_2 offspring is subject to the following probabilities:

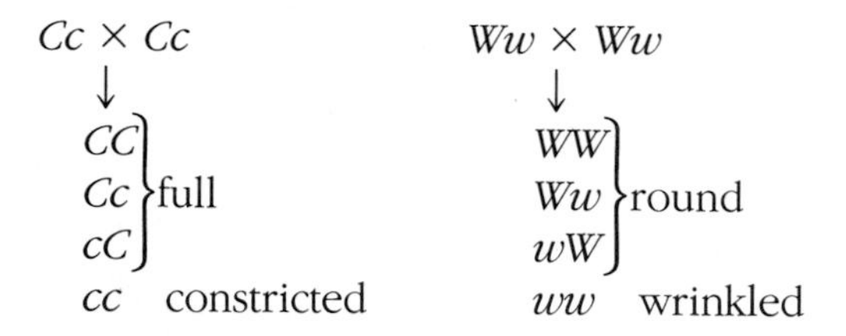

The forked-line method then allows us to confirm the 9:3:3:1 phenotypic ratio. Remember that this represents proportions of 9/16:3/16:3/16:1/16. Note that we are applying the product law as we compute the final probabilities:

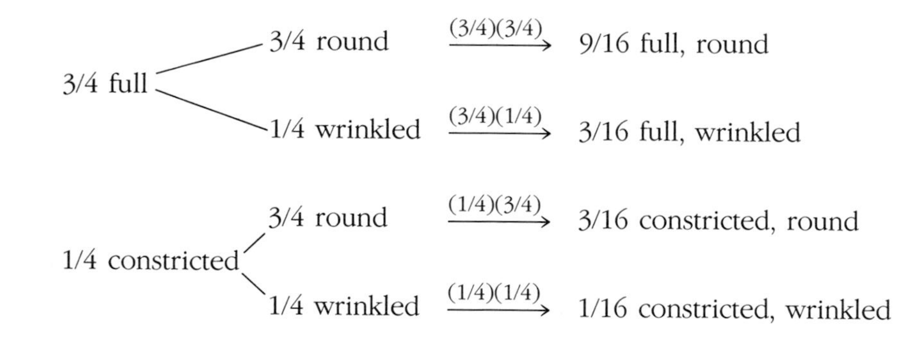

2 Determine the probability that a plant of genotype *CcWw* will be produced from parental plants of the genotypes *CcWw* and *Ccww*.

SOLUTION: Because the two gene pairs demonstrate straightforward dominance and recessiveness and independently assort during gamete formation, we need only calculate the individual probabilities of the two separate events (*Cc* and *Ww*) and apply the product law to calculate the final probability:

$$Cc \times Cc \rightarrow 1/4\ CC{:}1/2\ Cc{:}1/4\ cc$$

$$Ww \times ww \rightarrow 1/2\ Ww{:}1/2\ ww$$

$$p = (1/2\ Cc)(1/2\ Ww) = 1/4\ CcWw$$

PROBLEMS AND DISCUSSION QUESTIONS

When working genetics problems in this and succeeding chapters, always assume that members of the P_1 generation are homozygous, unless the information given indicates or requires otherwise.

1 In a cross between a black and a white guinea pig, all members of the F_1 generation are black. The F_2 generation is made up of approximately 3/4

black and 1/4 white guinea pigs.
Diagram this cross, showing the genotypes and phenotypes.

2 Albinism in humans is inherited as a simple recessive trait. For the following families, determine the genotypes of the parents and offspring. When two alternative genotypes are possible, list both.
(a) Two nonalbino (normal) parents have five children, four normal and one albino.
(b) A normal male and an albino female have six children, all normal.

3 What advantages were provided by Mendel's choice of the garden pea in his experiments?

4 Pigeons may exhibit a checkered or plain pattern. In a series of controlled matings, the following data were obtained:

	F_1 Progeny	
P_1 Cross	Checkered	Plain
(a) checkered × checkered	36	0
(b) checkered × plain	38	0
(c) plain × plain	0	35

How are the checkered and plain patterns inherited? Predict the results of the $F_1 \times F_1$ mating from cross (b).

5 Mendel crossed peas having round seeds and yellow cotyledons with peas having wrinkled seeds and green cotyledons. All the F_1 plants had round seeds with yellow cotyledons. Diagram this cross through the F_2 generation using both the Punnett square and forked-line, or branch diagram, methods.

6 Determine the genotypes of the F_2 plants given here by analyzing the phenotypes of the offspring of these crosses.

F_2 Plants	Offspring
(a) round, yellow × round, yellow	3/4 round, yellow 1/4 wrinkled, yellow
(b) wrinkled, yellow × round, yellow	6/16 wrinkled, yellow 2/16 wrinkled, green 6/16 round, yellow 2/16 round, green

7 Is either of the crosses in Problem 6 a test cross?

8 Which of Mendel's postulates can only be demonstrated in crosses involving at least two pairs of traits? Define it.

9 Correlate Mendel's four postulates with what is now known about homologous chromosomes, genes, alleles, and the process of meiosis.

10 What is the basis for homology among chromosomes?

11 Distinguish between homozygosity and heterozygosity.

12 In *Drosophila, gray* body color is dominant to *ebony* body color, while *long* wings are dominant to *vestigial* wings. Work the following crosses through the F_2 generation and determine the genotypic and phenotypic ratios for each generation. Assume the P_1 individuals are homozygous.
(a) gray, long × ebony, vestigial
(b) gray, vestigial × ebony, long
(c) gray, long × grey, vestigial

13 How many different types of gametes can be formed by individuals of the following genotypes: (a) *AaBb*, (b) *AaBB*, (c) *AaBbCc*, (d) *AaBBcc*, (e) *AaBbcc*, and (f) *AaBbCcDdEe*? What are they in each case?

14 Using the forked-line, or branch diagram, method, determine the genotypic and phenotypic ratios of the trihybrid crosses (a) *AaBbCc* × *AaBBCC*, (b) *AaBBCc* × *aaBBCc*, and (c) *AaBbCc* × *AaBbCc*.

15 Mendel crossed peas with round, green seeds to ones with wrinkled, yellow seeds. All F_1 plants had seeds that were round and yellow. Predict the results of test-crossing these F_1 plants.

16 Below are shown F_2 results of two of Mendel's monohybrid crosses. Calculate the χ^2 value and determine the p value for both. Which of the two crosses shows a greater amount of deviation?

(a)	Full pods	882
	Constricted pods	299
(b)	Violet flowers	705
	White flowers	224

17 In one of Mendel's dihybrid crosses, he observed 315 smooth, yellow, 108 smooth, green, 101 wrinkled, yellow, and 32 wrinkled, green F_2 plants. Analyze these data using the chi-square test to see if
(a) they fit a 9:3:3:1 ratio
(b) the smooth:wrinkled traits fit a 3:1 ratio
(c) the yellow:green traits fit a 3:1 ratio

18 For the following pedigree, predict the mode of inheritance and the resulting genotypes of each individual. Assume that the alleles *A* and *a* control the expression of the trait.

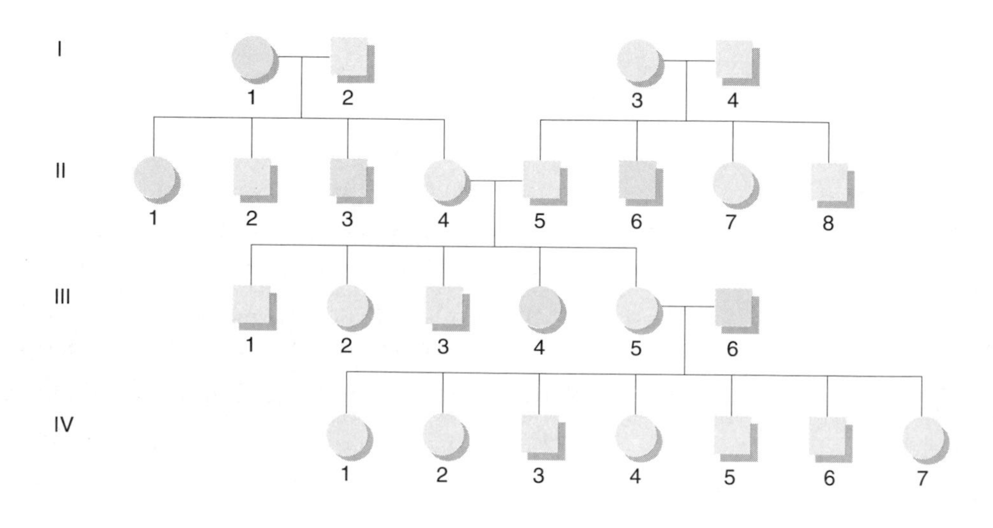

19 Which of Mendel's postulates are illustrated by the pedigree in Problem 18? List and define these postulates.

20 Thalassemia is an inherited anemic disorder in humans. Individuals can be completely normal, they can exhibit a "minor" anemia, or they can exhibit a "major" anemia. Assuming that only a single gene pair and two alleles are involved in the inheritance of these conditions, which phenotype is recessive?

21 A geneticist, in assessing data that fell into two phenotypic classes, observed values of 250:150. She decided to perform chi-square analysis using two different null hypotheses: (a) the data fit a 3:1 ratio; and (b) the data fit a 1:1 ratio. Calculate the χ^2 values for each hypothesis. What can be concluded about each hypothesis?

22 The basis for rejection of any null hypothesis is arbitrary. The researcher can set more or less stringent standards by deciding to raise or lower the p value used to reject or fail to reject the hypothesis. In the case of chi-square analysis of genetic crosses, would the use of a standard of $p = 0.10$ be more or less stringent in failing to reject the null hypothesis? Explain.

SELECTED READINGS

CUMMINGS, M. R. 1991. *Human heredity: principles and issues.* 2nd ed. St. Paul: West.

DUNN, L. C. 1965. *A short history of genetics.* New York: McGraw-Hill.

PETERS, J., ed. 1959. *Classic papers in genetics.* Englewood Cliffs, N.J.: Prentice-Hall.

SNEDECOR, G. W., and COCHRAN, W. G. 1980. *Statistical methods.* 7th ed. Ames, Iowa: Iowa State University Press.

SOKAL, R. R., and ROHLF, F. J. 1981. *Biometry—The principles and practice of statistics in biological research.* 2nd ed. New York: W. H. Freeman.

SOUDEK, D. 1984. Gregor Mendel and the people around him. *Am. J. Hum. Genet.* 36:495–98.

STURTEVANT, A. H. 1965. *A history of genetics.* New York: Harper & Row.

VOELLER, B. R., ed. 1968. *The chromosome theory of inheritance: Classical papers in development and heredity.* New York: Appleton-Century-Crofts.

4 Modification of Mendelian Ratios

Chapter Concepts

Specific phenotypes are often controlled by one or more gene pairs whose alleles exhibit modes of expression other than dominance and recessiveness. In all such cases, however, the Mendelian principles of segregation and independent assortment are operative during the distribution of the alleles into gametes.

In Chapter 3, we discussed the simplest principles of transmission genetics. We saw that genes are present on homologous chromosomes and that these chromosomes segregate from each other and independently assort with other segregating chromosomes during gamete formation. These two postulates are the fundamental principles of gene transmission from parent to offspring. However, if gene expression does not adhere to a simple dominant/recessive mode, or if more than one pair of genes influences the expression of a single trait, the classic 3:1 and 9:3:3:1 ratios may be modified. Although we will consider more complex modes of inheritance, the fundamental principles set down by Mendel hold true in these situations as well.

In this chapter, our discussion will initially be restricted to the inheritance of traits that are under the control of only one set of genes. In diploid organisms, where homologous pairs of chromosomes exist, two copies of each gene influence such traits. The copies need not be identical since alternate forms of genes, or **alleles,** occur within populations. How alleles act to influence a given phenotype will be our major consideration. Then we will proceed to consider how a single phenotype may be controlled by more than one set of genes. This general phenomenon is referred to as **gene interaction,** indicating that phenotypes are frequently under the influence of the expression of more than one gene pair.

We will conclude by examining cases where genes are present on the X chromosome, illustrating **sex-linkage.** Prior to that point, in this and preceding chapters, we have restricted our discussion to chromosomes other than the X and Y pair. All such chromosomes are called **autosomes** to distinguish them from the X and Y chromosomes. As we will see, sex-linkage provides yet another modification of Mendelian ratios.

POTENTIAL FUNCTION OF AN ALLELE

Following the rediscovery of Mendel's work in the early 1900s, research focused on the many ways in which genes can influence an individual's phenotype. This course of investigation, stemming from Mendel's findings, is called **neo-Mendelian genetics** (*neo* from the Greek word meaning since or new).

Each type of inheritance described in this chapter was investigated when observations of genetic data did not precisely conform to the expected Mendelian ratios. Hypotheses that modified and extended the Mendelian principles were proposed and tested with specifically designed crosses. The explanations for these observations were in accordance with the principle that a phenotype is under the control of one or more genes located at specific loci on one or more pairs of homologous chromosomes. If we adhere to the principles of segregation and independent assortment, we can predict accurately the transmission of any number of allele pairs.

To understand the various modes of inheritance, we must first examine the potential function of an allele. Alleles are alternate forms of the same gene; therefore, alleles contain modified genetic information and often specify an altered gene product. For example, in human populations there are well over 100 alleles of the genes that specify the protein portion of hemoglobin. Even though each allele specifies a modified component of hemoglobin, they all store information necessary for this molecule's chemical synthesis. Once manufactured, however, the product of an allele may or may not be functional.

The allele that occurs most frequently in a population, or the one that is arbitrarily designated as normal, is called **wild type.** This common allele is often dominant, such as the allele for tall plants in the garden pea, and its product is functional in the cell. Wild-type alleles are, of course, responsible for the corresponding wild type phenotype and thus serve as standards for comparison against all mutations occurring at a particular locus.

The process of **mutation** is the source of new alleles. Each allele may be recognized by a change in the phenotype. A new phenotype results from a change in functional activity of the cellular product controlled by that gene. Usually, the alteration or mutation is expressed as a loss of the specific wild-type function. For example, if a gene is responsible for the synthesis of a specific enzyme, a mutation in the gene may change the conformation of this enzyme, thus eliminating its affinity for the substrate. This mutation results in a total loss of function. On the other hand, another organism may have a different mutation in this gene, which results in an enzyme with a reduced or increased affinity for binding the substrate. This mutation may reduce or enhance rather than eliminate the functional capacity of the gene product. In either case, the overt phenotype of the organism may or may not be altered in a discernible way.

SYMBOLS FOR ALLELES

In Chapter 3, we learned to symbolize alleles for very simple Mendelian traits. We used the lowercase form of the initial letter of the name of a recessive trait to denote the recessive allele and the same letter in uppercase form to refer to the corresponding dominant allele. Thus, for *tall* and *dwarf,* where dwarf is recessive, *D* and *d* represent the alleles responsible for these respective traits.

As more complex inheritance patterns were investigated, another useful system was developed to discriminate between wild-type and mutant traits. In this system, the initial letter, or the first two letters, of the name of the mutant trait is selected. If the trait is recessive, the lowercase form of the letters is used; if it is dominant, the uppercase form of the initial letter is used. The contrasting wild-type trait is denoted by the same letter, but with a + as a superscript.

For example, *ebony* is a recessive body color mutation in the fruit fly, *Drosophila melanogaster.* The normal wild type body color is gray. Using the above system, *ebony* is denoted by the symbol *e* while gray is denoted by e^+. If we focus on the *ebony* mutation, the responsible locus may be occupied by either the wild type allele (e^+) or the mutant allele (e). A diploid fly may thus exhibit three possible genotypes:

e^+/e^+: gray homozygote (wild type)
e^+/e : gray heterozygote (wild type)
e/e : ebony homozygote (mutant)

The slash is used to indicate that the two allele designations represent the same locus on two homologous chromosomes. If we were instead considering a dominant mutation such as *Wrinkled (Wr)* wing in *Drosophila,* the three possible designations would be Wr^+/Wr^+, Wr^+/Wr, and Wr/Wr. The latter two genotypes express the wrinkled-wing phenotype.

One advantage of this system is that further abbreviation may be used when convenient: the wild type allele may simply be denoted by the + symbol. Using *ebony* as an example under consideration in a cross, the designations of the three possible genotypes become:

$+/+$: gray homozygote (wild type)
$+/e$: gray heterozygote (wild type)
e/e : ebony homozygote (mutant)

As we will see in Chapter 5, this abbreviation is particularly useful when two or three genes linked together on the same chromosome are considered simultaneously.

Still other allele designations are sometimes useful. The system just described works well with alleles that are either dominant or recessive to one another. However, if no dominance exists, we may simply use uppercase letters and superscripts to denote alleles, e.g., R^1 and R^2, L^M and L^N, I^A and I^B. Their use will become apparent in the ensuing three sections.

Finally, note that in each of the many crosses discussed in the next few chapters, only one or a few gene pairs are involved. It may be useful for you to remember that in each cross, all genes not under consideration are assumed to be normal, or wild type.

INCOMPLETE, OR PARTIAL, DOMINANCE

Incomplete, or **partial, dominance** in the offspring is based on the observation of intermediate phenotypes generated by a cross between parents with contrasting traits. For example, if plants such as four-o'clocks or snapdragons with red flowers are crossed with plants with white flowers, offspring may have pink flowers. It appears that neither red nor white flower color is dominant. Since some red pigment is produced in the F_1 intermediate-colored pink flowers, dominance appears to be incomplete or partial.

If this phenotype is under the control of a single pair of alleles where neither is dominant, the results of the F_1 (pink) × F_1 (pink) cross can be predicted. The resulting F_2 generation is shown in Figure 4.1, confirming the hypothesis that only one pair of alleles determines these phenotypes. The genotypic ratio (1:2:1) of the F_2 generation is identical to that of Mendel's monohybrid cross. Because there is no dominance, however, the phenotypic ratio is identical to the genotypic ratio. Note here that since neither of the alleles is recessive, we have chosen not to use upper and lowercase letters. Instead, we have chosen R^1 and R^2 to denote the red and white alleles. We could have chosen W^1 and W^2.

CODOMINANCE

If one pair of alleles is responsible for the production of two distinct and detectable gene products, a situation dif-

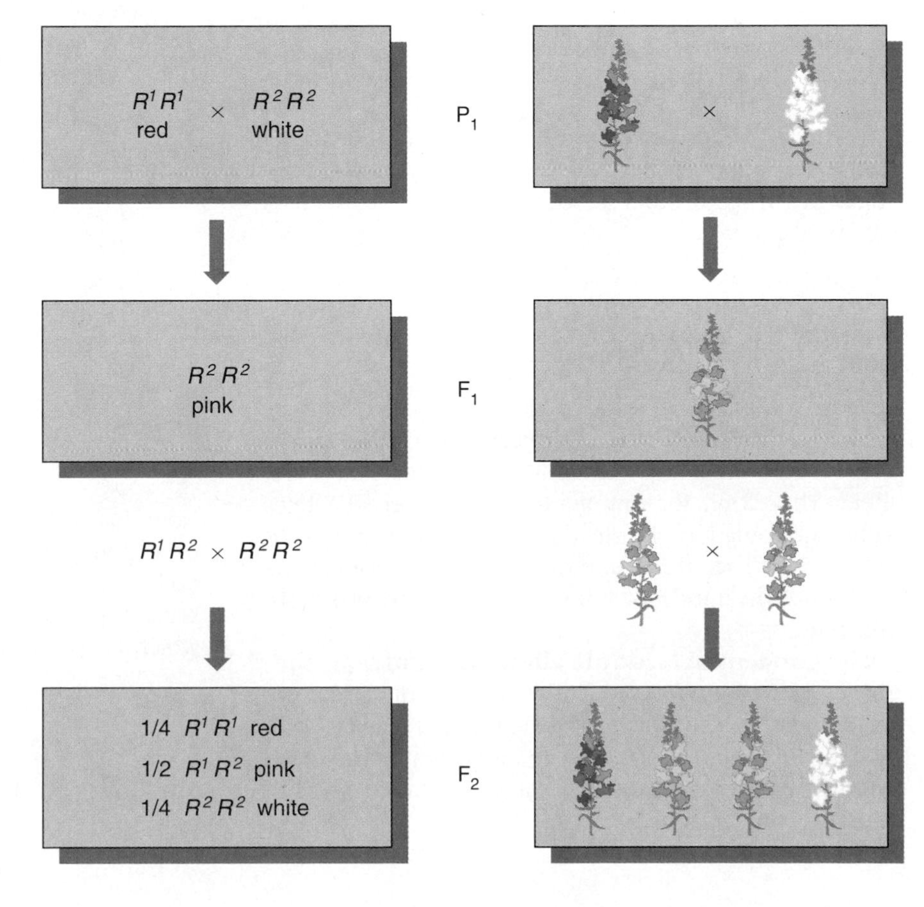

FIGURE 4.1
Incomplete dominance illustrated by flower color. The photograph illustrates not only red, white, and pink snapdragons, but other colors produced as a result of the cffcct of other genes.

ferent from incomplete dominance or dominance/recessive arises. The distinct genetic expression of two alleles is called **codominance.** The **MN blood group** in humans illustrates this phenomenon and is characterized by a molecule called a glycoprotein found on the surface of red blood cells. In the human population, two forms of this **glycoprotein** exist, designated M and N. An individual may exhibit either one or both of them.

The MN system is under the control of an autosomal locus found on chromosome 4 and two alleles designated L^M and L^N. Because humans are diploid, three combinations are possible, each resulting in a distinct blood type:

Genotype	Phenotype
L^ML^M	M
L^ML^N	MN
L^NL^N	N

As predicted, a mating between two MN parents may produce children of all three blood types:

$$L^ML^N \quad \times \quad L^ML^N$$
$$\downarrow$$
$$1/4\ L^ML^M$$
$$1/2\ L^ML^N$$
$$1/4\ L^NL^N$$

Codominant inheritance results in distinct evidence of the gene products of both alleles. This characteristic distinguishes it from other modes of inheritance, such as incomplete dominance.

MULTIPLE ALLELES

Because the information stored in any gene is extensive, mutations may modify this information in many ways. Each change has the potential for producing a different allele. Therefore, for any gene, the number of alleles within a population of individuals need not be restricted to only two. When three or more alleles of the same gene are found, the mode of inheritance is called **multiple allelism.**

The concept of multiple alleles can only be studied in populations. Any individual diploid organism has, at most, two homologous gene loci that may be occupied by different alleles of the same gene. However, among members of a species, many alternative forms of the same gene may exist.

The ABO Blood Group

The simplest possible case of multiple alleles is that in which there are three alleles of one gene. This situation exists in the inheritance of the **ABO blood group** in humans, discovered by Karl Landsteiner in the early 1900s. The ABO system, like the MN blood types, is characterized by the presence of native antigens on the surface of red blood cells. The A and B antigens are distinct from the MN antigens and are under the control of a different gene, located on chromosome 9. As in the MN system, one combination of alleles in the ABO system exhibits a codominant mode of inheritance.

When individuals are examined, four phenotypes are revealed. Each individual has either the A antigen (A phenotype), the B antigen (B phenotype), the A and B antigens (AB phenotype), or neither antigen (O phenotype). In 1924, it was hypothesized that these phenotypes were inherited as the result of three alleles of a single gene. This hypothesis was based on studies of the blood types of many different families.

Although different designations may be used, we will use the symbols I^A, I^B, and I^O for the three alleles. The I designation stands for **isoagglutinogen,** another term for antigen. If we assume that the I^A and I^B alleles are responsible for the production of their respective A and B antigens and that I^O is an allele that does not produce any detectable A or B antigens, the various genotypic possibilities can be listed and the appropriate phenotype assigned to each:

Genotype	Antigen	Phenotype
I^AI^A	A	A
I^AI^O	A	
I^BI^B	B	B
I^BI^O	B	
I^AI^B	A and B	AB
I^OI^O	Neither	O

Note that in these assignments the I^A and I^B alleles behave dominantly to the I^O allele, but codominantly to each other. Our knowledge of human blood types has several practical applications, the most important of which is compatible blood transfusions.

The Bombay Phenotype

The biochemical basis of the ABO blood type system has now been carefully worked out. The A and B antigens are actually carbohydrate groups (sugars) that are bound to lipid molecules (fatty acids) protruding from the membrane of the red blood cell. The specificity of the A and B antigens is based on the terminal sugar of the carbohydrate group.

Both the A and B antigens are derived from a precursor molecule called the **H substance,** to which one of two terminal sugars is added. In extremely rare instances, first recognized in a woman in Bombay, the H substance is incompletely formed. As a result, the individual cannot add either terminal sugar and exhibits blood type O. This condition called the **Bombay phenotype,** has been shown to be due to a rare recessive mutation, *h,* at a locus separate from that controlling the A and B antigens. Thus, an individual who has an *hh* genotype adds neither antigen, even though she or he may contain the I^A and/or I^B alleles. This information helped explain why the woman in Bombay was functionally type O even though one of her parents was type AB (and thus she should not have been type O), and she was able to donate the I^B allele to her children (Figure 4.2).

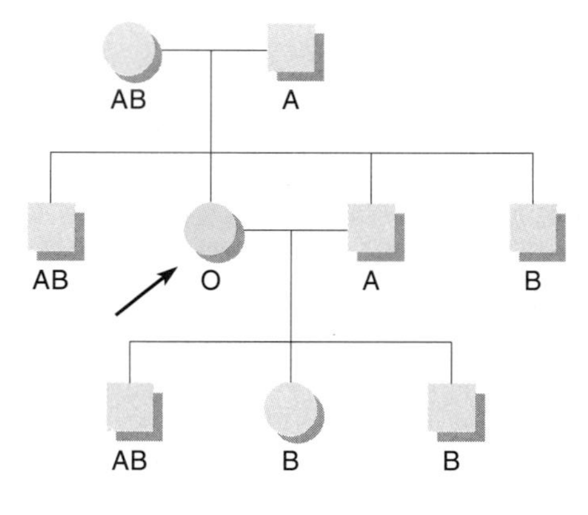

FIGURE 4.2
A partial pedigree of a woman displaying the Bombay phenotype. Functionally, her ABO blood group behaves as type O. Genetically, she is type B.

Multiple Alleles in *Drosophila:* The *white* Locus

Many other phenotypes in plants and animals are known to be controlled by multiple allelic inheritance. In *Drosophila,* for example, many alleles are known at practically every locus. The recessive *white* eye mutation, discovered by Thomas H. Morgan and Calvin Bridges in 1912, is only one of over 100 alleles that may occupy this locus. In this allelic series, eye colors range from complete absence of pigment in the *white* allele, to deep ruby in the *white-satsuma* allele, to orange in the *white-apricot* allele, to a buff color in the *white-buff* allele. These alleles are designated w, w^{sat}, w^a, and w^{bf}, respectively (Table 4.1). In each of these cases, the total amount of pigment in these mutant eyes is reduced to less than 20 percent of that found in the brick-red, wild-type eye.

TABLE 4.1
Some of the alleles present at the white locus of *Drosophila melanogaster* and their eye color phenotype

Allele	Name	Eye Color
w	*white*	pure white
w^a	*white-apricot*	yellowish orange
w^{bf}	*white-buff*	light buff
w^{bl}	*white-blood*	yellowish ruby
w^{cf}	*white-coffee*	deep ruby
w^e	*white-eosin*	yellowish pink
w^{mo}	*white-mottled orange*	light mottled orange
w^{sat}	*white-satsuma*	deep ruby
w^{sp}	*white-spotted*	fine grain, yellow mottling
w^t	*white-tinged*	light pink

LETHAL ALLELES

Many gene products are essential to an organism's survival. Mutations resulting in the synthesis of a gene product that is nonfunctional can sometimes be tolerated in the heterozygous state; that is, one wild-type allele may be sufficient to produce enough of the essential product to allow survival. However, such a mutation behaves as a **recessive lethal allele,** and homozygous recessive individuals will not survive. The time of death will depend upon when the product is needed during development or in adulthood.

In some cases, the allele responsible for a lethal effect when homozygous may result in a distinctive mutant phenotype when present heterozygously. Such an allele is behaving as a recessive lethal but is dominant with respect to the phenotype. For example, a mutation causing a yellow coat in mice was discovered in the early part of this century. The yellow coat varies from the normal wild type coat phenotype, as illustrated in Figure 4.3. Crosses between the various combinations of the two strains yields unusual results:

Crosses
A: normal × normal → all normal
B: yellow × yellow → 2/3 yellow: 1/3 normal
C: normal × yellow → 1/2 yellow: 1/2 normal

These results are explained on the basis of a single pair of alleles. With regard to coat color, the mutant *yellow* allele

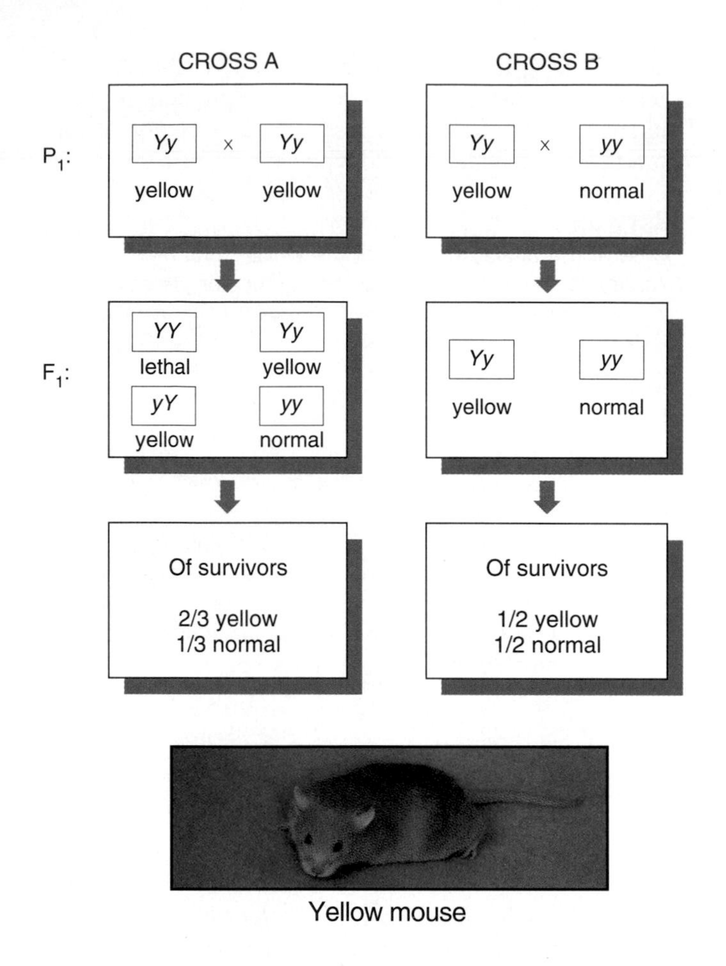

FIGURE 4.3
Inheritance patterns in two crosses involving the mutant *yellow* allele (*Y*) in the mouse. Note that the mutant allele behaves dominantly to the normal allele (*y*) in controlling coat color, but it also behaves as a recessive lethal allele. The genotype *YY* does not survive.

(*Y*) is dominant to the normal wild-type allele (*y*), so heterozygous mice will have yellow coats. However, the yellow allele also behaves as a recessive lethal. When present in two copies, the mice die before birth. Thus, no homozygous yellow mice are ever recovered. The genetic basis for these three crosses is provided in Figure 4.3.

Other alleles are known to behave as **dominant lethals.** In humans, a disorder called **Huntington disease** (also referred to as Huntington's chorea) is due to a dominant autosomal allele that behaves quite differently from the alleles just described. While individuals homozygous for the lethal gene apparently never survive, heterozygotes develop normally well into adulthood. Affected individuals then undergo gradual nervous and motor degeneration until they die. This lethal disorder is particularly tragic because it has such a late onset, typically about age 40. By that time, the affected individual may have produced a family. Each child has a 50% probability of also developing the disorder and passing the lethal allele to his or her offspring. The American folk singer and composer Woody Guthrie died from this disease.

Dominant lethal alleles are rarely observed on autosomes. In order for them to exist in a population, the affected individual must reproduce before the allele's lethality is expressed, as occurs in Huntington disease. If all affected individuals die before reaching the reproductive age, the mutant gene will not be passed to future generations. The mutation will thus disappear from the population unless it recurs as a result of mutation.

COMBINATIONS OF TWO GENE PAIRS

Each example discussed so far modifies Mendel's 3:1 F_2 monohybrid ratio. Therefore, combining any two of these modes of inheritance in a dihybrid cross will likewise modify the classical 9:3:3:1 ratio. Having established the foundation for the modes of inheritance of incomplete dominance, codominance, multiple alleles, and lethal genes, we can now deal with the situation of two modes of inheritance occurring simultaneously. Mendel's principle of independent assortment applies to these situations, provided that the genes controlling each character are not located on the same chromosome.

Suppose, for example, that a mating occurs between two humans who are both heterozygous for the autosomal recessive gene that causes albinism and who are both of blood type AB. What is the probability of any particular phenotypic combination occurring in each of their children? Albinism is inherited in the simple Mendelian fashion, and the blood types are determined by the series of three multiple alleles, I^A, I^B, and I^O. The solution to this problem is diagrammed in Figure 4.4, using the forked-line method.

Instead of this dihybrid cross yielding the classical four phenotypes in a 9:3:3:1 ratio, six phenotypes occur in a 3:6:3:1:2:1 ratio, establishing the expected probability for each phenotype.

GENE INTERACTION: DISCONTINUOUS VARIATION

Soon after the rediscovery of Mendel's work, experimentation revealed that phenotypic characters were often under the control of more than one gene pair. This was a significant discovery because it revealed for the first time that genetic influence on the phenotype is much more sophisticated than envisioned by Mendel. Instead of single genes controlling the development of individual parts of the plant and animal body, it soon became clear that

GENOMIC (PARENTAL) IMPRINTING

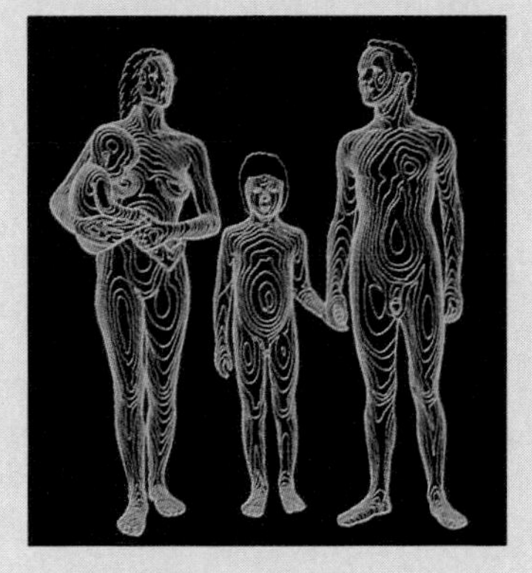

In humans there are some genetic disorders in which the severity of expression depends on whether the disease was inherited from the father or the mother. For example, Huntington disease is a fatal neurological disorder that usually has an age of onset of around 40. However, in certain individuals onset is much earlier, even in childhood. In 90 percent of early onset cases the disorder was inherited from the father. In another example, expression of a severe, early onset form of myotonic dystrophy depends on the disease being inherited from the mother.

This phenomenon of differential expression of a genetic trait depending on the parental origin of the trait results from a relatively recently discovered process called genomic imprinting. We do not understand how imprinting occurs, although it is probably related to processes that generate differential gene activity in cells, but we do know that it occurs differently in the cell lines leading to the production of male gametes (sperm) and female gametes (eggs). Imprinting alters the functioning of genes; for example, a gene that has been imprinted in the father may function poorly in his children, whereas essentially the same gene would be imprinted to function well as the mother formed eggs. The converse pattern can also occur. Since each child normally receives an allele of each gene from each parent, the well functioning and poorly functioning genes compensate for each other. The process is perfectly natural. It does not seem to apply to all genes. It is clearly different from mutation, because it can be reversed every generation as genes pass from mother to son to granddaughter and so on. Only in unusual cases, like the diseases mentioned previously, can we identify its effects.

For several years, a number of lines of evidence have implied the existence of genomic imprinting. However, it is only recently that specific genes have been identified as imprinted genes. In 1991, three mouse genes were shown to be imprinted; one of them is the gene encoding insulinlike growth factor II(*Igf2*). A mouse that carries two normal alleles of this gene is normal in size, whereas a mouse that carries two mutant alleles lacks a growth factor and is a dwarf. The size of a heterozygous mouse (one allele normal and one mutant) depends on the parent origin of the normal allele. It is normal in size if the normal allele came from the father, but dwarf if it came from the mother. From this, it can be deduced that the normal *Igf2* gene is imprinted to function poorly during the course of egg production in females and to function well during sperm production in males. We can see, then, that a normal-sized male heterozygote (which must have inherited the mutant allele from his mother) will pass the mutant allele to half his offspring. Even if their mother is a homozygote for the normal allele, these will be dwarf, because even a normal *Igf2* allele received from a mother functions poorly, due to her imprinting. Conversely, a dwarf female heterozygote (which received the mutant allele from her father) will have only normal sized offspring when crossed to a normal male. We can actually measure the activity of the *Igf2* gene in embryos. Imprinting in this case is just as strong as the effect of the mutant allele; the normal *Igf2* gene is effectively silent after imprinting in the mother. The remarkable thing is that, unlike mutation, the inactivation is completely reversible by re-imprinting in the male germ line.

The area of genomic imprinting is a relatively new field of research, and many questions are still unanswered. At this time it is not known how many genes are subject to imprinting, nor its developmental role. Further, it remains unclear whether or not this genetic process only affects function in some tissues but not in others. Finally, the molecular mechanism of imprinting is still a matter for conjecture, and so is the evolutionary significance of this phenomenon.

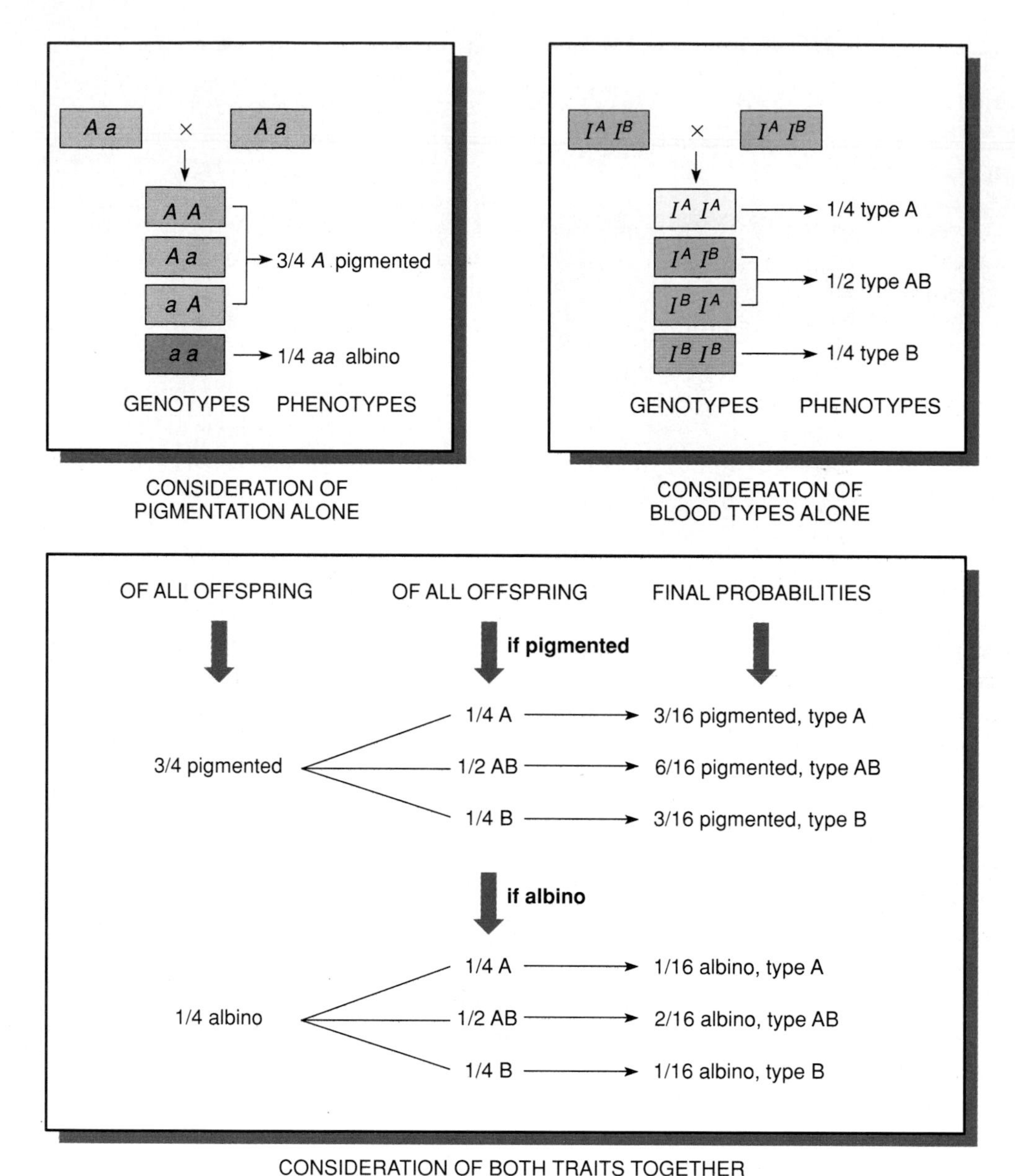

FIGURE 4.4
The calculation of the probabilities in the mating involving the ABO blood type and albinism in humans using the forked-line method.

each phenotypic character is influenced by many gene products.

The concept of **gene interaction** does not mean that two or more genes, or their products, necessarily interact directly to influence a particular phenotype. Instead, this concept implies that the cellular function of numerous gene products is related to the development of a common phenotype. To clarify this point, we will present several examples that directly illustrate gene interaction at the biochemical level.

In the following discussion, we make several assumptions and adopt certain conventions:

1. In each case, distinct phenotypic classes are produced, each clearly discernible from all others. Such traits illustrate **discontinuous variation,** where discrete phenotypic categories are qualitatively different from one another.

2. The genes considered in each cross are not linked on the same chromosome and therefore assort independently of one another during gamete formation.

3. If complete dominance exists between the alleles of any gene pair, such that *AA* and *Aa* or *BB* and *Bb* are equivalent in their genetic effects, the designations $A-$ or $B-$ will be used for both combinations. Therefore, the dash ($-$) indicates that either allele may be present, without consequence to the phenotype.

4. All P_1 crosses will involve homozygous individuals

(e.g., $AABB \times aabb$, $AAbb \times aaBB$, or $aaBB \times AAbb$). Therefore, each F_1 generation will consist of only heterozygotes of genotype *AaBb*.

5 In each example, the F_2 generation produced from these heterozygous parents will be the main focus of analysis. When two genes are involved (Figure 4.5), the F_2 genotypes fall into four categories: 9/16 $A{-}B{-}$, 3/16 $A{-}bb$, 3/16 $aaB{-}$, and 1/16 *aabb*. Because of dominance, all genotypes in each category are equivalent in their effect on the phenotype.

The study of gene interaction has revealed inheritance patterns in which these four categories are grouped together in various ways, such that each grouping yields a different phenotype. Thus, the 9:3:3:1 ratio, characteristic of Mendel's dihybrid cross, is modified in several ways (Figure 4.6). In the next several sections, we shall proceed to discuss a number of these examples.

Epistasis

Perhaps the best examples of gene interaction are those illustrating the phenomenon of **epistasis.** Derived from the Greek word meaning "stoppage," epistasis can in genetics be equated with the word *masking*. The phenomenon occurs when the expression of one gene or gene pair masks or modifies the expression of another gene or gene pair. This masking may occur under different conditions. For example, the presence of two recessive alleles at one locus may prevent or override the expression of the genes at a second locus (or several other loci). Or, a single dominant allele at the first locus may influence the expression of the alleles at a second gene locus. In a third case, two gene pairs may complement one another such that at least one dominant allele at each locus is required to express a particular phenotype. Each of these three forms of epistasis will be examined in more detail.

An example of the homozygous recessive condition at one locus masking the expression of a second locus was examined earlier in this chapter when we discussed the Bombay phenotype. There, the homozygous condition (*bb*) masked the expression of the I^A and I^B alleles. Another example is seen in the inheritance of coat color in mice (Case 1 of Figure 4.6). Normal wild-type coat color is agouti, a grayish pattern formed by alternating bands of

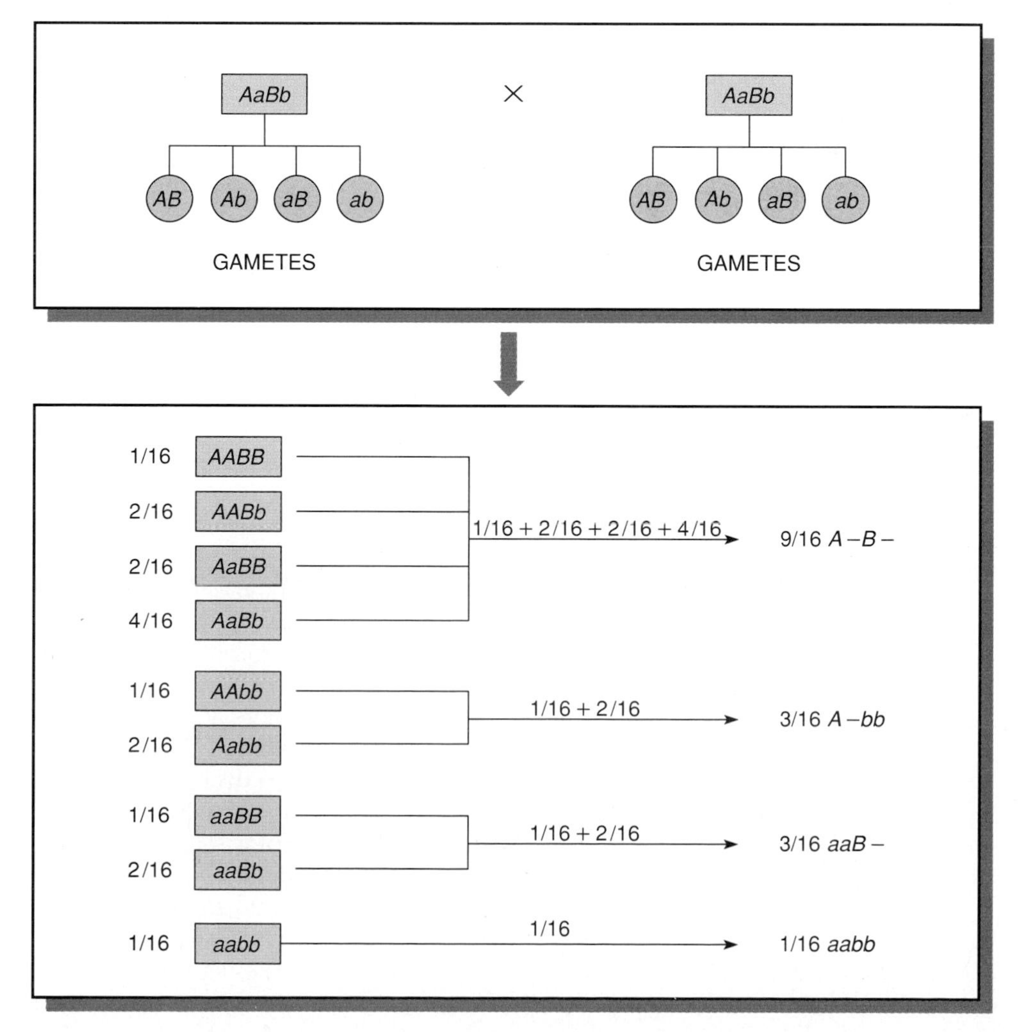

FIGURE 4.5
Generation of the four genotypic groupings from the nine genotypes produced in a cross between double heterozygotes.

<table>
<tr><th colspan="3">F1 : AaBb × AaBb →</th><th colspan="9">F2 Genotypes</th><th rowspan="2">Final Phenotypic Ratio</th></tr>
<tr><th>Case</th><th>Organism</th><th>Character</th><th>AABB 1/16</th><th>AABb 2/16</th><th>AaBB 2/16</th><th>AaBb 4/16</th><th>AAbb 1/16</th><th>Aabb 2/16</th><th>aaBB 1/16</th><th>aaBb 2/16</th><th>aabb 1/16</th></tr>
<tr><td></td><td>Pea</td><td>Mendel's dihybrid</td><td colspan="4">9/16</td><td colspan="2">3/16</td><td colspan="2">3/16</td><td>1/16</td><td>9:3:3:1</td></tr>
<tr><td>1</td><td>Mouse</td><td>Coat color</td><td colspan="4">Agouti</td><td colspan="2">Black</td><td colspan="3">Albino</td><td>9:3:4</td></tr>
<tr><td>2</td><td>Squash</td><td>Color</td><td colspan="6">White</td><td colspan="2">Yellow</td><td>Green</td><td>12:3:1</td></tr>
<tr><td>3</td><td>Pea</td><td>Flower color</td><td colspan="4">Purple</td><td colspan="5">White</td><td>9:7</td></tr>
<tr><td>4</td><td>Squash</td><td>Fruit shape</td><td colspan="4">Disc</td><td colspan="4">Sphere</td><td>Long</td><td>9:6:1</td></tr>
<tr><td>5</td><td>Chicken</td><td>Color</td><td colspan="6">White</td><td colspan="2">Colored</td><td>White</td><td>13:3</td></tr>
<tr><td>6</td><td>Mouse</td><td>Color</td><td colspan="4">White-spotted</td><td colspan="2">White</td><td colspan="2">Colored</td><td>White-spotted</td><td>10:3:3</td></tr>
<tr><td>7</td><td>Shepard's purse</td><td>Seed capsule</td><td colspan="8">Triangular</td><td>Ovoid</td><td>15:1</td></tr>
<tr><td>8</td><td>Flour beetle</td><td>Color</td><td>Red</td><td>Sooty</td><td>Red</td><td>Sooty</td><td colspan="2">Black</td><td colspan="2">Jet</td><td>Black</td><td>6:3:3:4</td></tr>
</table>

FIGURE 4.6
The basis of modified dihybrid F_2 phenotypic ratios, resulting from crosses between doubly heterozygous F_1 individuals. The four groupings shown in Figure 4.5 are combined in various ways to produce these ratios.

pigment on each hair. Agouti is dominant to nonagouti (black) hair caused by a recessive mutation, *b*. Thus, *B*− results in agouti, while *bb* yields black coat color. When homozygous, a recessive mutation, *a*, at a separate locus, eliminates pigmentation altogether, yielding albino mice (*aa*), regardless of the genotype at the other locus. The presence of at least one *A* allele simply allows pigmentation to occur.

In a cross between agouti (*AABB*) and albino (*aabb*), members of the F_1 are all *AaBb* and have agouti coat color. In the F_2 progeny of a cross between two F_1 heterozygotes, the following genotypes and phenotypes are observed:

F_1: *AaBb* × *AaBb*

↓

F_2 Ratio	Genotype	Phenotype	Final Phenotypic Ratio
9/16	*A*− *B*−	agouti	9/16 agouti
3/16	*A*− *bb*	black	3/16 black
3/16	*aa B*−	albino	4/16 albino
1/16	*aa bb*	albino	

Gene interaction yielding the observed 9:3:4 F_2 ratio can be envisioned as a two-step process:

$$\text{Precursor Molecule (colorless)} \xrightarrow[A-]{\text{Gene } A} \text{Black Pigment} \xrightarrow[B-]{\text{Gene } B} \text{Agouti Pattern}$$

In the presence of an *A* allele, black pigment can be made from a colorless substance. In the presence of a *B* allele, the black pigment is deposited during the development of hair in a pattern producing the agouti phenotype. If the *bb* genotype occurs, all of the hair remains black. If the *aa* genotype occurs in the presence of an *A* allele, no black pigment is produced, regardless of the presence of the *B* or *b* alleles, and the mouse is albino. Therefore, the *aa* genotype masks or suppresses the expression of the *B* gene, illustrating epistasis.

A second type of epistasis occurs when a dominant allele at one genetic locus masks the expression of the alleles of a second locus. For instance, Case 2 of Figure 4.6 deals with the inheritance of fruit color in summer squash. Here, the dominant allele *A* results in white fruit color regardless of the genotype at a second locus, *B*. In the absence of the dominant *A* allele (the *aa* genotype), *BB* or *Bb* results in yellow color, while *bb* results in green color. Therefore, if two white-colored double heterozygotes (*AaBb*) are crossed together, an interesting phenotypic ratio occurs because of this type of epistasis:

F_1: *AaBb* × *AaBb*

↓

F_2 Ratio	Genotype	Phenotype	Final Phenotypic Ratio
9/16	*A– B–*	white	12/16 white
3/16	*A– bb*	white	
3/16	*aa B–*	yellow	3/16 yellow
1/16	*aa bb*	green	1/16 green

Of the offspring, 9/16 are *A–B–* and are thus white. The 3/16 bearing the genotypes *A–bb* are also white. Of the remaining squash, 3/16 are yellow (*aaB–*), while 1/16 are green (*aabb*). When combined, the modified ratio of 12:3:1 occurs.

Our third example (Case 3 of Figure 4.6) is demonstrated in a cross between two strains of white-flowered sweet peas. Unexpectedly, the F_1 plants were all purple, and the F_2 occurred in a ratio of 9/16 purple to 7/16 white. The proposed explanation for these results suggests that the presence of at least one dominant allele of each of two gene pairs is essential in order for flowers to be purple. All other genotype combinations yield white flowers because the homozygous recessive condition at either locus masks the expression of the dominant allele at the other locus.

The cross is shown as follows:

P_1: *AAbb* × *aaBB*
white white

↓

F_1: All *AaBb* (purple)

↓

F_2 Ratio	Genotype	Phenotype	Final Phenotypic Ratio
9/16	*A– B–*	purple	9/16 purple
3/16	*A– bb*	white	7/16 white
3/16	*aa B–*	white	
1/16	*aa bb*	white	

We can envision the way in which two gene pairs might yield such results:

Precursor Substance (colorless)	Gene *A* ↓ ⟶ *A–*	Intermediate Product (colorless)	Gene *B* ↓ ⟶ *B–*	Final Product (purple)

At least one dominant allele from each pair of genes is necessary to ensure both biochemical conversions to the final product, yielding purple flowers. In the cross above, this will occur in 9/16 of the F_2 offspring. All other plants (7/16) have flowers that remain white.

Novel Phenotypes

Other cases of gene interaction yield novel, or new, phenotypes in the F_2 generation, in addition to producing modified dihybrid ratios. Case 4 in Figure 4.6 depicts the inheritance of fruit shape in the summer squash *Cucurbita pepo.* When plants with disc-shaped fruit (*AABB*) are crossed to plants with long fruit (*aabb*), the F_1 generation all have disc fruit. However, in the F_2 progeny, fruit with a novel shape—sphere—appear as well as fruit exhibiting the parental phenotypes. The F_2 generation, with a modified 9:6:1 ratio, is generated as follows:

F_1: *AaBb* × *AaBb*
disc disc

↓

F_2 Ratio	Genotype	Phenotype	Final Phenotypic Ratio
9/16	*A– B–*	disc	9/16 disc
3/16	*A– bb*	sphere	6/16 sphere
3/16	*aa B–*	sphere	
1/16	*aa bb*	long	1/16 long

In this example of gene interaction, both gene pairs influence fruit shape equivalently. A dominant allele at either locus ensures a sphere-shaped fruit. In the absence of dominant alleles, the fruit is long. However, if both dominant alleles (*A* and *B*) are present, the fruit is flattened into a disc shape.

Other Modified Dihybrid Ratios

The remaining cases (5–8) in Figure 4.6 illustrate additional modifications of the dihybrid ratio and provide still other examples of gene interactions. All cases (1–8) have two things in common. First, in arriving at a suitable explanation of the inheritance pattern of each one, we have not violated the principles of segregation and independent assortment. Therefore, the added complexity of inheritance in these examples does not detract from the validity of Mendel's conclusions. Second, the F_2 phenotypic ratio in each example has been expressed in sixteenths. When a similar observation is made in crosses where the inheritance pattern is unknown, it suggests to geneticists that two gene pairs are controlling the observed phenotypes. You should make the same inference in the analysis of genetics problems. Other insights into solving genetics problems are provided in the Insights and Solutions section at the conclusion of this chapter.

GENE INTERACTION: CONTINUOUS VARIATION

In the preceding section, examples in Figure 4.6 illustrated gene interaction leading to phenotypic variation that is easily classified into distinct traits. Pea plants may be tall or dwarf; squash shape may be spherical, disc-shaped, or elongated; and fruit-fly eye color red, brown, scarlet, or white. These phenotypes are examples of discontinuous variation where discrete phenotypic categories exist. There are many other traits in a population that demonstrate considerably more variation and are not easily categorized into distinct classes. Such phenotypes represent **continuous variation.** For example, in addition to his work with sweet peas, Mendel experimented with beans. In a cross between a purple- and a white-flowered strain, the F_1 was purple. However, the F_2 contained not only purple- and white-flowered bean plants, but ones with numerous intermediate shades. He was unable to explain these results satisfactorily, but recognized that they were inconsistent with most of his data derived from sweet peas. It was not until more than 50 years later that the inheritance of characters exhibiting continuous variation was explained.

It is now known that traits exhibiting continuous variation are often controlled by two or more genes that make additive contributions to the phenotype. Such traits are said to exhibit **continuous or quantitative variation** and are examples of **polygenic inheritance.** We will examine such patterns of inheritance in which a phenotypic trait is controlled by genes at two or more loci. Later in the text (Chapter 20), we will return to this general topic and outline the statistical tools used by geneticists to study traits that exhibit continuous variation.

Quantitative Inheritance: Polygenes

In the late eighteenth century, Josef Gottlieb Kolreuter showed that when tall and dwarf tobacco plants were crossed, the F_1 generation was not all tall, as Mendel found with garden peas. Instead, the individual plants were all intermediate in height. When the F_2 generation was examined, individuals showed continuous variation in height, ranging from tall to dwarf. The majority of the F_2 plants were intermediate like the F_1, but a few were as tall or dwarf as the P_1 parents. These distributions are depicted in histograms in Figure 4.7. Note that the F_2 data assume a normal distribution, as evidenced by the bell-shaped curve in the histogram.

At the beginning of the twentieth century, geneticists noted that many characters in different species had similar patterns of inheritance, such as height and stature in humans, seed size in the broad bean, grain color in wheat, and kernel number and ear length in corn. In each case, offspring in the succeeding generation seemed to be a blend of their parents' characteristics.

The issue of whether continuous variation could be accounted for in Mendelian terms caused considerable controversy in the early 1900s. Those such as William Bateson and Gudny Yule, who adhered to the Mendelian explanation of inheritance, suggested that a large number of factors or genes were responsible for the observed patterns. This proposal, called the **multiple-factor** or **multiple-gene hypothesis,** implied that many factors or genes contribute to the phenotype in a *cumulative* or *quantitative* way. However, some geneticists argued that Mendel's unit factors could not account for the blending of parental phenotypes characteristic of these patterns of inheritance and were thus skeptical of his ideas.

By 1920, the conclusions of several critical sets of experiments largely resolved the controversy and demonstrated that Mendelian factors could account for continuous variation. In one experiment, Edward M. East performed crosses between two strains of the tobacco plant *Nicotiana longiflora.* The fused inner petals of the flower, or corollas, of strain A were decidedly shorter than the corollas of strain B. With only minor variation, each strain was true breeding. Thus, the differences between them were clearly under genetic control.

When plants from the two strains were crossed, the F_1, F_2, and selected F_3 data (Figure 4.8) demonstrated a very distinct pattern. The F_1 generation displayed corollas that were intermediate in length, compared with the P_1 varieties, and showed only minor variability among individuals. While corolla lengths of the P_1 plants were about 40 mm and 94 mm, the F_1 generation contained plants with corollas that were all about 64 mm. In the F_2 generation, lengths varied much more, ranging from 52 mm to 82 mm. Most individuals were similar to their F_1 parents, and as the deviation from this average increased, fewer and fewer plants were observed. When the data are plotted graphically (number vs. length), a bell-shaped curve results.

East further experimented with this population by selecting F_2 plants of various corolla lengths and allowing them to produce separate F_3 generations. Several are illustrated in Figure 4.8. In each case, a bell-shaped distribution was observed with most individuals similar in height to the F_2 parents, but with considerable variation around this value.

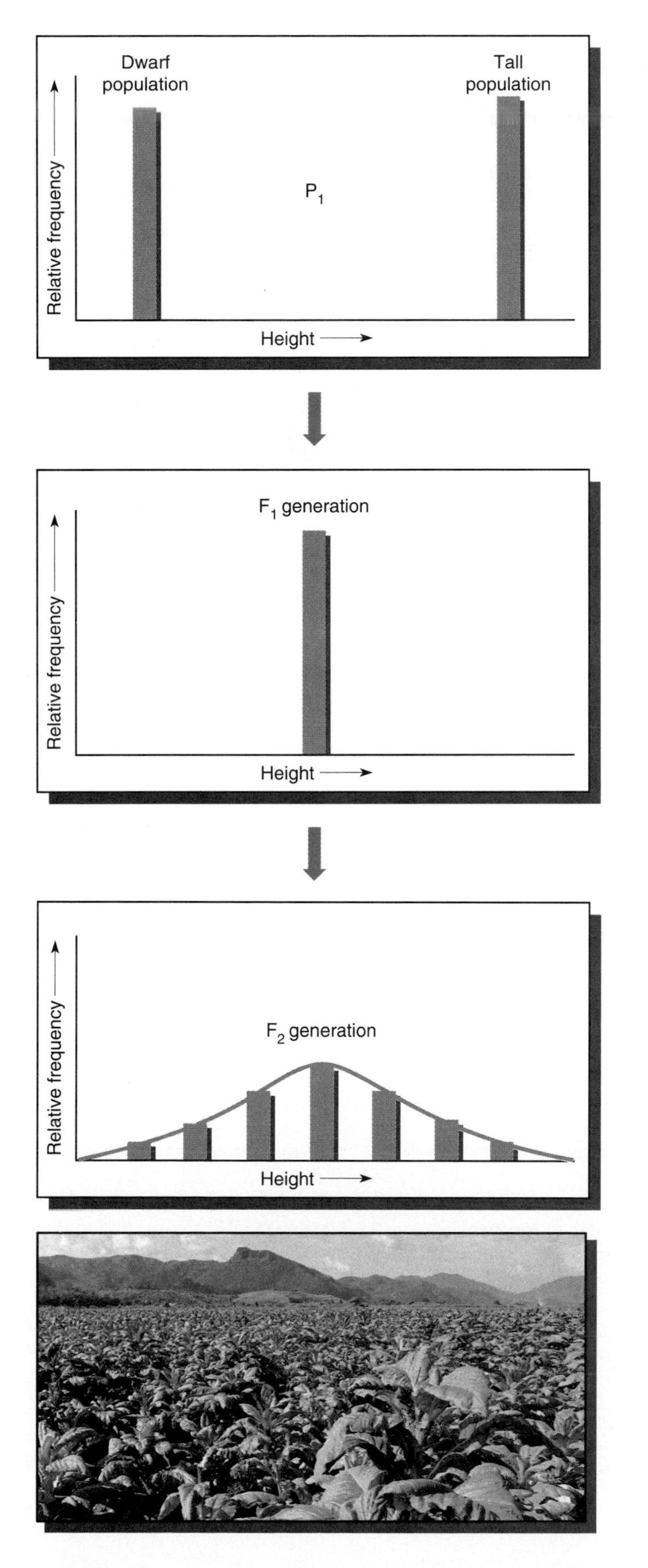

East's experiments thus demonstrated that although the variation in corolla length seemed continuous, experimental crosses resulted in the segregation of distinct phenotypic classes as observed in the three independent F_3 categories. This finding strongly suggested that the multiple-factor hypothesis could account for traits that deviate considerably in their expression.

The multiple-factor hypothesis, suggested by the observations of East and others, embodies the following major points:

1. Characters under such control can usually be quantified by measuring, weighing, counting, etc.
2. Two or more pairs of genes, located throughout the genome, account for the hereditary influence on the phenotype in an **additive way.** Because many genes may be involved, inheritance of this type is often called **polygenic.**
3. Each gene locus may be occupied by either an additive allele, which contributes a set amount to the phenotype, or by a nonadditive allele, which does not contribute quantitatively to the phenotype.
4. The total effect of each additive allele at each locus, while small, is approximately equivalent to all other additive alleles at other gene sites.
5. Together, the genes controlling a single character produce substantial phenotypic variation.
6. This genetic variation is affected by environmental factors. For example, despite their genetic basis, height in humans is influenced by individual diets, and plant characters are influenced by rainfall and soil nutrients.
7. Analysis of polygenic traits requires the study of large numbers of progeny from a population of organisms.

These points, as well as an explanation of the multiple-factor hypothesis, can be illustrated by examining Herman Nilsson-Ehle's experiments involving grain color in wheat performed in the early twentieth century. In one set of experiments, wheat with red grain was crossed to wheat with white grain (Figure 4.9). The F_1 generation demonstrated an intermediate color. In the F_2, approximately 15/16 of the plants showed some degree of red grain, while 1/16 of the plants showed white grain. Since the ratio occurred in sixteenths, we can hypothesize that two gene pairs control the phenotype and, if so, they

FIGURE 4.7
Histograms showing the relative frequency of individuals expressing various height phenotypes derived from Kolreuter's cross between dwarf and tall tobacco plants carried to an F_2 generation. The photograph shows a field of tobacco plants.

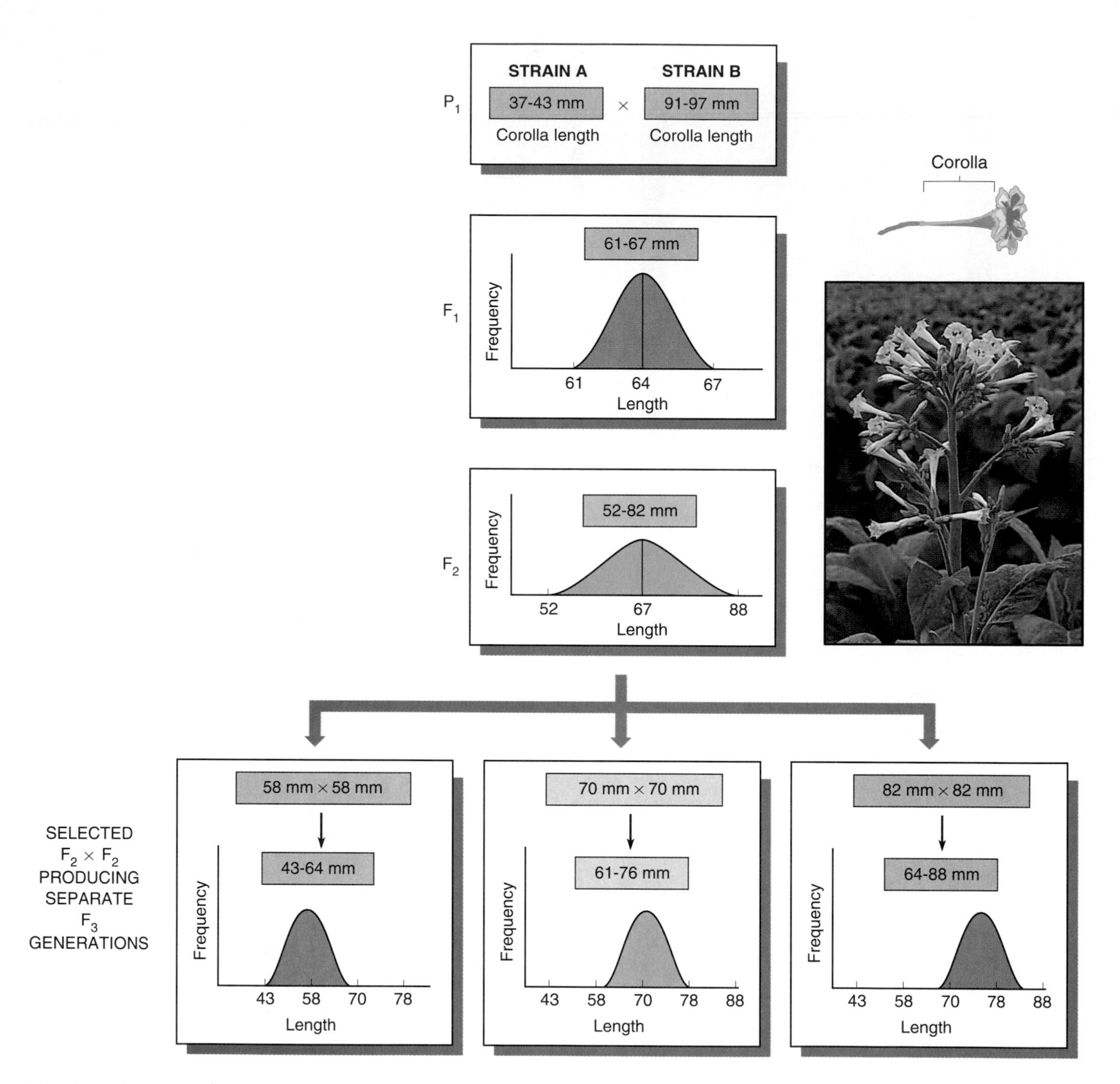

FIGURE 4.8
The F_1, F_2, and selected F_3 results of East's cross between two strains of *Nicotiana* with different corolla lengths. Plants of strain A vary from 37 to 43 mm, while plants of strain B vary from 91 to 97 mm. The photograph illustrates the flower and corolla of a tobacco plant.

segregate independently from one another in a Mendelian fashion.

Upon careful examination of the F_2, grain with color could be classified into four different shades of red. If two gene pairs were operating, each with one potential additive allele and one potential nonadditive allele, we can envision how the multiple factor hypothesis could account for this variation. In the P_1, both parents were homozygous; the red parent contains only additive alleles (uppercase), while the white parent contains only nonadditive alleles (lowercase). The F_1, being heterozygous, contains only two additive alleles and expresses an inter-

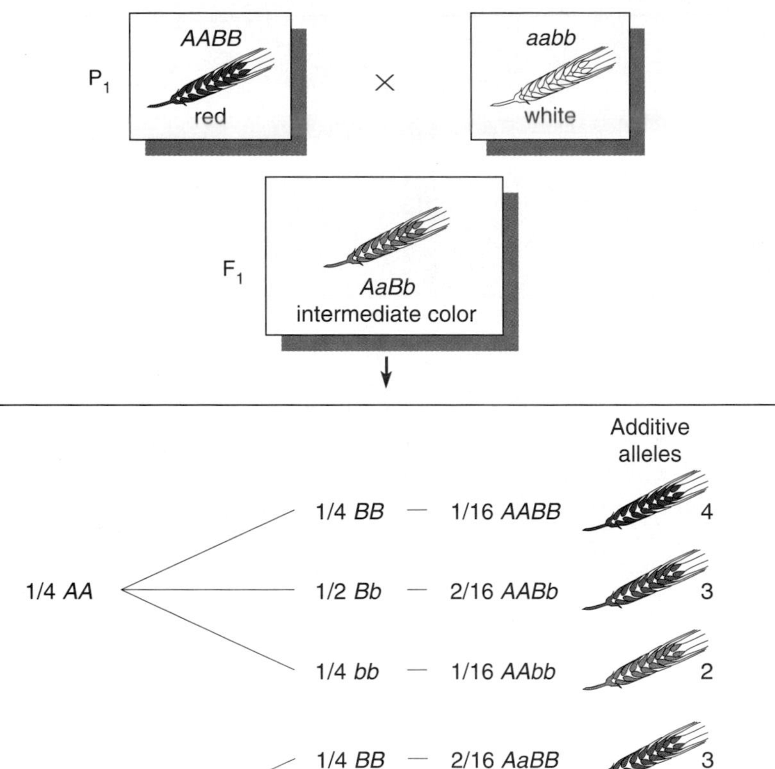

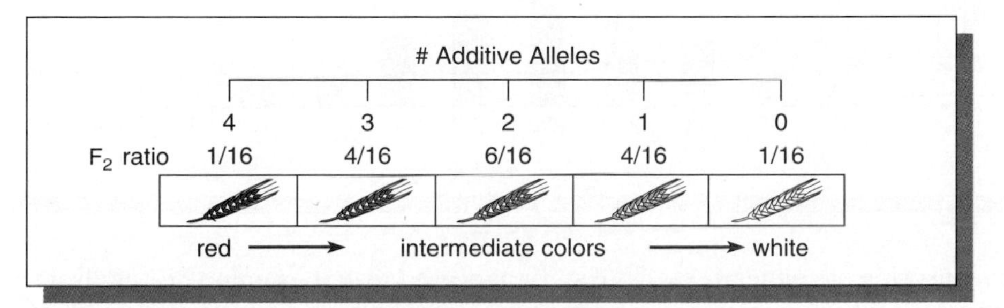

FIGURE 4.9
An illustration of how the multiple factor hypothesis can account for the 1:4:6:4:1 phenotype ratio of grain color when all alleles designated by an uppercase letter are additive and contribute an equal amount of pigment to the phenotype.

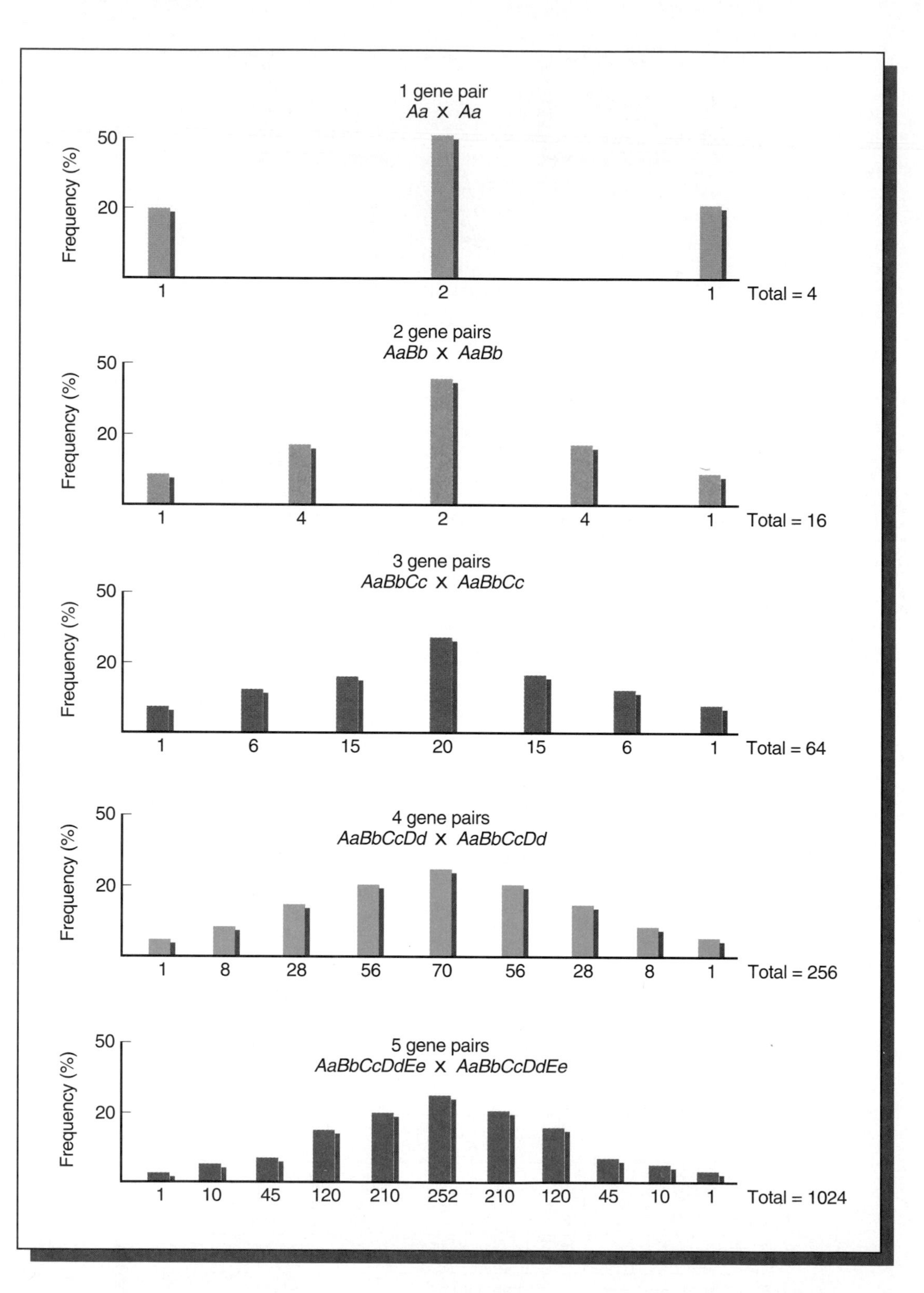

FIGURE 4.10
The results of crossing two heterozygotes where polygenic inheritance is in operation with one to five gene pairs. Each histogram bar indicates a distinct phenotypic class from one extreme (left end) to the other extreme (right end). Each phenotype results from a different number of additive alleles.

mediate phenotype. In the F_2, each offspring has either 4, 3, 2, 1, or 0 additive alleles (Figure 4.9). Wheat with no additive alleles (1/16) is white like one of the P_1 parents, while wheat with 4 additive alleles is red like the other P_1 parent. Plants with 3, 2, or 1 additive alleles constitute the other three categories of red color observed in the F_2, with most (6/16) having 2 additive alleles like the F_1 plants.

Thus, it appeared that, if examined carefully, even apparent continuous variation could be explained in a Mendelian fashion. This explanation fits nicely with the data. Multiple-factor inheritance as just described is now an accepted mechanism to account for phenotypes displaying continuous variation. This type of inheritance, in which alleles contribute additively to a phenotype, is different from any other mode of inheritance discussed thus far. Although the Nilsson-Ehle experiment involved two gene pairs, there is no reason why three, four, or more gene pairs cannot function in controlling various phenotypes. As we saw in Nilsson-Ehle's initial cross, if two gene pairs were involved, only five F_2 phenotypic categories, in a 1:4:6:4:1 ratio, would be expected. On the other hand, as three, four, five, or more gene pairs become involved, greater and greater numbers of classes would be expected to appear in more complex ratios. The number of phenotypes and the expected F_2 ratios of crosses involving up to five gene pairs are illustrated in Figure 4.10.

TABLE 4.2
Determination of number of gene pairs from polygenic inheritance patterns

n	Ratio from Cross between Heterozygous Parents	Number of Distinct F_2 Phenotypic Classes
1	1/4	3
2	1/16	5
3	1/64	7
4	1/256	9
5	1/1024	11

Calculating the Number of Polygenes

It is often of interest to determine the number of genes that are involved in the control of polygenic traits. If the ratio of F_2 individuals resembling *either* of the two most extreme phenotypes can be determined, then the number of gene pairs involved (n) may be calculated using the following simple formula:

$$\frac{1}{4^n} = \text{ratio of } F_2 \text{ individuals expressing either extreme phenotype}$$

In our past example, the P_1 phenotypes represent these two extremes. In Figure 4.9, 1/16 of the F_2 are either red *or* white like the P_1 classes; this ratio can be substituted on the right side of the equation prior to solving for n:

$$\frac{1}{4^n} = \frac{1}{16}$$
$$\frac{1}{4^2} = \frac{1}{16}$$
$$n = 2$$

Table 4.2 lists the ratio and the number of F_2 phenotypic classes produced in crosses involving up to five gene pairs.

For low numbers of gene pairs, it is sometimes easier to use the $(2n + 1)$ rule. If n equals the number of gene pairs, $2n + 1$ will determine the total number of categories of possible phenotypes. Where $n = 2$, $2n + 1 = 5$, since each phenotypic category could have 4, 3, 2, 1, or 0 additive alleles. Where $n = 3$, $2n + 1 = 7$, since each phenotypic category could have 6, 5, 4, 3, 2, 1, or 0 additive alleles, and so on.

Polygenic control is a significant concept because it is believed to be the mode of inheritance for a vast number of traits. For example, height, weight, and stature in all plants and animals, grain yield in crops, beef and milk production in cattle, and egg production in chickens are thought to be under polygenic control. Thus, knowledge of this mode of inheritance is of prime importance in animal breeding and agriculture. In humans, the degree of skin pigmentation is thought to be under similar genetic control. In most cases of polygenic inheritance, the genotype establishes the potential range in which a particular phenotype may fall, while environmental factors determine how much of the potential will be realized. In the crosses described in this section we have assumed an optimal environment, which minimizes variation resulting from that source.

GENES ON THE X CHROMOSOME: SEX LINKAGE

We near the conclusion of this chapter with a discussion of still one other mode of neo-Mendelian inheritance: **sex-linkage.** This phenomenon results from the fact that one of the sexes in many organisms contains a pair of *unlike* chromosomes, the X and Y, which are involved in sex determination. For example, in both fruit flies (*Drosophila*) and humans, males contain an X and a Y chromosome, while females contain two X chromosomes. As we will see below, the unique pattern of inheritance stems from the fact that the Y, while behaving as a homologue to the X in meiosis, is for the most part genetically blank, or void of genes. Sex-linkage, therefore, involves the transmission and expression of the normal complement of genes located on the X chromosome. To distinguish this pair of sex chromosomes, all other pairs are referred to as **autosomal chromosomes** or just **autosomes.**

Sex-Linkage in *Drosophila*

One of the first cases of sex linkage was documented by Thomas H. Morgan around 1920 during his studies of the *white* eye mutation in *Drosophila*. We will use this case to illustrate sex-linkage. The normal wild type eye color red is dominant to white.

Morgan's work established that the inheritance pattern of the white-eye trait was clearly related to the sex of the parent carrying the mutant allele. Unlike the outcome of the typical monohybrid cross, reciprocal crosses between white- and red-eyed flies did not yield identical results. In contrast, in all of Mendel's monohybrid crosses, F_1 and F_2 data were very similar regardless of which P_1 parent exhibited the recessive mutant trait. Morgan's analysis led to the conclusion that the *white* locus is present on the X chromosome rather than one of the autosomes. As such, both the gene and the trait are said to be **sex-linked** or **X-linked.**

Results of reciprocal crosses between white-eyed and red-eyed flies are shown in Figure 4.11. The obvious differences in phenotypic ratios in both the F_1 and F_2 generations are dependent on whether or not the P_1 white-eyed parent was male or female.

Morgan was able to correlate these observations with the difference found in the sex chromosome composition between male and female *Drosophila.* He hypothesized that the recessive allele for white eye is found on the X chromosome, but its corresponding locus is absent from the Y chromosome. Females thus have two available gene sites, one on each X chromosome, while males have only one available gene site on their single X chromosome.

Morgan's interpretation of sex-linked inheritance, shown in Figure 4.12 provides a suitable theoretical explanation for his results. Since the Y chromosome lacks homology with most genes on the X chromosome, what-

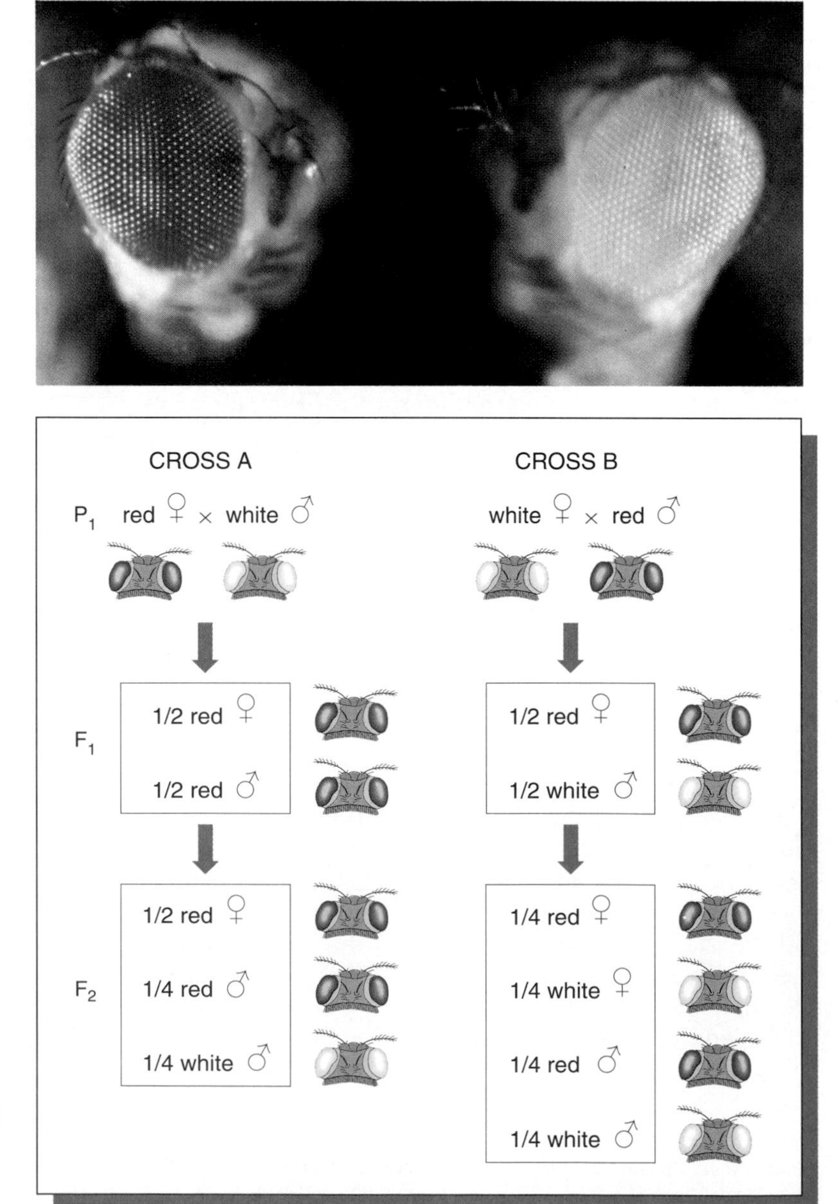

FIGURE 4.11
The F_1 and F_2 results of T. H. Morgan's reciprocal crosses involving the sex-linked white-eye mutation in *Drosophila melanogaster.* The photograph contrasts white eyes with the brick red wild-type eye color.

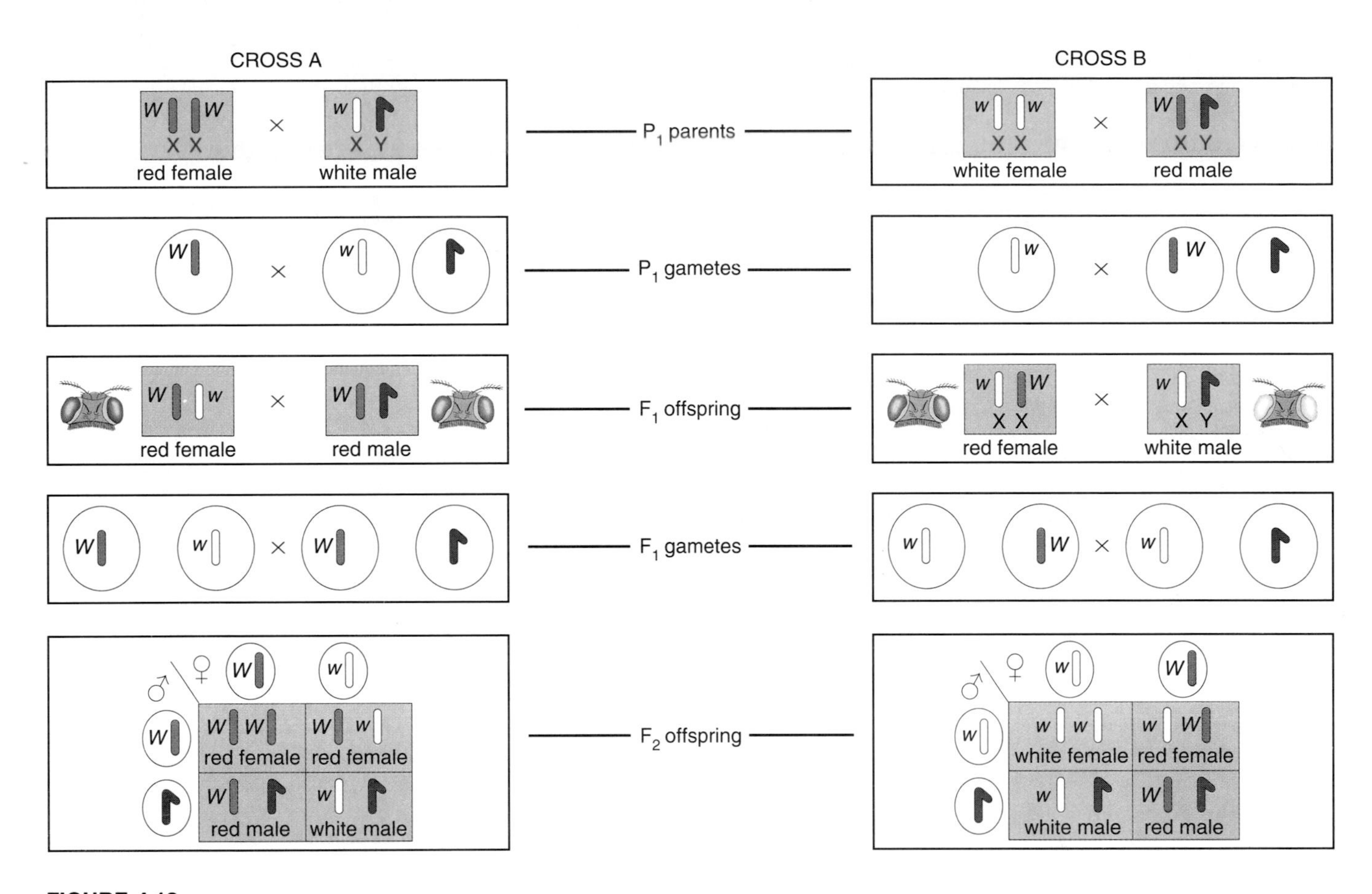

FIGURE 4.12
The chromosomal explanation of the results of the sex-linked crosses shown in Figure 4.11.

ever alleles are present on the X chromosome of the males will be directly expressed in the phenotype. Since males cannot be either homozygous or heterozygous for sex-linked genes, this condition is referred to as being **hemizygous.** In such cases, no alternative alleles are present, and the concept of dominance and recessiveness is irrelevant.

One result of sex linkage is the **crisscross pattern of inheritance,** whereby phenotypic traits controlled by recessive sex-linked genes are passed from mothers to all sons. This pattern occurs because females exhibiting a recessive trait must contain the mutant allele on both X chromosomes. Since male offspring receive one of their mother's two X chromosomes and are hemizygous for all alleles present on that X, all sons will express the same recessive sex-linked traits as their mother.

Sex-Linked Inheritance in Humans

In humans, many genes and the respective traits controlled by them are recognized as being linked to the X chromosome. These sex-linked traits may be easily identified in pedigrees because of the crisscross pattern of inheritance. A pedigree for one form of human color blindness is shown in Figure 4.13. The mother in generation I passes the trait to all her sons but to none of her daughters. If the offspring in generation II marry normal individuals, the color-blind sons will produce all normal male and female offspring (III-1, 2, and 3); the normal-visioned daughters will produce normal-visioned female offspring (III-4, 6, and 7), as well as color-blind (III-8) and normal-visioned (III-5) male offspring.

Many sex-linked human genes have now been identified, as shown in Table 4.3. For example, the genes controlling two forms of hemophilia and one form of muscular dystrophy are located on the X chromosome. Additionally, numerous genes whose expression yields enzymes are sex linked. Glucose-6-phosphate dehydrogenase and hypoxanthine-guanine-phosphoribosyl transferase are two examples. In the latter case, the **Lesch-Nyhan syndrome** results from the mutant form of the X-linked gene product.

Because of the way in which sex-linked genes are transmitted, unusual circumstances may be associated with recessive sex-linked disorders, in comparison to recessive autosomal disorders. For example, if a sex-linked disorder debilitates or is lethal to the affected individual prior to reproductive maturation, the disorder occurs exclusively in males. This is the case because the

FIGURE 4.13
(a) A human pedigree of the sex-linked color blindness trait. (b) The most probable genotypes of each individual in the pedigree.

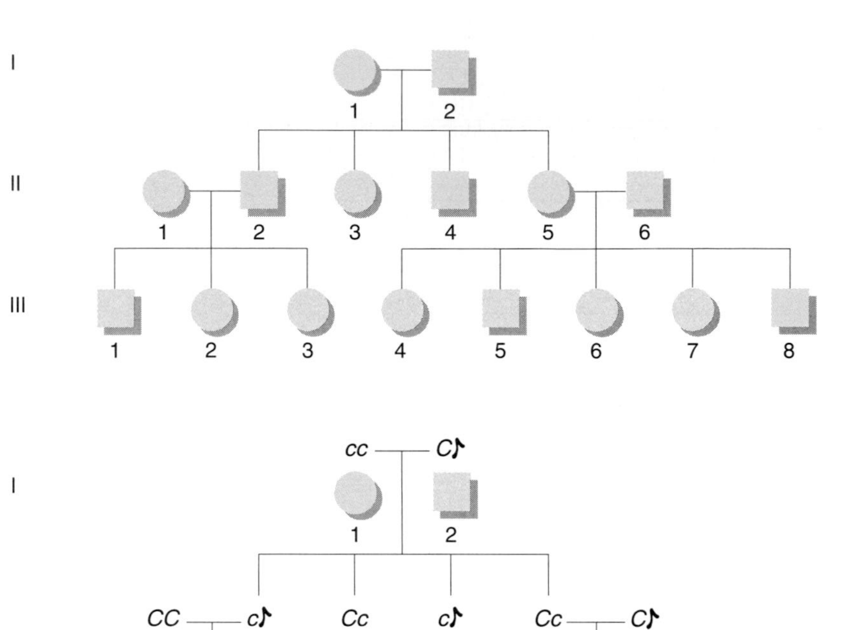

TABLE 4.3
Human sex-linked traits.

Condition	Characteristics
Color blindness, deutan type	Insensitivity to green light.
Color blindness, protan type	Insensitivity to red light.
Fabry's disease	Deficiency of galactosidase A; heart and kidney defects, early death.
G-6-PD deficiency	Deficiency of glucose-6-phosphate dehydrogenase; severe anemic reaction following intake of primaquines in drugs and certain foods including fava beans.
Hemophilia A	Classical form of clotting deficiency; lack of clotting factor VIII.
Hemophilia B	Christmas disease; deficiency of clotting factor IX.
Hunter syndrome	Mucopolysaccharide storage disease resulting from iduronate sulfatase enzyme deficiency; short stature, clawlike fingers, coarse facial features, slow mental deterioration, and deafness.
Ichthyosis	Deficiency of steroid sulfatase enzyme; scaly dry skin, particularly on extremities.
Lesch-Nyhan syndrome	Deficiency of hypoxanthine-guanine phosphoribosyl transferase enzyme (HGPRT), leading to motor and mental retardation, self-mutilation, and early death.
Muscular dystrophy (Duchenne type)	Progressive, life-shortening disorder characterized by muscle degeneration and weakness; sometimes associated with mental retardation. Deficiency of the protein dystrophin.

only sources of the lethal allele in the population are females who carry it heterozygously but do not express the disorder. They pass the allele to one-half of their sons, who develop the disorder because they are hemizygous, but who rarely, if ever, reproduce. Heterozygous females also pass the allele to one-half of their daughters, who become carriers but do not develop the disorder. An example of such a sex-linked disorder is the Duchenne form of muscular dystrophy. The disease has an onset prior to age 6 and is often lethal prior to age 20. It normally occurs only in males.

SEX-LIMITED AND SEX-INFLUENCED INHERITANCE

Our final topics involve inheritance affected by the sex of the organism, but not necessarily by genes on the X chromosome. There are numerous examples in different organisms where the sex of the individual plays a determining role in the expression of certain phenotypes. In some cases, the expression of a specific phenotype is absolutely limited to one sex; in others, the sex of an individual influences the expression of a phenotype that is not limited to one sex or the other. This distinction differentiates **sex-limited inheritance** from **sex-influenced inheritance.**

In domestic fowl, tail and neck plumage is often distinctly different in males and females (Figure 4.14), demonstrating sex-limited inheritance. Cock-feathering is longer, more curved, and pointed, while hen-feathering is shorter and more rounded. The inheritance of feather type is due to a single pair of autosomal alleles whose expression is modified by the individual's sex hormones.

As shown in the following chart, hen-feathering is due to a dominant allele, *H*; but regardless of the homozygous presence of the recessive *h* allele, all females remain hen-feathered. Only in males does the *hh* genotype result in cock-feathering.

	Phenotype	
Genotype	♀	♂
HH	Hen-feathered	Hen-feathered
Hh	Hen-feathered	Hen-feathered
hh	Hen-feathered	Cock-feathered

In the development of certain breeds of fowl, one allele or the other has become fixed in the population. In the Leghorn breed, all individuals are of the *hh* genotype, and thus all males and females show distinctive plumage. Sebright bantams are all *HH,* showing no sexual distinction in feathering.

Cases of sex-influenced inheritance include pattern baldness in humans, horn formation in sheep, and certain coat patterns in cattle. In such cases, autosomal genes are responsible for the contrasting phenotypes displayed by both males and females, but the expression of these

FIGURE 4.14
Hen-feathering (left) versus cock-feathering (right) in domestic fowl. The feathers in the hen are shorter and less curved.

genes is dependent on the hormone constitution of the individual. Thus, the heterozygous genotype may exhibit one phenotype in one sex and the contrasting one in the other. For example, **pattern baldness** in humans, where the hair is very thin on the top of the head (Figure 4.15) is inherited in the following way:

	Phenotype	
Genotype	♀	♂
BB	Bald	Bald
Bb	Not bald	Bald
bb	Not bald	Not bald

Even though females may display pattern baldness, this phenotype is much more prevalent in males. When females do inherit the *BB* genotype, the phenotype is much less pronounced than in males and is expressed later in life.

FIGURE 4.15
Pattern baldness, a sex-influenced autosomal trait in humans.

CHAPTER SUMMARY

1 Since Mendel's work was rediscovered, the study of transmission genetics has been expanded to include modes of inheritance controlling phenotypes that are often influenced by two or more genes in a variety of ways.

2 Incomplete or partial dominance is exhibited when an intermediate phenotypic expression of a trait occurs in an organism heterozygous for two alleles.

3 Codominance is exhibited when distinctive expression of two alleles occurs in a heterozygous organism.

4 The concept of multiple allelism applies to populations, since a diploid organism may host only two alleles at any given locus. Within a population, however, many alleles of the same gene often occur.

5 Lethal mutations usually result in the inactivation or the lack of synthesis of gene products essential during development. Such mutations may be recessive or dominant. Some lethal genes, such as that causing Huntington disease, are not expressed until adulthood.

6 Mendel's classic F_2 ratio is often modified in instances where gene interaction results in either continuous or discontinuous variation.

7 Epistasis involves discontinuous variation, where two or more genes influence a single characteristic. Usually, the expression of one of the genes masks the expression of the other gene or genes.

8 Polygenic (or quantitative) inheritance results in continuous variation and is illustrated when products of several genes make additive contributions to phenotypes.

9 Genes located on the X chromosome display a unique mode of inheritance referred to as sex-linkage.

10 Sex-limited and sex-influenced inheritance occur when the sex of the organism affects the phenotype.

KEY TERMS

ABO blood groups
additive alleles
alleles
antigen
autosomal chromosomes (autosomes)
Bombay phenotype
codominance
continuous variation
crisscross inheritance
discontinuous variation
dominance
epistasis
gene interaction
hemizygous
Huntington Disease
incomplete dominance
isoagglutinogen
Lesch-Nyhan syndrome
lethal allele
MN blood groups
multiple alleles
multiple-factor (multiple gene) inheritance
mutation
Neo-Mendelian genetics
pattern baldness
polygenic inheritance
sex-influenced inheritance
sex-limited inheritance
sex-linked (X-linked) inheritance
wild type allele

Genetic problems take on an added complexity if they involve two independent characters and multiple alleles, incomplete dominance, or epistasis. The most difficult types of problems are those that were faced by pioneering geneticists in the laboratory. In these problems they had to determine the mode of inheritance by working backwards from the observations of offspring to parents of unknown genotype. For example, consider the problem of comb shape inheritance in chickens, where walnut, rose, pea, and single are the observed distinct phenotypes. *How is comb shape inherited, and what are the genotypes of the P_1 generation of each cross? Use the following data to answer these questions.*

Cross 1: single × single ⟶ all single
Cross 2: walnut × walnut ⟶ all walnut
Cross 3: rose × pea ⟶ all walnut
Cross 4: $F_1 \times F_1$ of Cross 3
walnut × walnut ⟶ 93 walnut
28 rose
32 pea
10 single

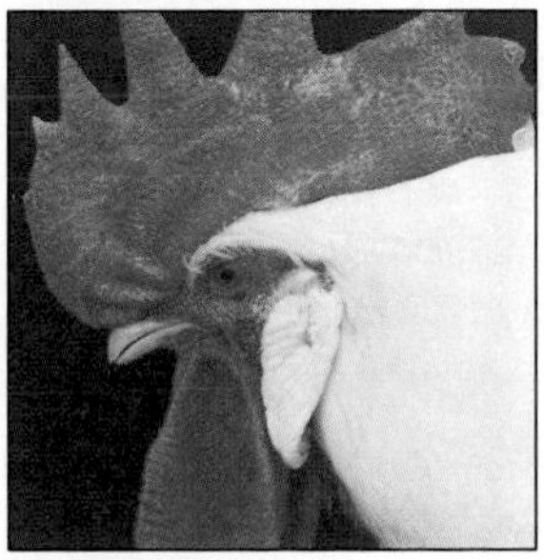

Single

Walnut

Rose

Pea

At first glance, this problem may appear quite difficult. The approach used in solving it must involve two steps. First, carefully analyze the data for any useful information. Once you determine something concrete, follow an empirical approach; that is, formulate a hypothesis and, in a sense, test it against the given data. Look for a pattern of inheritance that is consistent with all cases.

(a) In this problem there are two immediately useful facts. First, in Cross 1, P_1 singles breed true. Second, while P_1 walnut breeds true (Cross 2), a walnut phenotype is also produced in crosses between rose and pea (Cross 3). When these F_1 walnuts are crossed together (Cross 4), all four comb shapes are produced in a ratio that approximates a 9:3:3:1. This observation should immediately suggest a cross involving two gene pairs because the resulting data closely resemble the ratio of Mendel's dihybrid crosses. Since only one trait is involved (comb shape), epistasis must be occurring. This may serve as a working hypothesis, and we must now propose how the two gene pairs interact to produce each phenotype.

(b) If we call the allele pairs *A, a* and *B, b,* we might predict that since walnut represents 9/16 in Cross 4, $A-$ $B-$ will produce walnut. We might also hypothesize that in the case of Cross 2, the genotypes were $AABB \times AABB$ where walnut was seen to breed true. (Recall that $A-$ and $B-$ mean *AA* or *Aa* and *BB* or *Bb,* respectively.)

(c) Since single is the phenotype representing 1/16 of the offspring of Cross 4, we could predict that this phenotype is the result of the *aabb* genotype. This is consistent with Cross 1.

(d) Now we have only to determine the genotypes for rose and pea. The most logical prediction would be that at least one dominant *A* or *B* allele combined with the double recessive condition of the other allele pair can account for these phenotypes. For example,

$$A-bb \rightarrow \text{rose}$$
$$aa\,B- \rightarrow \text{pea}$$

If, in Cross 3, *AAbb* (rose) were crossed with *aaBB* (pea), all offspring would be *AaBb* (walnut). This is consistent with the data, and we must now only look at Cross 4. We predict these walnut genotypes to be *AaBb* (as above), and from the cross

$$\frac{AaBb}{\text{(walnut)}} \times \frac{AaBb}{\text{(walnut)}}$$

we expect

9/16 $A-B-$ (walnut)
3/16 $A-bb$ (rose)
3/16 $aa\,B-$ (pea)
1/16 $aa\,bb$ (single)

Our prediction is consistent with the information we were given. The initial hypothesis of the epistatic interaction of two gene pairs proves consistent throughout, and the problem has been solved.

This example illustrates the need to have a basic theoretical knowledge of transmission genetics. Then, you must search for the appropriate clues so that you can proceed in a stepwise fashion toward a solution. Mastering problem solving requires practice, but provides a great deal of satisfaction.

Apply this general approach to the following problems:

1 In radishes, flower color may be red, purple, or white. The edible portion of the radish may be long or oval. When only flower color is studied, red × white yields all purple. If these F_1 purples are interbred, no dominance is evident, and the F_2 generation consists of 1/4 red:1/2 purple:1/4 white. Regarding radish shape, long is dominant to oval in a normal Mendelian fashion.

(a) Determine the F_1 and F_2 phenotypes from a cross between a true-breeding red long radish and one that is white oval. Be sure to define all gene symbols initially.

SOLUTION: This is a modified dihybrid cross where the gene pair controlling color exhibits incomplete dominance. Shape is controlled conventionally. First, establish gene symbols:

RR = red $\quad Rr$ = purple $\quad rr$ = white
$O-$ = long $\quad oo$ = oval

P_1: $RROO$ × $rroo$
(red long) (white oval)

F_1: all $RrOo$ (purple long)

$F_1 \times F_1$: $RrOo \times RrOo$

F_2:			
1/4 RR	3/4 $O-$	3/16 $RR\ O-$	red long
	1/4 oo	1/16 $RR\ oo$	red oval
2/4 Rr	3/4 $O-$	6/16 $Rr\ O-$	purple long
	1/4 oo	2/16 $Rr\ oo$	purple oval
1/4 rr	3/4 $O-$	3/16 $rr\ O-$	white long
	1/4 oo	1/16 $rr\ oo$	white oval

Note above that to generate the F_2 results, we have used the forked-line method. First, the outcome of crossing F_1 parents for the color genes is considered ($Rr \times Rr$). Then the outcome of shape is considered ($Oo \times Oo$).

(b) A red oval plant was crossed with a plant of unknown genotype and phenotype, yielding the data shown below. Determine the genotype and phenotype of the unknown plant.

Offspring: 103 red long:101 red oval:98 purple long:100 purple oval

SOLUTION: Since the two characters are inherited independently, consider them separately. The data indicate a 1/4:1/4:1/4:1/4 proportion. First, consider color:

P_1: red × ??? (unknown)
F_1: 204 red (1/2)
198 purple (1/2)

Since the red parent *must* be *RR*, the unknown must have a genotype of *Rr* to produce these results. It is thus purple. Now, consider shape:

P_1: oval × ??? (unknown)
F_1: 201 long (1/2)
201 oval (1/2)

Since the oval plant *must* be *oo*, the unknown plant must have a genotype of *Oo* to produce these results. It is thus long. The unknown plant is thus:

RrOo purple long

2 In humans, color blindness is inherited as a sex-linked recessive trait. A woman with normal vision, but whose father is color-blind, marries a male who has normal vision. Predict the color vision of their male and female offspring.

SOLUTION: The female is heterozygous since she inherited an X chromosome with the mutant allele from her father. Her husband is normal. Therefore, the parental genotypes are

$$Cc \times C\upharpoonleft$$

All female offspring are normal (*CC* or *Cc*). One-half of the male children will be color-blind ($c\upharpoonleft$), and the other half will have normal vision ($C\upharpoonleft$).

PROBLEMS AND DISCUSSION QUESTIONS

1 In shorthorn cattle, coat color may be red, white, or roan. Roan is an intermediate phenotype expressed as a mixture of red and white hairs. The following data were obtained from various crosses:

red × red ⟶ all red
white × white ⟶ all white
red × white ⟶ all roan
roan × roan ⟶ 1/4 red : 1/2 roan : 1/4 white

How is coat color inherited? What are the genotypes of parents and offspring for each cross?

2 Contrast incomplete dominance and codominance.

3 With regard to the ABO blood types in humans, determine the genotypes of the male parent and female parent below:

Male Parent: Blood type B whose mother was type O
Female Parent: Blood type A whose father was type B

Predict the blood types of the offspring that this couple may have and the expected proportion of each.

4 Distinguish between epistasis and polygenic inheritance, and between discontinuous and continuous variation.

5 Three gene pairs located on separate autosomes determine flower color and shape as well as plant height. The first pair exhibits incomplete dominance where color can be red, pink (the heterozygote), or white. The second pair leads to personate (dominant) or peloric (recessive) flower shape, while the third gene pair produces either the dominant tall trait or the recessive dwarf trait. Homozygous plants that are red, personate, and tall are crossed to those that are white, peloric, and dwarf. Determine the F_1 genotype(s) and phenotype(s). If the F_1 plants are interbred, what proportion of the offspring will exhibit the same phenotype as the F_1 plants?

6 As in Problem 5, color may be *red, white,* or *pink,* and flower shape may be *personate* or *peloric.* For the following crosses, determine the P_1 and F_1 genotypes. What phenotype ratios would result from crossing the F_1 of (a) to the F_1 of (b)?

(a) red peloric × white personate ⟶ F_1 = all pink personate
(b) red personate × white peloric ⟶ F_1 = all pink personate
(c) pink personate × red peloric ⟶ F_1 = 1/4 red personate
1/4 red peloric
1/4 pink personate
1/4 pink peloric
(d) pink personate × white peloric ⟶ F_1 = 1/4 white personate
1/4 white peloric
1/4 pink personate
1/4 pink peloric

7 In some plants a red pigment, cyanidin, is synthesized from a colorless precursor. The addition of a hydroxyl group (OH^-) to the cyanidin molecule causes it to become purple. In a cross between two randomly selected purple plants, the following results were obtained:

94 purple
31 red
43 colorless

How many genes are involved in the determination of these flower colors? Which genotypic combinations produce which phenotypes? Diagram the purple × purple cross.

8 In rats, the following genotypes of two independently assorting autosomal genes determine coat color:

A– B– (gray)
A– bb (yellow)
aa B– (black)
aa bb (cream)

A third gene pair on a separate autosome determines whether or not any color will be produced. The *CC* and *Cc* genotypes allow color according to the expression of the *A* and *B* alleles. However, the *cc* genotype results in albino rats regardless of the *A* and *B* alleles present. Determine the F_1 phenotypic ratio of the following crosses:
(a) *AA bb CC* × *aa BB cc*
(b) *Aa BB CC* × *AA Bb cc*
(c) *Aa Bb Cc* × *Aa Bb cc*

9 Given the inheritance pattern of coat color in rats, as described in Problem 8, predict the genotype and phenotype of the parents who produced the following F_1 offspring:
(a) 9/16 gray:3/16 yellow:3/16 black:1/16 cream
(b) 9/16 gray:3/16 yellow:4/16 albino
(c) 27/64 gray:16/64 albino:9/64 yellow:9/64 black:3/64 cream

10 Assume that height in a plant is controlled by two gene pairs and that each additive allele contributes 5 cm to a base height of 20 cm (i.e., *aabb* is 20 cm).
(a) What is the height of an *AABB* plant?
(b) Predict the phenotypic ratios of F_1 and F_2 plants in a cross between *aabb* and *AABB*.
(c) List all genotypes that give rise to plants that are 25 and 35 cm in height.

11 An inbred strain of plants has a mean height of 24 cm. A second strain of the same species from a different geographical region also has a mean height of 24 cm. When plants from the two strains are crossed together, the F_1 plants are the same height as the parent plants. However, the F_2 generation shows a wide range of heights; the majority are like the P_1 and F_1 plants, but approximately 4 of 1000 are only 12 cm high, and about 4 of 1000 are 36 cm high.
(a) What mode of inheritance is occurring here?
(b) How many gene pairs are involved?
(c) How much does each gene contribute to plant height?
(d) Indicate one possible set of genotypes for the original P_1 parents and the F_1 plants that could account for these results.
(e) Indicate three possible genotypes that could account for F_2 plants that are 18 cm high and F_2 plants that are 33 cm high.

12 A husband and wife have normal vision, although both of their fathers are red-green color-blind, which is inherited as a sex-linked recessive condition. What is the probability that their first child will be
(a) a normal son?
(b) a normal daughter?
(c) a color-blind son?
(d) a color-blind daughter?

13 In humans, the ABO blood type is under the control of autosomal multiple alleles. Color blindness is a recessive sex-linked trait. If two parents who are both type A and have normal vision produce a son who is color-blind and is type O, what is the probability that their next child will be a female who is normal visioned and type O?

14 In spotted cattle, the colored regions may be mahogany or red. If a red female and a mahogany male, both derived from separate true-breeding lines, are mated and the cross carried to an F_2 generation, the following results are obtained:

F_1: 1/2 mahogany males
1/2 red females
F_2: 3/8 mahogany males
1/8 red males
1/8 mahogany females
3/8 red females

When the reciprocal of the initial cross is performed (mahogany female and red male), identical results are obtained. Explain these results by postulating how the color is genetically determined. Diagram the crosses.

15 In cats, yellow coat color is determined by the *b* allele and black coat color by the *B* allele. The heterozygous condition results in a color known as tortoise shell. These genes are sex-linked. What kinds of offspring would be expected from a cross of a black male and a tortoise-shell female? What are the chances of getting a tortoise-shell male?

16 In *Drosophila,* a sex-linked recessive mutation, *scalloped* (*sd*), causes irregular wing margins. Diagram the F_1 and F_2 results if:
(a) A *scalloped* female is crossed with a normal male.
(b) A *scalloped* male is crossed with a normal female.
Compare these results to those that would be obtained if *scalloped* were not sex-linked.

17 Another recessive mutation in *Drosophila, ebony* (*e*), is on an autosome (chromosome 3) and causes darkening of the body compared with wild-type flies. What phenotypic F_1 and F_2 male and female ratios will result if a *scalloped*-winged female with normal body color is crossed with a normal-winged *ebony* male? Work this problem by both the Punnett square method and the forked-line method.

18 While *vermilion* is sex linked and brightens the eye color, *brown* is an autosomal recessive mutation that darkens the eye. Flies carrying both mutations lose all pigmentation and are white eyed. Predict the F_1 and F_2 results of the following crosses:
(a) vermilion females × brown males
(b) brown females × vermilion males
(c) white females × wild males.

19 Erma and Harvey were a compatible barnyard pair, but a curious sight. Harvey's tail was only 6 cm while Erma's was 30 cm. Their F_1 piglet offspring all grew tails that were 18 cm. When inbred, an F_2 generation resulted in many piglets (Erma and Harvey's grandpigs) whose tails ranged in 4 cm intervals from 6 to 30 cm (6, 10, 14, 18, 22, 26, 30). Most had 18 cm tails, while 1/64 had 6 cm and 1/64 had 30 cm tails.
(a) Explain how tail length is inherited by describing the mode of inheritance, indicating how many gene pairs are at work, and designating the genotypes of Harvey, Erma, and their 18 cm offspring.
(b) If one of the 18 cm F_1 pigs were mated with the 6 cm F_2 pigs, what phenotypic ratio would be predicted if many offspring resulted? Diagram the cross.

SELECTED READINGS

DRAYNA, D., and WHITE, R. 1985. The genetic linkage map of the human X chromosome. *Science* 230: 753–58.
PETERS, J. A., ed. 1959. *Classic papers in genetics.* Englewood Cliffs, N.J.: Prentice-Hall.
RACE, R. R., and SANGER, R. 1975. *Blood groups in man.* 6th ed. Oxford, England: Blackwell Scientific Publishers.
VOGEL, F., and MOTULSKY, A. G. 1986. 2nd ed. *Human genetics: Problems and approaches.* New York: Springer-Verlag.
WATKINS, M. W. 1966. Blood group substances. *Science* 152:172–81.

5 Linkage and Chromosome Mapping

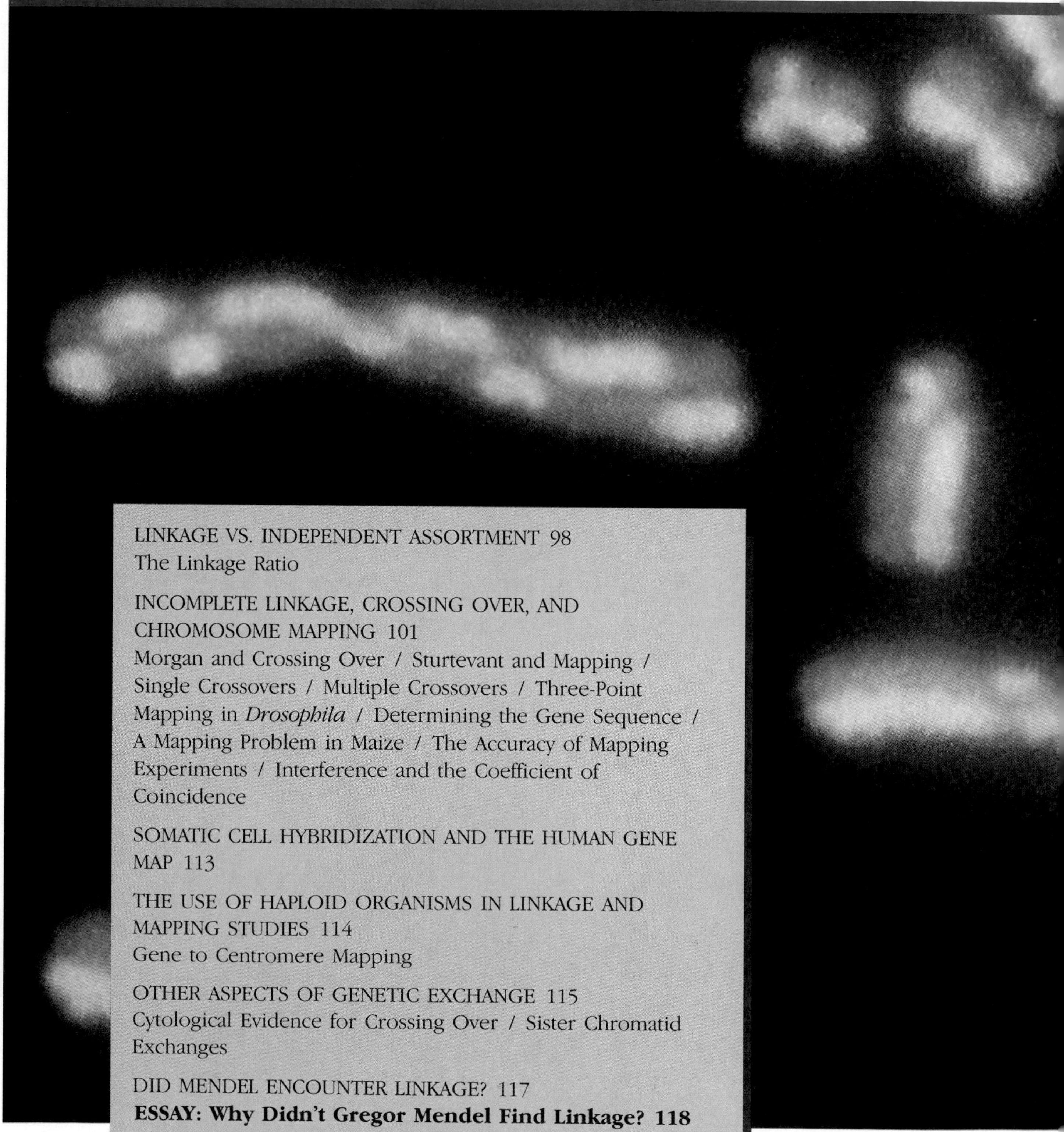

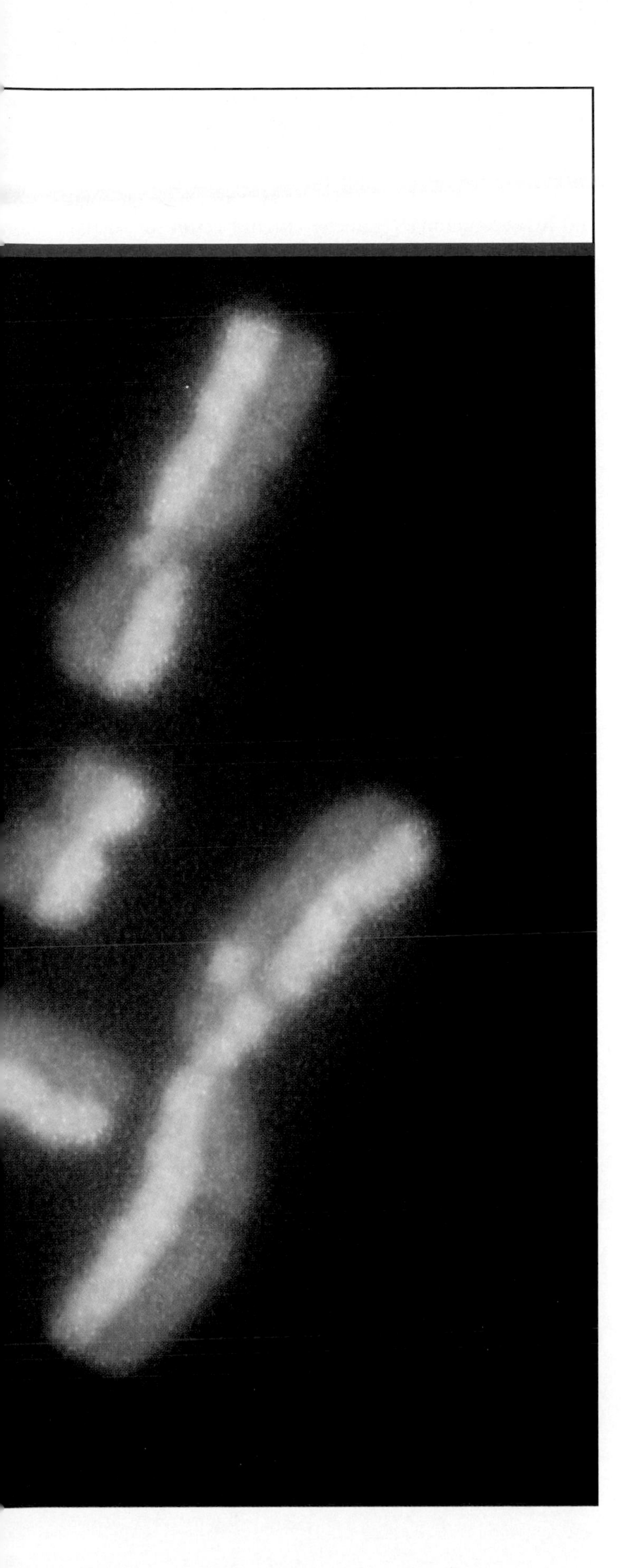

Chapter Concepts

Many genes reside on each chromosome. Unless separated by crossing over, alleles present at the many loci on each homologue segregate as a unit during gamete formation. Recombinant gametes resulting from crossing over enhance genetic variability within a species and serve as the basis for constructing chromosomal maps.

As early as 1903, Walter Sutton, who along with Theodor Boveri united the fields of cytology and genetics, pointed out the likelihood that in organisms there are many more "unit factors" than chromosomes. Soon thereafter, genetic investigations with several organisms revealed that certain genes segregate as if they were somehow joined or linked together. Further investigations showed that such genes are part of the same chromosome and are indeed transmitted as a single unit.

We now know that most chromosomes consist of very large numbers of genes and, in fact, contain sufficient DNA to encode thousands of these units. Genes that are part of the same chromosome are said to be **linked** and to demonstrate **linkage** in genetic crosses.

Since the chromosome, not the gene, is the unit of transmission during meiosis, linked genes are not free to undergo independent assortment. Instead, the alleles at all loci of one chromosome should, in theory, be transmitted as a unit during gamete formation. However, in many instances this does not occur. During the first meiotic prophase, when homologues are paired or synapsed, a reciprocal exchange of chromosome segments may take place. This event, called **crossing over,** results in the reshuffling or **recombination** of the alleles between homologues.

The degree of crossing over between any two loci on a single chromosome is proportionate to the distance between them. Thus, the percentage of recombinant gametes varies, depending on which loci are being considered. This correlation serves as the basis for the construction of **chromosome maps,** which provide the relative locations of genes on chromosomes.

In this chapter, we will discuss linkage, crossing over, and chromosome mapping. We will conclude the chapter by entertaining the rather intriguing question of why Mendel, who studied seven genes and seven chromosomes, did not encounter linkage. Or did he?

LINKAGE VS. INDEPENDENT ASSORTMENT

In order to provide a simplified overview of the major theme of this chapter, Figure 5.1 illustrates and contrasts the meiotic consequences of independent assortment, linkage without crossing over, and linkage with crossing over. Two homologous pairs of chromosomes are considered in the case of independent assortment and one homologous pair in the two cases of linkage.

Figure 5.1(a) illustrates the results of independent assortment of two pairs of nonhomologous chromosomes, each containing one heterozygous gene pair. Thus, no linkage is exhibited by the two genes. When a large number of meiotic events are observed, four genetically different gametes are formed in equal proportions.

We can compare these results with those that occur if the same genes are instead linked on the same chromosome. If no crossing over occurs between the two genes [Figure 5.1(b)], only two genetically different gametes are formed. Each gamete receives the alleles present on one homologue or the other, which has been transmitted intact as the result of segregation. This case illustrates **complete linkage,** which is said to result in the production of only **parental** or **noncrossover gametes.** The two parental gametes are formed in equal proportions. While complete linkage between two genes seldomly occurs, consideration of the theoretical consequences of this concept is useful when studying crossing over.

Figure 5.1(c) illustrates the results when crossing over occurs between two linked genes. As you will note, this crossover involves only two nonsister chromatids of the four chromatids present in the tetrad. This exchange generates two new allele combinations, called **recombinant** or **crossover gametes.** The two chromatids not involved in the exchange result in noncrossover gametes [as those in part (b) of this figure].

The frequency with which crossing over occurs between any two linked genes is proportional to the distance separating the respective loci along the chromosome. It is theoretically possible for two randomly selected genes to be so close to each other that crossover events are too infrequent to be detected. This circumstance, complete linkage, results in the production of only parental gametes, as shown in case (b). If a distinct but small distance separates two genes, few recombinant and many parental gametes will be formed. As the distance between two genes increases, the proportion of recombinant gametes increases and that of the parental gametes decreases. Thus, in case (c), the proportion of crossover gametes varies depending on the distance between the genes studied.

As will be explored again later in this chapter, when two linked genes whose loci are far apart are considered, the number of recombinant gametes approaches, but never exceeds 50 percent. If 50 percent recombinants occurred, a 1:1:1:1 ratio of the four types (two parental and two recombinant gametes) would result. In such a case, transmission of two linked genes would be indistinguishable from that of two unlinked, independently assorting genes. That is, the proportion of the four possible genotypes would be identical in Cases (a) and (c) of Figure 5.1.

The Linkage Ratio

If complete linkage exists between two genes because of their close proximity, as in case (b) of Figure 5.1, and organisms with mutant alleles representing these genes are mated, a unique F_2 phenotypic ratio results that is characteristic of linkage. To illustrate this ratio, we will consider a cross involving the closely linked recessive mutant genes *brown* (*bw*) eye and *heavy* (*hv*) wing vein in *Drosophila melanogaster* (Figure 5.2). The normal, wild-type alleles bw^+ and hv^+ are both dominant and result in red eyes and thin wing veins, respectively.

In this cross, flies with mutant brown eyes and normal thin veins are mated to flies with normal red eyes and mutant heavy veins. In more concise terms, brown-eyed flies are crossed with heavy-veined flies. If we extend the system of using genetic symbols established in Chapter 4, linked genes may be represented by placing their allele designations above and below a single or double horizontal line. Those placed above the line are located at loci on one homologue and those placed below at the homologous loci on the other homologue. Thus, we may represent the P_1 generation as follows:

$$P_1: \frac{bw\ hv^+}{bw\ hv^+} \times \frac{bw^+\ hv}{bw^+\ hv}$$

brown, thin   red, heavy

Since the genes are located on an autosome, no distinction for male and female is necessary.

In the F_1 generation each fly receives one chromosome of each pair from each parent; all flies are heterozygous for both gene pairs and exhibit the dominant traits of red eyes and thin veins:

$$F_1: \frac{bw\ \ hv^+}{bw^+\ hv}$$

red, thin

As shown in Figure 5.2, when the F_1 generation is interbred, because of complete linkage, each F_1 individual

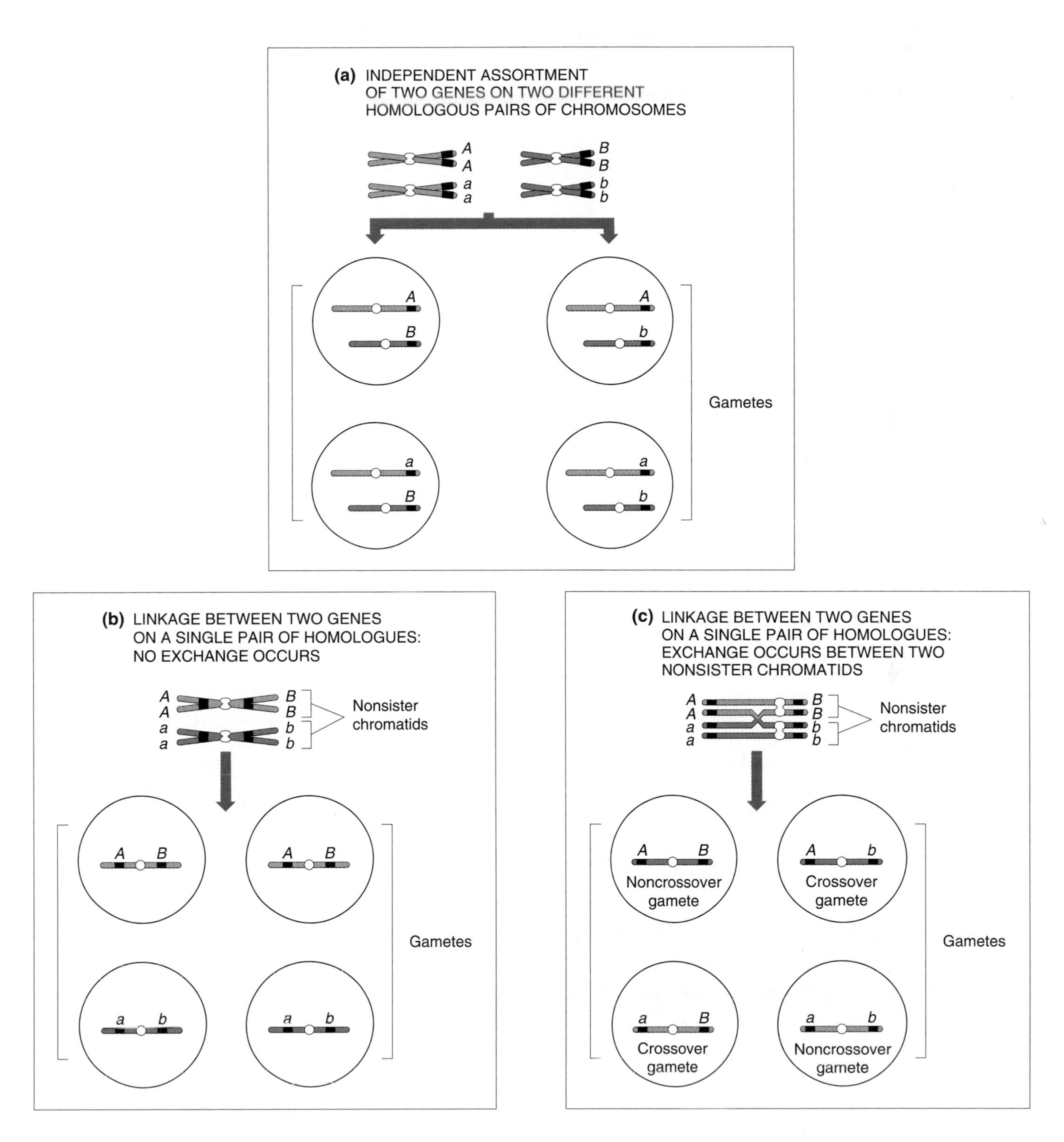

FIGURE 5.1
A comparison of the results of gamete formation where two heterozygous genes are (a) on two different pairs of chromosomes; (b) on the same pair of homologues, but with no exchange occurring between them; and (c) on the same pair of homologues, with an exchange between two nonsister chromatids.

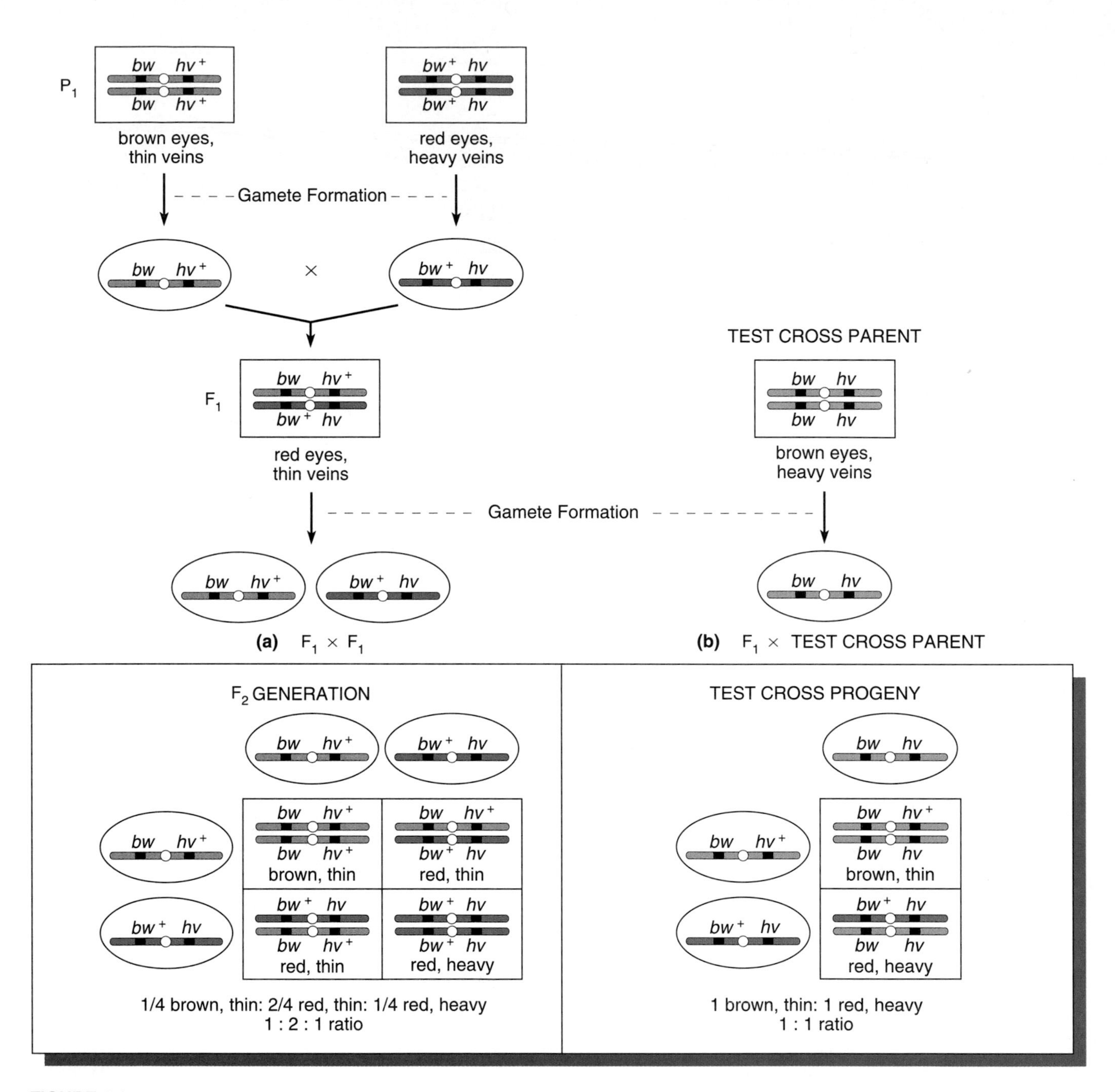

FIGURE 5.2
(a) The results of a cross between two linked genes demonstrating complete linkage. The cross is carried to the F_2 generation. (b) The results of a test cross with an F_1 individual are also illustrated.

forms only parental gametes. Following fertilization, the F_2 generation will be produced in a 1:2:1 phenotypic and genotypic ratio. One-fourth of this generation will show brown eyes and thin veins; one-half will show both wild-type traits, red eyes and thin veins; and one-fourth will show red eyes and thick veins. In more concise terms; the ratio is 1 brown:2 wild:1 thick. Such a ratio is characteristic of complete linkage. Even though it is extremely rare for two random genes to exhibit complete linkage, theoretically, it is possible.

Figure 5.2 also demonstrates the results of a test cross with the F_1 flies. Such a cross produces a 1:1 ratio of brown, thin and red, thick flies. Had the genes controlling these traits been incompletely linked or located on separate autosomes, four phenotypes rather than two would have been produced.

When large numbers of mutant genes present in any given species are investigated, genes located on the same chromosome will show evidence of linkage to one another. As a result, **linkage groups** can be established, one for each chromosome. Hence, the number of linkage groups should correspond to the haploid number of chromosomes. In organisms where large numbers of mutant genes are available for genetic study, this correlation has been upheld.

INCOMPLETE LINKAGE, CROSSING OVER, AND CHROMOSOME MAPPING

If two genes linked on the same chromosome are selected randomly it is unlikely that their respective loci will be contiguous to each other. Therefore, because crossing over will occur between them, complete linkage is rarely achieved. Instead, crosses involving two linked genes usually produce a percentage of offspring resulting from recombinant gametes. This percentage is variable, depending upon which two genes are involved in the cross. This phenomenon was first studied and explained by two *Drosophila* geneticists, Thomas H. Morgan and his undergraduate student, Alfred H. Sturtevant.

Morgan and Crossing Over

As you may recall from our earlier discussion of sex-linked genes in Chapter 4, Morgan first discovered the phenomenon of sex linkage. In his studies, he investigated numerous *Drosophila* mutations located on the X chromosome. When he analyzed crosses involving only one trait, he was able to deduce the mode of sex-linked inheritance. However, when he made crosses involving two sex-linked genes, his results were at first puzzling. For example, as shown in Cross A of Figure 5.3, he crossed mutant *yellow* body (y) and *white* eyes (w) females with wild-type males (gray bodies and red eyes). The F_1 females were wild-type, while the F_1 males expressed both mutant traits. In the F_2, 98.7 percent of the offspring showed the parental phenotypes—yellow-bodied, white-eyed flies and wild type flies. The remaining 1.3 percent of the flies were either yellow-bodied with red eyes or gray-bodied with white eyes. It was as if the genes had somehow separated from each other during gamete formation in the F_1 flies.

When he made crosses involving other sex-linked genes, the results were even more puzzling (Cross B of Figure 5.3). The same basic pattern was observed, but the proportion of F_2 phenotypes differed; for example, in a cross involving white-eye, miniature-wing mutants, only 62.8 percent of the F_2 showed the parental phenotypes, while 37.2 percent of the offspring appeared as if the mutant genes had been separated during gamete formation.

In 1911, Morgan was faced with two questions: (1) What was the source of gene separation? and (2) Why did the frequency of the apparent separation vary depending on the genes being studied? The proposed answer to the first question was based on his knowledge of earlier cytological observations made by F. A. Janssens and others. Janssens had observed that synapsed homologous chromosomes in meiosis wrapped around each other, forming crosslike areas called **chiasmata** [sing.: **chiasma**]. Morgan proposed that these chiasmata could represent points of genetic exchange.

In the crosses shown in Figure 5.3, Morgan postulated that if an exchange occurred between the mutant genes on the two X chromosomes of the F_1 females, it would lead to the observed results. He suggested that such exchanges led to 1.3 percent recombinant gametes in the *yellow-white* cross and 37.2 percent in the *white-miniature* cross. On the basis of this and other experimentation, Morgan concluded that linked genes exist in a linear order along the chromosome, and that a variable amount of exchange occurs between any two genes.

As an answer to the second question involving the frequency of gene separation, Morgan proposed that two genes located relatively close to each other along a chromosome are less likely to have a chiasma form between them than if the two genes are farther apart on the chromosome. Thus, the closer two genes are, the less likely it is that a genetic exchange will occur between them. Morgan proposed the term **crossing over** to describe the physical exchange leading to recombination.

Sturtevant and Mapping

Morgan's student, Alfred H. Sturtevant, was the first to realize that his mentor's proposal could be used to map the sequence of and distance between linked genes. For example, Sturtevant compiled further data on recombination between the genes represented by the *yellow, white,* and *miniature* mutants initially studied by Morgan. Frequencies of crossing over between each pair of these three genes were observed in separate crosses to be:

(1) *yellow, white*	0.5%
(2) *white, miniature*	34.5%
(3) *yellow, miniature*	35.4%

Since the sum of (1) and (2) is approximately equal to (3), Sturtevant suggested that the recombination frequencies between linked genes are additive. On this basis, he predicted that the order of the genes on the X chromosome was *yellow-white-miniature.* In arriving at this conclusion, he reasoned as follows: The *yellow* and *white* genes are apparently close to each other because the re-

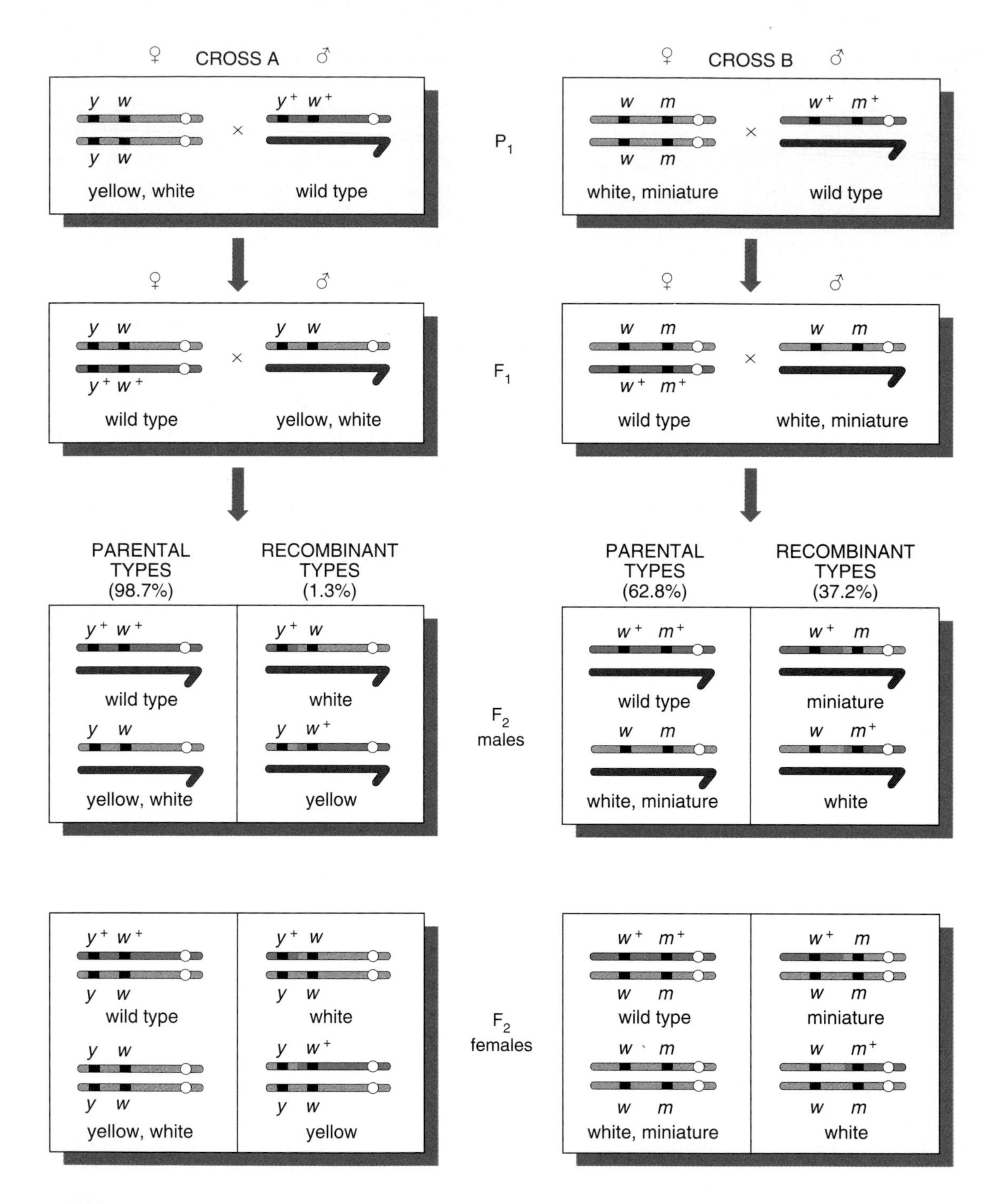

FIGURE 5.3
The F_1 and F_2 results of crosses involving the yellow-body, white-eye mutations and the white-eye, miniature-wing mutations. In the F_2 generation of Cross A, 1.3 percent of the flies demonstrate recombinant phenotypes, which are either *white* or *yellow*. In the F_2 generation of Cross B, 37.2 percent of the flies demonstrate recombinant phenotypes, which are either *miniature* or *white*.

combination frequency is low. However, both of these genes are quite far apart from *miniature* because the *white, miniature* and *yellow, miniature* combinations show large recombination frequencies. Since *miniature* shows more recombination with *yellow* than with *white* (35.4 vs. 34.5), it follows that *white* is between the other two genes, not outside of them.

Sturtevant suggested that the frequency of exchange could be taken as an estimate of the distance between two genes or loci along the chromosome. He constructed

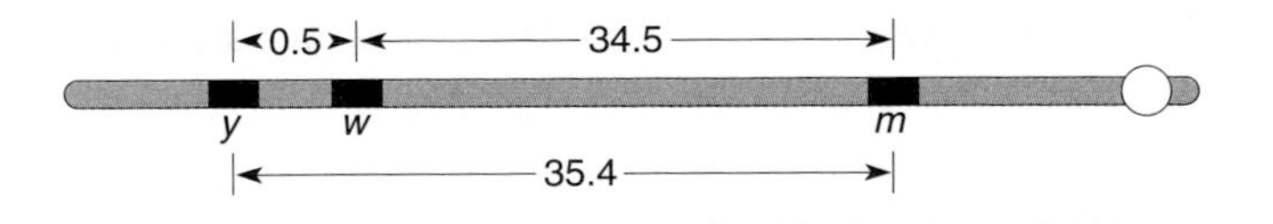

FIGURE 5.4
A simple map of the *yellow* (*y*), *white* (*w*), and *miniature* (*m*) genes on the X chromosome of *Drosophila melanogaster.* Each number represents the percentage of recombinant offspring produced in the three crosses, each involving two different genes.

a map of the three genes on the X chromosome, with one map unit being equated with one percent recombination between two genes.* In the preceding example, the distance between *yellow* and *white* would be 0.5 map unit, and between *yellow* and *miniature,* 35.4 map units. It follows that the distance between *white* and *miniature* should be (35.4 − 0.5) or 34.9. This estimate is close to the actual frequency of recombination between *white* and *miniature* (34.5). The simple map for these three genes is shown in Figure 5.4.

In addition to these three genes, Sturtevant considered three other genes on the X chromosome and produced a more extensive map including all six genes. He and a colleague, Calvin Bridges, soon began a search for autosomal linkage in *Drosophila.* By 1923, they had clearly shown that linkage and crossing over were not restricted to sex-linked genes. Rather, they had discovered linked genes, on autosomes, between which crossing over occurred.

During this work, they made another interesting observation. In *Drosophila,* crossing over was shown to occur only in females. The fact that no crossing over occurs in males made genetic mapping much less complex to analyze in *Drosophila.* However, crossing over does occur in both sexes in most other organisms.

Although many refinements in chromosome mapping have developed since Sturtevant's initial work, his basic principles are accepted as correct. They have been used to produce detailed chromosome maps of organisms for which large numbers of linked mutant genes are known. In addition to providing the basis for chromosome mapping, Sturtevant's findings were historically significant to the field of genetics. In 1910, the **chromosomal theory of inheritance** was still being widely disputed. Even Morgan was skeptical of this theory prior to the time he conducted the bulk of his experimentation. Research has now firmly established that chromosomes contain genes in a linear order and that these genes are the equivalent of Mendel's unit factors.

*In honor of Morgan's work, one map unit is often referred to as a centimorgan (cM).

Single Crossovers

Why should the relative distance between two loci influence the amount of recombination and crossing over observed between them? The basis for this variation is explained in the following analysis.

During meiosis, a limited number of crossover events occurs in each tetrad. These recombinant events occur randomly along the length of the tetrad. Therefore, the

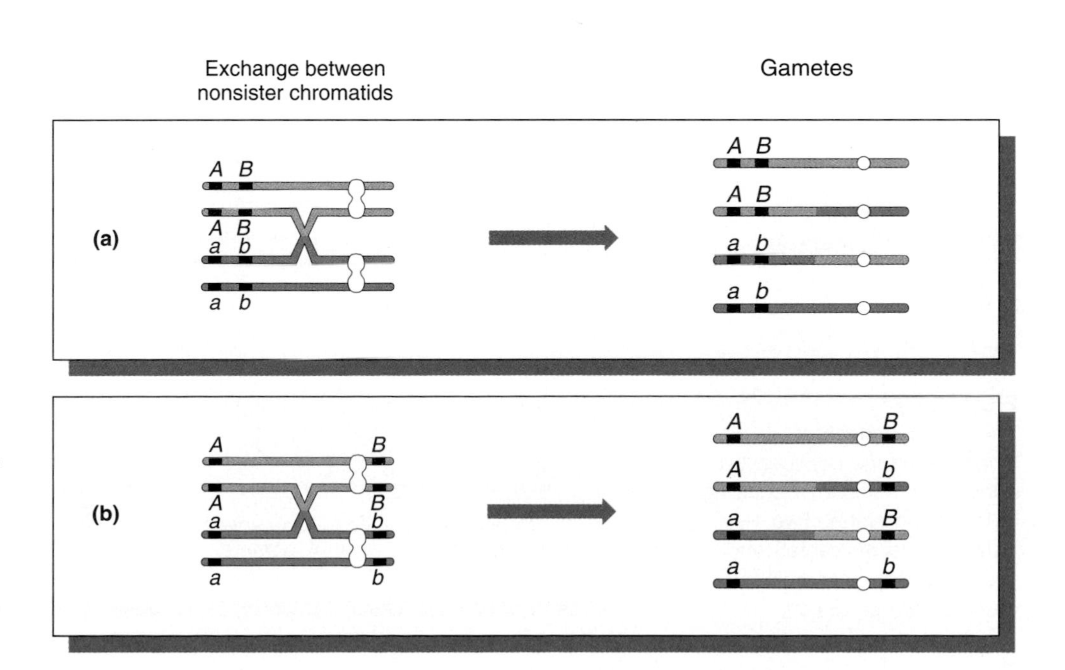

FIGURE 5.5
Two cases of exchange between nonsister chromatids and the gametes subsequently produced. In (a) the exchange does not separate the alleles of the two genes, only parental gametes are formed, and the exchange goes undetected. In (b) the exchange separates the alleles, resulting in recombinant gametes.

closer two loci reside along the axis of the chromosome, the less likely it is that any single crossover event will occur between them. The same reasoning suggests that the farther apart two linked loci are, the more likely it is that a random crossover event will occur between them.

In Figure 5.5(a), a single crossover occurs between two nonsister chromatids, but not between the two loci; therefore, the crossover goes undetected because no recombinant gametes are produced. In (b), where two loci are quite far apart, the crossover occurs between them, yielding recombinant gametes.

When a single crossover occurs between two nonsister chromatids, the other two strands of the tetrad will not be involved in this exchange and may enter a gamete unchanged. Thus, even if a single crossover occurs 100 percent of the time between two linked genes, this recombinant event will subsequently be observed in only 50 percent of the potential gametes formed. This concept is diagrammed in Figure 5.6. Theoretically, when only singles exchanges are considered, and when 20 percent recombinant gametes are observed, crossing over actually occurs between these two loci in 40 percent of the tetrads. The general rule is that under these conditions, the percentage of tetrads involved in an exchange between two genes is twice as great as the percentage of recombinant gametes produced. Therefore, the theoretical limit of crossing over is 50%.

When two linked genes are greater than 50 map units apart, a crossover can theoretically be expected to occur between them in 100 percent of the tetrads. If this prediction were to be achieved, each tetrad would yield equal proportions of the four gametes shown in Figure 5.6, just as if the genes were on different chromosomes and assorting independently. For a variety of reasons, this theoretical limit is never achieved.

Multiple Crossovers

It is possible that in a single tetrad, two, three, or more exchanges will occur between nonsister chromatids as a result of several crossing over events. Double exchanges of genetic material result from **double crossovers,** as shown in Figure 5.7. For a double exchange to be studied, three gene pairs must be investigated, each heterozygous for two alleles. Before we can determine the frequency of recombination between all three loci, we must review some simple probability calculations.

The probabilities of a single exchange occurring between the *A* and *B* or the *B* and *C* genes are directly related to the physical distance between each locus. The closer *A* is to *B* and *B* is to *C*, the less likely it is that a single exchange will occur between either of the two sets of loci. In the case of a double crossover, two separate and independent events or exchanges must occur simultaneously. The mathematical probability of two independent events occurring simultaneously is equal to the product of the individual probabilities. This is the product law previously introduced in Chapter 3.

Suppose that crossover gametes resulting from single exchanges between *A* and *B* are recovered 20 percent of the time ($p = 0.20$) and between *B* and *C* 30 percent of the time ($p = 0.30$). The probability of recovering a double-crossover gamete arising from two exchanges, between *A* and *B* and between *B* and *C*, is predicted to be $(0.20) \cdot (0.30) = 0.06$, or 6%. It is apparent from this calculation that the frequency of double-crossover gametes

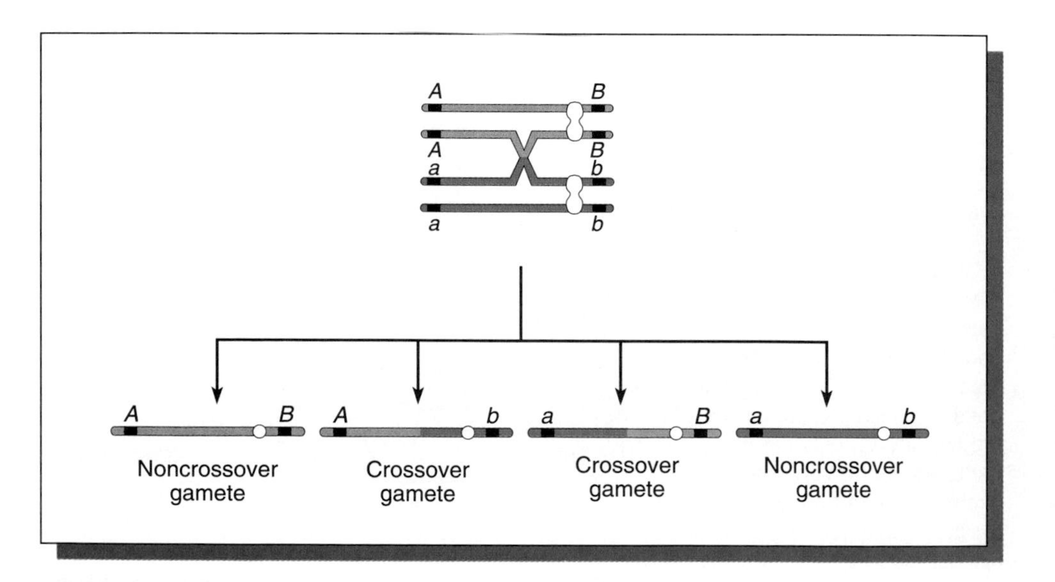

FIGURE 5.6
The consequences of a single exchange between two nonsister chromatids occurring in the tetrad stage. Two noncrossover (parental) and two crossover (recombinant) gametes are produced.

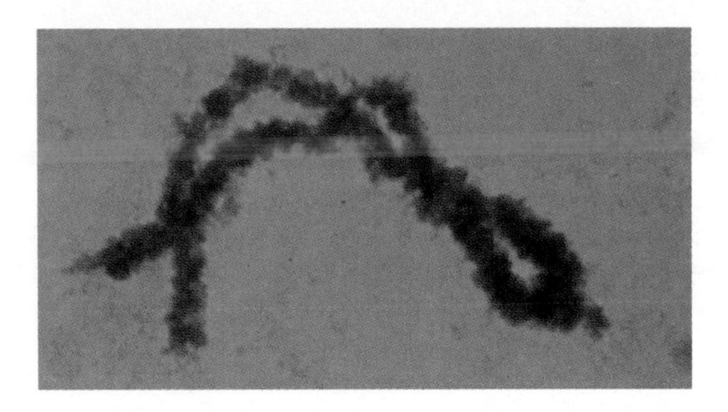

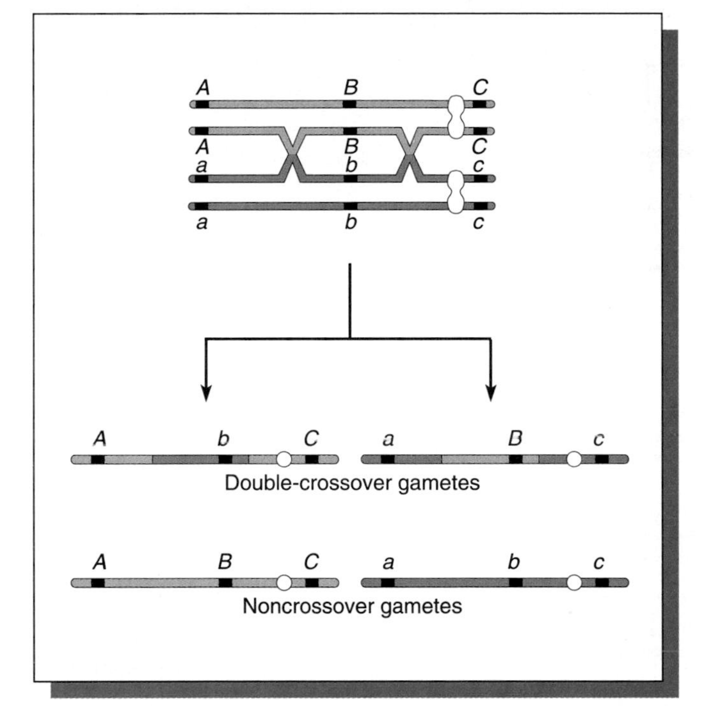

FIGURE 5.7
The results of a double exchange occurring between nonsister chromatids. Because the exchanges involve only two strands, two noncrossover gametes and two double crossover gametes are produced. The photograph illustrates chiasmata found in a tetrad isolated during the first meiotic prophase stage.

is always expected to be much lower than that of either single-crossover class of gametes.

If three genes are relatively close together along one chromosome, the expected frequency of double-crossover gametes is extremely low. For example, consider the $A - B$ distance in Figure 5.7 to be 3 map units and the $B - C$ distance in that figure to be 2 map units. The expected double-crossover frequency would be $(0.03) \cdot (0.02) = 0.0006$, or 0.06%. This translates to only 6 events in 10,000. Thus, in a mapping experiment involving closely linked genes, very large numbers of offspring are required in order to detect double-crossover events. In this example, it would be unlikely that a double crossover would be observed even if 1000 offspring were examined. If these probability considerations are extended, it is evident that if four or five genes were being mapped, even fewer triple and quadruple crossovers can be expected to occur.

Three-Point Mapping in *Drosophila*

The information presented in the previous section serves as the basis for mapping three or more linked genes in a single cross. To illustrate this, we will examine a situation involving three linked genes.

Three criteria must be met for a successful mapping cross:

1 The genotype of the organism producing the crossover gametes must be heterozygous at all loci under consideration.

2 The cross must be constructed so that genotypes of all gametes can be accurately determined by observing the phenotypes of the resulting offspring. This is necessary because the gametes and their genotypes can never be observed directly. Thus, each phenotypic class must reflect the genotype of the gametes of the parents producing it.

3 A sufficient number of offspring must be produced in the mapping experiment to recover a representative sample of all crossover classes.

These criteria are met in the three-point mapping cross from *Drosophila melanogaster* shown in Figure 5.8. In this cross, three sex-linked recessive mutant genes—*yellow* body color, *white* eye color, and *echinus* eye shape—are considered. In order to diagram the cross, we must assume some theoretical sequence, even though we do not yet know if it is correct. In Figure 5.8, we will initially assume the sequence of the three genes to be $y - w - ec$. If this is incorrect, our analysis will reveal the correct sequence.

In the P_1 generation, males hemizygous for all three wild-type alleles are crossed to females that are homozygous for all three recessive mutant alleles. Therefore, the P_1 males are wild type with respect to body color, eye color, and eye shape. They are said to have a wild-type phenotype. The females, on the other hand, exhibit the three mutant traits—*yellow* body color, *white* eyes, and *echinus* eye shape.

This cross produces an F_1 generation consisting of females heterozygous at all three loci and males that, because of the Y chromosome, are hemizygous for the three mutant alleles. Phenotypically, all F_1 females are wild type, while all F_1 males are *yellow, white,* and *echinus.* The genotype of the F_1 females fulfills the first crite-

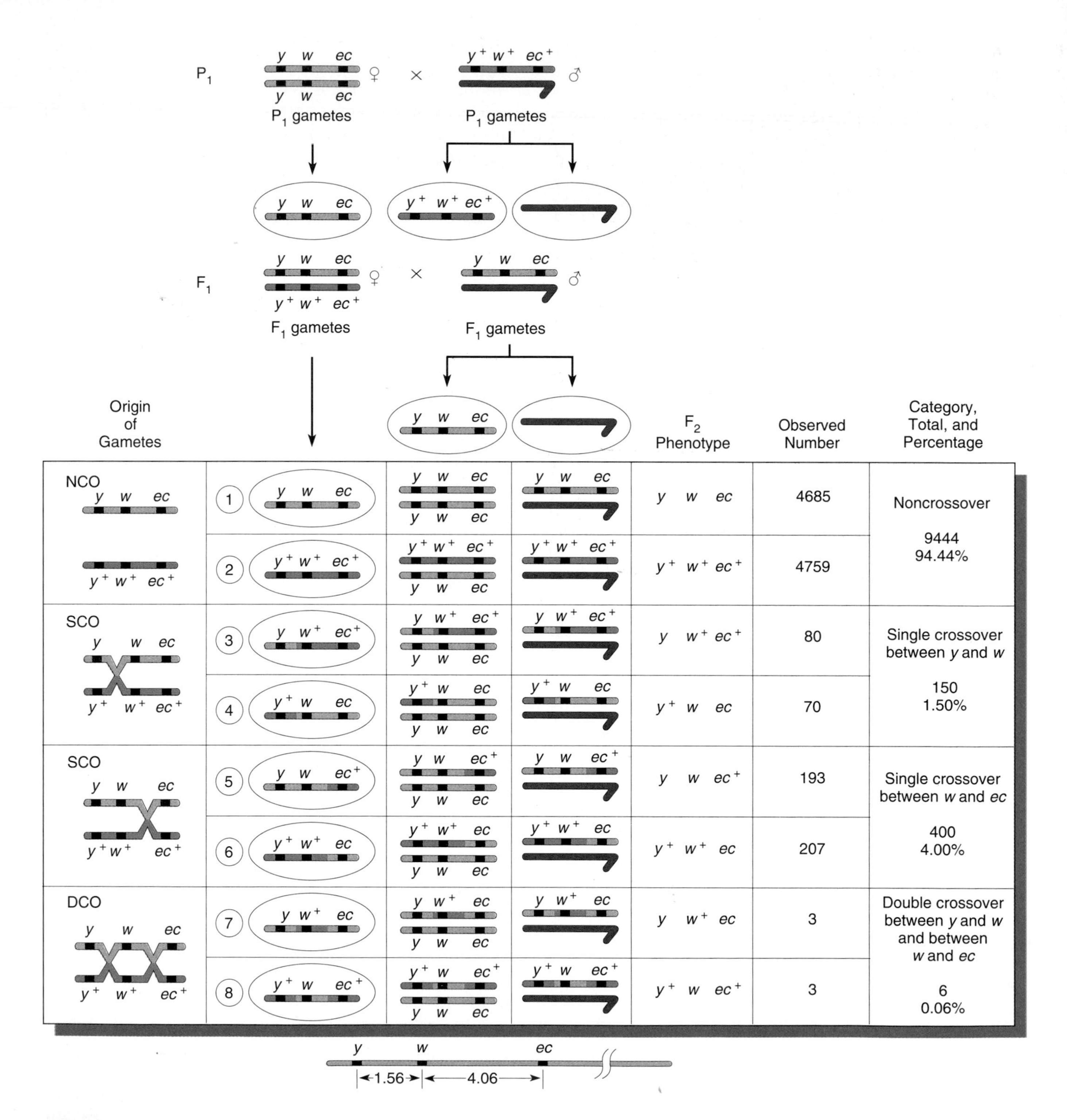

FIGURE 5.8
A three-point mapping cross involving the *yellow* (y or y^+), *white* (w or w^+), and *echinus* (ec or ec^+) genes in *Drosophila melanogaster*. NCO, SCO, and DCO refer to noncrossover, single crossover, and double crossover groups, respectively. Note that because of the complexity of this and several of the ensuing figures, centromeres have not been included on the chromosomes.

rion for constructing a map of the three linked genes; that is, it is heterozygous at the three loci and may serve as the source of recombinant gametes generated by crossing over. Note that because of the genotypes of the P_1 parents, all three mutant alleles are on one homologue, and all three wild-type alleles are on the other homologue.

In our cross, the second criterion is met by virtue of the gametes formed by the F_1 males. Every gamete will contain either an X chromosome bearing the three mutant alleles or a Y chromosome, which is genetically inert for the three loci being considered. Whichever type participates in fertilization, the genotype of the gamete produced by the F_1 female will be expressed phenotypically in the F_2 male and female offspring derived from it. As a result, all F_1 noncrossover and crossover gametes can be detected by observing the F_2 phenotypes.

With these two criteria met, we can construct a chromosome map from the crosses illustrated in Figure 5.8. First, we must determine which F_2 phenotypes correspond to the various noncrossovers and crossover categories. Two of these can be determined immediately.

The **noncrossover** F_2 phenotypes are determined by the combination of alleles present in the parental gametes formed by the F_1 female. Each such gamete contains one or the other of the X chromosomes unaffected by crossing over. As a result of segregation, approximately equal proportions of the two types of gametes, and subsequently the F_2 phenotypes, are produced. Because they are derived from a heterozygote, the genotypes of the two parental gametes and the phenotypes of the two F_2 phenotypes complement one another. For example, if one is wild type, the other is completely mutant. This is the case in the cross under consideration. In other situations, if one chromosome shows one mutant allele or trait, the other shows the other two mutant traits, and so on. They are therefore called **reciprocal classes** of gametes and phenotypes.

The two noncrossover phenotypes are most easily recognized because *they exist in the greatest proportion.* Figure 5.8 shows that classes (1) and (2) are present in the greatest numbers. Therefore, flies that are *yellow, white,* and *echinus* and those that are normal or wild type for all three characters constitute the noncrossover category and represent 94.44 percent of the F_2 offspring.

The second category that can be easily detected is represented by the double-crossover phenotypes. Because of their probability of occurrence, *they must be present in the least numbers.* Remember that this group represents two independent but simultaneous single-crossover events. Two reciprocal phenotypes can be identified: class (7), which shows the mutant traits *yellow* and *echinus* but normal eye color; and class (8), which shows the mutant trait *white* but normal body color and eye shape. Together these double-crossover phenotypes constitute only 0.06 percent of the F_2 offspring.

The remaining four phenotypic classes represent two categories resulting from single crossovers. Classes (3) and (4), reciprocal phenotypes produced by single-crossover events occurring between the *yellow* and *white* loci, are equal to 1.50 percent of the F_2 offspring. Classes (5) and (6), constituting 4.00 percent of the F_2 offspring, represent the reciprocal phenotypes resulting from single-crossover events occurring between the *white* and *echinus* loci.

The map distances between the three loci can now be calculated. The distance between *y* and *w*, or between *w* and *ec*, is equal to the percentage of all detectable exchanges occurring between them. For any two genes under consideration, this will include all appropriate single crossovers as well as all double crossovers. The latter are included because they represent two simultaneous single crossovers. For the *y* and *w* genes this includes classes (3), (4), (7), and (8), totaling 1.50% + 0.06%, or 1.56 map units. Similarly, the distance between *w* and *ec* is equal to the percentage of offspring resulting from an exchange between these two loci: classes (5), (6), (7), and (8), totaling 4.00% + 0.06%, or 4.06 map units. The map of these three loci on the X chromosome, based on these data, is shown at the bottom of Figure 5.8.

Determining the Gene Sequence

In the preceding example, the order or sequence of the three genes along the chromosome was assumed to be *y* – *w* – *ec*. Our analysis shows this sequence to be consistent with the data. In most mapping experiments, the gene sequence is not known, and this constitutes another variable in the analysis. Had the gene order been unknown in this example, it could have been determined using a straightforward method.

This method is based on the fact that there are only three possible orders, each containing one of the three genes in between the other two.

(I) *w-y-ec* (*y* is in the middle)
(II) *y-ec-w* (*ec* is in the middle)
(III) *y-w-ec* (*w* is in the middle)

If you use the following steps during your analysis, you will be able to determine the gene order:

1 Assuming any of the three orders, first determine the **arrangement** of alleles along each homologue of the heterozygous parent giving rise to noncrossover and crossover gametes (the F_1 female in our example).

2 Determine whether a double-crossover event occurring within that arrangement will produce the **observed** double-crossover phenotypes. Remember that these phenotypes occur least frequently and can be easily identified.

3 If this order does not produce the correct phenotypes, try each of the other two orders. One must work!

These steps will be illustrated for the cross shown in Figure 5.8. The three possible orders are labeled **I**, **II**, and **III**, as just shown. Either *y*, *ec*, or *w* must be in the middle.

1 Assuming *y* is between *w* and *ec*, the arrangement of alleles along the homologues of the F_1 heterozygote is

$$\textbf{I} \quad \frac{w \quad y \quad ec}{w^+ \; y^+ \; ec^+}$$

We know this because of the way in which the P_1 generation was crossed. The P_1 female contributed an X chromosome bearing the *w*, *y*, and *ec* alleles, while the P_1 male contributed an X chromosome bearing the w^+, y^+, and ec^+ alleles.

2 A double crossover within the above arrangement would yield the following gametes:

$$\underline{w \;\; y^+ \; ec} \quad \text{and} \quad \underline{w^+ \; y \;\; ec^+}$$

Following fertilization, if *y* is in the middle, the F_2 double-crossover phenotypes will correspond to the above gametic genotypes, yielding offspring that are *white, echinus* and offspring that are *yellow*. Instead, however, determination of the actual double crossovers reveals them to be *yellow, echinus* flies and *white* flies. Therefore, our assumed order is incorrect.

3 If we try the other orders, one with the ec/ec^+ alleles in the middle **II** or one with the w/w^+ alleles in the middle **III**:

$$\textbf{II} \quad \frac{y \quad ec \quad w}{y^+ \; ec^+ \; w^+} \quad \text{or} \quad \textbf{III} \quad \frac{y \quad w \quad ec}{y^+ \; w^+ \; ec^+}$$

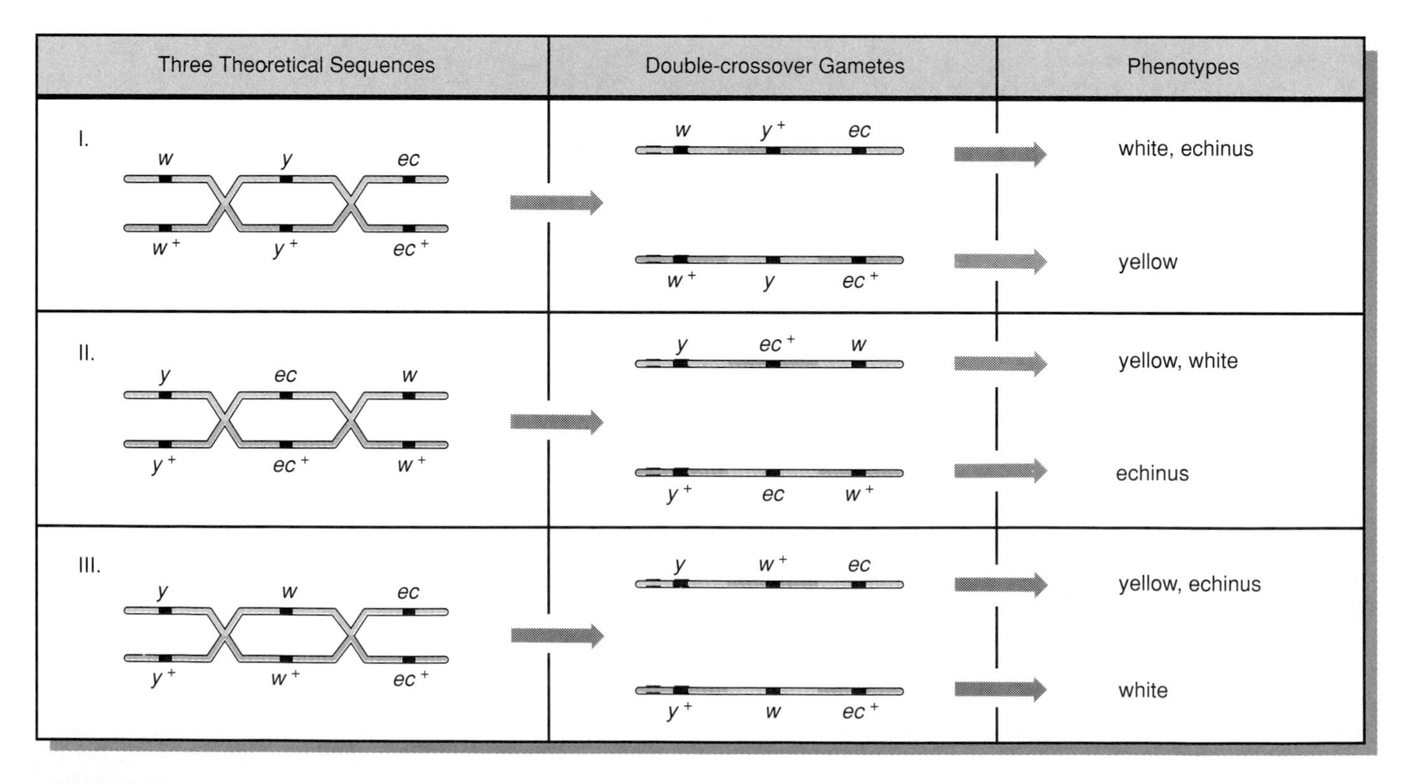

FIGURE 5.9
A summary of the three possible orders of the *white, yellow,* and *echinus* genes, the results of a double crossover in each case, and the resulting phenotypes produced.

we see that arrangement II again provides **predicted** double-crossover phenotypes that do not correspond to the **actual** double-crossover phenotypes. The predicted phenotypes are *yellow, white* flies and *echinus* flies in the F_2 generation. Therefore, this order is also incorrect. However, arrangement III **will** produce the **observed** phenotypes, *yellow, echinus* flies and *white* flies. Therefore, this order, where the *w* gene is in the middle, is correct.

To summarize, utilizing this method is rather straightforward. Determine the arrangement of alleles on the homologues of the heterozygote yielding the crossover, gametes. This is done by locating the reciprocal noncrossover phenotypes. Then, test each of three possible orders to determine which one yields the observed double-crossover phenotypes. Whichever of the three does so represents the correct order. Testing the three possibilities in our example is summarized in Figure 5.9.

A Mapping Problem in Maize

Having established the basic principles of chromosome mapping, we will now consider a related problem in maize (corn) where the gene sequence and interlocus distances are unknown.

This analysis differs from the preceding discussion in several ways and therefore will expand your knowledge of mapping procedures:

1 The previous mapping cross involved sex-linked genes. Here, autosomal genes are considered.

2 In the discussion of this cross we will make a transition in the use of symbols, as first suggested in Chapter 4. Instead of using the symbols bm^+, v^+, and pr^+, we will simply use + to denote each wild-type allele. This symbol is less complex to manipulate, but requires a better understanding of mapping procedures.

When we consider three autosomally linked genes in maize, the experimental cross must still meet the same three criteria established for the X-linked genes in *Drosophila:* (1) one parent must be heterozygous for all traits under consideration; (2) the gametic genotypes produced by the heterozygote must be apparent from observing the phenotypes of the offspring; and (3) a sufficient sample size must be available for complete analysis.

In maize, the recessive mutant genes *bm* (brown midrib), *v* (virescent seedling), and *pr* (purple aleurone) are linked on chromosome 5. Assume that a female plant is known to be heterozygous for all three traits. Nothing is known about: (1) the arrangement of the mutant alleles on the maternal and paternal homologues of this heterozygote; (2) the sequence of genes; or (3) the map distances between the genes. What genotype must the male plant have to allow successful mapping? In order to meet the second criterion, the male must be homozygous for all three recessive mutant alleles. Otherwise, offspring of this cross showing a given phenotype might represent more than one genotype, making accurate mapping impossible.

Figure 5.10 diagrams this cross. As shown, we do not know either the arrangement of alleles or the sequence of loci in the heterozygous female. Various possibilities are shown each followed by a question mark. In the test cross male parent, *we do not know the sequence,* but must write it down initially as a random selection. Note that we have chosen initially to place *v* in the middle. This may or may not be correct.

The offspring have been arranged in groups of two for each pair of reciprocal phenotypic classes. The two members of each reciprocal class are derived from either no crossing over **(NCO),** one of two possible single-crossover events **(SCO),** or a double crossover **(DCO).**

To solve this problem, consider the following questions. It will be helpful to refer to Figures 5.10 and 5.11 as you consider them.

1 *What is the actual heterozygous arrangement of alleles in the female parent?*

Determine the two noncrossover classes, those that occur in the highest frequency. In this case, they are $\underline{+\ v\ bm}$ and $\underline{pr\ +\ +}$. Therefore, the arrangement of alleles on the homologues of the female parent must be as shown in Figure 5.11(a). These homologues segregate into gametes, unaffected by any recombination event. Any other arrangement of alleles could not yield the observed noncrossover classes. (Remember that $\underline{+\ v\ bm}$ is equivalent to $\underline{pr^+\ v\ bm}$, and that $\underline{pr\ +\ +}$ is equivalent to $\underline{pr\ v^+\ bm^+}$.)

2 *What is the correct sequence of genes?*

We know that the arrangement of alleles is

$$\frac{+\ v\ bm}{pr\ +\ +}$$

But, is the assumed sequence correct? That is, will a double-crossover event yield the observed double-crossover phenotypes following fertilization? Simple observation shows that *it will not* [Figure 5.11(b)]. Try the other two orders [Figure 5.11(c) and (d)], *keeping the same arrangement:*

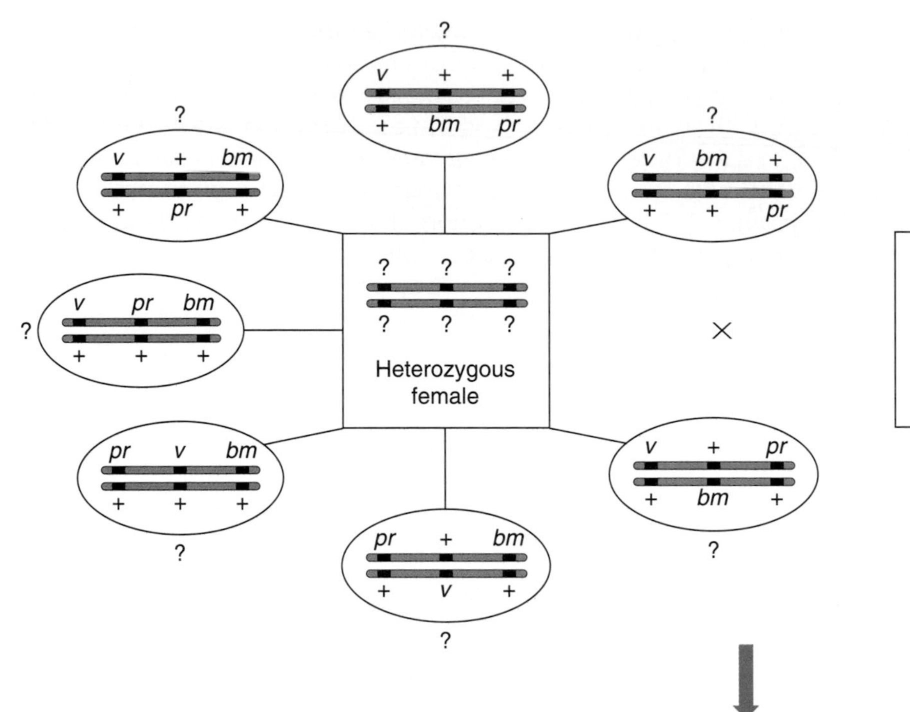

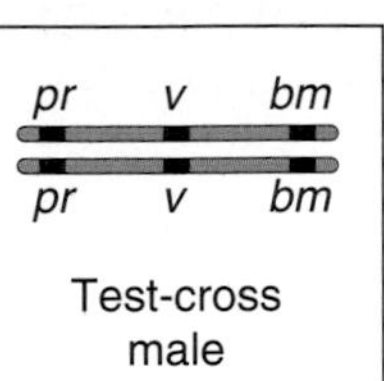

Offspring Phenotype	Number	Total and Percentage	Exchange Classification
+ *v* *bm*	230	467	Noncrossover
pr + +	237	42.1%	(NCO)
pr *v* +	82	161	Single crossover
+ + *bm*	79	14.5%	(SCO)
+ *v* +	200	395	Single crossover
pr + *bm*	195	35.6%	(SCO)
pr *v* *bm*	44	86	Double crossover
+ + +	42	7.8%	(DCO)

FIGURE 5.10
The results of a three-point mapping cross in maize where neither the arrangement of alleles nor the sequence of and distance between the genes are known. The results have been expressed with the *v* gene in the middle, although it is not known if this is correct. At the top left are shown various allele arrangements and gene sequences which are possible in the heterozygous female parent.

$$\frac{+\ bm\ v}{pr\ +\ +} \quad \text{and} \quad \frac{v\ +\ bm}{+\ pr\ +}$$

Only the latter case will yield the observed double-crossover classes (Figure 5.11(d)]. Therefore, the *pr* gene is in the middle. From this point on, work the problem using the above arrangement and sequence with the *pr* locus in the middle.

3 *What is the distance between each pair of genes?* Having established the sequence of loci as *v–pr–bm*, we can now determine the distance between *v* and *pr* and between *pr* and *bm*. Remember that the map distance between two genes is calculated on the basis of all detectable recombinational events occurring between them. This includes both the single- and double-crossover events involving the two genes being considered.

Figure 5.11(e) shows that the phenotypes $\underline{v\ pr\ +}$

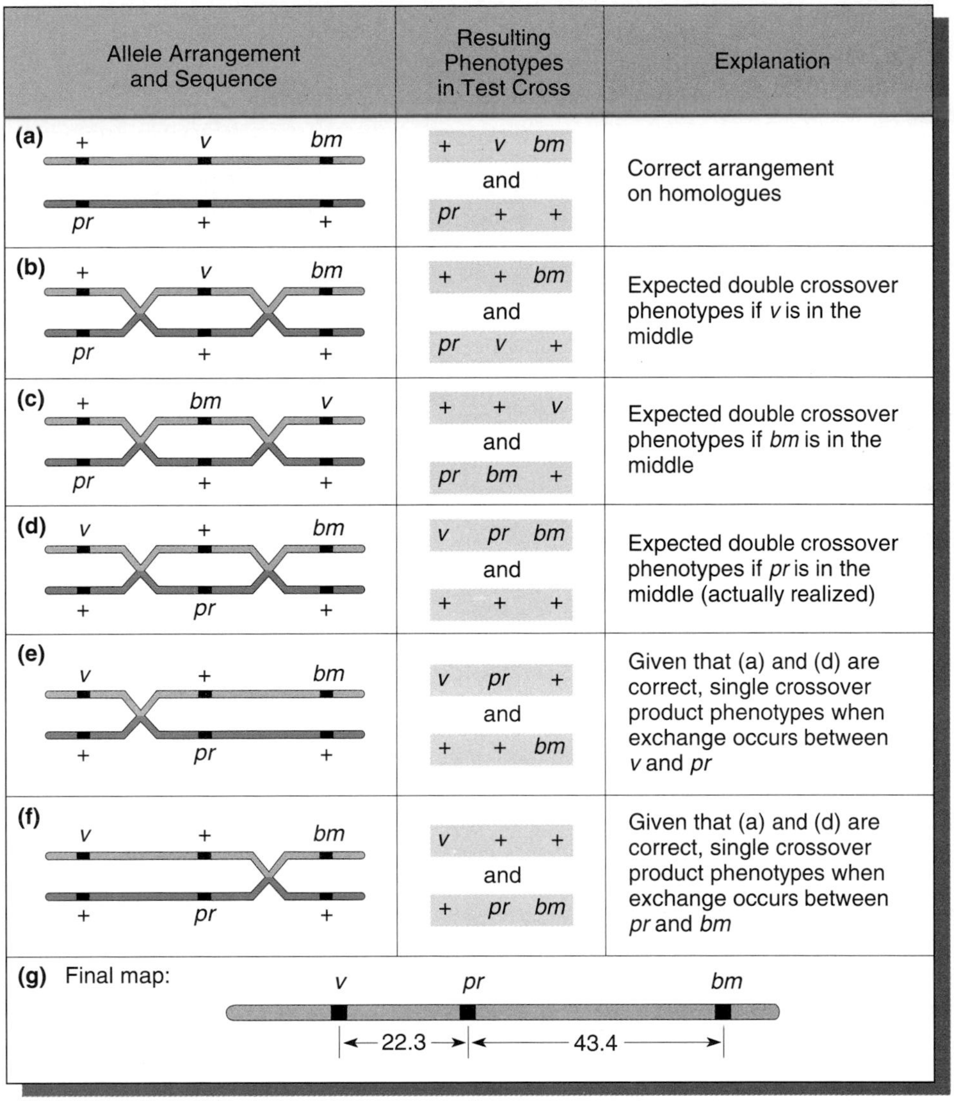

	Allele Arrangement and Sequence	Resulting Phenotypes in Test Cross	Explanation
(a)	+ *v* *bm* / *pr* + +	+ *v* *bm* and *pr* + +	Correct arrangement on homologues
(b)	+ *v* *bm* / *pr* + +	+ + *bm* and *pr* *v* +	Expected double crossover phenotypes if *v* is in the middle
(c)	+ *bm* *v* / *pr* + +	+ + *v* and *pr* *bm* +	Expected double crossover phenotypes if *bm* is in the middle
(d)	*v* + *bm* / + *pr* +	*v* *pr* *bm* and + + +	Expected double crossover phenotypes if *pr* is in the middle (actually realized)
(e)	*v* + *bm* / + *pr* +	*v* *pr* + and + + *bm*	Given that (a) and (d) are correct, single crossover product phenotypes when exchange occurs between *v* and *pr*
(f)	*v* + *bm* / + *pr* +	*v* + + and + *pr* *bm*	Given that (a) and (d) are correct, single crossover product phenotypes when exchange occurs between *pr* and *bm*
(g)	Final map: *v* — 22.3 — *pr* — 43.4 — *bm*		

FIGURE 5.11
The steps used in producing a map of the three genes involved in the cross shown in Figure 5.10, where neither the arrangement of alleles nor the sequence of genes are initially known in the heterozygous female parent.

and *+ + bm* result from single crossovers between the *v* and *pr* loci, accounting for 14.5 percent of the offspring. By adding the percentage of double crossovers (7.8%) to the number obtained for single crossovers, the total distance between the *v* and *pr* loci is calculated to be 22.3 map units.

Figure 5.11(f) shows that the phenotypes *v + +* and *+ pr bm* result from single crossovers between the *pr* and *bm* loci, totaling 35.6 percent. With the addition of the double-crossover classes (7.8%) the distance between *pr* and *bm* is calculated to be 43.4 map units.

The final map for all three genes in this example is shown in Figure 5.11(g).

The Accuracy of Mapping Experiments

Until now, we have considered crossover frequencies to be directly proportional to the distance between any two loci along the chromosome. However, it is not always possible to detect all crossover events. A case in point is a double exchange that occurs between the two loci in question. As shown in Figure 5.12(a), if a double exchange occurs, the original arrangement of alleles on each nonsister homologue is recovered. Therefore, even though crossing over occurs, it is impossible to detect in these cases. This factor holds for all even-numbered exchanges between two loci.

As a result of this and other types of multiple exchanges, mapping determinations will usually underestimate the actual distance between two genes. The farther apart two genes are, the greater the probability that undetected crossovers will occur. Therefore, the discrepancy is minimal for two genes that are relatively close together, but increases as the distance becomes greater, as noted in Figure 5.12(b), where the relationship between recombination frequency and map distance is graphed.

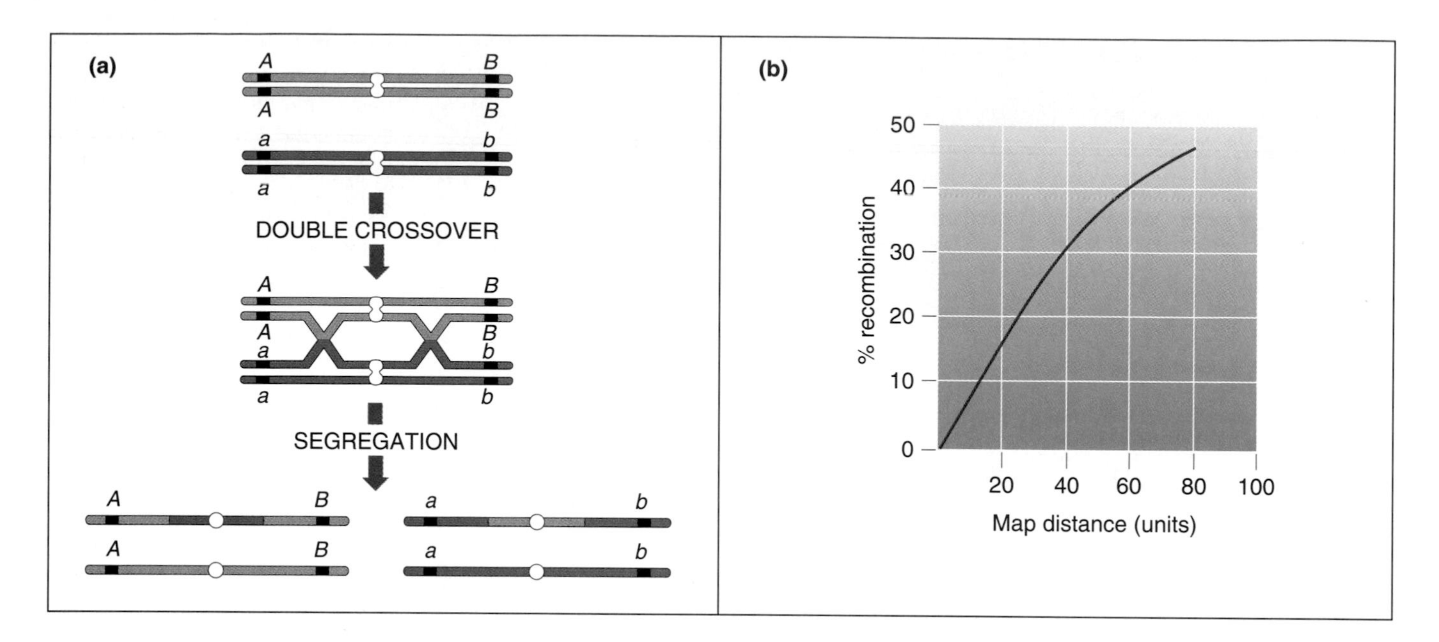

FIGURE 5.12
(a) An illustration of a double crossover that goes undetected because no rearrangement of alleles occurs. (b) The relationship between the frequency of recombination and map distance, as studied in *Drosophila, Neurospora,* and maize.

Interference and the Coefficient of Coincidence

Still another factor tends to limit the accuracy of mapping data. This factor involves the actual reduction of the number of expected double crossovers when genes are reasonably close to one another along the chromosome. This reduction, called **interference,** will be illustrated for three-point mapping.

We have already considered the probability relationships between single- and double-crossover events. In theory, the percentage of **expected double crossovers (DCOs)** is predicted by multiplying the percentage of the total crossovers between each pair of genes. Remember that a DCO represents two single-crossover events. For example, the expected double-crossover frequency of the cross illustrated in Figures 5.10 and 5.11 may be calculated in the following manner:

$$\text{DCO}_{\text{exp}} = (0.223) \times (0.434) = 0.097 = 9.7\%$$

Frequently, this predicted figure does not correspond precisely with the observed DCO frequency. Generally, there are fewer DCOs observed than predicted. In the maize cross, only 7.8% DCOs were observed. In some cases there may be more DCOs than expected. These disparities are explained by the concept of **interference,** which is quantified by calculating the **coefficient of coincidence (*C*)**:

$$C = \frac{\text{Observed DCO}}{\text{Expected DCO}}$$

In the maize cross, we have

$$C = \frac{0.078}{0.097} = 0.804$$

Once C is calculated, interference (I) may be quantified using the simple equation

$$I = 1 - C$$

In the maize cross, we have

$$I = 1.000 - 0.804 = 0.196$$

If interference is complete, and no double crossovers occur, $I = 1.0$. If fewer DCOs than expected occur, I is a positive number (as above) and **positive interference** has occurred. If more DCOs than expected occur, I is a negative number and **negative interference** has occurred.

In eukaryotic systems, positive interference is most often observed. It appears that a crossover event in one region of a chromosome inhibits a second crossover in neighboring regions of the chromosome. In general, the closer genes are to one another along the chromosome, the more positive interference is observed (and the lower the *C* value). Perhaps a mechanical stress is imposed on chromatids during crossing over such that one chiasma inhibits the formation of a second chiasma in the neighboring region.

SOMATIC CELL HYBRIDIZATION AND THE HUMAN GENE MAP

In humans, where neither designed matings nor large numbers of offspring are available, the earliest linkage studies were based on pedigree analysis. Attempts were made to establish whether a trait was sex-linked or autosomal. All of the traits shown to be sex-linked are linked to the X chromosome. For autosomal traits, geneticists tried to distinguish clearly whether pairs of traits demonstrated linkage or independent assortment. In this way, it was hoped that a human gene map could be created.

The results, even as recently as the 1960s, were discouraging because of the limitations of this approach and because of the relatively high haploid number of human chromosomes (23). Relatively few cases of either X-linkage or autosomal linkage were established, and almost no mapping information became available.

However, in the 1960s, a new technique, **somatic cell hybridization,** was developed that aided immensely in assigning human genes to their respective chromosomes. This technique, first discovered by Georges Barsky, relies on the fact that two cells in culture can be induced to fuse into a single hybrid cell. While Barsky used two mouse cell lines, it soon became evident that cells from different organisms will also fuse together. When this event occurs, an initial cell type called a **heterokaryon** is produced. The hybrid cell contains two nuclei in a common cytoplasm. Using the proper techniques, it is possible to fuse human and mouse cells, for example, and isolate the hybrids from the parental cells.

As the heterokaryons are cultured *in vitro,* two interesting changes occur. Eventually, the nuclei fuse together, creating what is termed a **synkaryon.** Then, as culturing is continued for many generations, chromosomes from one of the two parental species are gradually lost. In the case of the human-mouse hybrid, human chromosomes are lost randomly until eventually, the synkaryon has a full complement of mouse chromosomes and only a few human chromosomes. It is the result of this event that allows the assignment of human genes to the chromosomes upon which they reside.

The experimental rationale is straightforward. If a specific human gene product is synthesized in a synkaryon containing one to three human chromosomes, then the gene responsible for that product must reside on one of the human chromosomes remaining in the hybrid cell. Or, if the human gene product is absent, the responsible gene cannot be present on any of the remaining human chromosomes. Ideally, a panel of 23 hybrid cell lines, each with but one unique human chromosome, would allow the immediate assignment of any human gene for which the product could be characterized.

In practice, a panel of cell lines, each with several remaining human chromosomes is most often utilized. The correlation of the presence or absence of each chromosome with the presence or absence of each gene product is called **synteny testing.** Consider, for example, the hypothetical data provided in Figure 5.13, where four gene products (*A*, *B*, *C*, and *D*) are tested in relationship

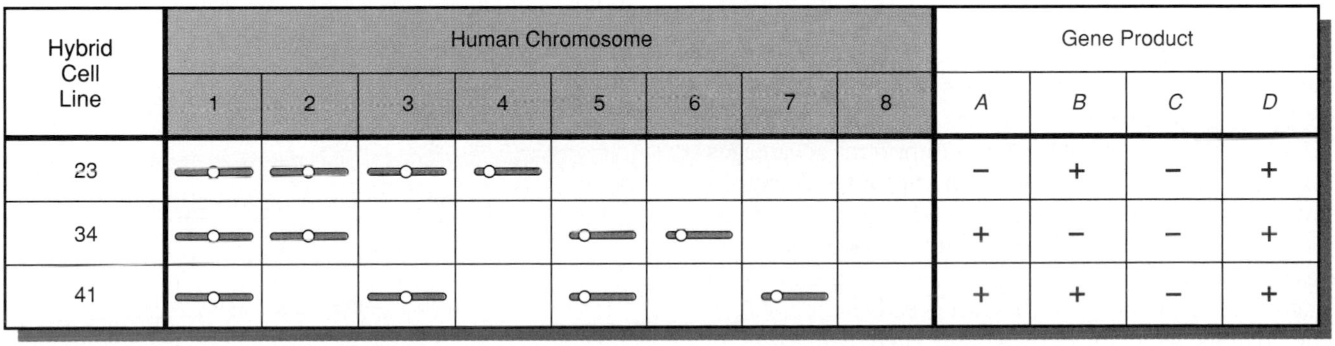

Hybrid Cell Line	Human Chromosome								Gene Product			
	1	2	3	4	5	6	7	8	*A*	*B*	*C*	*D*
23	━○━	━○━	━○━	━○━					−	+	−	+
34	━○━	━○━			━○━	━○━			+	−	−	+
41	━○━		━○━		━○━		━○━		+	+	−	+

FIGURE 5.13
A hypothetical grid of data used in synteny testing to assign genes to their appropriate human chromosomes. Three somatic hybrid cell lines, designated 23, 34, and 41, have each been scored for the presence or absence of human chromosomes 1 through 8, as well as for their ability to produce the hypothetical human gene products A through D.

to eight human chromosomes. Let us carefully analyze the gene that produces product *A*:

1. Product *A* is not produced by cell line 23, but chromosomes 1, 2, 3, and 4 are present in cell line 23. Therefore, we can rule out the presence of gene *A* on those four chromosomes and conclude that it must be on chromosome 5, 6, 7, or 8.

2. Product *A* is produced by cell line 34, which contains chromosomes 5 and 6, but not 7 and 8. Therefore, gene *A* is on chromosome 5 or 6, not 7 or 8.

3. Product *A* is also produced by cell line 41, which contains chromosome 5 but not chromosome 6. Therefore, gene *A* is on chromosome 5, according to this analysis.

Using a similar approach, gene *B* can be assigned to chromosome 3. You should perform this analysis to demonstrate for yourself that this is correct. Gene *C* presents a unique situation. The data indicate that it is not present on any of the first seven chromosomes (1–7). While it might be on chromosome 8, no direct evidence supports this conclusion. Other panels are needed. We shall leave gene *D* for you to analyze. Upon what chromosome does it reside? Using the approach described above, literally hundreds of human genes have been assigned to one chromosome or another.

THE USE OF HAPLOID ORGANISMS IN LINKAGE AND MAPPING STUDIES

Many of the single-celled eukaryotes are haploid during the vegetative stages of their life cycle. The alga *Chlamydomonas* and the mold *Neurospora* illustrate this genetic condition. These organisms do form reproductive cells that fuse during fertilization, producing a diploid zygote. However, this structure soon undergoes meiosis, resulting in haploid vegetative cells that are then propagated by mitotic divisions.

Haploid organisms have several important advantages in genetic studies compared with diploid eukaryotes. They can be cultured and manipulated in genetic crosses much more easily. In addition, a haploid organism contains only a single allele of each gene, which is expressed directly in the phenotype. This fact greatly simplifies genetic analysis. As a result, organisms such as *Chlamydomonas* and *Neurospora* have served as subjects of research investigations in many areas of genetics, including linkage and mapping studies.

In order to perform genetic experiments with such organisms, crosses are made, and following fertilization, the meiotic structures are isolated. Because in each of these structures all four meiotic products give rise to spores, each structure is called a **tetrad.** *This term has a different meaning here than when it was used earlier to describe a precise chromatid configuration in meiosis.* In any case, individual tetrads are isolated, and the resultant cells are grown and analyzed separately from those of other tetrads. In the results to be described, the data will reflect the proportion of tetrads that showed one combination of genotypes, the proportion that showed another combination, and so on. Thus, such experimentation is called **tetrad analysis.**

Gene to Centromere Mapping

When a single gene is analyzed in *Neurospora,* as diagrammed in Figure 5.14, the data can be used to calculate the map distance between the gene and the centromere. This process is sometimes referred to as **mapping the centromere.** It may be accomplished by experimentally determining the frequency of recombination using tetrad data.

If no crossover event occurs between the gene under study and the centromere, the pattern of ascospores in the ascus appears as shown in Figure 5.14(a) (*aaaa*++++). This pattern represents **first division segregation** because the two alleles are separated during the first meiotic division. However, a crossover event will alter this pattern, as shown in Figure 5.14(b) (*aa*++*aa*++) and 5.14(c) (++*aaaa*++). Actually, two other recombinant patterns may occur, depending on the chromatid orientation during the second meiotic division: ++*aa*++*aa* and *aa*++++*aa*. All four of the latter patterns reflect **second division segregation** because the two alleles are not separated until the second meiotic division. Usually, the ordered tetrad data are condensed to reflect the genotypes of the identical ascospore pairs. Thus, five combinations are possible:

First Division Segregation

a a + +

Second Division Segregation

a + *a* +
+ *a* + *a*
+ *a a* +
a + + *a*

In order to calculate the distance between the gene and the centromere, a large number of asci resulting from a controlled cross must be scored. Using these data, the distance (d) is calculated:

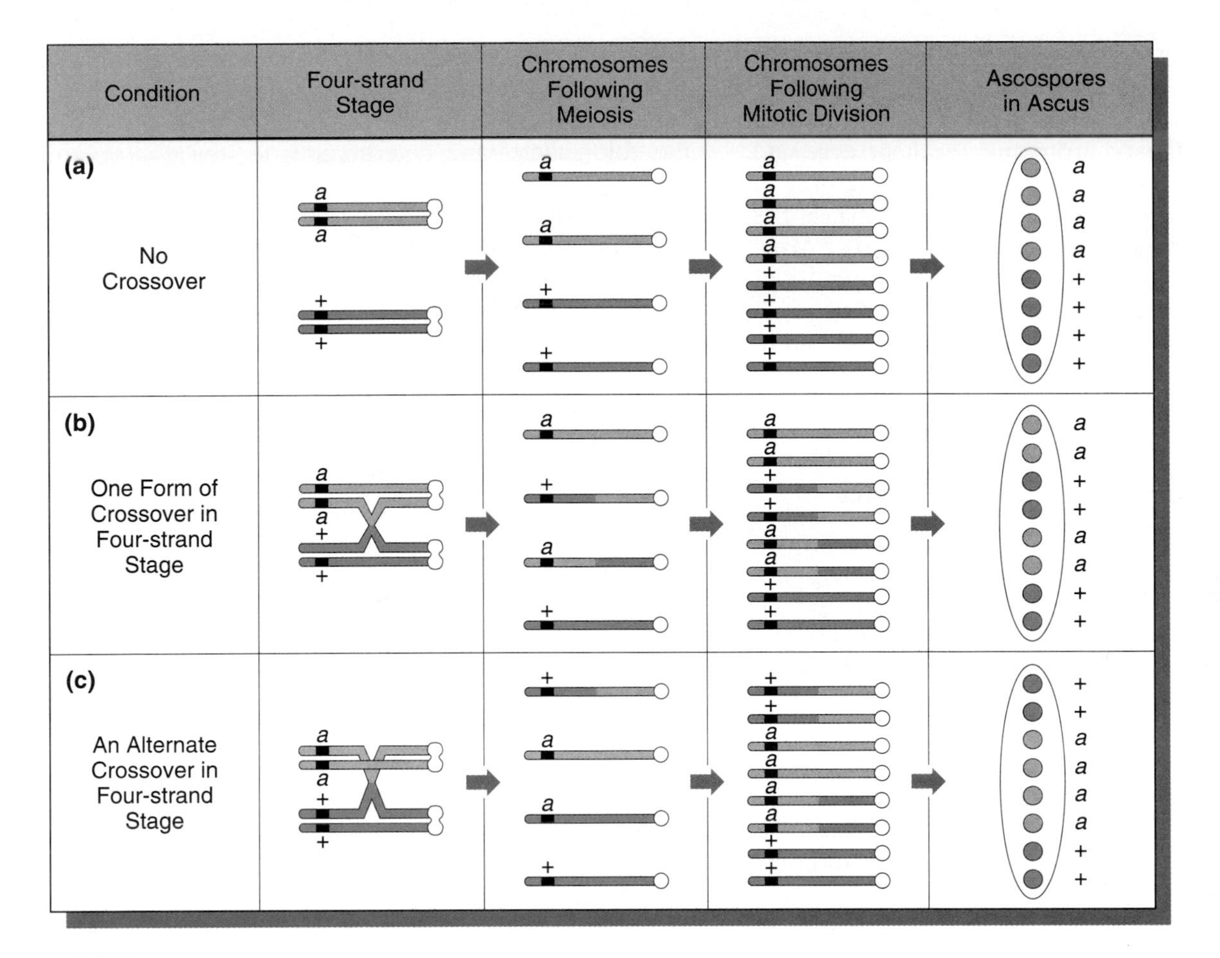

FIGURE 5.14
Three ways in which different ascospore patterns can be generated in *Neurospora.* Analysis of these patterns may serve as the basis of "gene to centromere" mapping, as described in the text.

$$d = 1/2\left(\frac{\text{second division segregant asci}}{\text{total asci scored}}\right)$$

The recombination percentage is only one-half the number of second division segregants because crossing over in each of them has occurred in only two of the four strands during meiosis.

To illustrate, assume that *a* represents albino and + represents wild type in *Neurospora.* In crosses between the two genetic types, suppose the following data were observed:

65 first-division segregants
70 second-division segregants

The distance between *a* and the centromere is thus:

$$d = 1/2\left(\frac{70}{135}\right) = 0.259$$

or about 26 map units.

As the distance increases up to 50 units, in theory, all asci should reflect second division segregation. However, numerous factors prevent this. As in diploid organisms, accuracy is greatest when the gene and centromere are relatively close together.

Analysis in haploid organisms can also be performed in order to distinguish between linkage and independent assortment of two genes. Mapping distances can be calculated between gene loci once linkage is established. As a result, detailed maps of organisms such as *Neurospora* and *Chlamydomonas* are now available.

OTHER ASPECTS OF GENETIC EXCHANGE

We have established that careful analysis of crossing over during gamete formation can serve as the basis for the construction of chromosome maps in both diploid and haploid organisms. We should not, however, lose sight of the real biological significance of the process, which is to generate genetic variation in gametes, and subsequently, in the offspring derived from the resultant eggs and sperm. Because of the critical role of crossing over in generating variation, the study of genetic exchange has remained an important topic for study in genetics. Many questions need to be addressed. For example, does crossing over involve an actual exchange of chromosome arms? Does exchange occur between paired sister chromatids during mitosis? We will briefly consider observations that attempt to answer these questions.

Cytological Evidence for Crossing Over

Visual proof that genetic crossing over in higher organisms is accompanied by an actual physical exchange between homologous chromosomes has been demonstrated independently by Curt Stern in *Drosophila* and by Harriet Creighton and Barbara McClintock in *Zea mays* (corn). Since the experiments are similar, we will consider only the work with corn. In Creighton and McClintock's work, two linked genes on chromosome 9 were studied. At one locus, the alleles *colorless* (*c*) and *colored* (*C*) control endosperm coloration. At the other locus, the alleles *starchy* (*Wx*) and waxy (*wx*) control the carbohydrate characteristics of the endosperm.

A corn plant was obtained that was heterozygous at both loci and contained two unique cytological markers on one of the homologues. The markers consisted of a densely stained knob at one end of the chromosome and a translocated piece of another chromosome (8) at the other end. The arrangement of alleles and cytological markers in this plant is shown in Figure 5.15. Creighton and McClintock crossed this plant to one homozygous for the color alleles and heterozygous for the endosperm alleles. They obtained several different phenotypes in the offspring, one of which could arise only as a result of crossing over. The chromosomes of this plant, with the colorless waxy phenotype (Case I in Figure 5.15) were examined for the presence of the cytological markers. As expected, if genetic crossing over is accompanied by a physical exchange between homologues, the translocated chromosome is still present, but the knob is not. In a second plant (Case II), the phenotype colored starchy is potentially the result of crossing over. If so, chromosomes from it should contain the dense knob but not the translocated chromosome. This was the case, and again the findings supported the conclusion that a physical exchange took place. Similar findings by Stern using *Drosophila* leave no doubt about this conclusion involving the cytological basis of crossing over.

Sister Chromatid Exchanges

While homologous chromosomes do not usually pair up or synapse in somatic cells, each individual chromosome in prophase and metaphase of mitosis consists of two identical sister chromatids, joined at a common centromere. Surprisingly, several experimental approaches have demonstrated that reciprocal exchanges similar to crossing over occur between sister chromatids. While these **sister chromatid exchanges (SCE)** do not pro-

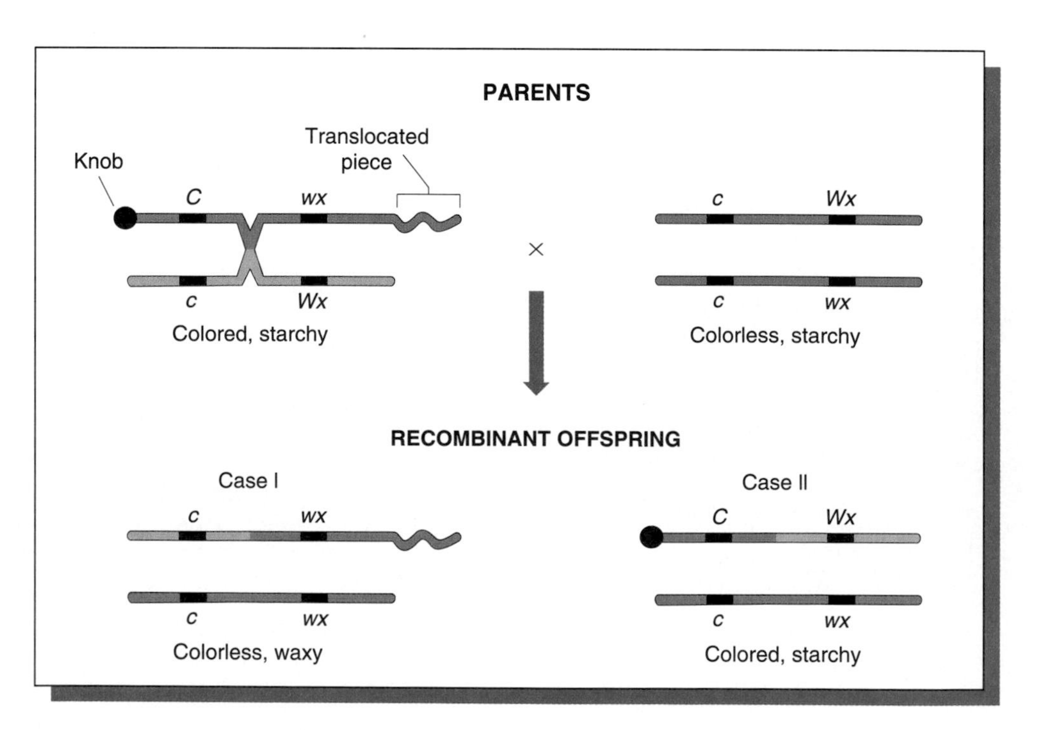

FIGURE 5.15
The phenotypes and chromosome compositions observed in Creighton and McClintock's demonstration in maize that crossing over involves a breakage and rejoining process.

duce new allelic combinations, evidence is accumulating that attaches significance to these events.

Identification and study of SCEs are facilitated by several modern staining techniques. In one approach, if cells are allowed to replicate for several generations in the presence of the thymidine analogue **bromodeoxyuridine (BUdR),** and the cells are then shifted to medium lacking the analogue, chromatids with or without BUdR are then distinguishable. Chromatids with both strands containing BUdR stain less brightly than chromatids with only one strand of the double helix containing the analogue. In Figure 5.16 numerous instances of SCE events may be detected.

While the significance of sister chromatid exchanges is still uncertain, several observations have led to great interest in this phenomenon. It is known, for example, that agents that induce chromosome damage (such as viruses, X-rays, ultraviolet light, and certain chemical mutagens) also increase the frequency of sister chromatid exchanges. This frequency of SCEs is also elevated in **Bloom syndrome,** an autosomal recessive disorder in humans. This rare genetic disease is characterized by retardation of growth, a great sensitivity of the facial skin to the sun, abnormal immunological function, and a predisposition to cancer. Increased breaks and rearrangements between nonhomologous chromosomes are observed in addition to excessive amounts of sister chromatid exchanges.

DID MENDEL ENCOUNTER LINKAGE?

We conclude this chapter by examining a modern-day interpretation of the experiments that serve as the cornerstone of transmission genetics—the crosses with garden peas performed by Mendel.

It has been said often that Mendel had extremely good fortune in his classical experiments with the garden pea. In none of his crosses did he encounter apparent linkage relationships between any of the seven mutant characters. Had Mendel obtained highly variable data characteristic of linkage and crossing over, these unorthodox ratios might have hindered his successful analysis and interpretation.

The accompanying boxed article by Stig Blixt, reprinted in its entirety, demonstrates the inadequacy of this hypothesis. As we shall see, some of Mendel's genes were indeed linked. We shall leave it to Stig Blixt to enlighten you as to why Mendel did not detect linkage.

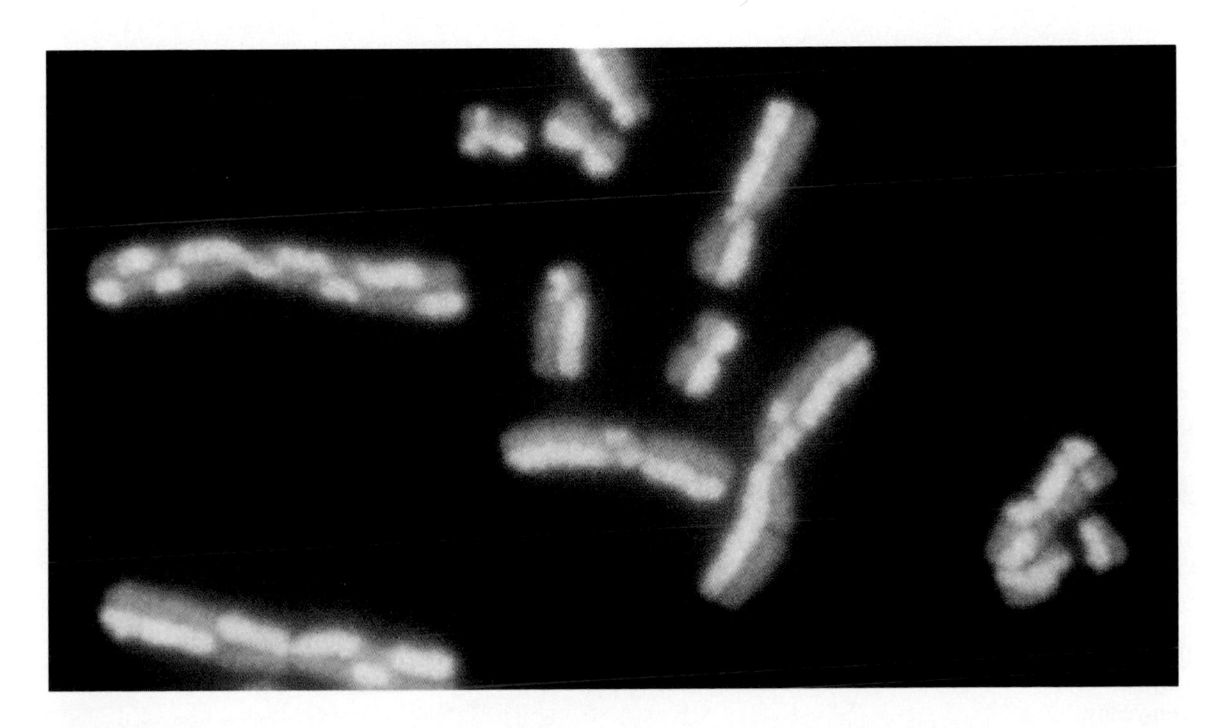

FIGURE 5.16
Demonstration of sister chromatid exchanges (SCEs) in mitotic chromosomes. Sometimes called harlequin chromosomes because of their patchlike appearance, chromatids containing the thymidine analogue BUdR in both DNA strands fluoresce *less* brightly than those with the analogue in only one strand. These chromosomes were stained with 33258-Hoechst reagent and viewed under fluorescence microscopy.

WHY DIDN'T GREGOR MENDEL FIND LINKAGE?

It is quite often said that Mendel was very fortunate not to run into the complication of linkage during his experiments. He used seven genes, and the pea has only seven chromosomes. Some have said that had he taken just one more, he would have had problems. This, however, is a gross oversimplification. The actual situation, most probably, is shown in Table 1. This shows that Mendel worked with three genes in chromosome 4, two genes in chromosome 1, and one gene in each of chromosome[s] 5 and 7. It seems at first glance that, out of the 21 dihybrid combinations Mendel theoretically could have studied, no less than four (that is, *a-i*, *v-fa*, *v-le*, *fa-le*) ought to have resulted in linkages. As found, however, in hundreds of crosses and shown by the genetic map of the pea[1], *a* and *i* in chromosome 1 are so distantly located on the chromosome that no linkage is normally detected. The same is true for *v* and *le* on the one hand, and *fa* on the other, in chromosome 4. This leaves *v-le*, which ought to have shown linkage.

Mendel, however, seems not to have published this particular combination and thus, presumably, never made the appropriate cross to obtain both genes segregating simultaneously. It is therefore not so astonishing that Mendel did not run into the complication of linkage, although he did not avoid it by choosing one gene from each chromosome.

Weibullsholm Plant Breeding Institute, Landskrona, Sweden and Centro Energia Nucleare na Agricultura, Piracicaba, SP, Brazil STIG BLIXT

Received March 5; accepted June 4, 1975.

[1]Blixt, S. 1974. In *Handbook of genetics,* ed. R. C. King. New York: Plenum Press.

TABLE 1
Relationship between modern genetic terminology and character pairs used by Mendel.

Character Pair Used by Mendel	Alleles in Modern Terminology	Located in Chromosome
Seed color, yellow-green	*I-i*	1
Seed coat and flowers, colored-white	*A-a*	1
Mature pods, smooth expanded-wrinkled indented	*V-v*	4
Inflorescences, from leaf axiis-umbellate in top of plant	*Fa-fa*	4
Plant height, 1m-around 0.5 m	*Le-le*	4
Unripe pods, green-yellow	*Gp-gp*	5
Mature seeds, smooth-wrinkled	*R-r*	7

CHAPTER SUMMARY

1 Genes located on the same chromosome are said to be linked. Alleles located on the same homologue, therefore, can be transmitted together during gamete formation. However, the mechanism of crossing over between homologues during meiosis results in the reshuffling of alleles, thereby contributing to genetic variability within gametes.

2 Early in this century, geneticists realized that crossing over could provide an experimental basis for mapping the location of linked genes relative to one another along the chromosome.

3 Somatic cell hybridization techniques have made possible linkage and mapping analysis of human genes.

4 Mapping studies are also possible in haploid organisms such as *Chlamydomonas* and *Neurospora*. Studies using *Neurospora* indicate that crossing over occurs during the four-strand stage of meiosis.

5 Cytological investigations of both corn and *Drosophila* have revealed that crossing over requires actual breaking and rejoining of nonsister chromatids.

6 An exchange of genetic material between sister chromatids may occur during mitosis as well. These events are referred to as sister chromatid exchanges (SCEs). An elevated frequency of such events is seen in the human disorder Bloom syndrome.

KEY TERMS

Bloom syndrome
chiasma
chiasmatype theory
chromosome maps
chromosomal theory of inheritance
coefficient of coincidence (C)
crossing over
heterokaryon
interference (I)
linkage
recombination
sister chromatid exchange (SCE)
somatic cell hybridization
synkaryon
synteny testing
tetrad analysis

1 In rabbits, *black* (B) is dominant to *brown* (b), while *full color* (C) is dominant to *chinchilla* (C^{ch}). The genes controlling these traits are linked. Rabbits that are heterozygous for both traits and express *black, full color,* were crossed to rabbits that are *brown, chinchilla,* with the following results:

31 *brown, chinchilla*
35 *black, full*
16 *brown, full*
19 *black, chinchilla*

Determine the arrangement of alleles in the heterozygous parents and the map distance between the two genes.

SOLUTION: This a two-point map problem, where the two reciprocal phenotypes most prevalent are the noncrossovers. The less frequent reciprocal phenotypes arise from a single crossover. The arrangement of alleles is derived from the noncrossover phenotypes because they enter gametes intact. The cross is shown as follows:

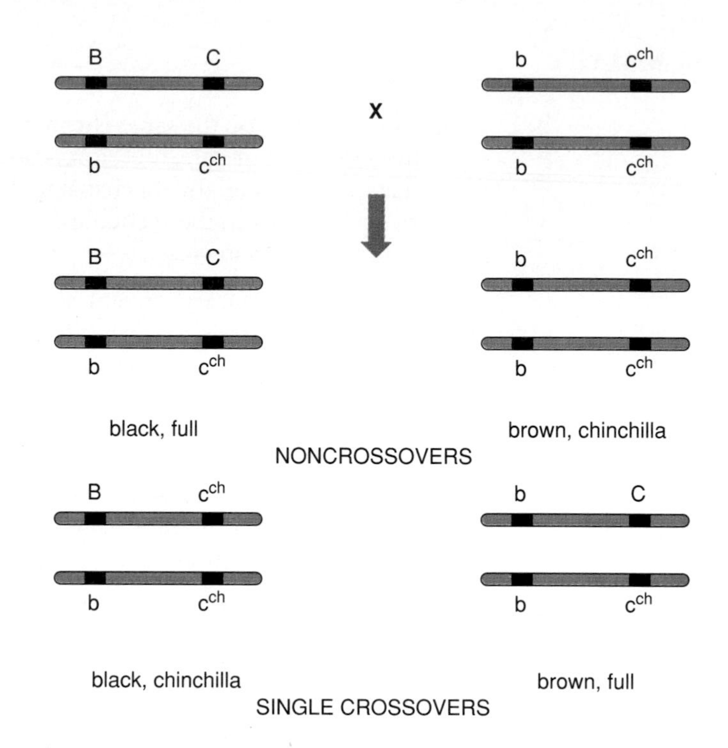

The single crossovers give rise to 35/100 offspring (35%). Therefore, the distance between the two genes is 35 map units (mu).

2 In *Drosophila, Lyra* (*Ly*) and *Stubble* (*Sb*) are dominant mutations located at locus 40 and 58, respectively, on chromosome 3. A recessive mutation with bright red eyes was discovered and shown also to be on chromosome 3. A map was obtained by crossing a female who was heterozygous for all three mutations to a male homozygous for the bright red mutation (which we will call *br*). The following data were obtained:

	Phenotype	Number
(1)	*Ly Sb br*	404
(2)	+ + +	422
(3)	*Ly* + +	18
(4)	+ *Sb br*	16
(5)	*Ly* + *br*	75
(6)	+ *Sb* +	59
(7)	*Ly Sb* +	4
(8)	+ + *br*	2
		1000

Determine the location of the bright red mutation on chromosome 3.

SOLUTION: First, determine the **arrangement** of the alleles on the homologues of the heterozygous crossover parent (the female in this case). This is done by locating the most frequent reciprocal phenotypes, which arise from the noncrossover gametes. These are phenotypes (1) and (2). Each one represents the arrangement of alleles on one of the homologues. Therefore, the arrangement is:

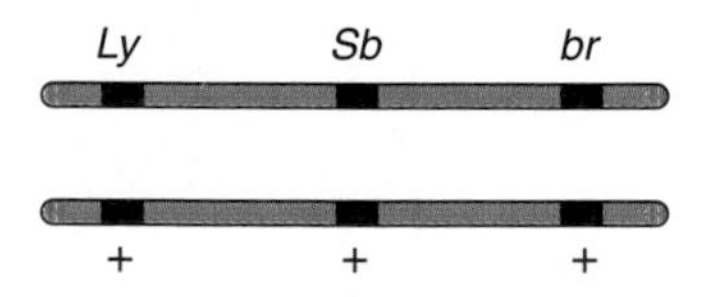

Second, determine the **correct sequence** of the three loci along the chromosome. This is done by determining which sequence will yield the observed double-crossover phenotypes, which are the least frequent reciprocal phenotypes (7 and 8).

If the sequence is correct as written, then a double crossover

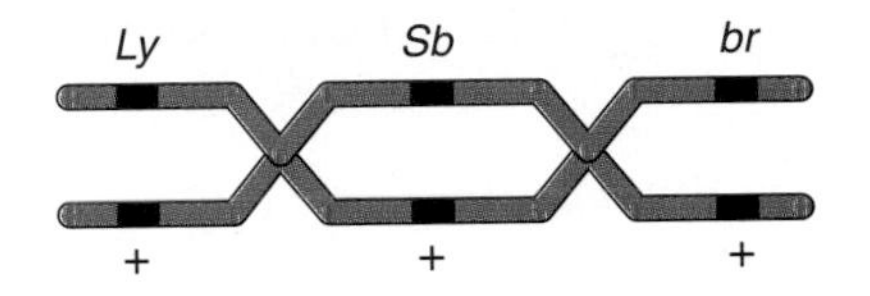

will yield *Ly+br* and *+Sb+* as phenotypes. Inspection shows that these categories (5 and 6) are actually single crossovers, not double crossovers. Therefore, the sequence, as written, is incorrect. There are only two other possible sequences. The *br* gene is either to the left of *Ly*, or it is between *Ly* and *Sb*:

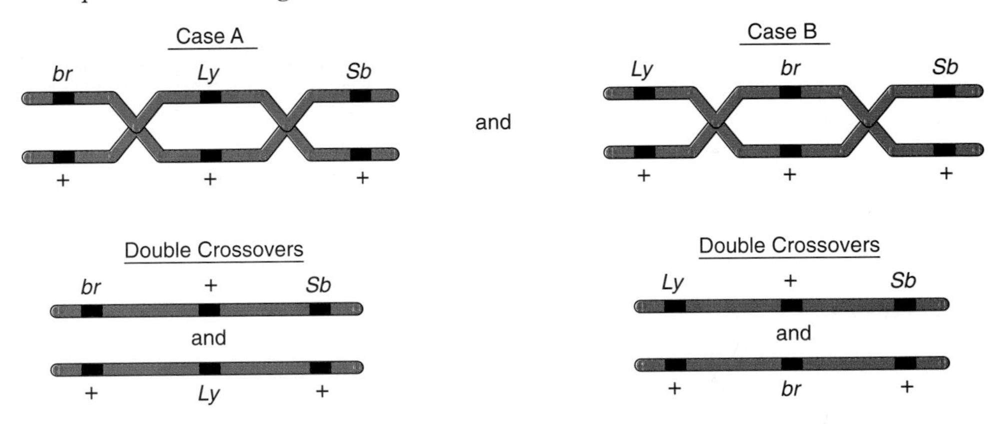

Comparison with the actual data shows that Case B is correct. The double-crossover gametes yield flies that express *Ly* and *Sb*, but not *br*, or express *br*, but not *Ly* and *Sb*. Therefore, the correct arrangement *and* sequence is:

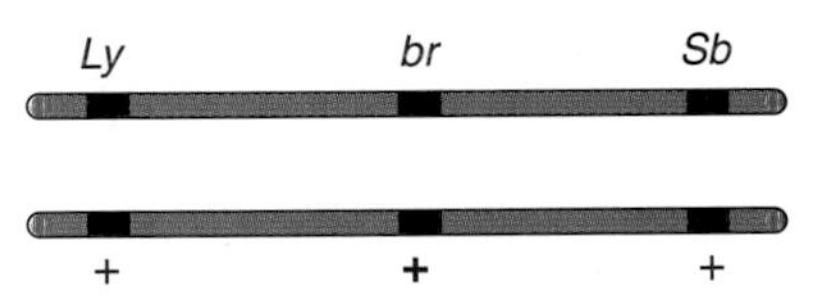

Once this has been determined, it is possible to determine the location of *br* relative to *Ly* and *Sb*. A single crossover between *Ly* and *br*

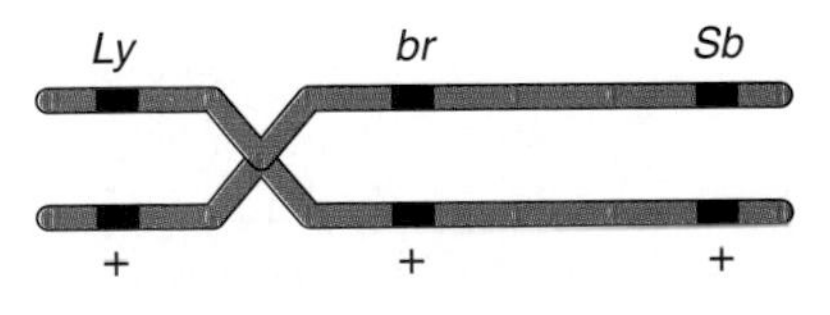

yields flies that are *Ly* + + and + *br Sb* (categories 3 and 4). Therefore, the distance between the *Ly* and *br* loci is equal to:

$$\frac{18 + 16 + 4 + 2}{1000} = \frac{40}{1000} = 0.04 = 4 \text{ mu}$$

Remember that we must add in the double crossovers since they represent two single crossovers occurring simultaneously. Because we need to know the frequency of all crossovers between *Ly* and *br*, they must be included. Similarly, the distance between the *br* and *Sb* loci is derived mainly from single crossovers between them:

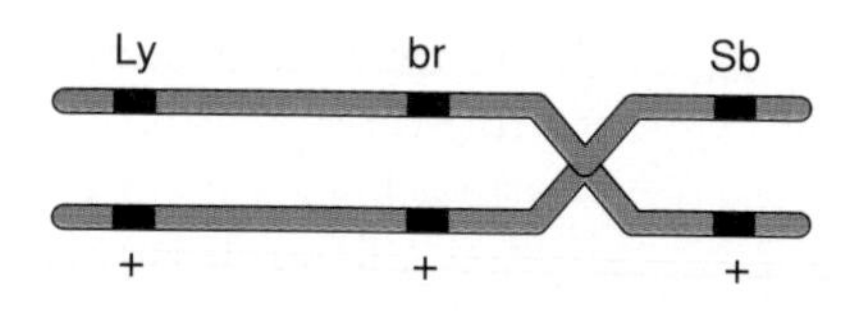

This event yields *Ly br + and + + Sb* phenotypes (categories 5 and 6). Therefore, the distance equals

$$\frac{75 + 59 + 4 + 2}{1000} = \frac{140}{1000} = 0.14 = 14 \text{ mu}$$

The final map shows that *br* is located at locus 44, since *Lyra* and *Stubble* are known:

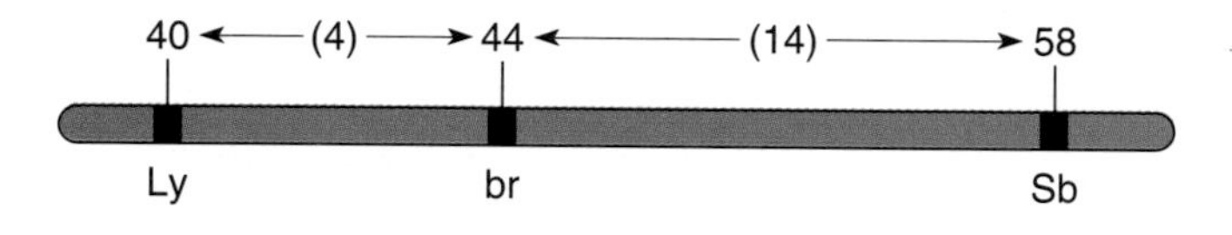

PROBLEMS AND DISCUSSION QUESTIONS

1 Why does more crossing over occur between two distantly linked genes than between two genes that are very close together on the same chromosome?

2 Why are double-crossover events expected in lower frequency than single-crossover events?

3 What essential criteria must be met in order to execute a successful mapping cross?

4 The genes *dumpy* wing (*dp*), *clot* eye (*cl*), and *apterous* wing (*ap*) are linked on chromosome 2 of *Drosophila.* In a series of two-point mapping crosses, the following genetic distances were determined:

dp-ap	42
dp-cl	3
ap-cl	39

What is the sequence of the three genes?

5 In corn, colored aleurone (in the kernels) is due to the dominant allele *R*. The recessive allele *r*, when homozygous, produces colorless aleurone. The plant color (not the kernel color) is controlled by the gene pair *Y* and *y*. The dominant *Y* gene results in green color, while the homozygous presence of the recessive *y* gene causes the plant to appear yellow. In a test cross between a plant of unknown genotype and phenotype and a plant that is homozygous recessive for both traits, the following progeny were obtained:

Colored green	88
Colored yellow	12
Colorless green	8
Colorless yellow	92

Explain how these results were obtained by determining the exact genotype and phenotype of the unknown plant, including the precise association of the two genes on the homologues, including the distance between them.

6 Two different female *Drosophila* were isolated, each heterozygous for the autosomally linked genes *b* (*black* body), *d* (*dachs* tarsus), and *c* (*curved* wings). These genes are in the order *d—b—c*, with *b* being closer to *d* than to *c*. Shown below is the genotypic arrangement for each female along with the various gametes formed by both. Identify which categories are noncrossovers (NCO), single crossovers (SCO), and double crossovers (DCO) in each case. Which category is likely to have the fewest gametes? Which will have the most gametes?

Female A		Female B	
$\frac{d\ b\ +}{+\ +\ c}$		$\frac{d\ +\ +}{+\ d\ c}$	
------Gametes------			
(1) d b c	(5) d + +	(1) *d b* +	(5) *d b* c
(2) + + +	(6) + *b* c	(2) + + c	(6) + + +
(3) + + c	(7) *d* + c	(3) *d* + c	(7) *d* + +
(4) *d b* +	(8) + *b* +	(4) + *b* +	(8) + *b* c

7 In *Drosophila,* a cross was made between females expressing the three sex-linked recessive traits, *scute* (sc) bristles, *sable* body(*s*), and *vermilion* eyes (*v*), and wild-type males. In the F_1, all females were wild type, while all males expressed all three mutant traits. The cross was carried to the F_2 generation, and 1000 offspring were counted with the results shown below. No determination of sex was made in the F_2 data.

Phenotype	Offspring
sc s v	314
+ + +	280
+ *s v*	150
sc + +	156
sc + *v*	46
+ *s* +	30
sc s +	10
+ + *v*	14

(a) Determine the genotypes of the P_1 and F_1 parents, using proper nomenclature.
(b) Determine the sequence of the three genes and the map distance between them.
(c) Are there more or fewer double crossovers than expected? Calculate the coefficient of coincidence. Does this represent positive or negative interference?

8 In *Drosophila, Dichaete* (*D*) is a chromosome 3 mutation with a dominant effect on wing shape. It is lethal when homozygous. The genes *ebony* (*e*) and *pink* (*p*) are chromosome 3 recessive mutations affecting the body and eye color, respectively.
Flies from a *Dichaete* stock were crossed to homozygous *ebony, pink* flies, and the F_1 progeny, with a *Dichaete* phenotype, were back-crossed to the *ebony, pink* homozygotes. The results of this back cross were:

Phenotype	Number
Dichaete	401
ebony, pink	389
Dichaete, ebony	84
pink	96
Dichaete, pink	2
ebony	3
Dichaete, ebony, pink	12
wild type	13

(a) Diagram this cross, showing the genotypes of the parents and offspring of both crosses.

(b) What is the sequence and interlocus distance between these three genes?

9 In a cross in *Neurospora,* involving two alleles *B* and *b*, the following tetrad patterns were observed. Calculate the distance between the locus and the centromere.

Tetrad Pattern	Number
BBbb	36
bbBB	44
BbBb	4
bBbB	6
BbbB	3
bBBb	7

10 A female of genotype

$$\frac{a \quad b \quad c}{+ \quad + \quad +}$$

produces 100 meiotic tetrads. Of these, 68 show no crossover events. Of the remaining 32, 20 show a crossover between *a* and *b*, 10 show a crossover between *b* and *c*, and 2 show a double-crossover between *a* and *b* and between *b* and *c*. Of the 400 gametes produced, how many of each of the 8 different genotypes will be produced? Assuming the order $a - b - c$ and the allele arrangement shown above, what is the map distance between these loci?

11 With two pairs of genes involved (*P*/*p* and *Z*/*z*), a test cross (*ppzz*) with an organism of unknown genotype indicated that the gametes produced were in the following proportions:

PZ, 42.4%; *Pz*, 6.9%; *pZ*, 7.1%; and *pz*, 43.6%

Draw all possible conclusions from these data.

12 In *Drosophila,* two mutations, *Stubble* (*Sb*) and *curled* (*cu*), are linked on chromosome 3. *Stubble* is a dominant gene that is lethal in a homozygous state, and *curled* is a recessive gene. If a female of the genotype

$$\frac{Sb \quad cu}{+ \quad +}$$

is to be mated to detect recombinants among her offspring, what male genotype would you choose as a mate?

13 Another cross in *Drosophila* involved the recessive, sex-linked genes *yellow* (*y*), *white* (*w*), and *cut* (*ct*). A female that was yellow-bodied and white-eyed with normal wings was crossed to a male whose eyes and body were normal but whose wings were cut. The F_1 females were wild type for all three traits, while the F_1 males expressed the yellow-body, white-eye traits. The cross was carried to an F_2, and only male offspring were tallied. On the basis of the data shown below, a genetic map was constructed.

Phenotype	Male Offspring
y + *ct*	9
+ *w* +	6
y w ct	90
+ + +	95
+ + *ct*	424
y w +	376
y + +	0
+ *w ct*	0

(a) Diagram the genotypes of the F_1 parents.
(b) Construct a map, assuming that *white* is at locus 1.5 on the X chromosome.
(c) Were any double-crossover offspring expected?
(d) Could the F_2 female offspring be used to construct the map? Why or why not?

SELECTED READINGS

ALLEN, G. E. 1978. *Thomas Hunt Morgan: The man and his science.* Princeton, NJ: Princeton University Press.

CHAGANTI, R., SCHONBERG, S., and GERMAN, J. 1974. A manyfold increase in sister chromatid exchange in Bloom's syndrome lymphocytes. *Proc. Natl. Acad. Sci.* 71: 4508–12.

CREIGHTON, H. S., and McCLINTOCK, B. 1931. A correlation of cytological and genetical crossing over in *Zea mays. Proc. Natl. Acad. Sci.* 17: 492–97.

DOUGLAS, L., and NOVITSKI, E. 1977. What chance did Mendel's experiments give him of noticing linkage? *Heredity* 38: 253–57.

EPHRUSSI, B., and WEISS, M. C. 1969. Hybrid somatic cells. *Scient. Amer.* (April) 220: 26–35.

MORGAN, T. H. 1911. An attempt to analyze the constitution of the chromosomes on the basis of sex-linked inheritance in *Drosophila. J. Exp. Zool.* 11: 365–414.

NEUFFER, M. G., JONES, L., and ZOBER, M. 1968. *The mutants of maize.* Madison, WI: Crop Science Society of America.

PERKINS, D. 1962. Crossing-over and interference in a multiply marked chromosome arm of *Neurospora.* Genetics 47: 1253–74.

RUDDLE, F. H., and KUCHERLAPATI, R. S. 1974. Hybrid cells and human genes. *Scient. Amer.* (July) 231: 36–49.

STAHL, F. W. 1979. *Genetic recombination.* New York: W. H. Freeman.

STURTEVANT, A. H. 1913. The linear arrangement of six sex-linked factors in *Drosophila,* as shown by their mode of association. *J. Exp. Zool.* 14: 43–59.

———. 1965. *A history of genetics.* New York: Harper & Row.

VOELLER, B. R., ed. 1968. *The chromosome theory of inheritance: Classical papers in development and heredity.* New York: Appleton-Century-Crofts.

WOLFF, S., ed. 1982. *Sister chromatid exchange.* New York: Wiley-Interscience.

6 Chromosome Variations and Sex Determination

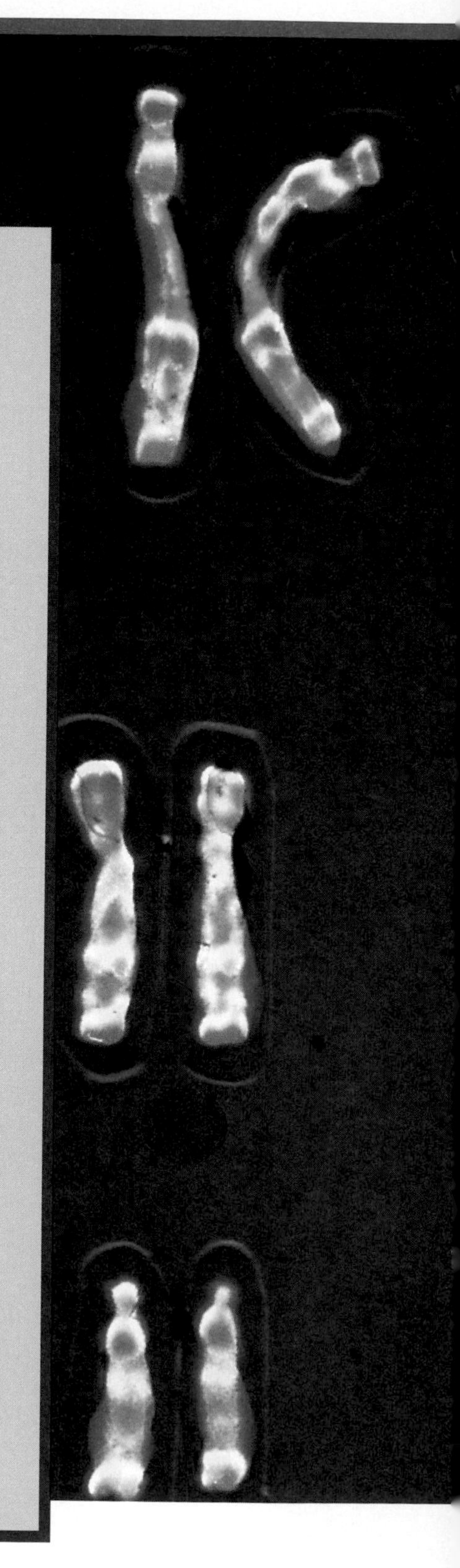

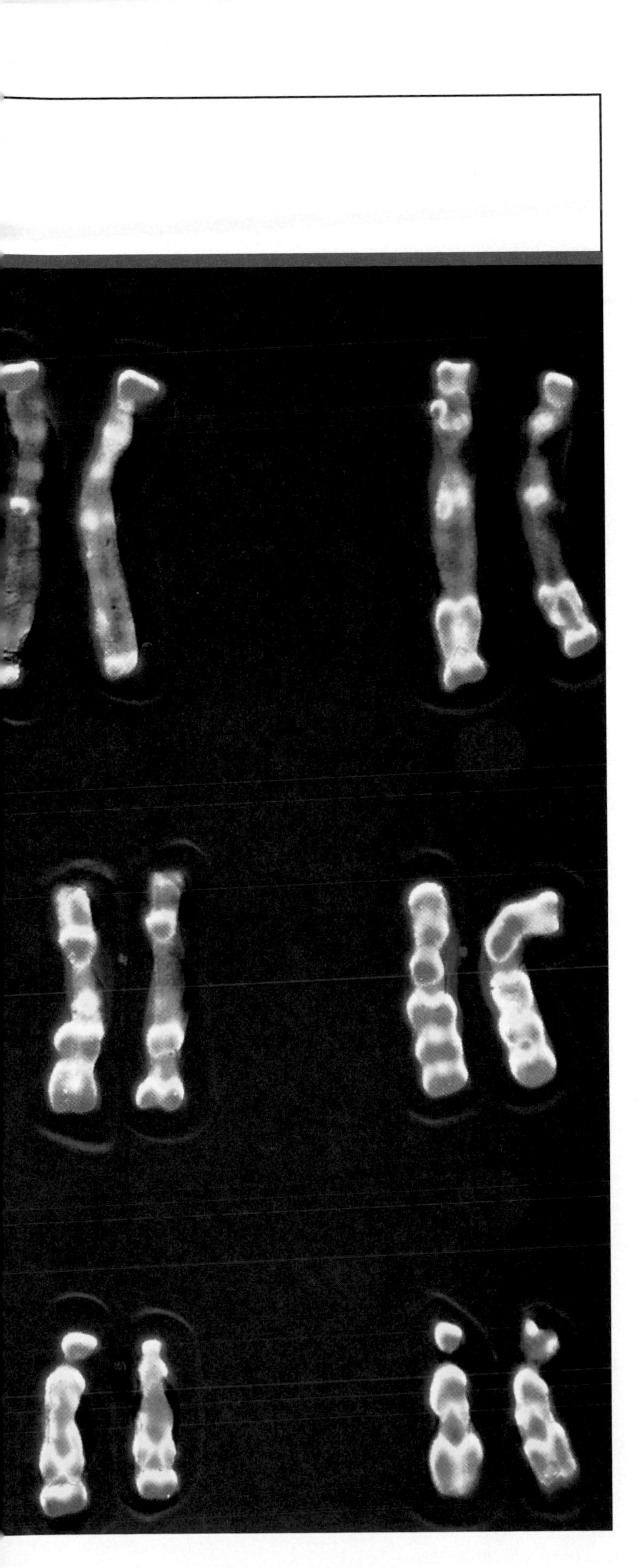

CHAPTER CONCEPTS

Genetic information of a diploid organism is delicately balanced in both content and location within the genome. A change in chromosome number or in the arrangement of a chromosome region results often in phenotypic variation or disruption of development of an organism. Because the chromosome is the unit of transmission in meiosis, such variations are passed to offspring in a predictable manner, resulting in many interesting genetic situations. As a result of studying variation in the sex chromosomes, valuable insights have been gained into the mode of sex determination in many organisms.

Up to this point in the text, we have emphasized how mutations and the resulting alleles affect an organism's phenotype, and how traits are passed from parents to offspring according to Mendelian principles. In this chapter, we shall look at phenotypic variation occurring as a result of changes in the genetic material that are more substantial than alterations of individual genes. These involve modifications at the level of the chromosome.

Although members of diploid species normally contain precisely two haploid chromosome sets, many cases are known in which some variation from this pattern occurs. Modifications include variations in the number of individual chromosomes as well as rearrangements of the genetic material either within or among chromosomes. Taken together, such changes are called **chromosome mutations** or **chromosome aberrations,** to distinguish such genetic alterations from gene mutations. Since it is the chromosome, and not the gene, that is transmitted according to Mendelian laws, chromosome aberrations are transmitted to offspring in a predictable manner, resulting in many interesting examples of heritable phenotypic variation.

As a result of studying variations in the composition of sex chromosomes, valuable insights have been provided into how sex is determined. As we discuss such variations, we will contrast the cases of humans with the fruit fly *Drosophila*.

VARIATION IN CHROMOSOME NUMBER: AN OVERVIEW

Before embarking on a discussion of variations involving the number of chromosomes in organisms and sex determination, it is useful to establish the terminology used that describes such changes. Variation in chromosome number ranges from the addition or loss of one or more chromosomes to the addition of one or more haploid sets of chromosomes. When an organism gains or loses one or more chromosomes, but not a complete set, the condition of **aneuploidy** is created. This is contrasted with the condition of **euploidy,** where complete haploid sets of chromosomes are found. If there are three or more sets, the more general term **polyploidy** is applicable. Those with three sets are specifically **triploid;** those with four sets are **tetraploid,** and so on. Table 6.1 provides a useful organizational framework for you to follow as we discuss each of these categories and the subsets within them.

CHROMOSOME COMPOSITION AND SEX DETERMINATION IN HUMANS

In our discussion of sex linkage in Chapter 4, we pointed out that, as part of the diploid chromosome composition of both humans and *Drosophila,* females possess two X chromosomes while males possess one X and one Y chromosome. This observation might lead us to conclude that the Y chromosome causes maleness in both species; however, this is not necessarily the case. Perhaps the lack of a second X chromosome somehow causes maleness while the Y plays no part whatsoever in sex determination. Perhaps the presence of two X chromosomes causes femaleness while the Y plays no role. The evidence that clarified which explanation was correct awaited the study of variations in the sex chromosome composition of both humans and flies. As we shall see, the first explanation, where the Y determines maleness, is valid in humans but not in *Drosophila.*

Klinefelter and Turner Syndromes

Around the 1940 it was observed that two human abnormalities, the **Klinefelter** and **Turner syndromes,*** are characterized by aberrant sexual development. Individuals with Klinefelter syndrome [Figure 6.1(a)] have genitalia and internal ducts that are usually male, but their testes are underdeveloped and fail to produce sperm. Although masculine development occurs, feminine sexual development is not entirely suppressed. Slight enlargement of the breasts is common, for example. Intersexuality may lead to abnormal social development.

In Turner syndrome [Figure 6.1(b)], the affected individual has female external genitalia and internal ducts, but the ovaries are rudimentary. Other characteristic abnormalities include short stature (usually under five feet); a short, webbed neck; and a broad, shieldlike chest.

In 1959, the karyotypes of individuals with these syndromes were independently determined to be abnormal with respect to the sex chromosomes. Individuals with Klinefelter syndrome most often have an XXY complement in addition to 44 autosomes. People with this karyotype are designated **47,XXY.** Individuals with Turner syndrome have only 45 chromosomes, including just a single X chromosome and are designated **45,X.** Both condi-

*Although the possessive form of the names of most syndromes (eponyms) is often used (e.g., Klinefelter's), the current preference is to use the nonpossessive form for syndromes.

TABLE 6.1
Terminology for variation in chromosome numbers

Term	Explanation
Aneuploidy	$2n$ plus or minus chromosomes
Monosomy	$2n - 1$
Trisomy	$2n + 1$
Tetrasomy, pentasomy, etc.	$2n + 2$, $2n + 3$, etc.
Euploidy	Multiples of n
Diploidy	$2n$
Polyploidy	$3n$, $4n$, $5n$, . . .
Triploidy	$3n$
Tetraploidy, pentaploidy, etc.	$4n$, $5n$, etc.
Autopolyploidy	Multiples of the same genome
Allopolyploidy	Multiples of different genomes

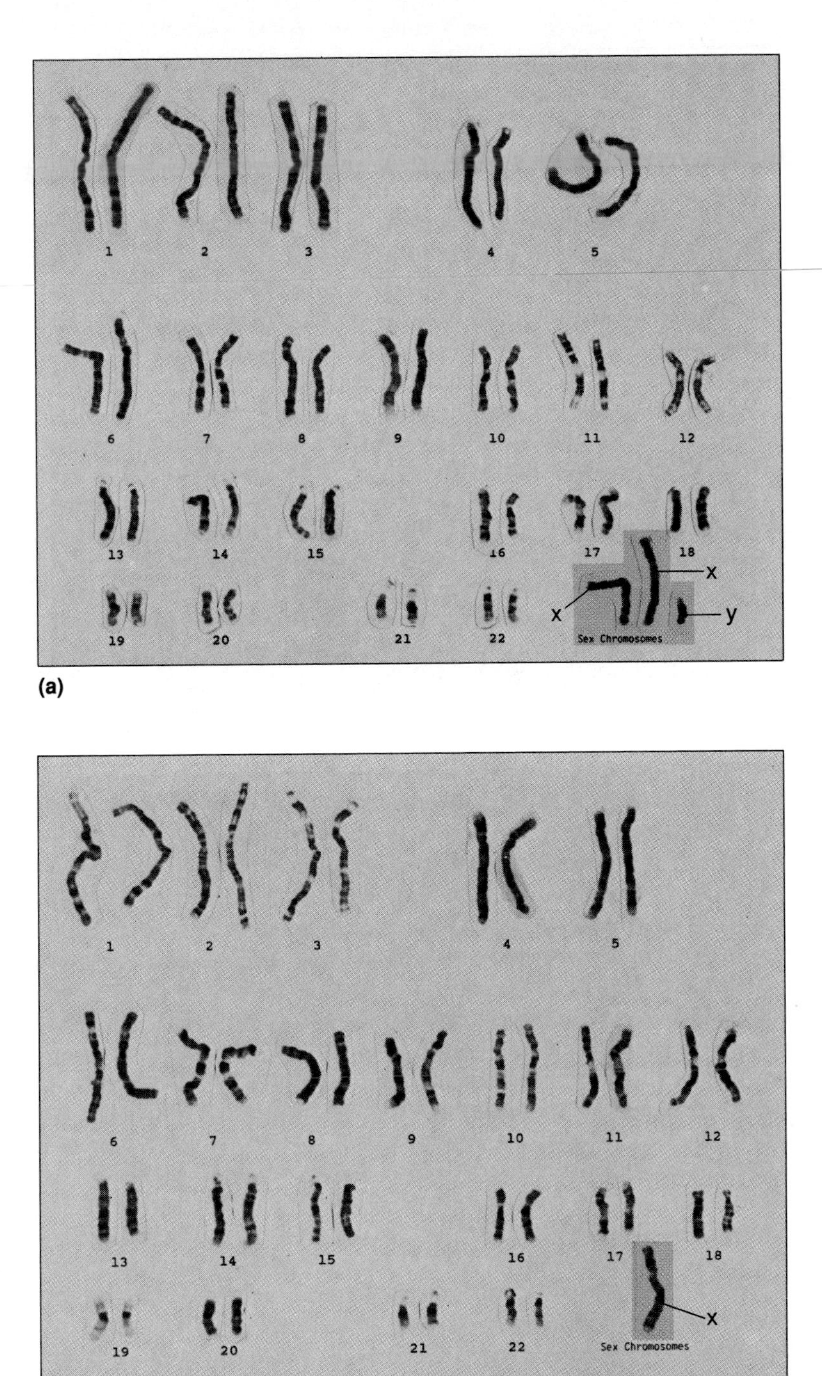

FIGURE 6.1
(a) The karyotype of an individual with Klinefelter syndrome. In the karyotype, two X chromosomes and one Y chromosome are visible, creating the 47,XXY condition. (b) The karyotype of an individual with Turner syndrome. In the karyotype, only one X chromosome is visible, creating the 45,X condition. The shading identifies the affected chromosomes.

tions result from nondisjunction of the sex chromosomes during meiosis. Note the convention used in designating chromosome compositions. The number indicates how many chromosomes are present, and the information after the comma designates the deviation from the normal diploid content.

These karyotypes and their corresponding sexual phenotypes allow us to conclude that the Y chromosome determines maleness in humans. In its absence, the sex of the individual is female, even if only a single X chromosome is present. The presence of the Y chromosome in the individual with Klinefelter syndrome is sufficient to determine maleness, even though its expression is not complete. Similarly, in the absence of a Y chromosome, as in case of individuals with Turner syndrome, no masculinization occurs.

Klinefelter syndrome occurs in about 2 of every 1000 male births. The karyotypes **48,XXXY; 48,XXYY;**

49,XXXXY; and **49,XXXYY** are similar phenotypically to **47,XXY,** but manifestations are often more severe in individuals with a greater number of X chromosomes.

Karyotypes other than 45,X also lead to Turner syndrome. These include mosaic individuals with two apparent cell lines, the most common chromosome combinations being **45,X/46,XY** and **45,X/46,XX.** Turner syndrome is observed in only about 1 in 3000 female births, a frequency much lower than that for Klinefelter syndrome. One explanation for this difference is the observation that a substantial majority of **45,X** fetuses die *in utero* and are aborted spontaneously.

47,XXX Syndrome

The presence of three X chromosomes along with a normal set of autosomes **(47,XXX)** results in female differentiation. This syndrome, which is estimated to occur in about 1 of 1200 female births, is highly variable in expression. Frequently, 47,XXX women are perfectly normal. In other cases, underdeveloped secondary sex characteristics, sterility, and mental retardation may occur. In rare instances, **48,XXXX** and **49,XXXXX** karyotypes have been reported. The syndromes associated with these karyotypes are similar to but more pronounced than the 47,XXX. Thus, the presence of additional X chromosomes appears to disrupt the delicate balance of genetic information essential to normal female development.

47,XYY Condition

A third human trisomy involving the sex chromosomes has been discovered and intensively investigated. This case involves males with one X and two Y chromosomes in addition to the normal complement of 44 autosomes **(47,XYY).** Many studies have attempted to clarify the effect of this chromosome composition.

In 1965, Patricia Jacobs discovered 9 of 315 males in a Scottish maximum security prison to have the 47,XYY karyotype. These males were significantly above average in height and were involved in criminal acts of serious social consequence. Of the 9 males studied, 7 were of subnormal intelligence, and all suffered personality disorders. In several other studies, similar findings were obtained.

Because of these investigations, the phenotype and frequency of the 47,XYY condition in criminal and noncriminal populations have been examined more extensively. Above-average height and subnormal intelligence have been generally substantiated, and the frequency of males displaying this karyotype is indeed higher in penal and mental institutions compared with unincarcerated males.

The possible correlation between this chromosome composition and antisocial and criminal behavior has been of considerable interest. A particularly relevant question involves the characteristics displayed by XYY males who are not incarcerated. The only nearly constant association is that such individuals are over 6 feet tall! Since it is now clear that many XYY males do not exhibit any form of antisocial behavior and lead normal lives, we must conclude that there is no direct correlation between the extra Y chromosome and behavior.

The Human Y Chromosomes and Male Development

Before turning to other types of chromosome variation and their effects in humans and other organisms, it is appropriate to review information available concerning *how* the Y chromosome results in male development in humans. We have previously alluded to the fact that this chromosome, unlike the X, is nearly genetically blank. Although it shares only limited homology with loci on the X chromosome, it does carry genetic information that controls sexual development.

Therefore, some region of the Y chromosome exists, presumably a gene, that is responsible for the **testis-determining factor (TDF),** a product that somehow triggers the undifferentiated gonadal tissue of the embryo to form testes. In its absence, female development occurs. Research has focused on just what constitutes the region and the product.

It is now clear that a small part of the human Y chromosome contains a gene called *SRY* (*sex-determining region Y*). In rare individuals sex chromosome compositions do not match their expected sexual phenotype. Evidence proving that SRY is indeed the responsible gene has relied on the molecular geneticist's ability to identify the presence or absence of DNA sequences in these particular individuals.

There are human males who demonstrate two X and no Y chromosomes. They have attached to one of their Xs the region of the Y containing *SRY.* There are also females who have only one X but also have one Y chromosome. Their Y is missing the *SRY* region. These observations argue strongly in favor of the role of *SRY* in male development.

The final proof that this gene is responsible for causing maleness involves an experiment using **transgenic mice,** where a similar region *Sry,* has been identified. Such animals arise from fertilized eggs that have foreign DNA injected into them and incorporated into the genetic composition of the developing embryo. When DNA containing only *Sry* is injected into normal XX eggs, most mice develop into males!

Thus, the relevant gene has been identified. It appears to be present in all mammals and thus to have been conserved throughout evolution. How the product of this gene triggers the embryonic gonadal tissue to develop into testes rather than ovaries is now a reasonable question to ask and one that is amenable to investigation.

THE X CHROMOSOME AND DOSAGE COMPENSATION

The presence of two X chromosomes in normal human females and only one X in normal human males is unique compared with the equal numbers of autosomes present in the cells of both sexes. On theoretical grounds alone, it is possible to speculate that this situation should create a genetic dosage problem between males and females for all X-linked genes. Females have two copies and males only one. The additional X chromosomes in both males and females exhibiting the various syndromes discussed earlier in this chapter should compound this dosage problem. In this section, we will describe certain research findings regarding X-linked gene expression which demonstrate that a **genetic dosage compensation mechanism** does indeed exist.

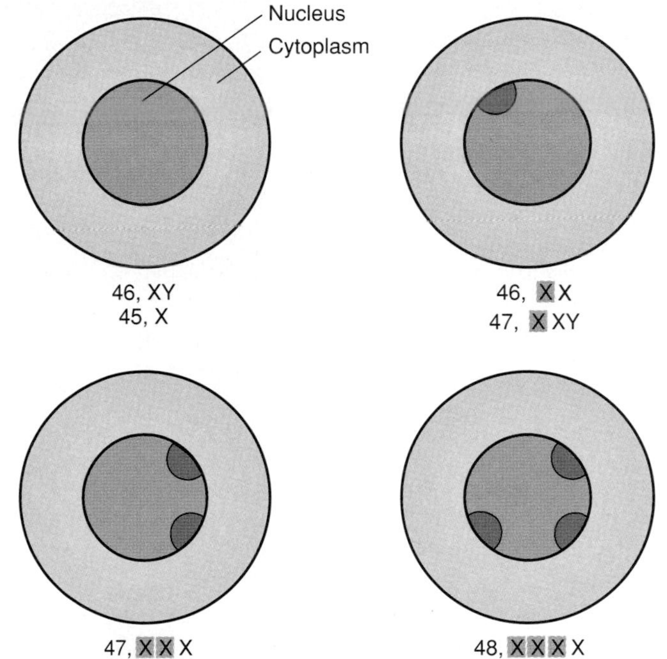

FIGURE 6.2
Diagrammatic representation of Barr body occurrence in various human karyotypes.

Barr Bodies and Dosage Compensation

Murray L. Barr and Ewart G. Bertram's experiments with female cats, and Keith Moore and Barr's subsequent study with humans, demonstrate a genetic mechanism in mammals that compensates for X chromosome dosage disparities. Barr and Bertram observed a darkly staining body in interphase nerve cells of female cats. They found that this structure was absent in similar cells of males. In human females, this body can be easily demonstrated in cells derived from the buccal mucosa or in fibroblasts but not in similar male cells. This highly condensed structure, about 1 μm in diameter, lies against the nuclear envelope of interphase cells. It stains positively in the Feulgen reaction for DNA.

Current experimental evidence strongly suggests that this body, called a **sex chromatin body** or simply a **Barr body,** is an inactivated X chromosome. Ohno was the first to suggest that the Barr body arises from one of the two X chromosomes. This hypothesis is attractive because it provides a mechanism for dosage compensation. If one of the two X chromosomes is inactive in the cells of females, the dosage of genetic information that may be expressed in males and females is equivalent. Convincing but indirect evidence for this hypothesis comes from the study of the sex chromosome syndromes described earlier in this chapter. Regardless of how many X chromosomes exist, all but one of them appear to be inactivated and can be seen as Barr bodies. For example, none is seen in Turner 45,X females; one is seen in Klinefelter 47,XXY males; two in 47,XXX females; three in 48,XXXX females; and so on (Figure 6.2). Therefore, the number of Barr bodies follows an $N - 1$ rule, where N is the total number of X chromosomes present.

The Lyon Hypothesis

In mammalian females one X chromosome is of maternal origin, and the other is of paternal origin. Which one is inactivated? Is the inactivation random? Is the same chromosome inactive in all somatic cells? In 1961, Mary Lyon and Liane Russell independently proposed a hypothesis that answers these questions. They postulated that the inactivation of X chromosomes occurs randomly in somatic cells at a point early in embryonic development, and that once inactivation has occurred, all progeny cells have the same X chromosome inactivated.

This explanation, which has come to be called the **Lyon hypothesis,** was initially based on observations of female mice heterozygous for sex-linked coat color genes. The pigmentation of these heterozygous females was mottled, with large patches of skin expressing the color allele on one X and other patches expressing the allele on the other X. Indeed, if one or the other of the two X chromosomes was inactive in adjacent patches of cells, such a phenotypic pattern would result. Similar mottling occurs in the black and orange patches of female tortoise-shell (or calico) cats. Such sex-linked coat color patterns do not occur in male cats since all cells are hemizygous for only one sex-linked coat color allele. Two such cats are compared in Figure 6.3.

FIGURE 6.3
A male and female cat from the same litter, illustrating the Lyon hypothesis. The female (on the left) is a calico with orange and black patches resulting from the inactivation of one or the other X chromosome early in development. The male (on the right) is hemizygous for the orange allele. The white patches are due to still another gene.

The Lyon hypothesis is generally accepted as valid. One extension of the hypothesis is that mammalian females are mosaics for all heterozygous X-linked alleles. Some areas of the body express only the maternally derived alleles, and others express only the paternally derived alleles. Two especially interesting examples involve **red-green color blindness** and **anhidrotic ectodermal dysplasia,** both X-linked recessive disorders. In the former case, hemizygous males are fully color-blind in all retinal cells. However, heterozygous females display mosaic retinas with patches of defective color perception and surrounding areas with normal color perception. In the latter disorder, hemizygous males show absence of teeth, sparse hair growth, and lack of sweat glands. The skin of heterozygous females reveals patterns of tissue with and without sweat glands. In both examples, random inactivation of one or the other X chromosome early in the development of heterozygous females has led to these occurrences.

The least understood aspect of the Lyon hypothesis concerns how inactivation occurs. While studies on the mechanisms that inactivate an X chromosome are just beginning, evidence suggests that DNA in inactivated chromosome regions has been chemically modified. One line of work has shown an association between the addition of methyl groups to cytosine residues and the condensation of chromosomes. Perhaps this process of methylation is part of the mechanism responsible for X chromosome inactivation.

CHROMOSOME COMPOSITION AND SEX DETERMINATION IN *DROSOPHILA*

Because males and females in *Drosophila* have the identical sex chromosome composition as humans, we might assume that the Y also causes maleness in these flies. However, the elegant work of Calvin Bridges in 1916 showed this not to be true. He studied flies with quite varied chromosome compositions leading him to the conclusion that the Y chromosome is not involved in sex determination in this organism. Instead, Bridges proposed that both the X chromosomes and autosomes together play a critical role in sex determination.

His work can be divided into two phases: a study of offspring resulting from **nondisjunction** of the X chromosomes during meiosis in females, and subsequent work with progeny of triploid ($3n$) females. Nondisjunction is the failure of paired chromosomes to segregate or separate during the anaphase stage of the first or second meiotic divisions. The result is the production of two abnormal gametes, one of which contains an extra chromosome ($n+1$) and the other that lacks a chromosome ($n-1$). Fertilization of such gametes produces ($2n+1$)

or $(2n - 1)$ aneuploid zygotes. As a result, in addition to the normal complement of six autosomes, the resulting flies had either an XXY or an XO sex chromosome composition. (The zero signifies that the second chromosome is absent.) The XXY flies were normal females, and the XO flies were sterile males. The presence of the Y chromosome in the XXY flies did not cause maleness, and its absence in the XO flies did not produce femaleness. From these data Bridges concluded that the Y chromosome in *Drosophila* lacks male-determining factors, but apparently contains genetic information essential to male fertility since the XO males were sterile.

Bridges was able to clarify the mode of sex determination in *Drosophila* by studying the progeny of triploid females ($3n$), which have *three* copies each of the haploid complement of chromosomes. These females apparently originate from rare diploid eggs fertilized by normal haploid sperm. Triploid females have heavy-set bodies, coarse bristles, coarse eyes, and may be fertile. During meiosis a wide range of chromosome complements is distributed into gametes that give rise to offspring with a variety of abnormal chromosome constitutions. A correlation between the sexual morphology, chromosome composition, and Bridges' interpretation is shown in Figure 6.4. *Drosophila* has a haploid number of four, thereby displaying three pairs of autosomes in addition to its sex chromosomes.

Bridges realized that the critical factor in determining sex is the **ratio of X chromosomes to the number of haploid sets of autosomes present.** Normal (2X:2A) and triploid (3X:3A) females each have a ratio equal to 1.0, and both are fertile. As the ratio exceeds unity (3X:2A, or 1.5, for example), what was originally called a

CHROMOSOME COMPOSITION	CHROMOSOME FORMULATION	RATIO OF X CHROMOSOMES TO AUTOSOME SETS	SEX
	3X/2A	1.5	Metafemale
	3X/3A	1.0	Female
	2X/2A	1.0	Female
	2X/3A	0.67	Intersex
	3X/4A	0.75	Intersex
	X/2A	0.50	Male
	XY/2A	0.50	Male
	XY/3A	0.33	Metamale

NORMAL DIPLOID MALE

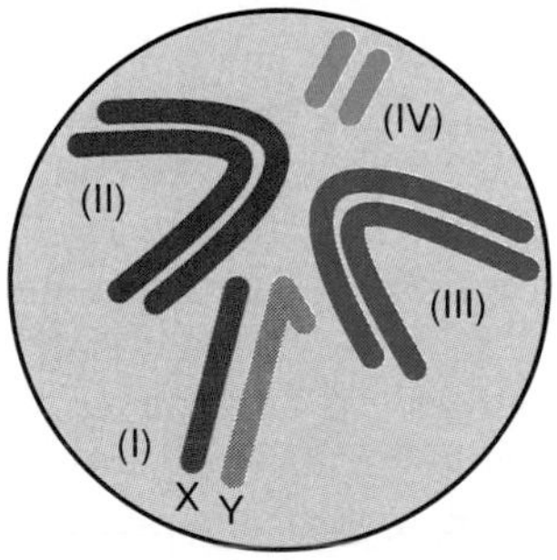

(IV)
(II)
(III)
(I)
X Y

6 autosomes
+
X Y

FIGURE 6.4
Chromosome compositions, the ratios of X chromosomes to sets of autosomes, and the resultant sexes in *Drosophila melanogaster.* The normal diploid male chromosome composition is shown as a reference in the circle below.

superfemale is produced. Since this female is rather weak and infertile and has lowered viability, this type is now more appropriately called a **metafemale.**

Normal (XY: 2A) and sterile (X0: 2A) males each have a ratio of 1:2, or 0.5. When the ratio decreases to 1:3, or 0.33, as in the case of an XY:3A male, infertile **metamales** result. Other flies recovered by Bridges in these studies contained an X:A ratio intermediate between 0.5 and 1.0. These flies were generally larger, and they exhibited a variety of morphological abnormalities and rudimentary bisexual gonads and genitalia. They were invariably sterile and were designated as **intersexes.**

These results indicate that in *Drosophila,* male-determining factors are not localized on the sex chromosomes, but are instead found on the autosomes. Some female-determining factors, however, are localized on the X chromosomes. Thus, with respect to primary sex determination, male gametes containing one of each autosome plus a Y chromosome result in male offspring *not* because of the presence of the Y chromosome but because of the lack of an X chromosome. This mode of sex determination is explained by the **genic balance theory.** Bridges proposed that a threshold for maleness is reached when the X:A ratio is 1:2 (X:2A), but that the presence of an additional X (XX:2A) alters this balance and results in female differentiation.

ANEUPLOIDY

We turn now to a consideration of variations in the number of autosomes and the genetic consequence of such changes. The most common examples of **aneuploidy,** where an organism has a chromosome number other than an exact multiple of the haploid set, are cases in which a single chromosome is either added to or lost from a normal diploid set. Such circumstances can arise as a result of primary or secondary nondisjunction (Figure 6.5). The loss of one chromosome produces a $2n - 1$ complement and is called **monosomy;** the gain of one chromosome produces a $2n + 1$ complement and is described as **trisomy.** The $2n + 2$ and $2n + 3$ conditions are called **tetrasomy** and **pentasomy,** indicating that an individual has four or five copies of a chromosome in an otherwise diploid genome.

Monosomy

Although monosomy for one of the sex chromosomes is fairly common, monosomy for one of the autosomes is not easily tolerated in animals. In *Drosophila,* flies monosomic for the very small chromosome 4—a condition referred to as **Haplo-IV**—develop more slowly, exhibit a reduced body size, and have impaired viability. Monosomy for the larger chromosomes 2 and 3 is apparently lethal because such flies have never been recovered.

The failure of monosomic individuals to survive in many animal species is at first quite puzzling, since at least a single copy of every gene is present in the remaining homologue. However, if just one of those genes is represented by a lethal allele, the unpaired chromosome condition leads to the death of the organism. This occurs because monosomy unmasks recessive lethals that are tolerated in heterozygotes carrying the corresponding wild-type alleles.

Aneuploidy is tolerated better in the plant kingdom. Monosomy for autosomal chromosomes has been observed in maize, tobacco, the evening primrose *Oenothera,* and the Jimson weed *Datura,* among other plants. Nevertheless, such monosomic plants are less viable usually than their diploid derivatives. Haploid pollen grains, which undergo extensive development before participating in fertilization, are particularly sensitive to the lack of one chromosome and are often sterile.

Partial Monosomy: Cri-du-Chat Syndrome

In humans, autosomal monosomy has not been reported beyond birth. Individuals with such chromosome complements are undoubtedly conceived, but none apparently survive embryonic and fetal development. There are, however, examples of survivors with **partial monosomy,** where only part of one chromosome is lost. These cases are also referred to as **segmental deletions.** One such case was first reported by Jérôme LeJeune in 1963 when he described the clinical symptoms of the **cri-du-chat syndrome.** This syndrome is associated with the loss of about one-half the short arm of chromosome 5 (Figure 6.6). Thus, the genetic constitituion may be designated as **46,5p−,** meaning that such an individual has all 46 chromosomes but that some of the p arm (the short arm) is missing. Infants with this syndrome may exhibit anatomic malformations, including gastrointestinal and cardiac complications, and are often mentally retarded. Abnormal development of the glottis and larynx is characteristic of individuals with this syndrome. As a result, the infant has a cry similar to that of the meowing of a cat, thus giving the syndrome its name.

Since 1963, over 300 cases of cri-du-chat syndrome have been reported worldwide. An incidence of 1 in 50,000 live births has been estimated. The size of the deletion appears to influence the physical, psychomotor, and mental skill levels of those children who survive. Although the effects of the syndrome are severe, many individuals achieve a level of social development in the

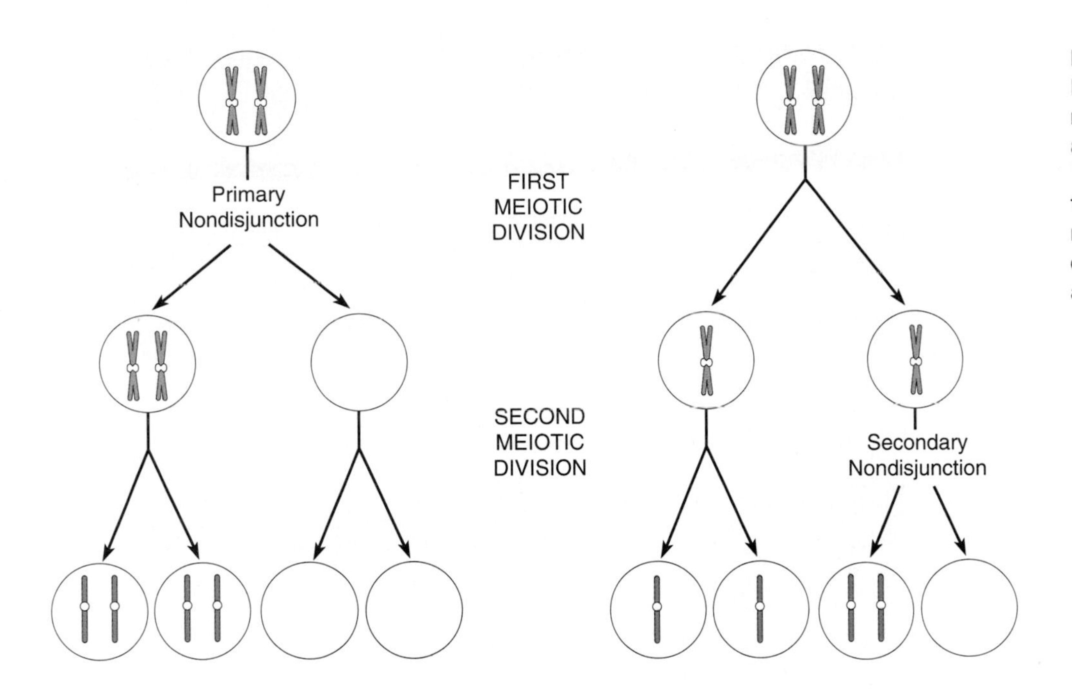

FIGURE 6.5
Diagram illustrating nondisjunction during the first and second meiotic divisions. In both cases, gametes are formed either containing two members of a specific chromosome or lacking it altogether.

trainable range. Those who receive home care and early special schooling are ambulatory, develop self-care skills, and learn to communicate verbally.

Trisomy

In general, the effects of trisomy parallel those of monosomy. However, the addition of an extra chromosome produces somewhat more viable individuals in both animal and plant species than does the loss of a chromosome. In animals, this is often true, providing that the involved chromosome is relatively small. However, the addition of a large autosome to the diploid complement in both *Drosophila* and humans has severe effects and is usually lethal during development.

In plants, trisomic individuals are viable, but their phenotype may be altered. A classical example involves the Jimson weed *Datura,* whose diploid number is 24.

(a)

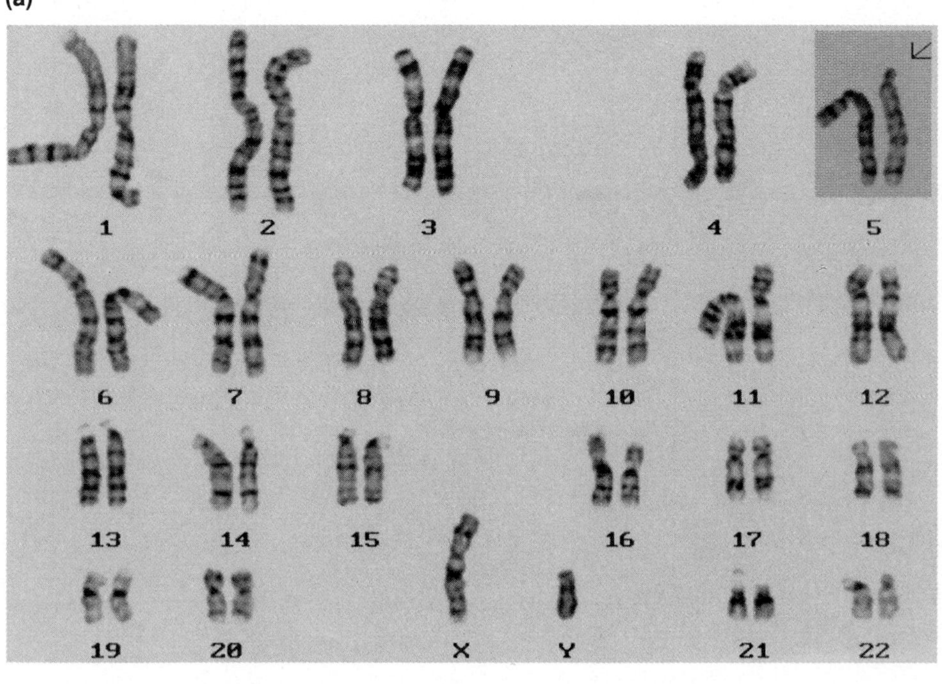

(b)

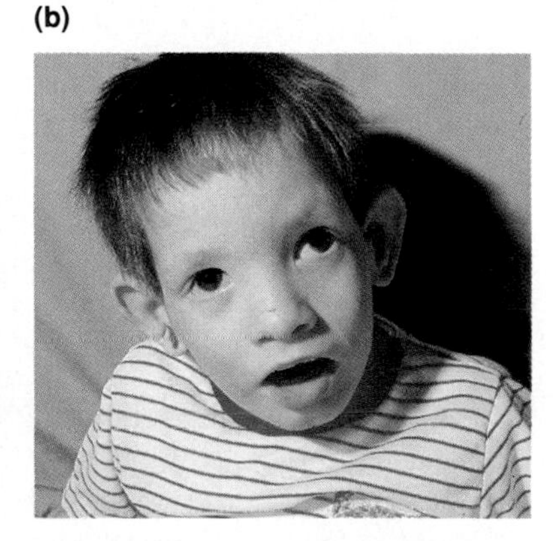

FIGURE 6.6
(a) A representative karyotype and (b) photograph of a child exhibiting cri-du-chat syndrome (46,5p−). In the karyotype, the shading identifies the nearly complete absence of the short arm of one member of the chromosome 5 homologues.

FIGURE 6.7
Drawings of capsule phenotypes of the fruits of *Datura stramonium.* In comparison with wild type, each phenotype is the result of trisomy of one of the twelve chromosomes characteristic of the haploid genome. The photograph illustrates the entire plant.

Twelve different primary trisomic conditions are possible, and examples of each one have been recovered. Each trisomy alters the phenotype of the plant's capsule sufficiently (Figure 6.7) to produce a unique phenotype. These capsule phenotypes were first thought to be caused by point mutations.

Down Syndrome

The only human autosomal trisomy in which a significant number of individuals survive longer than a year past birth was discovered in 1866 by Langdon Down. The condition is now known to result from trisomy of chromosome 21, one of the G group* (Figure 6.8), and is called **Down syndrome** or simply **trisomy 21** (and is designated **47, 21+**). This trisomy is found in approximately 3 infants in every 2000 live births.

The external phenotype of these individuals is similar so that they bear a striking resemblance to one another.

*On the basis of size and centromere placement, human chromosomes are divided into seven groups: A (1–3); B (4–5); C (6–12); D (13–15); E (16–18); F (19–20); and G (21–22).

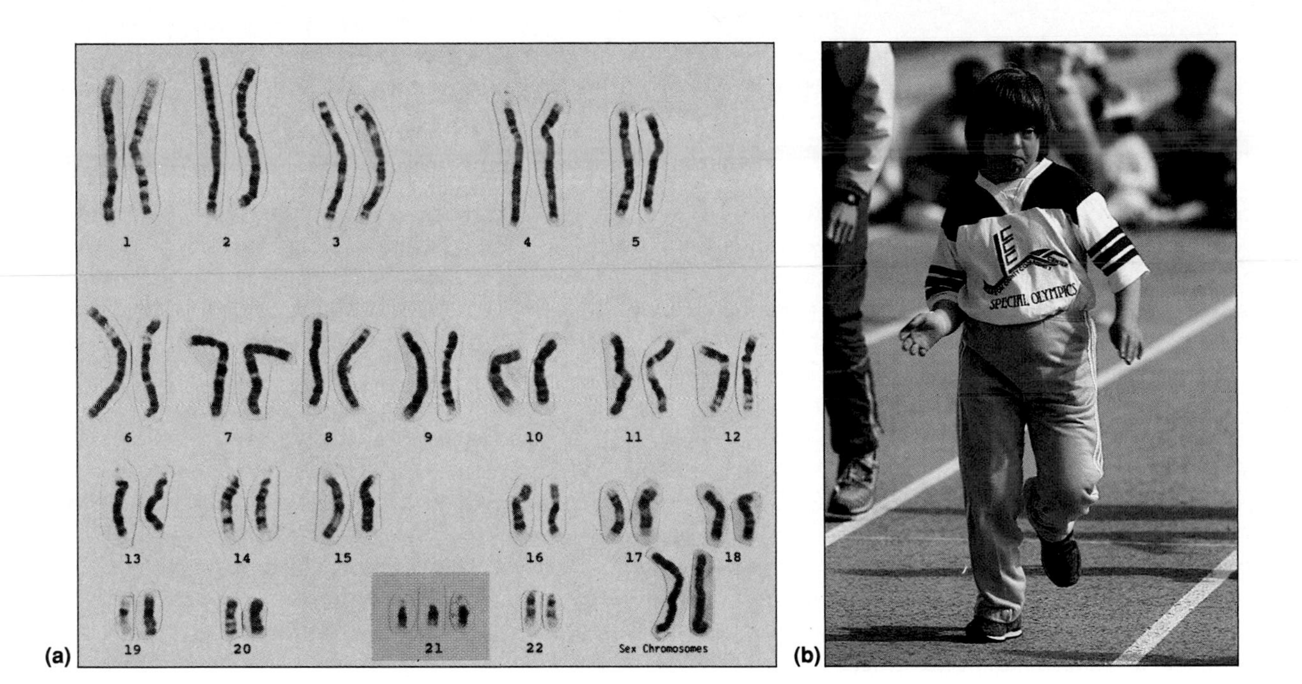

FIGURE 6.8
(a) The karyotype and (b) a photograph of a child with Down syndrome. In the karyotype, three members of the G-group chromosome 21 are present, creating the 47,21+ condition.

They display a prominent epicanthic fold in the corner of the eye and are characteristically short. They may have small, round heads; protruding, furrowed tongues, which cause the mouth to remain partially open; and short, broad hands with fingers showing characteristic palm and fingerprint patterns. Physical, psychomotor, and mental development is retarded, and the IQ is seldom above 70. Their life expectancy is shortened, and few individuals survive to be 50.

Down children are prone to respiratory disease and heart malformations and show an incidence of leukemia approximately 15 times higher than that of the normal population. However, careful medical scrutiny and treatment throughout their lives has extended their survival significantly.

One way in which this trisomic condition may originate is through nondisjunction of chromosome 21 during meiosis. Failure of paired homologues to disjoin during anaphase I or II can result in male or female gametes with the $n + 1$ chromosome composition. Following fertilization with a normal gamete, the trisomic condition is created. Chromosome analysis has shown that while the additional chromosome may be derived from either the mother or father, the ovum is most often the source.

Before the development of techniques that distinguish paternal from maternal homologues, this conclusion was supported by other indirect evidence derived from studies of the age of mothers giving birth to Down infants. Figure 6.9 shows the analysis of the distribution of maternal age and the incidence of Down syndrome newborns.

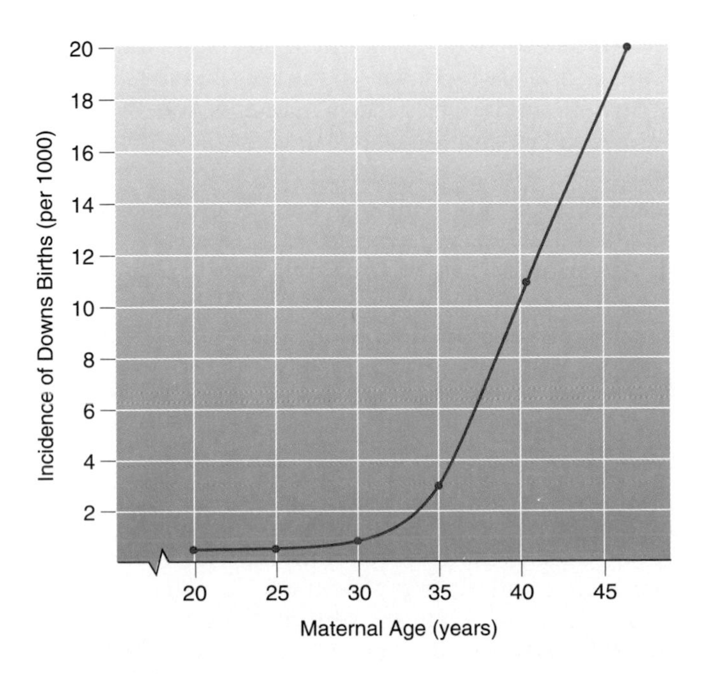

FIGURE 6.9
Incidence of Down syndrome births contrasted with maternal age.

The frequency of Down births increases dramatically as the age of the mother increases. While the frequency is about 1 in 1000 at maternal age 30, a tenfold increase to a frequency of 1 in 100 is noted at age 40. The frequency increases still further to about 1 in 50 at age 45.

While the nondisjunctional event that produces Down syndrome seems more likely to occur during oogenesis in women between the ages of 35 and 45, we do not know with certainty why this is so. However, one observation may be relevant. In human females, all primary oocytes have been formed by birth. Therefore, once ovulation begins, each succeeding ovum has been arrested in meiosis for about a month longer than the one preceding it. Thus, women 30 or 40 years old produce ova that are significantly older and arrested longer than those they ovulated 10 or 20 years previously. However, it is not yet known whether ovum age is the cause of the increased incidence of nondisjunction leading to Down syndrome.

These statistics pose a serious problem for the woman who becomes pregnant late in her reproductive years. Genetic counseling early in such pregnancies serves two purposes. First, it informs the parents about the probability that their child will be affected and educates them about Down syndrome. Although some individuals with Down syndrome must be institutionalized, others are trainable and may be cared for at home. Further, they are noted as being affectionate, loving children. Second, a genetic counselor may recommend a prenatal diagnostic technique such as **amniocentesis** or **chorionic villus sampling (CVS).** These techniques require the removal and culture of fetal cells. The karyotype of the fetus may then be determined by cytogenetic analysis. If the fetus is diagnosed as having Down syndrome, a therapeutic abortion is one option the parents may consider.

Since Down syndrome appears to be caused by a random error—nondisjunction of chromosome 21 during maternal or paternal meiosis—the disorder is not expected to be inherited. Nevertheless, a similar condition called **familial Down syndrome** runs in families and has been reported many times. Familial Down syndrome involves a **translocation** of chromosome 21, another type of chromosomal aberration, which we will discuss later in this chapter.

Viability in Human Aneuploid Conditions

The reduced viability of individuals with recognized monosomic and trisomic conditions is evident. Only two other trisomies in humans survive to term. Both **Patau** and **Edwards syndromes, 47, 13+** and **47,18+,** respectively, result in severe malformations and early lethality. Such observations lead us to believe that many other aneuploid conditions arise, but that the affected fetuses do not survive to term. This observation has been confirmed by karyotypic analysis of spontaneously aborted fetuses.

These studies have revealed some striking statistics. At least 15 to 20 percent of all conceptions are terminated in spontaneous abortion (some estimates are considerably higher)! About 30 percent of all spontaneous abortuses demonstrate some form of chromosomal anomaly, and approximately 90 percent of all chromosomal anomalies are terminated prior to birth as a result of spontaneous abortion.

A large percentage of spontaneous abortuses demonstrating chromosomal abnormalities are aneuploids. The aneuploid with highest incidence among abortuses is the 45,X condition, which produces an infant with Turner syndrome if the fetus survives to term.

An extensive review of this subject by David H. Carr also reveals that a significant percentage of abortuses are trisomic for one of the chromosome groups. Trisomies for every human chromosome have been recovered. Monosomies were seldomly found, however, even though nondisjunction should produce $n - 1$ gametes with a frequency equal to $n + 1$ gametes. This finding suggests that *gametes* lacking a single chromosome are functionally impaired to a serious degree or that the embryo dies so early in its development that recovery occurs infrequently. Various forms of polyploidy and other miscellaneous chromosomal anomalies were also found in Carr's study.

These observations support the hypothesis that normal embryonic development requires a precise diploid complement of chromosomes to maintain a delicate equilibrium in the expression of genetic information. The prenatal mortality of most aneuploids provides a barrier against the introduction of a general form of genetic anomaly into the human population.

POLYPLOIDY AND ITS ORIGINS

The term **polyploidy** describes instances where more than two multiples of the haploid chromosome set are found. The naming of polyploids is based on the number of sets of chromosomes found: a **triploid** has $3n$ chromosomes; a **tetraploid** has $4n$; a **pentaploid,** $5n$; and so forth. Several general statements may be made about polyploidy. This condition is relatively infrequent in most animal species, but is well known in lizards, amphibians, and fish. It is much more common in plant species. Odd numbers of chromosomes sets are not usually maintained reliably from generation to generation because a polyploid organism with an uneven number of homologues usually does not produce genetically balanced gametes. For this reason, triploids, pentaploids, and so on, are not usually found in species that depend solely upon sexual reproduction for propagation.

Polyploidy can originate in two ways: (1) the addition of one or more extra sets of chromosomes, identical to the normal haploid complement of the same species, results in **autopolyploidy;** and (2) the combination of chromosome sets from different species may occur as a consequence of hybridization and results in **allopolyploidy.** Thus, the distinction between auto- and allopolyploidy is based on the genetic origin of the extra chromosome sets.

In our discussion of polyploidy, we will use the following symbols to clarify the origin of additional chromosome sets. For example, if A represents the haploid set of chromosomes of any organism, then

$$A = a_1 + a_2 + a_3 + a_4 \cdots a_n$$

where a_1, a_2, and so on represent individual chromosomes, and where n is the haploid number. Thus, a normal diploid organism would be represented simply as *AA*.

Autopolyploidy

In **autopolyploidy,** each additional set of chromosomes is identical to the parent species. Therefore, **triploids** are represented as *AAA,* **tetraploids** are *AAAA,* and so forth.

Autotriploids may arise in several ways. A failure of all chromosomes to segregate during meiotic divisions may produce a diploid gamete. If such a gamete is fertilized by a haploid gamete, a zygote with three sets of chromosomes is produced. Or, occasionally two sperm may fertilize an ovum, resulting in a triploid zygote. Triploids may also be produced under experimental conditions by crossing diploids with tetraploids. Diploid organisms produce gametes with n chromosomes while tetraploids produce $2n$ gametes. Upon fertilization, the desired triploid is produced.

Tetraploid cells may be produced experimentally from diploid cells by applying cold or heat shock during meiosis or by applying colchicine to somatic cells undergoing mitosis. **Colchicine,** an alkyloid derived from the autumn crocus, interferes with spindle formation, and thus, replicated chromosomes cannot enter anaphase and migrate to the poles. When colchicine is removed, the cell may reenter interphase. When the paired sister chromatids separate and uncoil, the nucleus will contain twice the diploid number of chromosomes and is effectively $4n$. This process is illustrated in Figure 6.10.

In general, autopolyploid plants are larger than their diploid relatives. Often, the flower and fruit are increased in size. This increase seems to be due to larger cell size rather than greater cell number. Although autopolyploids do not contain new or unique information compared with the diploid relative, such varieties may be of greater commercial value. Economically important triploid plants include several potato species of the genus *Solanum,* Winesap apples, commercial bananas, seedless watermelons, and the cultivated tiger lily *Lilium tigrinum.* These plants are propagated asexually. Diploid bananas contain hard seeds, but the commercial, triploid, "seedless" variety has edible seeds. Tetraploid alfalfa, coffee, peanuts, and McIntosh apples are also of economic value because they are either larger or grow more vigorously than their diploid or triploid counterparts. The commercial strawberry is an octoploid. These observations attest to the importance of autopolyploidy in domesticated plants.

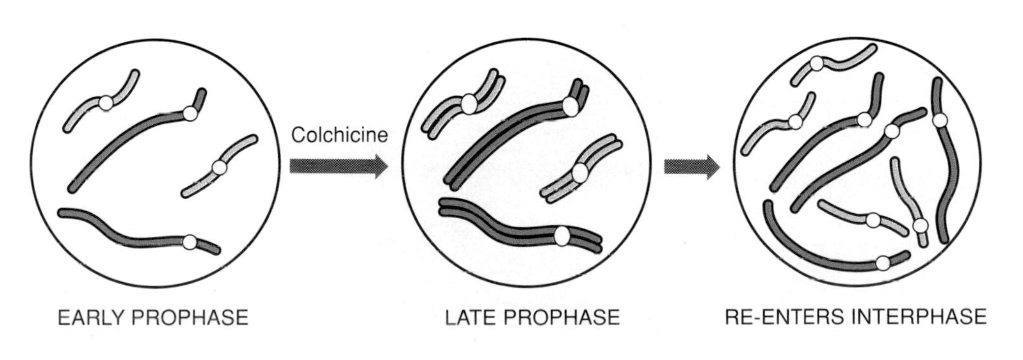

FIGURE 6.10
Illustration of the potential involvement of colchicine in the production of an autotetraploid. The figure follows two pairs of homologous chromosomes. Colchicine interferes with the formation of spindle fibers and thus inhibits the movement of chromosomes to the poles during anaphase. Paired sister chromatids may then separate, effectively doubling the number of chromosomes. (See Figure 2.7 for an overview of mitosis.)

Allopolyploidy

Polyploidy may also result from hybridization of two closely related species. Thus, if a haploid ovum from a species with chromosome sets *AA* is fertilized by sperm from a species with sets *BB,* the resulting hybrid is *AB,* where $A = a_1, a_2, a_3, \cdots, a_n$ and $B = b_1, b_2, b_3, \cdots, b_n$. The hybrid may be sterile because of its inability to produce viable gametes. This occurs because, most often, some or all of the *a* and *b* chromosomes cannot synapse in meiosis, and unbalanced genetic conditions result. If, however, the new *AB* genetic combination undergoes a natural or induced chromosomal doubling, a fertile *AABB* tetraploid is produced. These events are illustrated in Figure 6.11. Since this polyploid contains the equivalent of four haploid genomes and since the hybrid contains unique genetic information compared with either parent, such an organism is called an **allotetraploid** (from the Greek word *allo,* meaning other or different). An equivalent term, **amphidiploid,** is also used to describe this situation in which a hybrid organism contains two complete diploid genomes. This term is preferred when the original species are known.

A classical example of allotetraploidy in plants is the cultivated species of American cotton, *Gossypium* (Figure 6.11). This species has 26 pairs of chromosomes: 13 are large and 13 are much smaller. When it was discovered that Old World cotton had only 13 pairs of large chromosomes, allopolyploidy was suspected. After an examination of wild American cotton revealed 13 pairs of small chromosomes, this speculation was strengthened. J. O. Beasley was able to reconstruct the origin of cultivated cotton experimentally. He crossed the Old World strain with the wild American strain and then treated the hybrid with colchicine to double the chromosome number. The result of these treatments was a fertile allotetraploid variety of cotton. It contained 26 pairs of chromosomes and characteristics similar to the cultivated variety.

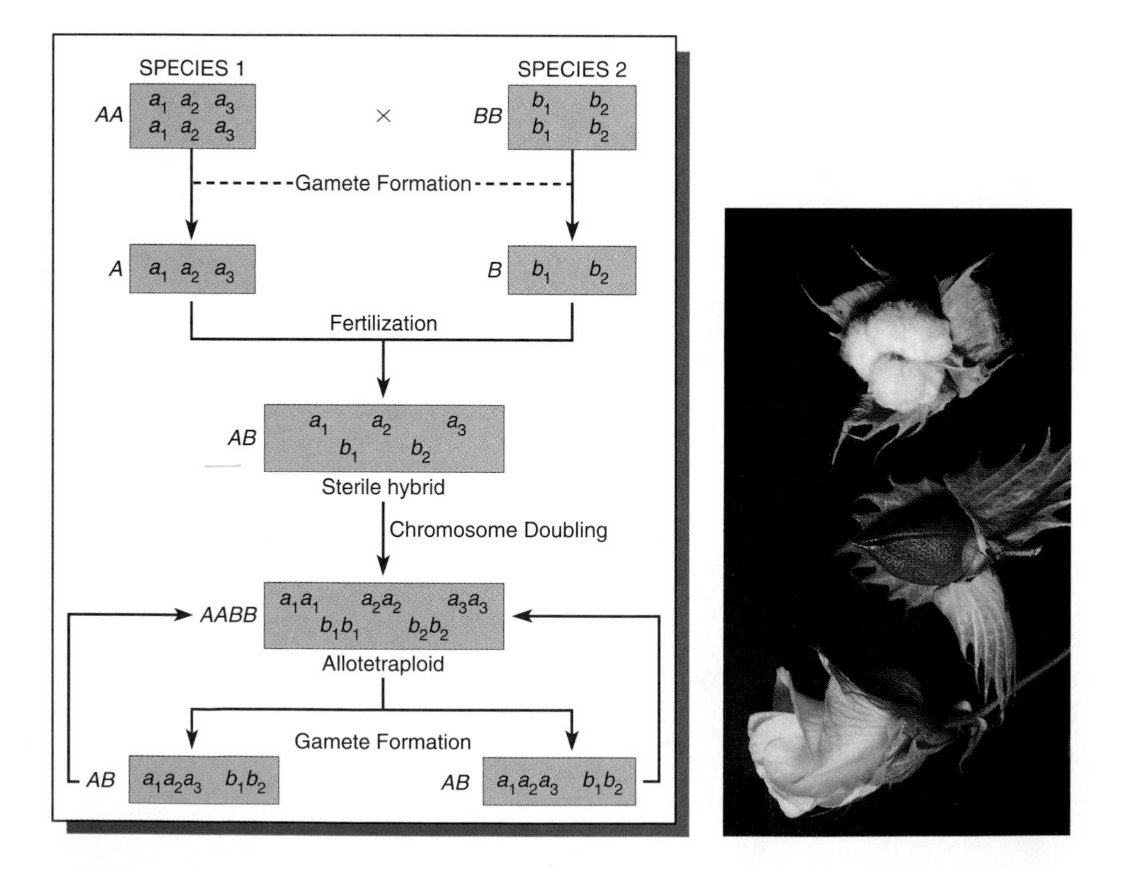

FIGURE 6.11
The origin and propagation of an allotetrapolid. Species 1 contains genome (*A*) consisting of three distinct chromosomes, a_1, a_2, and a_3. Species 2 contains genome (*B*) consisting of two distinct chromosomes, b_1 and b_2. Following fertilization between members of the two species and chromosome doubling, a fertile allotetraploid containing four complete haploid genomes (*AABB*) is formed. The allotetraploid form of *Gossypium,* which is believed to have arisen in this way, is illustrated.

Allotetraploids often exhibit characteristics of both parental species. Successful commercial hybridization has been performed using the grasses wheat and rye. Wheat (genus *Triticum*) has a basic haploid genome of 7 chromosomes. Cultivated autopolyploids exist, including tetraploid ($4n = 28$) and hexaploid ($6n = 42$) varieties. Rye (genus *Secale*) also has a genome consisting of 7 chromosomes. The only cultivated species is diploid ($2n = 14$).

Using the technique outlined in Figure 6.11 and 6.12, geneticists have produced various allopolyploid hybrids. When tetraploid wheat is crossed with diploid rye, and the F_1 treated with colchicine, a hexaploid variety ($6n = 42$) is derived. The hybrid, designated *Triticale* (see Figure 1.8), represents a new genus. Fertile hybrid varieties derived from various wheat and rye species may be crossed together or back-crossed. These crosses have created many variations of the genus *Triticale*.

The hybrid plants demonstrate characteristics of both wheat and rye. For example, certain hybrids combine the high protein content of wheat with the high content of the amino acid lysine of rye. The lysine content is low in wheat and thus is a limiting nutritional factor. Wheat is considered a high-yielding grain, while rye is noted for its versatility of growth in unfavorable environments. *Triticale* species, combining both traits, have the potential of significantly increasing grain production. Programs designed to improve crops through hybridization have long been underway in several underdeveloped countries of the world, as discussed in Chapter 1.

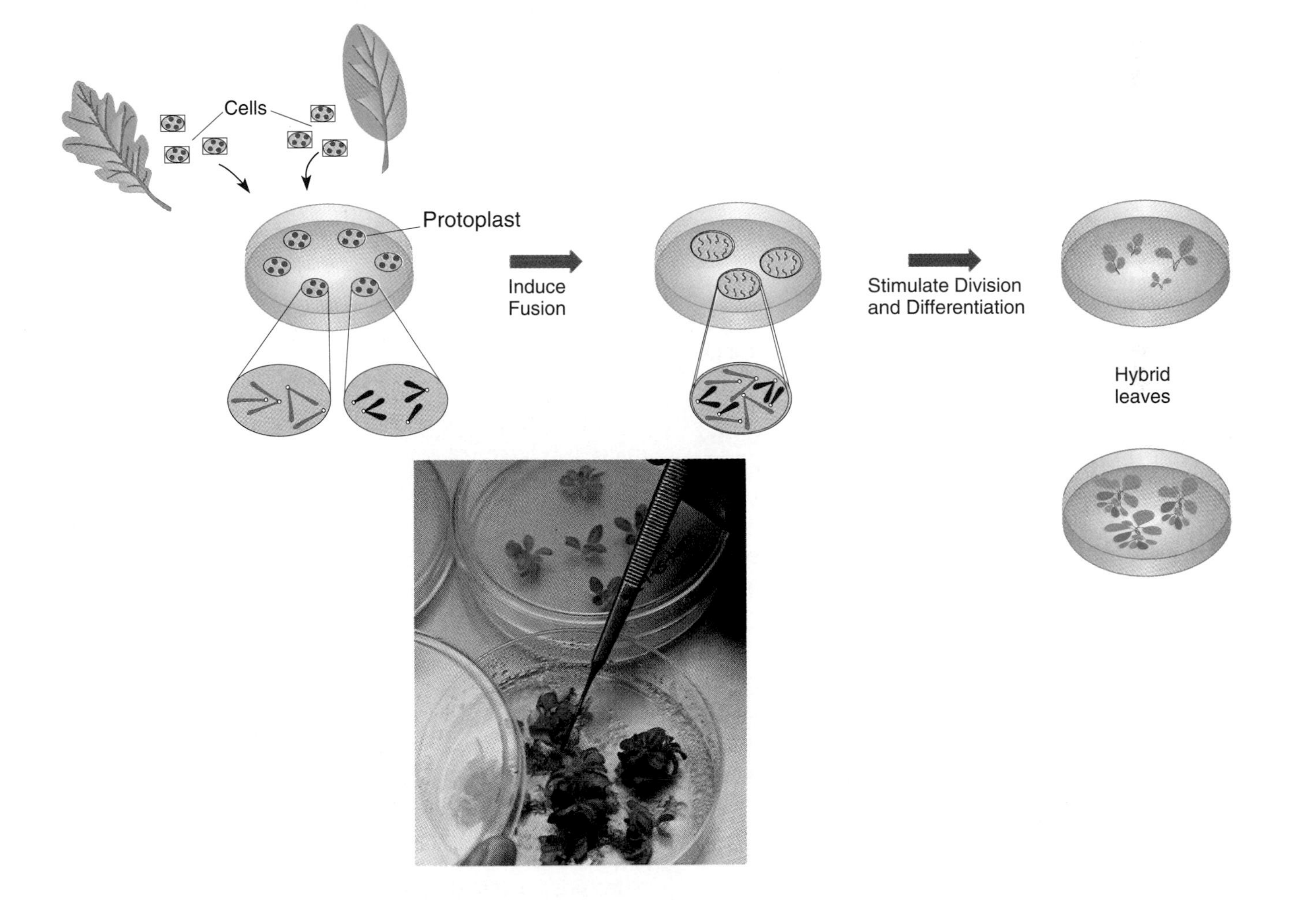

FIGURE 6.12
Application of somatic cell hybridization technique in the production of an amphidiploid. Cells from the leaves of two species of plants are removed and cultured. The cell walls are digested away and the resultant protoplasts are induced to undergo cell fusion. The hybrid cell is selected and stimulated to divide and differentiate, as illustrated in the photograph. An amphidiploid has a complete set of chromosomes from each parental cell type and displays phenotypic characteristics of each. Two pairs of chromosomes from each species are depicted.

Recall that in the previous chapter, we discussed the use of **somatic cell hybrids** to map human genes. This technique has also been applied to the production of allopolyploid plants. Cells from the developing leaves of plants can be isolated and treated to remove their cell wall, resulting in **protoplasts.** These altered cells can be maintained in culture and stimulated to fuse with other protoplasts from other plants, producing somatic cell hybrids.

When cells from different plant species are fused in this way, hybrid cells are produced with unique chromosome combinations. Since protoplasts can be induced to divide and differentiate into stems that develop leaves (Figure 6.12), the potential for producing allopolyploids is available in the research laboratory. In some cases, entire plants may be derived from cultured protoplasts. If only stems and leaves are produced, these may be grafted onto the stem of another plant. If flowers are formed, fertilization may yield mature seeds, which, upon germination, yield an allopolyploid plant.

VARIATION IN CHROMOSOME STRUCTURE AND ARRANGEMENT: AN OVERVIEW

The second general class of chromosome aberrations includes structural changes that delete, add, or rearrange substantial portions of one or more chromosomes. Included in this category are **deletions** and **duplications** of genes or part of a chromosome and **rearrangements** of genetic material in which a chromosome segment is either inverted, exchanged with a segment of nonhomologous chromosome, or merely transferred to one, leading to **translocations.** Before discussing these aberrations, we will present several general statements pertaining to them.

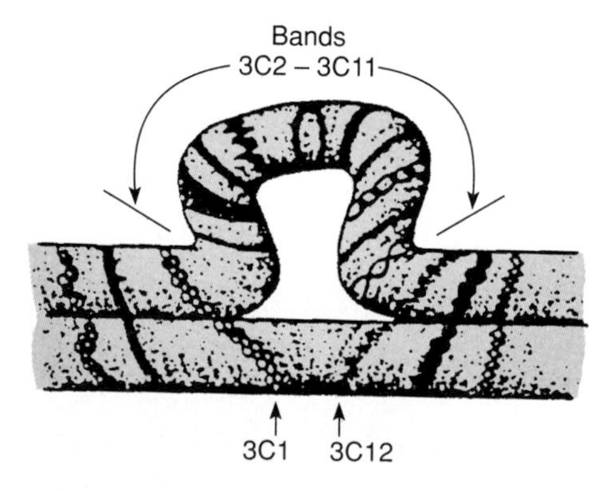

FIGURE 6.13
Deletion loop formed in salivary chromosomes heterozygous for the *Notch* deletion on the X chromosome of *Drosophila melanogaster.* The deficiency encompasses bands 3C2 through 3C11.

In most instances, these changes are due to one or more breaks along the axis of a chromosome, followed by either the loss or rearrangement of some genetic material. Chromosomes can break spontaneously, but the rate of breakage may increase in cells exposed to chemicals or radiation. Although the ends of chromosomes, the **telomeres,** do not readily fuse with ends of "broken" chromosomes or with other telomeres, the ends produced at points of breakage are "sticky" and can rejoin other such ends. If breakage-rejoining does not merely reestablish the original relationship, and if the alteration occurs in germ plasm, the gametes will contain the structural rearrangement, which will be heritable.

If the aberration is found in one homologue but not the other, unusual but characteristic pairing configurations are formed during meiotic synapsis. These patterns are useful in identifying the type of change that has occurred. If no loss or gain of genetic material occurs, individuals bearing the aberration "heterozygously," will likely to be unaffected phenotypically. However, the unusual pairing arrangements often lead to gametes that are duplicated or deficient for chromosomal regions. Thus, the offspring of "carriers" of certain aberrations often have an increased probabilty of demonstrating phenotypic manifestations of the aberration.

DELETIONS

When a portion of a chromosome is lost, the missing piece is referred to as a **deletion** (or a **deficiency**). The deletion may occur either at one end or from the interior of the chromosome. These are called **terminal** or **intercalary deletions,** respectively. Both result from one or more breaks in the chromosome. During synapsis between a chromosome with a large intercalary deficiency and a normal complete homologue, the unpaired region of the normal homologue must "buckle out." Such a configuration is called a **deficiency loop** or **compensation loop.** Such a loop can be visualized in the large polytene chromosomes derived from insect salivary gland cells. Polytene chromosomes, discussed in detail in Chapter 2, display a banding pattern characteristic of each chromosome, making them useful in cytological studies. The formation of such a loop in *Drosophila* polytene chromosomes is diagrammed in Figure 6.13.

DUPLICATIONS

When any part of the genetic material—a single locus or a large piece of a chromosome—is present more than once in the genome, it is called a **duplication.** As in deletions, pairing in heterozygotes may produce a compensation loop. Duplications may arise as the result of

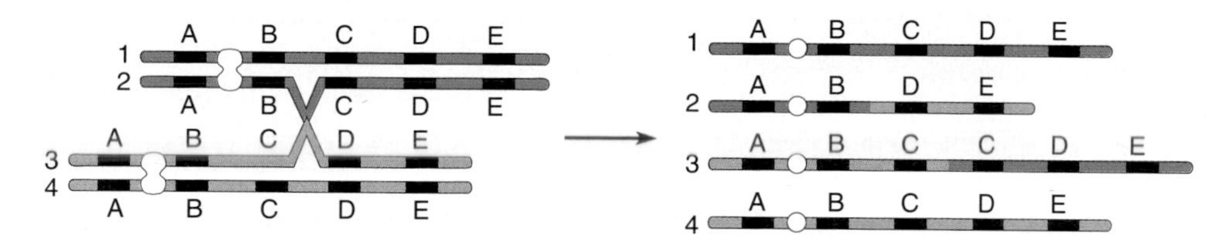

FIGURE 6.14
The origin of duplicated and deficient chromosomes as a result of unequal crossing over. The tetrad at the left is mispaired during synapsis. A single crossover between chromatids 2 and 3 results in a deficient and duplicated chromosome (see chromosomes 2 and 3 respectively, on the right). The two chromosomes uninvolved in the crossover event remain normal in their gene sequence and content.

unequal crossing over between synapsed chromosomes during meiosis (Figure 6.14) or through a replication error prior to meiosis. In the former case, both a duplication and a deficiency are produced.

Three interesting aspects of duplications will be considered. First, they may result in gene redundancy. Second, as with deletions, duplications may produce phenotypic variation. Third, according to one convincing theory, duplications have also been an important source of genetic variability during evolution.

Gene Redundancy and Amplification—Ribosomal RNA Genes

Although many gene products are not needed in every cell of an organism, other gene products are known to be essential components of all cells. For example, ribosomal RNA must be present in abundance in order to support protein synthesis. The more metabolically active a cell is, the higher is the demand for this molecule. In theory, a single copy of the gene encoding rRNA may be inadequate in many cells. Studies using the technique of molecular hybridization, which allows the determination of the percentage of the genome coding for specific RNA sequences, show that indeed there are multiple copies of the genes coding for rRNA. Such DNA is called **rDNA,** and the general phenomenon is called **gene redundancy.** For example, in the common intestinal bacterium *Escherichia coli* (*E. coli*) about 0.4 percent of the haploid genome consists of rDNA. This is equivalent to 5 to 10 copies of the gene. In *Drosophila melanogaster,* 0.3 percent of the haploid genome, equivalent to 130 copies, consists of rDNA. Although the presence of multiple copies of the same gene is not restricted to those coding for rRNA, we will focus on them in this section.

In some cells, particularly oocytes, even the normal redundancy of rDNA may be insufficient to provide adequate amounts of rRNA and ribosomes. Oocytes store abundant nutrients in the ooplasm for use by the embryo during early development. In fact, more ribosomes are included in the oocytes than in any other cell type. By considering how the amphibian *Xenopus laevis* acquires this abundance of ribosomes, we will see a second way in which the amount of rRNA is increased. This phenomenon is referred to as **gene amplification.**

The genes that code for rRNA are located in an area of the chromosome known as the **nucleolar organizer region (NOR).** The NOR is intimately associated with the nucleolus, which is a processing center for ribosome production. Molecular hybridization analysis has shown that each NOR in the frog *Xenopus* contains the equivalent of 400 redundant gene copies coding for rRNA. Even this number of genes is apparently inadequate to synthesize the vast amount of ribosomes that must accumulate in the amphibian oocyte to support development following fertilization. To further amplify the number of rRNA genes, the rDNA is selectively replicated, and each new set of genes is released from its template. Since each new copy is equivalent to the NOR, multiple small nucleoli are formed around each NOR in the oocyte. As many as 1500 of these "micronucleoli" have been observed in a single oocyte. If we multiply the number of micronucleoli (1500) by the number of gene copies in each NOR (400), we see that amplification in *Xenopus* oocytes can result in over half a million gene copies! If each copy is transcribed only 20 times during the maturation of the oocyte, in theory, sufficient copies of rRNA may be produced to result in well over one million ribosomes.

The *Bar* Eye Mutation in *Drosophila*

Duplications may cause phenotypic variation that at first might appear to be caused by a simple gene mutation. The *Bar* eye phenotype in *Drosophila* is a classical example. Instead of the normal oval eye shape, *Bar*-eyed flies have narrow, slitlike eyes. This phenotype *appears* to be inherited as a dominant sex-linked mutation.

In the early 1920s, Alfred H. Sturtevant and Thomas H. Morgan discovered and investigated this "mutation." As illustrated in Figure 6.15, normal wild-type females (B^+/B^+) have about 800 facets in each eye. Heterozygous

FIGURE 6.15
Summary of the *Bar* duplication in *Drosophila melanogaster.*

Genotype	Facet Number	Phenotype	Normal = 16A segments
B^+/B^+	779		B^+ B^+
B/B^+	358		B B^+
B/B	68		B B
B^+/B^D	45		B^+ B^D

females (B/B^+) have about 350 facets, while homozygous females (B/B) average only about 50 facets. Females are occasionally recovered with even fewer facets and are designated as *double Bar* (B^+/B^D).

About ten years later, Calvin Bridges and Herman J. Muller compared the polytene X chromosome banding pattern of the *Bar* fly with that of the wild-type fly. Such chromosomes contain specific banding patterns that have been well categorized into regions. Their studies reveal that one copy of region 16A of the X chromosome is present in wild-type flies and that this region is duplicated in *Bar* flies and triplicated in *double Bar* flies. These observations provide evidence that the *Bar* phenotype is not the result of a simple chemical change in the gene, but is instead a duplication.

The Role of Gene Duplication in Evolution

One of the most intriguing aspects of the study of evolution is the consideration of the mechanisms for genetic variation. The origin of unique gene products present in phylogenetically advanced organisms but absent in less advanced, ancestral forms is a topic of particular interest. In other words, how do "new" genes arise?

In 1970, Susumo Ohno published a provocative monograph *Evolution by Gene Duplication* in which he suggests that gene duplication is essential to the origin of new genes during evolution. Ohno's thesis is based on the supposition that the gene products of unique genes, present as only a single copy in the genome, are indispensable to the survival of members of any species during evolution. Therefore, unique genes are not free to accumulate mutations sufficient to alter their primary function and give rise to new genes.

However, if an essential gene were to become duplicated in a germ cell the new copy would be inherited in all future generations. Since it is an extra copy, major mutational changes in it are tolerated because the original gene still provides the genetic information for its essential function. The duplicated copy is now free to undergo large numbers of mutational changes within the organisms of any species over long periods of time. Over short intervals, the new genetic information may be of no practical advantage. However, over long periods of evolutionary time, the gene may change sufficiently so that its product assumes a divergent role in the cell. The new function may impart an adaptability and survival advantage to organisms carrying such unique genetic information. Thus, Ohno has outlined a mechanism through which sustained genetic variability may have originated.

Ohno's thesis is supported by the discovery of gene families, groups of contiguous genes that are related, whose products are similar in function, but whose DNA sequence is different. The various human hemoglobins are examples (see Chapter 12). Other genes have been discovered with sequence homology whose products are even more distinctive. These include trypsin and chymotrypsin, myoglobin, and hemoglobin, and the light and heavy immunoglobulin chains. As more and more genes and protein molecules are sequenced, the list grows.

INVERSIONS

The **inversion,** another class of structural variation, is a type of chromosomal aberration in which a segment of a chromosome is turned around 180° within a chromosome. An inversion does not involve a loss of genetic information, but simply rearranges the linear gene sequence. An inversion requires two breaks along the length of the chromosome and subsequent reinsertion of the inverted segment. Figure 6.16 illustrates how an inversion might arise. By forming a chromosomal loop prior to breakage, the newly created "sticky ends" are brought close together and rejoined.

The inverted segment may be short or quite long and may or may not include the centromere. If the centromere is not part of the rearranged chromosome segment, the inversion is said to be **paracentric.** If the centromere is part of the inverted segment, the term **pericentric** describes the inversion, as in Figure 6.14.

Although inversions may seem to have a minimal impact on the genetic material, their consequences are of interest to geneticists. Organisms heterozygous for inversions may produce aberrant gametes.

Consequences of Inversions During Gamete Formation

If only one member of a homologous pair of chromosomes has an inverted segment, normal linear synapsis during meiosis is not possible. Organisms with one inverted chromosome and one noninverted homologue are called **inversion heterozygotes.** Pairing between two such chromosomes in meiosis may be accomplished only if they form an **inversion loop.**

If crossing over does not occur within the inverted segment of the inversion heterozygote, the homologues will segregate and result in two normal and two inverted chromatids that are distributed into gametes. However, if crossing over occurs within the inversion loop, abnormal chromatids are produced. The effects of single exchange events within a paracentric inversion is diagrammed in Figure 6.17.

As in any meiotic tetrad, a single crossover between nonsister chromatids produces two parental chromatids and two recombinant chromatids. One recombinant chromatid is **dicentric** (two centromeres), and one recombinant chromatid is **acentric** (lacking a centromere). Both contain duplications and deletions of chromosome segments as well. During anaphase, an acentric chromatid moves randomly to one pole or the other or may be lost, while a dicentric chromatid is pulled in two directions. This polarized movement produces **dicentric bridges** that are cytologically recognizable. A dicentric chromatid will usually break at some point so that part of the chromatid goes into one gamete and part into another gamete during the reduction divisions. Therefore, gametes containing either recombinant chromatid are deficient in genetic material. When such a gamete participates in fertilization, the zygote most often develops abnormally, if at all.

A similar chromosomal imbalance is produced as a result of a crossover event between a chromatid bearing a pericentric inversion and its noninverted homologue. The recombinant chromatids that are directly involved in the exchange have duplications and deletions. However, no acentric or dicentric chromatids are produced. Gametes receiving these chromatids also produce inviable embryos following their participation in fertilization.

Because fertilization events involving these aberrant chromosomes do not produce viable offspring, it appears as if the inversion suppresses crossing over since crossover gametes are not recovered in the offspring. Actually, in inversion heterozygotes, the inversion has the effect of **suppressing the recovery of crossover products when chromosome exchange occurs within the inverted region.** If crossing over always occurred within a paracentric or pericentric inversion, 50 percent of the gametes would be ineffective. The viability of the resulting zygotes is therefore greatly diminished. Furthermore, up to one-half of the viable gametes have the

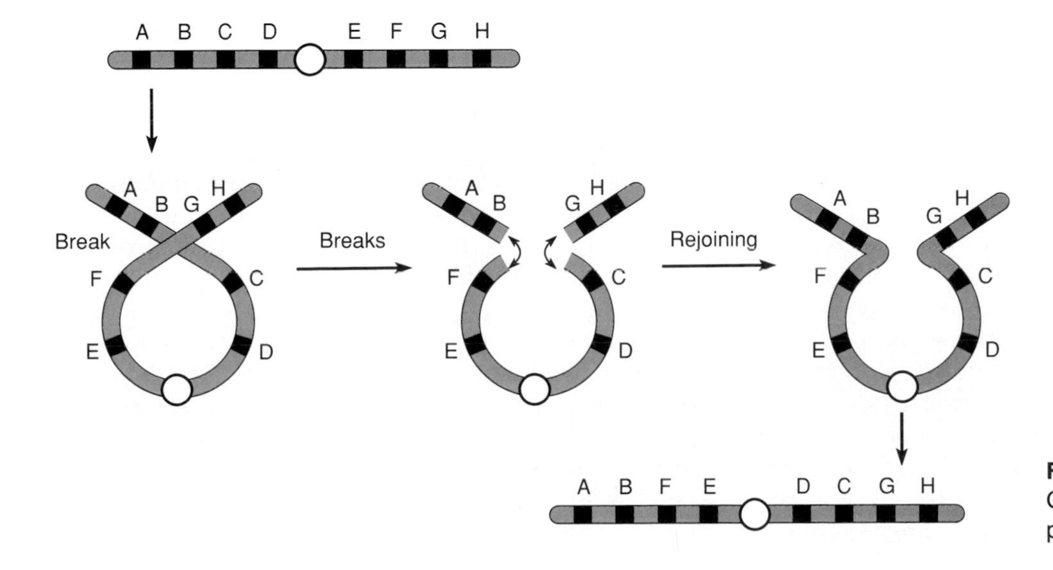

FIGURE 6.16
One possible origin of a pericentric inversion.

FIGURE 6.17
The effects of a single-crossover event between nonsister chromatids at a point within a paracentric inversion loop. Two altered gametes are produced, one which is acentric and one which is dicentric. Both altered forms are duplicated and deleted.

(a) Paracentric Inversion Heterozygote

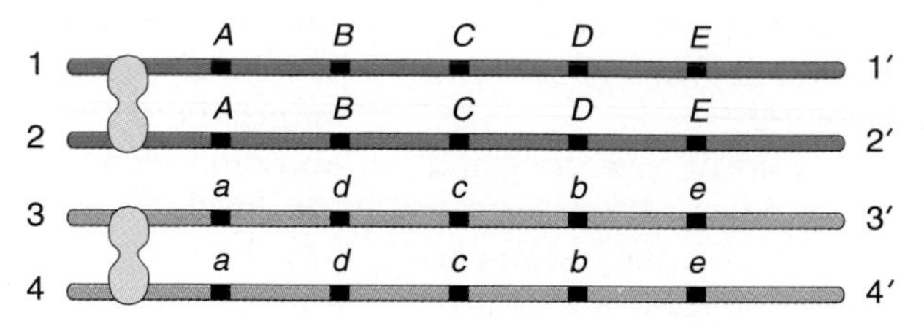

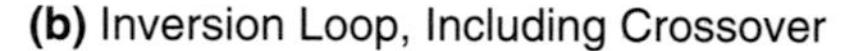

(b) Inversion Loop, Including Crossover

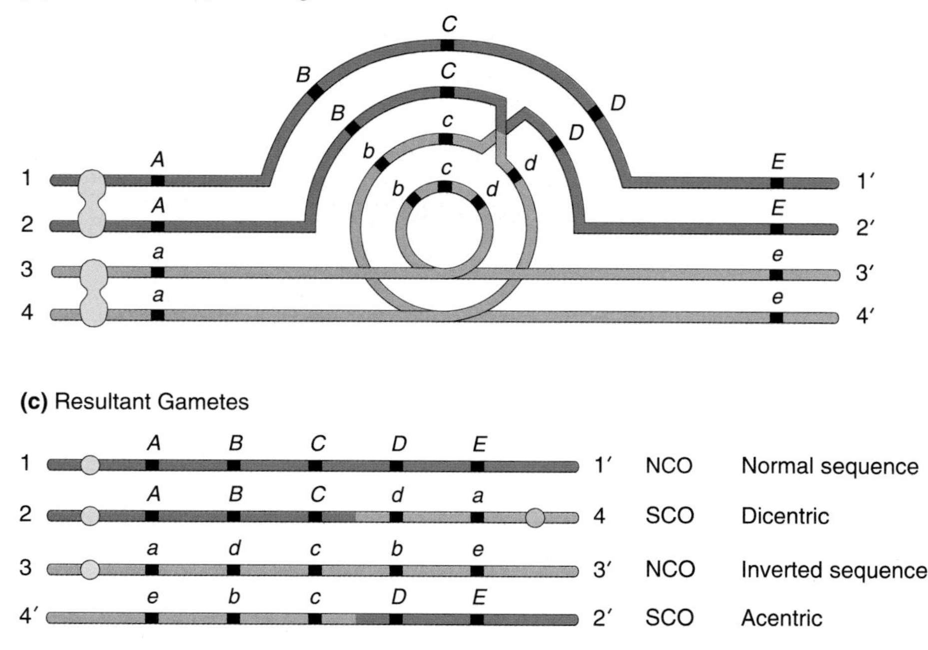

(c) Resultant Gametes

inverted chromosome, and the inversion will be perpetuated within the species. The cycle will be repeated continuously during meiosis in future generations.

TRANSLOCATIONS

Translocation, as the name implies, involves the movement of a segment of a chromosome to a new place in the genome. Translocation may occur within a single chromosome or between nonhomologous chromosomes. The exchange of segments between two nonhomologous chromosomes is a type of structural variation called a **reciprocal translocation.** The origin of a relatively simple reciprocal exchange is illustrated in Figure 6.18(a). The least complex way for this event to occur is for two nonhomologous chromosome arms to come close to each other so that an exchange is facilitated. For this type of translocation, only two breaks are required. If the exchange includes internal chromosome segments, four breaks are required, two on each chromosome.

The genetic consequences of reciprocal translocations are, in several instances, similar to those of inversions. For example, genetic information is not lost or gained. Rather, there is only a rearrangement of genetic material. The presence of a translocation does not, therefore, directly alter viability of individuals bearing it.

Homologues heterozygous for a reciprocal translocation undergo unorthodox synapsis during meiosis. As shown in Figure 6.18(b), pairing results in a crosslike configuration. As with inversions, genetically unbalanced gametes are also produced as a result of this unusual alignment during meiosis. In the case of translocations, however, aberrant gametes are not necessarily the result of crossing over. To see how unbalanced gametes are produced, focus on the homologous centromeres in Figure 6.18(b) and (c). According to the principle of independent assortment, the chromosome containing centromere 1 will migrate randomly toward one pole of the

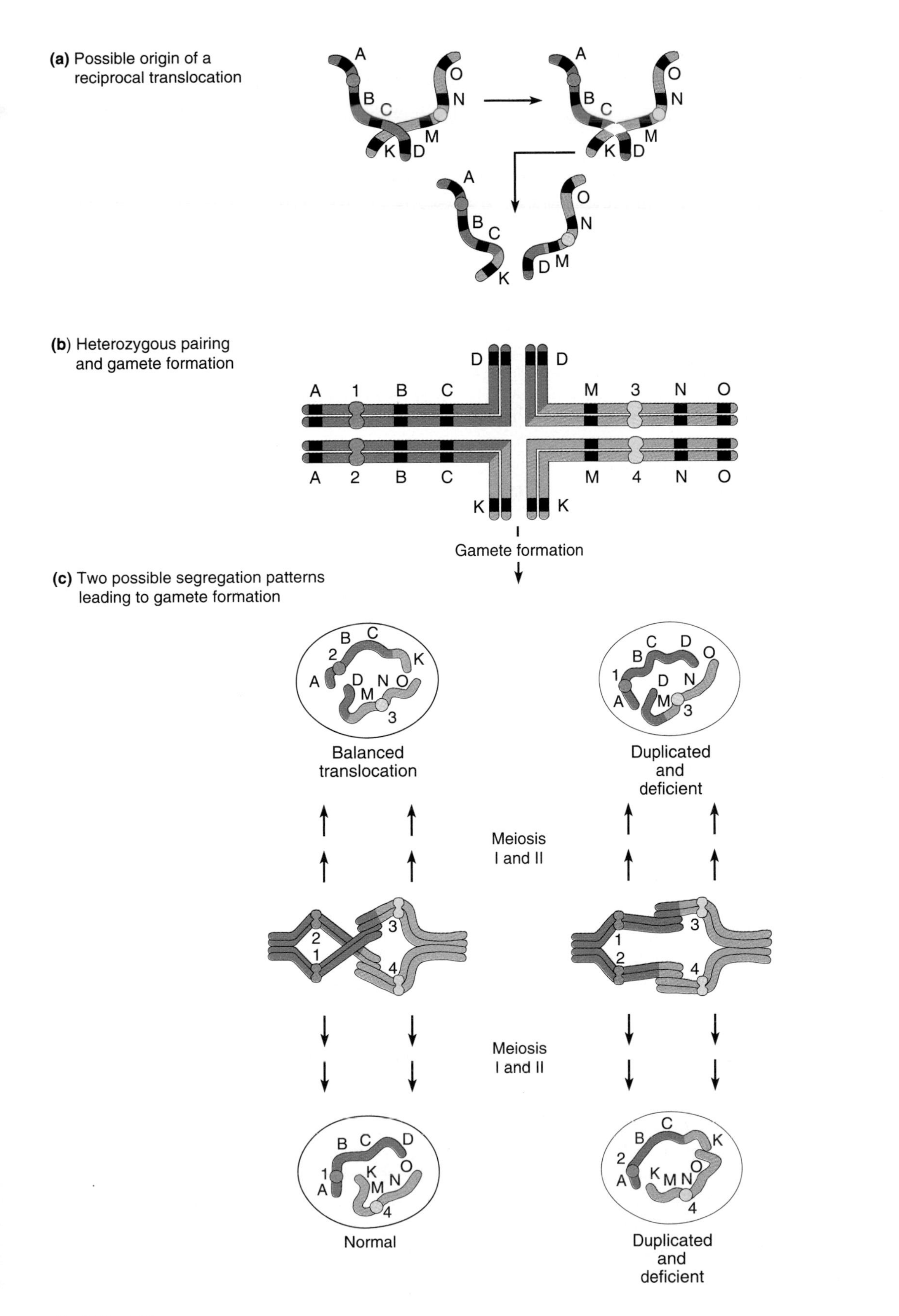

FIGURE 6.18
Origin, synapsis, and gamete formation of a simple reciprocal translocation.

spindle during the first meiotic anaphase; it will travel with *either* the chromosome having centromere 3 *or* centromere 4. The chromosome with centromere 2 will move to the other pole along with *either* the chromosome containing centromere 3 *or* centromere 4. This results in four potential meiotic products. The 1,4 combination contains chromosomes uninvolved in the translocation. The 2,3 combination, however, contains translocated chromosomes. These contain a complete complement of genetic information and are balanced. The other two potential products, the 1,3 and 2,4 combinations, will contain chromosomes displaying duplicated and deleted segments.

When incorporated into gametes, the resultant meiotic products are genetically unbalanced. If they participate in fertilization, lethality often results. As few as 50 percent of the progeny of parents heterozygous for a reciprocal translocation may survive. This condition, called **semisterility,** has an impact on the reproductive fitness of organisms, thus playing a role in evolution. Furthermore, in humans, such an unbalanced condition results in partial monosomy or trisomy, leading to a variety of birth defects.

Translocations in Humans: Familial Down Syndrome

Research performed since 1959 reveals numerous translocations in members of the human population. One common type of translocation involves breaks at the extreme ends of the short arms of two nonhomologous acrocentric chromosomes. These small segments are lost, and the larger segments fuse at their centromeric region. This type of translocation produces a new, large submetacentric or metacentric chromosome, often called a **Robertsonian translocation.**

Just such a translocation accounts for cases in which Down syndrome is inherited or familial. Earlier in this chapter we pointed out that most instances of Down syndrome are due to trisomy 21. This chromosome composition results from nondisjunction during meiosis of one parent. Trisomy accounts for over 95 percent of all cases of Down syndrome. In such cases, the chance of the same parents producing a second afflicted child is extremely low. However, in the remaining families with a Down child, the syndrome occurs in a much higher frequency over several generations.

Cytogenetic studies of the parents and their offspring from these unusual cases explain the cause of **familial Down syndrome.** Analysis reveals that one of the parents contains a **14/21 D/G translocation** (Figure 6.19). That is, one parent has the majority of the G-group chromosome 21 translocated to one end of the D-group chromosome 14. This individual has only 45 chromosomes, but apparently has a nearly complete diploid content of genetic information. During meiosis, one-fourth of the individual's gametes will have two copies of chromosome 21: a normal chromosome and a second copy translocated to chromosome 14. When such a gamete is fertilized by a standard haploid gamete, the resulting zygote has 46 chromosomes but three copies of chromosome 21. These individuals exhibit Down syndrome. Other potential surviving offspring contain either the standard diploid genome without a translocation or the balanced translocation like the parent. Both cases result in normal individuals. Knowledge of translocations has allowed geneticists to resolve the seeming paradox of an inherited trisomic phenotype in an individual with an apparent diploid number of chromosomes.

FRAGILE SITES IN HUMANS

We conclude this chapter by briefly discussing the results of an interesting discovery made around 1970 during observations of metaphase chromosomes prepared following human cell culture. In cells derived from certain individuals, a specific area along one of the chromosomes failed to stain, giving the appearance of a gap. In other individuals whose chromosomes displayed or expressed such morphology, the gap appeared in different positions within the set of chromosomes. Such areas eventually become known as **fragile sites,** since they were considered susceptible to chromosome breakage when cultured in the absence of certain chemicals such as folic acid, which is normally present in the medium. Fragile sites were at first considered curiosities until a strong association was subsequently shown to exist between one of the sites and a form of mental retardation.

The molecular nature of fragile sites is unknown. Because they appear to represent points along the chromosome susceptible to breakage, these sites may indicate regions that are not tightly coiled or compacted. It should be noted that almost all studies of fragile sites have been carried out in mitotically dividing cells, and it is unknown whether such sites are also expressed in meiotic cells.

Most fragile sites do not appear to be associated with any clinical syndrome except for a folate-sensitive site on the X chromosome (Figure 6.20). Those bearing this site exhibit **fragile X syndrome** (or **Martin-Bell syndrome**), the most common form of inherited mental retardation, affecting about 1 in 1250 males and 1 in 2500 females. Because it is a dominant trait, females carrying only one fragile X chromosome can be mentally retarded. The trait is not fully expressed, since only about 30 percent of fragile X females are retarded, while about 20 percent of fragile X-bearing males are phenotypically

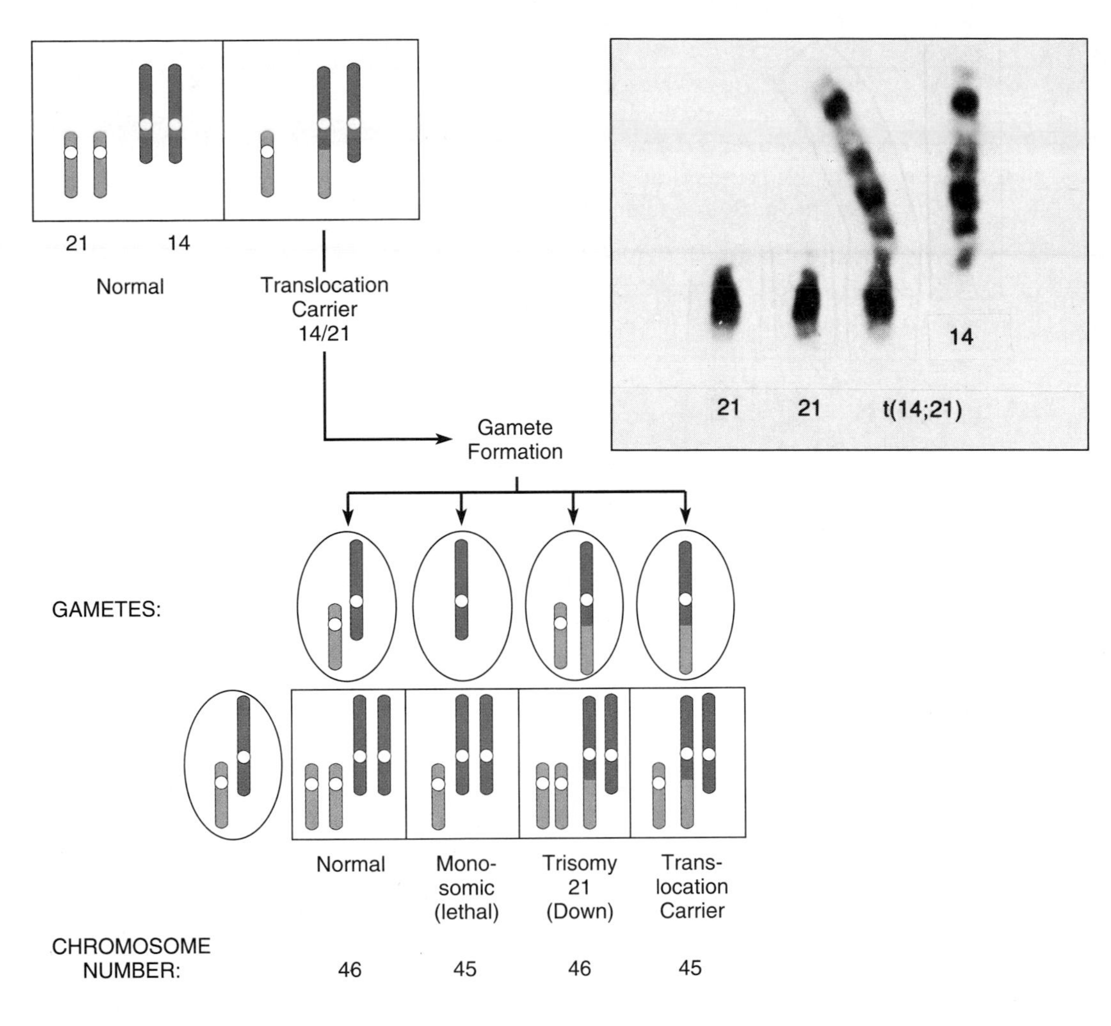

FIGURE 6.19
The origin of a 14/21 D/G translocation leading to familial Down syndrome.

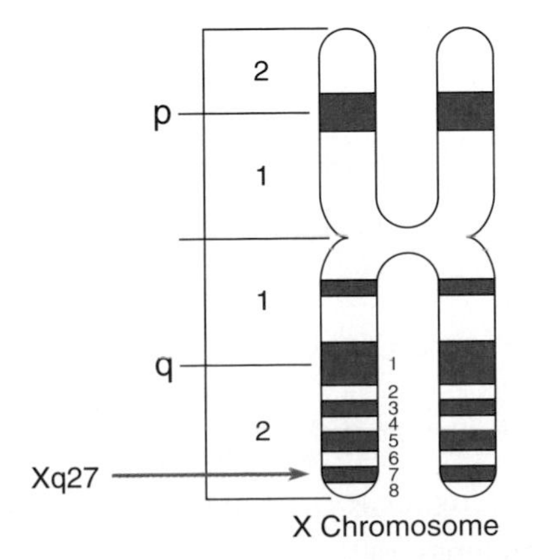

FIGURE 6.20
An illustration of the region of the human X chromosome associated with fragile X-linked mental retardation. The arrow identifies the fragile site (Xq27) of the chromosome.

normal. In addition to mental retardation, affected males have characteristic long, narrow faces with protruding chins, enlarged ears, and increased testicular size.

A gene that spans the fragile site may be responsible for this syndrome. This gene, known as *FMR-1,* contains a region that is increased in size in fragile X chromosomes. The fragile X breakpoint is contained in this variable length region, and full expression of the phenotype depends on both an increase in length and chemical modifications to the DNA nucleotides in this gene segment. The *FMR-1* gene is known to be expressed in the brain, but the exact nature and function of the encoded protein awaits further research. The gene may encode a DNA-binding protein based on its sequence. Although other genes may turn out to be involved in fragile X syndrome, the structure and function of the FMR-1 gene is currently the leading candidate to provide insights into the molecular basis of this common form of mental retardation.

CHAPTER SUMMARY

1 Investigations into the uniqueness of each organism's chromosomal constitution have further enhanced our understanding of genetic variation. Alterations of the precise diploid content of chromosomes are referred to as chromosomal aberrations or chromosomal mutations.

2 In humans, the study of individuals with altered sex chromosome compositions has established that the Y chromosome is responsible for male differentiation. The absence of the Y leads to female differentiation. Similar studies in *Drosophila* have excluded the Y in such a role, instead demonstrating that a balance between the number of X chromosomes and sets of autosomes is the critical factor.

3 Deviations from the expected chromosomal number, or mutations in the structure of the chromosome, are inherited in predictable Mendelian fashion; they often result in inviable organisms or substantial changes in the phenotype.

4 Aneuploidy is the gain or loss of one or more chromosomes from the diploid content, resulting in conditions of monosomy, trisomy, tetrasomy, etc. Studies of monosomy in Turner syndrome, and trisomic disorders in the Down syndrome have increased our understanding of the delicate genetic balance that must exist in order for normal development to occur.

5 When complete sets of chromosomes are added to the diploid genome, polyploidy is created. These sets may have identical or diverse genetic origin, creating either autopolyploidy or allopolyploidy, respectively.

6 Large segments of the chromosome may be modified by deletions or duplications. Deletions may produce serious conditions such as the cri-du-chat syndrome in humans, while duplications may be particularly important as a source of redundant or unique genes.

7 Inversions and translocations, while altering the gene order along chromosomes, cause no loss of genetic information or deleterious effects initially. However, heterozygous combinations may cause genetically abnormal gametes following meiosis, often causing lethality.

8 Fragile sites in human mitotic chromosomes have sparked research interest because one such site on the X chromosome is associated with the most common form of inherited mental retardation.

KEY TERMS

acentric chromosome
allopolyploidy
allotetraploid
amniocentesis
amphidiploid
aneuploidy
autopolyploidy
Barr body
chorionic villus sampling (CVS)
colchicine
compensation loop
cri-du-chat syndrome
deficiency
deletion
dicentric chromosome
dosage compensation
Down syndrome
duplication
Edwards syndrome
euploidy
fragile sites
fragile X syndrome (Martin-Bell syndrome)
gene amplification
gene redundancy
intersex
inversion
inversion loop
Klinefelter syndrome

Lyon hypothesis
metafemale
metamale
monosomy
nondisjunction
nucleolar organizer region (NOR)
paracentric inversion
partial monosomy
Patau syndrome
pentasomy
pericentric inversion
polyploidy
protoplast
segmental deletion
semisterility
sex chromatin body
sex-determining region Y (SRY)
somatic cell hybrid
telomere
testis determining factor (TDF)
tetraploid
tetrasomy
transgenic mice
translocation
triploid
trisomy
Turner syndrome
Barr body
sex chromatin body

1 In a cross using maize involving three genes, *a, b,* and *c,* a heterozygote (*abc*/+ + +) was test-crossed (to *abc*/*abc*). Even though the three genes were separated by at best five map units, only two phenotypes were recovered: *abc* and + + +. Additionally, the cross produced significantly fewer viable plants than expected. Can you propose why no other phenotypes were recovered and why the viability was reduced?

SOLUTION: One of the two chromosomes may contain an inversion that overlaps all three genes, effectively precluding the recovery of any "crossover" offspring. If this is a paracentric inversion and the genes are clearly separated (assuring that a significant number of crossovers will occur between them), then numerous acentric and dicentric chromosomes will be formed, resulting in the observed reduction in viability.

2 If a haplo-IV female *Drosophila* with normal bristles is crossed with a male with a diploid set of chromosomes, but who is homozygous for the recessive chromosome 4 mutation, *shaven* (*sv*), what F_1 phenotypic ratio might be expected?

SOLUTION: The bristle phenotypes will be governed by the fact that the normal-bristled P_1 female produces gametes, one-half of which contain a chromosome 4 (sv^+) and one-half that have no chromosome 4. Following fertilization by sperm from the *shaven* male, one-half of the offspring will receive two members of chromosome 4 and be heterozygous for *sv,* expressing normal bristles. The other half will have only one copy of chromosome 4 (indicated below by $sv^+/-$ or by $sv/-$). Since its origin is from the male parent, where the chromosome bears the *sv* allele, these flies will express *shaven* since there is no wild-type allele present to mask this recessive allele.

P_1: $sv^+/- \times sv/sv$
normal females shaven male

F_1: $1/2\ sv^+/sv$ males and females → 1/2 wild type
$1/2\ sv/-$ males and females → 1/2 shaven

PROBLEMS AND DISCUSSION QUESTIONS

1 Considering the information presented in Chapter 4 about lethal genes, speculate as to why the vast majority of human 45,X conceptions fail to survive to birth.

2 It has been suggested that any male-determing genes contained on the Y chromosome in humans should not be located in the limited region that synapses with the X chromosome during meiosis. What might be the outcome if such genes were located in this region?

3 Define and distinguish between the following pairs of terms:

aneuploidy/euploidy
monosomy/trisomy
autopolyploidy/allopolyploidy
paracentric inversion/pericentric inversion

4 What evidence suggests that Down syndrome is more often the result of nondisjunction during oogenesis rather than during spermatogenesis?

5 What conclusions have been drawn about human aneuploidy as a result of karyotypic analyses of abortuses?

6 Discuss the role of polyploidy in evolution. How is its role different in animal and plant species?

7 What experimental techniques may be used to convert diploid plants to tetraploids?

8 Discuss the origin of cultivated American cotton.

9 For a species with a diploid number of 18, indicate how many chromosomes will be present in the somatic nuclei of individuals who are haploid, triploid, tetraploid, trisomic, and monosomic.

10 Discuss the possible mechanisms involved in the production of deletions, duplications, inversions, and translocations.

11 Contrast the synaptic configurations of homologous pairs of chromosomes in which one member is normal and the other member has sustained a deletion, a duplication, or an inversion.

12 In a cross in *Drosophila,* a female heterozygous for the autosomally linked genes *a, b, c, d,* and *e* (*abcde*/+++++) was test-crossed to a male homozygous for all recessive alleles. Even though the distance between each of the above loci was at least 3 map units, only four phenotypes were recovered:

Phenotype	No. of Flies
+ + + + +	440
a b c d e	460
+ + + + *e*	48
a b c d +	52
Total	1000

Why are many expected crossover phenotypes missing? Can any of these loci be mapped from the data given here? If so, determine map distances.

13 Discuss Ohno's hypothesis on the role of gene duplication in the process of evolution

14 A human female with Turner syndrome also expresses the recessive sex-linked trait hemophilia as her father did. Which parent underwent nondisjunction during meiosis, giving rise to the gamete responsible for the syndrome?

15 The primrose *Primula kewensis* has 36 chromosomes that are similar in appearance to the chromosomes in two related species, *Primula floribunda* ($2n = 18$) and *Primula verticillata* ($2n = 18$). How could *P. kewensis* arise from these species? How would you describe *P. kewensis* in genetic terms?

16 *Drosophila* may be trisomic for chromosome 4 and remain fertile. Predict the F_1 results of crossing:

trisomic bent ($b/b/b$) $\times$ normal disomic (B/B)

SELECTED READINGS

BARR, M. L. 1966. The significance of sex chromatin. *Int. Rev. Cytol.* 19:35–39. LYON, M.F. 1962. Sex chromatin and gene action in the mammalian X chromosome. *Am. J. Hum. Genet.* 14:135–48.

———. 1988 X-chromosome inactivation and the location and expression of X-linked genes. *Am. J. Hum. Genet.* 42:8–16.

BLAKESLEE, A. F. 1934. New jimson weeds from old chromosomes. *J. Hered.* 25:80–108.

BORGAONKER, D. S. 1989. *Chromosome variation in man: A catalogue of chromosomal variants and anomalies.* 5th ed. New York: Alan R. Liss.

BOUE, A. 1985. Cytogenetics of pregnancy wastage. *Adv. Hum. Genet.* 14:1–58.

BURGIO, G. R., et al., Eds. 1981. *Trisomy* 21. New York; Springer-Verlag.

CARR, D. H. 1971. Genetic basis of abortion. *Ann. Rev. Genet.* 5:65–80.

COURT-BROWN, W. M. 1968. Males with an XYY sex chromosome complement. *J. Med. Genet.* 5:341–59.

CUMMINGS, M. R. 1991. *Human heredity: Principles and issues.* 2nd ed. St. Paul: West.

DeARCE, M. A., and KEARNS, A. 1984. The fragile X syndrome: The patients and their chromosomes. *J. Med. Genet.* 21:84–91.

ERICKSON, J. D. 1979. Paternal age and Down syndrome. *Amer. J. Hum. Genet.* 31:489–97.

GUPTA, P. K., and PRIYADARSHAN, P. M. 1982. *Triticale:* Present status and future prospects. *Adv. in Genet.* 21:256–346.

HECHT, F. 1988. Enigmatic fragile sites on human chromosomes. *Trends in Genetics* 4:121–22.

HOOK, E. B. 1973. Behavioral implications of the humans XYY genotype. *Science* 179:139–50.

HULSE, J. H., and SPURGEON, D. 1974 Triticale. *Scient. Amer.* (Aug.) 231:72–81.

JACOBS, P. A., et al. 1974. A cytogenetic survey of 11,680 newborn infants. *Ann. Hum. Genet.* 37:359–76.

KOOPMAN. P., et al. 1991. *Male development of chromosomally female mice transgenic for Sry. Nature* 351:117–121.

LEWIS, W. H., ed. 1980. *Polyploidy: Biological relevance.* New York: Plenum Press.

MANTELL, S. H., MATHEWS, J. A., and McKEE, R. A. 1985. *Principles of plant biotechnology: An introduction to genetic engineering in plants.* Oxford: Blackwell.

McLAREN, A. 1988. Sex determination in mammals. *Trends in Genetics* 4:153–57.

OBE, G., and BASLER, A. 1987. *Cytogenetics: Basic and applied aspects.* New York: Springer-Verlag.

OHNO, S. 1970. *Evolution by gene duplication.* New York: Springer-Verlag.

SHEPARD, J., et al. 1983. Genetic transfer in plants through interspecifc protoplast fusion. *Science* 21:683–88.

SUTHERLAND, G. 1985. The enigma of the fragile X chromosome. *Trends in Genetics* 1:108–11.

SWANSON, C. P., MERZ, T., and YOUNG, W. J. 1981. *Cytogenetics: The chromosome in division, inheritance, and evolution.* 2nd ed. Englewood Cliffs, NJ: Prentice-Hall.

TJIO, J. H., and LEVAN, A. 1956. The chromosome number of man. *Hereditas* 42:1–6

WILKINS, L. E., BROWN, J. A., and WOLF, B. 1980. Psychomotor development in 65 home-reared children with cri-du-chat syndrome. *J. Pediatr.* 97:401–5

YUNIS, J. J., ed. 1977. *New chromosomal syndromes.* Orlando: Academic Press.

7 DNA—The Physical Basis of Life

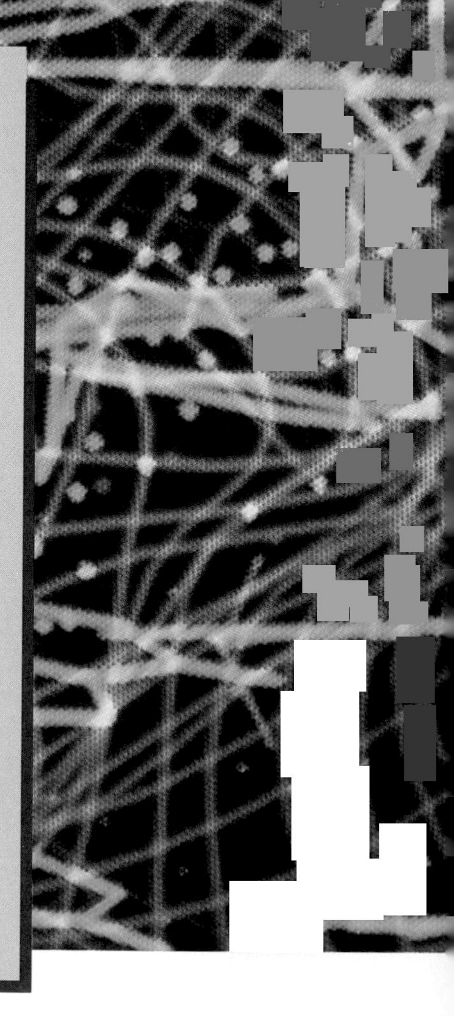

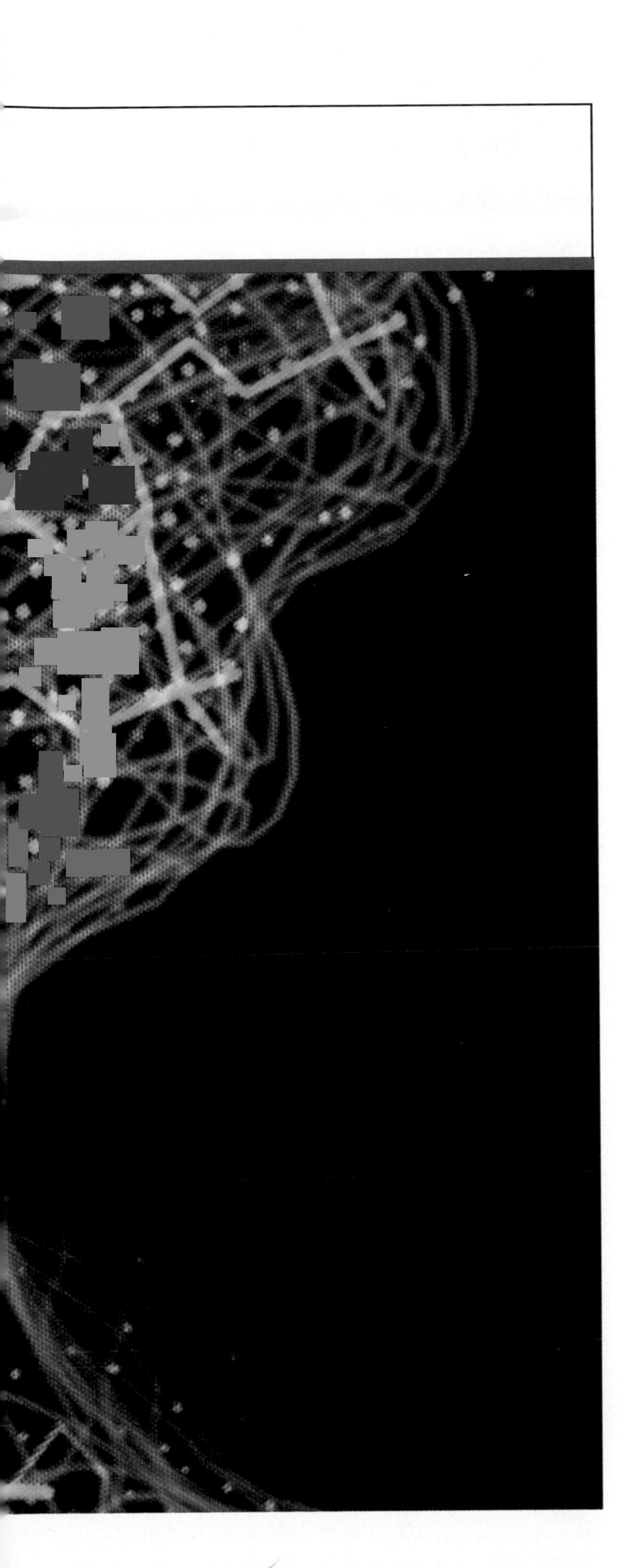

Chapter Concepts

With few exceptions, the nucleic acid DNA serves as the genetic material in every living thing. The structure of DNA provides the chemical basis for storing and expressing genetic information within the cells, as well as transmitting it to future generations. The molecule takes the form of a double-stranded helix united by hydrogen bonds formed between complementary nucleotides. In some viruses, RNA serves as the genetic material.

Earlier in the text, we discussed the presence of genes on chromosomes that control phenotypic traits and the way in which the chromosomes are transmitted through gametes to future offspring. Logically, there must be some form of information contained in genes, which, when passed to a new generation, influences the form and characteristics of the offspring; this is called the **genetic information.** We might also conclude that this same information in some way directs the many complex processes leading to the adult form.

Until 1944 it was not clear what chemical component of the chromosome makes up genes and constitutes the genetic material. Since chromosomes were known to have both a nucleic acid and a protein component, both were considered candidates. In 1944, however, there emerged direct experimental evidence that the nucleic acid, DNA, serves as the informational basis for the process of heredity.

Once the importance of DNA in genetic processes was realized, work was intensified with the hope of discerning not only the structural basis of this molecule but also the relationship of its structure to its function. Between 1944 and 1953, many scientists sought information that might answer the most significant and intriguing question in the history of biology: How does DNA serve as the genetic basis for the living process? The answer was believed to depend strongly on the chemical structure of the DNA molecule, given the complex but orderly functions ascribed to it.

These efforts were rewarded in 1953 when James Watson and Francis Crick set forth their hypothesis for the double-helical nature of DNA. The assumption that the molecule's functions would be clarified more easily once its general structure was determined proved to be correct. This chapter initially reviews the evidence that DNA is the genetic material and then discusses the elucidation of its structure.

CHARACTERISTICS OF THE GENETIC MATERIAL

The genetic material has several characteristics: **replication, storage of information, expression of that information,** and **variation by mutation.** "Replication" of the genetic material is one facet of cell division, a fundamental property of all living organisms. Once the genetic material of cells has been replicated, it must then be partitioned equally into daughter cells.

The characteristic of "storage" may be viewed as genetic information that is present, but that may or may not be expressed. It is clear that while most cells contain a complete complement of DNA, at any point in time they express only a part of this genetic potential. For example, bacteria turn many genes on only in response to specific environmental conditions, only to turn them off when such conditions change. In vertebrates, skin cells may display active melanin genes but never activate their hemoglobin genes; digestive cells activate many genes specific to their function, but do not activate their melanin genes.

"Expression" of the stored genetic information is a complex process and is the basis for the concept of **information flow** within the cell. Figure 7.1 shows a simplified illustration of this concept. The initial event is the **transcription** of DNA, resulting in the synthesis of three types of RNA molecules: **messenger RNA (mRNA), transfer RNA (tRNA),** and **ribosomal RNA (rRNA).** Of these, mRNAs are translated into proteins. Each type of mRNA is the product of a specific gene and leads to the synthesis of a different protein. **Translation** occurs in conjunction with ribosomes and involves tRNA, which adapts the chemical information in mRNA to the amino acids that make up proteins. Collectively, these processes serve as the foundation for the **central dogma of molecular genetics:** "DNA makes RNA, which makes proteins."

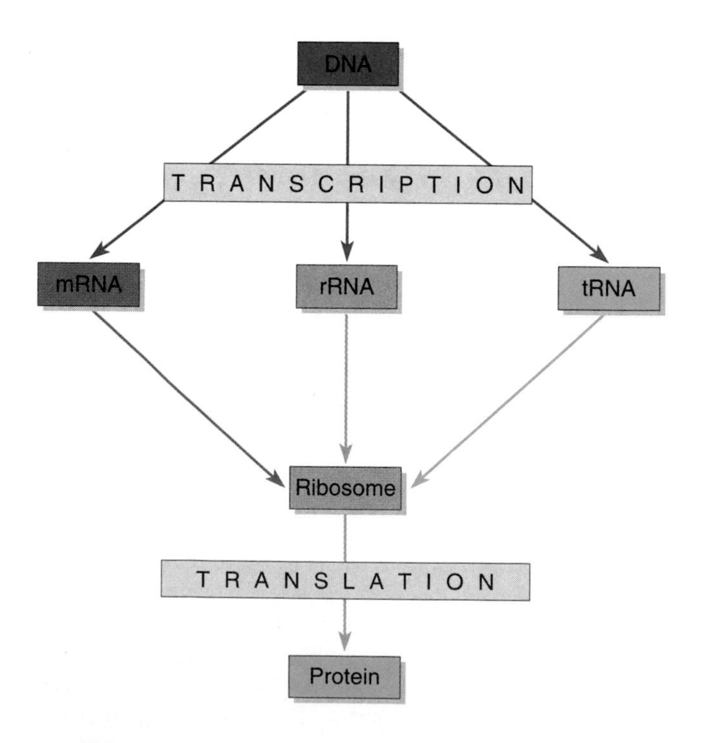

FIGURE 7.1
A simplified view of information flow involving DNA, RNA, and proteins within cells.

The genetic material is also the source of newly arising "variability" among organisms through the process of mutation. If a change in the chemical composition of DNA occurs, the alteration will be reflected during transcription and translation, affecting the specified protein. If a mutation is present in gametes, it will be passed to future generations and, with time, may become distributed in the population. Genetic variation, which also includes rearrangements within and between chromosomes, provides the raw material for the process of evolution.

THE GENETIC MATERIAL: EARLY STUDIES

The idea that genetic material is transmitted physically from parent to offspring has been accepted for as long as the concept of inheritance has existed. Beginning in the late nineteenth century, research into the structure biomolecules progressed considerably, setting the stage for the description of the genetic material in chemical terms. Although proteins and nucleic acid were both considered major candidates for the role of the genetic material, many geneticists, until the 1940s, favored proteins. This is not surprising, since a diversity of proteins was known to be abundant in cells and much more was known about protein chemistry.

DNA was first studied in 1868 by a Swiss chemist, Friedrick Miescher. He was able to isolate cell nuclei and derive an acid substance containing DNA that he called **nuclein.** As investigations progressed, however, DNA, which was shown to be present in chromosomes, seemed to lack the chemical diversity necessary to store extensive genetic information. This conclusion was based largely on Phoebus A. Levene's observations in 1910 that DNA contained approximately equal amounts of four quite similar molecules called nucleotides. Levene postulated incorrectly that identical groups of these four components were repeated over and over, which was the basis of his **tetranucleotide hypothesis** for DNA structure. Attention was thus directed away from DNA, favoring proteins, much by default. However, in the 1940s, Erwin Chargaff showed that Levene's proposal was incorrect when he demonstrated that an *unequal* proportion of the four nucleotides is the rule for most organisms studied. We will see later that the structure of DNA accounts for Chargaff's observations.

EVIDENCE FAVORING DNA IN BACTERIA AND BACTERIOPHAGES

The 1944 publication by Oswald Avery, Colin MacLeod, and Maclyn McCarty concerning the chemical identity of a "transforming principle" in bacteria marked the initial event leading to the acceptance of DNA as the genetic material. Along with the subsequent findings of other research teams, this work constituted direct experimental proof that, in the organisms studied, DNA, and not protein, is the biomolecule responsible for heredity. This period marked the beginning of an era of discovery in biology that has revolutionized our understanding of life on earth. The impact on biology parallels the work that followed the publication of Darwin's theory of evolution and that following the rediscovery of Mendel's postulates of transmission genetics. Together, these constituted three great revolutions in biology.

The initial evidence implicating DNA as the genetic material was derived from studies of prokaryotic bacteria and the viruses that infect them, called bacteriophages. The reasons for their use will become apparent as the experiments are studied. Primarily, bacteria and their viruses are capable of rapid growth because they complete life cycles in hours. They may also be experimentally manipulated and mutations may be easily induced and selected.

Transformation Studies

The research that provided the foundation for Avery, MacLeod, and McCarty's work was initiated in 1927 by Frederick Griffith, a medical officer in the British Ministry of Health. He performed experiments with several different strains of the bacterium *Diplococcus pneumoniae*.* Some were **virulent strains,** which cause pneumonia in certain vertebrates (notably humans and mice), while others were **avirulent strains,** which do not cause illness.

The difference in virulence is related to the polysaccharide capsule of the bacterium. Virulent strains have this capsule, whereas avirulent strains do not. The nonencapsulated bacteria are readily engulfed and destroyed by phagocytic cells in the animal's circulatory system. Virulent bacteria, which possess the polysaccharide coat, are not easily engulfed; they multiply and cause pneumonia.

The presence or absence of the capsule is the basis for a visible difference between colonies of virulent and avirulent strains. Encapsulated bacteria form a **smooth,** shiny-surfaced colony **(*S*)** when grown on an agar culture plate; nonencapsulated strains produce **rough** colonies **(*R*)** (Figure 7.2). Thus, virulent and avirulent strains may be distinguished easily by standard microbiological culture techniques.

Each strain of *Diplococcus* may be one of dozens of different types called **serotypes.** The specificity of the serotype is due to the detailed chemical structure of the polysaccharide constituent of the thick, slimy capsule. Serotypes are identified by immunological techniques and are usually designated by Roman numerals. Griffith used the virulent type II*R* and the avirulent type III*S* in his critical experiments. Table 7.1 summarizes the characteristics of these strains.

TABLE 7.1
Strains of *Diplococcus pneumoniae* used by Frederick Griffith in his original transformation experiments.

Serotype	Colony Morphology	Capsule	Virulence
II*R*	Rough	Absent	Avirulent
III*S*	Smooth	Present	Virulent

Griffith knew from the work of others that only living virulent cells would produce pneumonia in mice. If heat-killed virulent bacteria are injected into mice, no pneumonia results, just as living avirulent bacteria fail to produce the disease. Griffith's critical experiment (Figure 7.2) involved an injection into mice of living II*R* (avirulent) cells combined with heat-killed III*S* (virulent) cells. Neither cell type caused death in mice when injected alone. Griffith expected that the double injection would not kill the mice, but after five days all mice receiving double injections were dead. Analysis of the blood of the dead mice revealed large numbers of living type III*S* bacteria! As far as could be determined, these III*S* bacteria were identical to the III*S* strain from which the heat-killed cell preparation had been made. Control mice, injected only with living avirulent II*R* bacteria, did not develop pneumonia and remained healthy. This finding suggested strongly that the occurrence of the living III*S* bacteria in the dead mice was not caused by faulty technique or contamination, but that some interaction between the two types of injected bacteria had occurred.

*Note that this organism is now designated *streptococcus pnemonia*.

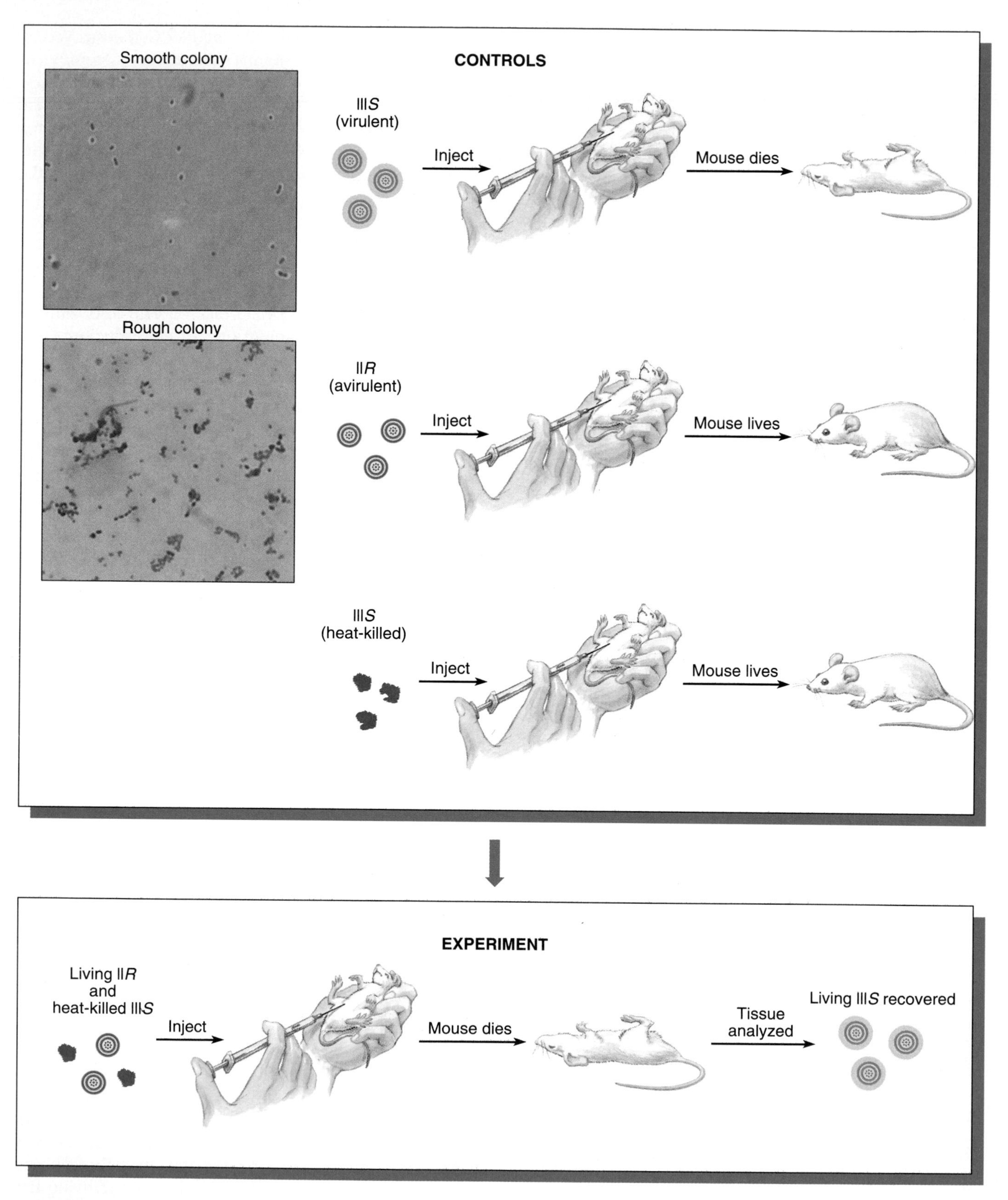

FIGURE 7.2
Summary of Griffith's transformation experiment. The photographs show bacterial colonies containing cells with capsules (type III*S*) and without capsules (type II*R*).

Griffith suggested that the heat-killed III*S* bacteria were responsible for converting live avirulent II*R* cells into virulent III*S* cells. Calling the phenomenon **transformation,** he suggested that the **transforming principle** might be some part of the polysaccharide capsule *or* some compound required for capsule synthesis, although the capsule alone did not cause pneumonia. To use Griffith's term, the transforming principle from the dead III*S* cells served as a "pabulum" for the II*R* cells.

Griffith's work led other physicians and bacteriologists to explore the phenomenon of transformation. By 1931, Henry Dawson and his coworkers showed that transformation could occur *in vitro* (in a test tube containing only bacterial cells); that is, injection into mice was not necessary for transformation to occur. By 1933, Lionel J. Alloway had refined the *in vitro* experiments by using extracts from *S* cells added to living *R* cells. The soluble filtrate from the heat-killed *S* cells was as effective in inducing transformation as were the intact cells! Alloway and others did not view transformation as a genetic event, but rather as a physiological modification of some sort. Nevertheless, the experimental evidence that a chemical substance was responsible for transformation was quite convincing.

Then, in 1944, after ten years of work, Avery, MacLeod, and McCarty published their results in what is now regarded as a classic paper in the field of molecular genetics. They reported that they had obtained the transforming principle in a highly purified state, and that beyond reasonable doubt it was DNA. The details of their work are outlined in Figure 7.3.

These researchers began their isolation procedure with large quantities (50–75 liters) of liquid cultures of type III*S* virulent cells. The cells were centrifuged, collected, and heat-killed. Following various chemical treatments a soluble filtrate was derived from these cells that still retains the ability to induce transformation of type II*R* avirulent cells.

Further testing established beyond doubt that the transforming principle was DNA. Treatment was first performed with **proteolytic** (protein-digesting) **enzymes** and an RNA-digesting enzyme, **ribonuclease.** Such treatment destroyed the activity of any remaining protein and RNA. Nevertheless, transforming activity was not diminished. The conclusion was drawn that neither protein nor RNA was responsible for transformation. The final confirmation came with experiments using crude samples of the DNA-digesting enzyme **deoxyribonuclease,** which was isolated from dog and rabbit sera. This digestion of the soluble filtrate was shown to destroy transforming activity. There was no doubt that the active transforming principle in these experiments was DNA!

The great amount of work, the confirmation and reconfirmation of the conclusions drawn, and the brilliant logic involved in the research of these three scientists are truly impressive. The conclusion of their 1944 publication, however, can be very simply stated: "The evidence presented supports the belief that a nucleic acid of the deoxyribose type is the fundamental unit of the transforming principle."

Avery and his coworkers recognized the genetic and biochemical implications of their work. They suggested that the transforming principle interacts with the II*R* cell and gives rise to a coordinated series of enzymatic reactions that culminates in the synthesis of the type III*S* capsular polysaccharide. They emphasized that, once transformation occurs, the capsular polysaccharide is produced in successive generations. Transformation is therefore heritable, and the process affects the genetic material.

Transformation has now been shown to occur in *Hemophilus influenzae, Bacillus subtilis, Shigella paradysenteriae,* and *Escherichia coli,* among many other microorganisms. Transformation of numerous genetic traits other than colony morphology has been demonstrated, including ones involving resistance to antibiotics and the ability to metabolize various nutrients. Thus, geneticists believed that most, if not all, traits could be transformed under suitable experimental conditions.

The Hershey-Chase Experiment

The second major piece of evidence supporting DNA as the genetic material was provided by the study of the bacterial virus T2. This virus, also called a **bacteriophage** or just a **phage,** has as its host the bacterium *Escherichia coli* and consists of a protein coat surrounding a core of DNA. The phage's external structure is composed of a hexagonal head plus a tail. The life cycle of bacteriophages such as T2 is shown in Figure 7.4.

In 1952, Alfred Hershey and Martha Chase published the results of experiments designed to clarify the events leading to phage reproduction. Several of the experiments clearly established the independent functions of phage protein and nucleic acid in the reproduction process associated with the bacterial cell. Hershey and Chase knew from existing data that:

1 T2 phages consist of approximately 50 percent protein and 50 percent DNA.

2 Infection is initiated by adsorption of the phage by its tail fibers to the bacterial cell.

3 The production of new viruses occurs within the bacterial cell.

It would appear that some molecular component of the phage, DNA and/or protein, enters the bacterial cell and directs viral reproduction. Which was it?

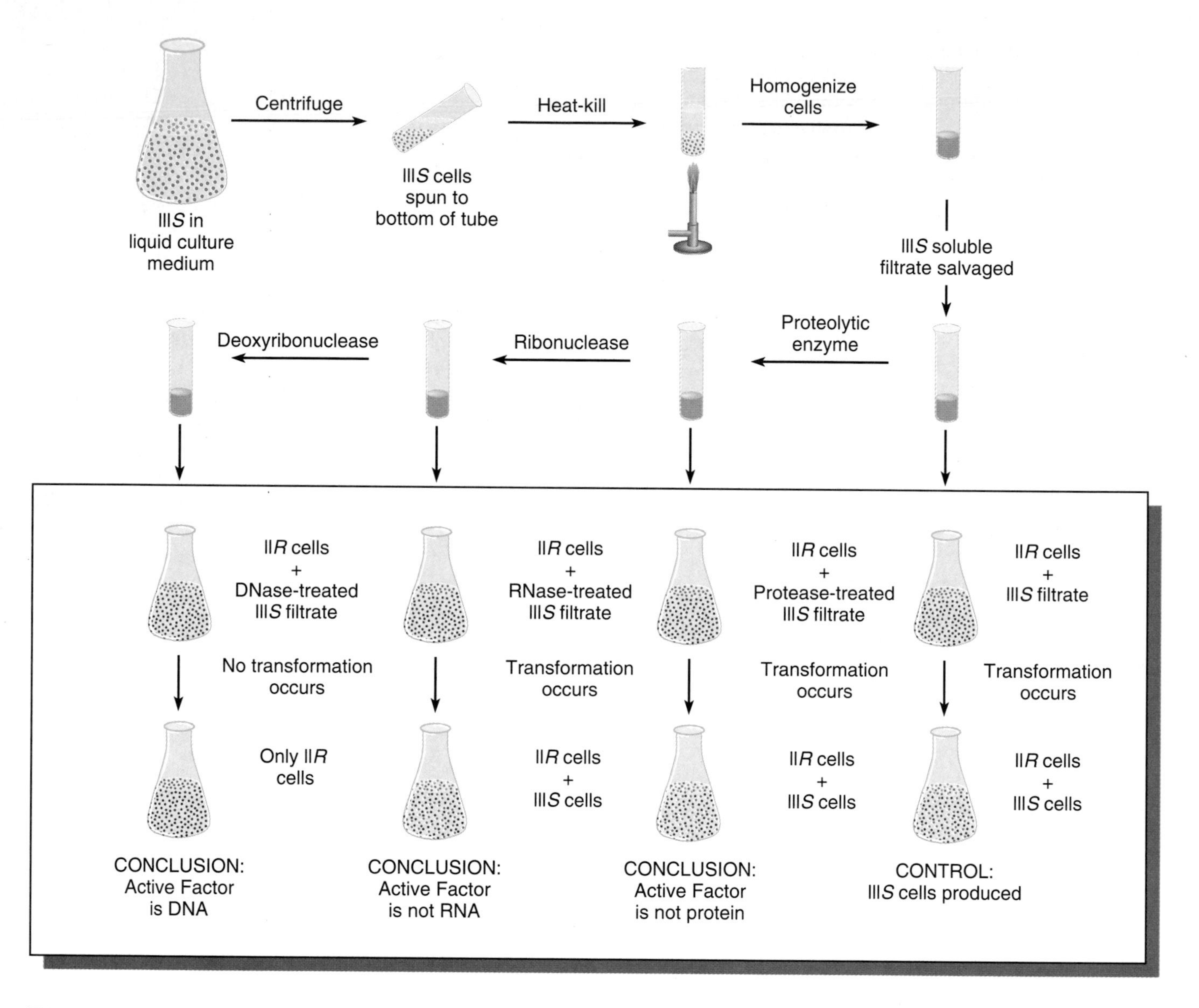

FIGURE 7.3
Summary of Avery, MacLeod, and McCarty's experiment demonstrating that DNA is the transforming principle.

Hershey and Chase used radioisotopes to follow the molecular components of phages during infection. Both ^{32}P and ^{35}S, radioactive forms of phosphorus and sulfur, were used. Since DNA contains phosphorus but not sulfur, ^{32}P effectively labels DNA. Since proteins contain sulfur but not phosphorus, ^{35}S labels protein. **This is a key point in the experiment.** If *E. coli* cells are first grown in the presence of ^{32}P **or** ^{35}S and then infected with T2 viruses, the progeny phage will have **either** a labeled DNA core **or** a labeled protein coat, respectively. These radioactive phages may be isolated and used to infect unlabeled bacteria (Figure 7.5).

When labeled phage and unlabeled bacteria are mixed, an adsorption complex is formed as the phages attach their tail fibers to the bacterial wall. These complexes were isolated and subjected to a high shear force by placing them in a blender. This force strips off the attached phages, which may then be analyzed separately (Figure 7.5). By tracing the radioisotopes, Hershey and Chase were able to demonstrate that most of the ^{32}P-labeled DNA had been transferred into the bacterial cell following adsorption; on the other hand, most of the ^{35}S-labeled protein remained outside the bacterial cell and was recovered in the phage "ghosts" (empty phage coats) after the blender treatment. Following this separation, the bacterial cells, which now contained viral DNA, were eventually lysed as new phages were produced.

Hershey and Chase interpreted these results to indicate that the protein of the phage coat remains outside the host cell and is not involved in the production of new phages. On the other hand, and most importantly, phage DNA enters the host cell and directs phage multiplication. Thus they had demonstrated that in phage T2, DNA, not protein, is the genetic material.

This experimental work, along with that of Avery and his colleagues, provided convincing evidence to most geneticists that DNA was the molecule responsible for heredity. Since then, many significant findings have been based on this conclusion. These many findings, constituting the field of molecular genetics.

Transfection Experiments

During the eight years following the publication of the Hershey–Chase experiment, additional research provided even more solid proof that DNA is the genetic material. These studies involved the same organisms used by Hershey and Chase.

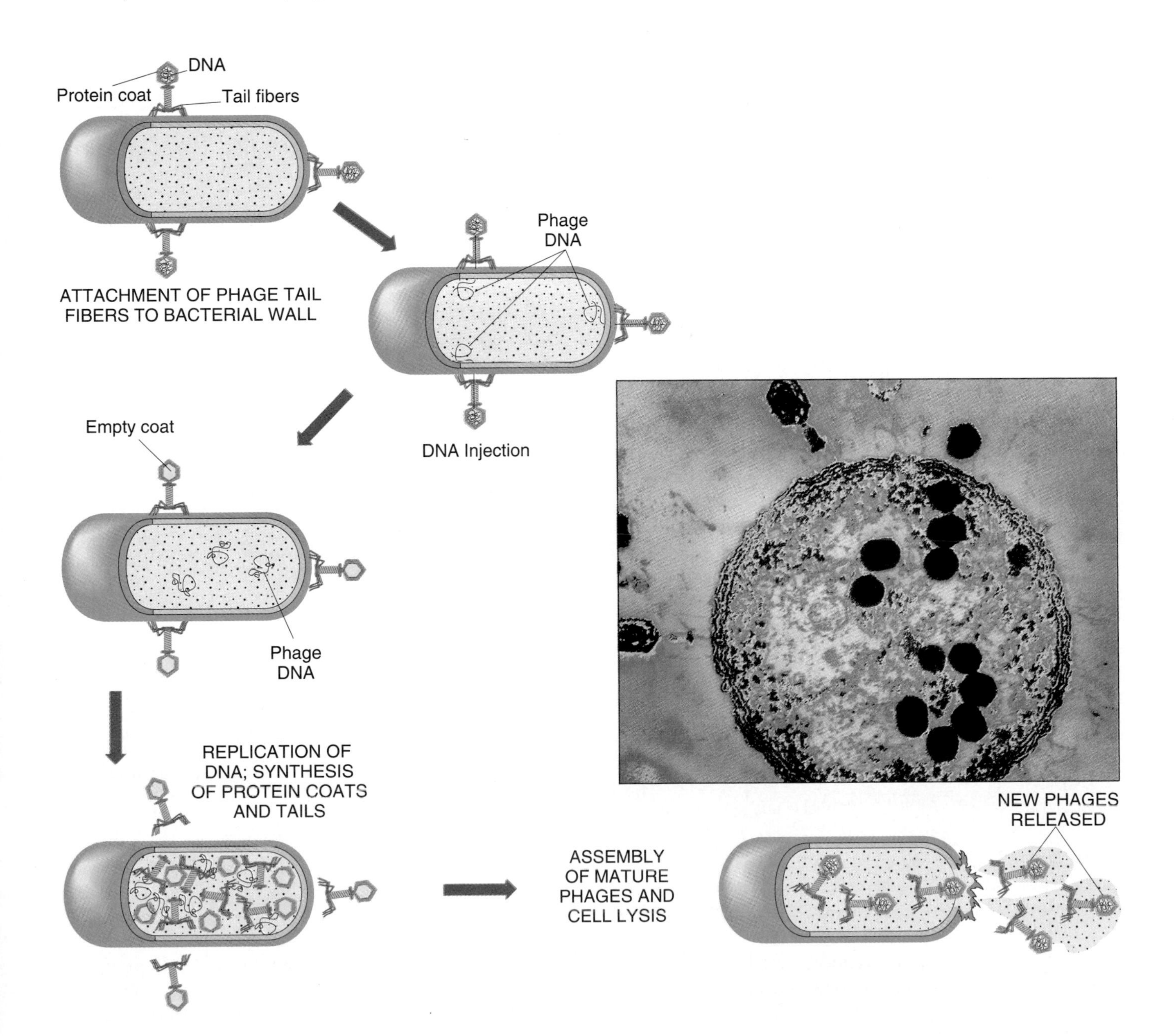

FIGURE 7.4
The life cycle of a T-even bacteriophage. The electron micrograph shows *E. coli* during infection by phage T2.

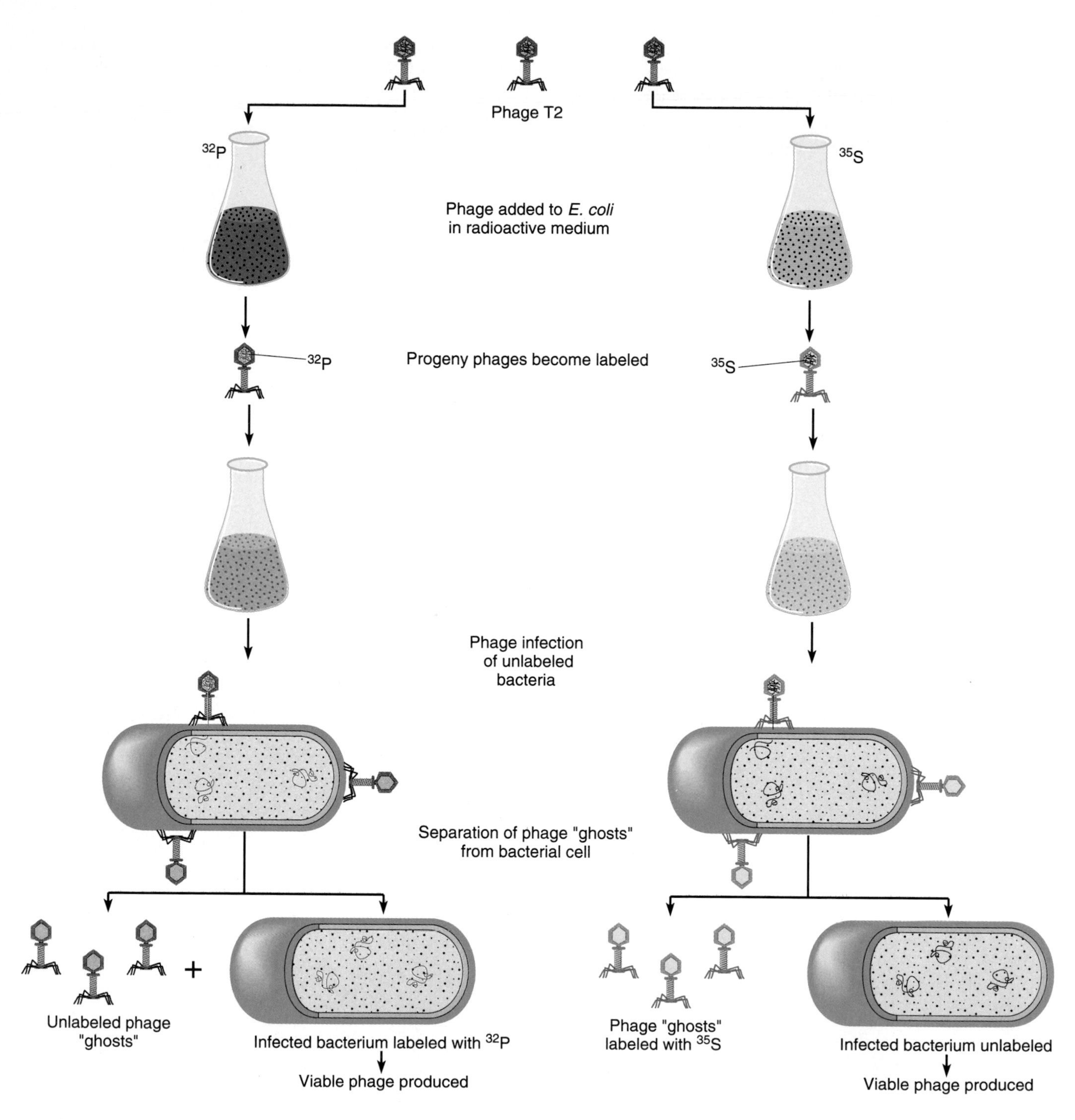

FIGURE 7.5
Summary of the Hershey-Chase experiment demonstrating that DNA and not protein is responsible for directing the reproduction of phage T2 during the infection of *E. coli.*

In 1957, several reports demonstrated that if *E. coli* were treated with the enzyme **lysozyme,** the outer wall of the cell could be removed without destroying the bacterium. Enzymatically treated cells are naked, so to speak, and contain only the cell membrane as the outer boundary of the cell. Such structures are called **protoplasts** (or **spheroplasts**). John Spizizen and Dean Fraser independently reported that by using protoplasts, they were able to initiate phage multiplication with disrupted T2 particles. That is, provided protoplasts are used, it is not necessary for a virus to be intact in order for infection to occur.

Similar but refined experiments were reported in 1960 using only the DNA purified from bacteriophage. This process of infection by only the viral nucleic acid, called **transfection,** proves conclusively that phage DNA alone contains all the necessary information for production of mature viruses. Thus, the evidence that DNA serves as the genetic material was further strengthened, even though all direct evidence had been obtained from bacterial and viral studies.

INDIRECT EVIDENCE FAVORING DNA IN EUKARYOTES

Most eukaryotic organisms are not amenable to the types of experiments that demonstrate DNA is the genetic material in bacteria and viruses. Therefore, support for this concept in eukaryotes was initially based only on circumstantial evidence. While such evidence is indirect, many diverse observations, all pointing to the same conclusion, together provided support that favored DNA. We shall look at several such observations.

Distribution of DNA

The genetic material should be found where it functions—in the nucleus as part of chromosomes. Both DNA and protein fit this criterion. However, protein is also abundant in the cytoplasm, while DNA is not. Both mitochondria and chloroplasts are known to perform genetic functions, and DNA is also present in these organelles. Thus, DNA is found only where primary genetic function is known to occur. Protein, however, is found in all parts of the cell. These observations are consistent with an interpretation favoring DNA over protein as the genetic material.

Since it is generally accepted that chromosomes within the nucleus contain the genetic material, a correlation should exist between the ploidy of cells and the amount of the molecule that functions as the genetic material. Meaningful comparisons may be made between gametes (sperm and eggs) and somatic or body cells. The latter are recognized as being diploid ($2n$) and containing twice the number of chromosomes as gametes, which are haploid (n).

Table 7.2 compares the amount of DNA found in haploid sperm and diploid reticulocytes (the nucleated precursors of red blood cells) from a variety of organisms. There is a close correlation between the amount of DNA and the number of sets of chromosomes. No such correlation can be observed between gametes and diploid cells for the other major classes of organic molecules, the proteins, lipids, and carbohydrates.

TABLE 7.2
DNA content of sperm and reticulocytes of various species (in picograms).

Species	Sperm (n)	Reticulocyte ($2n$)
Human	3.25	7.30
Chicken	1.26	2.49
Trout	2.67	5.79
Carp	1.65	3.49
Shad	0.91	1.97

Mutagenesis

Ultraviolet light (UV) is one of a number of agents capable of inducing mutations in the genetic material. Bacteria and other simple organisms may be irradiated with various wavelengths of ultraviolet light, and the effectiveness of each wavelength measured by the number of mutations it induces. When the data are plotted, an **action spectrum** of ultraviolet light as a mutagenic agent is obtained. This action spectrum may then be compared with the **absorption spectrum** of any molecule suspected to be the genetic material (Figure 7.6). The molecule serving as the genetic material is expected to absorb at the wavelengths found to be mutagenic.

UV light is most mutagenic at the wavelength (λ) of 260 nanometers (nm). Both DNA and RNA absorb UV light most strongly at 260 nm. On the other hand, protein absorbs most strongly at 280 nm, yet no significant muta-

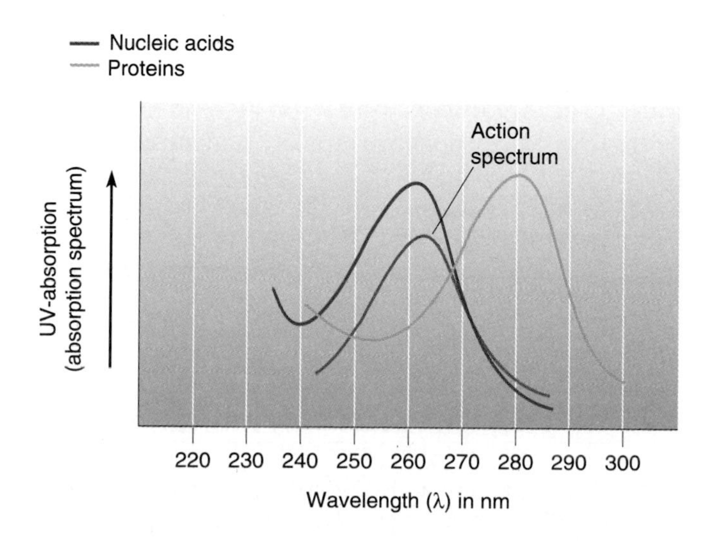

FIGURE 7.6
The absorption spectrum of nucleic acids and proteins when subjected to ultraviolet light compared to the action spectrum (the effectiveness of inducing mutation) of ultraviolet light.

genic effects are observed at this wavelength. Thus, this indirect evidence supports the idea that a nucleic acid is the genetic material and tends to exclude protein.

DIRECT EVIDENCE FAVORING DNA IN EUKARYOTES

While the circumstantial evidence just described does not constitute direct proof that DNA is the genetic material in eukaryotes, these observations spurred researchers to forge ahead under this assumption. Today, there is no doubt of the validity of this conclusion.

The strongest evidence that DNA is the genetic material is provided by a current experimental procedure called **recombinant DNA technology.** In this procedure, segments of eukaryotic DNA corresponding to specific genes are isolated and literally spliced into bacterial DNA. Such a complex can be inserted into a bacterial cell and its genetic expression monitored. If a eukaryotic gene is chosen that is foreign to the bacterial genetic information, the presence of the corresponding foreign protein product demonstrates directly that this DNA is functional in expressing genetic information. This has been shown to be the case in numerous instances. For example, the products of the human genes specifying the hormone insulin and the immunologically important molecule interferon are made by bacteria following recombinant DNA procedures.

Work in the laboratory of Beatrice Mintz and others has further strengthened this evidence. This research has demonstrated that DNA encoding the human β-globin gene, when microinjected into a fertilized mouse egg, is later found in the adult mouse tissue and can be transmitted to that mouse's progeny! Such mice are examples of **transgenic organisms.** More recent work has introduced DNA representing the growth hormone gene from rats into fertilized mouse eggs. About one-third of the resultant animals grew to twice the size of normal mice, indicating that the foreign DNA was present *and* functional. Subsequent generations receive the gene and display the trait that it governs.

RNA AS THE GENETIC MATERIAL

Some viruses contain an RNA core rather than one composed of DNA. In these viruses, it would thus appear that RNA must serve as the genetic material—an exception to the general rule that DNA performs this function. In 1956, it was demonstrated that when purified RNA from **tobacco mosaic virus (TMV)** was spread on tobacco leaves, the characteristic lesions caused by viral infection would appear later on the leaves. It was concluded that RNA is the genetic material of this virus.

In 1965 and 1966, Norman Pace and Sol Spiegelman demonstrated further that RNA from the phage Qβ could be isolated and replicated *in vitro.* Replication was dependent on an enzyme, **Qβ RNA replicase,** which was isolated from host *E. coli* cells following normal infection. When the RNA replicated *in vitro* was added to *E. coli* protoplasts, infection and viral multiplication (transfection) occurred. Thus, RNA synthesized in a test tube can amply serve as the genetic material in these phages.

Finally, one other group of RNA-containing viruses bears mentioning. These are the **retroviruses,** which replicate in an unusual way. Their RNA serves as a template for the synthesis of the complementary DNA molecule! The process, designated as reverse transcription, occurs under the direction of an RNA-dependent DNA polymerase enzyme called **reverse transcriptase.** Because the genetic material can be represented by this DNA intermediate, it may be incorporated into the genome of the host cell. Once present, if the DNA is expressed, transcription yields retroviral RNA chromosomes. Retroviruses include the human immunodeficiency virus (HIV) that causes AIDS and many RNA tumor viruses.

THE STRUCTURE OF DNA

Having established that DNA is the genetic material in most living things, we turn now to a consideration of the structure of this nucleic acid. In 1953, James Watson and Francis Crick proposed that the structure of DNA is in the form of a double helix. Their proposal was published in a short paper in *Nature.* In a sense, this publication constituted the finish line in a highly competitive scientific race to obtain what some consider to be the most significant finding in the history of biology. This "race," as recounted in Watson's book *The Double Helix,* demonstrates the human interaction, genius, frailty, and intensity involved in the scientific effort that eventually led to the elucidation of DNA structure.

The data available to Watson and Crick, crucial to the development of their proposal, came primarily from two sources: base composition analysis of hydrolyzed samples of DNA, and X-ray diffraction studies of DNA. The analytical success of Watson and Crick may be attributed to model building that conformed to the existing data. If the structure of DNA may be analogized by a puzzle, Watson and Crick, working in the Cavendish Laboratory in Cambridge, England, were the first to successfully put together all of the pieces.

Nucleic Acid Chemistry

Before turning to this work, a brief introduction to nucleic acid chemistry is in order. DNA is a nucleic acid, and **nucleotides** are the building blocks of all nucleic acid molecules. Sometimes called mononucleotides, these structural units consist of three essential components: a **nitrogenous base,** a **pentose sugar** (5 carbons), and a **phosphate group.** There are two kinds of nitrogenous bases: the nine-membered, double-ringed **purines** and the six-membered, single-ringed **pyrimidines.** Two types of purines and three types of pyrimidines are found commonly in nucleic acids. The two purines are **adenine** and **guanine,** abbreviated **A** and **G.** The three pyrimidines are **cytosine, thymine,** and **uracil,** abbreviated **C, T,** and **U.** The chemical structures of A, G, C, T, and U are shown in Figure 7.7(a). Both DNA and RNA contain A, C, and G; only DNA contains the base T, whereas only RNA contains the base U. Each nitrogen or carbon atom of the ring structures of purines and pyrimi-

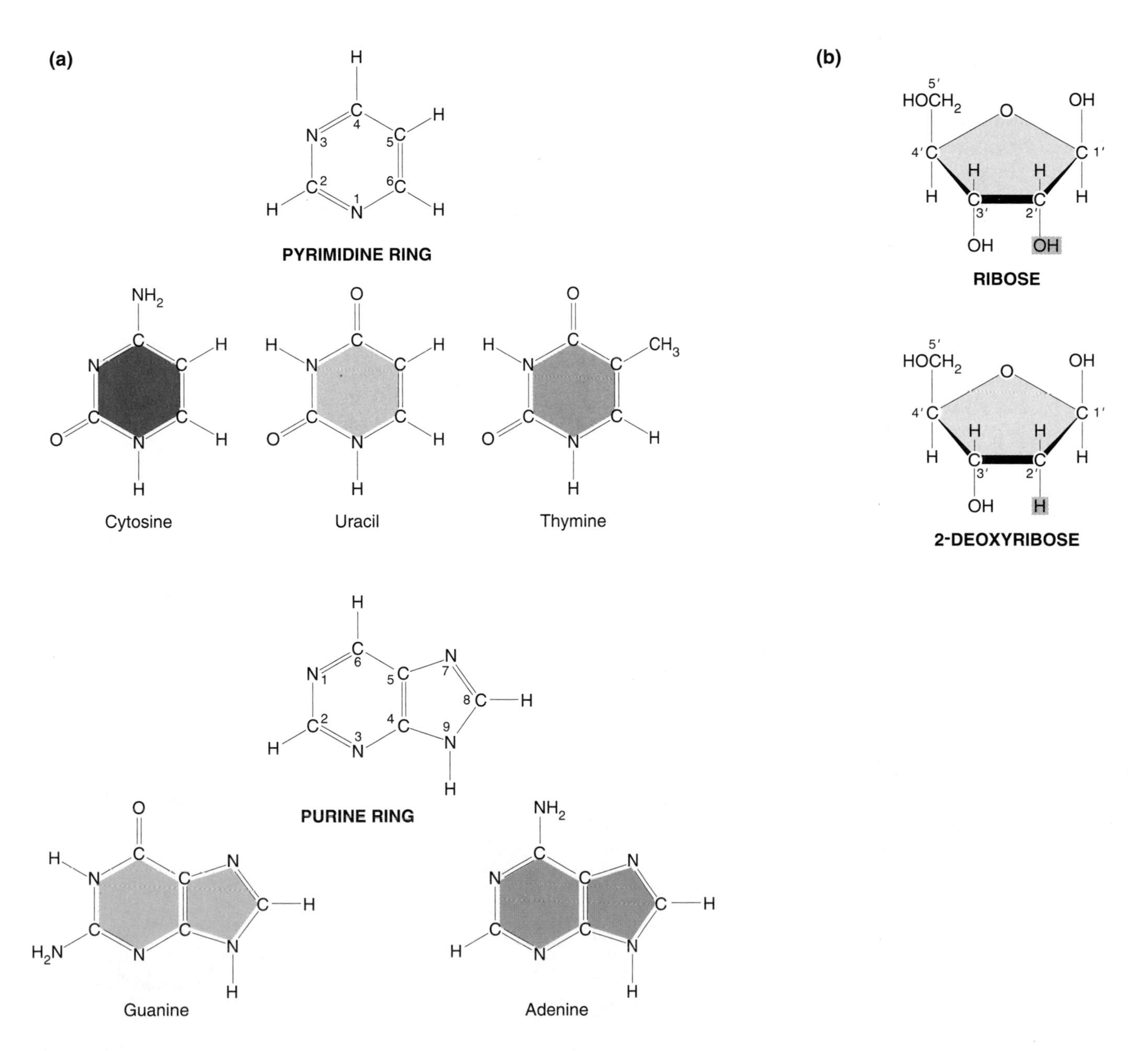

FIGURE 7.7
(a) Chemical structures of the pyrimidines and purines that serve as the nitrogenous bases in RNA and DNA. (b) Chemical ring structures of ribose and 2-deoxyribose, which serve as the pentose sugars in RNA and DNA, respectively.

dines is designated by a number without a prime sign. Note that corresponding atoms in the two rings are numbered differently in most cases.

The pentose sugars found in nucleic acids give them their names. **Ribonucleic acids (RNA)** contain **ribose,** while **deoxyribonucleic acids (DNA)** contain **deoxyribose.** Figure 7.7(b) shows the structures for these two pentose sugars. Each carbon atom is distinguished by a number with a prime sign (e.g., C-1′, C-2′, etc.). As you can see, deoxyribose has a hydrogen atom rather than a hydroxyl group at the C-2′ position compared with ribose. This lack of an oxygen atom is the only difference between the two sugars.

If a molecule is composed of a purine or pyrimidine base and a ribose or deoxyribose sugar, the chemical unit is called a **nucleoside.** If a phosphate group is added to the nucleoside, the molecule is now called a **nucleotide.** Nucleosides and nucleotides are named according to the specific nitrogenous base (A, T, G, C, or U) that is part of the molecule. The structure of a nucleotide and the nomenclature used in naming DNA nucleosides and nucleotides are shown in Figure 7.8.

The bonding between the three components of a nucleotide is highly specific. The C-1′ atom of the sugar is involved in the chemical linkage to the nitrogenous base. If the base is a purine, the N-9 atom is covalently bonded to the sugar. If the base is a pyrimidine, the bonding involves the N-1 atom. In a nucleotide, the phosphate group may be bonded to the C-2′, C-3′, or C-5′ atom of the sugar. The C-5′—phosphate configuration is shown in Figure 7.8. It is by far the most prevalent one in biological systems.

Mononucleotides may also be described by the term **nucleoside monophosphate (NMP).** The addition of one or two phosphate groups results in **nucleoside diphosphates (NDP)**and **triphosphates. (NTP)** The triphosphate form (Figure 7.9) is significant because it serves as the precursor molecule during nucleic acid synthesis within the cell. Additionally, the triphosphates **adenosine triphosphate (ATP)** and **guanosine triphosphate (GTP)** are important in the cell's bioenergetics because of the large amount of energy involved in the addition or removal of the terminal phosphate group. The hydrolysis of ATP or GTP to ADP or GDP and inorganic phosphate (P_i) is accompanied by the release of a large amount of energy in the cell. When these chemical conversions are coupled to energy-required reactions, the energy produced may be used to drive them. As a result, ATP and GTP are involved in many cellular activities.

The linkage between two mononucleotides consists of a phosphate group linked to two sugars. A **phosphodiester** bond is formed, because phosphoric acid has been joined to two alcohols (the hydroxyl groups on the two sugars) by an ester linkage on both sides. Figure 7.10

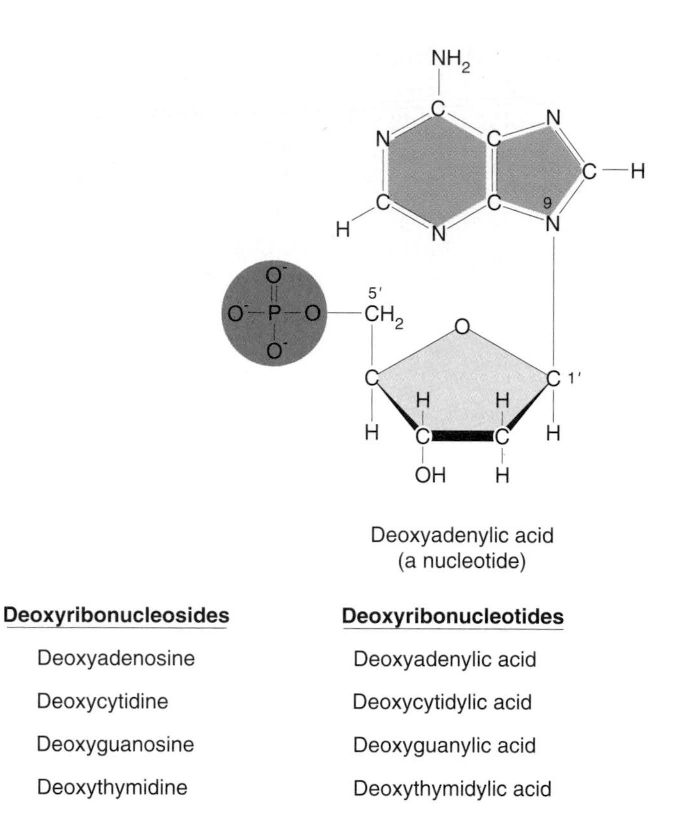

Deoxyribonucleosides	Deoxyribonucleotides
Deoxyadenosine	Deoxyadenylic acid
Deoxycytidine	Deoxycytidylic acid
Deoxyguanosine	Deoxyguanylic acid
Deoxythymidine	Deoxythymidylic acid

FIGURE 7.8
The structure of a nucleotide, deoxyadenylic acid. The nomenclature used in naming the nucleosides and nucleotides of DNA is also shown.

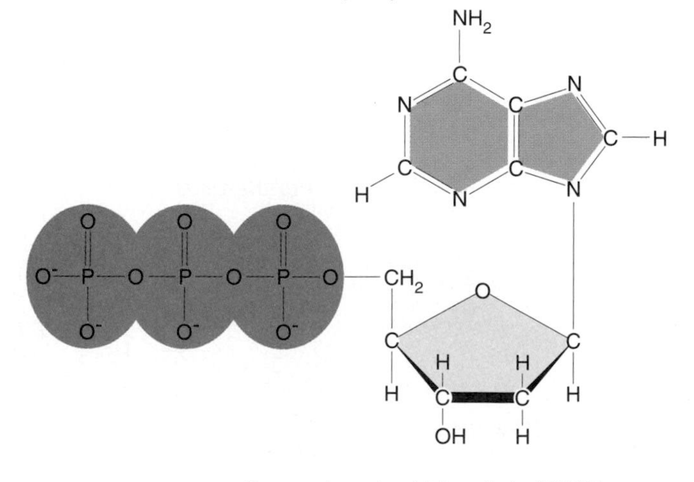

FIGURE 7.9
The basic structure of a nucleoside triphosphate, as illustrated by deoxyadenosine triphosphate.

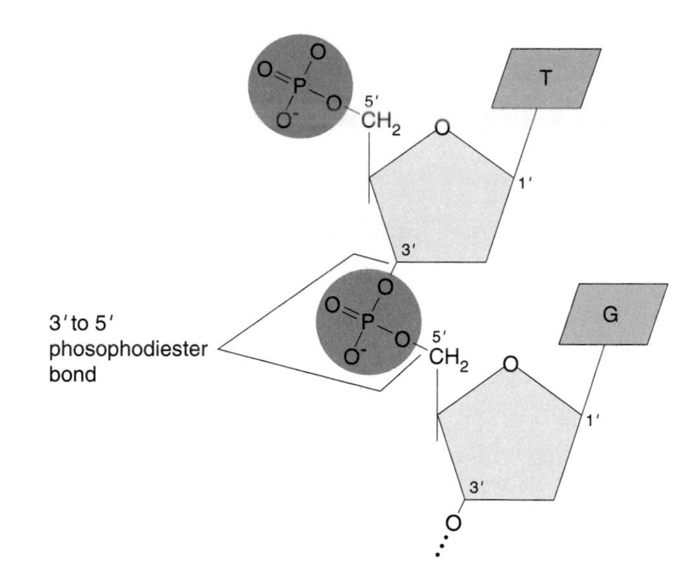

FIGURE 7.10
The linkage of two nucleotides by the formation of C-3′–C-5′ (3′–5′) phosphodiester bond, producing a dinucleotide.

shows the resultant phosphodiester bond in DNA. Each bond has a C-5′ end and a C-3′ end. The joining of two nucleotides forms a dinucleotide; of three nucleotides, a trinucleotide; and so forth. When long chains of nucleotides are formed, the structure is called a **polynucleotide.**

Long polynucleotide chains account for the observed molecular weight and explain the most important property of DNA—storing enough information to account for the enormous amount of genetic variation that we observe. If each nucleotide position in this long chain may be occupied by any one of four nucleotides, extraordinary variation is possible. For example, a polynucleotide that is only 1000 nucleotides in length may be arranged 4^{1000} different ways, each one different from all other possible sequences. This potential variation in molecular structure is essential if DNA is to serve the function of storing the vast amounts of chemical information necessary to direct cellular activities.

Base Composition Studies

Between 1949 and 1953, Erwin Chargaff and his colleagues used chromatographic methods to separate and quantitate nitrogenous bases in DNA samples from various organisms. Table 7.3(a) provides some of Chargaff's original data. Part (b) provides still other base composition data. As we shall see, interpretation of Chargaff's data was critical to the successful model of DNA put forward by Watson and Crick.

On the basis of these data, the following conclusions may be drawn:

TABLE 7.3
DNA base composition data.
(a) Chargaff's data.

Source	Approximate Amount*			
	A	T	G	C
Ox thymus	26	25	21	16
Ox spleen	25	24	20	15
Yeast	24	25	14	13
Avian tubercle bacilli	12	11	28	26
Human sperm	29	31	18	18

SOURCE: From Chargaff, 1950.
*Moles of nitrogenous constituent per mole of P (often, the recovery was less than 100 percent).

(b) Base compositions of DNAs from various sources.

Source	Base Composition and Base Ratios					
	A	T	G	C	A/T	G/C
Human	30.9	29.4	19.9	19.8	1.05	1.00
Sea urchin	32.8	32.1	17.7	17.3	1.02	1.02
E coli	24.7	23.6	26.0	25.7	1.04	1.01
Sarcina lutea	13.4	12.4	37.1	37.1	1.08	1.00
T 7 phage	26.0	26.0	24.0	24.0	1.00	1.00

1 The number of adenine residues approximately equals the number of thymine residues in the DNA of any species. Also, the number of guanine residues is approximately equivalent to the number of cytosine residues.

2 The sum of the purines (A + G) approximately equals the sum of the pyrimidines (C + T).

3 The percentage of G + C does not necessarily equal the percentage of A + T. As we can see, this ratio varies greatly among species [Table 7.3(b)].

These conclusions indicate a definite pattern of base composition of DNA molecules. These data served as the initial clue to "the puzzle." Additionally, they directly refute the tetranucleotide hypothesis, which stated that all four bases are present in equal amounts.

X-Ray Diffraction Analysis

When fibers of a DNA molecule are subjected to X-ray bombardment, these rays are scattered according to the molecule's atomic structure. The pattern of scatter may be captured as spots on photographic film and analyzed, particularly for the overall shape of and regularities within the molecule. This process, X-ray diffraction analysis, was successfully applied to the study of protein structure by Linus Pauling and other chemists. The technique had been attempted on DNA as early as 1938 by William Astbury. By 1947, he had detected a periodicity within the structure of the molecule of 3.4 Å that suggested to him that the bases were stacked like coins on top of one another.

Between 1950 and 1953, Rosalind Franklin, working in the laboratory of Maurice Wilkins, obtained improved X-ray data from more purified samples of DNA (Figure 7.11). Her work confirmed the 3.4 Å periodicity seen by Astbury and suggested that the structure of DNA was some sort of helix. However, she did not propose a definitive model. Pauling had analyzed the work of Astbury and others and incorrectly proposed that DNA was a triple helix.

The Watson–Crick Model

Watson and Crick published their analysis of DNA structure in 1953. By building models under the constraints of the information just discussed, they proposed the double-helical form of DNA as shown in Figure 7.12. This model has the following major features:

1 Two right-handed helical polynucleotide chains are coiled around a central axis.

2 The double helix measures 20 Å (2.0 nm) in diameter.

3 Each complete turn of the helix is 34 Å (3.4 nm) long; thus, 10 bases exist in each chain per turn.

4 In any segment of the molecule, alternating larger **major grooves** and smaller **minor grooves** are apparent along the axis.

5 The two chains are **antiparallel;** that is, their C-5′-to-C-3′ orientations run in opposite directions.

6 The bases of both chains are flat structures, lying perpendicular to the axis; they are "stacked" on one another, 3.4 Å (0.34 nm) apart and are located on the inside of the structure.

7 The nitrogenous bases of opposite chains are paired to one another as the result of the formation of **hydrogen bonds** (described in the following discussion); specifically, only A-T and G-C pairs are allowed.

A more recent and accurate analysis of the form of DNA that served as the basis for the Watson–Crick model has revealed a minor structural difference. A precise measurement of the number of base pairs (bp) per turn has demonstrated a value of 10.4 rather than the 10.0 predicted by Watson and Crick.

One of the most significant features of the structure proposed by Watson and Crick is the specificity of base pairing. Chargaff's data had suggested that the amounts of A equaled T and that G equaled C. Watson and Crick realized that when placed opposite each other in the model, the members of each such base pair form hydrogen bonds, providing the chemical stability necessary to hold the two chains together. Arranged in this way, both major

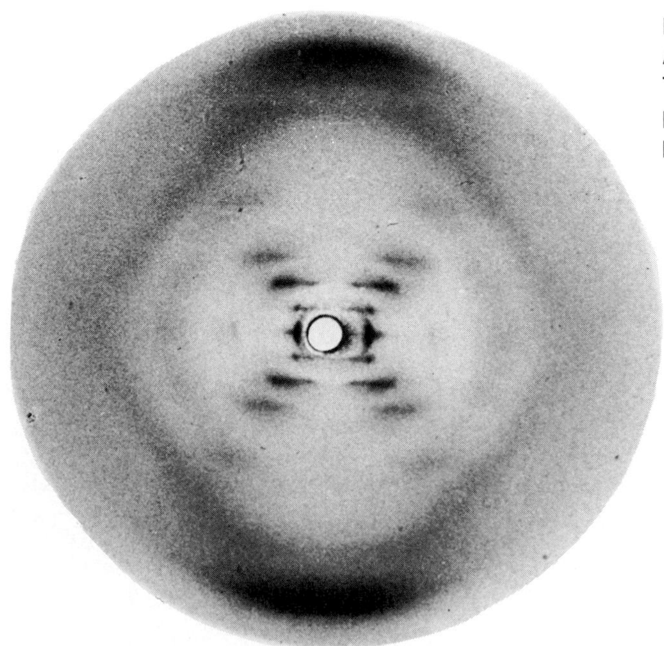

FIGURE 7.11
An X-ray diffraction photograph of the B form of crystallized DNA. The dark patterns at the top and bottom provide an estimate of the periodicity of nitrogenous bases, which are 3.4 Å apart. The central pattern is indicative of the molecule's helical structure.

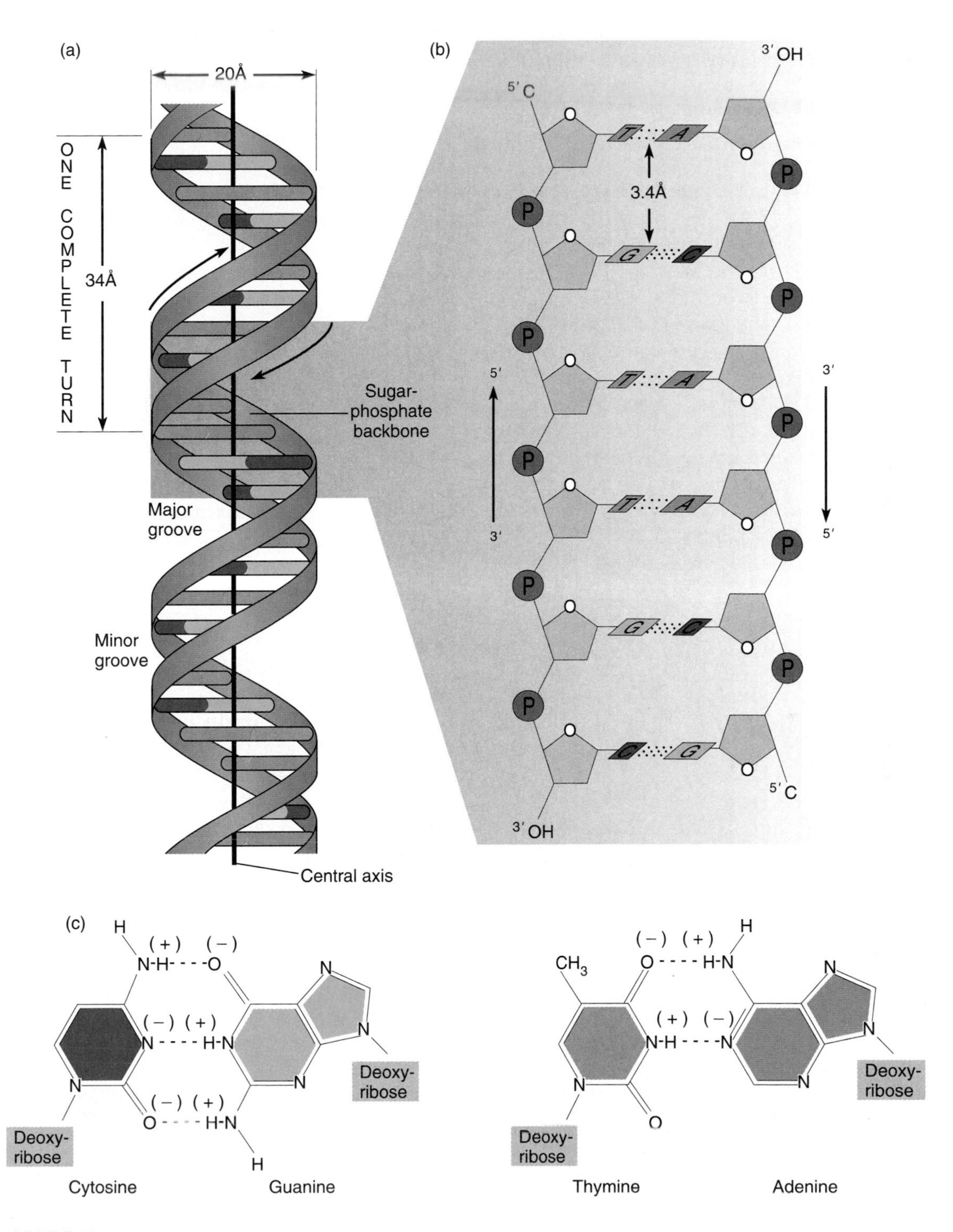

FIGURE 7.12
(a) A schematic representation of the DNA double helix as proposed by Watson and Crick. The ribbonlike strands constitute the sugar-phosphate backbones, and the horizontal rungs constitute the nitrogenous base pairs, of which there are 10 per complete turn. The major and minor grooves are apparent. A solid vertical line representing the central axis has been placed through the center of the helix. (b) A representation of the antiparallel nature of the two strands of the helix. (c) The hydrogen bonds formed between cytosine and guanine and between thymine and adenine.

and minor grooves become apparent along the axis. Further, with one purine (A or G) opposite one pyrimidine (T or C) as each "rung of the spiral staircase" of the proposed helix, the Watson-Crick model conforms to the 20 Å (2 nm) diameter suggested by X-ray diffraction studies.

The specific A-T and G-C base pairing is the basis for the important genetic concept of **complementarity.** This term is used to describe the chemical affinity provided by the hydrogen bonds between the bases. As we will see, this concept is vital to our understanding of how DNA replicates and how gene expression occurs.

It is appropriate to inquire into the nature of a hydrogen bond, and whether is it strong enough to stabilize the helix. A hydrogen bond is a very weak electrostatic attraction between a covalently bonded hydrogen atom and an atom with an unshared electron pair. The hydrogen atom assumes a partial positive charge, while the unshared electron pair—characteristic of covalently bonded oxygen and nitrogen atoms—assumes a partial negative charge. These opposite charges are responsible for the weak chemical attractions. As oriented in the double helix, adenine forms two hydrogen bonds with thymine, and guanine forms three hydrogen bonds with cytosine. Although two or three hydrogen bonds taken alone are very weak, two or three thousand bonds in tandem (which would be found in two long polynucleotide chains) are capable of providing great stability to the helix.

Still another stabilizing factor is the arrangement of sugars and bases along the axis. In the Watson–Crick model, the hydrophobic or "water-fearing" nitrogenous bases are stacked almost horizontally on the interior of the axis, thus shielded from water. The hydrophilic sugar-phosphate backbone is on the outside of the axis, where both components may interact with water. These molecular arrangements provide significant chemical stabilization to the helix.

Point 5 in the model requires special emphasis. Antiparallelity means that the two chains run in opposite directions. One chain is in the 5′-to-3′ orientation while the other chain is in the 3′-to-5′ orientation. Given the constraints of bond angles, a double helix of the nature described by Watson and Crick could not easily be constructed if both chains were in the same orientation.

The Watson–Crick model had an immediate effect on the emerging discipline of molecular biology. Even in their initial 1953 article, the authors noted, "It has not escaped our notice that the specific pairing we have postulated immediately suggests a possible copying mechanism for the genetic material." Two months later, in a second article in *Nature,* Watson and Crick pursued this idea, suggesting a specific mode of replication of DNA—the semiconservative model. The second article also alluded to two new concepts: the storage of genetic information in the sequence of the bases, and the mutation or genetic change that would result from alteration of the bases. These ideas have received vast amounts of experimental support since 1953 and are now universally accepted. Thus, the "synthesis" of ideas by Watson and Crick was a remarkable feat and highly significant in the history of genetics and biology. For their work, they, along with Wilkins, received the Nobel Prize in Physiology and Medicine in 1962. This was to be one of many such awards bestowed for work in the field of molecular genetics.

OTHER FORMS OF DNA

Under different conditions of isolation, several conformational forms of DNA have been recognized. At the time Watson and Crick performed their analysis, two forms—**A-DNA** and **B-DNA**—were known. Watson and Crick's analysis was based on X-ray studies of the B form by Franklin, which is present under aqueous, low-salt conditions and is believed to be the biologically significant conformation.

While DNA studies around 1950 relied on the use of X-ray diffraction, more recent investigations have been performed using **single crystal X-ray analysis.** The earlier studies achieved limited resolution of about 5 Å, but single crystals diffract X-rays at about 1 Å, near atomic resolution. As a result, every atom is "visible," and much greater structural detail is available during analysis.

Using these modern techniques, the A form of DNA has now been scrutinized. A-DNA is prevalent under high-salt or dehydration conditions. In comparison to B-DNA (Figure 7.13), A-DNA is slightly more compact, with 11 base pairs in each complete turn of the helix, which is 23 Å in diameter. While it is also a right-handed helix, the orientation of the bases is somewhat different. They are tilted and displaced laterally in relation to the axis of the helix. As a result of these differences, the appearance of the major and minor grooves is modified compared with those in B-DNA. It seems doubtful that A-DNA occurs under biological conditions.

Still another form of DNA, called **Z-DNA,** was discovered by Andrew Wang, Alexander Rich, and their colleagues in 1979 when they examined a small synthetic DNA fragment containing only C-G base pairs. Z-DNA takes on the rather remarkable configuration of a left-handed double helix (Figure 7.13). Like A- and B-DNA, Z-DNA consists of two antiparallel chains held together by Watson–Crick base pairs. Beyond these characteristics, Z-DNA is quite different. The left-handed helix is 18 Å (1.8 nm) in diameter, contains 12 base pairs per turn, and assumes a zigzag conformation (hence its name). The

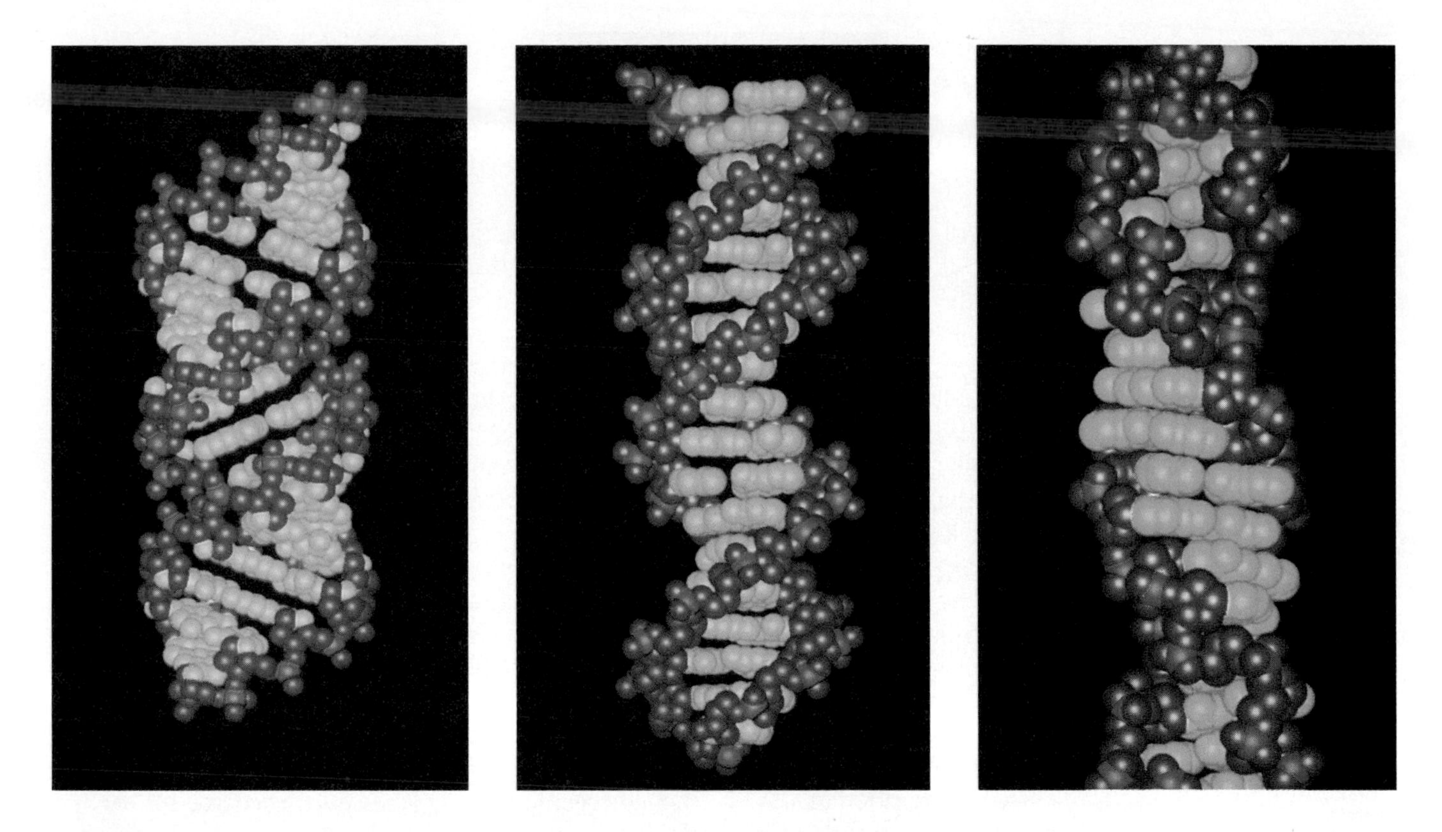

FIGURE 7.13
Space-filling models comparing the structure of the helix of A-DNA (left) and B-DNA (center). Both are right-handed helices. On the right is a space-filling model of the Z form of DNA, which is a left-handed helix. Note the near elimination of the grooves in this molecule compared with B-DNA.

major groove present in B-DNA is nearly eliminated in Z-DNA.

Speculation has abounded over the possibility that regions of Z-DNA exist in the chromosomes of living organisms. The unique helical arrangement could provide an important recognition point for the interaction with other molecules. However, it is still not clear whether Z-DNA occurs *in vivo*.

THE STRUCTURE OF RNA

The second category of nucleic acids, introduced at the beginning of this chapter, is the ribonucleic acids, or RNA. The structure of these molecules is similar to DNA, with several important exceptions. While RNA also has as its building blocks nucleotides linked into polynucleotide chains, ribose replaces deoxyribose and uracil replaces thymine. Another important difference is that most RNA is thought of as being single stranded. However, RNA molecules sometimes fold back on themselves following their synthesis; such a configuration results when regions of complementarity occur in positions that allow stretches of base pairs to form. Furthermore, some animal viruses that have RNA as their genetic material contain it in the form of double-stranded helices. Thus, there are several instances where RNA does not exist strictly as a linear, single-stranded molecule.

At least three classes of cellular RNA molecules function during the expression of genetic information: **ribosomal RNA (rRNA), messenger RNA (mRNA),** and **transfer RNA (tRNA).** These molecules all originate as complementary copies of one of the two strands of DNA segments during the process of transcription. That is, their nucleotide sequence is complementary to the deoxyribonucleotide sequence of DNA, which serves as the template for their synthesis. Uracil replaces thymine in RNA and is complementary to adenine during transcription.

Each class of RNA may be characterized by its size, sedimentation behavior when centrifuged, and genetic function. Sedimentation behavior depends on a molecule's density, mass, and shape, and its measure is called the **Svedberg coefficient (*S*).** Table 7.4 relates the *S* values, molecular weights, and approximate number of nucleotides of the major forms of RNA. As you can see, there is a wide variation in size of the three classes of RNA.

SCREENING YOUR DNA BLUEPRINT

The genetic blueprint of an individual is determined by the unique set of genes we each carry. Within the next 15 years scientists expect to decode the DNA and base pair sequence of our 50,000 to 100,000 genes. As a result it will be possible to screen individuals for thousands of different genetic traits, including genetic disorders and predispositions to diseases that are not exclusively due to inheritance. Genetic screening coupled with cures or treatment will have enormous medical benefits; however, it must be kept in mind that the results of genetic tests could also lead to discrimination based entirely on individual genotype.

Genetic screening for a limited number of disorders has been carried out in North America for several years. The following contrasting examples indicate the benefits and problems encountered in genetic screening programs. Tay-Sachs disease is an untreatable genetic disease that results in death in early childhood. Individuals who carry one copy of the deleterious gene (carriers) are unaffected, but have a one in four chance of having a child with the disorder if they marry another carrier. This genetic disorder has a relatively high frequency in the Ashkenazi Jewish population and, accordingly, screening programs were set up in this population to detect carriers. The programs were voluntary and were coupled with education and genetic counseling. Reproduction choices, including prenatal diagnosis with the option of abortion, were explained to couples at risk. The success of this program is indicated by a dramatic decrease in the incidence of this disease; the frequency has dropped by at least 80 percent.

In contrast, a program to screen North American blacks for sickle cell trait backfired. These individuals are carriers (heterozygotes) who are not affected, but had a risk of having a child with sickle cell disease, a severe and often fatal blood disorder. In this case, though, the carriers were often mistakenly labeled as having the disease. As a result, many carriers suffered employment, educational, and insurance discrimination. The program, mandatory in some American states, caused significant damage to many people and has since been discontinued. It should serve to remind us of the importance of community education in screening programs.

As knowledge of the human genome expands, the scope of genetic testing will increase. Social and ethical problems will also increase. Individuals who are discovered to be heterozygous for deleterious genes, and it is estimated that on average we all carry four or five deleterious genes, could face discrimination. Many insurance companies believe that genetic screening could be used to provide an accurate assessment of disease risk. In other instances, genetic screening may be used to determine an individual's suitability for employment. Companies may decide that employees with one gene combination or another will do better on the job.

Viewed at a naive level, the argument in favor of aggressive action based on genetic tests often seems to make social and economic sense. However, it must be kept in mind that many of our traits, including behavioral disorders and predispositions to certain medical problems, are determined by complex interactions between genes and the environment. Genetic screening may reveal susceptibility to a problem, but predictions are far from certainties in the great majority of cases. It must be remembered, too, that some of the most outrageous of Nazi programs were based upon unfounded confidence in the validity of their views on heredity. Many geneticists and lawyers agree that legislation is needed to regulate the collection, use, and potential abuse of genetic test results. The areas of confidentiality, informed consent, and insurance and employment controls must also be considered. Such an approach reflects the growing public awareness of social and ethical problems that are being created by the rapid advances in genetic technology.

TABLE 7.4
Sedimentation coefficients, molecular weights, and number of nucleotides for various RNAs.

RNA Type	Abbreviation	Sedimentation Coefficient	Molecular Weight	Number of Nucleotides
Ribosomal RNA	rRNA	5*S*	35,000	105
		18*S*	700,000	1740
		28*S*	1,800,000	4850
Transfer RNA	tRNA	4*S*	23,000–30,000	75–90
Messenger RNA	mRNA	6–50*S*	25,000–1,000,000	100–10,000

Ribosomal RNA is the largest of these molecules and usually constitutes about 80 percent of all RNA in the cell. The various forms of rRNA found in prokaryotes and eukaryotes differ distinctly in size. These molecules constitute an important structural component of **ribosomes,** which function in the synthesis of proteins, the process called **translation.**

Messenger RNA molecules carry genetic information from the DNA of the gene to the ribosome, where translation into protein occurs. They vary considerably in length, which is partially a reflection of the variation in the size of the gene serving as the template for transcription of mRNA species.

Transfer RNA, the smallest class of RNA molecules, carries amino acids to the ribosome during translation. More than one tRNA molecule interacts simultaneously with the ribosome, and the molecule's smaller size facilitates these interactions.

HYDROGEN BONDS AND THE ANALYSIS OF NUCLEIC ACIDS

The unique nature of the hydrogen bond imparts an interesting and important set of qualities to the chemical behavior of nucleic acids under both laboratory and physiological conditions. For example, if DNA is isolated and subjected to slow heating, the double helix is denatured and unwinds. If a mixture of single strands complementary to each other are slowly cooled, they will reassociate, reforming the helix. In the laboratory, these transformations may be "tracked" by monitoring the absorption of UV light (or optical density) at 260 nm (OD_{260}), using a spectrophotometer.

During unwinding, the viscosity of DNA decreases and UV absorption increases. A melting profile, where OD_{260} is plotted against temperature, is shown for two DNA molecules in Figure 7.14. The midpoint of each curve is called the **melting temperature (T_m),** where 50 percent of the strands are unwound. The molecule with a higher T_m has a higher percentage of G ≡ C base pairs than A=T base pairs compared to the molecule with the lower T_m, since G ≡ C pairs share three hydrogen bonds compared to two present between A=T pairs.

Molecular Hybridization Techniques

The property of denaturation/renaturation of nucleic acids is the basis for one of the most powerful and useful techniques in molecular genetics—**molecular hybridization**. Provided that a reasonable degree of base complementarity exists and under the proper temperature conditions, two nucleic acid strands from different sources will rejoin, as described above. As a result, mo-

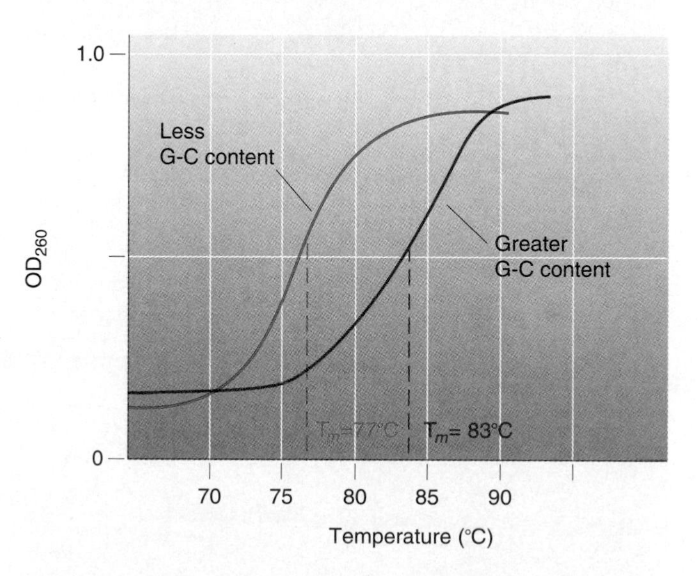

FIGURE 7.14
Comparison of the increase in UV absorbance with an increase in temperature for two DNA molecules with differing G-C contents. The molecule with a melting point (T_m) of 83°C has a greater G-C content than the molecule with a T_m of 77°C.

lecular hybridization is possible between DNA strands from different species and between DNA and RNA strands. For example, an RNA molecule will hybridize with the segment of DNA from which it was transcribed or with a DNA molecule from a different species, providing its nucleotide sequence is nearly the same.

The technique can even be performed using the DNA present in cytological preparations as the "target" for hybrid formation. This process is called ***in situ* molecular hybridization.** Mitotic or interphase cells are first fixed to slides and then subjected to hybridization conditions. Labeled single-stranded DNA or RNA is added, and hybridization is monitored. The nucleic acid that is added may be either radioactive or contain a fluorescent label to allow its detection. In the former case, the technique of autoradiography may be used. In Figure 7.15 the use of fluorescence is illustrated. A "probe," consisting of a short fragment of DNA complementary to DNA present in the centromere regions of chromosomes, has been hybridized. Fluorescence occurs only in the centromere regions, thus identifying each one along its chromosome. The use of this technique in identification of chromosomal locations housing other specific genetic information has been a valuable addition to the repertoire of experimental geneticists.

Reassociation Kinetics and Repetitive DNA

One extension of molecular hybridization procedures is the analysis of the *rate* of reassociation of complementary single strands of DNA. This technique, called **reassociation kinetics,** was first refined and studied by Roy Brit-

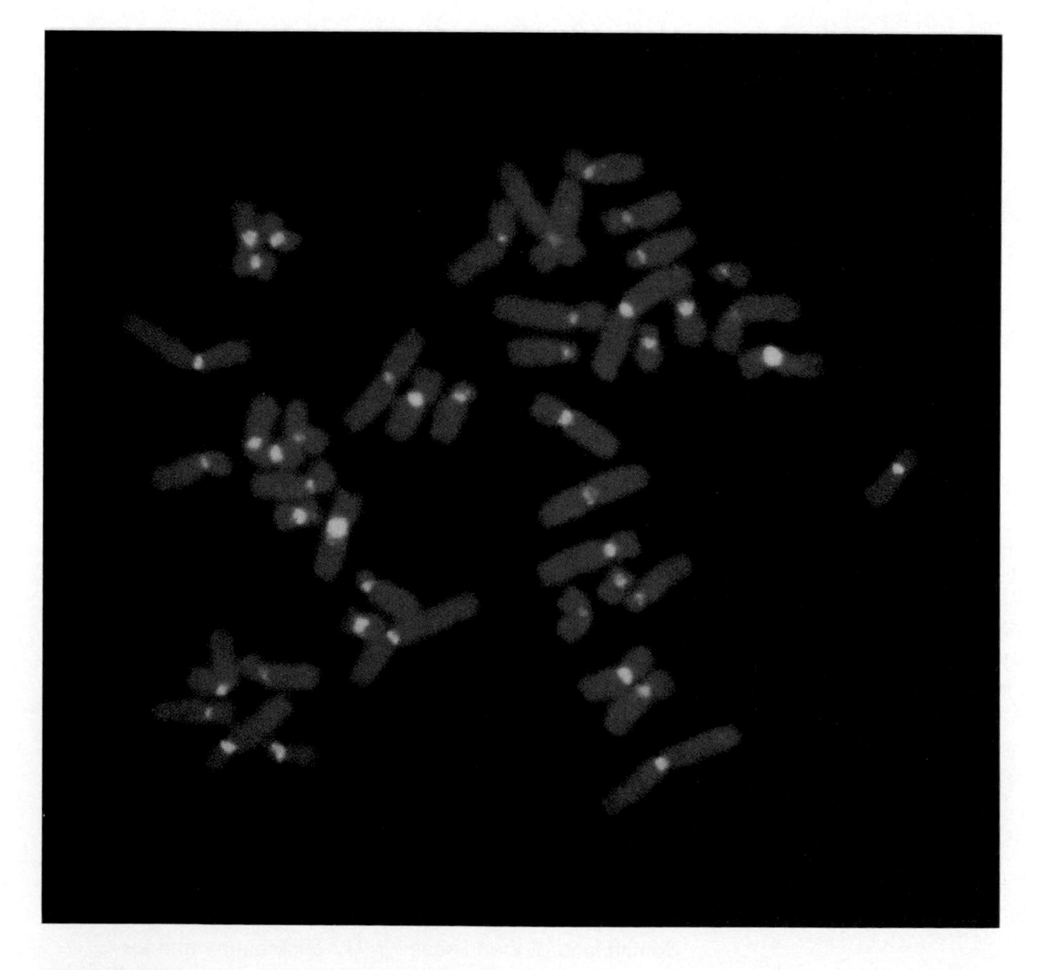

FIGURE 7.15
In situ hybridization of human metaphase chromosomes using a DNA probe. The probe, specific to centromeric DNA, produces a yellow fluorescence signal indicating hybridization. The red fluorescence is produced by counterstaining chromosomal DNA.

ten and David Kohne. As we shall see, many insights may be derived from such analysis.

The DNA used in such studies is first fragmented into small pieces several hundred base pairs long. After the DNA has been dissociated into single strands by heating, the temperature is lowered, and reassociation is monitored. During reassociation, pieces of single-stranded DNA collide randomly. If they are complementary, a stable double strand is formed; if not, they are free to encounter other DNA fragments. The process continues until all possible matches are made.

The results of such an experiment are presented in Figure 7.16. The percentage of reassociation of DNA fragments is plotted against a logarithmic scale of normalized time, a function referred to as $\mathbf{C_0t}$. In this term, $\mathbf{C_0}$ is equal to the initial DNA concentration and $\mathbf{t}$ is equal to time in seconds.

A great deal of information can be obtained from studies comparing the reassociation of DNA of different organisms. For example, we may compare the point in the reaction when one-half of the DNA is present as double-stranded fragments. This point is called the **half reaction time.** Provided that all DNA fragments contain unique nucleotide sequences, and all are about the same size, $\mathbf{C_0t_{1/2}}$ varies directly with the total length of the DNA.

In Figure 7.16 DNAs from three phage or bacterial sources are compared, each with a different genome size. As genome size increases, the curves obtained have a similar shape but are shifted farther and farther to the right, indicative of an extended reassociation time. Reassociation occurs more slowly in larger genomes because it takes longer for initial matches to be made if there are greater numbers of unique DNA fragments. This is so because collisions are random; more sequences present will result in greater numbers of mismatches before each correct match is made.

When reassociation kinetics of DNA from eukaryotic organisms with much larger genome sizes were first studied, a surprising observation was made. The data showed that *some* of the DNA segments reassociate even more rapidly than those derived from *E. coli*! The remainder of the DNA, as expected because of its greater size and complexity, takes longer to reassociate.

For example, Britten and Kohne examined DNA derived from calf thymus tissue (Figure 7.17). Based on these observations, they hypothesized correctly that the rapidly reassociating fraction might represent repetitive sequences present many times in the calf genome. This interpretation would explain why these DNA segments reassociate so rapidly. Multiple copies of the same sequence are much more likely to make initial matches, thus reassociating more quickly than single copies. On the other hand, the remaining DNA segments consist of unique nucleotide sequences present only once in the genome; because there are many more of these unique sequences in calf thymus compared with *E. coli,* their reassociation takes longer.

The copies present many times in the genome are collectively referred to as **repetitive DNA.** Repetitive DNA is characteristic of eukaryotic genomes and is important to our understanding of how genetic information is organized in chromosomes. Careful study has shown that there are two major categories of repetitive DNA in eukaryotes (Figure 7.18).

The first category reassociates very rapidly (much more rapidly than *E. coli* DNA) and is called **highly re-**

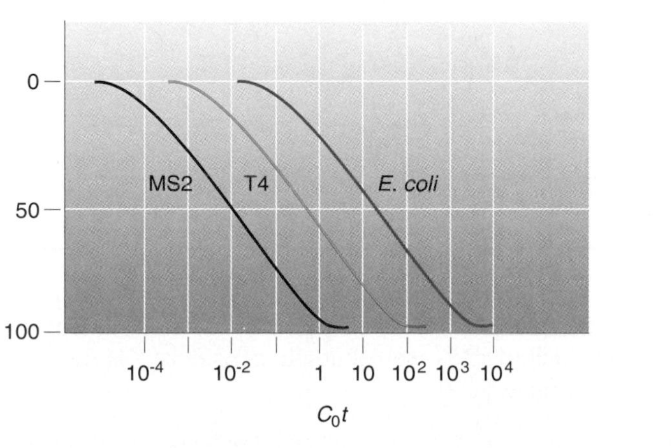

FIGURE 7.16
Comparison of the reassociation rate (C_0t) of DNA derived from phage MS-2, phage T4, and *E. coli.* The genome of T4 is larger than MS-2, and that of *E. coli* is larger than T4.

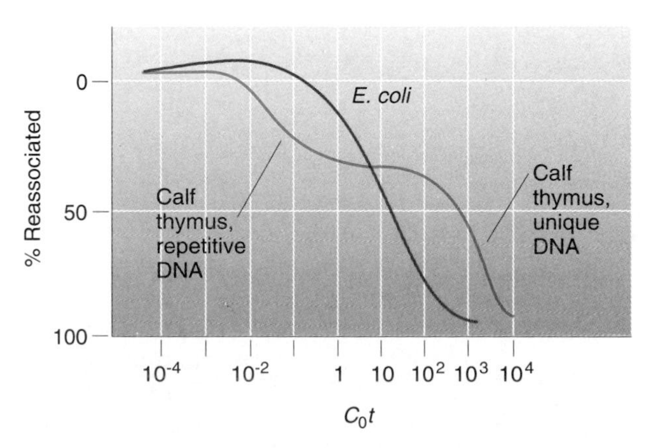

FIGURE 7.17
C_0t curve of calf thymus DNA compared with *E. coli.* The repetitive fraction of calf DNA reassociates more quickly than that of *E. coli,* while the more complex unique calf DNA takes longer to reassociate than that of *E. coli.*

petitive DNA. This fraction constitutes 10 to 15 percent of the genome of mammals, and even more in seed plants. It reassociates so rapidly because it consists of many sets of *short* repeating nucleotide sequences. Most of these sequences are 5 to 10 base pairs (bp) in length, and they may be present up to a million times in the genome. Most organisms have various types of simple sequence DNA, which are usually similar but not identical in sequence. In humans, for example, at least ten varieties exist.

These rapidly reassociating DNA sequences are thought not to be transcribed into RNA and are most often clustered in the regions of centromeres and sometimes in the telomeres of chromosomes. Evidence revealing this has been derived from *in situ* hybridization studies.

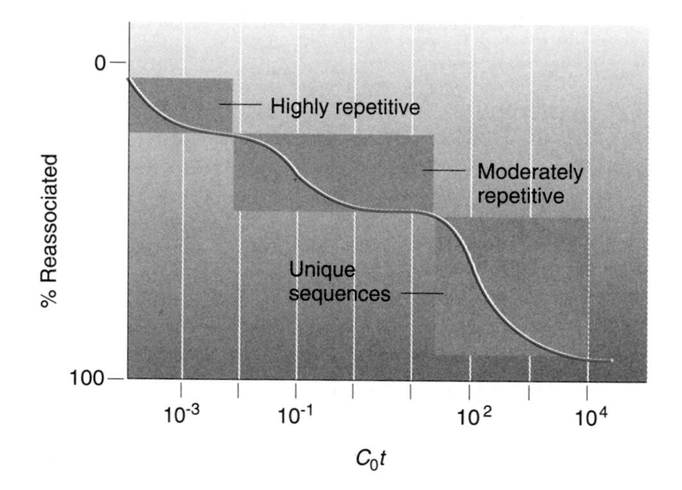

FIGURE 7.18
Reassociation analysis of eukaryotic DNA containing highly repetitive, moderately repetitive, and unique sequences. This pattern is similar to that found in mammals, including humans.

The second category of repetitive DNA does not reassociate as quickly as the simple sequence DNA. Called **moderately repetitive** or **intermediate repeat DNA,** this fraction consists of *longer* sequences that are present *fewer* times than highly repetitive DNA. These sequences are interspersed throughout the genome and make up approximately 20 percent of the total DNA in invertebrates and up to 40 percent in mammals.

Among the prominently studied classes in this fraction are the ***Alu* sequences** in mammals. Varying between 150 and 300 base pairs in length, they are characterized by a sequence that is recognized as a substrate by the highly specific nuclease called *Alu* I. In humans, *Alu* sequences are 300 base pairs long and present in the genome about one-half million times. This group of sequences constitutes up to 10 percent of the human genome.

Although we will explore the significance of repetitive DNA sequences in Chapter 12, a short discussion is appropriate here. Highly repetitive sequences are usually much too short to serve as genes. Much of moderately repetitive DNA is also nongenic and not transcribed. However, some of this fraction consists of sequences that *are* transcribed and constitute repeat or duplicate copies of various genes. Included in this fraction are genes coding for ribosomal RNA, ribosomal proteins, and histones (positively charged proteins bound to DNA in eukaryotes). On the other hand, most genes are present only once within the genome and are included as part of what is called **single copy** or **unique sequence DNA.** Estimates are that most eukaryotic organisms contain between 5,000 and 50,000 genes. *Drosophila* is likely to be closer to the lower estimate while humans are thought to be closer to the higher estimate.

CHAPTER SUMMARY

1 The existence of a genetic material capable of replication, storage, expression, and mutation is deducible from the observed patterns of inheritance in organisms. Both proteins and nucleic acids were initially considered as the possible candidates for the genetic material.

2 Transformation studies, as well as experiments using bacteria infected with bacteriophages, strongly suggested that DNA was the genetic material for bacteria and most viruses.

3 Initially, only circumstantial evidence supported the concept of DNA controlling inheritance in eukaryotes. This included the distribution of DNA in the cell, quantitative analysis of DNA, and the UV-induced mutagenesis. Recent recombinant DNA techniques, as well as experiments with transgenic mice, have provided direct experimental evidence that the eukaryote genetic material is DNA.

4 Numerous viruses provide an important exception to this general rule, because many of them use RNA as their genetic material. These include bacteriophages as well as some plant and animal viruses.

5 By the 1950s, many scientists sought to determine the structure of DNA. These efforts culminated in 1953 with Watson and Crick's proposal. Based on base-pairing information and X-ray diffraction data, they constructed a model, the key features of which include two antiparallel polynucleotide chains held together in a right-handed double helix by the hydrogen bonds formed between complementary bases. To date, the basic tenets of this double helix hold true.

6 RNA, in comparison, varies from DNA by virtue of being single stranded, containing uracil rather than thymine, and having ribose rather than deoxyribose.

7 The unique nature of the hydrogen bond allows double-stranded nucleic acids to be dissociated and reassociated experimentally, leading to the analytical techniques of molecular hybridization and the study of reassociation kinetics.

8 The latter investigation has led to the discovery of repetitive DNA sequences characteristic of eukaryotes in contrast to unique DNA sequences.

KEY TERMS

absorption spectrum
action spectrum
adenine
adenosine triphosphate (ATP)
A-DNA
Alu sequences
antiparallel chains
bacteriophage
B-DNA
central dogma of molecular genetics
complementarity
cytosine
deoxyribose
genetic expression
genetic information
guanine
guanosine triphosphate (GTP)
hydrogen bond
information flow
information storage
in situ molecular hybridization
major groove
melting temperature (T_m)
minor groove
messenger RNA (mRNA)
molecular hybridization
nitrogenous base
nuclein
nucleoside diphosphate (NDP)
nucleoside monophosphate (NMP)
nucleoside triphosphate (NTP)
nucleotide
pentose sugar
phosphate group
phosphodiester bond
polynucleotide
purine
pyrimidine
$Q\beta$ RNA replicase
reassociation kinetics

recombinant DNA technology
repetitive DNA
replication
retroviruses
reverse transcriptase
ribose
ribosomal RNA (rRNA)
ribosome
serotypes
single copy DNA
single crystal X-ray analysis
spheroplast
Svedberg coefficient (*S*)
tetranucleotide hypothesis
thymine
transfer RNA (tRNA)
transformation
transgenic organisms
translation
tobacco mosaic virus (TMV)
transcription
transfection
uracil
variation by mutation
Z-DNA

In contrast to the preceding chapters, this chapter does not emphasize genetic problem solving. Instead, it recounts some of the initial experimental analysis that served as the cornerstone of modern genetics. Quite fittingly, then, our Insights and Solutions section shifts its emphasis to experimental rationale and analytical thinking, an approach that will continue through the remainder of the text whenever appropriate.

1 Based strictly on the transformation analysis of Avery, MacLeod, and McCarty, what objection might be made to the conclusion that DNA is the genetic material? What other conclusion might be considered?

ANSWER: Based solely on their results, it may be concluded that DNA is essential for transformation. However, DNA might have been a substance that caused capsular formation by *directly* converting nonencapsulated cells to ones with a capsule. That is, DNA may simply have played a catalytic role in capsular synthesis, leading to cells displaying smooth type III colonies.

2 What observations argue against this objection?

ANSWER: First, transformed cells pass the trait on to their progeny cells, thus supporting the conclusion that DNA is responsible for heredity, not for the direct production of polysaccharide coats. Second, subsequent transformation studies over the next five years showed that other traits, such as antibiotic resistance, could be transformed. Therefore, the transforming factor has a broad general effect, not one specific to polysaccharide synthesis.

3 If RNA was the universal genetic material, how would this have affected the Avery experiment and the Hershey-Chase experiment?

ANSWER: In the Avery experiment, RNase rather than DNase would have eliminated transformation. Had this occurred, Avery and his colleagues would have concluded that RNA was the transforming factor. Hershey and Chase would have received identical results, since ^{32}P would also label RNA, but not protein.

4 Sea urchin DNA, which is double-stranded, was shown to contain 17.5 percent of its bases in the form of cytosine (C). What percentages of the other three bases are present in this DNA?

SOLUTION: The amount of C equals G, so guanine is also present as 17.5 percent. The remaining bases, A and T, are present in equal amounts and together they represent the rest of the bases (100 − 35). Therefore, A = T = 65/2 = 32.5 percent.

PROBLEMS AND DISCUSSION QUESTIONS

1 The functions ascribed to the genetic material are replication, expression, storage, and mutation. What does each of these terms mean?

2 Discuss the reasons why proteins were generally favored over DNA as the genetic material before 1940. What was the role of the tetranucleotide hypothesis in this controversy?

3 Contrast the various contributions made to an understanding of transformation by Griffith with those of Avery and his coworkers.

4 Why were ^{32}P and ^{35}S chosen for use in the Hershey–Chase experiment? Discuss the rationale and conclusions of this experiment.

5 Why is the early evidence that DNA serves as the genetic material in eukaryotes called circumstantial? List and discuss these evidences. What direct evidence exists?

6 What are the exceptions to the general rule that DNA is the genetic material in all organisms? What evidence supports these exceptions?

7 Draw the chemical structure of the three components of a nucleotide and then link the three together. What atoms are removed from the structures when the linkages are formed?

8 Adenine may also be named 6-amino purine. How would you name the other four nitrogenous bases using this alternate system? (= O is oxy, and $-CH_3$ is methyl.)

9 Describe the various characteristics of the Watson–Crick double-helix model for DNA.

10 What evidence did Watson and Crick have at their disposal in 1953? What was their approach in arriving at the structure of DNA?

11 Had Chargaff's data from a single source indicated the following, what might Watson and Crick have concluded?

	A	T	C	G
%	29	19	21	31

Why would this conclusion be contradictory to Wilkins and Franklin's data?

12 List three main differences between DNA and RNA.

13 What is the chemical basis of molecular hybridization?

14 Why does unique sequence DNA take longer to reassociate than repetitive DNA sequences?

15 A genetics student was asked to draw the chemical structure of an adenine- and thymine-containing dinucleotide derived from DNA. The student made more than six major errors. His answer is shown below. One of them is circled, numbered (1), and explained. Find five others. Circle them, number them (2)—(6), and briefly explain each, following the example below.

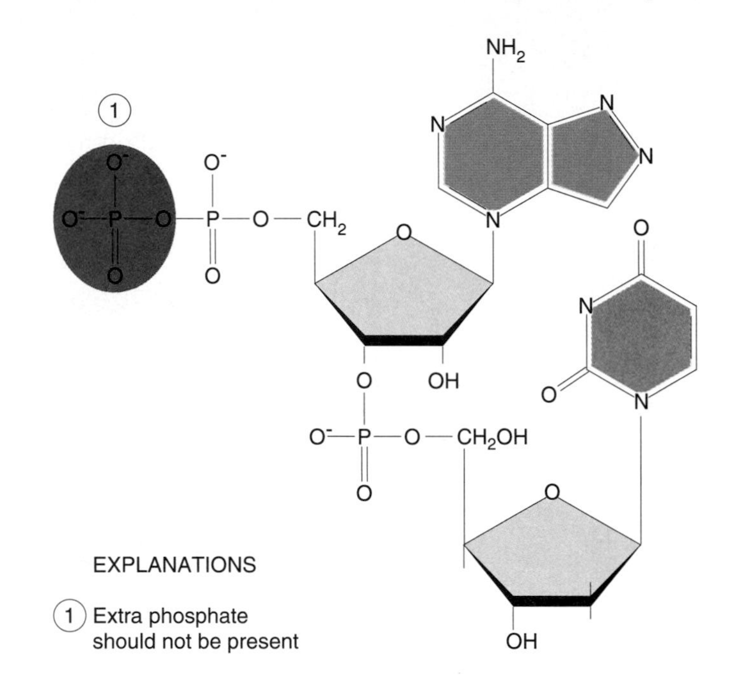

16 Draw the predicted C_0t curve if the following mixture of nucleic acid molecules were analyzed in a single experiment: 10^6 copies of a short, identical sequence; 10^3 copies of a much longer DNA sequence; and 10^2 unique sequences of DNA.

17 The DNA of the bacterial virus T4 produces a $C_0t_{1/2}$ of about 0.5 and contains 10^5 nucleotide pairs in its genome. How many nucleotide pairs are present in the genome of the virus MS2 and the bacterium *E. coli,* whose respective DNAs produce $C_0t_{1/2}$ values of 0.001 and 10.0?

18 *Newdate: March 1, 2005.* A unique creature has been discovered during exploration of outer space. Recently, its genetic material has been isolated and analyzed. This material is similar in some ways to DNA in its chemical makeup. It contains in abundance the 4-carbon sugar erythrose and a molar equivalent of phosphate groups. Additionally, it contains six nitrogenous bases: adenine (A), guanine (G), thymine (T), cytosine (C), hypoxanthine (H), and xanthine (X). These bases exist in the following relative proportions.

$$A = T = H \quad \text{and} \quad C = G = X$$

X-ray diffraction studies have established a regularity to the molecule and a constant diameter of about 30 Å.

Together, these data have suggested a model for the structure of this molecule.

(a) Propose a general model of this molecule. Describe it briefly.
(b) What base-pairing properties must exist for H and for X in the model?
(c) Given the constant diameter of 30 Å, do you think that H and X are *either* (1) both purines or both pyrimidines, *or* (2) one is a purine and one is a pyrimidine?

SELECTED READINGS

AVERY, O. T., MacLEOD, C. M., and McCarty, M. 1944. Studies on the chemical nature of the substance inducing transformation of pneumococcal types. Induction of transformation by a desoxyribonucleic acid fraction isolated from pneumococcus type III. *J. Exp. Med.* 79:137–58. (Reprinted in Taylor, J. H. 1965. *Selected papers in molecular genetics.* Orlando: Academic Press.)

BRITTEN, R. J., and KOHNE, D.E. 1970. Repeated segments of DNA. *Scient. Amer.* (April) 222:24–31.

DeROBERTIS, E. M., and GURDON, J. B. 1979. Gene transplantation and the analysis of development. *Scient. Amer.* (Dec.) 241:74–82.

DICKERSON, R. E., et al. 1982. The anatomy of A-, B-, and Z-DNA. *Science* 216:475–85.

DICKERSON, R. E. 1983. The DNA helix and how it is read. *Scient. Amer.* (June) 249:94–111.

DUBOS, R. J. 1976. *The professor, the Institute and DNA: Osward T. Avery, his life and scientific achievements.* New York: Rockefeller University Press.

FELSENFELD, G. 1985. DNA. *Scient. Amer.* (Oct.) 253:58–78.

FRANKLIN, R. E., and GOSLING, R. G. 1953. Molecular configuration in sodium thymonucleate. *Nature* 171:740–41.

GRIFFITH, F. 1928. The significance of pneumococcal types. *J. Hyg.* 27:113–59.

HERSHEY, A. D., and CHASE, M. 1952. Independent functions of viral protein and nucleic acid in growth of bacteriophage. *J. Gen. Phys.* 36:39–56. (Reprinted in Taylor, J. H. 1965. *Selected papers in molecular genetics.* Orlando: Academic Press.)

JUDSON, H. 1979. *The eighth day of creation: Makers of the revolution in biology.* New York: Simon & Schuster.

LEVENE, P. A., and SIMMS, H. S. 1926. Nucleic acid structure as determined by electrometric titration data. *J. Biol. Chem.* 70:327–41.

McCARTY, M. 1985. *The transforming principle: Discovering that genes are made of DNA.* New York: W. W. Norton.

PALMITER, R. D., and BRINSTER, R. L. 1985. Transgenic mice. *Cell* 41:343–45.

STENT, G. S., ed. 1981. *The double helix: Text, commentary, review, and original papers.* New York: W. W. Norton.

STEWART, T. A., WAGNER, E. F., and MINTZ, B. 1982. Human β-globin gene sequences injected into mouse eggs, retained in adults, and transmitted to progeny. *Science* 217:1046–48.

VARMUS, H. 1988. Retroviruses. *Science* 240:1427–35.

WATSON, J. D. 1968. *The double helix.* New York: Atheneum.

WATSON, J. D., and CRICK, F. C. 1953a. Molecular structure of nucleic acids. A structure for deoxyribose nucleic acids. *Nature* 171:737–38.

———. 1953b. Genetic implications of the structure of deoxyribose nucleic acid. *Nature* 171:964.

WEINBERG, R. A. 1985. The molecules of life. *Scient. Amer.* (Oct.) 253:48–57.

8 Replication and Synthesis of DNA

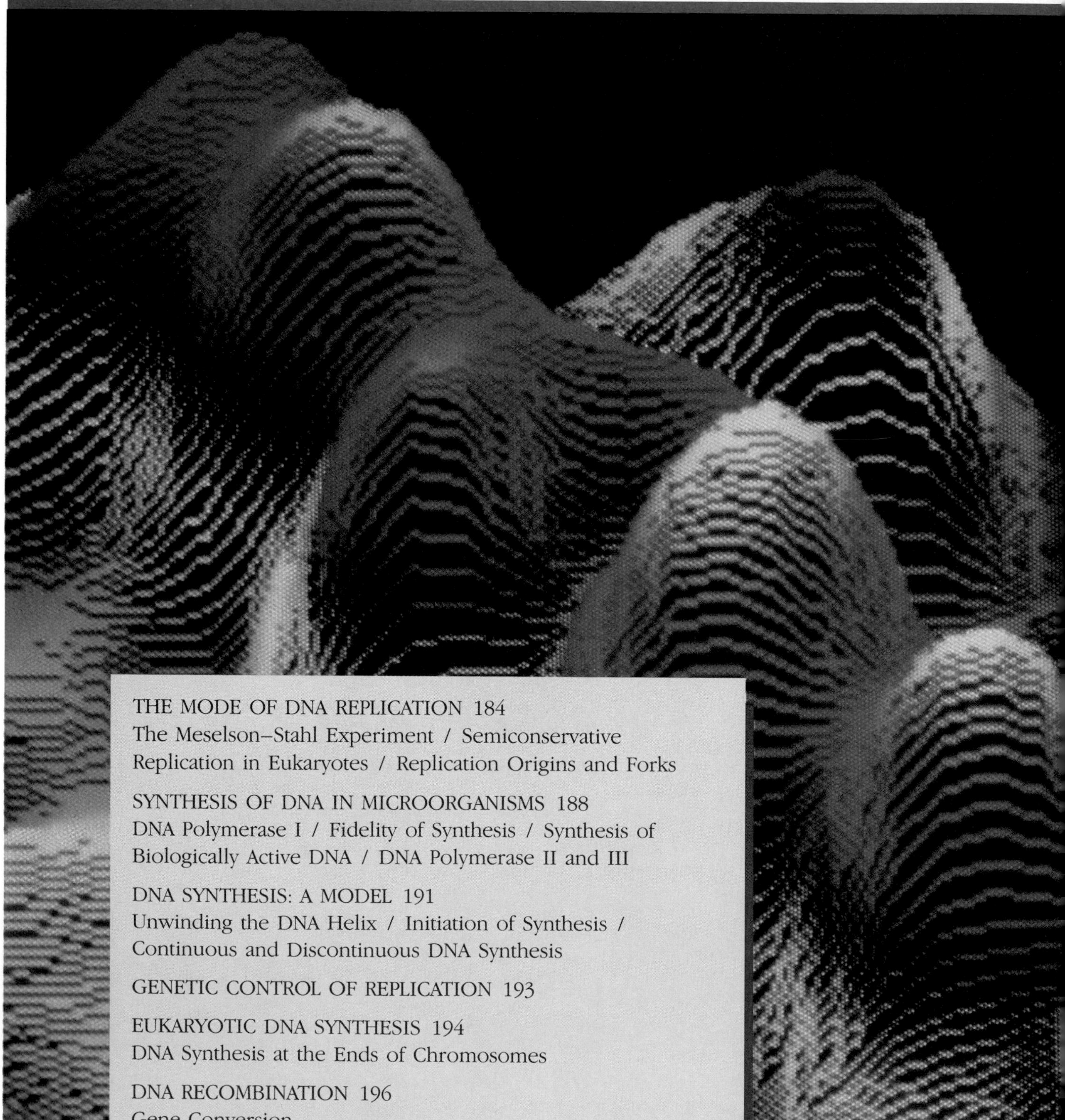

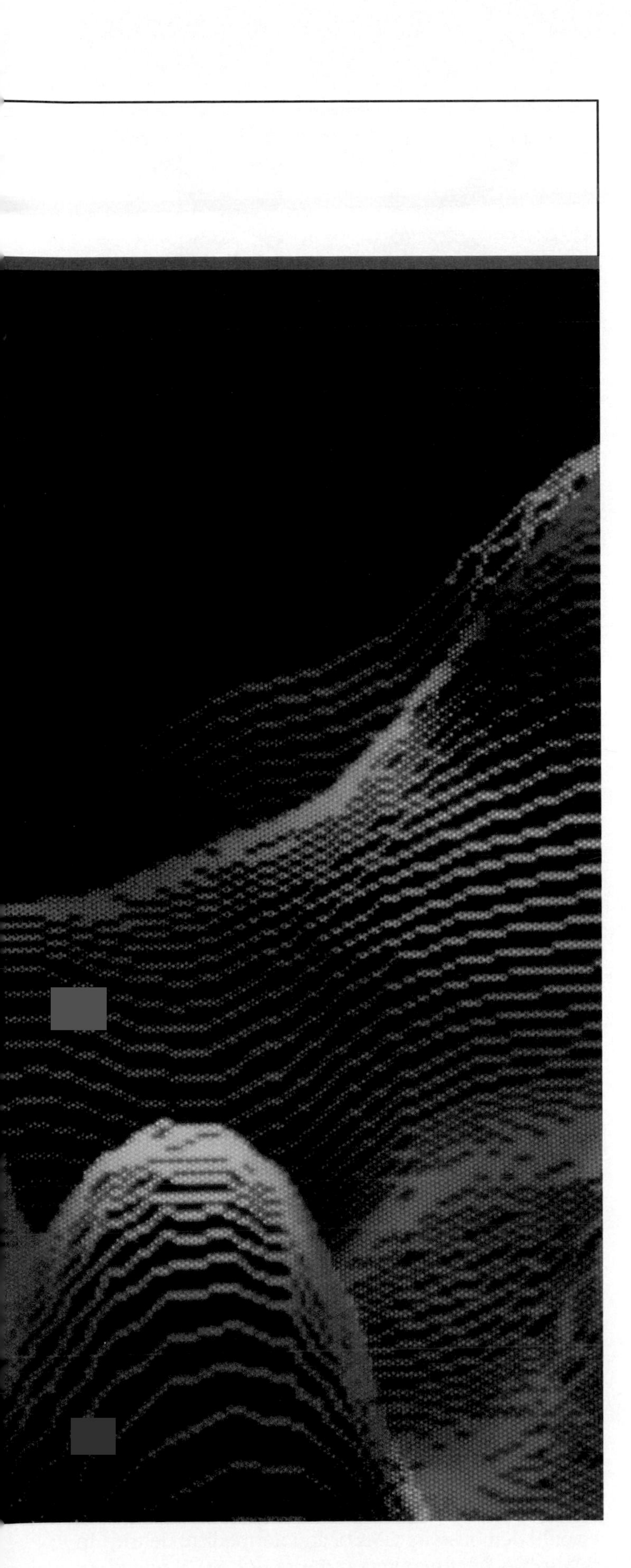

Chapter Concepts

Genetic continuity between parental and progeny cells is made possible by semiconservative replication of DNA, as predicted by the Watson–Crick model. Each strand of the parent helix serves as a template for the production of its complement. Synthesis of DNA is a complex but orderly process orchestrated by a myriad of enzymes and other molecules. Together, they function with great fidelity to polymerize nucleotides into polynucleotide chains. Other enzymes interact with DNA, leading to genetic recombination.

Following Watson and Crick's proposal for the structure of DNA, scientists focused their attention on how this molecule is replicated. This process is an essential function of the genetic material and must be executed precisely if genetic continuity between cells is to be maintained following cell division. This is an enormous, complex task. Consider for a second that in the human genome, some three billion base pairs exist. To faithfully duplicate a molecule of this size requires a mechanism of extreme precision. Even an error rate of 10^{-6} (one in a million) will create an excessive number of errors (3000) during each replication cycle. While not error free, an extremely accurate system of DNA replication has evolved in all organisms.

As Watson and Crick suggested in 1953, the model of the double helix gave them the initial insight into how replication could occur. This mode, called **semiconservative replication,** has since received strong experimental support from studies of viruses, prokaryotes, and eukaryotes.

Once the general mode of replication was made clear, research was intensified to determine the precise details of DNA synthesis. What has since been discovered is that numerous enzymes and other proteins are needed to copy a DNA helix. Because of the complexity of the chemical events during synthesis, this subject remains an extremely active area of research.

In this chapter, we will discuss the general mode of replication as well as the specific details of the synthesis of DNA. The research leading to this knowledge is still another link in our understanding of life processes at the molecular level.

THE MODE OF DNA REPLICATION

It was apparent to Watson and Crick that because of the arrangement and nature of the nitrogenous bases, each strand of a DNA double helix could serve as a template for the synthesis of its complement. They proposed that if the helix were unwound, each nucleotide along the two parent strands would have an affinity for its complementary nucleotide. As we learned in Chapter 7, the complementarity is due to the potential hydrogen bonds that can be formed. If thymidylic acid (T) were present, it would "attract" adenylic acid (A); if guanidylic acid (G) were present, it would "attract" cytidylic acid (C); and so on. If these nucleotides were then covalently linked into polynucleotide chains along both templates, the result would be the production of two new but identical double strands of DNA. This concept is illustrated in Figure 8.1. Each replicated DNA molecule would consist of one "old" and one "new" strand; hence the reason for the name semiconservative replication.

There are two other possible modes of replication that also rely on the parental strands as a template. In **conservative replication,** synthesis of complementary polynucleotide chains occurs as described above. Following synthesis, however, the two newly created strands are brought back together, and the parental strands reassociate. The original helix is thus "conserved."

In the second alternative mode, called **dispersive replication,** the parental strands are seen to be dispersed into two new double helices following replication. Thus, each strand consists of both old and new DNA. This mode would involve cleavage of the parental strands during replication. Therefore, it is the most complex of the three possibilities and is least likely, as such.

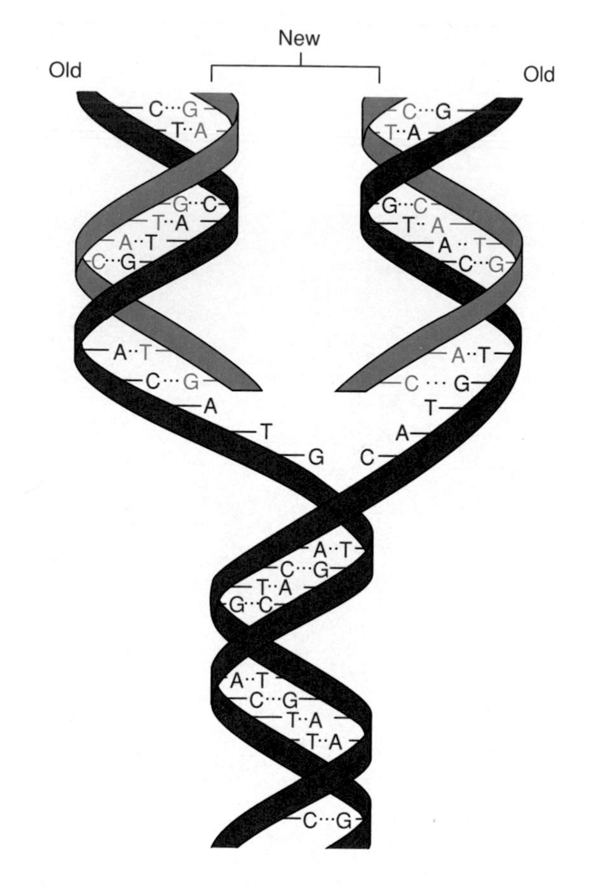

FIGURE 8.1
General model of semiconservative replication of DNA.

The Meselson–Stahl Experiment

In 1958, Matthew Meselson and Franklin Stahl published the results of an experiment providing strong evidence that semiconservative replication is the mode used by cells to produce new DNA molecules. *E. coli* cells were grown for many generations in a medium where $^{15}NH_4Cl$ (ammonium chloride) was the only nitrogen source. A "heavy" isotope of nitrogen, ^{15}N contains one more neutron than the naturally occurring ^{14}N isotope. Unlike "radioactive" isotopes, ^{15}N is stable and is thus not radioactive. After many generations, all nitrogen-containing molecules, including the nitrogenous bases of DNA, contained the isotope in the *E. coli* cells. DNA containing ^{15}N may be distinguished from ^{14}N-containing DNA by the use of **sedimentation equilibrium centrifugation,** where samples are "forced" by centrifugation through a density gradient of a heavy metal salt such as cesium chloride. The more dense ^{15}N-DNA will reach equilibrium in the gradient at a point closer to the bottom (where the density is greater) than will ^{14}N-DNA.

In this experiment, uniformly labeled ^{15}N cells were transferred to a medium containing only $^{14}NH_4Cl$. Thus, all subsequent synthesis of DNA during replication contained the "lighter" isotope of nitrogen. The time of transfer was taken as time zero ($t = 0$). The *E. coli* cells were allowed to replicate during several generations with cell samples removed at various intervals. From each sample, DNA was isolated and subjected to sedimentation equilibrium centrifugation. The results are depicted in Figure 8.2.

After one generation, the isolated DNA was all present in a single band of intermediate density—the expected result for semiconservative replication. Each replicated molecule would be composed of one new ^{14}N-strand and one old ^{15}N-strand, as seen in Figure 8.3. This result was not consistent with the conservative replication mode, in which two distinct bands would have been predicted to occur.

After two cell divisions, DNA samples showed two density bands: one was intermediate and the other was lighter, corresponding to the ^{14}N position in the gradient. Similar results occurred after a third generation, except that the proportion of the ^{14}N-band increased. If replication were dispersive, all subsequent generations after $t = 0$ would demonstrate DNA of an intermediate density. In each subsequent generation the ratio $^{14}N/^{15}N$ would in-

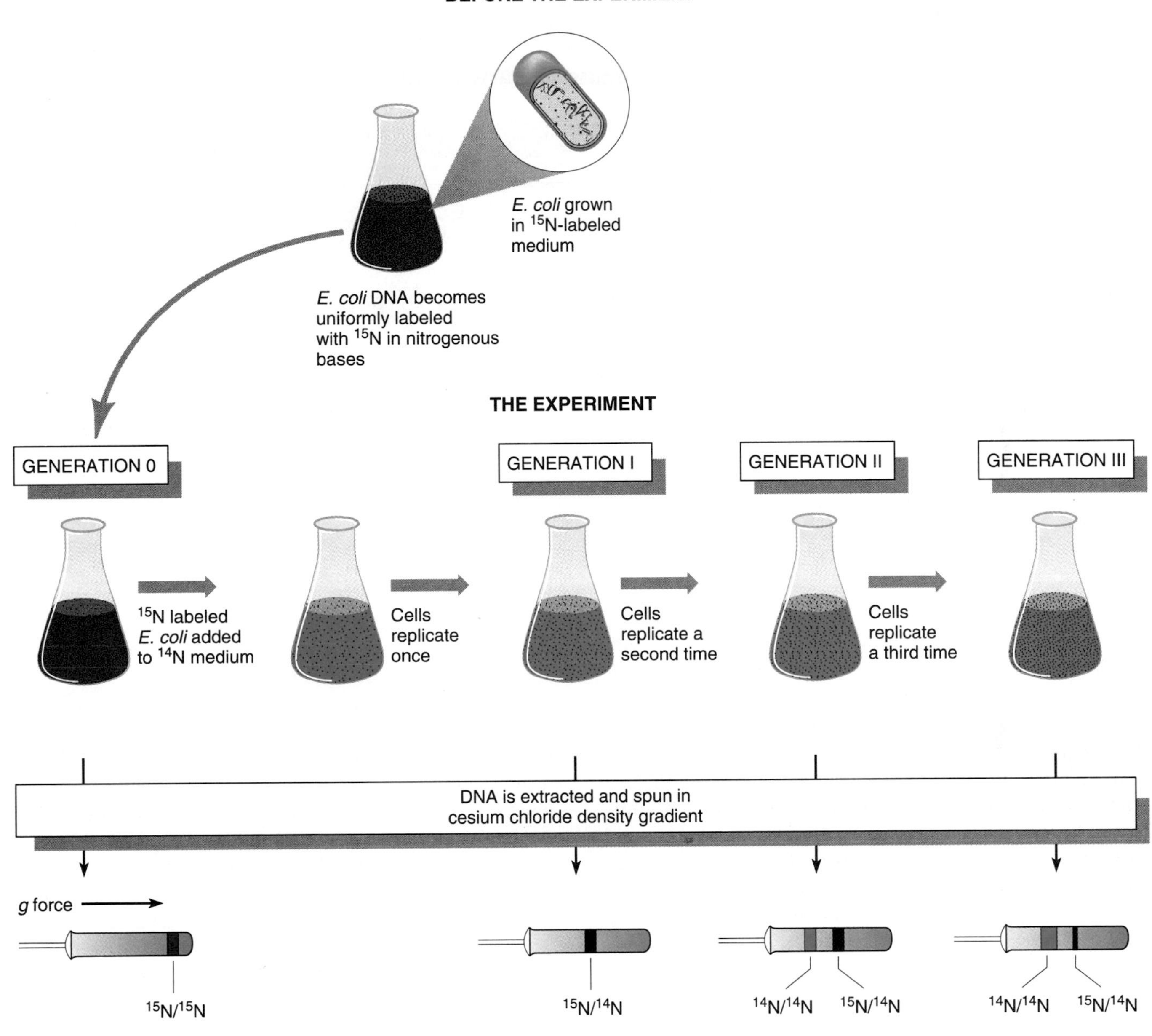

FIGURE 8.2
The Meselson–Stahl experiment.

crease, and the hybrid band would become lighter and lighter, eventually approaching the ^{14}N-band. This result was not observed. Thus, the results of the Meselson–Stahl experiment provided strong support for the semiconservative mode of DNA replication, as postulated by Watson and Crick.

Semiconservative Replication in Eukaryotes

In 1957, the year before the work of Meselson and his colleagues was published, evidence was also presented that supported semiconservative replication in a eukaryotic organism. J. Herbert Taylor, Philip Woods, and Walter Hughes experimented with root tips of the broad bean *Vicia faba*. Root tips are an excellent source of dividing cells. These researchers examined the chromosomes of these cells following replication of DNA. They were able to monitor the process of replication by labeling DNA with ^{3}H-thymidine, a radioactive precursor of DNA, and performing autoradiography.

The technique of autoradiography is a cytological procedure that allows the isotope to be localized within the cell. In this procedure, a photographic emulsion is placed over a section of cellular material (root tips in this

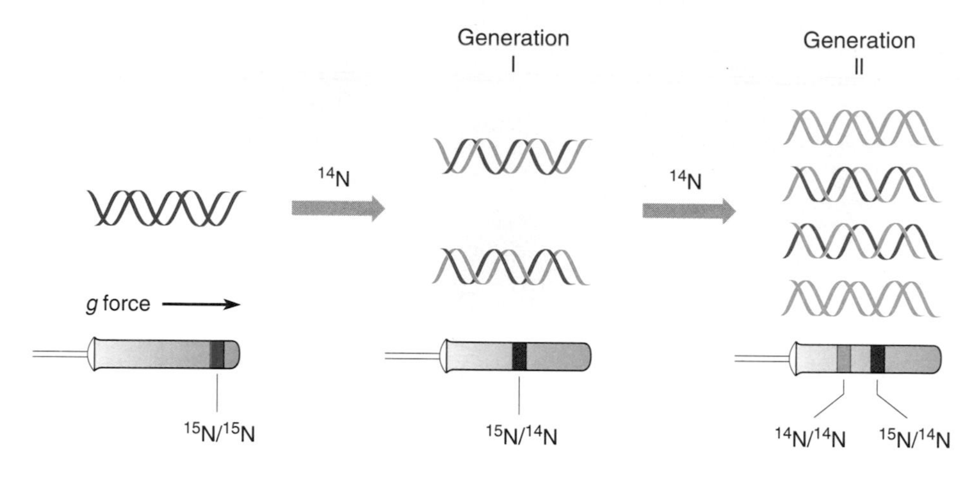

FIGURE 8.3
The expected results of two generations of semiconservative replication in the Meselson–Stahl experiment.

experiment), and the preparation is stored in the dark. The slide is then developed, much as photographic film is processed. Since the radioisotope emits energy, the emulsion turns black at the approximate point of emission following development. The end result is the presence of dark spots or "grains" on the surface of the section, localizing newly synthesized DNA within the cell.

Root tips were grown for approximately one generation in the presence of the radioisotope and then placed in unlabeled medium, where cell division continued. At the conclusion of each generation, cultures were arrested at metaphase by the addition of colchicine, and chromosomes were examined by autoradiography. Figure 8.4 illustrates replication of a single chromosome over two division cycles as well as the distribution of grains. In this experiment, labeled thymidine is found only in association with chromosomes that contain newly synthesized DNA.

The results are compatible with the semiconservative mode of replication. After the first replication cycle, radioactivity is detected over both sister chromatids. This finding is expected because each chromatid will contain one "new" radioactive DNA strand and one "old" unlabeled strand. After the second replication cycle, which also takes place in unlabeled medium, only one of the two new sister chromatids should be radioactive because half of the parent strands are unlabeled. With only minor exceptions this result was observed.

Together, the Meselson–Stahl experiment and the experiment by Taylor, Woods, and Hughes soon led to the general acceptance of the semiconservative mode of replication. The same conclusion has been reached in studies with other organisms. Since this mode is suggested by the double-helix model of DNA, these experiments also strongly supported Watson and Crick's proposal for DNA structure.

Replication Origins and Forks

The mode of replication just established represents the general pattern by which DNA is duplicated. Before turning to the details of the biosynthesis of DNA, we will mention several other topics relevant to the complete description of semiconservative replication. The first concerns the **origin of replication** along any chromosome. Is there but a single origin or more than one point where DNA synthesis begins? And, is each point of origin random or located in a specific region along the chromosome? As we address these questions, we shall define the length of DNA that is replicated following the initiation of synthesis at a single origin as a unit called the **replicon.**

A second topic involves the direction of replication. Once it begins, does it move in a single direction or in both directions away from the origin? This consideration distinguishes between **unidirectional** and **bidirectional replication,** respectively. At each point of replication, the strands of the helix must unwind, creating what is called a **replication fork.** Bidirectional replication creates two such forks, which move apart in opposite directions away from the origin.

The evidence is reasonably clear regarding these topics. In bacteria and most bacterial viruses, which have only a single circular chromosome, there is one specific region where replication is initiated. In *E. coli* this region, called ***ori,*** has been located. It consists of 245 base pairs, but only a small number are actually essential to the initiation of DNA synthesis. Since there is but a single point of origin of DNA synthesis, in bacteriophages and bacteria, the entire chromosome constitutes one replicon. In *E. coli,* replication is bidirectional from *ori* and proceeds until the entire chromosome is replicated, as illustrated in Figure 8.5. Both strands of the parent helix are copied at the two replication forks that are estab-

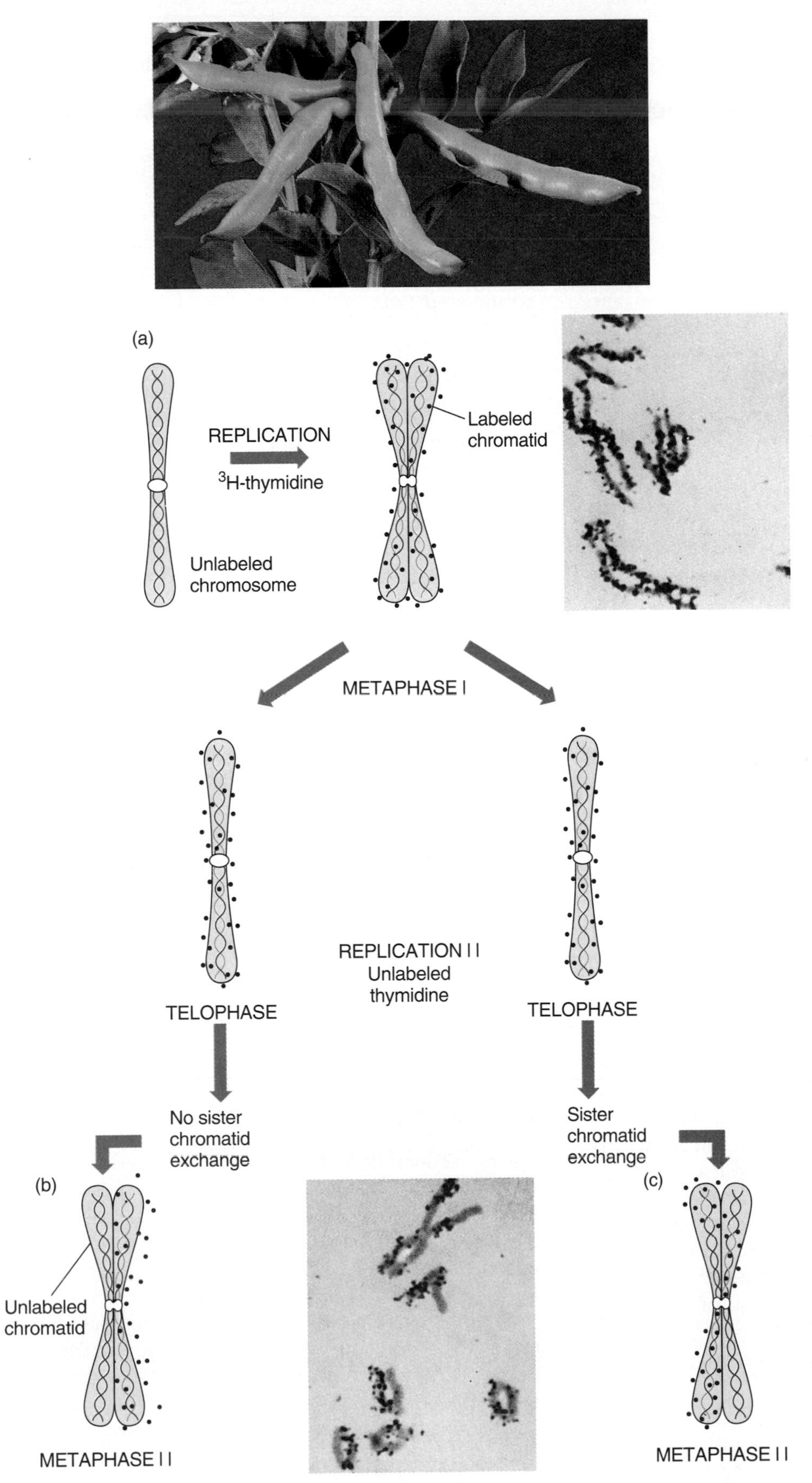

FIGURE 8.4
Depiction of the experiment by Taylor, Woods, and Hughes demonstrating the semiconservative mode of replication of DNA in root tips of *Vicia faba.* The plant is shown in the photograph. In (a) an unlabeled chromatid proceeds through the cell cycle in the presence of ^{3}H-thymidine. As it enters mitosis, both sister chromatids of each chromosome are labeled, as shown by autoradiography. After a second generation of replication, this time in the absence of ^{3}H-thymidine, only one chromatid of each chromosome is expected to be surrounded by grains (b). In all cases, except where a reciprocal exchange has occurred between sister chromatids (c), the expectation was upheld.

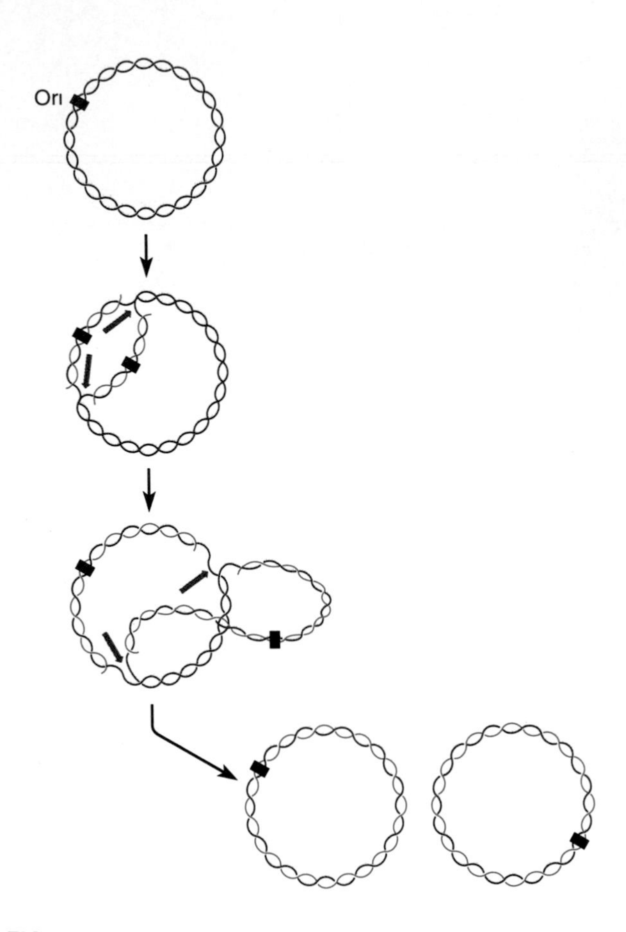

FIGURE 8.5
Bidirectional replication of the *E. coli* chromosome.

lished. These forks move away from the origin in opposite directions and eventually merge as semiconservative replication of the entire chromosome is completed.

Compared to bacteria, as described above, a major difference in replication exists in eukaryotes. While replication is bidirectional, creating two replication forks during each replication cycle, there are multiple origins along each chromosome. As a result, during the S phase of interphase, there are numerous replicating events occurring along each chromosome. Eventually, the numerous replication forks merge, completing replication of the entire chromosome. The presence of multiple replicons is undoubtedly related to the much greater length and complexity of a single eukaryotic chromosome compared to one from a bacterium. As a result, replication can be completed in a reasonable period of time.

SYNTHESIS OF DNA IN MICROORGANISMS

The determination that replication is semiconservative and bidirectional indicates only the pattern of DNA duplication and the association of finished strands with one another once synthesis is completed. A much more complex issue is how the **actual synthesis** of long complementary polynucleotide chains occurs on a DNA template. As in most studies of molecular biology, this question was first approached by using microorganisms. Research began about the same time as the Meselson–Stahl work, and even today this topic is an active area of investigation. What is most apparent in this research is the tremendous chemical complexity of the biological synthesis of DNA.

DNA Polymerase I

Studies of the enzymology of DNA replication were first reported by Arthur Kornberg and colleagues in 1957. They isolated an enzyme from *E. coli* that was able to direct DNA synthesis in a cell-free (*in vitro*) system. The enzyme is now called **DNA polymerase I,** since it was the first of several to be isolated. Kornberg determined the following major requirements for *in vitro* DNA synthesis under the direction of the enzyme:

1 All four deoxyribonucleoside triphosphates (dATP, dCTP, dGTP, dTTP = dNTP)*

2 Template DNA

If any one of the four deoxyribonucleoside triphosphates was omitted from the reaction, no synthesis occurred. If derivatives of these precursor molecules other than the nucleoside triphosphate were used (nucleotides or nucleoside diphosphates), synthesis did not occur. If no template DNA was added, synthesis of DNA occurred but was reduced greatly. Thus, most synthesis directed by Kornberg's enzyme appeared to be exactly the type required for semiconservative replication. The reaction is summarized in Figure 8.6. The enzyme has since been shown to consist of a single polypeptide containing 928 amino acids.

Fidelity of Synthesis

Having shown how DNA was synthesized, Kornberg sought to demonstrate the accuracy, or fidelity, with which the enzyme had replicated the DNA template. Since the nucleotide sequences of the template and the product could not be determined in 1957, he had to rely initially on several indirect methods.

One of Kornberg's approaches was to compare the nitrogenous base compositions of the DNA template with those of the recovered DNA product. Table 8.1 shows Kornberg's base composition analysis of three different DNA templates. These may be compared with the DNA

*dNTP designates the deoxyribose forms of the four nucleoside triphosphates; in a similar way, dNMP refers to the monophosphate forms.

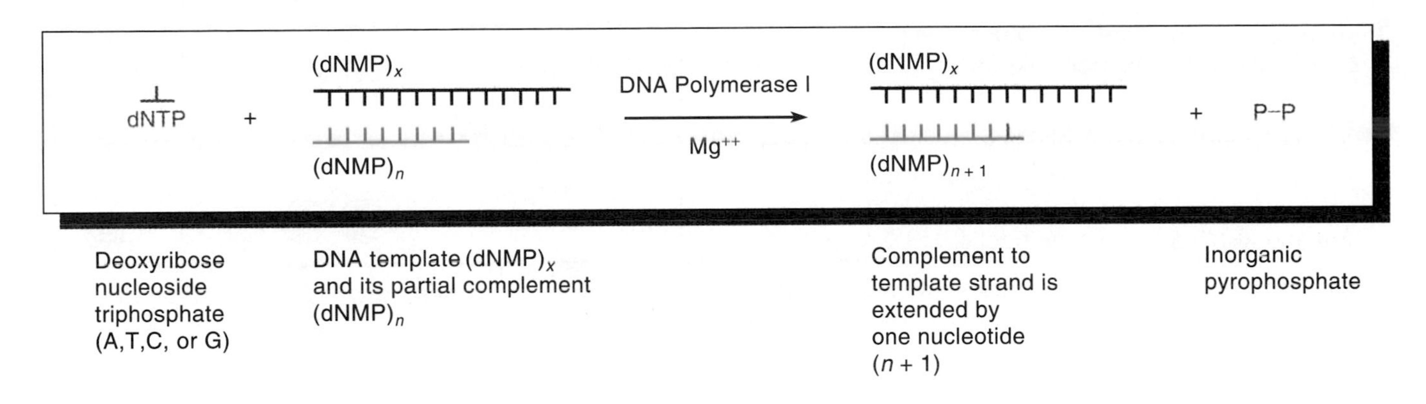

FIGURE 8.6
Chemical reaction catalyzed by DNA polymerase I. During each step, a single nucleotide is added to the growing complement of the DNA template, using a nucleoside triphosphate as the substrate. The release of inorganic pyrophosphate drives the reaction energetically.

product synthesized in each case. Within experimental error, the base composition of each product agreed with the template DNAs used. These data, along with other types of comparisons of template and product, suggested that the templates were replicated faithfully.

Synthesis of Biologically Active DNA

Despite Kornberg's extensive work, not all researchers were convinced that DNA polymerase I was the enzyme that replicates DNA within cells (*in vivo*). The primary reservations involved observations that the *in vitro* rate of synthesis was very slow, that the enzyme was much more effective replicating single-stranded DNA than double-stranded DNA, and that the enzyme appeared to be able to *degrade* DNA as well as to *synthesize* it.

Faced with the uncertainty of the true cellular function of DNA polymerase I, Kornberg pursued another approach. He reasoned that if the enzyme could be used to synthesize **biologically active DNA** *in vitro,* then DNA polymerase I must be the major catalyzing force for DNA synthesis within the cell. The term *biological activity* means that the DNA synthesized is capable of supporting metabolic activities and directing reproduction of the organism from which it was originally duplicated.

TABLE 8.1
Base composition of the DNA template and the product of replication in Kornberg's early work.

Organism	Template or Product	%A	%T	%G	%C
T2	Template	32.7	33.0	16.8	17.5
	Product	33.2	32.1	17.2	17.5
E. coli	Template	25.0	24.3	24.5	26.2
	Product	26.1	25.1	24.3	24.5
Calf	Template	28.9	26.7	22.8	21.6
	Product	28.7	27.7	21.8	21.8

SOURCE: From Kornberg, 1960. Copyright 1960 by the American Association for the Advancement of Science.

In 1967, Mehran Goulian, Kornberg, and Robert Sinsheimer showed that the DNA of the small bacteriophage ϕX174 could be completely copied by DNA polymerase I *in vitro,* and that the new product could be isolated and used to successfully infect *E. coli.* This resulted in the production of mature phages under the direction of the synthetic DNA, thus demonstrating biological activity!

This demonstration of biological activity was viewed as a precise assessment of faithful copying. If even a single error in one of the 5386 nucleotides constituting the ϕX174 chromosome had occurred, the change would most likely have constituted a lethal mutation, precluding the production of viable phages.

DNA Polymerase II and III

Although DNA synthesized under the direction of polymerase I demonstrated biological activity, a more serious reservation about the enzyme's true biological role was raised in 1969. Peter DeLucia and John Cairns reported the discovery of a mutant strain of *E. coli* that was deficient in polymerase I activity. The mutation was designated *polA1*. In the absence of the functional enzyme, this mutant strain of *E. coli* still duplicated its DNA and successfully reproduced! Other properties of the mutation led DeLucia and Cairns to conclude that in the absence of polymerase I, these cells are highly deficient in their ability to "repair" DNA. For example, the mutant strain is highly sensitive to ultraviolet light and radiation, both of which damage DNA and are mutagenic. Nonmutant bacteria are able to repair a great deal of UV-induced damage.

These observations led to two conclusions.

1 There must be at least one other enzyme present in *E. coli* cells that is responsible for replicating DNA *in vivo*.

2 DNA polymerase I may serve only a secondary function *in vivo*. This function is now believed by Kornberg and others to occur during the normal course of replication and to be critical to the fidelity of DNA synthesis.

To date, two unique DNA polymerases have been isolated from cells lacking polymerase I activity. These two enzymes have also been isolated from normal cells that also contain polymerase I.

The characteristics of these two enzymes, called **DNA polymerase II and III,** are contrasted with DNA polymerase I in Table 8.2. As is evident from that information, all three share several characteristics. While none can *initiate* DNA synthesis on a template, all can *elongate* an existing DNA strand, called a **primer.** As we shall see, RNA is also an adequate primer and is, in fact, what is utilized! Elongation occurs by polymerizing nucleotides in the 5′-to-3′ direction. That is, each new nucleotide is added at its 5′-phosphate end to the 3′-C of the growing polynucleotide. Each addition creates a new exposed 3′-OH group on the sugar, which can then participate in the next reaction. Synthesis occurring in this way is illustrated in Figure 8.7

These enzymes are all large, complex proteins exhibiting a molecular weight in excess of 100,000 daltons. All three possess 3′-to-5′ exonuclease activity. This means that they can polymerize in one direction and then reverse directions and excise nucleotides just added. As we will see, this activity provides a capacity to proofread and remove an incorrect nucleotide.

What then are the roles of the three polymerases *in vivo*? Polymerase III is considered to be the enzyme responsible for the polymerization essential to replication. Its 3′-to-5′ exonuclease activity also allows it to proofread, excise, and then correct base pairs created in error during polymerization. As we will soon see, gaps are a natural occurrence on one of the two strands during replication as RNA primers are removed. It is believed that polymerase I, originally studied by Kornberg, is responsible for removing the primer as well as for the synthesis that fills these gaps. Its exonuclease activity also allows proofreading to occur during this process.

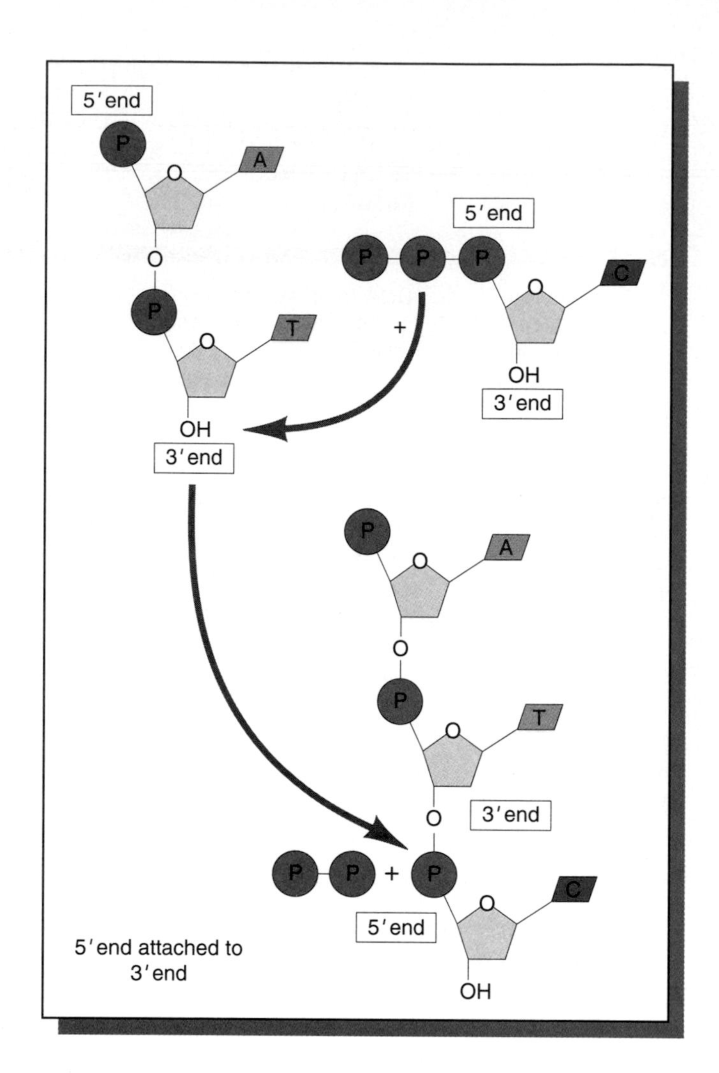

FIGURE 8.7
Demonstration of 5′-to-3′ synthesis of DNA.

TABLE 8.2
Comparative properties of the three bacterial DNA polymerases.

Properties	I	II	III
Initiation of chain synthesis	–	–	–
5′ to 3′ polymerization	+	+	+
3′ to 5′ exonuclease activity	+	+	+
5′ to 3′ exonuclease activity	+	–	–
Molecular weight	103,000	90,000	167,500
Molecules of polymerase/cell	400	?	15

Polymerase II is suspected to be involved in repair synthesis of DNA that has been damaged by external forces such as ultraviolet light. If so, its role is also important to survival.

We end this section by emphasizing the complexity of the DNA polymerase III molecule. Its active form, called a **holoenzyme,** consists of seven separate polypeptide chains. The largest, the α subunit, has a molecular weight of 140,000 daltons and is responsible for the polymerization activity of the holoenzyme. The ϵ subunit possesses the 3′-to-5′ exonuclease activity. While the function of the other five subunits is not precisely clear, one binds ATP, a step essential to the initial polymerization step. It is also believed that two holoenzymes may function together as a dimer at the replication fork. Together with several other proteins at the replication fork, a complex nearly as large as a ribosome is present. It has been called a **replisome.**

DNA SYNTHESIS: A MODEL

We have thus far established that replication is semiconservative and bidirectional along a single replicon in bacteria and many viruses. And we know that synthesis is in the 5′-to-3′ direction primarily under the direction of DNA polymerase III, creating two replication forks. These move in opposite directions away from the origin of synthesis. We are now ready to pursue several other aspects of DNA synthesis. We will combine this new information with that previously established into a coherent model. It takes into account the following general points.

1 A mechanism must exist by which the helix is initially unwound (or denatured) and stabilized in this "open" configuration.

2 As unwinding proceeds, a tension is created farther down the helix that must be reduced.

3 A primer of some sort must be synthesized so that polymerization can commence under the direction of DNA polymerase III. The primer, surprisingly, is RNA!

4 Once the RNA primers have been synthesized, DNA polymerase III commences synthesis of the complement of both strands of the parent molecule. Since the strands are antiparallel, continuous synthesis in the direction that the replication fork moves is possible along only one of the two strands. On the other strand, synthesis is discontinuous in the opposite direction.

5 The RNA primers must be removed prior to completion of replication. The gaps that are temporarily created must be filled with DNA complementary to the template at each location.

6 The newly synthesized DNA strand that fills each temporary gap must be ligated to the adjacent strand of DNA synthesized by DNA polymerase III.

As we consider the above points, two figures (Figures 8.8 and 8.9) will be used to illustrate how each issue is resolved.

Unwinding the DNA Helix

As discussed earlier, there is but a single origin along the circular chromosome of most bacteria and viruses where DNA synthesis is initiated. This region of the *E. coli* chromosome has been particularly well studied. Called ***oriC,*** the origin of replication consists of about 250 base pairs, which are characterized by the presence of repeating sequences of 9 and 13 bases (called **9mers** and **13mers**). One particular protein, called **dnaA** (because it is encoded by the gene called *dnaA,*) is responsible for the initial step in unwinding the helix. A number of subunits of the dnaA protein bind to each of several 9mers. This step is essential in facilitating the subsequent binding of **dnaB** and **dnaC proteins** that further open and destabilize the helix (Figure 8.8). Proteins such as these that require the energy normally supplied by the hydrolysis of ATP in order to break hydrogen bonds and denature the double helix are called **helicases.**

As unwinding proceeds, another group of proteins, called **single-stranded binding proteins (SSBPs)** stabilize this conformation. As unwinding continues, tighter coiling occurs farther down the helix. **DNA gyrase,** a member of a large group of enzymes called **DNA topoisomerases,** can function to reduce the tension produced (Fig. 8.9). The enzyme accomplishes this by causing a break in both strands and a topological manipulation of the DNA. Then, in a more relaxed conformation, the gaps created in both strands of the DNA are resealed. These reactions also require the hydrolysis of ATP to drive them.

In combination with the polymerase complex, these proteins create an array of molecules that participate in DNA synthesis and are part of what we have previously called the replisome.

Initiation of Synthesis

Once a small portion of the helix is unwound, initiation of synthesis may occur. As previously mentioned, DNA polymerase III requires a free 3′ end in order to elongate a polynucleotide chain. This prompted researchers to investigate how the first nucleotide can be added, since no free 3′-hydroxyl group is initially present. There is now evidence that, at least on one strand, RNA is involved in initiating DNA synthesis. It is thought that a short segment of RNA, complementary to DNA, is first synthesized on the DNA template. The RNA is made under the direction of a form of the enzyme RNA polymerase called **RNA primase.** The short segment of RNA that is synthesized serves as a primer, and it is to this primer that DNA polymerase III begins to add 5′-deoxyribonucleotides (Figure 8.9). RNA primase does not require a free 3′ end to initiate synthesis. After DNA synthesis occurs at an area adjacent to the RNA primer, the RNA segment is clipped off and replaced with DNA. Both steps are thought to be performed by DNA polymerase I. RNA priming has been recognized in viruses, bacteria, and several eukaryotic organisms and is thought to be a universal phenomenon.

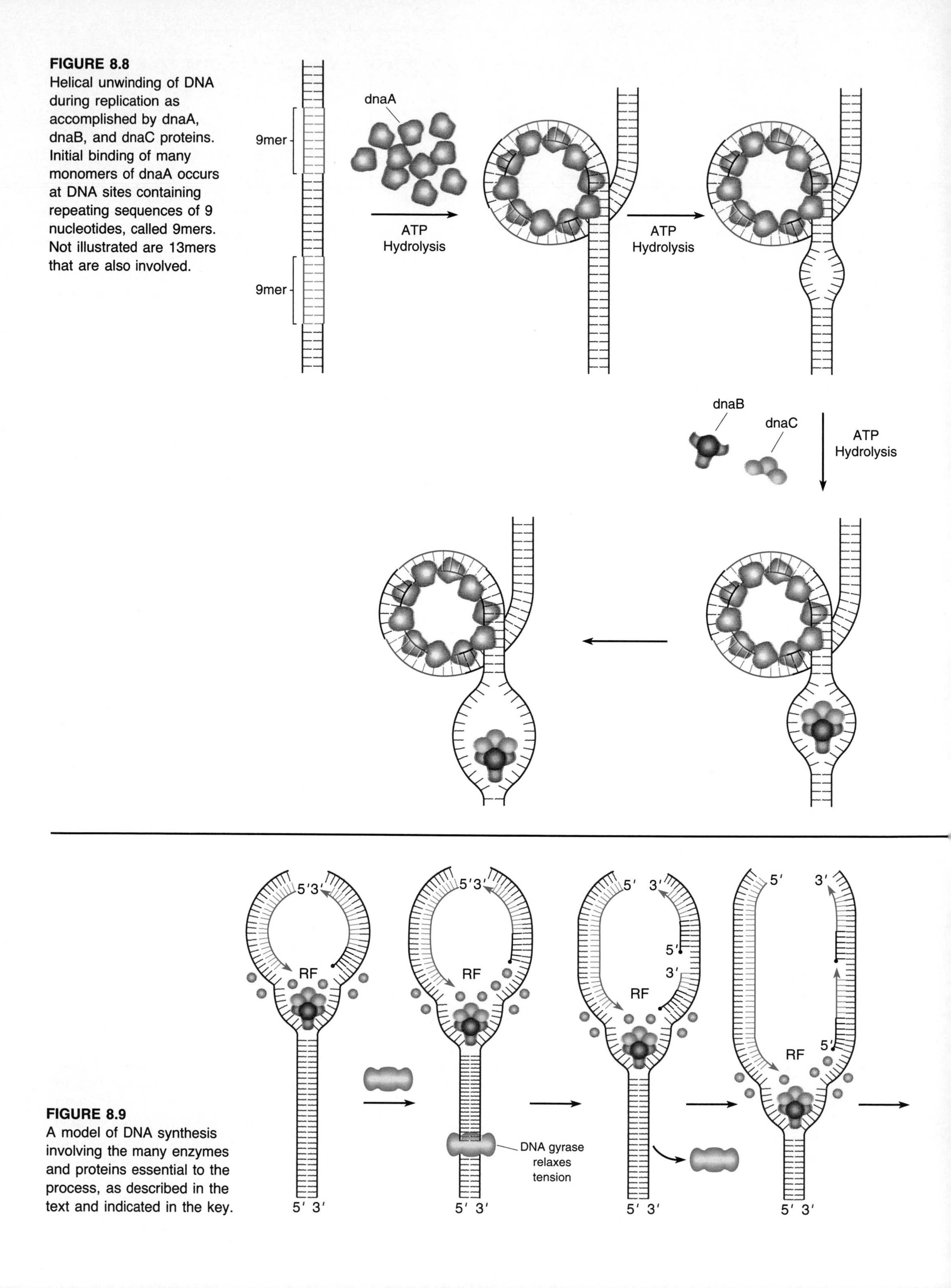

FIGURE 8.8
Helical unwinding of DNA during replication as accomplished by dnaA, dnaB, and dnaC proteins. Initial binding of many monomers of dnaA occurs at DNA sites containing repeating sequences of 9 nucleotides, called 9mers. Not illustrated are 13mers that are also involved.

FIGURE 8.9
A model of DNA synthesis involving the many enzymes and proteins essential to the process, as described in the text and indicated in the key.

Continuous and Discontinuous DNA Synthesis

We must now reconsider the fact that the two strands of a double helix are antiparallel to each other. One runs in the 5′-to-3′ direction, while the other has the opposite 3′-to-5′ polarity. Since DNA polymerase III synthesizes DNA in only the 5′-to-3′ direction, simultaneous synthesis of antiparallel strands along an advancing replication fork must occur in one direction along one strand and in the opposite direction on the other.

Thus, as the strands unwind and the replication fork progresses down the helix, only one strand can serve as a template for **continuous synthesis.** This strand is called the **leading strand.** As the fork progresses, many points of initiation are necessary on the opposite, or **lagging strand** resulting in **discontinuous synthesis** (Figure 8.10).

Evidence in support of discontinuous synthesis was first provided by Reiji and Tuneko Okazaki and their colleagues. They discovered that when bacteriophage DNA is replicated in *E. coli,* some of the newly formed DNA is found as small fragments containing 1000 to 2000 nucleotides. RNA primers are part of each such fragment. These pieces, called **Okazaki fragments,** must then be enzymatically joined to create longer DNA strands. As synthesis proceeds, the Okazaki fragments of low molecular weight are indeed converted into longer and longer DNA strands of higher molecular weight.

Discontinuous synthesis of DNA requires the removal of the RNA primer as well as an enzyme that can unite the smaller products into the longer continuous molecule. **DNA ligase** has been shown to be capable of performing this latter function. The evidence that DNA ligase does perform this function during DNA synthesis is strengthened by the observation of a **ligase-deficient mutant strain** of *E. coli.* In this strain, Okazaki fragments accumulate in particularly large amounts. Apparently, they are not joined adequately.

Discontinuous synthesis on the lagging strand, as described above, is characteristic of bacteria as well as eukaryotic cells.

GENETIC CONTROL OF REPLICATION

Much of what we know and have outlined in the previous section concerning the details of DNA replication in viruses and bacteria has been based on the genetic analysis of the process. For example, we have already discussed the *polA1* mutation, the study of which revealed that DNA polymerase I is not the major enzyme responsible for replication. Many other mutations have been isolated that interrupt or seriously impair some aspect of replication, such as the ligase-deficient mutation mentioned above. These can be studied if they are **temperature-sensitive mutations** (one type of **conditional mutation**) that express the mutant condition at a restrictive temperature but function normally at a permissive temperature. Investigation of such temperature-sensitive mutants provides insights into the product and the associated function of the normal, nonmutant gene.

As shown in Table 8.3, the enzyme product or its general role in replication has been ascertained for a variety of genes in *E. coli.* For example, numerous mutations in

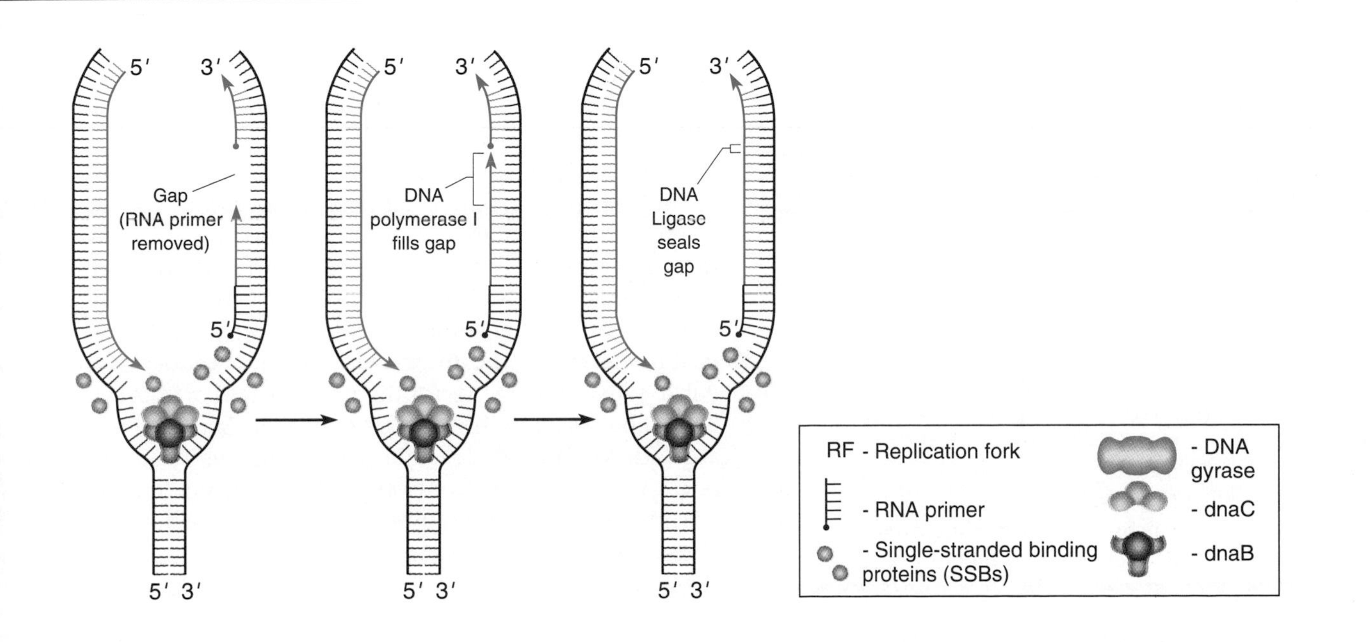

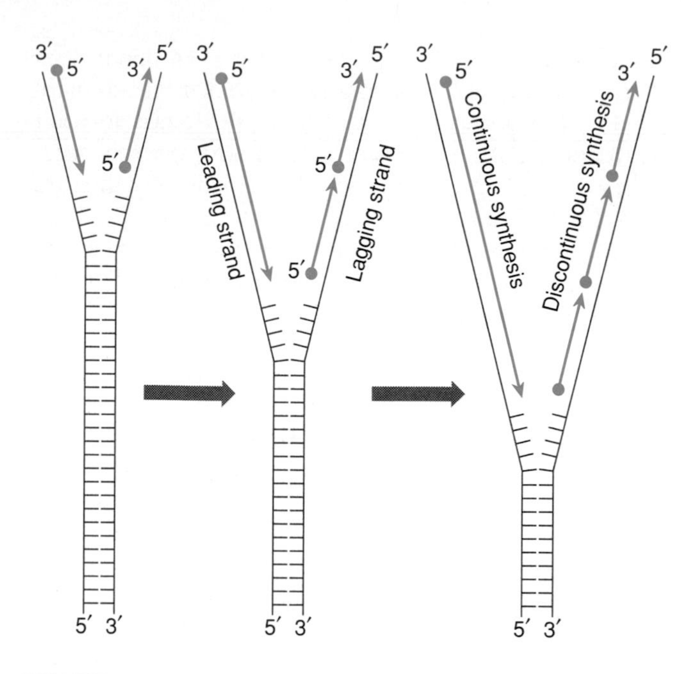

FIGURE 8.10
Illustration of the opposite polarity of DNA synthesis along the two strands necessitated by the requirement of 5′-to-3′ synthesis of DNA polymerase III. On the lagging strand, synthesis must be discontinuous, resulting in the production of Okazaki fragments. On the leading strand, synthesis is continuous.

genes specifying the subunits of polymerases I, II, and III have been isolated. Genes have also been identified that encode products involved in specification of the origin of synthesis, helix-unwinding and stabilization, initiation and priming, relaxation of supercoiling, repair, and ligation. The discovery of such a large group of genes attests to the complexity of the process of replication, even in the relatively simple prokaryote. This complexity is not unexpected, given the enormous quantity of DNA that must be unerringly replicated in a very brief time. As we will see, the process is even more difficult to investigate in eukaryotes.

TABLE 8.3
A list of various *E. coli* mutant genes and their products or functions.

Mutant Gene	Enzyme or Role in Replication
polA	DNA polymerase I
polB	DNA polymerase II
dnaN,Q,X,Z	DNA polymerase III subunits
dnaG	Primase
dnaA,I,P	Initiation
dnaB,C	Helicase at *oriC*
oriC	Origin of replication
gyrA,B	Gyrase subunits
lig	Ligase
rep	Helicase
ssb	Single-stranded binding proteins
rpoB	RNA polymerase subunit

EUKARYOTIC DNA SYNTHESIS

Most features of DNA synthesis found in microorganisms are now thought to apply also to eukaryotic systems. However, the DNA of eukaryotes is complexed with a variety of proteins, some of which maintain the molecule's structural integrity. The presence of these proteins and the general complexity of the eukaryotic cell compared with the prokaryotic cell have made analysis much more difficult.

Eukaryotic cells contain four different kinds of DNA polymerases, called $\boldsymbol{\alpha}$, $\boldsymbol{\beta}$, $\boldsymbol{\gamma}$, and $\boldsymbol{\delta}$. It appears the α form is the major enzyme involved in the replication of nuclear DNA. The β form may be involved in DNA repair, while the γ form is the only DNA polymerase found in mitochondria. Presumably, it is unique in its function within that organelle. The δ form, like α, appears to function in nuclear DNA replication. DNA polymerases of eukaryotes have the same fundamental requirements for DNA synthesis as bacterial and viral systems: four deoxyribonucleoside triphosphates, a template, and a primer.

Data and observations derived from autoradiographic and electron microscopic studies have provided many other insights. As mentioned earlier, eukaryotic DNA synthesis is bidirectional, creating two replication forks from each point of origin. As would be expected because of the increased amount of DNA in eukaryotes compared to prokaryotes, many more points of origin are present. In mammals, there are about 25,000 replicons present in the genome, each consisting of an average of 100,000 to 200,000 base pairs (100–200 kb). In *Drosophila,* there are about 3500 replicons per genome of an average size of 40 kb. The many points of origin are thought not to be random along each chromosome, but instead to represent specific points of initiation of DNA synthesis. Not all origins are activated at the same time during the S phase of the cell cycle. The presence of smaller replicons in eukaryotes compensates for their slower rate of DNA synthesis compared to that in prokaryotes. In *E. coli,* 100 kb are added to a growing chain per minute, while eukaryotic synthesis ranges from only 0.5 to 5 kb per minute. Nevertheless, *E. coli* requires 40 minutes to replicate its chromosome, while *Drosophila,* with 40 times more DNA, accomplishes the same task in only 3 minutes!

Even though DNA synthesis is slower in eukaryotes, most general aspects of chain elongation are thought to be similar. The actions of a variety of proteins—DNA helicase and SSBPs—are believed to modulate strand separation that precedes RNA priming by a primase enzyme. On the lagging strand, synthesis is discontinuous, resulting in Okazaki fragments that are linked together by DNA ligase. These fragments are about 10 times smaller (100–150 nucleotides) than in prokaryotes. While synthesis may occur continuously on the leading strand, it is not clear whether or not it can proceed uninterrupted for the full length of a replicon. Thus, synthesis on the leading strand may be **semidiscontinuous.**

DNA Synthesis at the Ends of Chromosomes

One final aspect of eukaryotic versus prokaryotic DNA synthesis involves the fact that eukaryotic chromosomes are linear compared to the circular forms displayed by bacteria, as well as most bacteriophages. A special problem is encountered at the "ends" of linear molecules of DNA during replication. Such open ends are part of each telomeric region of each chromosome.

While synthesis can proceed normally to the end of the leading strand, a problem is encountered on the **lagging strand.** The 3′ end of the lagging strand is an inadequate template for the synthesis of the RNA primer essential to replication. The result is that normal replication procedures will leave a gap on the end of the complement of the lagging strand following each replication cycle (Figure 8.11a).

In bacteria and viruses that contain circular DNA molecules this is not a problem, since no free ends are encountered during replication. In eukaryotes, the discovery of a unique enzyme, **telomerase,** has helped us understand how some organisms solve this problem. In the ciliated protozoan, *Tetrahymena,* the many telomeres terminate in the sequence 5′-TTGGGG-3′. The enzyme is capable of adding repeats of TTGGGG to the ends of molecules already containing this sequence. This process, now known to occur in other organisms under the direction of a similar enzyme, prevents the telomeric ends from shortening following each replication (Figure 8.11b).

Further investigation of the *Tetrahymena* telomerase enzyme, isolated and studied extensively by Elizabeth Blackburn and Carol Greider, has yielded an extraordinary finding. This enzyme adds the same TTGGGG sequence to other DNA termini. Thus, it is not using a specific terminus as a template to somehow repeat the sequence. Blackburn and Greider have now established how the enzyme works. They have discovered that the enzyme contains a short piece of RNA that is essential to its catalytic activity! The functional enzyme is thus a ribonucleoprotein. The RNA component encodes the sequences added by the enzyme, serving as its own template. The RNA contains 129 bases, including the sequence 5′-CCCCAA-3′, which is complementary to the sequence whose synthesis it directs. Analogous enzyme functions have now been found in still other organisms, including the ciliate *Euplotes* and cultured human cells. In both cases, the RNA-containing telomerase enzyme behaves like the retroviral enzyme, reverse transcriptase, in that it synthesizes the DNA complement of an RNA template. In this case, the template is much shorter, and the enzyme supplies its own!

The analysis of telomeric DNA sequences, discussed in more detail in Chapter 12, has shown them to be highly conserved throughout evolution. Such conservation re-

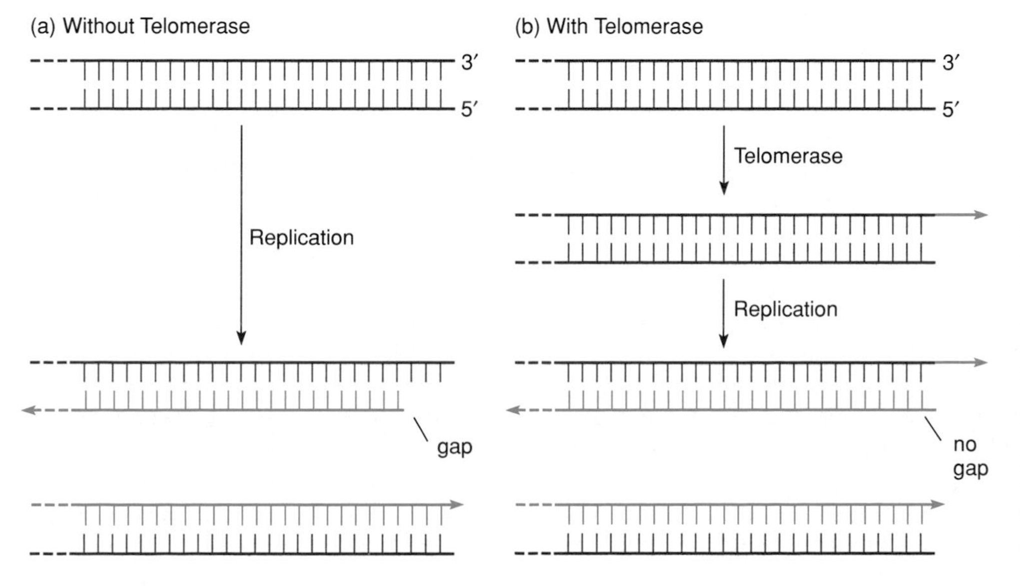

FIGURE 8.11
Replication of the telomere with and without the telomerase enzyme. New DNA synthesis is shown in blue. The dashed lines indicate the nonterminus portion of the chromosome, while the 3′ and 5′ designations identify the terminus of the chromosome occupied by the telomere. Without the new synthesis provided by telomerase, a gap is created on the lagging strand following each round of replication.

flects not only the critical function of telomeres, but also the unusual nature of their DNA replication.

DNA RECOMBINATION

We conclude this chapter by returning to a topic discussed in Chapter 5—**genetic recombination.** There, it was pointed out that the process of crossing over depends on breakage and rejoining of the DNA strands between homologues. Now that we have discussed the chemistry and replication of DNA, it is appropriate to consider how recombination occurs at the molecular level. In general, the following information pertains to genetic exchange between any two homologous double-stranded DNA molecules, whether they be viral or bacterial chromosomes or eukaryotic homologues during meiosis.

While there are several models available to explain crossing over, they all share certain common features. First, all are based on the initial proposals put forth independently by Robin Holliday and Harold L. K. Whitehouse in 1964. They also depend on the complementarity between DNA strands for their precision of exchange. Finally, each model relies on a series of enzymatic processes in order to accomplish genetic recombination.

One such model is illustrated in Figure 8.12. It begins with two paired DNA duplexes or homologues (a), each of which has a single-stranded nick introduced (b) at an identical position by an endonuclease. The ends of the strands produced by these cuts are then displaced and subsequently pair with their complements on the opposite duplex (c). A ligase then seals the loose ends (d), creating hybrid duplexes called **heteroduplex DNA molecules.** The exchange creates a cross-bridged or **Holliday structure.** The position of this cross-bridge can then move down the chromosomes by branch migration (e). This occurs as a result of a zipperlike action as hydrogen bonds are broken and then reformed between complementary bases of the displaced strands of each duplex. This migration yields an increased length of heteroduplex DNA on both homologues.

If the duplexes now separate (f), and the bottom portions rotate 180° (g), an intermediate planar structure called a **chi form** is created. If the two strands on opposite homologues previously uninvolved in the exchange are now nicked by an endonuclease (h), and ligation occurs (i), recombinant duplexes are created. Note that the arrangement of alleles is altered as a result of the crossover that occurs.

Evidence supporting the above model includes the electron microscopic visualization of chi-form planar molecules from bacteria where four duplex arms are joined at a single point of exchange (Figure 8.12). Additionally, the discovery in *E. coli* of the ***recA protein*** provides important evidence. This molecule promotes the exchange of reciprocal single-stranded DNA molecules as must occur in step (c) of the model. Further, *recA* enhances the hydrogen bond formation during strand displacement, thus initiating heteroduplex formation. Finally, all enzymes essential to the nicking and ligation process have been discovered and investigated. Mutations that prevent genetic recombination have been found in a number of genes in viruses and bacteria. These are thought to represent genes, the products of which play an essential role in this process.

Gene Conversion

A modification of the above model has helped us to better understand a unique genetic phenomenon known as **gene conversion.** Initially found in yeast by Carl Lindegren and in *Neurospora* by Mary Mitchell, gene conversion is characterized by a genetic exchange ratio involving two closely linked genes that is nonreciprocal. If one were to cross two *Neurospora* strains each bearing a separate mutation ($a+ \times +b$), a reciprocal recombination event between the genes would yield spore pairs of the $++$ and *ab* genotypes. However, a nonreciprocal exchange yields one pair without the other. Working with pyridoxine mutants, Mitchell observed such a ratio. She found several asci with the $++$ genotype, but not the reciprocal product (*ab*). Since the frequency of these events was higher than the predicted mutation rate, and thus could not be accounted for by that phenomenon, they were called *gene conversions.* They were so-named because it appeared that one allele had somehow been "converted" to another during an event where genetic exchange also occurred. Similar findings are apparent in the study of other fungi as well.

This phenomenon is now considered to be a consequence of the process of genetic recombination, as discussed in the previous section. During this process, heteroduplex formation is most often accompanied by mismatched bases. These appear as "errors" to the enzyme system that is capable of correcting these mismatches. It is during this repair process that the conversion step occurs. We will return to the topic of DNA repair (such as this) in Chapter 11.

Gene conversion events have helped to explain other puzzling genetic phenomena in fungi. For example, when mutant and wild type alleles of a single gene are studied in a cross, asci should yield equal numbers of mutant and wild type spores. However, exceptional asci with 3:1 or 1:3 ratios are sometimes observed. These ratios are more easily understood when interpreted in terms of gene conversion. The phenomenon has also been detected during mitotic events in fungi as well as during the study of unique compound chromosomes in *Drosophila.*

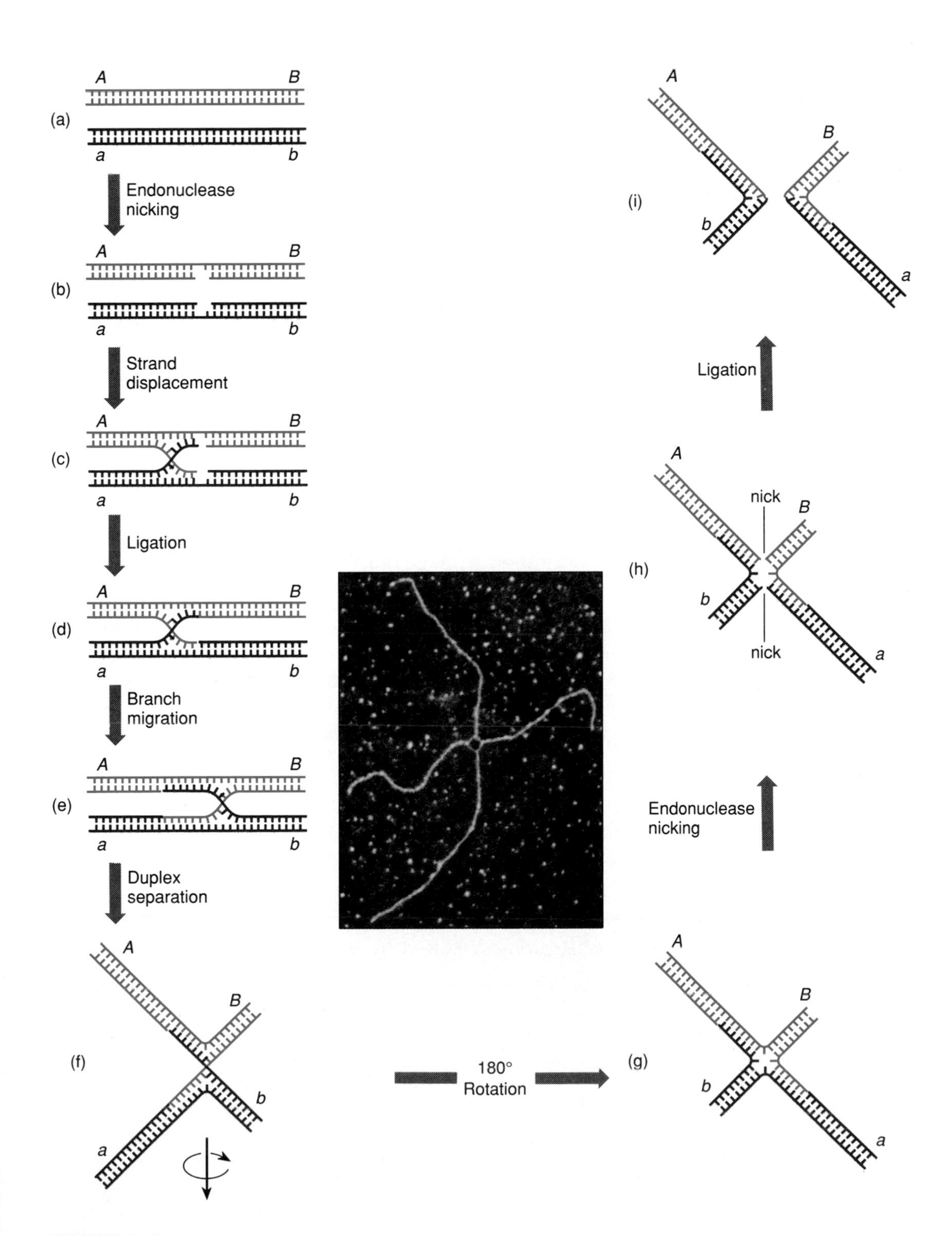

FIGURE 8.12
A possible molecular sequence depicting how genetic recombination occurs as a result of the breakage and rejoining of heterologous DNA strands. Each stage is described in the text. The electron micrograph shows DNA in a chi-form structure similar to the diagram in (g). The DNA is from the Col E1 plasmid of *E. coli*.

CHAPTER SUMMARY

1 In theory, three modes of DNA replication are possible: semiconservative, conservative, and dispersive. Though all three rely on base complementarity, semiconservative replication is the most straightforward and was predicted.

2 In 1958, Meselson and Stahl resolved this problem in favor of semiconservative replication in *E. coli,* showing that newly synthesized DNA consists of one old strand and one new strand. Taylor, Woods, and Hughes used root tips of the broad bean to demonstrate semiconservative replication in eukaryotes.

3 During the same period, Kornberg isolated DNA polymerase I from *E. coli* and demonstrated it to be an enzyme capable of *in vitro* DNA synthesis, provided that a template and precursor nucleoside triphosphates were supplied.

4 The subsequent discovery of the *polA1* mutant strain of *E. coli,* capable of DNA replication in spite of its lack of polymerase I activity, cast doubt on this enzyme's *in vivo* replicative function. DNA polymerases II and III were then isolated. Polymerase III has been identified as the enzyme responsible for DNA replication *in vivo.*

5 During the process of DNA synthesis, the double helix unwinds, forming a replication fork where synthesis begins. Proteins stabilize the unwound helix and assist in relaxing the coiling tension created ahead of the replication activity.

6 Synthesis is initiated at specific sites along each template strand by RNA primase, which results in a short segment of RNA that provides a suitable 3′ end, upon which DNA polymerase III can begin polymerization.

7 Due to the antiparallel nature of the double helix, polymerase III synthesizes DNA continuously on the leading strand in a 5′-to-3′ direction. On the opposite strand, called the lagging strand, synthesis results in short Okazaki fragments that are later joined by DNA ligase.

8 DNA polymerase I removes and replaces the RNA primer with DNA, which is joined to the adjacent polynucleotide by DNA ligase.

9 The isolation of numerous phage and bacterial mutant genes affecting many of the molecules involved in the replication of DNA has helped to define the complex genetic control of the entire process.

10 DNA replication in eukaryotes is similar to but more complex than replication in prokaryotes. For example, replication at the ends (telomeres) of linear molecules poses a special problem, solved by a unique RNA-containing enzyme called telomerase.

11 Recombination between genetic molecules relies on a series of enzymes that can cut, realign, and reseal DNA strands. The phenomenon of gene conversion, where one allele appears to be converted to another, relies on repair enzymes correcting mismatches of bases during the process of recombination.

KEY TERMS

bidirectional replication
biologically active DNA
chi form
conditional mutation
conservative replication
continuous DNA synthesis
discontinuous DNA synthesis
dispersive replication
dnaA proteins
dnaB proteins
dnaC proteins
DNA gyrase
DNA helicases
DNA ligase
DNA polymerase I, II, III
DNA topoisomerases
gene conversion
genetic recombination
heteroduplex DNA molecules
Holliday structure
holoenzyme
lagging strand
leading strand
Okazaki fragments
oriC
origin of replication
primer
recA protein
replication fork
replicon
replisome
RNA primase
sedimentation equilibrium centrifugation
semiconservative replication
semidiscontinuous DNA synthesis
single-stranded binding proteins (SSBP)
temperature-sensitive mutation
9mers
13mers

1 Predict the theoretical results of conservative and dispersive models of DNA synthesis using the conditions of the Meselson–Stahl experiment. Follow the results through two generations of replication after cells have been shifted to ^{14}N-containing medium, using the following sedimentation pattern:

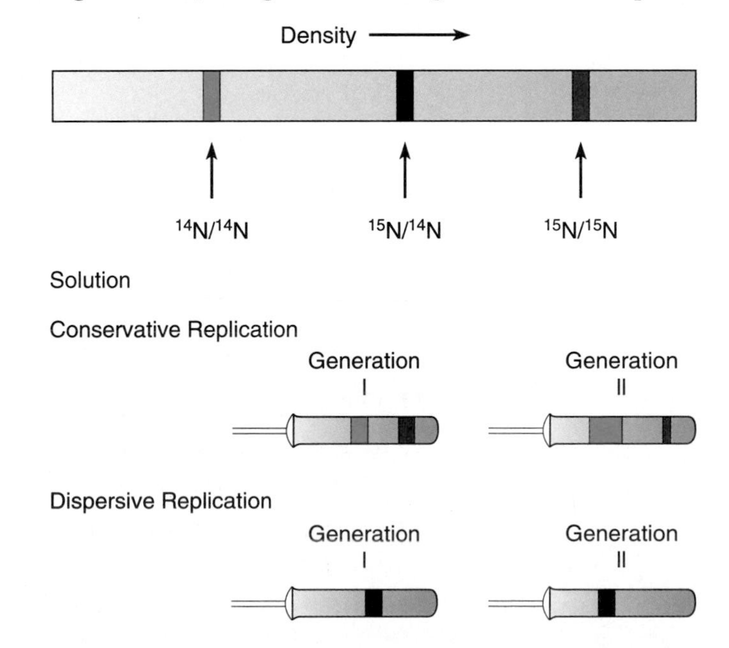

2 Mutations in the *dnaA* gene of *E. coli* are lethal and can only be studied following the isolation of conditional, temperature-sensitive mutations. Such mutant strains grow nicely and replicate their DNA at the permissive temperature of 18°C, but they do not grow or replicate their DNA at the restrictive temperature of 37°C. Two observations were useful in determining the function of the *dnaA* gene product. First, *in vitro* studies using DNA templates that have been nicked (opened) do not require the *dnaA* protein. Second, if intact cells are grown at 18°C and then shifted to 37°C, DNA synthesis continues at this temperature until one round of replication is completed, and DNA synthesis stops. What do these observations suggest about the role of the *dnaA* gene product?

ANSWER: These observations suggest that *in vivo* the *dnaA* protein is essential to the initiation of DNA synthesis. At 18°C (the permissive temperature) the mutation is not expressed and DNA synthesis begins. Following the shift to the restrictive temperature, DNA synthesis already initiated continues, but no new synthesis can begin. Since the *dnaA* protein is not required for synthesis of "nicked" DNA, this observation suggests that the protein functions during initiation by interacting with the intact helix and somehow facilitating the localized denaturing necessary for synthesis to proceed. In fact, both conclusions are valid.

PROBLEMS AND DISCUSSION QUESTIONS

1 Compare conservative, semiconservative, and dispersive modes of DNA replication.

2 In the Meselson–Stahl experiment, which of the three modes of replication could be ruled out after one round of replication? After two rounds?

3 Predict the results of the experiment by Taylor, Woods, and Hughes if replication were (a) conservative and (b) dispersive.

4 What are the requirements for the *in vitro* synthesis of DNA under the direction of DNA polymerase I?

5 What is meant by "biologically active" DNA?

6 Why was the phage ϕX174 chosen for the experiment demonstrating biological activity?

7 What was the significance of the *polA1* mutation?

8 Summarize the properties of polymerase I, II, and III.

9 List the proteins that unwind DNA during *in vivo* DNA synthesis. How do they function?

10 Define and indicate the significance of (a) Okazaki fragments, (b) DNA ligase, and (c) primer RNA.

11 Outline the current model for DNA synthesis.

12 Why should DNA synthesis be more complex in eukaryotes than in bacteria? How is DNA synthesis similar in the two types of organisms?

13 Describe why DNA synthesis is "problematic" at the telomeres. What is unique about the enzyme that makes synthesis possible?

14 Define gene conversion. How does it occur?

SELECTED READINGS

DARNELL, J., LODISH, H., and BALTIMORE, D. 1990. *Molecular cell biology.* 2nd ed. New York: Scient. Amer. Books.

DeLUCIA, P., and CAIRNS, J. 1969. Isolation of an *E. coli* strain with a mutation affecting DNA polymerase. *Nature* 224:1164–66.

DENHARDT, D. T., and FAUST, E. A. 1985. Eukaryotic DNA replication. *BioEssays* 2:148–53.

DRESLER, D., and POTTER, H. 1982. Molecular mechanisms in genetic recombination. *Ann. Rev. Biochem.* 51:727–61.

HOLLIDAY, R. 1964. A mechanism for gene conversion in fungi. *Genet. Res.* 5:282–304.

HUBERMAN, J. C. 1987. Eukaryotic DNA replication: A complex picture partially clarified. *Cell* 48:7–8.

KORNBERG, A. 1960. Biological synthesis of DNA. *Science* 131:1503–8.

———. 1974. *DNA synthesis.* New York: W. H. Freeman.

———. 1979. Aspects of DNA replication. *Cold Spr. Harb. Symp.* 43:1–10.

KORNBERG, A., and BAKER, T. A. 1992. *DNA replication.* 2nd ed. New York: W. H. Freeman.

LINDEGREN, C. C. 1953. Gene conversion in *Saccharomyces. J. Genet.* 51:625–37.

MESELSON, M., and STAHL, F. W. 1958. The replication of DNA in *Escherichia coli. Proc. Natl. Acad. Sci.* 44:671–82.

MITCHELL, M. B. 1955. Aberrant recombination of pyridoxine mutants of *Neurospora. Proc. Natl. Acad. Sci.* 41:215–20.

RADMAN, M., and WAGNER, R. 1988. The high fidelity of DNA duplication. *Scient. Amer.* (August) 259:40–46.

STAHL, F. W. 1979. *Genetic recombination: Thinking about it in phage and fungi.* New York: W. H. Freeman.

———. 1987. Genetic recombination. *Scient. Amer.* (Feb.) 256:90–101.

TAYLOR, J. H., WOODS, P. S., and HUGHES, W. C. 1957. The organization and duplication of chromosomes revealed by autoradiographic studies using tritium-labeled thymidine. *Proc. Natl. Acad. Sci.* 48:122–28.

WANG, J. C. 1982. DNA topoisomerases. *Scient. Amer.* (July) 247:94–108.

WATSON, J. D., et al. 1987. *Molecular biology of the gene.* Vol. 1. *General principles.* 4th ed. Menlo Park, CA: Benjamin/Cummings.

WHITEHOUSE, H. L. K. 1982. *Genetic recombination: Understanding the mechanisms.* New York: Wiley.

Storage and Expression of Genetic Information

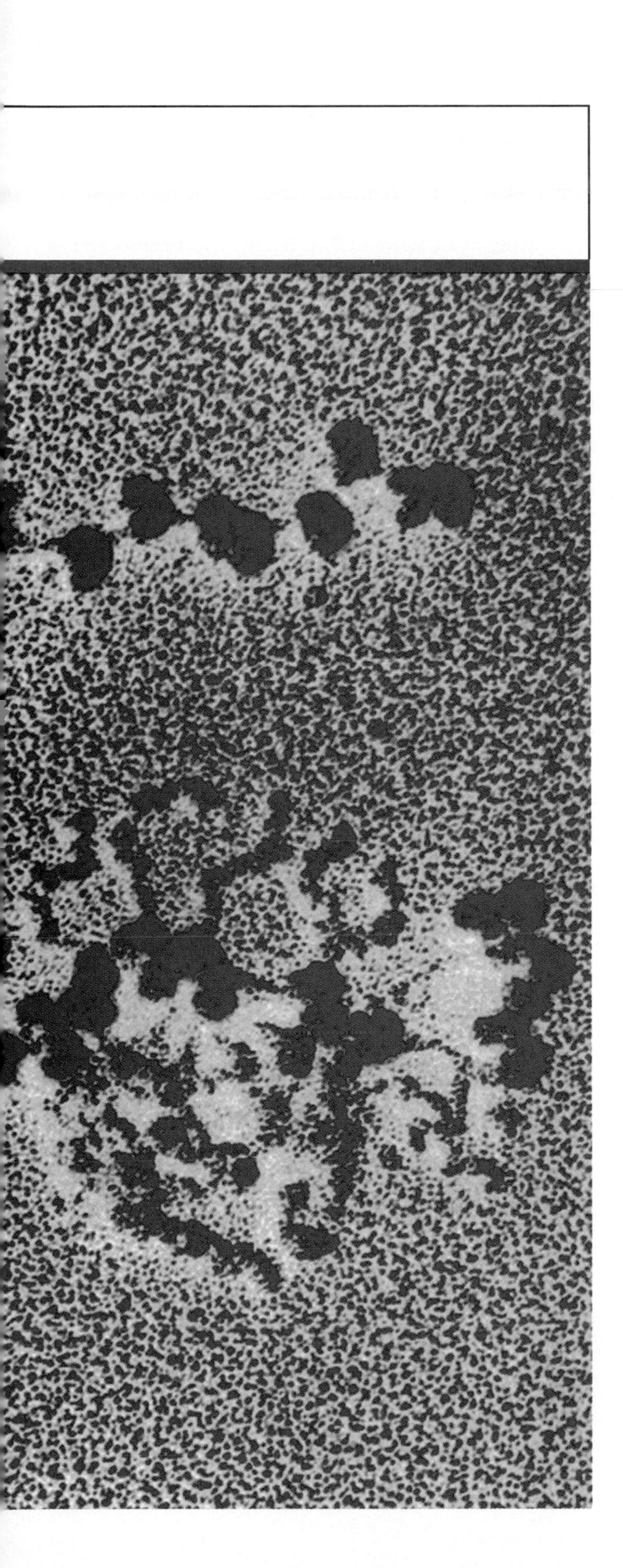

Chapter Concepts

Genetic information, stored in DNA and transferred to RNA during the process of transcription, is present as three-letter code words. Using four different letters, corresponding to the four ribonucleotides in RNA, the 64 possible triplet codons specify the 20 different amino acids found in proteins, as well as provide signals that initiate and terminate protein synthesis. This unique language is the basis of life as we know it.

In previous chapters, we have established that DNA is the genetic material and elucidated the molecular structure of DNA. This molecule provides the chemical basis for storage of genetic information. Although we have yet to discuss the details of the processes, we know that the DNA of an active gene is transcribed into an RNA complement that is translated ultimately into the end product of the gene—a protein. Thus, the sequence of deoxyribonucleotides in DNA is first transferred to the complementary sequence of ribonucleotides in RNA, which then specifies the insertion of amino acids into protein. The end products of various genes are responsible for the normal and mutant phenotypes of organisms.

A fundamental question, then, is how an RNA molecule consisting of only four different types of nucelotides (A, U, C, and G) can specify 20 different amino acids. This question poses an intriguing theoretical problem. Although the earliest proposals were imaginative, it was not until ingenious analytical research was applied that the code was deciphered. It was established that the code is triplet in nature. Code words, or **codons,** consisting of three ribonucleotides direct the insertion of amino acids into a polypeptide chain during its synthesis.

The work leading to these discoveries occurred most intensively in the late 1950s and early 1960s, one of the most exciting periods in the study of molecular genetics. This research revealed the intricacies of the specific chemical language that serves as the basis of all life on earth.

THE GENETIC CODE: AN OVERVIEW

Before considering the various analytical approaches used in arriving at our current understanding of the genetic code, we shall provide a summary of the general features that characterize it.

1 The code is written in linear form using the ribonucleotide bases that compose mRNA molecules as the letters. The ribonucleotide sequence is, of course, derived from its complement in DNA.

2 Each code word within the mRNA contains three letters. Thus, the code is a **triplet.** Called a **codon,** each group of *three* ribonucleotides specifies *one* amino acid.

3 The code is **unambiguous,** meaning that each triplet specifies only a single amino acid.

4 The code is **degenerate,** meaning that more than one triplet may specify a given amino acid. This is the case for 18 of the 20 amino acids.

5 The code contains "start" and "stop" punctuation signals. Certain triplets are necessary to initiate and to terminate translation.

6 No commas (or internal punctuation) are used in the code. Thus, the code is said to be **commaless.** Once translation of mRNA begins, each triplet codon is read in turn, one after the other.

7 The code is **nonoverlapping.** Once translation commences, any single ribonucleotide at a specific location within the mRNA is part of only one triplet.

8 The code is nearly **universal.** With only minor exceptions, a single coding dictionary is used by almost all viruses, prokaryotes, and eukaryotes.

EARLY THINKING ABOUT THE CODE

Before it became clear that mRNA serves as an intermediate in transferring genetic information from DNA to proteins, it was thought that DNA itself might directly encode proteins during their synthesis. The central question, whether DNA or RNA houses the code, was how only four letters—the four nucleotides—could specify 20 words—the amino acids. Once mRNA was discovered, it was clear that even though genetic information is stored in DNA, the code that is translated into proteins resides in RNA.

As to the size of the code, Sidney Brenner argued on theoretical grounds that it must be a triplet since three-letter words represent the minimal use of four letters to specify 20 amino acids. For example, four nucleotides, taken two at a time provide only 16 unique code words (4^2). While a triplet code provides 64 words (4^3)—clearly more than the 20 needed—it is much simpler than a four-letter code, where 256 words (4^4) would be specified. Brenner also argued that the code was nonoverlapping.

By 1960, many significant questions had been posed, and the scene was clearly set for the design of experimentation to decipher the code. In 1961, several noteworthy events occurred. First, Francois Jacob and Jacques Monod formally postulated the existence of an intermediate messenger RNA (mRNA), suggesting how investigators must design experiments aimed at deciphering the code.

Second, research by Francis Crick and colleagues provided some of the earliest experimental evidence concerning the nature of the code. Studying phage T4 and the unique type of mutation called a **frameshift,** they provided support for the triplet nature of the code.

Frameshift mutations occur as a result of the addition or deletion of one or more nucleotides in the gene and subsequently the mRNA transcribed by it. The gain or the loss of one or more letters shifts the frame of reading during translation. They found that the gain *or* loss of one *or* two nucleotides caused such a mutation, but that when three nucleotides were involved, the frame of reading was reestablished (Figure 9.1). This would not occur if

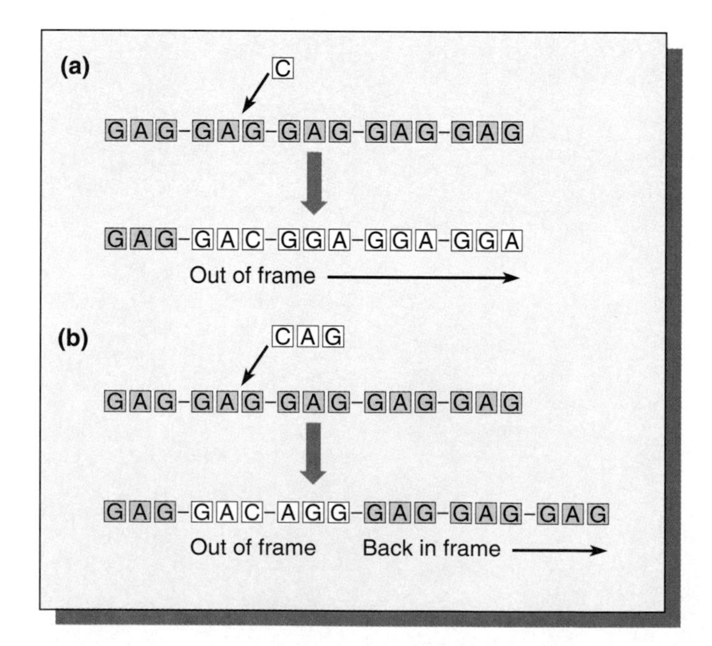

FIGURE 9.1
Schematic diagram of the effect of frameshift mutations on a DNA sequence repeating the triplet sequence GAG. In part (a) the insertion of a single nucleotide has shifted all subsequent reading frames. In part (b) the insertion of three nucleotides changes only two reading frames, but all remaining frames are retained in their original sequence.

the code consisted of anything other than a triplet. Their work also suggested that most all triplet codes were *not* blank, but instead encoded amino acids, supporting the concept of a degenerate code.

DECIPHERING THE CODE: INITIAL STUDIES

In 1961 Marshall Nirenberg and J. Heinrich Matthaei published results characterizing the first specific coding sequences. These results served as a cornerstone for the complete analysis of the code. Their success was dependent on the use of two experimental tools: **a cell-free protein-synthesizing system,** and an enzyme, **polynucleotide phosphorylase,** which allowed the production of synthetic mRNAs. These mRNAs served as templates for polypeptide synthesis in the cell-free system.

In the cell-free (*in vitro*) system, amino acids can be incorporated into polypeptide chains. This *in vitro* mixture, as might be expected, must contain the essential factors for proteins synthesis in the cell: ribosomes, tRNAs, amino acids, and other molecules essential to translation. In order to follow protein synthesis, one or more of the amino acids must be radioactive. Finally, an mRNA must be added, which serves as the template to be translated.

In 1961 mRNA had yet to be isolated. However, the use of the enzyme polynucleotide phosphorylase allowed artificial synthesis of RNA templates, which could be added to the cell-free system. This enzyme, isolated from bacteria, catalyzes the reaction shown in Figure 9.2. Discovered in 1955 by Marianne Grunberg-Manago and Severo Ochoa, the enzyme functions metabolically in bacterial cells to degrade RNA. However, *in vitro,* with high concentrations of ribonucleoside diphosphates, the reaction can be "forced" in the opposite direction to synthesize RNA, as illustrated.

In contrast to RNA polymerase, polynucleotide phosphorylase requires no DNA template. As a result, ribonucleotides are assembled at random, according to the relative concentration of the four ribonucleoside diphosphates added to the reaction mixtures. **This point is absolutely critical to understanding the work of Nirenberg and others in the ensuing discussion.**

Taken together, the cell-free system for protein synthesis and the availability of synthetic mRNAs provides a means of deciphering the ribonucleotide composition of various triplets encoding specific amino acids.

Nirenberg and Matthaei's Homopolymer Codes

In their initial experiments, Nirenberg and Matthaei synthesized **RNA homopolymers,** each consisting of only one ribonucleotide. Therefore, the mRNA added to the *in vitro* system was either UUUUUU . . . , AAAAAA . . . , CCCCCC . . . , or GGGGGG. . . . In testing each mRNA, they were able to determine which, if any, amino acids were incorporated into newly synthesized proteins. They determined this by labeling one of the 20 amino acids added to the *in vitro* system and conducting a series of experiments, each with a different amino acid made radioactive.

For example, consider one of the initial experiments, where ^{14}C-phenylalanine was used (Table 9.1). From these and related experiments, Nirenberg and Matthaei concluded that the message poly U (polyuridylic acid) directs the incorporation of only phenylalanine into the homopolymer polyphenylalanine. Assuming a triplet code, they had determined the first specific codon assignment! UUU codes for phenylalanine.

In the same way, they quickly found that AAA codes for lysine and CCC codes for proline. Poly G did not serve as an adequate template, probably because the molecule folds back on itself. Thus, the assignment for GGG had to await other approaches. Note that the specific triplet codon assignments were possible only because of the use of homopolymers. This method yields only the composi-

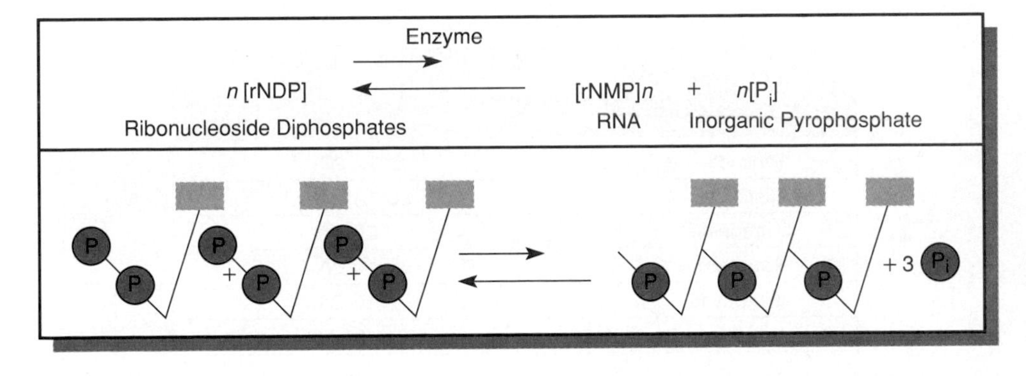

FIGURE 9.2
The reaction catalyzed by the enzyme polynucleotide phosphorylase. Note that the equilibrium of the reaction favors the degradation of RNA but can be "pushed" in favor of synthesis.

TABLE 9.1
Incorporation of ^{14}C-phenylalanine into protein.

Artificial mRNA	Radioactivity (counts/min)
None	44
Poly U	39,800
Poly A	50
Poly C	38

SOURCE: After Nirenberg and Matthaei, 1961, p. 1595.

tion of triplets, not the sequence. However, three U's, C's, or A's can have only one possible sequence (i.e., UUU, CCC, and AAA).

The Use of Mixed Copolymers

With these techniques in hand, Nirenberg and Matthaei, and Ochoa and coworkers turned to the use of **RNA heteropolymers.** In this technique, two or more different ribonucleoside diphosphates are added in combination to form the message. These researchers reasoned that if the relative proportion of each type of ribonucleoside diphosphate is known, the frequency of any particular triplet codon occurring in the synthetic mRNA can be predicted. If the mRNA is then added to the cell-free system and the percentage of any particular amino acid present in the new protein is ascertained, correlations may be made and composition assignments predicted.

This concept is illustrated in Figure 9.3. Suppose that A and C are added in a ratio of 1A:5C. Now, the insertion of a ribonucleotide at any position along the RNA molecule during its synthesis is determined by the ratio of A:C. Therefore, there is a 1/6 possibility for an A and a 5/6 chance for a C to occupy each position. On this basis, we can calculate the frequency of any given triplet appearing in the message.

For AAA, the frequency is $(1/6)^3$, or about 0.4 percent. For AAC, ACA, and CAA, the frequencies are identical—that is, $(1/6)^2(5/6)$, or about 2.3 percent. Together, all

FIGURE 9.3
Results and interpretation of a mixed copolymer experiment where a ratio of 1A:5C is used (1/6A:5/6C).

Possible Compositions	Probability of Occurrence of Any Triplet	Possible Triplets	Final %
3A	$(1/6)^3$ = 1/216 = 0.4%	AAA	0.4
2A:1C	$(1/6)^2(5/6)$ = 5/216 = 2.3%	AAC ACA CAA	3 × 2.3 = 6.9
1A:2C	$(1/6)(5/6)^2$ = 25/216 = 11.6%	ACC CAC CCA	3 × 11.6 = 34.8
3C	$(5/6)^3$ = 128/216 = 57.9%	CCC	57.9
			100.0%

Transcription of Synthetic Message

CCCCCCCCACCCCCCAACCACCCCCACCCCCACCCAAACCCCCACCCCCC RNA

Translation of Message

Percentage of Amino Acids in Protein		Probable Base Composition Assignments
Proline	69	CCC, 2C:1A
Histidine	14	2C:1A, 1C:2A
Threonine	12	2C:1A
Asparagine	2	1C:2A
Glutamine	2	1C:2A
Lysine	1	AAA

three 2A:1C triplets account for 6.9 percent of the total three-letter sequences. In the same way, each of three 1A:2C triplets accounts for $(1/6)(5/6)^2$, or 11.6 percent (or a total of 34.8 percent). CCC is represented by $(5/6)^3$, or 57.9 percent of the triplets.

By examining the percentages of any given amino acid incorporated into the protein synthesized under the direction of this message, we may make tentative composition assignments (Figure 9.3). Since proline appears 69 percent of the time and since 69 percent is close to 57.9 percent + 11.6 percent, we can deduce that proline is coded by CCC and by one triplet of the 2C:1A variety. Histidine, at 14 percent, is probably coded by one 2C:1A (11 percent) and one 1C:2A (2 percent). Threonine, at 12 percent, is likely coded by only one 2C:1A. Asparagine and glutamine appear each to be coded by one of the 1A:2C triplets, and lysine appears to be coded by AAA.

In similar experiments, some using as many as all four ribonucleotides to construct the mRNA, many possible combinations were determined for all amino acids. Although this represented a very significant breakthrough, specific sequences of triplets were still unknown. Their determination awaited still other approaches.

The Triplet Binding Technique

It was not long before more advanced techniques were developed. In 1964 Nirenberg and Philip Leder developed the **triplet binding assay,** which led to specific assignments of triplets. The technique took advantage of the observation that ribosomes, when presented with an RNA sequence as short as three ribonucleotides, will bind to it and attract the correct charged tRNA. For example, if ribosomes are presented with an RNA triplet UUU and $tRNA^{phe}$, a complex will form that is similar to what actually occurs *in vivo.* The triplet acts like a codon in mRNA and it attracts the complementary sequence called the **anticodon** found within tRNA (Figure 9.4). Anticodons are those triplet sequences in tRNAs that are complementary to the codons of mRNA. Although it was not yet feasible to chemically synthesize long stretches of RNA, triplets of known sequence could be constructed in the laboratory to serve as templates.

All that was needed now was a method to determine which tRNA-amino acid was bound to the triplet RNA-ribosome complex. The test system devised was quite simple. The amino acid to be tested is made radioactive, and a charged tRNA (tRNA bonded to its amino acid) is produced. Since code compositions were known, it was possible to narrow the decision as to which amino acids should be tested for each specific triplet.

The radioactive charged tRNA, the RNA triplet, and ribosomes are incubated together on a nitrocellulose filter, which will retain the ribosomes but not the other components, such as charged tRNA. If radioactivity is not retained on the filter, an incorrect amino acid has been tested. If radioactivity remains on the filter, it is retained because the charged tRNA has bound to the triplet associated with the ribosome. In such a case, a specific codon assignment may be made.

Work proceeded in several laboratories, and in many cases clearcut, unambiguous results were obtained. Table 9.2, for example, shows 25 triplets assigned to 10 amino acids. However, in some cases the degree of binding was insufficient, and unambiguous assignments were not possible. Eventually, about 50 of the 64 triplets were as-

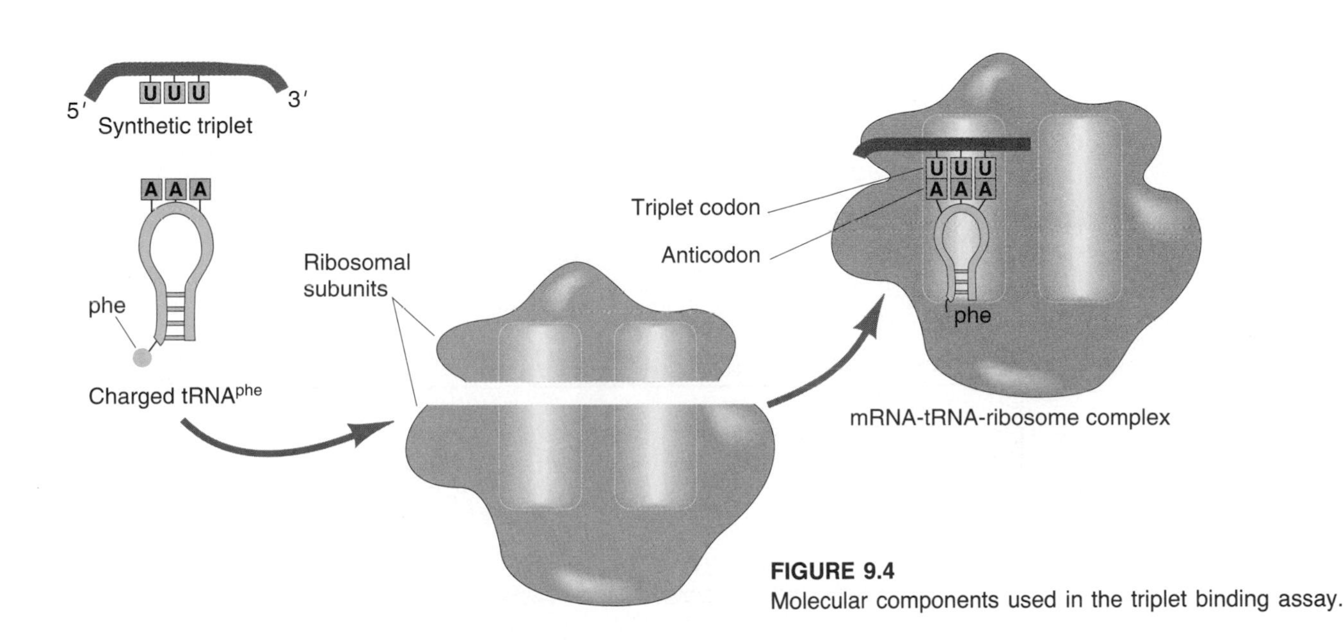

FIGURE 9.4
Molecular components used in the triplet binding assay.

TABLE 9.2
Amino acid assignments to specific trinucleotides derived from the triplet binding assay.

Trinucleotide	Amino Acid
UGU UGC	Cysteine
GAA GAG	Glutamic acid
AAU AUC AUA	Isoleucine
UUA UUG CUU	Leucine
CUC CUA CUG	Leucine
AAA AAG	Lysine
AUG	Methionine
UUU UUC	Phenylalanine
CCU CCC	Proline
CCG CCA	Proline
UCU UCC	Serine
UCA UCG	Serine
UCU UCC	Valine
UCA UCG	Valine

signed. The binding technique was a major innovation in deciphering the code. Based on these specific assignments of triplets to amino acids, two major conclusions were drawn. The genetic code is **degenerate;** that is, one amino acid may be specified by more than one triplet. The code is also **unambiguous;** that is, a single triplet specifies only one amino acid. As we shall see later in this chapter, these conclusions have been upheld with only minor exceptions.

The Use of Repeating Copolymers

Still another innovative technique used to decipher the genetic code was developed by Gobind Khorana. He was able to chemically synthesize long repeating RNA sequences, which could be used in the cell-free protein synthesizing system. First, he created shorter sequences (e.g., di-, tri-, and tetranucleotides), which were then replicated many times and finally joined enzymatically to form the long polynucleotides.

As illustrated in Figure 9.5, a dinucleotide made in this way is converted to a message with two repeating triplets. A trinucleotide is converted to one with three potential triplets, depending on the point at which initiation occurs. Similarly, a tetranucleotide creates four repeating triplets.

When these synthetic mRNAs are added to a cell-free system, the predicted number of different amino acids incorporated is upheld. Several examples are shown in Table 9.3. When such data are combined with conclusions drawn from other approaches (composition assignment, triplet bindings), specific assignments become possible.

From such interpretations, Khorana reaffirmed triplets already deciphered and filled in gaps left from other approaches. For example, the use of two tetranucleotide sequences, GAUA and GUAA, suggested to him that at least two triplets are termination signals. This conclusion was drawn because neither of these sequences directed the incorporation of any amino acids into a polypeptide

Repeating sequence	Polynucleotides	Repeating triplets
Dinucleotide UG	U G U G U G U G U G U G U G U G U . . . (Initiation)	UGU GUG
Trinucleotide UUG	U U G U U G U U G U U G U U G U U G U (Initiation)	UUG UGU GUU
Tetranucleotide UAUC	U A U C U A U C U A U C U A U C U A U C U (Initiation)	UAU CUA UCU AUC

FIGURE 9.5
The conversion of di-, tri-, and tetranucleotides into repeating copolymers. The triplet codons produced in each case are shown.

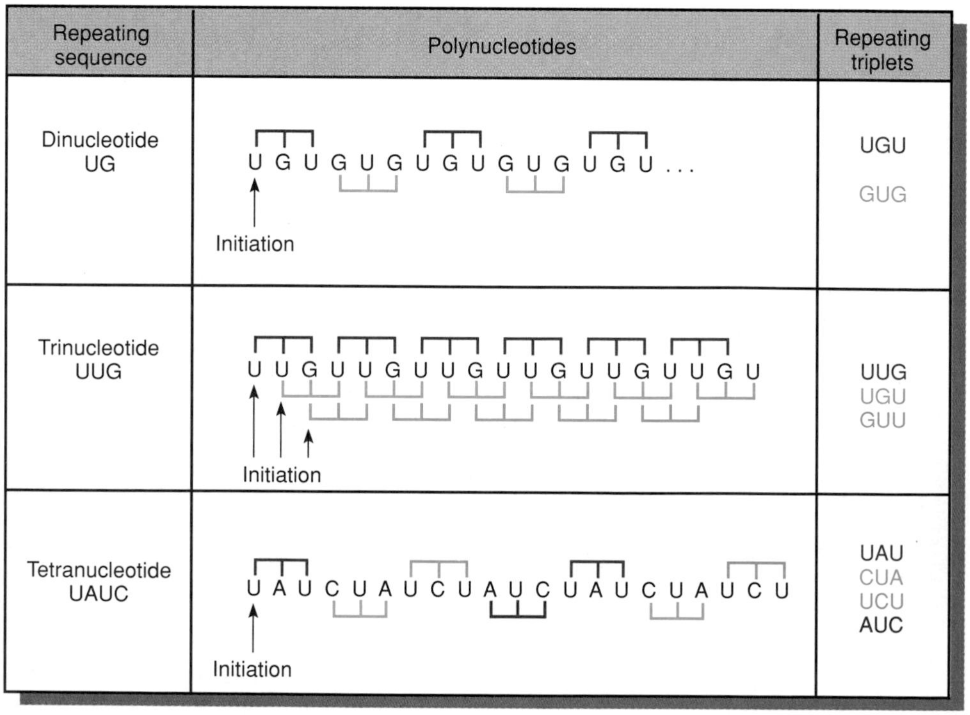

TABLE 9.3
Amino acids incorporated using repeating synthetic copolymers of RNA.

Repeating Copolymer	Codons Produced	Amino Acids in Polypeptide
UG	UGU	Cysteine
	GUG	Valine
AC	ACA	Threonine
	CAC	Histidine
UUC	UUC	Phenylalanine
	UCU	Serine
	CUU	Leucine
AUC	AUC	Isoleucine
	UCA	Serine
	CAU	Histidine
UAUC	UAU	Tyrosine
	CAU	Leucine
	UCU	Serine
	AUC	Isoleucine
GAUA	GAU	
	AGA	
	UAG	None
	AUA	

(Table 9.3). Since there are no triplets common to both messages, it was predicted that each repeating sequence contains at least one triplet that terminates protein synthesis. As we shall see, there are three such triplets, two of which are included in poly-(GAUA) and poly-(GUAA).

THE CODING DICTIONARY

The various techniques applied to decipher the genetic code have yielded a dictionary of 61 triplet codon-amino acid assignments. The remaining three triplets are termination signals, not specifying any amino acid. Figure 9.6 designates the assignments in a particularly illustrative form first suggested by Crick.

Degeneracy and Wobble

The degenerate nature of the code becomes apparent when we inspect this presentation of the genetic code. While the amino acids tryptophan and methionine are encoded by only single triplets, most amino acids are specified by two, three, or four triplets. Three amino acids (serine, arginine, and leucine) are coded by six triplets each.

What also becomes evident is the pattern of degeneracy. Most often in a set of codons specifying the same amino acid, the first two letters are the same, with only the third differing. This interesting pattern prompted Crick in 1966 to postulate the **wobble hypothesis.**

Crick concluded that the first two ribonucleotides of triplet codes are more critical than the third member in attracting the correct tRNA. Thus, Crick proposed that hydrogen bonding at the third position of the codon: anticodon interactions need not adhere as specifically as the first two members to the established base pairing rules. After examining the coding dictionary carefully, he proposed a new set of base pairing rules at the third position of the codon.

This relaxed base pairing requirement, or "wobble," allows the anticodon of a single tRNA species to pair with more than one triplet in mRNA. The code's degeneracy would often allow this to occur without changing the amino acid. Consistent with the coding assignments, it appears that U at the third position of the anticodon of tRNA may pair with A or G at the third position of the triplet in mRNA, and that G may likewise pair with U or C. Inosine, one of the modified bases found in tRNA, may pair with C, U, or A. Applying these wobble rules, a minimum of about 30 different tRNA species is necessary to accommodate the 61 triplets specifying an amino acid. If nothing more, wobble can be considered an economy measure, provided that fidelity of translation is not compromised.

Second Position

First Position (5′ end)	U	C	A	G	Third Position (3′ end)
U	UUU phe	UCU ser	UAU tyr	UGU cys	U
	UUC phe	UCC ser	UAC tyr	UGC cys	C
	UUA leu	UCA ser	UAA Stop	UGA Stop	A
	UUG leu	UCG ser	UAG Stop	UGG trp	G
C	CUU leu	CCU pro	CAU his	CGU arg	U
	CUC leu	CCC pro	CAC his	CGC arg	C
	CUA leu	CCA pro	CAA gln	CGA arg	A
	CUG leu	CCG pro	CAG gln	CGG arg	G
A	AUU ile	ACU thr	AAU asn	AGU ser	U
	AUC ile	ACC thr	AAC asn	AGC ser	C
	AUA ile	ACA thr	AAA lys	AGA arg	A
	AUG met	ACG thr	AAG lys	AGG arg	G
G	GUU val	GCU ala	GAU asp	GGU gly	U
	GUC val	GCC ala	GAC asp	GGC gly	C
	GUA val	GCA ala	GAA glu	GGA gly	A
	GUG val	GCG ala	GAG glu	GGG gly	G

FIGURE 9.6
The coding dictionary. AUG encodes methionine, which initiates most polypeptide chains. All other amino acids except tryptophan, which is encoded by UGG, are represented by two to six triplets.

Initiation and Termination

Initiation of protein synthesis using an mRNA template is a highly specific process. In bacteria, the initial amino acid inserted into all polypeptide chains is a modified form of methionine—**N-formylmethionine (fmet).** Only one codon, AUG, codes for methionine, and it is sometimes called the **initiator codon.** However, when AUG appears internally in mRNA, unformylated methionine is inserted into the polypeptide chain. Rarely, still another triplet, GUG, specifies methionine during initiation. It is not clear why this occurs, since GUG normally encodes valine.

In bacteria, either the formyl group is removed from the initial methionine upon completion of synthesis of a protein, or the entire formylmethionine residue is removed. In eukaryotes, methionine is also the initial amino acid during polypeptide synthesis. However, it is not formylated.

As mentioned in the preceding section, three other triplets (UAA, UAG, and UGA) serve as punctuation signals and do not code for any amino acid. They are not recognized by a tRNA molecule, and termination of translation occurs when they are encountered. Mutations occurring internally in a gene that produce any of the three triplets also result in premature termination. As a result, only a partial polypeptide is synthesized, since it is prematurely released from the ribosome. When such a change occurs, it is called a **nonsense mutation.** The terms **amber** (UAG), **ochre** (UAA), and **opal** (UGA) have been used to distinguish these three triplet codons.

CONFIRMATION OF CODE STUDIES: PHAGE MS2

All aspects of the genetic code discussed so far yield a fairly complete picture. The code is triplet in nature, degenerate, unambiguous, and commaless, but contains punctuation with respect to start and stop signals. These individual principles have been confirmed by the detailed analysis of the RNA-containing bacteriophage MS2 by Walter Fiers and his coworkers.

MS2 is a bacteriophage that infects *E. coli.* Its nucleic acid (RNA) contains only about 3500 ribonucleotides, making up only three genes. These genes specify a coat protein, an RNA-directed replicase, and a maturation protein (the A protein). This simple system of a small genome and few gene products allowed Fiers and his colleagues to sequence the genes and their products. The amino acid sequence of the coat protein was completed in 1970, and the nucleotide sequence of the gene and a number of nucleotides on each end of it were reported in 1972.

The coat protein contains 129 amino acids, and the gene contains 387 nucleotides, as expected for a triplet code. Each amino acid and triplet corresponds in linear sequence to the correct codon in the RNA code word dictionary, providing direct proof of the **colinear relationship** between nucleotide sequence and amino acid sequence. The codon for the first amino acid is preceded by AUG, the common initiator codon; and the codon for the last amino acid is succeeded by two consecutive termination codons, UAA and UAG.

By 1976, the other two genes and their protein products were sequenced, providing similar confirmation. The analysis clearly shows that the genetic code as established in bacterial systems is identical in this virus. We shall now briefly consider other evidence suggesting that the code is also identical in eukaryotes.

UNIVERSALITY OF THE CODE?

Between 1960 and 1978, it was generally assumed that the genetic code would be found to be universal, applying equally to viruses, bacteria, and eukaryotes. Certainly, the nature of mRNA and the translation machinery seemed to be very similar in these organisms. For example, cell-free systems derived from bacteria could translate eukaryotic mRNAs. Poly U was shown to stimulate translation of polyphenylalanine in cell-free systems when the components were derived from eukaryotes. Many recent studies involving recombinant DNA technology (see Chapter 15) have revealed that eukaryotic genes can be inserted into bacterial cells and transcribed and translated. Within eukaryotes, mRNAs from mice and rabbits have been injected into amphibian eggs and efficiently translated. For the many eukaryotic genes that have been sequenced, notably those for hemoglobin molecules, the amino acid sequence of the encoded proteins adheres to the coding dictionary established from bacterial studies.

However, several 1979 reports on the coding properties of DNA derived from yeast and human mitochondria (mtDNA) altered the principle of universality of the genetic language. Since then, mtDNA has been examined in many other organisms.

Not only do mitochondria contain DNA, but transcription and translation occur within these organelles. Cloned mtDNA fragments were sequenced and compared with the amino acid sequences of various mitochondrial proteins, revealing several exceptions to the coding dictionary (Table 9.4). Most surprising was that the codon UGA, normally causing termination, specifies the insertion of tryptophan during translation in yeast and human mitochondria. In human mitochondria, AUA, which normally specifies isoleucine, directs the internal insertion of methionine. In yeast mitochondria, threonine is inserted instead of leucine when CUA is encountered in mRNA.

TABLE 9.4
Exceptions to the universal code.

Triplet	Normal Code Word	Altered Code Word	Source
UGA	Termination	trp	Human and yeast mitochondria Mycoplasma
CUA	leu	thr	Yeast mitochondria
AUA	ile	met	Human mitochondria
AGA AGG	arg	Termination	Human mitochondria
UAA	Termination	gln	*Paramecium* *Tetrahymena* *Stylonychia*
UAG	Termination	gln	*Paramecium*

More recently, in 1985, several other exceptions to the standard coding dictionary have been discovered. These and prior aberrant codes are also summarized in Table 9.4. Such changes have been observed in the bacterium *Mycoplasma capricolum,* and in the protozoan ciliates *Paramecium, Tetrahymena,* and *Stylonychia.* As shown, each change converts one of the termination codons to glutamine or tryptophan. These changes are significant because both a prokaryote and several eukaryotes are involved, representing distinct species that have evolved over a long period of time.

Note the apparent pattern in several of the altered codon assignments. The change in coding capacity involves only a shift in recognition of the third, or wobble, position. For example, AUA specifies isoleucine during translation in the cytoplasm and methionine in the mitochondrion. In cytoplasmic translation, methionine is specified by AUG. In a similar way, UGA calls for termination in the cytoplasmic system but tryptophan in the mitochondrion. In the cytoplasm, tryptophan is specified by UGG. Although it has been suggested that such changes in codon recognition may represent an evolutionary trend toward reducing the number of tRNAs needed in mitochondria, the significance of these findings is not yet clear. It is known that only 22 tRNA species are encoded in human mitochondria. However, until still other examples are revealed, the differences must be considered as exceptions to the previously established general coding rules.

EXPRESSION OF GENETIC INFORMATION: AN OVERVIEW

Even while the genetic code was being studied, it was quite clear that proteins were the end products of many genes. Thus, while some geneticists were attempting to elucidate the code, other research efforts were directed toward the nature of genetic expression. The central question was how DNA, a nucleic acid, is able to specify a protein composed of amino acids. Put still another way, How is information transferred between DNA and protein? We shall return in the next chapter to examine the evidence supporting the conclusion that proteins are specified by genes, as well as to discuss protein structure and function. Here we shall emphasize the concept of **information flow** as it occurs between DNA and protein. This topic was extremely interesting and exciting in the early 1960s, and it remains no less so in the 1990s!

Genetic information, stored in DNA, was shown to be transferred to RNA during the initial stage of gene expression. The process by which RNA molecules are synthesized on a DNA template is called **transcription.** The ribonucleotide sequence of RNA, written in a genetic code, is then capable of directing the process of **translation.** During translation, polypeptide chains—the precursors of proteins—are synthesized. Protein synthesis is dependent on a series of **transfer RNA (tRNA)** molecules, which serve as adaptors between the codons of mRNA and the amino acids specified by them. In addition, the process occurs only in conjunction with an intricate cellular organelle, the **ribosome.**

The processes of transcription and translation are complex molecular events. Like the replication of DNA, both rely heavily on base-pairing affinities between complementary nucleotides. The initial transfer from DNA to mRNA produces a molecule complementary to the gene sequence of one of the two strands of the double helix. Then, each triplet codon is complementary to the anticodon region of tRNA as the corresponding amino acid is correctly inserted into the polypeptide chain during translation. In the following sections, we will describe in detail how these processes were discovered and how they are executed.

TRANSCRIPTION: RNA SYNTHESIS

The idea that RNA is involved as an intermediate molecule in the process of information flow between DNA and protein is suggested by the following observations:

1. DNA is, for the most part, associated with chromosomes in the nucleus of the eukaryotic cell. However, protein synthesis occurs in association with ribosomes located outside the nucleus in the cytoplasm. Therefore, DNA does not participate directly in protein synthesis.
2. RNA is synthesized in the nucleus of eukaryotic cells, where DNA is found, and is chemically similar to DNA.
3. Following its synthesis, most RNA migrates to the cytoplasm, where protein synthesis occurs.
4. The amount of RNA is generally proportional to the amount of protein in a cell.

Collectively, these observations suggest that genetic information, stored in DNA, is transferred to an RNA intermediate, which directs the synthesis of proteins. As with most new ideas in molecular genetics, the initial supporting experimental evidence was based on studies of bacteria and their phages. Elliot Volkin and his colleagues established that during initial infection RNA synthesis preceded phage protein synthesis. Work by Sol Spiegelman confirms that the RNA is complementary to phage DNA.

The results of all these experiments agree with the concept of a **messenger RNA (mRNA)** being made on a DNA template and then directing the synthesis of specific proteins. This concept was formally proposed by François Jacob and Jacques Monod in 1961 as part of a model for gene regulation in bacteria. Since then, mRNA has been isolated and thoroughly studied. There is no longer any question about its role in genetic processes.

RNA Polymerase

In order to prove that RNA may be synthesized on a DNA template, it was necessary to demonstrate that there is an enzyme capable of directing this synthesis. By 1959, several investigators, including Samuel Weiss, had independently discovered such a molecule from rat liver. Called **RNA polymerase,** it has the same general requirements as DNA polymerase, the major exception being that the ribose rather than the deoxyribose form of the sugar is present in each nucleotide. The overall reaction that summarizes the synthesis of RNA on a DNA template may be expressed as:

$$n(\mathrm{NTP}) \xrightarrow[\text{enzyme}]{\text{DNA}} (\mathrm{NMP})_n + n(\mathrm{PP}_i)$$

As the equation reveals, nucleoside triphosphates (NTPs) serve as substrates for the enzyme. It catalyzes the polymerization of nucleoside monophosphates (NMPs), or nucleotides, into a polynucleotide chain $(\mathrm{NMP})_n$. Nucleotides are linked during synthesis by $3'$-$5'$ phosphodiester bonds. The energy created by cleaving the triphosphate precursor into the monophosphate form drives the reaction and inorganic phosphates (PP_i) are produced.

A second equation summarizes the sequential addition of each ribonucleotide as the process of transcription progresses:

$$(\mathrm{NMP})_n + \mathrm{NTP} \xrightarrow[\text{enzyme}]{\text{DNA}} (\mathrm{NMP})_{n+1} + \mathrm{PP}_i$$

As this equation shows, each step of transcription involves the addition of one ribonucleotide (NMP) to the growing polyribonucleotide chain $(\mathrm{NMP})_{n+1}$, using a nucleoside triphosphate (NTP) as the precursor.

RNA polymerase from *E. coli* has been extensively characterized and shown to consist of subunits designated α, β, β', and σ. The active form of the enzyme $\alpha_2\beta\beta'\sigma$ has a molecular weight of almost 500,000 daltons. Of these various subunits, it is the **β polypeptide** that provides the catalytic basis and active site for transcription. As we will see, the **σ (sigma) subunit** plays a regulatory function involving the initiation of RNA transcription.

While there is but a single form of the enzyme in *E. coli,* separate forms of RNA polymerase are involved in the transcription of the three types of RNA in eukaryotes. The nomenclature used in describing these is summarized in Table 9.5. The three eukaryotic polymerases all consist of a greater number of polypeptide subunits than the bacterial form of the enzyme.

TABLE 9.5
RNA polymerases in eukaryotes.

Type	Product	Location
I	rRNA	Nucleolus
II	mRNAs	Nucleoplasm
III	5S RNA	Nucleoplasm
	tRNAs	Nucleoplasm

The Sigma Subunit, Promoters, and Template Binding

Transcription results in the synthesis of a single-stranded RNA molecule complementary to a region along one of the two strands of the DNA double helix. The initial step is referred to as **template binding,** where RNA polymerase interacts physically with DNA [Figure 9.7 (a)]. The accuracy of this initial binding in bacteria is achieved as a result of the recognition of specific DNA sequences

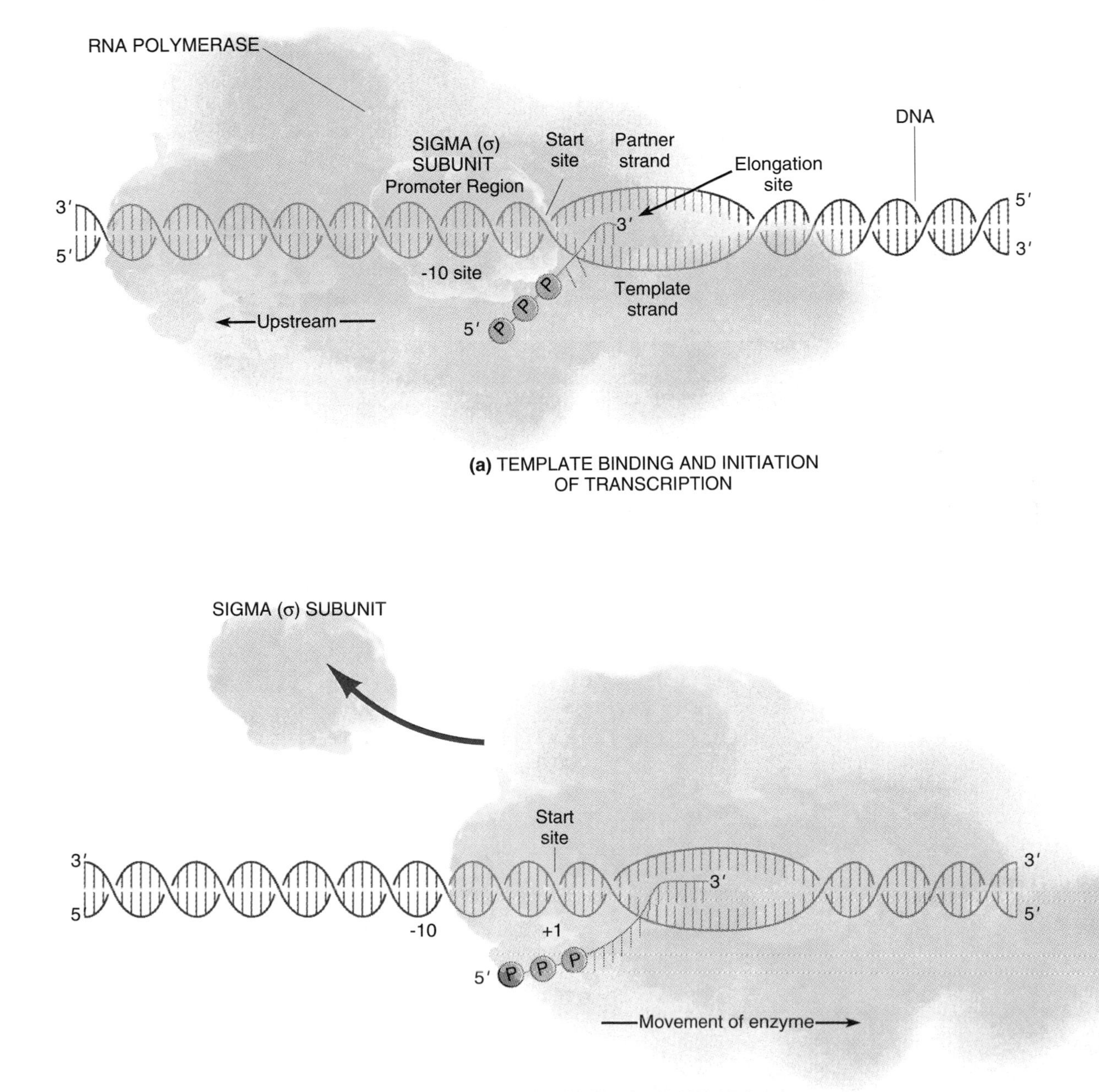

FIGURE 9.7
Schematic representation of the early stages of transcription, including: (a) template binding and initiation involving the sigma subunit of RNA polymerase; and (b) chain elongation, after sigma has dissociated from the transcription complex.

called **promoters** by the sigma subunit (σ) of the enzyme.

These regions are located upstream (to the left in the illustration) from the point of initial transcription of a gene. Most likely, the enzyme "explores" a length of DNA until the promoter region is recognized, and a tightly bound complex results. Once this occurs, the helix is denatured or unwound locally, making the DNA template accessible to the action of the enzyme.

A great deal is now known about promoters and template binding. The enzyme is a complex molecule, large enough to bind about 60 nucleotide pairs of the helix, 40 of which are upstream from the point of initial transcription. The promoter regions for many genes have been isolated and their binding capacities analyzed.

What has been revealed is extremely interesting. Two sequences, each six nucleotides long, are strikingly similar in over 100 bacterial promoters thus far analyzed. The initial sequence, first discovered by David Pribnow, occurs 10 nucleotides upstream from the start of transcription; it is thus referred to as the **−10 site,** or **Pribnow box.** While it is not invariant, the sequence represented by each nucleotide present most often in all promoters analyzed is called the **consensus sequence.** By convention this is always depicted as the sequence present on the nontemplate (or partner) strand:

(−10)
5′ . . . TATAAT . . . 3′

The importance of promoter sequences cannot be overemphasized. They govern the efficiency of initiation of transcription by directing RNA polymerase to the proper starting point of transcription. Both strong promoters and weak promoters are recognized, leading to a variation of initiation from once every 1 to 2 seconds to only once every 10 to 20 minutes. In fact, mutations in the two consesus sequences described above may have the effect of severely reducing the initiation of gene expression. And, as we shall see later in this chapter, a consensus sequence rich in adenine and thymine residues (called the TATA box) is also present, but farther upstream, in almost all eukaryotic genes.

The high degree of conservation through evolution attests to the critical nature of these consensus sequences and the important role they play in transcription.

The Synthesis of RNA

Once the promoter has been recognized and bound by the enzyme complex, RNA polymerase catalyzes the insertion of the first 5′-ribonucleoside triphosphate, which is complementary to the first nucleotide at the start site of the DNA template strand. Subsequent ribonucleotide complements are inserted and linked together by phosphodiester bonds as RNA polymerization proceeds. This process, called **chain elongation** [Figure 9.7(a)], continues in the 5′-to-3′ direction, creating a temporary DNA/RNA duplex whose chains run antiparallel to one another.

After a few ribonucleotides have been added to the growing RNA chain, the σ subunit is dissociated from the holoenzyme, and elongation proceeds under the direction of the core enzyme [Figure 9.7(b)]. In *E. coli* this process proceeds at the rate of about 50 nucleotides/second at 37°C.

Eventually, the enzyme traverses the entire gene and encounters a termination signal, a specific nucleotide sequence. Synthesis is completed usually in conjunction with the **termination factor, rho (ρ).** At that point, the transcribed RNA molecule is released from the DNA template, and the core enzyme dissociates. Under the direction of RNA polymerase, an RNA molecule is synthesized that is precisely complementary to a DNA sequence representing the template strand of a gene. Wherever an A, T, C, or G residue existed, a corresponding U, A, G, or C residue has been incorporated into the RNA molecule, respectively. The significance of this synthesis is enormous, for it is the initial step in the process of information flow within the cell.

TRANSLATION COMPONENTS NECESSARY FOR PROTEIN SYNTHESIS

Translation of mRNA is the biological polymerization of amino acids into polypeptide chains. The process, alluded to in our earlier discussion of the genetic code, occurs only in association with ribosomes. The central question in translation is how triplet ribonucleotides of mRNA direct specific amino acids into their correct position in the polypeptide. This question was answered once transfer RNA (tRNA) was discovered. This class of molecules adapts specific triplet codons in mRNA to their correct amino acids. This adaptor role of tRNA was postulated by Crick in 1957.

In association with a ribosome, mRNA presents a triplet codon that calls for a specific amino acid. Because a specific tRNA molecule contains within its composition three consecutive ribonucleotides complementary to the **codon,** they are called the **anticodon** and can base pair with the codon. This tRNA is covalently bonded to the amino acid called for. As this process occurs over and over as mRNA runs through the ribosome, amino acids are polymerized into a polypeptide.

In our discussion of translation, we will first consider the structure of the ribosome and transfer RNA, two of the major components essential for protein synthesis.

Ribosomal Structure

Because of its essential role in the expression of genetic information, the ribosome has been extensively analyzed. One bacterial cell contains about 10,000 of these structures, while a eukaryotic cell contains many times more. Electron microscopy has revealed that the bacterial ribosome is about 250 Å in its largest diameter and consists of a larger and a smaller subunit. Both subunits consist of one or more molecules of rRNA and an array of **ribosomal proteins.**

The specific differences between prokaryotic and eukaryotic ribosomes are summarized in Figure 9.8. The subunit and rRNA components are most easily isolated and characterized on the basis of their sedimentation behavior (their rate of migration) in sucrose gradients (see Chapter 7). The two subunits associated with each other constitute a **monosome.** In prokaryotes the monosome is a 70*S* particle, and in eukaryotes it is approximately 80*S*. Sedimentation coefficients, which reflect the variable rate of migration of different-sized particles and molecules, are not additive. For example, the 70*S* monosome consists of a 50*S* and a 30*S* subunit, and the 80*S* monosome consists of a 60*S* and a 40*S* subunit.

The larger subunit in prokaryotes consists of a 23*S* RNA molecule, a 5*S* rRNA molecule, and 31 ribosomal proteins. In the eukaryotic equivalent, a 28*S* rRNA molecule is accompanied by a 5.8*S* and 5*S* rRNA molecule and many more than 34 proteins. In the smaller prokaryotic subunits, a 16*S* rRNA component and 21 proteins are found. In the eukaryotic equivalent, a 18*S* rRNA component and many more than 21 proteins are found. The approximate molecular weights and number of nucleotides of these components are shown in Figure 9.8.

Molecular hybridization studies have established the degree of redundancy of the genes coding for the rRNA

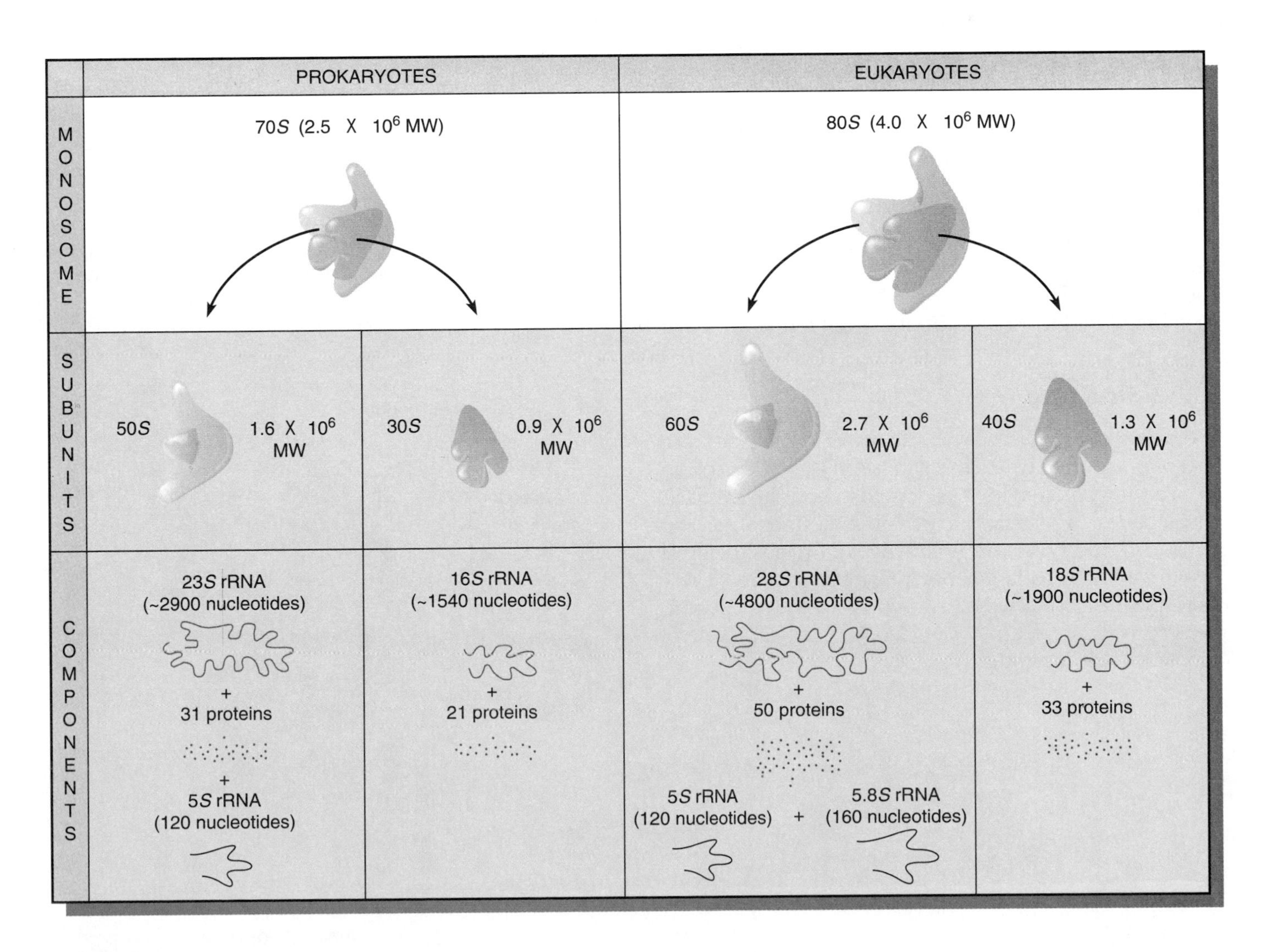

FIGURE 9.8
A comparison of the components in the prokaryotic and eukaryotic ribosomes.

components. The *E. coli* genome contains seven copies of a single sequence that codes for all three components—23*S*, 16*S*, and 5*S*. The initial transcript of these genes produces a 30*S* RNA molecule that is enzymatically cleaved into these smaller components.

In eukaryotes, many more copies of a sequence encoding the 28*S* and 18*S* components are present. In *Drosophila,* approximately 120 copies per haploid genome are each transcribed into a molecule of about 34*S*. This is processed to the 28*S*, 18*S*, and 5.8*S* rRNA species. In *X. laevis,* over 500 copies per haploid genome are present. In mammalian cells, the initial transcript is 45*S*. The rRNA genes are part of the moderately repetitive DNA fraction and are present in clusters at various chromosomal sites. Each cluster consists of **tandem repeats,** with each unit separated by a noncoding **spacer DNA** sequence. In humans, these gene clusters have been localized on the ends of chromosomes 13, 14, 15, 21, and 22.

The 5*S* rRNA component of eukaryotes is not part of the larger transcript, as it is in *E. coli.* Instead genes coding for this ribosomal component are distinct and located separately. In humans, a gene cluster encoding them has been located on chromosome 1.

In spite of the detailed knowledge available on the structure and genetic origin of the ribosomal components, a complete understanding of the function of these components has eluded geneticists. This is not surprising, because the ribosome is perhaps the most intricate of all cellular organelles. In bacteria, the monosome has a combined molecular weight of 2.5 million daltons!

tRNA Structure

Because of its small size and stability in the cell, tRNA has been investigated extensively. In fact, it is the best characterized RNA molecule. It is composed of only 75 to 90 nucleotides, displaying a nearly identical structure in bacteria and eukaryotes. In both types of organisms, tRNAs are transcribed as larger precursors, which are cleaved into mature 4S tRNA molecules. In *E. coli,* for example, $tRNA^{tyr}$ is composed of 77 nucleotides, yet its precursor contains 126 nucleotides.

In 1965, Robert Holley and his coworkers reported the complete sequence of $tRNA^{ala}$ (the superscript identifies the specific amino acid that binds a particular tRNA) isolated from yeast. Of great interest was the finding that there are a number of nucleotides unique to tRNA. Each is a modification of one of the four nitrogenous bases expected in RNA (G, C, A, and U). These include **inosinic acid,** which contains the purine **hypoxanthine, ribothymidylic acid,** and **pseudouridine** among others. These modified structures, sometimes referred to as *unusual, rare,* or *odd bases,* are created **posttranscriptionally.** That is, the unmodified base is produced during transcription, and then enzymatic reactions catalyze the modifications.

Holley's sequence analysis led him to propose the two-dimensional **cloverleaf model** of transfer RNA. It had been known that tRNA demonstrates secondary structure due to base pairing. Holley discovered that he could arrange the linear model in such a way that several stretches of base pairing would result. This arrangement created a series of paired stems and unpaired loops resembling the shape of a four-leaf clover. Loops consistently contained modified bases, which do not generally form base pairs. Holley's model is shown in Figure 9.9.

Since the triplets GCU, GCC, and GCA specify alanine, Holley looked for the theoretical anticodon of his $tRNA^{ala}$ molecule. He found it in the form of CGI, at the bottom loop of the cloverleaf. Recall from Crick's wobble hypothesis that I (inosinic acid) is predicted to pair with U, C, or A. Thus, the anticodon loop was established.

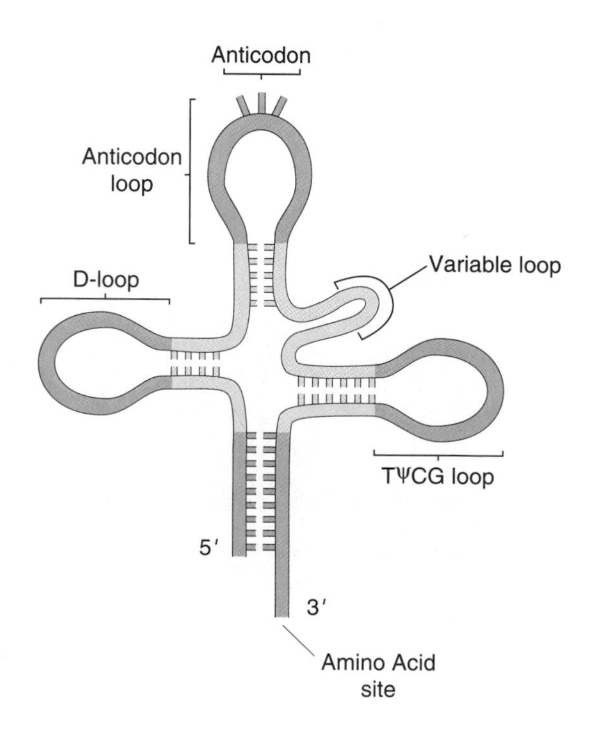

FIGURE 9.9
The two-dimensional cloverleaf model of transfer RNA.

As other tRNA species were examined, numerous constant features were observed. First, at the 3′ end, all tRNAs contain the sequence . . . . **pCpCpA.** It is to the terminal adenosine residue that the amino acid is joined covalently during charging. At the 5′ terminus, all tRNAs contain . . . **pG.**

Additionally, the lengths of various stems and loops are very similar. All tRNAs examined also contain an anticodon complementary to the known amino acid code for which it is specific. These anticodon loops are present in the same position of the cloverleaf as well.

The cloverleaf model was predicted strictly on the basis of nucleotide sequence. Thus, there was great interest in X-ray crystallographic examination of tRNA, which reveals three-dimensional structure. By 1974, Alexander Rich and his coworkers in the United States, and J. Roberts, B. Clark, Aaron Klug, and their colleagues in England had been successful in crystallizing tRNA and performing X-ray crystallography at a resolution of 3 Å. At such resolution, the pattern formed by individual nucleotides is discernible.

As a result of these studies, a complete three-dimensional model is now available (Figure 9.10). The model reveals tRNA to be L-shaped. At one end of the L is the anticodon loop and stem, and at the other end is the acceptor region where the amino acid is bound. It has been speculated that the shapes of the intervening loops may be recognized by specific **aminoacyl tRNA synthetases** (see the next section), the enzymes responsible for adding the amino acid to tRNA.

Charging tRNA

Before translation can proceed, the tRNA molecules must be chemically linked to their respective amino acids. This activation process, called **charging,** occurs under the direction of enzymes called **aminoacyl synthetases.** Because there are 20 different amino acids, there must be at least 20 different tRNA molecules and as many different enzymes. In theory, since there are 61 triplet codes, there could be the same number of specific tRNAs and enzymes. Because of the ability of the third member of a triplet code to wobble, however, it is now thought that there are about 30 different tRNAs; it is also believed that there are only 20 synthetases, one for each amino acid, regardless of the number of corresponding tRNAs.

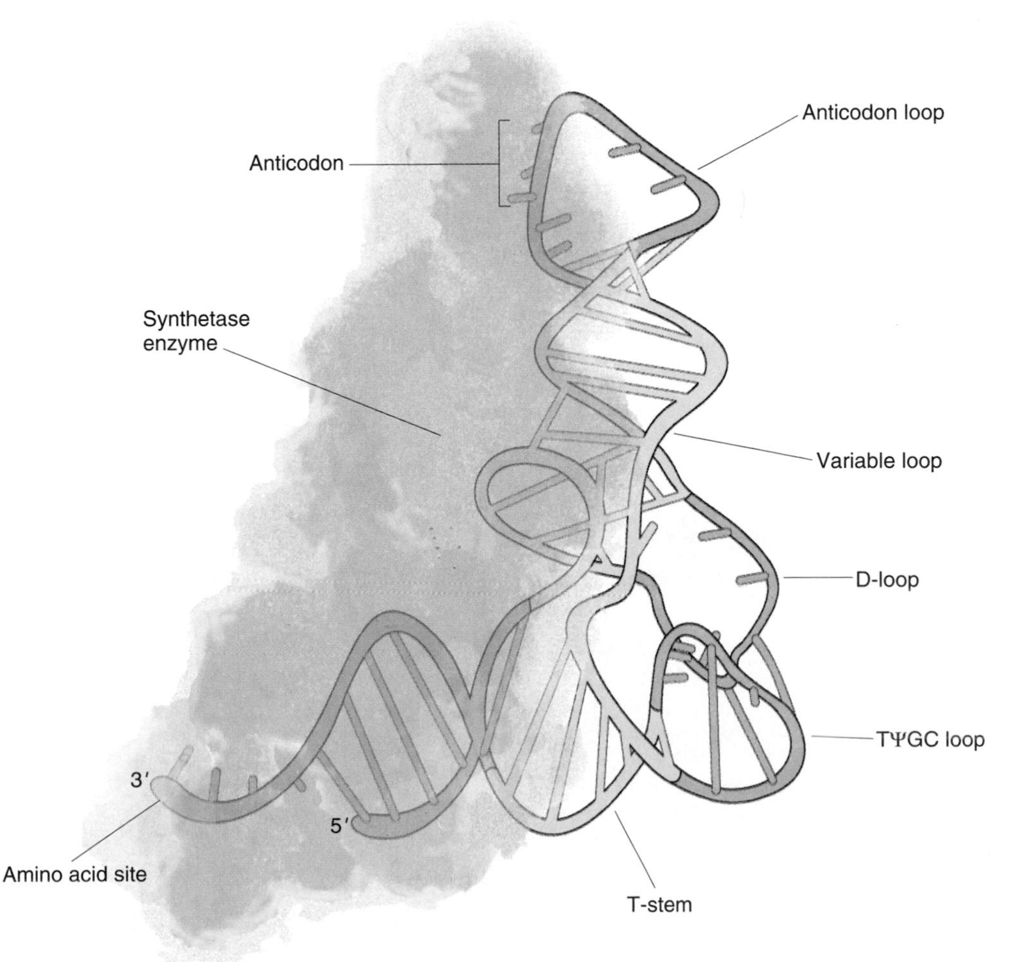

FIGURE 9.10
A three-dimensional model of transfer RNA as it associates with its corresponding aminoacyl synthetase, which catalyzes the addition of the appropriate amino acid.

The charging process is outlined in Figure 9.11. In the initial step, the amino acid is converted to an **activated form,** reacting with ATP to form an **aminoacyladenylic acid.** A covalent linkage is formed between the 5′ phosphate group and the carboxyl end of the amino acid. This molecule remains associated with the enzyme, forming an activated complex, which then reacts with a specific tRNA molecule. In this second step, the amino acid is transferred to the appropriate tRNA and bonded covalently. The charged tRNA may participate directly in protein synthesis. Aminoacyl-tRNA synthetases are highly specific enzymes because they recognize only one amino acid and only a corresponding tRNA. This is a crucial point if fidelity of translation is to be maintained. The basis for this binding has sometimes been referred to as the **second genetic code,** although this is not a particularly apt descriptor.

TRANSLATION: THE PROCESS

In a way similar to transcription, the process of translation can be best described by breaking it into discrete steps. Be aware, however, that translation is a dynamic, ongoing process. Correlate the following discussion with the step-by-step characterization of the process in Figure 9.12.

Initiation (Steps 1–3)

Recall that ribosomes serve as a nonspecific work-bench for the translation process. Most ribosomes, while uninvolved in translation, are dissociated into their large and small subunits. Initiation of translation of *E. coli* involves these subunits, an mRNA molecule, a specific initiator tRNA, GTP, Mg^{++}, and at least three **initiation factors (IFs).** The initiator molecules are not part of the ribosome, but are required to enhance the binding affinity of the various translational components. In prokaryotes, the initiation code of mRNA, AUG, calls for the modified amino acid formylmethionine.

The small ribosomal subunit binds to several initiation proteins, and this complex in turn binds to mRNA. In bacteria, this binding involves a sequence of up to six ribonucleotides, which precedes the initial AUG codon. It is this sequence (which is purine-rich and called the **Shine-Dalgarno sequence)** that base pairs with a region of the 16*S* rRNA of the small ribosomal subunit.

Another initiation protein then facilitates the binding of charged formylmethionyl-tRNA to the small subunit in response to the AUG triplet. This step "sets" the reading frame so that all subsequent groups of three ribonucleotides are translated accurately. This aggregate represents the **initiation complex,** which then combines with the large ribosomal subunit. In this process, a molecule of GTP is hydrolyzed, providing the required energy, and the initiation factors are released.

Elongation (Steps 4–9)

The ribosomal subunits contain two binding sites for charged tRNA molecules; these are labeled the **P,** or **peptidyl,** and the **A,** or **aminoacyl, sites.** The initiation tRNA binds to the P site, provided the AUG triplet is in the corresponding position of the small subunit. The sequence of the second triplet in mRNA dictates which

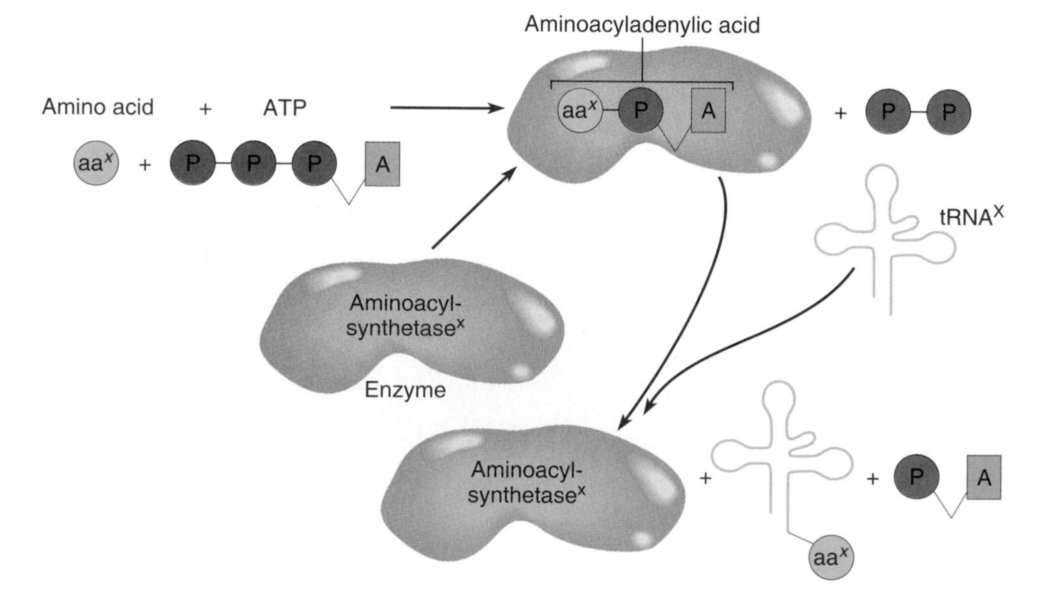

FIGURE 9.11
Steps involved in charging tRNA. The X denotes that for each amino acid, only a specific tRNA and a specific aminoacyl synthetase enzyme are involved in the charging process.

charged tRNA molecule will become positioned at the A site. Once it is present, **peptidyl transferase** catalyzes the formation of the peptide bond, which links the two amino acids together. This enzyme is part of the large subunit of the ribosome. At the same time, the covalent bond between the amino acid and the tRNA occupying the P site is hydrolyzed (or broken). The product of this reaction is a dipeptide, which is attached to the tRNA at the A site. The step in which the growing polypeptide chain has increased in length by one amino acid is called **elongation.**

Before elongation can be repeated, the tRNA attached to the P site, which is now uncharged, must be released from the large subunit. The entire **mRNA–tRNA–aa_2–aa_1** complex now shifts in the direction of the P site by a distance of three nucleotides. This event requires several protein **elongation factors (EFs)** as well as the energy derived from hydrolysis of GTP. The result is that the third triplet of mRNA is now in a position to direct another specific charged tRNA into the A site. One simple way to distinguish the two sites in your mind is to remember that, *following the shift,* the P site contains a tRNA attached to a peptide chain (*P* for peptide), while the A site contains a tRNA with an amino acid attached (*A* for amino acid).

The sequence of elongation is repeated over and over. An additional amino acid is added to the growing polypeptide chain each time the mRNA advances through the ribosome. The efficiency of the process is remarkably high; the observed error rate is only 10^{-4}. An incorrect amino acid will thus occur once in every 20 polypeptides of an average length of 500 amino acids! In *E. coli,* elongation occurs at a rate of about 15 amino acids per second at 37°C. The process can be likened to a tape moving through a tape recorder. As the tape moves, sequential sound is emitted from the recorder. Likewise, as mRNA moves, a growing polypeptide is produced by the ribosome.

Termination (Steps 10–11)

The termination of protein synthesis is signaled by one or more of three triplet codes: UAG, UAA, or UGA. These codons do not specify an amino acid, nor do they direct tRNA into the A site. The finished polypeptide is therefore still attached to the terminal tRNA at the P site. The termination codon signals the action of **GTP-dependent release factors** (Table 9.6), which cleave the polypeptide chain from the terminal tRNA. Once this cleavage occurs, the tRNA is released from the ribosome, which then dissociates into its subunits. If a termination codon should appear in the middle of an mRNA molecule as a result of mutation, the same process occurs, and the polypeptide chain is prematurely terminated. Note that the specific proteins that serve as initiation, elongation, and termination factors are characterized in Table 9.6.

TABLE 9.6
Various protein factors involved during translation in *E. coli.*

Process	Factor	Role
Initiation	IF1	Stabilizes 30*S* subunit
	IF2	Binds fmet-tRNA to 30*S*-mRNA complex; Binds to and stimulates GTP hydrolysis
	IF3	Binds 30*S* subunit to mRNA; Dissociates monosomes into subunits following termination
Elongation	EF-Tu	Binds GTP; Brings aminoacyl-tRNA to the A site
	EF-Ts	Generates active EF-Tu
	EF-G	Stimulates translocation; GTP-dependent
Termination	RF1	Catalyzes release of the polypeptide chain from tRNA and dissociation of the translation complex; Specific for UAA and UAG termination codons.
	RF2	Behaves like RF1; Specific for UGA and UAA codons
	RF3	Stimulates RF1 and RF2

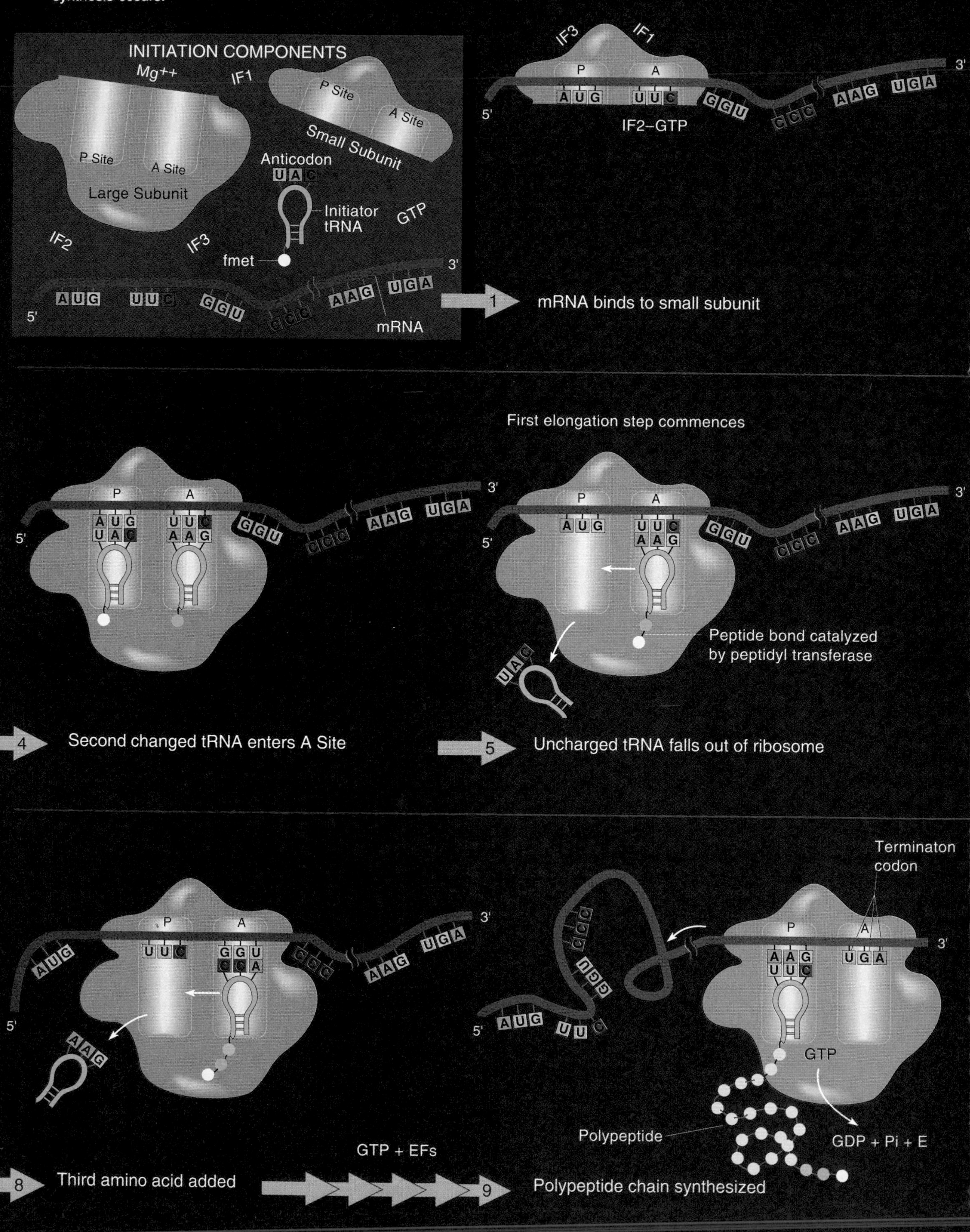

FIGURE 9.12
Schematic representation of the process of translation, depicting how protein synthesis occurs.

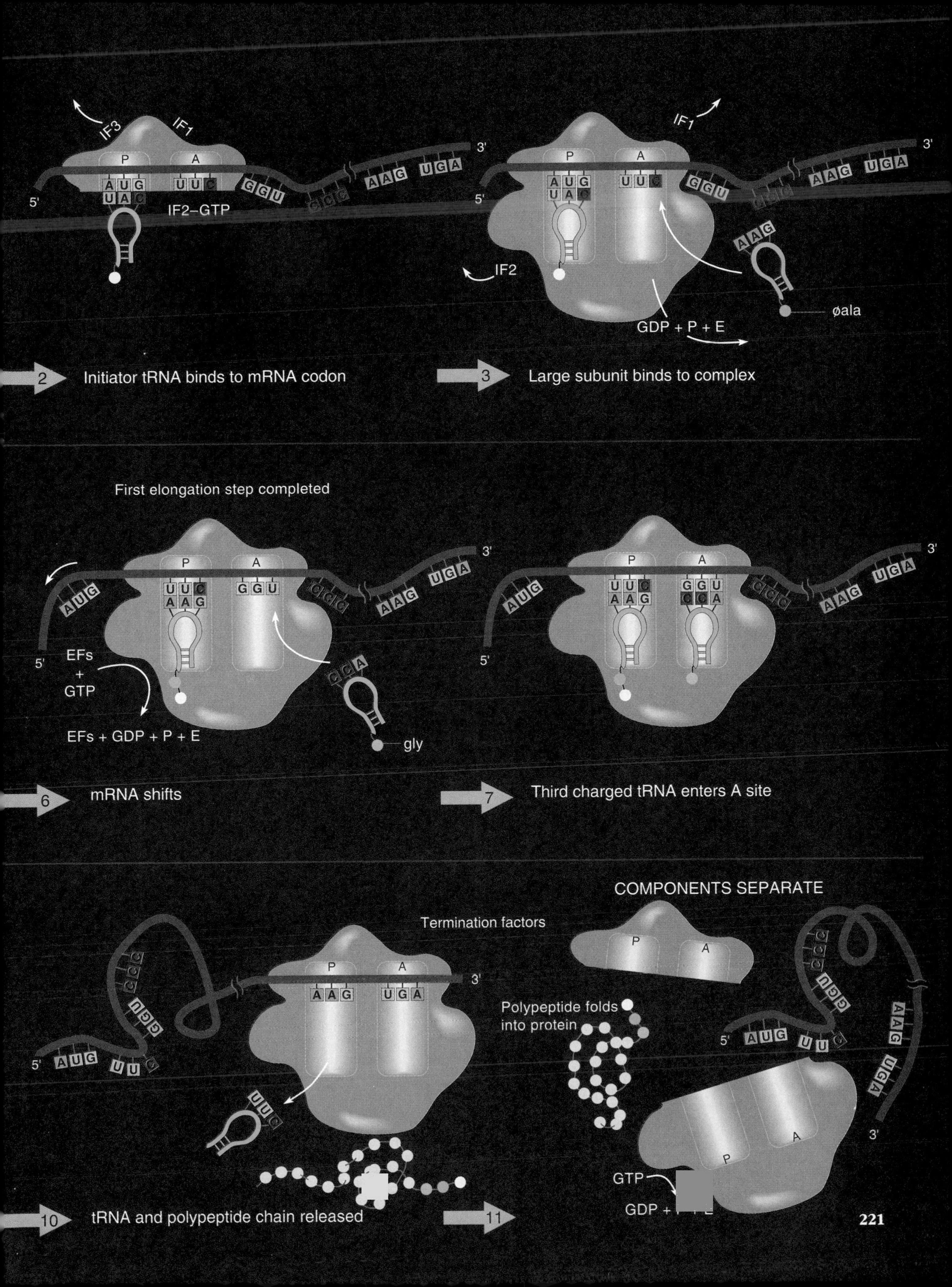
IF3
IF1
P
A
3'
5'
IF2–GTP
IF1
IF2
GDP + P + E
øala
2 Initiator tRNA binds to mRNA codon
3 Large subunit binds to complex
First elongation step completed
EFs + GTP
EFs + GDP + P + E
gly
6 mRNA shifts
7 Third charged tRNA enters A site
COMPONENTS SEPARATE
Termination factors
Polypeptide folds into protein
GTP
10 tRNA and polypeptide chain released
11

Polyribosomes

As elongation proceeds and the initial portion of mRNA has passed through the ribosome, the message is free to associate with another small subunit to form another initiation complex. This process can be repeated several times with a single mRNA and results in what are called **polyribosomes** or just **polysomes.**

Polyribosomes can be isolated and analyzed following a gentle lysis of cells. Figure 9.13(a) and (b) illustrates these complexes as seen under the electron microscope; note the presence in Part (a) of this figure of mRNA between the individual ribosomes. The micrograph in Part (b) is even more remarkable since it shows the polypeptide chains emerging from the ribosomes during translation. The formation of polysome complexes represents an efficient use of the components available for protein synthesis during a unit time.

To complete the analogy with tapes (mRNA) and tape recorders (ribosomes), in polysome complexes one tape would be played simultaneously, but at any given moment, the transcripts (polypeptides) would all be at different stages of completion.

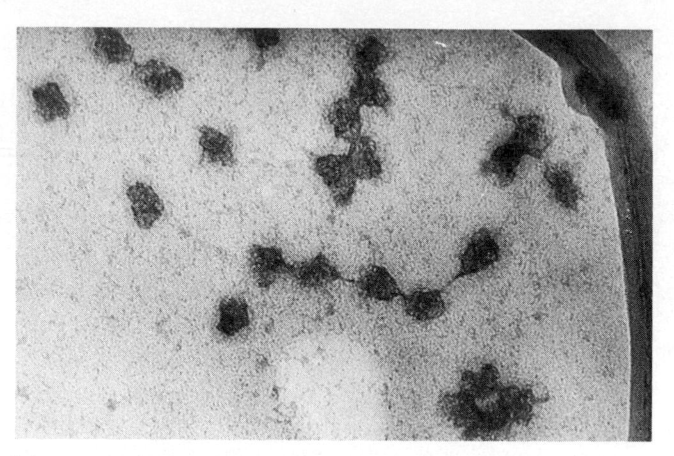

(a)

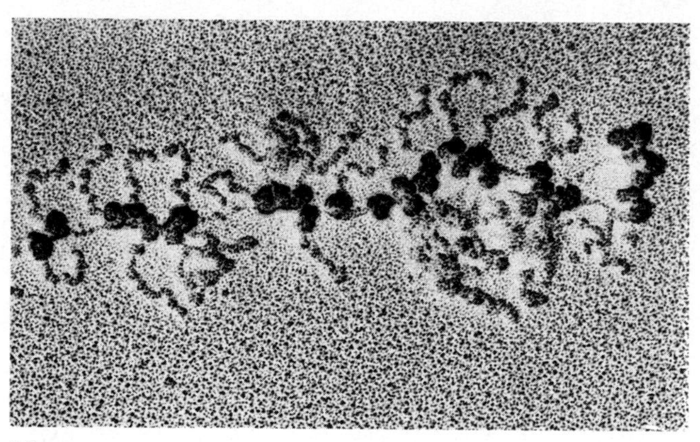

(b)

FIGURE 9.13
Polyribosomes visualized under the electron microscope. Those viewed by electron microscopy were derived in Part (a) from rabbit reticulocytes engaged in the translation of hemoglobin mRNA and in Part (b) from giant salivary gland cells of the midgefly, *Chironomus thummi.* In Part (b), the nascent polypeptide chain is apparent as it emerges from each ribosome. Its length increases as translation proceeds from left (5′) to right (3′) along the mRNA.

TRANSCRIPTION AND TRANSLATION IN EUKARYOTES

Much of our knowledge of transcription and translation has been derived from studies of prokaryotes. The general aspects of the mechanics of these processes are similar in eukaryotes. There are, however, numerous notable differences, several of which have been discussed earlier in this chapter. We will first summarize the differences and then expand on several of these in greater detail.

1 Transcription in eukaryotes occurs within the nucleus under the direction of three separate forms of RNA polymerase (Table 9.5); for the mRNA to be translated, it must move out into the cytoplasm.

2 The initiation and regulation of transcription are under the control of more extensive nucleotide sequences found in DNA upstream from the point of initial transcription. There are, in addition to promoters, other control units called enhancers.

3 Translation occurs on ribosomes that are larger and whose rRNA and proteins are more complex than those present in prokaryotes.

4 Protein factors similar to those in prokaryotes guide initiation, elongation, and termination of translation in eukaryotes. However, there appear to be more factors required during each of these steps.

5 Initiation of eukaryotic translation does not require the amino acid formylmethionine. However, the AUG triplet is essential to the formation of the translational complex, and a unique transfer RNA ($tRNA_i^{met}$) is used for its initiation.

6 Most notably, extensive modifications occur to eukaryotic RNA transcripts that eventually serve as mRNAs. The initial transcripts are much larger than those that are eventually translated. Thus, they are called **pre-mRNAs** and are thought to constitute a group of molecules found only in the nucleus—a group referred to generally as **heterogeneous RNA (hnRNA).** Only about 25 percent of hnRNA molecules are converted to mRNA. Those that are have substantial amounts of their ribonucleotide sequence excised, while the remaining segments are spliced back together prior to transla-

tion. This phenomenon has given rise to the concept of **split genes** in eukaryotes.

7 Prior to the processing of an mRNA transcript, a cap and a tail are added to the molecule. These modifications are essential to efficient processing and subsequently to translation.

8 Eukaryotic mRNAs are much longer lived than their bacterial counterparts. Most exist for hours, rather than minutes, prior to their degradation in the cell.

Eukaryotic Promoters, Enhancers, and Transcription Factors

The recognition of certain highly specific DNA regions by RNA polymerase is at the heart of orderly genetic function in all cells, but particularly in those of differentiated eukaryotes. Thus it comes as no surprise that the nature of both the polymerases and promoters leading to template binding were found to be more complex in these more advanced organisms. RNA polymerase of eukaryotes, as earlier discussed, exists in three forms and each is more compelx than the prokaryotic counterpart of the enzyme. In regard to the initial template binding step and promoter regions, most is known about polymerase II, which transcribes all mRNAs in eukaryotes.

There are at least three areas of DNA associated with most genes that are essential to the efficient initiation of transcription. The first is called the **Goldberg-Hogness** or **TATA box,** which is found about 25 nucleotide pairs upstream (−25) from the start point of transcription. The consensus sequence is a heptanucleotide consisting solely of A and T residues. The sequence and function are analogous to that found in the −10 region of prokaryotic genes. While mutations in this sequence reduce or eliminate transcription of a variety of genes, the TATA box may be fairly nonspecific and simply have the responsibility for fixing the site of initiation of transcription by facilitating the denaturation of the helix. Such a conclusion is supported by the fact that A = T base pairs are less stable than G ≡ C pairs.

The second set of noncoding regions of interest are found farther upstream from the TATA box. In different genes studied, regions anywhere from 50 to 500 nucleotides upstream appear also to modulate transcription. These regions have been located on the basis of the effects of their deletion on transcription. The loss of various noncoding regions appears to drastically reduce *in vivo* transcription. Some noncoding regions, such as those associated with globin and the viral SV40 genes, are about 50 to 100 nucleotides from the TATA sequence. Others, such as those associated with the sea urchin H2A gene and the *Drosophila* glue protein gene, are 200 to 500 nucleotides upstream. Frequently, GC-rich sequences as well as the sequence *CCAAT* are part of this noncoding region.

The third noncoding region is represented by elements called **enhancers.** While their location may vary, they often may be found even farther upstream. They have the effect of modulating transcription from a distance. Thus, they may not participate directly in template binding, even though they are essential to it. We will return to a discussion of these elements in Chapter 16.

Aside from the more extensive regulatory sequences present upstream in eukaryotes, a second major difference exists in the facilitation of template binding: the presence of protein transcription factors. It is generally believed that RNA polymerase II, cannot bind directly to promoter sites and initiate transcription without the presence of such factors.

It appears that the enzyme recognizes a complex of DNA bound by specific proteins. For example, there is a **TATA-factor** (also called **TFIID**), which binds to that promoter sequence and is critical to transcription. Still other factors specific to GC boxes and the CCAAT box have been discovered and studied. It may be that these transcription factors supplant the role of the sigma factor in the prokaryotic enzyme, thus providing a much greater degree of specific control of transcription. We will return in Chapter 16 to a consideration of the role of some of these factors in eukaryotic gene regulation.

Heterogeneous RNA and its Processing

Still other insights into regions of DNA that do not directly encode proteins have come from the study of RNA. This research has provided detailed knowledge of eukaryotic gene structure. The genetic code is written in the ribonucleotide sequence of mRNA. This information originated, of course, in the template strand of DNA, where complementary sequences of deoxyribonucleotides exist. In bacteria, the relationship between DNA and RNA appears to be quite direct. The DNA base sequence is transcribed into an mRNA sequence, which is then translated into an amino acid sequence according to the genetic code.

However, in eukaryotes the situation is much more complex than in bacteria. It has been found that many internal base sequences of a gene may never appear in the mature mRNA that is translated. Other modifications occur at the beginning and the end of the mRNA prior to translation. These findings have made it clear that in eukaryotes, complex processing of mRNA occurs before it participates in translation.

By 1970, accumulating evidence showed that eukaryotic mRNA is transcribed initially as a much larger precursor molecule than that which is translated. This notion was based on the observation by James Darnell and his coworkers of **heterogeneous RNA (hnRNA)** in mammalian cells. Heterogeneous RNA is of large but variable size (up to 10^7 daltons), is found only in the nucleus, and is rapidly degraded. Nevertheless, hnRNA was found to contain nucleotide sequences common to the smaller mRNA molecules of the cytoplasm. Thus, it was proposed that the initial transcript of a gene results in a large RNA molecule that must first be processed in the nucleus before it appears in the cytoplasm as a mature mRNA molecule.

A subsequent discovery provided further evidence for this proposal. Both hnRNAs and mRNAs were found to contain at the 3′ end a stretch of up to 200 adenylic acid residues. Such **poly A** sequences are added **posttranscriptionally** to the initial gene transcript. In higher eukaryotes, for example, transcription is terminated, and the 3′ end of the initial transcript is reduced in length close to an AAUAAA sequence and then polyadenylated. Subsequent investigation has shown poly A at the 3′ end of almost all mRNAs studied in a variety of eukaryotic organisms. The exceptions seem to be the products of histone genes and some yeast genes.

Poly A is added to the initial RNA transcript, which is then processed before its transport to the cytoplasm. The majority of the RNA transcript is cleaved from the 3′ poly A fragment and then rapidly degraded by nucleases. A further posttranscriptional modification of eukaryotic mRNA involves the 5′ end of these molecules, where a **7-methyl guanosine (7mG) residue,** or **cap** is added. The cap appears to be important to the successful translation of the message, perhaps by serving as a recognition point for ribosome binding. These various modifications are summarized in Figure 9.14.

Intervening Sequences and Split Genes

One of the most exciting discoveries in the history of molecular genetics occurred in 1977. At this time, direct evidence was provided by Susan Berget, Philip Sharp, and others that the genes of animal viruses as well as

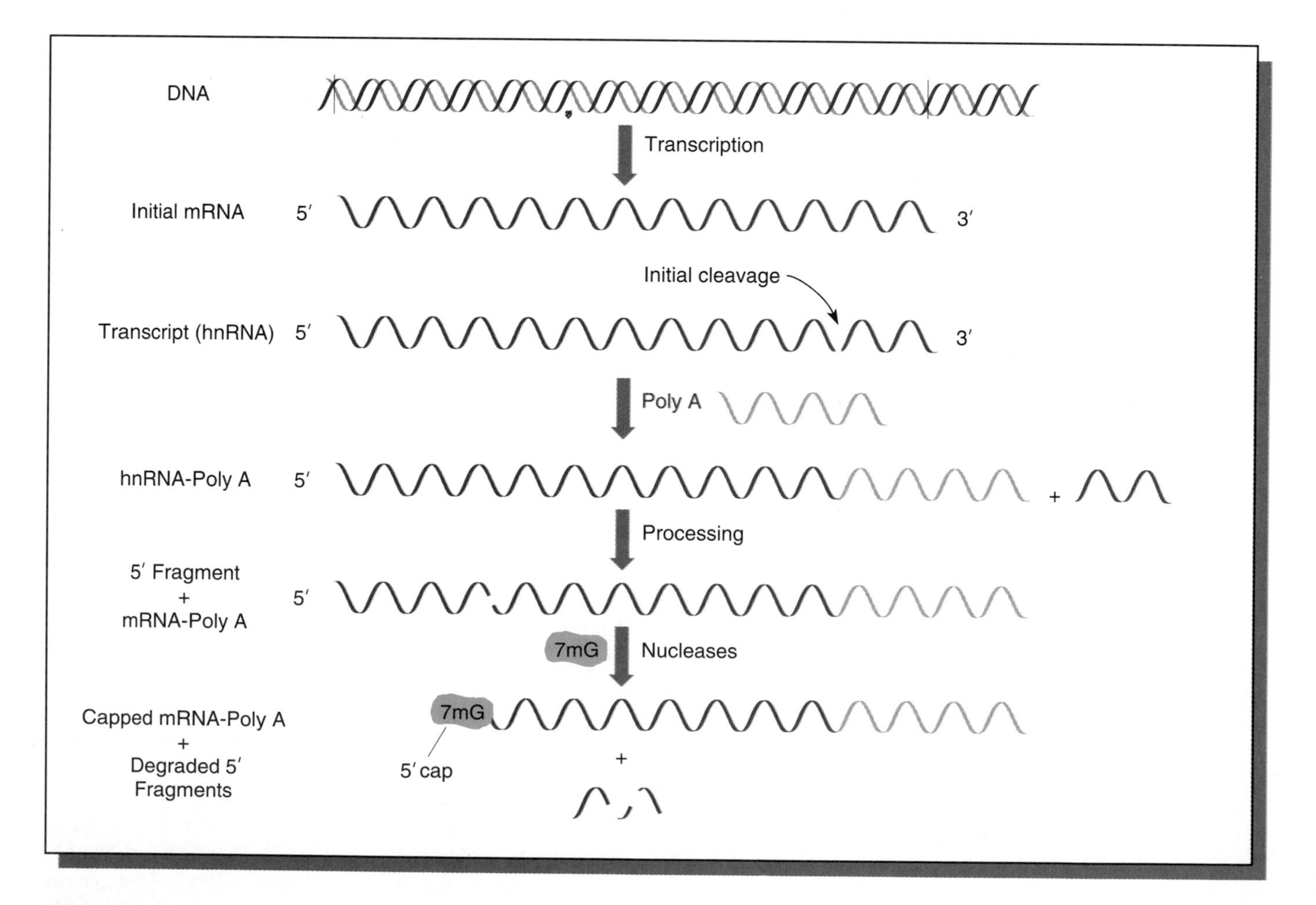

FIGURE 9.14
The conversion in eukaryotes of heterogeneous RNA (hnRNA) to messenger (mRNA), which contains a 5′ cap and a 3′ polyA tail.

eukaryotes contain internal nucleotide sequences that are not expressed in the amino acid sequence of the proteins that they encode. That is, certain internal sequences in DNA do not always appear in the mature mRNA that is translated into a protein.

Such nucleotide segments have been called **intervening sequences,** contained within **split genes.** Those DNA sequences not present in the final mRNA product are also called **introns** ("int" for intervening), and those retained and expressed are called **exons** ("ex" for expressed). Removal of introns occurs as a result of an excision and rejoining process referred to as **splicing.**

Similar discoveries were soon to be made in eukaryotes. Two approaches have been most fruitful. The first involves molecular hybridization of purified, functionally mature mRNAs with DNA containing the gene specifying that message. When hybridization occurs between nucleic acids that are not perfectly complementary, **heteroduplexes** are formed that may be visualized with the electron microscope. As illustrated in Figure 9.15 introns present in DNA but absent in mRNA must loop out and remain unpaired. This figure shows an electron micrograph and an interpretive drawing derived from hybridization between the template strand of the gene encoding chicken ovalbumin and the mature RNA prior to its translation into that protein. There are seven introns (A–G) whose sequences are present in DNA but not in the final mRNA.

The second approach provides more specific information. It involves a comparison of nucleotide sequences of DNA with those of mRNA and amino acid sequences. Such an approach allows the precise identification of all intervening sequences.

So far, a large number of genes from diverse eukaryotes have been shown to contain introns. One of the first so identified was the **beta-globin gene** in mice and rabbits, as studied independently by Philip Leder and Richard Flavell. The mouse gene contains an intron 550 nucleotides long, beginning immediately after the codon specifying the 104th amino acid. In the rabbit (Figure 9.16), there is an intron of 580 base pairs near the codon for the 110th amino acid. Additionally, a second intron of about 120 nucleotides exists earlier in both genes. Similar introns have been found in the beta-globin gene in all mammals examined.

Several genes, notably those coding for histones and interferon, appear to contain no introns. However, intervening sequences have been identified in immunoglobulin genes of the mouse, some tRNA genes in yeast, and rRNA genes in *Drosophila,* among many others. The case of tRNA is interesting. Four different $tRNA^{tyr}$ genes have been sequenced. In each case, an intron of 14 or 15 nucleotide pairs has been found immediately adjacent to the anticodon sequence.

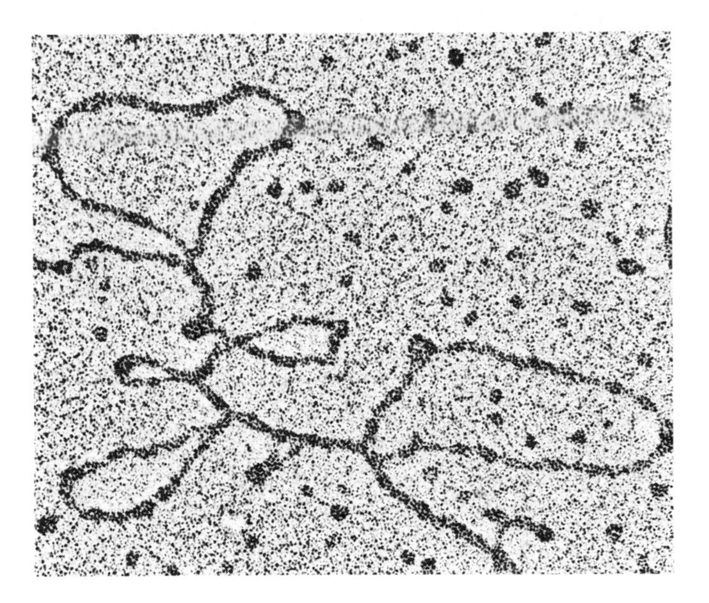

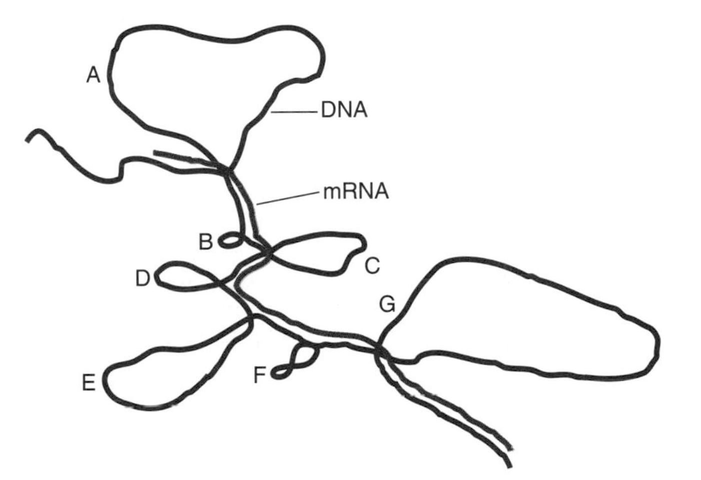

FIGURE 9.15
An electron micrograph and an interpretive drawing of the hybrid molecule formed between the template DNA strand of the ovalbumin gene and the mature ovalbumin mRNA. Seven DNA introns, A–G, produce unpaired loops.

As pointed out above, a more extensive set of introns has been located in the **ovalbumin gene** of chickens. The gene has been extensively characterized by Bert O'Malley in the United States and Pierre Chambon in France. As shown in Figure 9.16 the gene contains seven introns. Notice that the majority of the gene's DNA sequence is "silent," being composed of introns. The initial RNA transcript is four times the length of the mature mRNA. You should compare the information on the ovalbumin gene presented in Figures 9.15 and 9.16. Can you match the unpaired loops in 9.15 with the sequence of introns specified in 9.16?

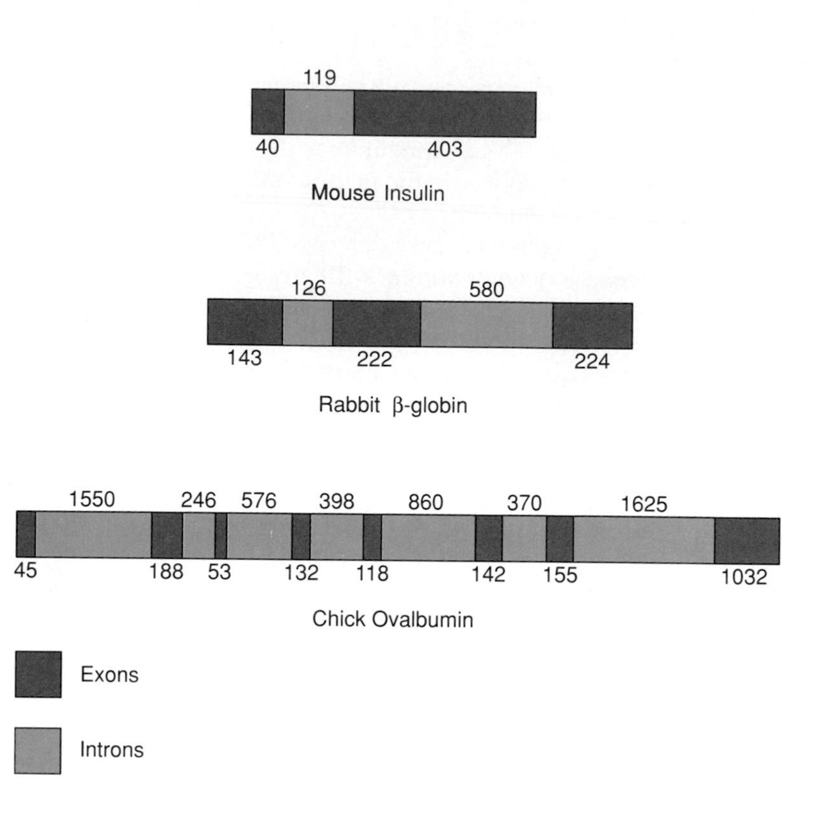

FIGURE 9.16
Intervening sequences in various eukaryotic genes. The numbers indicate nucleotides present in various intron and exon regions.

The list of genes containing intervening sequences is growing rapidly. In fact, few eukaryote genes seem to be without introns. An extreme example of the number of introns in a single gene is that found in one of the chicken genes, *pro-α-2(1) collagen.* One of several genes coding for a subunit of this connective tissue protein, *pro-α-2(l) collagen* contains about 50 introns. The precision with which cutting and splicing occur must be extraordinary if errors are not to be introduced into the mature mRNA.

The discovery of split genes represents one of the most exciting genetic findings in recent years. As a result, intensive investigation is now in progress to elucidate the mechanism by which introns of RNA are excised and exons are spliced back together. A great deal of progress has already been made. Interestingly, it appears that somewhat different mechanisms are utilized for each of the three types of RNA as well as for RNAs produced in mitochondria. The finding of "genes in pieces," as split genes have been described, raises many interesting questions and has provided great insights into the organization of eukaryotic genes. Thus, we will return to this topic again in Chapter 12.

OVERLAPPING GENES

In this chapter we established that the genetic code is nonoverlapping. This means that each ribonucleotide of an mRNA which specifies a polypeptide chain is part of only one triplet. However, this characteristic of the code does not rule out the possibility that a single mRNA may have multiple initiation points for translation. If so, these points could theoretically create several different frames of reading within the same mRNA, thus specifying more than one polypeptide. This concept, which would create **overlapping genes,** is illustrated in Figure 9.17(a).

That this might actually occur in some viruses was suspected when phage ϕX174 was carefully investigated. The circular DNA chromosome consists of 5386 nucleotides, which should encode a maximum of 1795 amino acids, sufficient for five or six proteins. However, it was realized that this small virus in fact synthesizes 11 proteins consisting of more than 2300 amino acids! Comparison of the nucleotide sequence of the DNA and the amino acid sequences of the polypeptides synthesized has clarified this paradox. At least four cases of multiple initiation have been discovered, creating seven overlapping genes [Figure 9.17(b)].

The use of overlapping reading frames optimizes the use of a limited amount of DNA present in this and other small viruses. However, such an approach to storing information has the distinct disadvantage that a single mutation may affect more than one protein and thus increase the chances that the change will be deleterious or lethal. In the case discussed above, a single mutation at the junction of genes *A* and *C* could affect three proteins (the A, C, and K proteins). It may be for this reason that overlapping genes have not become common in other organisms.

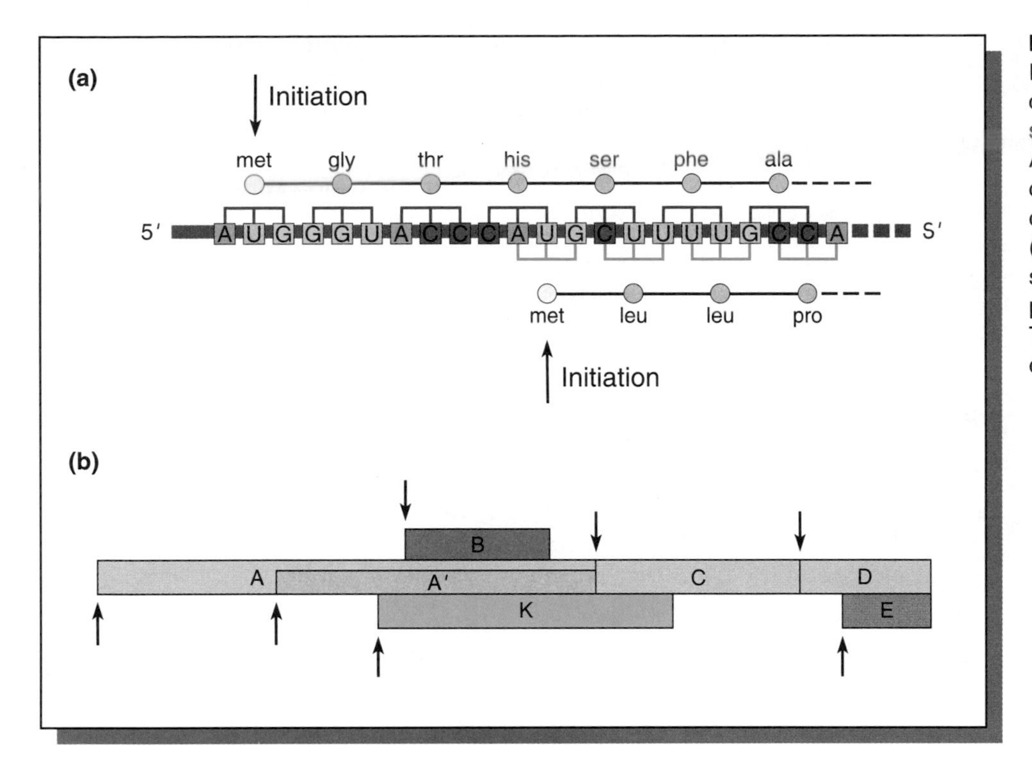

FIGURE 9.17
Illustration of the concept of overlapping genes. (a) An mRNA sequence initiated at two different AUG positions out of frame with one another will give rise to two distinct amino acid sequences. (b) The relative positions of the sequences encoding seven polypeptides of the phage ϕX174. Those encoding B, K, and E are out of frame.

CHAPTER SUMMARY

1 The genetic code, stored in DNA, is transferred to RNA, where it is used to direct the synthesis of polypeptide chains. It is degenerate, unambiguous, nonoverlapping, and comma-free.

2 The complete coding dictionary determined using various experimental approaches, reveals that of the 64 possible codons, 61 encode the 20 amino acids found in proteins, while three triplets terminate translation. One of these 61 is the initiation codon and specifies methionine.

3 The observed pattern of degeneracy often involves only the third letter of a triplet series and led Crick to propose the wobble hypothesis.

4 Confirmation for the coding dictionary, including codons for intiation and termination, was obtained by comparing the complete nucleotide sequence of phage MS2 with the amino acid sequence of a corresponding protein. Other findings support the belief that, with only minor exceptions, the code is universal for all organisms.

5 Transcription and translation—RNA and protein biosynthesis, respectively—are the fundamental processes essential to the expression of genetic information.

6 The processes of transcription and translation, like DNA replication, can be subdivided into the stages of initiation, elongation, and termination. Both processes rely on base-pairing affinities between complementary nucleotides.

7 Transcription describes the synthesis, under the direction of RNA polymerase, of a strand of RNA complementary to a DNA template.

8 The complex energy-requiring process of translation involves charged tRNA molecules, numerous proteins, ribosomes, and mRNA. Transfer RNA (tRNA) serves as the adaptor molecule between an mRNA triplet and the appropriate amino acid.

9 The processes of transcription and translation are more complex in eukaryotes than in prokaryotes. The primary transcript in eukaryotes must be modified in various ways, including the addition of a cap and poly A, and the removal, through splicing, of intervening sequences, or introns.

10 In some bacteriophages multiple initiation points may occur during the translation of RNA, resulting in multiple reading frames and overlapping genes.

KEY TERMS

aminoacyl site (A site)
aminoacyl tRNA synthetase
anticodon
chain elongation
charging (tRNA)
cloverleaf model
codon
consensus sequence
degenerate code
elongation
enhancer
exon
frameshift mutation
Goldberg – Hogness (TATA) box
heteroduplex
heterogeneous RNA (hnRNA)
information flow
initiation complex
initiation factors
initiator codon
intervening sequence
intron
messenger RNA (mRNA)
monosome
N-formylmethionine (fmet)
nonoverlapping code
nonsense mutation
overlapping genes
peptidyl site (P site)
poly A
polynucleotide phosphorylase
polyribosome
posttranscriptional modification
Pribnow box
promoter
release factors
ribosomal protein
ribosome
RNA polymerase
second genetic code
Shine-Dalgarno sequence
sigma subunit
spacer DNA
splicing
split genes
tandem repeats
template binding
termination factors
transcription
transfer RNA (tRNA)
translation
triplet
triplet binding assay
unambiguous code
universal code
wobble hypothesis

1 Had evolution seized on 6 bases (three complementary base pairs) rather than 4 bases within the structure of DNA, calculate how many triplet codons would be possible. Would 6 bases accommodate a two-letter code, assuming 20 amino acids and start and stop codons?

SOLUTION: Six things taken three at a time will produce $(6)^3$ or 216 triplet codes. If the code was a doublet, there would be $(6)^2$ or 36 two-letter codes, more than enough to accommodate 20 amino acids and start–stop punctuation.

2 In a heteropolymer experiment using 1/2C:1/4A:1/4G, how many different triplets will occur in the synthetic RNA molecule? How frequently will the most frequent triplet occur?

SOLUTION: There will be $(3)^3$ or 27 triplets produced. The most frequent will be CCC, present $(1/2)^3$ or 1/8 of the time.

3 In a regular copolymer experiment, where UUAC is repeated over and over, how many different triplets will occur in the synthetic RNA, and how many amino acids will occur in the polypeptide when this RNA is translated? Be sure to consult Figure 9.6.

SOLUTION: The synthetic RNA will repeat four triplets—UUA, UAC, ACU, and CUU—over and over.

Since both UUA and CUU encode leucine, while ACU and UAC encode threonine and tyrosine, respectively, polypeptides synthesized under the directions of such an RNA contain three amino acids in the repeating sequence leu-leu-thr-tyr.

4 Actinomycin D inhibits DNA-dependent RNA synthesis. This antibiotic is added to a bacterial culture where a specific protein is being monitored. Compared to a control culture, translation of the protein declines over a period of 20 minutes, until no further protein is made. Explain these results.

ANSWER: The mRNA, which is the basis for the translation of the protein, has a lifetime of about 20 minutes. When actinomycin D is added, transcription is inhibited and no new mRNAs are made. Those already present support the translation of the protein for up to 20 minutes.

PROBLEMS AND DISCUSSION QUESTIONS

1 Crick, Barnett, Brenner, and Watts-Tobin, in their studies of frameshift mutations, found that either 3 (+)'s or 3 (−)'s restored the correct reading frame. If the code were a sextuplet (consisting of six nucleotides), would the reading frame be restored by either of the above combinations?

2 In a mixed copolymer experiment using polynucleotide phosphorylase, 3/4 G:1/4 C was added to form the synthetic message. The resulting amino acid composition of the ensuing protein was determined:

Glycine	36/64 (56%)
Alanine	12/64 (19%)
Argininc	12/64 (19%)
Proline	4/64 (7%)

From this information

(a) Indicate the percentage (or fraction) of the time each possible triplet will occur in the message.

(b) Determine one consistent base composition assignment for the amino acids present.

(c) Considering the wobble hypothesis, predict as many specific triplet assignments as possible.

3 In a mixed copolymer experiment, messengers were created with either 4/5 C:1/5 A or 4/5 A:1/5 C. These messages yielded proteins with the following amino acid compositions. Using these data, predict the most specific coding composition for each amino acid.

4/5 C:1/5 A	Proline	63.0%	4/5 A:1/5 C	Proline	3.5%
	Histidine	13.0%		Histidine	3.0%
	Threonine	16.0%		Threonine	16.0%
	Glutamine	3.0%		Glutamine	13.0%
	Asparagine	3.0%		Asparagine	13.0%
	Lysine	0.5%		Lysine	50.0%
		98.5%			98.5%

4 In a coding experiment using repeating copolymers (as shown in Table 9.3), the following data were obtained:

Copolymer	Codons Produced	Amino Acids in Polypeptide
AG	AGA, GAG	Arg, Glu
AAG	AGA, AAG, GAA	Lys, Arg, Glu

AGG is known to code for arginine. Taking into account the wobble hypothesis, assign each of the four remaining different triplet codes to its correct amino acid.

5 In the triplet binding technique, radioactivity remains on the filter when the amino acid corresponding to the triplet is labeled. Explain the basis of this technique.

6 In studies of the amino acid sequence of wild-type and mutant forms of tryptophan synthetase in *E. coli,* the following changes have been observed:

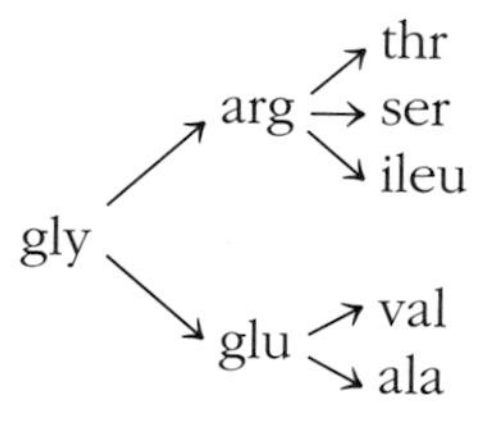

Determine a set of triplet codes in which a single nucleotide change produces each amino acid change.

7 Define and differentiate between transcription and translation. Where do these processes fit into the central dogma of molecular genetics?

8 List all of the molecular constituents present in a functional polyribosome.

9 Contrast the roles of tRNA and mRNA during translation and list all enzymes that participate in the transcription and translation process.

10 Francis Crick proposed the "adaptor hypothesis" for the function of tRNA. Why did he choose that description?

11 What molecule bears the codon? the anticodon?

12 The α chain of eukaryotic hemoglobin is composed of 141 amino acids. What is the minimum number of nucleotides in an mRNA coding for this protein chain? Assuming that each nucleotide is 0.34 nm long in the mRNA, how many triplet codes can at one time occupy space in a ribosome that is 20 nm in diameter?

13 Summarize the steps involved in charging tRNAs with their appropriate amino acids.

14 In 1962, F. Chapeville and others reported an experiment where they isolated radioactive ^{14}C-cysteinyl-tRNA^{cys} (charged tRNA^{cys} + cysteine). They then removed the sulfur group from the cysteine, creating alanyl-tRNA^{cys} (charged tRNA^{cys} + alanine). When alanyl-tRNA^{cys} was added to a synthetic mRNA calling for cysteine but not alanine, a polypeptide chain was synthesized containing alanine. What can you conclude from this experiment?

15 A short RNA molecule was isolated that demonstrated a hyperchrome shift indicating secondary structure. Its sequence was determined to be AGGCGCCGACUCUACU.

(a) Predict a two-dimensional model for this molecule.

(b) What DNA sequence would give rise to this RNA molecule through transcription?

(c) If the molecule were a tRNA fragment containing a CGA anticodon, what would the corresponding codon be?

(d) If the molecule were an internal part of a message, what amino acid sequence would result from it following translation? (Refer to the code chart in Figure 9.6.)

SELECTED READINGS

ALBERTS, B., et al. 1989. *Molecular biology of the cell.* 2nd ed. New York: Garland.

BARRELL, B. G., AIR, G., and HUTCHINSON, C. 1976. Overlapping genes in bacteriophage ϕX174. *Nature* 264:34–40.

BARRELL, B. G., BANKER, A. T., and DROUIN, J. 1979. A different genetic code in human mitochondria. *Nature* 282:189–94.

BRENNER, S. 1989. *Molecular Biology: A selection of papers.* Orlando: Academic Press.

BRENNER, S., JACOB, F., and MESELSON, M. 1961. An unstable intermediate carrying information from genes to ribosomes for protein synthesis. *Nature* 190:575–80.

CECH, T. R. 1986. RNA as an enzyme. *Scient. Amer.* (Nov.) 255, 5: 64–75.

CHAMBON, P. 1981. Split genes. *Scient. Amer.* (May) 244:60–71.

COLD SPRING HARBOR LABORATORY. 1966. The genetic code. *Cold Spr. Harb. Symp.,* Vol. 31.

CRICK, F. H. C. 1962. The genetic code. *Scient. Amer.* (Oct.) 207:66–77.

———. 1966. The genetic code: III. *Scient. Amer.* (Oct.) 215:55–63.

———. 1979. Split genes and RNA splicing. *Science* 204:264–71.

DARNELL, J. E. 1983. The processing of RNA. *Scient. Amer.* (Oct.) 249:90–100.

———. 1985. RNA. *Scient. Amer.* (Oct.) 253:68–87.

DICKERSON, R. E. 1983. The DNA helix and how it is read. *Scient. Amer.* (Dec.) 249:94–111.

FIERS, W., et al. 1976. Complete nucleotide sequence of bacteriophage MS2 RNA: Primary and secondary structure of the replicase gene. *Nature* 260:500–507.

HAMKALO, B. 1985. Visualizing transcription in chromosomes. *Trends in Genet.* 1:255–60.

HOLLEY, R. W., et al. 1965. Structure of a ribonucleic acid. *Science* 147:1462–65.

JUDSON, H. F. 1979. *The eighth day of creation.* New York: Simon and Schuster.

LAKE, J. A. 1981. The ribosome. *Scient. Amer.* (Aug.) 245:84–97.

MILLER, O. L., HAMKALO, B., and THOMAS, C. 1970. Visualization of bacterial genes in action. *Science* 169:392–95.

NIRENBERG, M. W. 1963. The genetic code: II. *Scient. Amer.* (March) 190:80–94.

RICH, A., and HOUKIM, S. 1978. The three-dimensional structure of transfer RNA. *Scient. Amer.* (Jan.) 238:52–62.

STEITZ, J. A. 1988. Snurps. *Scient. Amer.* (June) 258(6):56–63.

STRYER, L. 1988. *Biochemistry.* 3rd ed. New York: W.H. Freeman.

VOLKIN, E., ASTRACHAN, L., and COUNTRYMAN, J. L. 1958. Metabolism of RNA phosphorus in *E. coli* infected with bacteriophage T7. *Virology* 6:545–55.

WATSON, J. D., HOPKINS, N. H., ROBERTS, J. W., STEITZ, J. A., and WEINER, A. M. 1987. *Molecular Biology of the Gene.* 4th ed. Menlo Park, CA: Benjamin-Cummings.

ZUBAY, G. L., and MARMUR, J. 1973. *Papers in biochemical genetics.* 2nd ed. New York: Holt, Rinehart and Winston.

Proteins: The End Products of Genes

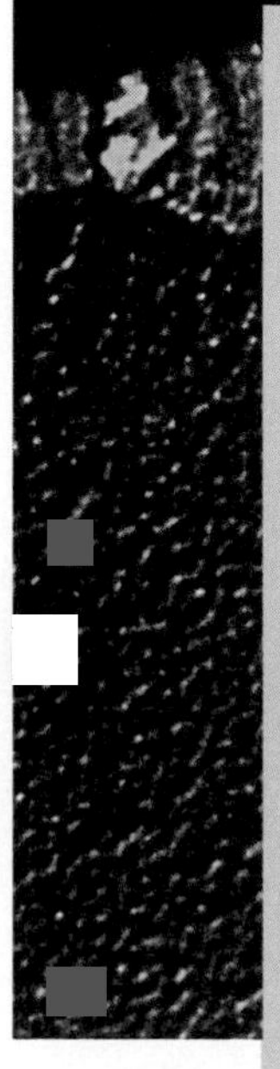

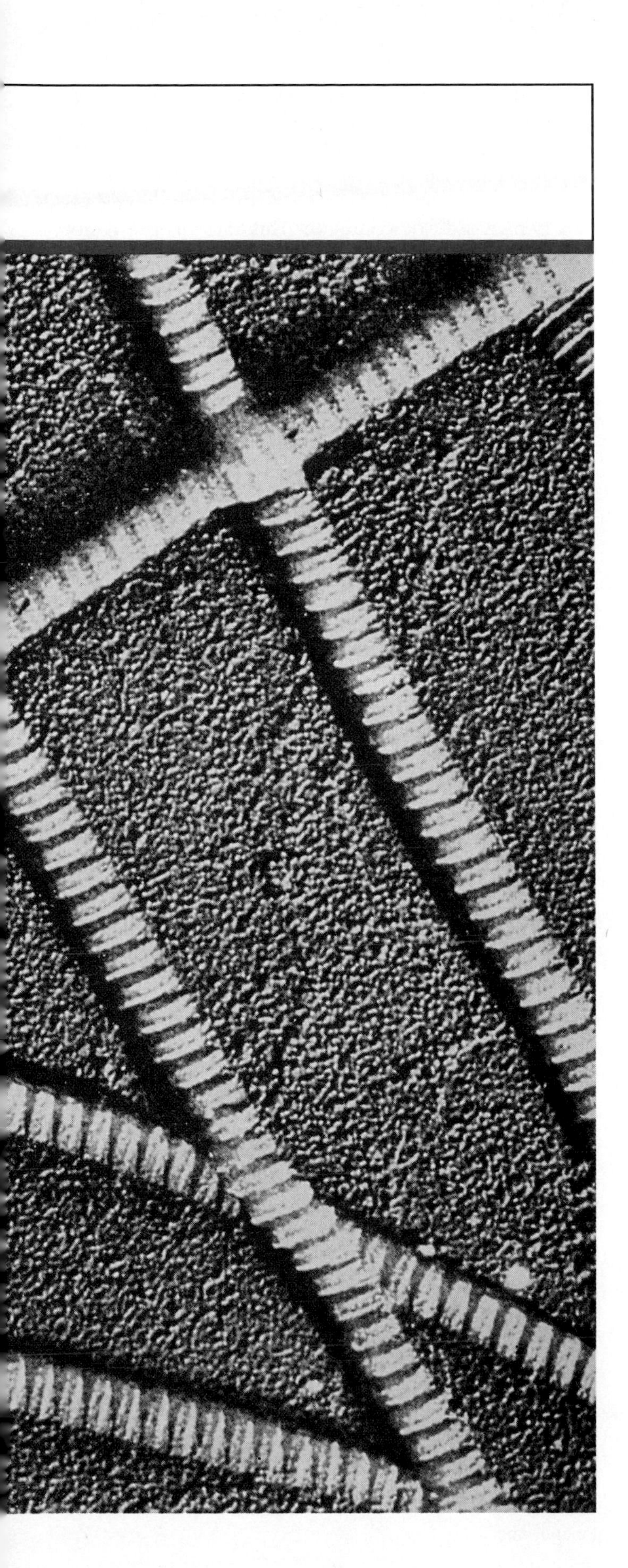

CHAPTER CONCEPTS

The end products of most genes are polypeptide chains. They achieve a three-dimensional conformation based on their primary amino acid sequences, often interacting with other such chains to create functional protein molecules. The function of any protein is closely tied to its structure, which can be disrupted by mutation, leading to a distinctive phenotypic effect.

In the previous chapter we established that there is a genetic code that stores information in the form of triplet nucleotides in DNA, and that this information can be expressed through the orderly processes of transcription and translation. The final product of gene expression, in almost all instances, is a polypeptide chain consisting of a linear series of amino acids, whose sequence has been prescribed by the genetic code. In this chapter, we will review the evidence that confirmed that proteins are the end products of genes. Then we will discuss briefly the various levels of protein structure, diversity, and function. This information provides an important foundation in our understanding of how mutations, which arise in DNA, can result in the variety of phenotypic effects observed in organisms.

GARROD AND BATESON: INBORN ERRORS OF METABOLISM

The first insight into the role of proteins in genetic processes was provided by observations made by Sir Archibald Garrod and William Bateson early in this century. Garrod was born into an English family of medical scientists. His father was a physician with a strong interest in the chemical basis of rheumatoid arthritis, and his eldest brother was a leading zoologist in London. Thus, it is not surprising that, as a practicing physician, Garrod became interested in several human disorders that seemed to be inherited. Although he also studied albinism and cystinuria, we will describe his investigation of the disorder **alkaptonuria.** Individuals afflicted with this disorder cannot metabolize the alkapton 2,5-dihydroxyphenylacetic acid, also known as **homogentisic acid.** As a result, an important metabolic pathway (Figure 10.1) is blocked. Homogentisic acid accumulates in cells and tissues and is excreted in the urine. The molecule's oxidation products are black and thus are easily detectable in the diapers of newborns and the urine of individuals. The products tend to accumulate in cartilaginous areas, causing a darkening of the ears and nose. In joints, this deposition leads to a benign arthritic condition. This rare disease is not serious, but it persists throughout an individual's life.

Garrod studied alkaptonuria by increasing dietary protein or adding to the diet the amino acids phenylalanine or tyrosine, which are chemically related to homogentisic acid. Under such conditions, homogentisic acid levels increase in the urine of alkaptonurics but not in unaffected individuals. He concluded that normal individuals can break down, or catabolize, this alkapton but that afflicted individuals cannot. By studying the patterns of inheritance of the disorder, Garrod further concluded that alkaptonuria was inherited as a simple recessive trait.

On the basis of these conclusions, Garrod hypothesized that the hereditary information controls chemical reactions in the body and that the inherited disorders he

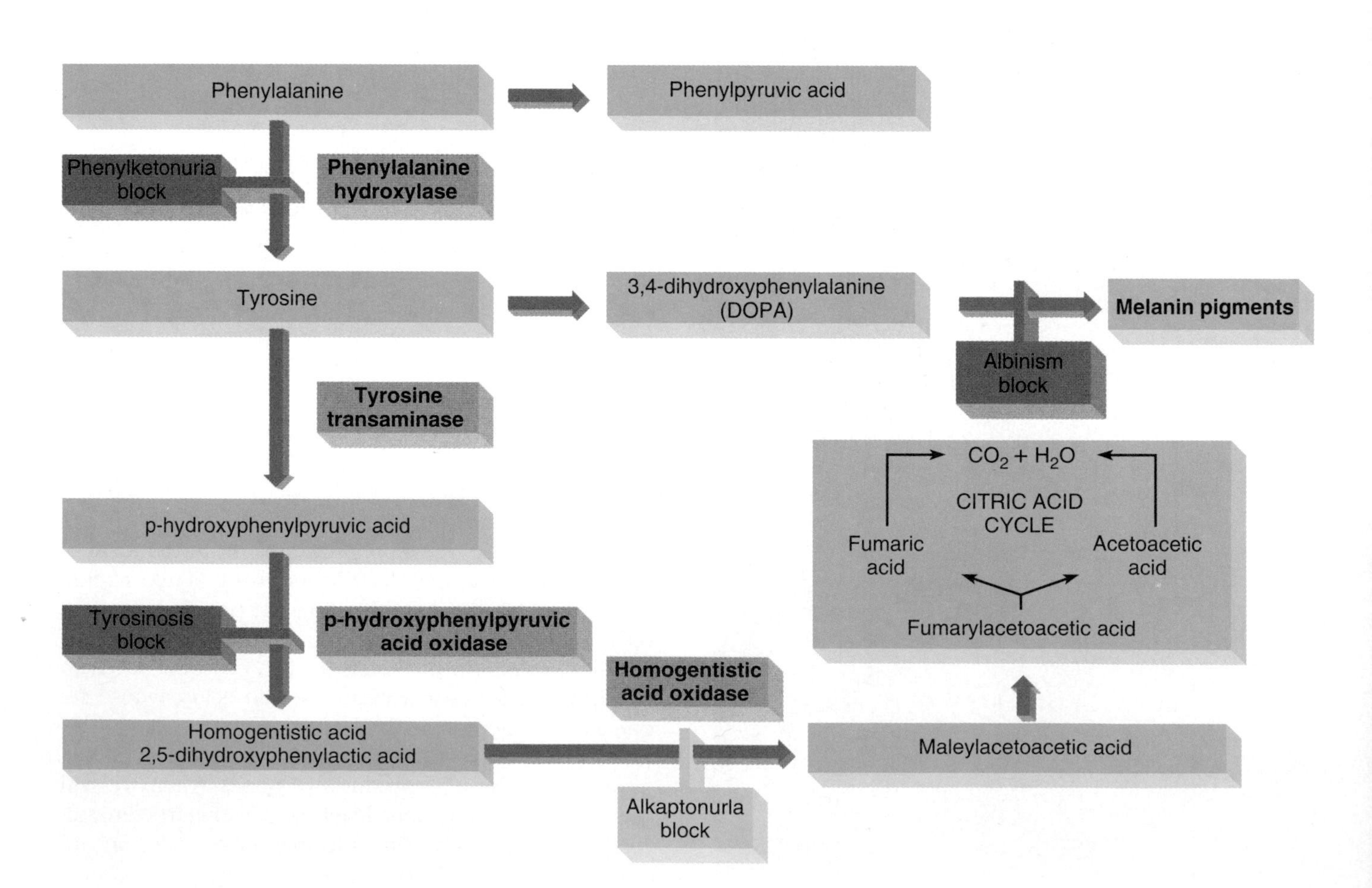

FIGURE 10.1
Metabolic pathway involving phenylalanine and tyrosine. Various metabolic blocks resulting from mutations lead to the disorders phenylketonuria, alkaptonuria, and albinism.

studied are the result of alternate modes of metabolism. While the terms *genes* and *enzymes* were not familiar during Garrod's time, he used the corresponding concepts of *unit factors* and *ferments*. Garrod published his initial observations in 1902.

Only a few geneticists, including Bateson, were familiar with or referred to Garrod's work. His ideas fit nicely with Bateson's belief that inherited conditions were caused by the lack of some critical substance. In 1909, Bateson published *Mendel's Principles of Heredity*, in which he linked ferments with heredity. However, for almost thirty years, most geneticists failed to see the relationship between genes and enzymes. Garrod and Bateson, like Mendel, were ahead of their time.

Phenylketonuria

Described first in 1934, **phenylketonuria (PKU)** may result in mental retardation and is inherited as an autosomal recessive disease. Afflicted individuals have still another step blocked in the metabolic pathway just discussed. They are unable to convert the amino acid phenylalanine to the amino acid tyrosine (see Figure 10.1). These molecules differ by only a single hydroxl group (—OH), present in tyrosine, but absent in phenylalanine. The reaction is catalyzed by the enzyme **phenylalanine hydroxlase,** which is inactive in affected individuals and active at about a 30 percent level in heterozygotes. The enzyme functions in the liver. While the normal blood level of phenylalanine is about 1 mg/100 ml, phenylketonurics show a level as high as 50 mg/100 ml.

As phenylalanine accumulates, it may be converted to phenylpyruvic acid and, subsequently, other derivatives. These are less efficiently resorbed by the kidney and tend to spill into the urine more quickly than phenylalanine. Both phenylalanine and derivatives enter the cerebrospinal fluid, resulting in elevated levels in the brain. The presence of these substances during early development is thought to be responsible for retardation.

Retardation can be prevented by PKU screening of newborns. When the condition is detected in the analysis of an infant's blood, a strict dietary regimen is instituted. A low phenylalanine diet can reduce such byproducts as phenylpyruvic acid, and abnormalities characterizing the disease can be diminished. Screening of newborns occurs routinely in almost every state in this country. Phenylketonuria occurs in approximately 1 in 11,000 births.

Knowledge of inherited metabolic disorders such as alkaptonuria and phenylketonuria has caused a revolution in medical thinking and practice. No longer is human disease attributed solely to the action of invading microorganisms, viruses, or parasites. We know now that literally thousands of abnormal physiological conditions are caused by errors in metabolism that are the result of mutant genes. These human biochemical disorders are far-ranging and include all classes of organic biomolecules.

THE ONE-GENE:ONE-ENZYME HYPOTHESIS

In two separate investigations beginning in 1933, George Beadle was to provide the first convincing experimental evidence that genes are directly responsible for the synthesis of enzymes. The first investigation, conducted in collaboration with Boris Ephrussi, involved *Drosophila* eye pigments. Encouraged by these findings, Beadle then joined with Edward Tatum to investigate nutritional mutations in the pink bread mold *Neurospora crassa*. The latter investigation led to the **one-gene:one-enzyme hypothesis.**

Beadle and Tatum: *Neurospora* Mutants

In the early 1940s, Beadle and Tatum chose to work with the organism *Neurospora crassa* because much was known about its biochemistry, and mutations could be induced and isolated with relative ease. By inducing mutations, they produced strains that had genetic blocks of reactions essential to the growth of the organism.

Beadle and Tatum knew that *Neurospora* could manufacture nearly everything necessary for normal development. For example, using rudimentary carbon and nitrogen sources, this organism can synthesize 9 water-soluble vitamins, 20 amino acids, numerous carotenoid pigments, and purines and pyrimidines. Beadle and Tatum irradiated asexual spores with X rays to increase the frequency of mutations and allowed them to progress through a sexual cycle, forming ascospores. These were grown on "complete" medium containing all necessary growth factors (vitamins, amino acids, etc.). Under such growth conditions, a mutant strain would be able to grow by virtue of supplements present in the enriched complete medium. All cultures were then transferred to minimal medium. If growth occurred on minimal medium, the organisms were able to synthesize all necessary growth factors themselves, and it was concluded that the culture did not contain a mutation. If no growth occurred, then it contained a nutritional mutation, and the only task remaining was to determine its type. Both cases are illustrated in Figure 10.2(a).

Literally thousands of individual spores derived by this procedure were isolated and grown on complete medium. In subsequent tests on minimal medium, many cultures failed to grow, indicating that a nutritional mutation had been induced. To identify the mutant type, the mutant strains were tested on a series of different minimal media, each containing a single vitamin, amino acid, purine, or pyrimidine until one supplement that permitted growth was found. Beadle and Tatum reasoned that the supplement that restores growth is the molecule that the mutant strain could not synthesize. Details of how these mutants are isolated are presented in Figure 10.2

The first mutant strain isolated required vitamin B-6

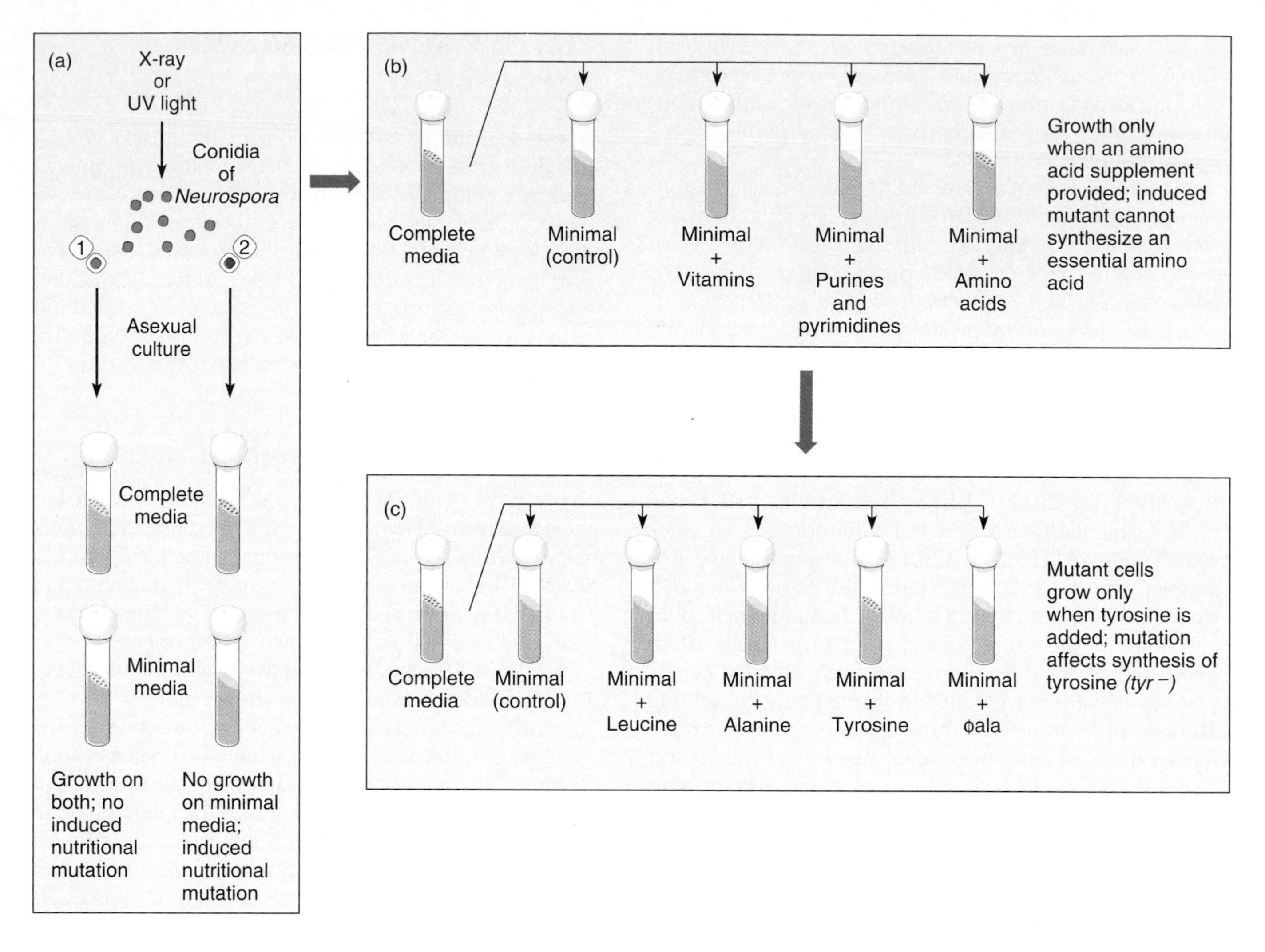

FIGURE 10.2
Induction, isolation, and characterization of a nutritional auxotrophic mutation in *Neurospora.* In (a), conidia 1 is not affected, but conidia 2 contains such a mutation. In (b) and (c), the precise nature of the mutation is determined to involve the biosynthesis of tyrosine.

(pyridoxin) in the medium, and the second one required vitamin B-1 (thiamine). Using the same procedure, Beadle and Tatum eventually isolated and studied hundreds of mutants.

The findings derived from testing over 80,000 spores convinced Beadle and Tatum that genetics and biochemistry have much in common. It seemed likely that each nutritional mutation caused the loss of the enzymatic activity that facilitates an essential reaction in wild-type organisms. It also appeared that a mutation could be found for nearly any enzymatically controlled reaction. Beadle and Tatum had thus provided sound experimental evidence that **one gene specifies one enzyme,** a hypothesis alluded to over thirty years earlier by Garrod and Bateson. With modifications, this concept was to become a major principle of genetics.

Genes and Enzymes: Analysis of Biochemical Pathways

The one-gene:one-enzyme concept and its attendant methods have been used over the years to work out many details of metabolism in *Neurospora, Escherichia coli,* and a number of other microorganisms. One of the first metabolic pathways to be investigated in detail was that leading to the synthesis of the amino acid arginine in *Neurospora.* By studying seven mutant strains, each requiring arginine for growth, Adrian Srb and Norman Horowitz were able to ascertain a partial biochemical pathway leading to the synthesis of this molecule. The rationale followed in their work illustrates how genetic analysis can be used to establish biochemical information.

Srb and Horowitz tested each mutant strain's ability to reestablish growth if either **citrulline** or **ornithine,** two compounds with close chemical similarity to arginine, was used as a supplement to minimal medium. If either was able to substitute for arginine, they reasoned that it must be involved in the biosynthetic pathway of arginine. They found that both molecules could be substituted in one or more strains.

Of the seven mutant strains, four of them (*arg 1–4*) grew if supplied with either citrulline, ornithine, or arginine. Two of them (*arg 5* and *6*) grew if supplied with citrulline or arginine. One strain (*arg 7*) would grow only if arginine was supplied. Neither citrulline nor ornithine could substitute for it. From these experimental observations, the following pathway and metabolic blocks for each mutation were deduced:

	arg 1–4		*arg 5–6*		*arg 7*	
Precursor	——+→	Ornithine	——+→	Citrulline	——+→	Arginine
	Enzyme A		Enzyme B		Enzyme C	

The reasoning supporting these conclusions is based on the following logic. If mutants *arg 1* through *4* can grow regardless of which of the three molecules is supplied as a supplement to minimal medium, the mutations preventing growth must cause a metabolic block that occurs *prior to* the involvement of ornithine, citrulline, or arginine in the pathway. When any of these three molecules is added, its presence bypasses the block. As a result, it can be concluded that both citrulline and ornithine are involved in the biosynthesis of arginine. However, the sequence of their participation in the pathway cannot be determined on the basis of these data.

On the other hand, the *arg 5* and *6* mutations grow if supplied citrulline but not ornithine. Therefore, ornithine must occur in the pathway *prior to* the block. Its presence will not overcome the block. Citrulline, however, does overcome the block, so it must be involved beyond the point of blockage. Therefore, the conversion of ornithine to citrulline represents the correct sequence in the pathway.

Finally, it can be concluded that *arg 7* represents a mutation preventing the conversion of citrulline to arginine. Neither ornithine nor citrulline can overcome the metabolic block because both participate earlier in the pathway.

Taken together, these reasons support the sequence of biosynthesis outlined here. Since Srb and Horowitz's work in 1944, the detailed pathway has been worked out and the enzymes controlling each step characterized. The chemical pathway is shown in Figure 10.3.

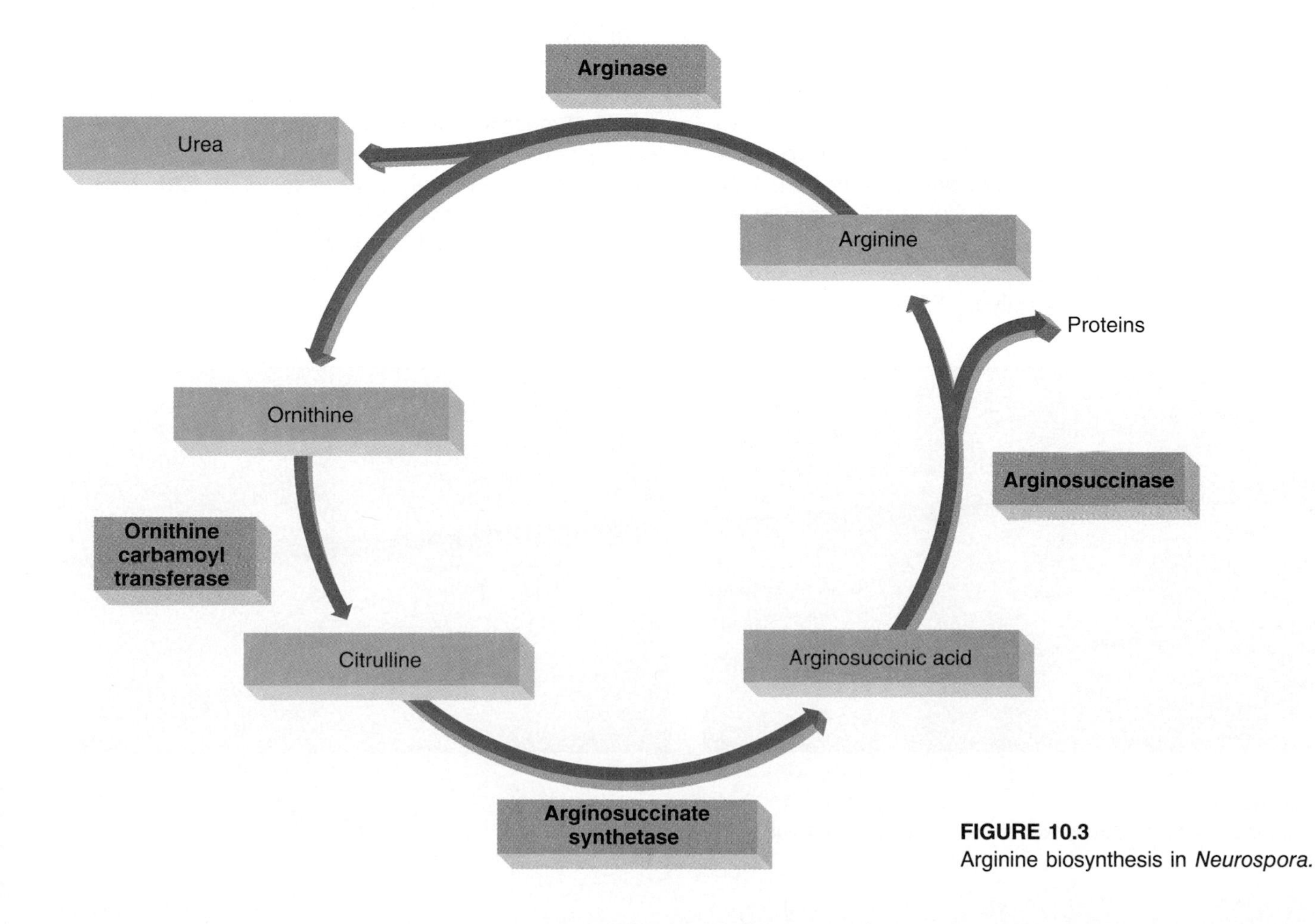

FIGURE 10.3
Arginine biosynthesis in *Neurospora.*

The concept of one-gene:one-enzyme developed in the early 1940s was not accepted immediately by all geneticists. This is not surprising, because it was not yet clear how mutant enzymes could cause variation in many phenotypic traits. For example, *Drosophila* mutants demonstrated altered eye size, wing shape, wing vein pattern, and so on. Plants exhibited mutant varieties of seed texture, height, and fruit size. How an inactive, mutant enzyme could result in such phenotypes was puzzling to many geneticists. Another reason for their reluctance to accept this concept was the paucity of information then available in molecular genetics. It was not until 1944 that Avery, MacLeod, and McCarty showed DNA to be the transforming factor and not until the early 1950s that most geneticists believed that DNA serves as the genetic material. However, by this time, the evidence in support of the concept of enzymes as gene products was overwhelming, and the concept was accepted as valid. Nevertheless, the question of how DNA specifies the structure of enzymes remained unanswered.

ONE-GENE: ONE-PROTEIN/ ONE-GENE: ONE-POLYPEPTIDE

Two factors soon modified the one-gene:one-enzyme hypothesis. First, while nearly all enzymes are proteins, not all proteins are enzymes. As the study of biochemical genetics proceeded, it became clear that all proteins are specified by the information stored in genes, leading to the more accurate phraseology **one-gene:one-protein.** Second, proteins were often shown to have a subunit structure consisting of two or more **polypeptide chains.** This is the basis of the **quaternary structure** of proteins, which we will discuss later in this chapter. Because each distinct polypeptide chain is encoded by a separate gene, a more modern statement of Beadle and Tatum's basic principle is **one-gene:one-polypeptide chain.** These modifications of the original hypothesis became apparent during the analysis of hemoglobin structure in individuals afflicted with sickle-cell anemia.

Sickle-Cell Anemia

The first direct evidence that genes specify proteins other than enzymes came from the work on mutant hemoglobin molecules derived from humans afflicted with the disorder **sickle-cell anemia.** Affected individuals contain erythrocytes, which, under low oxygen tension, become elongated and curved because of the polymerization of hemoglobin. The "sickle" shape of these erythrocytes is in contrast to the biconcave disc shape characteristic of normal individuals (Figure 10.4). Individuals with the disease suffer attacks when red blood cells aggregate in the venous side of capillary systems, where oxygen tension is very low. As a result, a variety of tissues may be deprived of oxygen and suffer severe damage. When this occurs, an individual is said to experience a sickle-cell crisis. If untreated, a crisis may be fatal. The kidneys, muscles, joints, brain, gastrointestinal tract, and lungs may be affected.

In addition to suffering crises, these individuals are anemic because their erythrocytes are destroyed more rapidly than normal red blood cells. Compensatory physiological mechanisms include increased red cell production by bone marrow and accentuated heart action. These mechanisms lead to abnormal bone size and shape as well as dilation of the heart.

In 1949, James Neel and E. A. Beet demonstrated that the disease is inherited as a Mendelian trait. Pedigree analysis revealed three genotypes and phenotypes con-

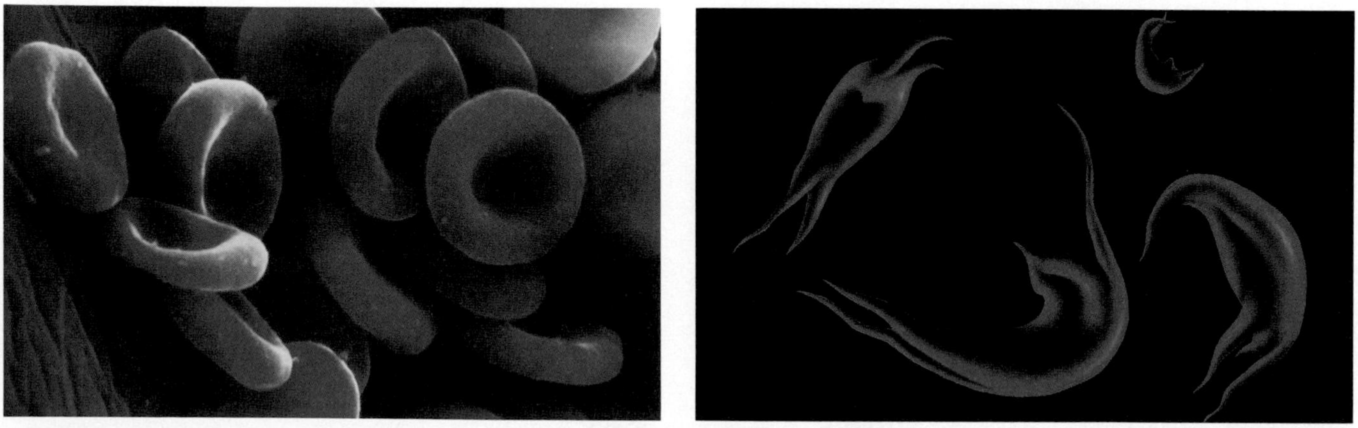

FIGURE 10.4
A comparison of erythrocytes derived from HbA (left) and HbS (right).

trolled by a single pair of alleles, *A* and *S*. Normal and affected individuals result from the homozygous genotypes *AA* and *SS*, respectively. The red blood cells of the *AS* heterozygote, which exhibits the **sickle-cell trait** but not the disease, undergo much less sickling since over half of their hemoglobin is normal. Although largely unaffected, such persons are carriers of the disorder.

In the same year, Linus Pauling and his coworkers provided the first insight into the molecular basis of the disease. They showed that hemoglobins isolated from diseased and normal individuals differ in their rates of electrophoretic migration. In this technique, charged molecules migrate in an electric field. If the net charge of two molecules is different, the rate of migration will vary. On this basis, Pauling and his colleagues concluded that a chemical difference exists between the two types of hemoglobin. The two molecules are now designated **HbA** and **HbS.**

Figure 10.5(a) illustrates the migration pattern of hemoglobin derived from individuals of all three possible genotypes when subjected to **starch gel electrophoresis.** The gel provides the supporting medium for the molecules during migration. In this experiment, samples are placed at a point of origin between the cathode (−) and the anode (+), and an electric field is applied. The migration pattern reveals that all molecules move toward the anode, indicating a net negative charge. However, HbA migrates farther than HbS, suggesting that its net negative charge is greater. The electrophoretic pattern of hemoglobin derived from carriers reveals the presence

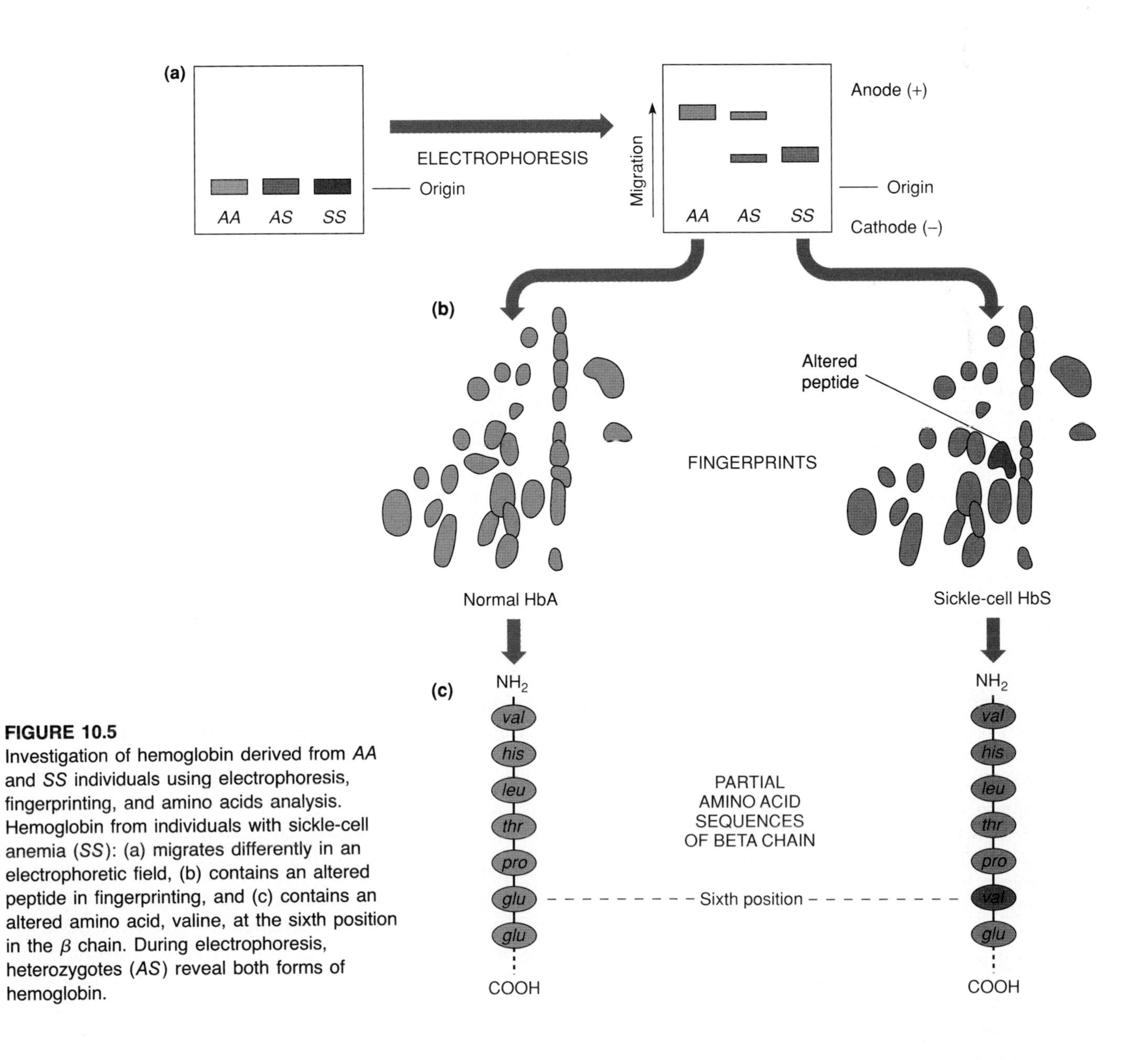

FIGURE 10.5
Investigation of hemoglobin derived from *AA* and *SS* individuals using electrophoresis, fingerprinting, and amino acids analysis. Hemoglobin from individuals with sickle-cell anemia (*SS*): (a) migrates differently in an electrophoretic field, (b) contains an altered peptide in fingerprinting, and (c) contains an altered amino acid, valine, at the sixth position in the β chain. During electrophoresis, heterozygotes (*AS*) reveal both forms of hemoglobin.

of both HbA and HbS, confirming their heterozygous genotype.

Pauling's findings suggested two possibilities. It was known that hemoglobin consists of four nonproteinaceous, iron-containing **heme groups** and a **globin portion** that contains four polypeptide chains. The alteration in net charge in HbS could be due, theoretically, to a chemical change in either component.

Work carried out between 1954 and 1957 by Vernon Ingram resolved this question. He demonstrated that the chemical change occurs in the primary structure of the globin portion of the hemoglobin molecule. Using the **fingerprinting technique,** Ingram showed that HbS differs in amino acid composition compared to HbA. Human adult hemoglobin contains two identical alpha (α) chains of 141 amino acids and two identical beta (β) chains of 146 amino acids in its quaternary structure.

The fingerprinting technique involves enzymatic digestion of the protein into peptide fragments. The mixture is then placed on absorbent paper and exposed to an electric field, where migration occurs according to net charge. The paper is then turned at a right angle and placed in a solvent, where chromatographic action causes migration of the peptides in the second direction. The end result is a two-dimensional separation of the peptide fragments into a distinctive pattern of spots or "fingerprints." Ingram's work revealed that HbS and HbA differed by only a single peptide fragment [Figure 10.5(b)]. Further analysis then revealed a single amino acid change: Valine was substituted for glutamic acid at the sixth position of the β chain, accounting for the peptide difference. [Figure 10.5(c)].

The significance of this discovery has been multifaceted. It clearly establishes that a single gene provides the genetic information for a single polypeptide chain. Studies of HbS also demonstrate that a mutation can affect the phenotype by directing a single amino acid substitution. Also, by providing the explanation for sickle-cell anemia, the concept of **molecular disease** was firmly established. Finally, this work led to a thorough study of human hemoglobins, which has provided valuable genetic insights.

In the United States, sickle-cell anemia is found almost exclusively in the black population. Extensive research is now underway to determine modes of treatment for those with the disease. It affects about one in every 625 black infants born in this country. Currently, about 50,000 to 75,000 individuals are afflicted. In about one of every 145 black married couples, both partners are heterozygous carriers, where each of their children has a 25 percent chance of having the disease.

Human Hemoglobins

Molecular analysis reveals that a variety of hemoglobin molecules are produced in humans. All are tetramers consisting of numerous combinations of seven distinct polypeptide chains, each encoded by a separate gene. In our discussion of sickle-cell anemia, we learned that **HbA** contains two **alpha (α)** and two **beta (β) chains.** HbA represents about 98 percent of all hemoglobin found in an individual's erythrocytes after the age of six months. The remaining 2 percent consists of **HbA_2,** a minor adult component. This molecule contains two alpha and two **delta (δ) chains.** The latter chain is very similar to the beta chain, consisting of 146 amino acids.

TABLE 10.1
Chain compositions of human hemoglobins from conception to adulthood.

Hemoglobin Type	Chain Composition
Embryonic	$\zeta_2\epsilon_2$
Fetal	$\alpha_2{}^A\gamma_2$ $\alpha_2{}^G\gamma_2$
Adult	$\alpha_2\beta_2$
Minor adult	$\alpha_2\delta_2$

During embryonic and fetal development, quite a different set of hemoglobins is found. The earliest set to develop is called **Gower 1** and contains two **zeta (ζ) chains,** which are alphalike, and two **epsilon (ϵ) chains,** which are betalike. By eight weeks of gestation, this embryonic form is gradually replaced by still another hemoglobin molecule with different chains. This molecule is called **HbF,** or **fetal hemoglobin,** and consists of two alpha chains and two **gamma (γ) chains.** These gamma chains are of nearly identical types and are designated **$^G\gamma$** and **$^A\gamma$.** Both are betalike and differ from each other by only a single amino acid.

The nomenclature and sequence of appearance of the firve tetramers described so far are summarized in Table 10.1. The genes coding for each of the seven chains have been mapped. Those coding for α and ζ are located on chromosome 16, while those coding for ϵ, $^G\gamma$, $^A\gamma$, δ, and β are located on chromosome 11. In each case, they are clustered together, constituting **gene families.**

COLINEARITY

Once it was established that genes specify the synthesis of polypeptide chains, the next logical question was how genetic information contained in the nucleotide sequence of a gene can be transferred to the amino acid sequence of a polypeptide chain. It seemed most likely that a **colinear relationship** would exist between the two molecules. That is, the order of nucleotides in the

DNA of a gene would correlate directly with the order of amino acids in the corresponding polypeptide.

While we have introduced this concept earlier in Chapter 9, it is useful to examine experimental evidence in its support. In studies of the A subunit of the enzyme **tryptophan synthetase** in *E. coli,* Charles Yanofsky sought to demonstrate colinearity. He isolated many independent mutants that had lost activity of the enzyme. He was able to map these mutations and establish their location with respect to one another within the gene. Then, he determined where the amino acid substitution had occurred in each mutant protein. When the two sets of data were compared, the colinear relationship was apparent. The location of each mutation in the *trp* A gene correlates with the position of the altered amino acid in the A polypeptide of tryptophan synthetase. This comparison is illustrated in Figure 10.6. Recall that we have already discussed the details of how information is transferred from DNA to protein (transcription and translation) in Chapter 9.

PROTEIN STRUCTURE AND FUNCTION

Having established that the genetic information is stored in DNA and influences cellular activities through the proteins it encodes, we turn now to a discussion of protein structure and function. How is it that these molecules play such a critical role in determining the complexity of cellular activities? As we will see, the structure of proteins is intimately related to the functional diversity of these molecules.

Protein Structure

First, we should differentiate between the terms **polypeptides** and **proteins.** Both describe a molecule composed of amino acids. The molecules differ, however, in their state of assembly and functional capacity.

Polypeptides are the precursors of proteins. As assembled on the ribosome during translation, the molecule is called a polypeptide. When released from the ribosome following translation, a polypeptide folds up and assumes a higher order of structure. When this occurs, a three-dimensional conformation in space is produced. In many cases, several polypeptides interact to produce this conformation. Whether or not several polypeptides interact, the three-dimensional conformation is essential to the function of the molecule. When the functional state is achieved, the molecule is appropriately called a protein.

The polypeptide chains of proteins, like nucleic acids, are linear nonbranched polymers. There are 20 amino acids that serve as the building blocks or subunits of proteins. Each amino acid has a **carboxyl group,** an **amino group,** and an **R (radical) group** (or side chain) bound covalently to a central carbon atom. The R group gives each amino acid its chemical identity. Figure 10.7 illustrates the 20 different R groups, which show a variety of configurations and may be divided into four main classes: (1) **nonpolar** or **hydrophobic;** (2) **polar** or **hydrophilic;** (3) **negatively charged;** and (4) **positively charged.** Because polypeptides are often long polymers, and because each position may be occupied by any one of 20 amino acids with unique chemical properties, an enormous variation in chemical activity is possible. For example, if an average polypeptide is composed of 200 amino acids (molecular weight of about 20,000 daltons), 20^{200} different molecules, each with a unique sequence, can be created using 20 different building blocks.

Around 1900, a German chemist Emil Fischer determined the manner in which the amino acids are bonded together. He showed that the amino group of one amino acid can react with the carboxyl group of another amino acid during a dehydration reaction, releasing a molecule of H_2O. The resulting covalent bond is known as a **pep-**

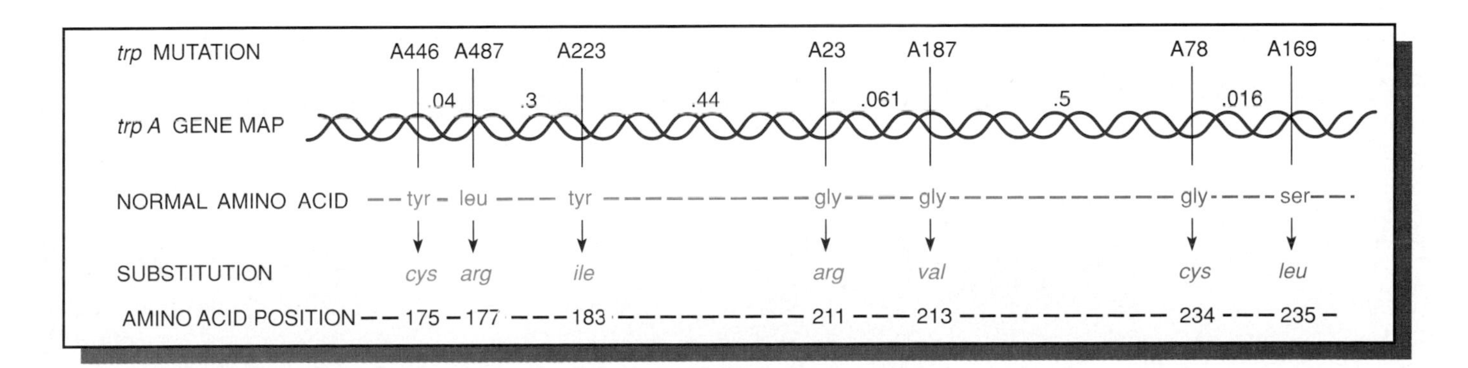

FIGURE 10.6
Demonstration of colinearity between the genetic map of various *trp A* mutations in *E. coli* and the affected amino acids in the protein product. The values shown between mutations represent linkage distances.

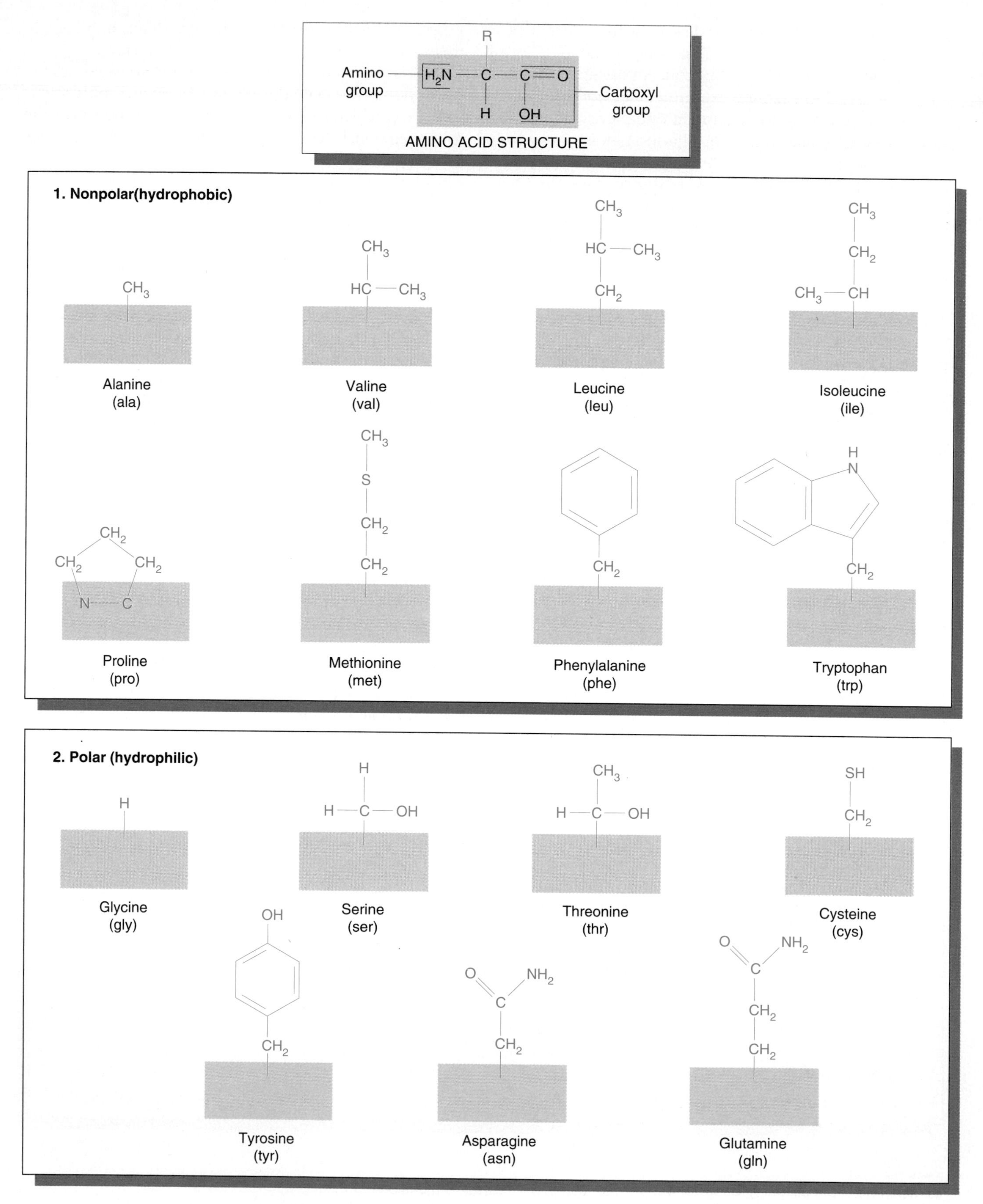
R
Amino group
H_2N — C — C = O
H OH
Carboxyl group
AMINO ACID STRUCTURE
1. Nonpolar(hydrophobic)
CH_3
Alanine (ala)
CH_3
HC — CH_3
Valine (val)
CH_3
HC — CH_3
CH_2
Leucine (leu)
CH_3
CH_2
CH_3 — CH
Isoleucine (ile)
CH_2
CH_2 CH_2
N — C
Proline (pro)
CH_3
S
CH_2
CH_2
Methionine (met)
CH_2
Phenylalanine (phe)
H
N
CH_2
Tryptophan (trp)
2. Polar (hydrophilic)
H
Glycine (gly)
H
H — C — OH
Serine (ser)
CH_3
H — C — OH
Threonine (thr)
SH
CH_2
Cysteine (cys)
OH
CH_2
Tyrosine (tyr)
O
NH_2
C
CH_2
Asparagine (asn)
O
NH_2
C
CH_2
CH_2
Glutamine (gln)

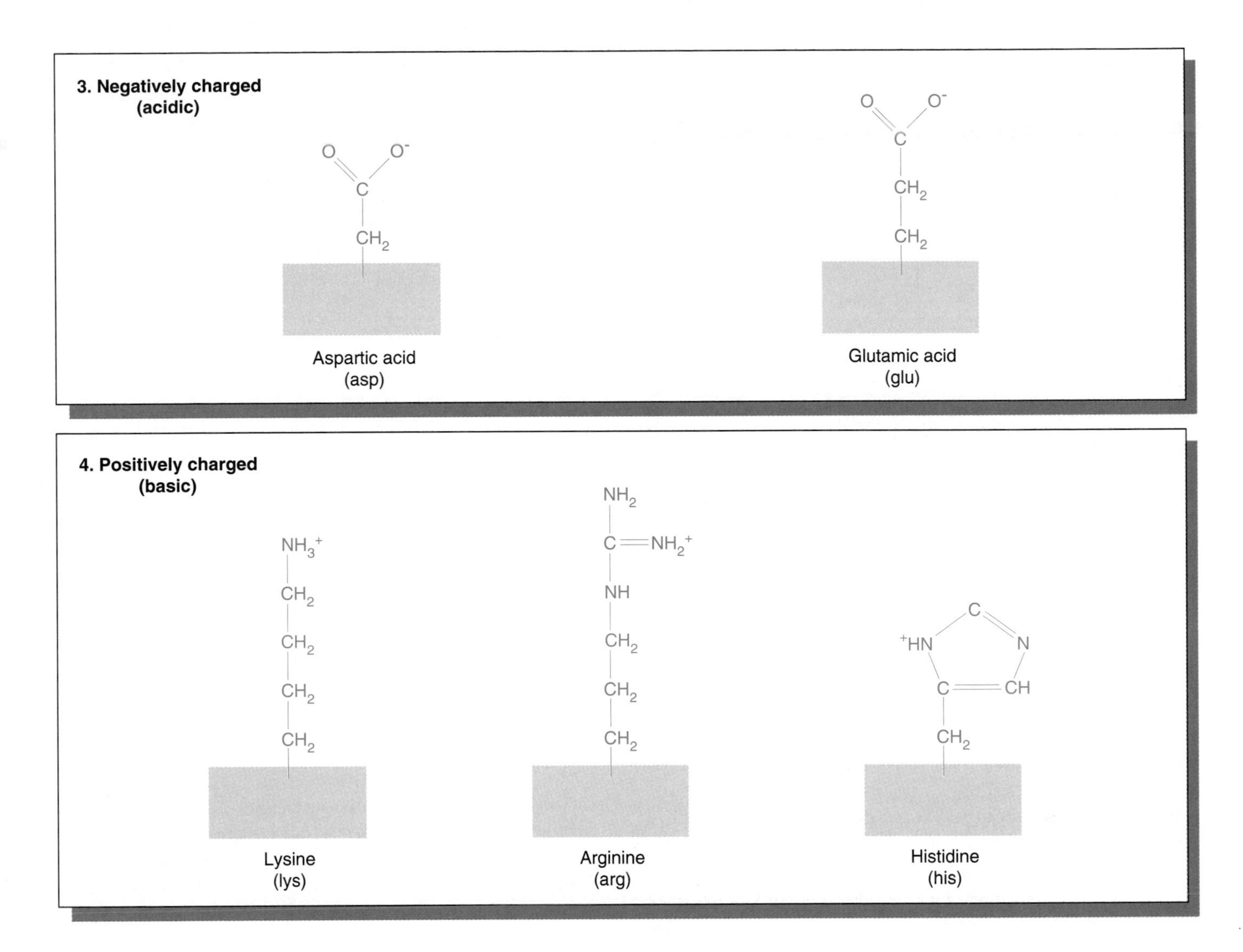

FIGURE 10.7
Chemical structures of the 20 amino acids found in living organisms, divided into four major categories.

tide bond (Figure 10.8). Two amino acids linked together constitute a **dipeptide,** three a **tripeptide,** etc. When more than ten amino acids are linked by peptide bonds, the chain is referred to as a **polypeptide.** Generally, no matter how long a polypeptide is, it will contain a free amino group at one end (the **N-terminus**) and a free carboxyl group at the other end (the **C-terminus**).

Four levels of protein structure are recognized: **primary (I°); secondary (II°); tertiary (III°);** and **quaternary (IV°).** The sequence of amino acids in the linear backbone of the polypeptide constitutes its **primary structure.** This sequence is specified by the sequence of deoxyribonucleotides in DNA via an mRNA intermediate. The primary structure of a polypeptide determines the specific characteristics of the higher orders of structure as a protein is formed.

The **secondary structure** refers to a regular or repeating configuration in space assumed by amino acids aligned closely to one another in the polypeptide chain. In 1951, Linus Pauling and Robert Corey predicted, on theoretical grounds, an **α (alpha) helix** as one type of secondary structure. The α-helix model [Figure 10.9(a)] has since been confirmed by X-ray crystallographic studies. It is rodlike and has the greatest possible theoretical stability. The helix is composed of a spiral chain of amino acids stabilized by hydrogen bonds.

The side chains of amino acids extend outward from the helix, and each amino acid residue occupies a vertical

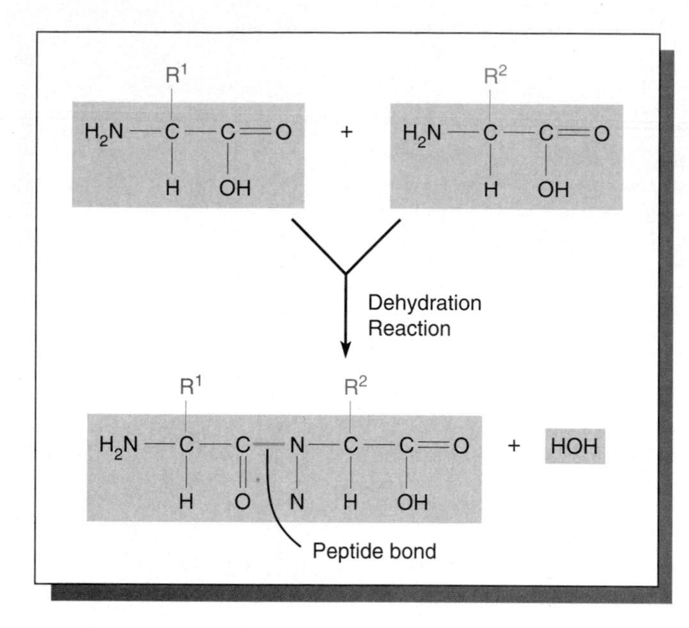

FIGURE 10.8
Peptide bond formation between two amino acids, resulting from a dehydration reaction.

distance of 1.5 Å in the helix. There are 3.6 residues per turn. While left-handed helices are theoretically possible, all proteins demonstrating an α helix are right-handed.

Also in 1951, Pauling and Corey proposed a second structure, the **β-pleated-sheet configuration.** In this model, a single polypeptide chain folds back on itself, or several chains run in either parallel or antiparallel fashion next to one another. Each such structure is stabilized by hydrogen bonds formed between atoms present on adjacent chains [Figure 10.9(b)]. A single zigzagging plane is formed in space with adjacent amino acids 3.5 Å apart.

As a general rule, most proteins demonstrate a mixture of α and β structure. Globular proteins, most of which are round in shape and water soluble, usually contain a core of β-pleated-sheet structure as well as many areas demonstrating α helical structure. The more rigid structural proteins, many of which are water insoluble, rely on more extensive β-pleated-sheet regions for their rigidity. For example, **fibroin,** the protein made by the silk moth, depends extensively on this form of secondary structure.

While the secondary structure describes the arrangement of amino acids within certain areas of a polypeptide chain, **tertiary protein structure** defines the three-dimensional conformation of the entire chain in space. The molecule twists and turns and loops around itself in a very specific fashion, characteristic of the specific protein. Three aspects of this III° structure are most important in determining this conformation and in stabilizing the molecule.

1. Covalent disulfide bonds form between closely aligned cysteine residues to form the unique amino acid cystine.
2. Nearly all of the polar, hydrophilic R groups are located on the surface, where they interact with water.
3. The nonpolar, hydrophobic R groups are usually located on the inside of the molecule, where they interact with one another, avoiding interaction with water.

It is important to emphasize that the three-dimensional conformation achieved by any protein is the direct result of the primary (I°) structure of the polypeptide. Thus, the genetic code need only specify the sequence of amino acids in order to encode information leading to the complete structure of proteins. The three stabilizing factors listed above depend on the location of each amino acid relative to all others in the chain. As folding occurs, the most thermodynamically stable conformation possible results.

The three-dimensional structures of the globular protein **myoglobin** is diagrammed in Figure 10.10. This III° level of organization is extremely important because the specific function of any protein is the direct result of its three-dimensional conformation.

The **quaternary level** of organization is characteristic of proteins composed of more than one polypeptide chain. The IV° structure indicates the conformation of the various chains in relation to one another. This type of protein is called **oligomeric,** and each chain is called a **protomer.** The individual protomers have native conformations that fit together in a specific complementary fashion. Hemoglobin, an oligomeric protein of four polypeptide chains, has been studied in great detail. Its IV° structure is shown in Figure 10.11. Many enzymes, including DNA and RNA polymerase, demonstrate IV° structure.

Protein Function

Proteins are the most abundant macromolecules found in cells. As the end products of genes, they play many diverse roles. For example, the respiratory pigments **hemoglobin** and **myoglobin** transport oxygen, which is essential for cellular metabolism. **Collagen** and **keratin** are examples of structural proteins associated with the skin, connective tissue, and hair of organisms. **Actin** and **myosin** are contractile proteins, found in abundance in muscle tissue. Still other examples are the **immunoglobins,** which function in the immune system of vertebrates; **transport proteins,** involved in movement of molecules across membranes; **hormones,** which regulate various types of chemical activity; and **histones,** which bind to DNA in eukaryotic organisms.

FIGURE 10.9
(a) The right-handed α helix, which represents one form of secondary structure of a polypeptide chain. (b) The β-pleated-sheet configuration, an alternate form of secondary structure of polypeptide chains. Some atoms are not shown for the sake of clarity.

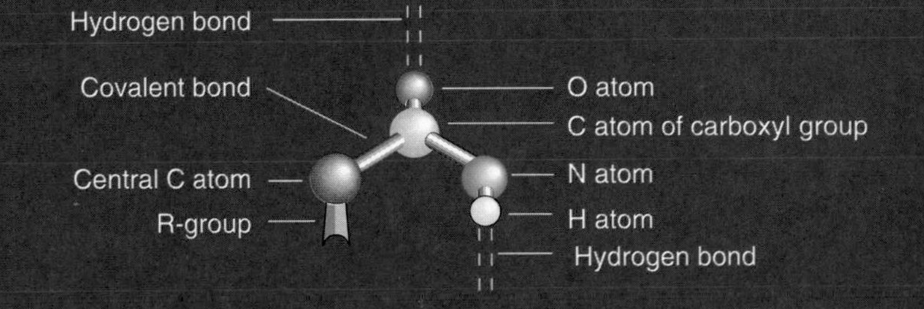

(a) alpha helix

KEY

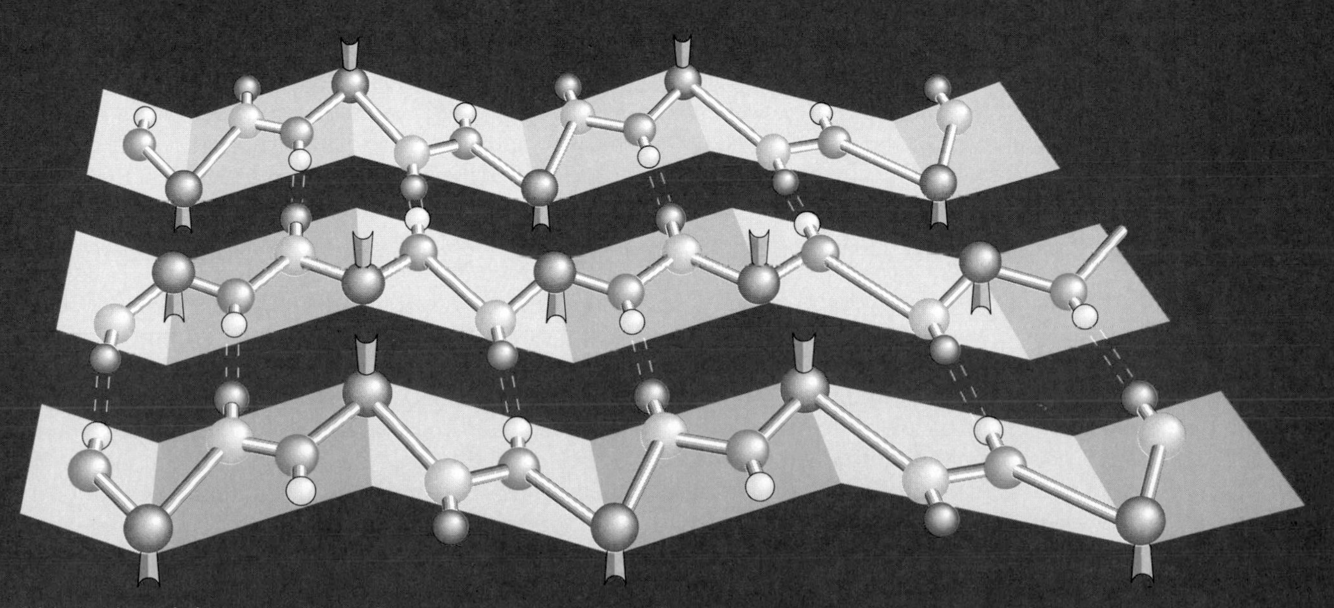

(b) beta pleated sheet

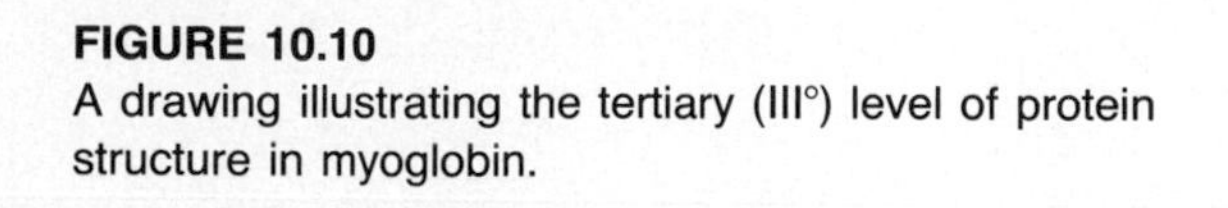

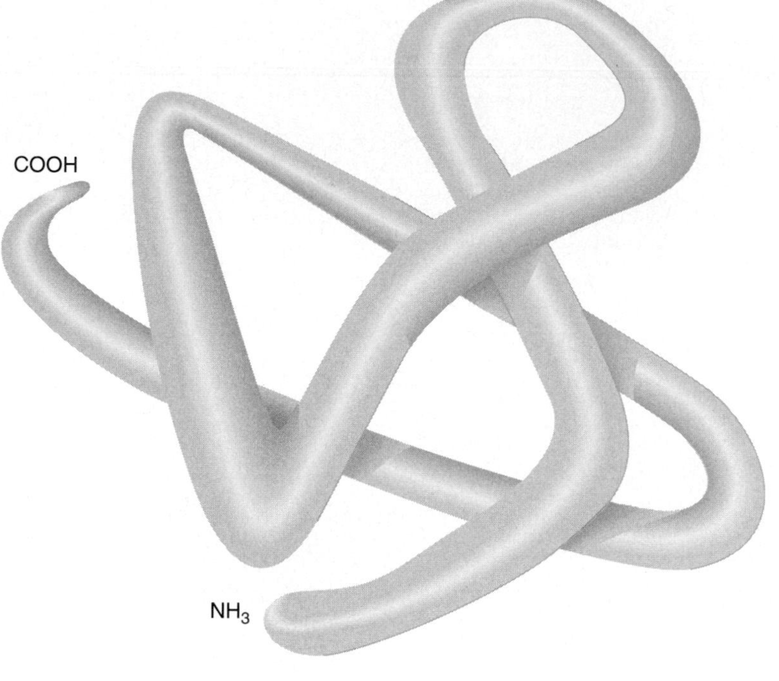

FIGURE 10.10
A drawing illustrating the tertiary (III°) level of protein structure in myoglobin.

The largest group of proteins with a related function are the **enzymes.** These molecules specialize in catalyzing biological reactions. Enzymes increase the rate at which a chemical reaction reaches equilibrium, but they do not alter the end point of the chemical equilibrium. Their remarkable, highly specific catalytic properties largely determine the biomolecular nature of any cell type. The specific functions of many enzymes involved in the genetic and cellular processes of cells are described throughout the text.

Catalysis is a process whereby the **energy of activation** for a given reaction is lowered. The energy of activation is the increased kinetic energy state that molecules must usually reach before they react with one another. While this state can be attained as a result of elevated temperatures, enzymes allow biological reactions to occur at lower physiological temperatures. In this way, enzymes make possible life as we know it.

The catalytic properties and specificity of an enzyme are determined by the chemical configuration of the molecule's **active site.** This site is associated with a crevice, a cleft, or a pit on the surface of the enzyme, which binds the reactants, or substrates, facilitating their interaction. Enzymatically catalyzed reactions control metabolic activities in the cell. Each reaction is either **catabolic** or **anabolic.** Catabolism is the degradation of large molecules into smaller, simpler ones with the release of chemical energy. Anabolism is the synthetic phase of metabolism, yielding nucleic acids, proteins, lipids, and carbohydrates. Metabolic pathways that serve the dual function of anabolism and catabolism are **amphibolic.**

Protein Structure and Function: The Collagen Fiber

In order to provide a more in-depth example of the relationship between protein structure and function, we shall consider one of the most interesting proteins found in vertebrates: **collagen.** In mammals, it is the most abundant protein, constituting up to 25 percent of the total in an individual. It has been extensively studied and clearly illustrates this relationship.

Collagen is found in many places in the body, including tendons, ligaments, bone, connective tissue, skin, blood vessels, teeth, and the lens and cornea of the eye. In each case, the presence of collagen increases the tensile strength of, and provides support to, the tissue. An inspection of the above list makes it evident that collagen must be a versatile protein, providing a variable degree of flexibility and support to these tissues.

While there are several types of collagen produced in vertebrates, we will restrict our discussion to just the major one, called *type I.* Its main subunit or building block is called **tropocollagen.** This consists of three polypeptide chains wrapped around one another in a triple helix. This unit is extremely large, being 15 Å in diameter and 3000 Å long. Each polypeptide contains about 1000 amino acids, and the three interacting chains of the helix are stabilized by hydrogen bonds between them. There is no hydrogen bonding between amino acids within a single chain. Thus, no α-helical or β-pleated-sheet secondary structure exists in collagen.

The maturation of the polypeptides into tropocollagen

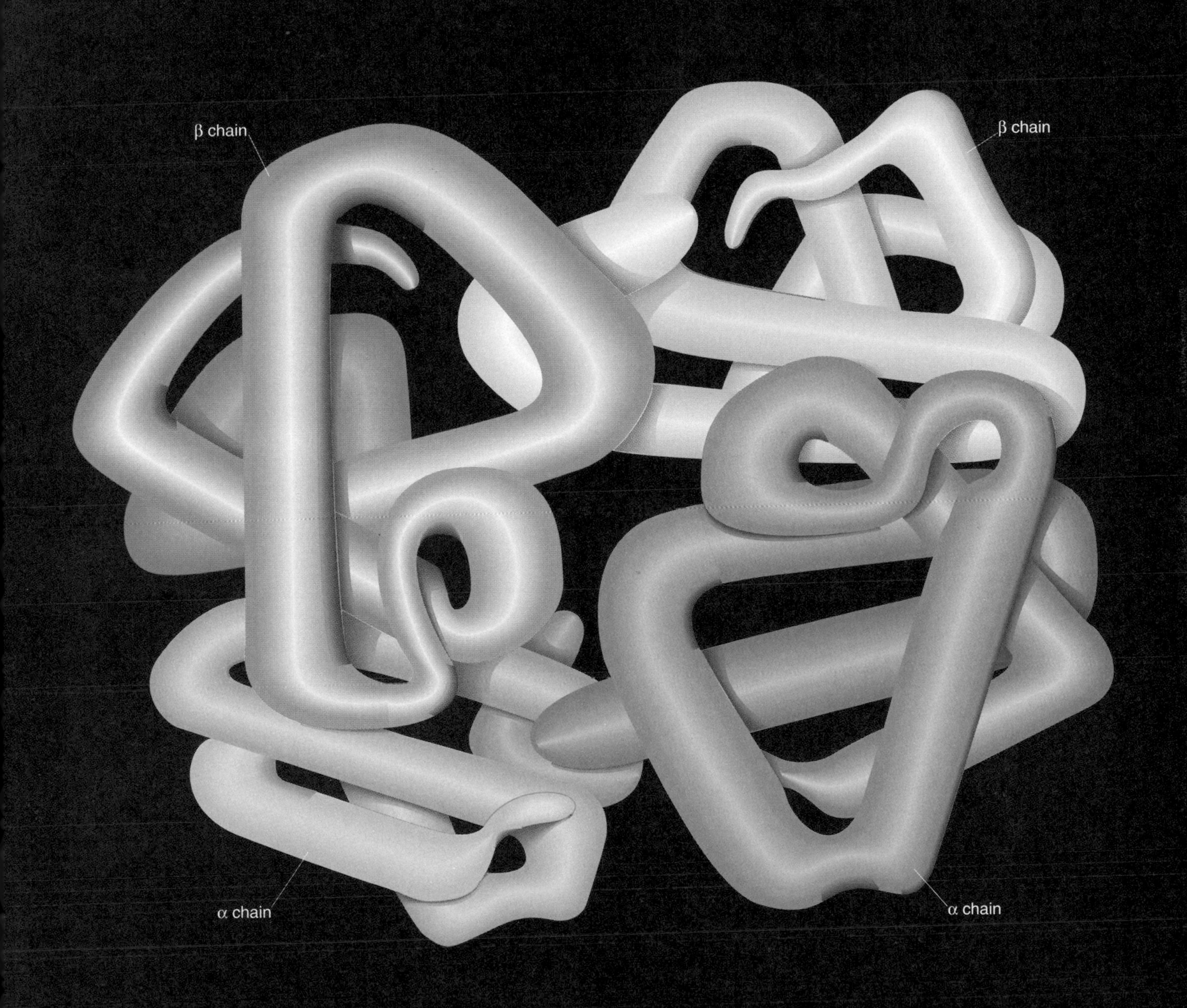

FIGURE 10.11
Schematic drawing illustrating the quaternary level (IV°) of protein structure as seen in hemoglobin. Four chains (2 alpha and 2 beta) interact with four heme groups to form the functional molecule.

and the subsequent condensation of these helical structures into the more densely coiled collagen fibers is a fascinating tale. The polypeptides are first synthesized by fibroblasts as even longer units called procollagen. These are secreted into extracellular spaces, where they are cleaved and shortened at both N-terminus and C-terminus by specific enzymes called **procollagen peptidases** [Figure 10.12(a)].

Once tropocollagen has been formed, these triple-helical rods associate spontaneously to form dense collagen fibers. The association follows an orderly pattern, where rows of end-to-end molecules line up in a staggered fashion next to one another [Figure 10.12(b)]. In each row a gap of approximately 400 Å exists between each tropocollagen unit. This complex structural arrangement creates protein fibers that strengthen and support a variety of tissues.

Although we have concentrated our discussion on type I collagen, there are about 20 different collagen genes, each encoding a slightly different primary chain. At least ten different combinations of chains have been discovered, each constituting a variant collagen molecule. It is likely that each imparts slightly different properties and characteristics to the different forms of collagen.

The Genetics of Collagen

We began this chapter with a discussion of an inherited biochemical disorder first studied early in this century by Garrod. It is appropriate that we conclude this chapter with a short discussion of more modern discoveries of inherited biochemical disorders, this time related to collagen.

At least two inherited disorders known to involve defects in collagen synthesis and assembly are fairly well characterized genetically. For example, the inability to adequately transform procollagen to collagen leads to the human connective tissue disorder **Ehlers-Danlos syndrome.** Individuals exhibiting this disorder have fragile, stretchable skin and are loose-jointed, providing hypermobility. Some forms of the syndrome render individuals susceptible to arterial and colon ruptures as well as periodontal problems. They exhibit elevated levels of procollagen and decreased activity of the enzyme procollagen peptidase.

Our second example involves the lethal human disorder **osteogenesis imperfecta.** There are many variations, but all affect long bone formation and result from some defect in collagen structure. In its most severe form (*type II*), widespread bone defects occur in the fetus and newborn; fractures occur spontaneously and death often occurs soon after birth.

Type I, in comparison, is fairly mild, with an onset sometimes as late as age 35 to 45. However, at that point, declining health is associated with the degeneration of collagen-rich tissues: Blood vessels weaken, bones become fragile, and they fracture frequently. A stroke or heart attack usually causes premature death.

In one instance, the specific genetic defect has been uncovered and found to involve but a single glycine residue in the primary procollagen chain. At position 988, a glycine residue has been altered to the amino acid cysteine by a mutation of a single base change in the gene. This alteration of the primary structure inhibits the formation of highly ordered fibers that make up the bundles of collagen and leads to all of the aberrant phenotypic effects associated with the disease.

Collagen disorders are quite prevalent in the human population. It is estimated that over 100,000 individuals worldwide suffer from osteogenesis imperfecta. While Ehlers-Danlos syndrome is much rarer, there are about 20,000 individuals who suffer from still another collagen-based disorder called **Marfan syndrome.** Long, thin arms, legs, and digits of the hands and feet characterize the appearance of affected individuals. Aortic heart disease most frequently is the cause of premature death. Abraham Lincoln is suspected to have suffered from this disorder. Given the complexity of the biochemistry of collagen and its wide distribution throughout the body, it is not surprising that these and other genetic disorders exist.

CHAPTER SUMMARY

1 Basic to Garrod's studies in the early 1900s of inborn, human metabolic disorders such as cystinuria, albinism, and alkaptonuria, was the concept that genes control the synthesis of specific metabolic products.

2 The investigation of nutritional requirements in *Neurospora* by Beadle and Tatum made it clear that mutations cause the loss of enzyme activity. Their work led to the concept of one gene : one enzyme.

3 The concept of the one gene : one enzyme hypothesis was later revised. Pauling and Ingram's investigations of hemoglobins from patients with sickle-cell anemia led to the discovery that one enzyme directs the synthesis of only one polypeptide chain.

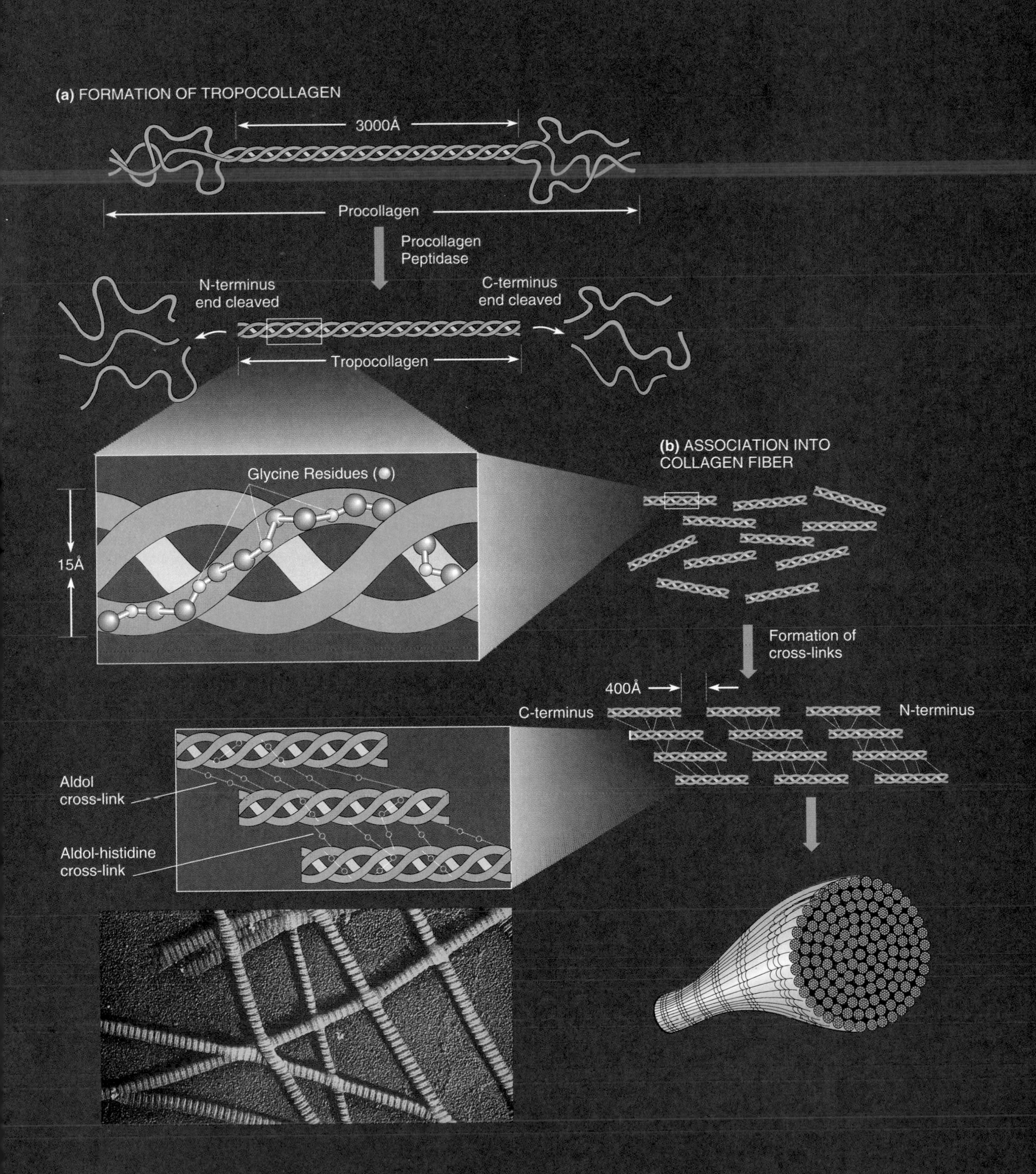

FIGURE 10.12
Various steps and intermediate structures involved in the formation of collagen. (a) The process of post-translational modification of polypeptides during the formation of tropocollagen, the precursor of collagen. (b) The formation of cross-links leading to the assembly of the collagen fiber. An electron micrograph illustrates such fibers.

4 Thorough investigations have revealed the existence of several major types of human hemoglobin molecules, found in the embryo, fetus, and the adult. Specific genes control each polypeptide chain constituting these various hemoglobin molecules.

5 The proposal suggesting that a gene's nucleotide sequence specifies in a colinear way the sequence of amino acids in a polypeptide chain was confirmed by Yanofsky's experiments involving mutations in the tryptophan synthetase gene in *E. coli.*

6 Proteins, the end products of genes, demonstrate four levels of structural organization that together provide the chemical basis for their three-dimensional conformation.

7 Of the myriad functions performed by proteins, the most influential role is assumed by enzymes. These highly specific, cellular catalysts play a central role in the production of all classes of molecules in living systems.

8 Collagen is an abundant, specialized protein in vertebrates, serving a structural role in a variety of tissues. Ehlers-Danlos syndrome, osteogenesis imperfecta, and Marfan syndrome are examples of inherited human disorders resulting from mutations that alter collagen structure and its function.

KEY TERMS

active site
alkaptonuria
alpha helix
amino acid
β-pleated-sheet
colinear relationship
collagen
Ehlers-Danlos syndrome
energy of activation
enzymes
fingerprinting technique
Gower 1 hemoglobin
HbA
HbA_2
HbF
HbS
hemoglobin
homogentisic acid
Marfan syndrome
myoglobin
one-gene:one-enzyme hypothesis
one gene:one polypeptide chain
osteogenesis imperfecta
peptide bond
phenylalanine hydroxlase
phenylketonuria
polypeptide chain
primary (I°) protein structure
quaternary (IV°) protein structure
secondary (II°) protein structure
sickle-cell anemia
starch gel electrophoresis
tertiary (III°) protein structure

1 The following growth responses were obtained using four mutant strains of *Neurospora* and the related compounds A, B, C, and D. None of the mutations grow on minimal medium. Draw all possible conclusions.

Mutation	Growth Product C	D	B	A
1	−	−	−	−
2	−	+	+	+
3	−	−	+	+
4	−	−	+	−

SOLUTION: First, nothing can be concluded about mutation 1, except that it is lacking some essential growth factor, perhaps even unrelated to the biochemical pathway represented by mutations 2–4; nor can anything be concluded about compound C. If it is involved in the pathway, it is a product synthesized prior to the synthesis of A, B, and D.

We must now analyze these three compounds and the control of their synthesis by the enzymes encoded by genes 2, 3, and 4. Since product B allows growth in all three cases, it may be considered the "end product." It bypasses the block in all three instances. Using similar reasoning, product A precedes B in the pathway since it allows a bypass in two or the three steps. Product D precedes B, yielding the more complete solution:

$$C(?) \longrightarrow D \longrightarrow A \longrightarrow B$$

Now determine which mutations control which steps. Since mutation 2 can be alleviated by products D, B, and A, it must control a step prior to all three products, perhaps the direct conversion to D, although we can't be certain. Mutation 3 is alleviated by B and A, so its effect must precede them in the pathway. Thus, we will assign it as controlling the conversion of D to A. Likewise, we can assign mutation 4 to the conversion of A to B, leading to the more complete solution:

$$C(?) \xrightarrow{2(?)} D \xrightarrow{3} A \xrightarrow{4} B$$

PROBLEMS AND DISCUSSION QUESTIONS

1 Discuss the potential difficulties involved in designing a diet to alleviate the symptoms of phenylketonuria.

2 The synthesis of flower pigments is known to be dependent upon enzymatically controlled biosynthetic pathways. In the crosses shown below, postulate the role of mutant genes and their products in producing the observed phenotypes.

(a) P_1: white strain A × white strain B
F_1: all purple
F_2: 9/16 purple: 7/16 white

(b) P_1: white × pink
F_1: all purple
F_2: 9/16 purple: 3/16 pink: 4/16 white

3 A series of mutations in the bacterium *Salmonella typhimurium* result in the requirement of either tryptophan or some related molecule in order for growth to occur. From the data shown below, suggest a biosynthetic pathway for tryptophan.

	Growth Response Supplement				
Mutation	Minimal Medium	Anthranilic Acid	Indole Glycerol Phosphate	Indole	Tryptophan
trp-8	−	+	+	+	+
trp-2	−	−	+	+	+
trp-3	−	−	−	+	+
trp-1	−	−	−	−	+

4 The study of biochemical mutants in organisms such as *Neurospora* has demonstrated that some pathways are branched. The data shown below illustrate the branched nature of the pathway resulting in the synthesis of thiamine. Why don't the data support a linear pathway? Can you postulate a branched pathway for the synthesis of thiamine in *Neurospora*?

	Growth Response Supplement			
Mutation	Minimal Medium	Pyrimidine	Thiazole	Thiamine
thi-1	−	−	+	+
thi-2	−	+	−	+
thi-3	−	−	−	+

5 Explain why the one-gene:one-enzyme concept is not considered completely accurate.

6 Using sickle-cell anemia as a basis, describe what is meant by a molecular or genetic disease. What are the similarities and dissimilarities between this type of a disorder and a disease caused by an invading microorganism?

7 Describe what *colinearity* means.

8 Define and compare the four levels of protein organization.

9 List as many different protein functions as you can, with an example of each.

10 How does an enzyme function? Why are enzymes essential for living organisms on earth?

SELECTED READINGS

BARTHOLOME, K. 1979. Genetics and biochemistry of phenylketonuria—Present state. *Hum. Genet.* 51:241–45.

BATESON, W. 1909. *Mendel's principles of heredity.* Cambridge, England: Cambridge University Press.

BEADLE, G. W. 1946. Genes and the chemistry of the organism. *Amer. Scient.* 34:31–53.

BEADLE, G. W., and TATUM, E. L. 1941. Genetic control of biochemical reactions in *Neurospora. Proc. Natl. Acad. Sci.* 27:499–506.

BRENNER, S. 1955. Tryptophan biosynthesis in *Salmonella typhimurium. Proc. Natl. Acad. Sci.* 41:862–63.

BYERS, P. H. 1989. Inherited disorders of collagen gene structure and expression. *Am. J. Med. Genet.* 34:72–80.

DICKERSON, R. E., and GEIS, I. 1983. *Hemoglobin: Structure, function, evolution, and pathology.* Menlo Park, CA: Benjamin/Cummings.

DOOLITTLE, R. F. 1985. Proteins. *Scient. Amer.* (Oct.) 253:88–99.

GARROD, A. E. 1902. The incidence of alkaptonuria: A study in chemical individuality. *Lancet* 2:1616–20.

———. 1909. *Inborn errors of metabolism.* London: Oxford University Press. (Reprinted 1963, Oxford University Press, London.)

INGRAM, V. M. 1957. Gene mutations in human hemoglobin: The chemical difference between normal and sickle cell hemoglobin. *Nature* 180:326–28.

KOSHLAND, D. E. 1973. Protein shape and control. *Scient. Amer.* (Oct.) 229:52–64.

NEEL, J. V. 1949. The inheritance of sickle-cell anemia. *Science* 110:64–66.

PAULING, L., ITANO, H. A., SINGER, S. J., and WELLS, I. C. 1949. Sickle cell anemia, a molecular disease. *Science* 110:543–48.

PROCKOP, D. J., and KIVIRIKKO, K. I. 1984. Heritable diseases of collagen. *New Engl. J. Med.* 311:376–86.

STRYER, L. 1988. *Biochemistry.* 3rd ed. New York: W.H. Freeman.

SYKES, B. 1985. The molecular genetics of collagen. *BioEssays* 3:112–17.

YANOFSKY, C. DRAPEAU, G., GUEST, J., and CARLTON, B. 1967. The complete amino acid sequence of the tryptophan synthetase A protein and its colinear relationship with the genetic map of the *A* gene. *Proc. Natl. Acad. Sci.* 57:296–98.

11

Mutation and Mutagenesis

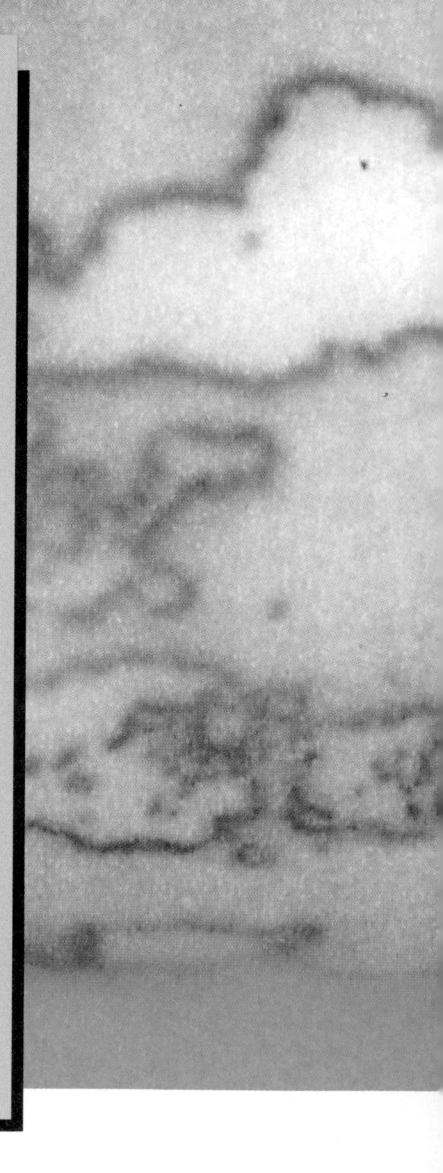

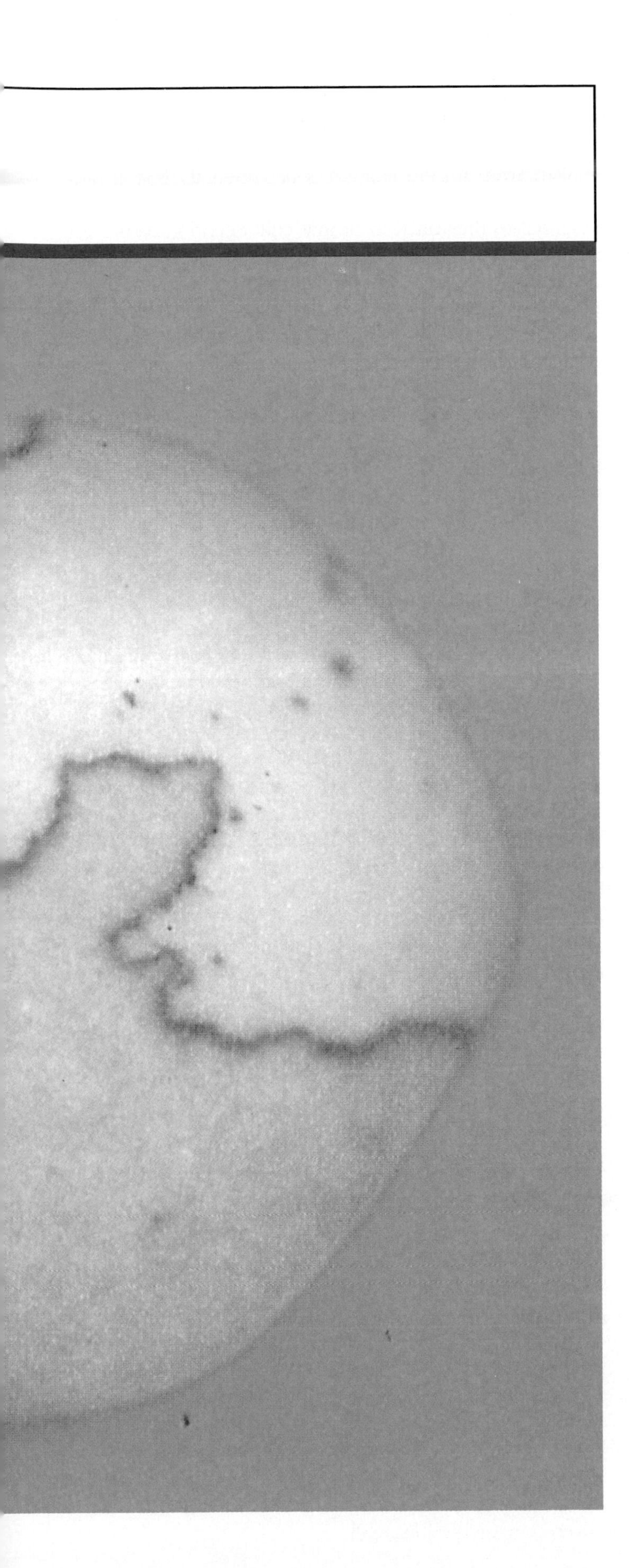

Chapter Concepts

Aside from chromosomal mutations, the major basis of diversity among organisms, even those that are closely related, is genetic variation at the level of the gene. The origin of such variation is gene mutation, whereby the coding sequence is altered as a result of base substitution or the addition or deletion of one or more bases within those sequences. Such mutations make the study of genetics possible and are the raw material upon which evolution relies.

In Chapter 7, we defined the four characteristics or functions ascribed to the genetic information: **replication, storage, expression, and variation by mutation.** In a sense, *mutation* is a failure to store the genetic information faithfully. If a change occurs in the stored information, it may be reflected in the expression of that information and will be propagated following replication. Historically, the term **mutation** includes both chromosomal changes and changes within single genes. We have discussed the former alterations in Chapter 6, referring to them collectively as **chromosomal aberrations.** In this chapter, we will be concerned with **gene mutations.** A change may be a simple substitution of one nucleotide or involve the insertion or deletion of one or more nucleotides within the normal sequence of DNA.

Mutations provide the basis for genetic studies. The resulting phenotypic variability allows the geneticist to identify and study the genes that control the traits that have been modified. Without the phenotypic variability that mutations provide, all genetic crosses would be meaningless. For example, if all pea plants displayed a uniform phenotype, Mendel would have had no basis for his experimentation. Because of the importance of mutations, great attention has been given to their origin, induction, and classification.

Certain organisms lend themselves to induction of mutations that can be detected easily and studied throughout reasonably short life cycles. Viruses, bacteria, fungi, fruit flies, certain plants, and mice variably fit these criteria. Thus, these organisms have often been used in studying mutation and mutagenesis, and through other studies have also contributed to more general aspects of genetic knowledge.

CLASSIFICATION OF MUTATIONS

There are various schemes by which mutations are classified. They are not mutually exclusive, but instead depend simply on which aspects of mutation are being investigated or discussed. In this section, we will describe three sets of distinctions concerning mutations.

Spontaneous versus Induced Mutations

All mutations are described as either **spontaneous** or **induced.** While these two categories overlap to some degree, **spontaneous mutations** are those that arise in nature. No specific agent—other than natural forces—is associated with their occurrence, and they are assumed to arise randomly as changes in the nucleotide sequences of genes.

What causes spontaneous mutations? Although their origin is not fully understood, many such mutations can be linked to normal chemical processes or phenomena that result in rare errors. Such errors may cause alterations in the chemical structure of nitrogenous bases that are part of existing genes. They occur more frequently, however, during the enzymatic process of DNA replication, which we will explore later in this chapter. It is generally agreed that any natural phenomenon that heightens chemical reactivity in cells will lead to more errors. For example, background radiation from cosmic and mineral sources and ultraviolet light from the sun are energy sources to which most organisms are exposed. As such, they may be factors leading to spontaneous mutations. Once an error is present in the genetic code, it may be reflected in the amino acid composition of the specified protein. If the changed amino acid is present in a part of the molecule critical to its biochemical activity, a functional alteration may result.

In contrast to such spontaneous events, those that arise as a result of the influence of any artificial factor are considered to be **induced mutations.** The earliest demonstration of the induction of mutation occurred in 1927, when Herman J. Muller reported that X rays could cause mutations in *Drosophila.* In 1928, Lewis J. Stadler reported the same finding in barley. In addition to various forms of radiation, a wide spectrum of chemical agents is also known to be mutagenic, the aspects of which we will examine later in this chapter.

Gametic versus Somatic Mutations

When considering the effects of mutation in eukaryotic organisms, it is important to distinguish whether the change occurs in somatic cells or gametes. Mutations arising in somatic cells are not transmitted to future generations. In tissues of an adult organism, thousands upon thousands of cells may be performing a similar function. Thus, a mutation in a single cell may not impair the organism, even if the mutation is detrimental. First, a random mutation might occur in a gene that is not active or essential to the function of that cell. Second, even if an active gene is affected, there are still thousands of unaffected cells to perform the function of the tissue in question.

Somatic mutations will have a greater impact if they are dominant or if they are sex-linked recessives, since such mutations are more likely to be immediately expressed. Similarly, the impact will be more noticeable if a somatic mutation occurs early in development, where an undifferentiated cell will give rise to several differentiated tissues or organs.

There is a very important exception to the above description. If a mutation occurs that disrupts the control of cell proliferation, such a mutation may cause cells to divide uncontrollably, causing a cancerous condition. This topic will be pursued extensively in Chapter 19.

Mutations in gametes or gamete-forming tissue are part of the germ line and are of greater concern because they have the potential of being expressed in all cells of an offspring. Such a mutation may also be transmitted to future generations, gradually increasing in frequency within the population. **Dominant autosomal mutations** will be expressed phenotypically in the first generation. **Sex-linked recessive mutations** arising in the gametes of a heterogametic female may be expressed in hemizygous male offspring. This will occur provided the male offspring receives the affected X chromosome. Because of heterozygosity, the occurrence of an **autosomal recessive mutation** in the gametes of either males or females (even one resulting in a lethal allele) may go unnoticed for many generations until it has become widespread in the population. The new allele will become evident only when a chance mating brings two copies of it together in the homozygous condition.

Categories of Mutation

Various types of mutations are classified on the basis of their effect on the organism. Even though some of these categories were introduced earlier, we will briefly review several of them again. A single mutation may well fall into more than one category.

The most obvious mutations are those affecting a **morphological trait.** For example, all of Mendel's pea characters and most genetic variations encountered in the study of *Drosophila* fit this designation. All such morphological variations deviate from the normal, wild-type phenotype.

A second broad category includes mutations that are **nutritional** or **biochemical variations** from the norm. In bacteria and fungi, the inability to synthesize an amino acid or vitamin is an example of a typical nutritional mutation. In humans, sickle-cell anemia and hemophilia are examples of biochemical mutations. While such mutations in these organisms are not visible and do not usually affect specific morphological characters, they can have a more general effect on the well-being and survival of the affected individual.

A third category consists of mutations that affect behavior patterns of an organism. For example, mating behavior or circadian rhythms of animals may be altered. The primary effect of **behavior mutations** is often difficult to discern. For example, the mating behavior of a fruit fly may be impaired if it cannot beat its wings. However, the defect may be in: (1) the flight muscles, (2) the nerves leading to them, or (3) the brain, where the nerve impulses that initiate wing movements originate. The study of behavior and the genetic factors influencing it has benefited immensely from investigations of behavior mutations.

Still another type of mutation may affect the regulation of genes. A regulatory gene may produce a product that controls the transcription of another gene. In other instances, a region of DNA either close to or far away from a gene may modulate its activity. In either case, **regulatory mutations** may disrupt this process and permanently activate or inactivate a gene. Our knowledge of genetic regulation depends on the study of mutations that disrupt this process.

Another group consists of **lethal mutations.** Nutritional and biochemical mutations may also fall into this category. A mutant bacterium that cannot synthesize a specific amino acid it needs will cease to grow if plated on a medium lacking that amino acid. Various human biochemical disorders, such as Tay-Sachs disease and Huntington disease, are lethal at different points in the life cycle of humans.

Finally, any of the above groups can exist as **conditional mutations.** Even though a mutation is present in the genome of an organism, it may not be evident under certain conditions. The best examples are the **temperature-sensitive mutations,** found in a variety of organisms. At certain permissive temperatures, a mutant gene product functions normally, only to lose its functional capability at a different restrictive temperature. When shifted to this temperature, the impact of the mutation becomes apparent, even lethal, and is amenable to investigation. The study of conditional mutations has been extremely important in experimental genetics, particularly in understanding the function of genes essential to the viability of organisms.

DETECTION OF MUTATION

Before geneticists can study directly the mutational process or obtain mutant organisms for genetic investigations, they must be able to detect mutations. The ease and efficiency of detecting mutations in a particular organism has generally determined the organism's usefulness in genetic studies. In this section, we will use several examples to illustrate how mutations are detected.

Detection in Bacteria and Fungi

Detection of mutations is most efficient in haploid microorganisms such as bacteria and fungi. Detection depends on a selection system where mutant cells are isolated easily from nonmutant cells. The general principles are similar in bacteria and fungi. To illustrate, we will describe how nutritional mutations in the fungus *Neurospora crassa* are detected, a discussion that we entertained in the previous chapter.

Neurospora is a pink mold that normally grows on bread. It may also be cultured in the laboratory. This eukaryotic mold is haploid in the vegetative phase of its life cycle. Thus, mutations may be detected without the complications generated by heterozygosity in diploid organisms. Wild-type *Neurospora* grows on a **minimal culture medium** of glucose, a few inorganic acids and salts, a nitrogen source such as ammonium nitrate, and the vitamin biotin. Induced nutritional mutants will not grow on minimal medium, but will grow on a supplemented or **complete medium** that also contains numerous amino acids, vitamins, nucleic acid derivatives, and so forth. Microorganisms that are nutritional wild types (requiring only minimal medium) are called **prototrophs,** while those mutants that require a specific supplement to the minimal medium are called **auxotrophs.**

The detection of nutritional mutants is illustrated in Figure 10.2. Nutritional mutants may be detected and isolated by their failure to grow on minimal medium and their ability to grow on complete medium. The mutant cells can not synthesize some essential compound absent in minimal medium but present in complete medium. Once a nutritional mutant is detected and isolated, the missing compound is determined by attempts to grow the mutant strain in a series of tubes, each containing minimal medium supplemented with a single compound. The auxotrophic mutation is defined in this way.

In Chapter 14, where the topics of bacterial and viral genetics are introduced, we will return to the topic of mutation. Techniques for the detection and study of such mutations are very similar to that just described. Analysis of mutations in microorganisms has been particularly important to the study of molecular genetics.

Detection in *Drosophila*

Muller, in his studies demonstrating that X rays are mutagenic, developed a number of detection systems in *Drosophila melanogaster.* These systems can be used to estimate the spontaneous and induced rates of sex-linked and autosomal recessive lethal mutations. We will consider the **attached-X procedure,** which assesses sex-linked mutations.

Attached-X females have two X chromosomes attached to a single centromere and one Y chromosome, in addition to the normal diploid complement of autosomes. When attached-X females are mated to males with normal sex chromosomes (XY), four types of progeny result: triplo-X females that die; viable attached-X females; YY males that also die; and viable XY males. Figure 11.1 illustrates how P_1 males that have been treated with a mutagenic agent produce F_1 male offspring that express any induced sex-linked recessive mutation.

The same approach works equally well if no mutagenic agent has been used. In such a case, spontaneous mutations are expressed in the first generation. In detection techniques devised for recessive autosomal lethals in *Drosophila,* dominant marker mutations are followed through a series of three generations. Although more cumbersome to perform, these techniques are also fairly efficient.

Detection in Plants

Genetic variation in plants is extensive. Mendel's peas, for example, were the basis for the fundamental postulates of transmission genetics. Studies of plants have also enhanced our understanding of gene interaction, polygenic inheritance, linkage, sex determination, chromosome rearrangements, and polyploidy. Most variations are detected visually. However, there are also techniques for the detection of biochemical mutations in plants. The first is the analysis of biochemical composition of plants. For example, the isolation of proteins from maize endosperm, hydrolysis of the proteins, and determination of the amino acid composition have revealed that the ***opaque-2* mutant strain** contains significantly more lysine than do other, nonmutant lines. As a result of this discovery, plant geneticists and other specialists have begun to analyze the amino acid compositions of various strains of several grain crops, including corn, rice, wheat, barley, and millet. When the results of these analyses are completed and catalogued, the information will be useful in combating malnutrition diseases resulting from inadequate protein or the lack of the essential amino acids in the diet.

The second detection technique involves tissue culture of plant cell lines in defined medium. The plant cells are handled as microorganisms, and biochemical requirements may be determined by adding or deleting nutrients in the culture medium. There are other advantages to this method. Techniques associated with conditional lethal mutants can be used on plant cells in tissue culture and then applied to the genetics of higher plants. Also, this method provides a detection system that is generally not useful with the intact plant. Temperature-sensitive mutations in plants are beginning to be explored, particularly in tobacco. These studies may add significantly to our understanding of plant growth, metabolism, and genetics.

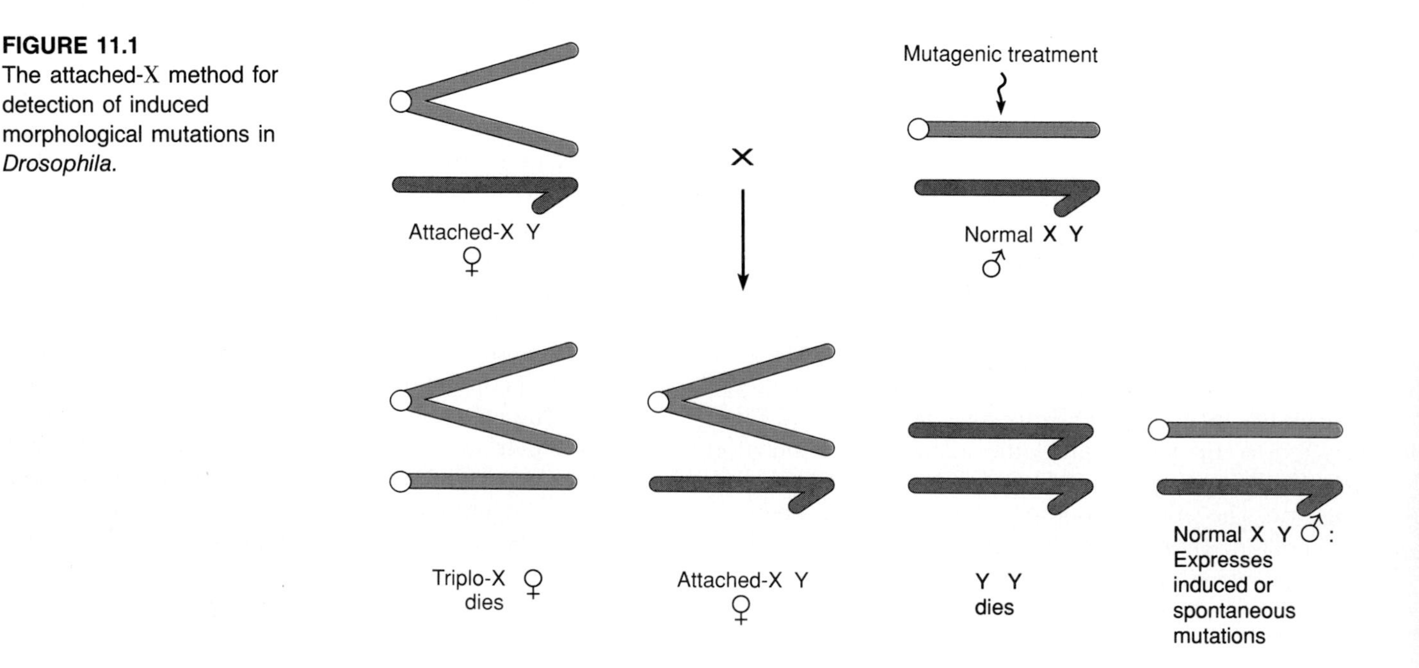

FIGURE 11.1
The attached-X method for detection of induced morphological mutations in *Drosophila.*

Detection in Humans

Since humans are obviously not suitable experimental organisms, the techniques that have been developed for the detection of mutations in organisms such as *Drosophila* are not available to human geneticists. To determine the mutational basis for any human characteristic or disorder, geneticists must first analyze a pedigree that traces the family history as far back as possible. Once any trait has been shown to be inherited, it is possible to predict whether the mutant allele is behaving as a dominant or a recessive and whether it is sex-linked or autosomal.

Dominant mutations are the simplest to detect. If they are present on the X chromosome, affected fathers pass the phenotypic trait to all their daughters. If dominant mutations are autosomal, approximately 50 percent of the offspring of an affected heterozygous individual are expected to show the trait. Figure 11.2 shows a hypothetical pedigree illustrating the initial occurrence of an autosomal dominant allele for cataracts of the eye. The parents in generation I were unaffected, but one of three offspring (Generation II) developed cataracts. This female, the proband, produced two children, of which the male child was affected. Of his six offspring (IV), four of six were affected (Generation III). These observations are consistent with, but do not prove, an autosomal dominant mode of inheritance. However, the high percentage of affected offspring in generation IV favors this conclusion. Also, the unaffected daughter in this generation argues against sex linkage because she received her X chromosome from her affected father. This conclusion is sound, provided that the mutant allele is completely penetrant.

Sex-linked recessive mutations may also be detected by pedigree analysis, as discussed in Chapter 4. The most famous case of a sex-linked mutation in humans is that of **hemophilia,** which was found in the descendants of Queen Victoria. The recessive mutation for hemophilia has occurred many times in human populations, but the political consequences of the mutation that occurred in the royal family have been sweeping. Inspection of the pedigree in Figure 11.3 leaves little doubt that Victoria was heterozygous (*Hh*) for the trait. Her father was not affected, and there is no reason to believe that her mother was a carrier, as was Victoria. Robert Massie's *Nicholas and Alexandra* and Robert and Suzanne Massie's *Journey* (see this chapter's selected readings) provide fascinating reading on the topic of hemophilia.

In a similar manner, it is possible to detect autosomal recessive alleles. Because this type of mutation is "hidden" when heterozygous, it is not unusual for the trait to appear only intermittently through a number of generations. An affected individual and a homozygous normal individual will produce unaffected carrier children. Matings between two carriers will produce, on the average, one-fourth affected offspring.

In addition to pedigree analysis, human cells may now be routinely cultured *in vitro*. This procedure has allowed the detection of many more mutations than any other form of analysis. Analysis of enzyme activity, protein migration in electrophoretic fields, and direct sequencing of DNA and proteins are among the techniques that have demonstrated wide genetic variation between individuals in human populations.

SPONTANEOUS MUTATION RATE

The types of detection systems just described allow geneticists to estimate mutation rates. It is of considerable interest to determine the rate of spontaneous mutation. Such information provides insights into evolution, and

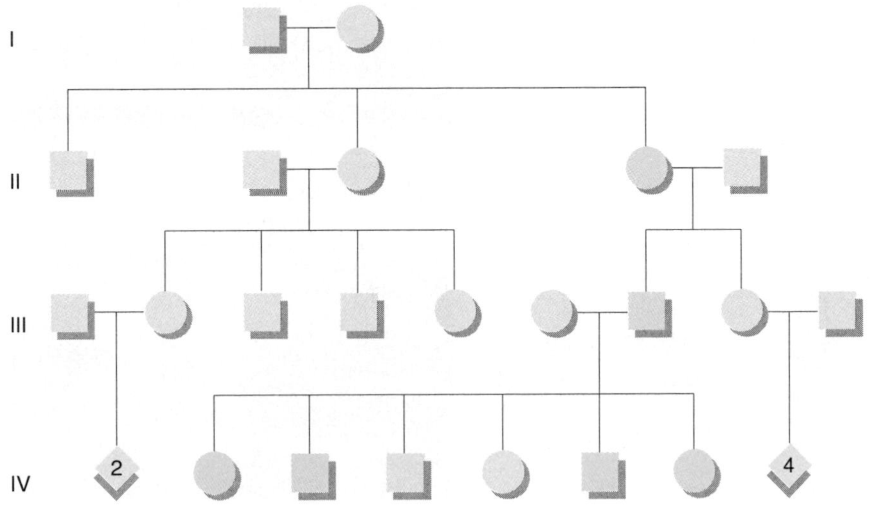

FIGURE 11.2
A hypothetical pedigree of inherited cataract of the eye in humans.

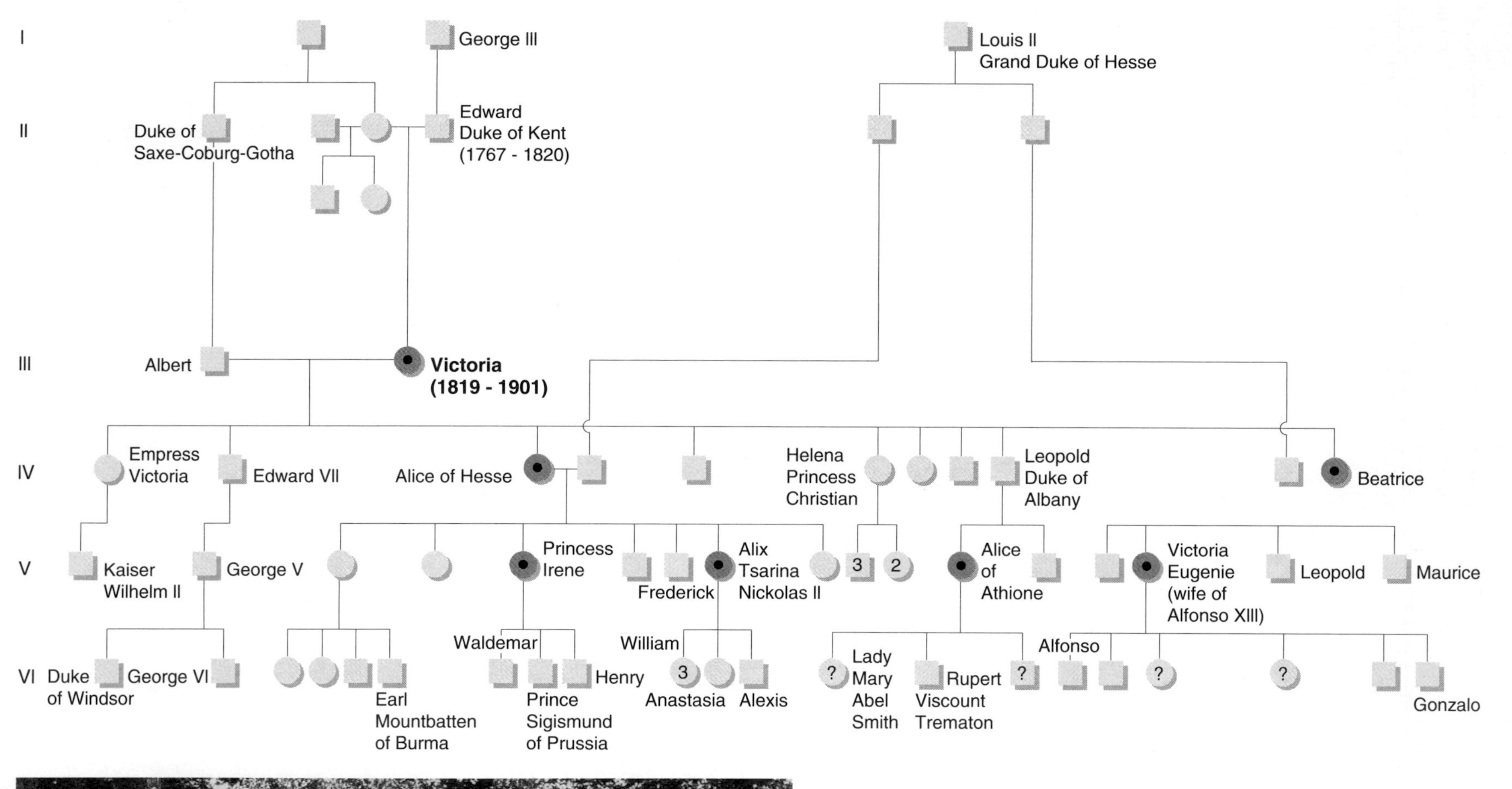

FIGURE 11.3
Pedigree of hemophilia in the royal family descended from Queen Victoria. The pedigree is typical of the transmission of sex-linked recessive traits. Circles with a dot in them indicate presumed female carriers heterozygous for the trait. Circles with a question mark indicate females whose status is uncertain. The photograph shows Queen Victoria (seated front center) and some of her immediate family.

provides the baseline for measuring the rate of experimentally induced mutation. Induction of mutation can only be ascertained when the induced rate exceeds clearly the spontaneous rate for the organism under study.

Examination of this rate in a variety of organisms reveals many interesting points. First, the rate is exceedingly low for all organisms studied. Second, the rate is seen to vary considerably in different organisms. Third, even within the same species, the spontaneous mutation rate varies from gene to gene.

Viral and bacterial genes undergo spontaneous mutation on an average of about 1 in 100 million (10^{-8}) cell divisions. While *Neurospora* exhibits a similar rate, maize, *Drosophila,* and humans demonstrate a rate several orders of magnitude higher. The genes studied in these groups average between 1/1,000,000 and 1/100,000 (10^{-6} and 10^{-5}) mutations per gamete formed. Mouse genes are still another order of magnitude higher in their spontaneous mutation rate, 1/100,000 to 1/10,000 (10^{-5} to 10^{-4}). It is not clear why such a large variation occurs in mutation rate. The variation might reflect the relative efficiency of enzyme systems whose function is to repair errors created during replication. Repair systems will be discussed later in this chapter.

THE MOLECULAR BASIS OF MUTATION

We will use a *simplified* definition of a gene in the description of the molecular basis of mutation. In this context, it is easiest to consider a gene as a linear sequence of nucleotide pairs representing stored chemical information. Since the genetic code is a triplet, each sequence of three nucleotides specifies a single amino acid in the corresponding polypeptide. Any change that disrupts the coded information provides sufficient basis for a mutation. The simplest change would be the substitution of a single nucleotide. In Figure 11.4, such a change is compared with our own written language, using three-letter words to be consistent with the genetic code. A single change can alter obviously the meaning of the sentence "THE CAT SAW THE DOG," creating what is called *missense.* These are analogies to what are most appropriately referred to as **base substitutions** or **point mutations.** The mutation has turned information that makes sense into various forms of missense.

Two terms are often used to describe nucleotide substitutions. If a pyrimidine replaces a pyrimidine or vice versa, a **transition** has occurred. If a purine and a pyrimidine are interchanged, a **transversion** has occurred.

A second type of change that could occur is the insertion or deletion of a single nucleotide at any point along the gene. As also illustrated in Figure 11.4, the remainder of the three-letter (code) words becomes garbled, creating much more extensive missense. These examples are called **frameshift mutations** because the frame of reading has become altered.

The analogy in Figure 11.4 demonstrates that duplications and deletions change all subsequent triplets in a gene. It is likely that one of many new triplets will be either UAA, UAG, or UGA. These are termination codons (see Chapter 9). When one is encountered during translation, polypeptide synthesis is terminated. Obviously, frameshift mutations can be very severe!

Tautomeric Shifts

In 1953, after they had proposed a molecular structure of DNA, Watson and Crick published a paper in which they discussed the genetic implications of this structure. They recognized that the purines and pyrimidines found in DNA could exist in **tautomeric forms;** that is, each can exist in several chemical forms, differing by only a single proton shift in the molecule. Watson and Crick suggested

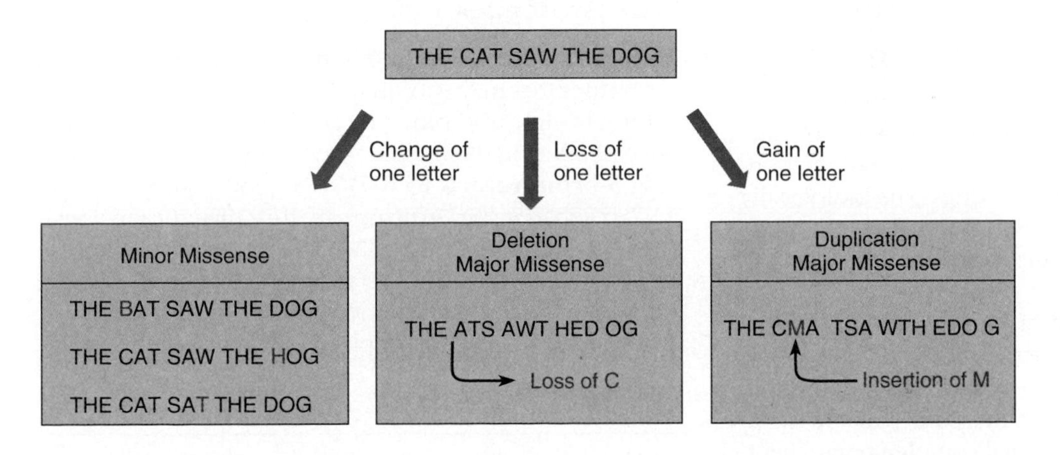

FIGURE 11.4
The impact of the substitution, addition, or deletion of one letter in a sentence composed of three-letter words, creating various levels of missense.

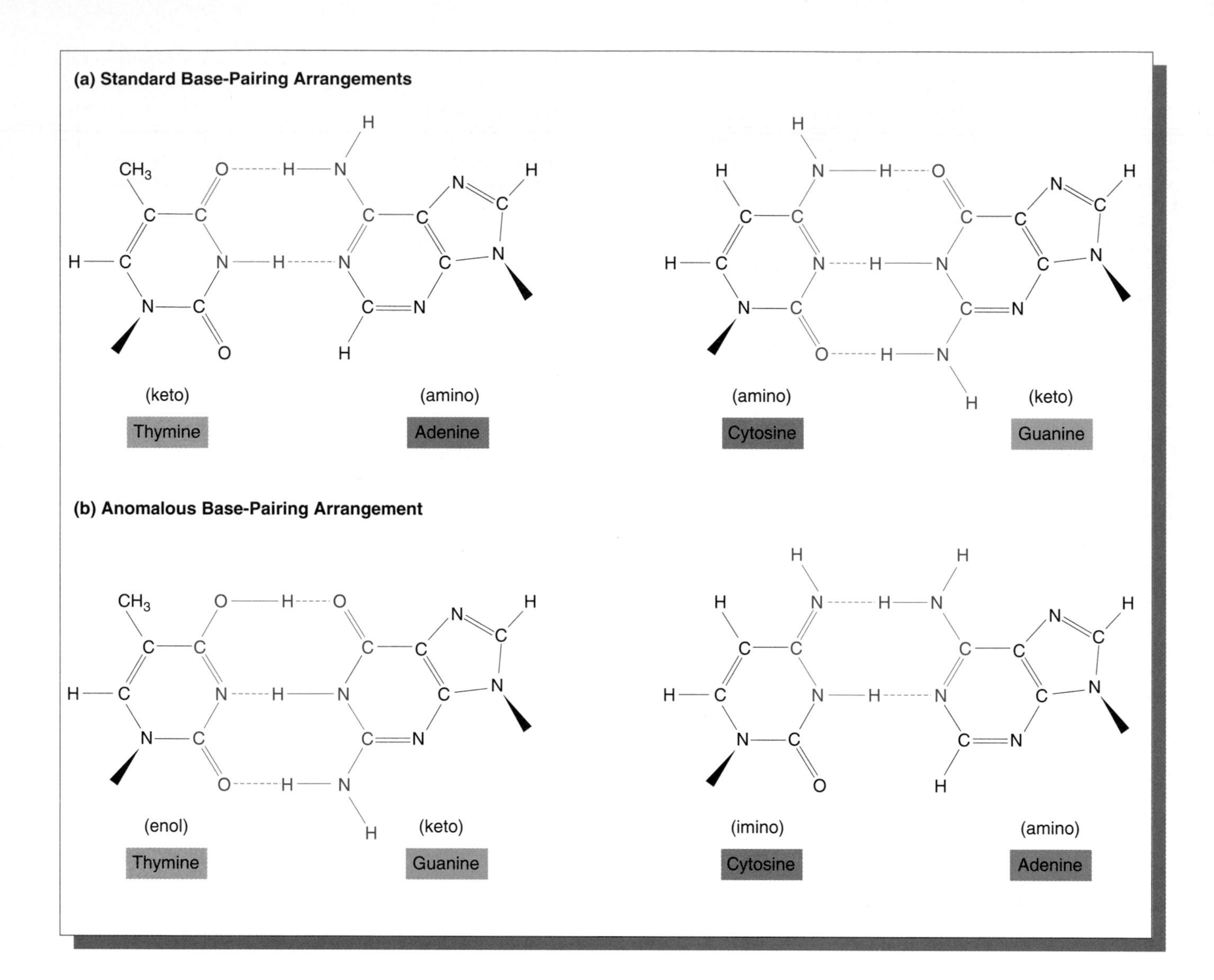

FIGURE 11.5
The standard base-pairing relationships compared with anomalous arrangements occurring as a result of tautomeric shifts. The dense arrow indicates the point of bonding to the pentose sugar.

that **tautomeric shifts** could result in base pair changes or mutations.

The *infrequent* tautomer is capable of hydrogen bonding with a normally noncomplementary base. However, the pairing is always between a pyrimidine and a purine. Figure 11.5 compares the normal base-pairing relationships with the rare unorthodox pairings. The biologically important tautomers involve keto-enol pairs for thymine and guanine, and amino-imino pairs for cytosine and adenine.

The effect occurs at a time of DNA replication when a rare tautomer in the template strand matches with a noncomplementary base. In the next round of replication, the "mismatched" members of the base pair are separated, and each specifies its normal complementary base. The end result is a mutation (see Figure 11.6).

Base Analogues

Base analogues, which are mutagenic chemicals, are molecules that may substitute for purines or pyrimidines during nucleic acid biosynthesis. The halogenated derivative of uracil in the number-5 position of the pyrimidine ring **5-bromouracil (5 BU)*** is a good example. Figure 11.7 compares the structure of this thymine analogue with the structure of thymine. The presence of the bromine atom in place of the methyl group increases the probability that a tautomeric shift will occur. If 5-BU is incorporated into DNA in place of thymine and a tauto-

*If 5-BU is chemically linked to d-ribose, the nucleoside analogue bromodeoxyuridine (BUdR) is formed.

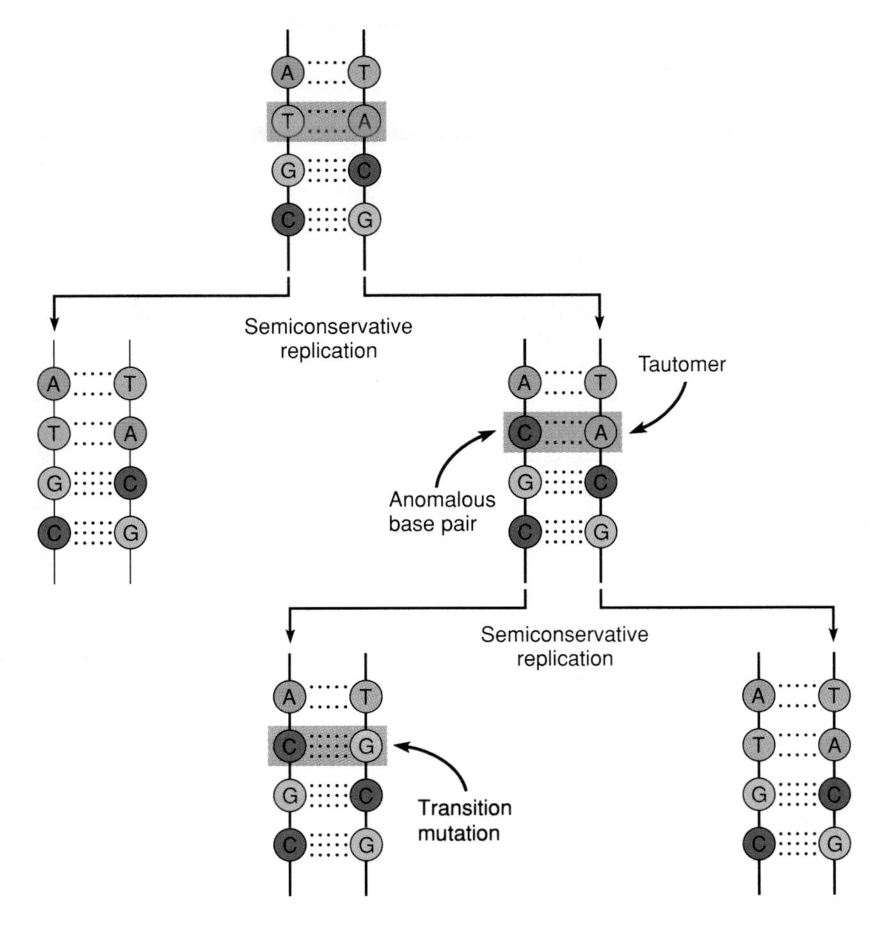

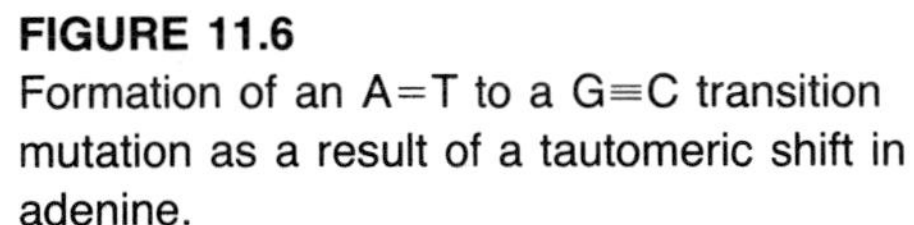
FIGURE 11.6
Formation of an A=T to a G≡C transition mutation as a result of a tautomeric shift in adenine.

meric shift occurs, the result is an A = T to G ≡ C transition (see Figure 11.6). If the tautomeric shift to the enol form occurs before the analogue is incorporated into DNA, and 5-BU is mistaken for C, the transition is from G ≡ C to A = T.

There are other base analogues that are mutagenic. One, **2-amino purine (2-AP),** can serve successfully as an analogue of adenine. In addition to its base-pairing affinity with thymine, 2-AP can also base pair with cytosine. As such, transitions from A = T to G ≡ C may result following replication.

Because of the specificity by which base analogues such as 2-AP induce transition mutations, base analogues may also be used to induce reversion to the wild-type nucleotide sequence. This alteration is called **reverse mutation.** The process can occur spontaneously, but at a much lower rate.

Alkylating Agents

The sulfur-containing **mustard gases** were one of the first groups of chemical mutagens discovered. This discovery was made as a result of studies concerned with chemical warfare during World War II. Mustard gases are **alkylating agents;** that is, they donate an alkyl group such as CH_3— or CH_3—CH_2— to amino or keto groups in nucleotides. **Ethylmethane sulfonate (EMS),** for example, alkylates the keto groups in the number-6 position of guanine and in the number-4 position of thymine (Figure 11.8). As with base analogues, base-pairing affinities are altered and transition mutations result. In the case of **6-ethyl guanine,** this molecule acts like a base analogue of adenine, causing it to pair with thymine. Table 11.1 lists the chemical names and structures of several frequently used alkylating agents known to be mutagenic.

Acridine Dyes and Frameshift Mutations

Other chemical mutagens cause **frameshift mutations.** These result from the addition or removal of one or more base pairs in the polynucleotide sequence of the gene. Inductions of frameshift mutations have been studied in detail with a group of aromatic molecules known as **acridine dyes.** The structures of **proflavin,** the most widely studied acridine mutagen, and **acridine orange** are shown in Figure 11.9. Acridine dyes are of about the same dimension as a nitrogenous base pair and are

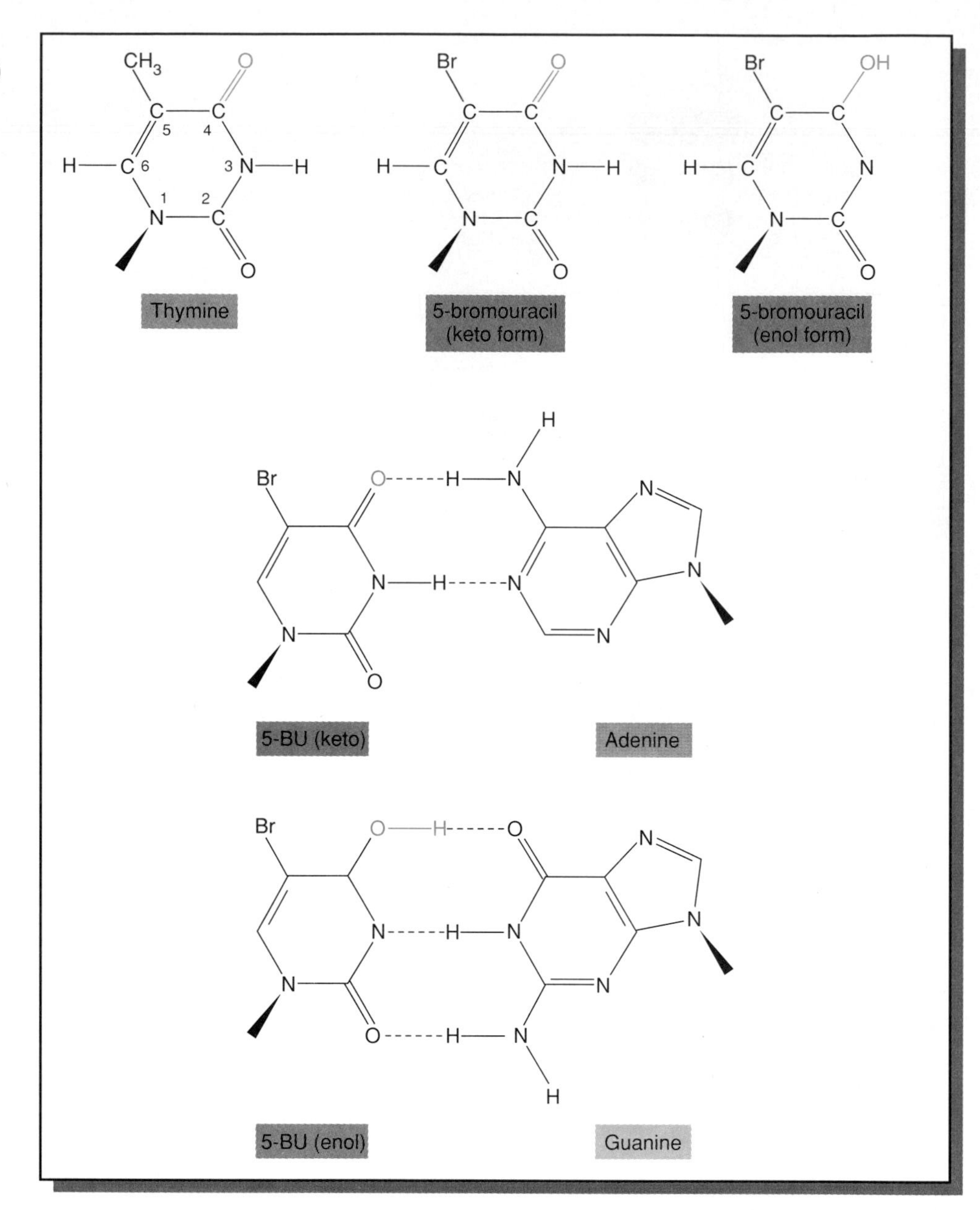

FIGURE 11.7
Similarity of 5-bromouracil (5-BU) structure to thymine structure. In the common keto form, 5-BU pairs normally with adenine, behaving as an analogue. In the rare enol form, it pairs anomalously with guanine.

TABLE 11.1
Alkylating agents.

Common Name or Symbol	Chemical Name	Chemical Structure
Mustard gas (sulfur)	Di-(2-chloroethyl)sulfide	$Cl—CH_2—CH_2—S—CH_2—CH_2—Cl$
EMS	Ethylmethane sulfonate	$CH_3—CH_2—O—S(=O)(O)—CH_3$
EES	Ethylethane sulfonate	$CH_3—CH_2—O—S(=O)(O)—CH_2—CH_3$

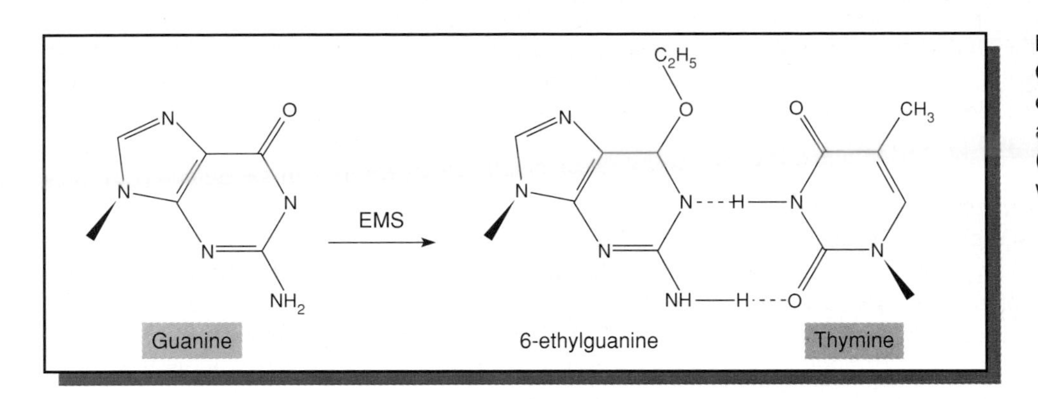

FIGURE 11.8
Conversion of guanine to 6-ethylguanine by the alkylating agent ethylmethane sulfonate (EMS). 6-ethylguanine base pairs with thymine.

known to intercalate or wedge between purines and pyrimidines of intact DNA. Intercalation of acridine dyes induces contortions in the DNA helix, causing deletions and additions.

One model suggests that the resultant frameshift mutations are generated at gaps produced in DNA during replication, repair, or recombination. During these events, there is the possibility of slippage and improper base pairing of one strand with the other. The model suggests that intercalation of the acridine into an improperly base-paired region can extend the existence of these slippage structures. If so, the probability increases that the mispaired configuration will exist when synthesis and rejoining occurs, thereby resulting in an addition or deletion of one or more bases from one of the strands.

MUTATIONS IN HUMANS: TWO CASE STUDIES

The preceding section summarizes the molecular basis of mutation. The information presented relies almost exclusively on the study of nucleic acid chemistry and has described numerous ways in which one nucleotide pair can be converted to another, either spontaneously or as a result of an induced change. We know that such mutations actually occur, primarily because analysis of amino acid sequences of many proteins within populations of a species show substantial diversity. This diversity has arisen during evolution and is a reflection of changes in the triplet codons following substitution of one or more nucleotides in the DNA sequences constituting genes.

As our ability to analyze DNA more directly increases we are better able to look specifically at the actual nucleotide sequence of genes and gain greater insights into mutation. Several techniques capable of accurate, rapid sequencing of DNA are available. These and other approaches involving the analysis of DNA have extended greatly our knowledge in molecular genetics.

In this section, we examine the results of two studies that have investigated the actual gene sequence of various mutations affecting humans. The first provides an interesting insight into the molecular basis of the **ABO antigens,** originally presented in Chapter 4 as an example of multiple alleles. The second case involves the nature of

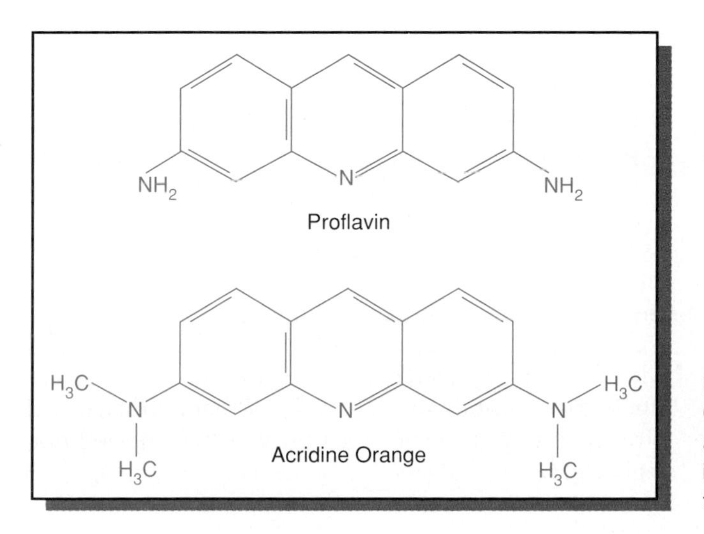

FIGURE 11.9
Chemical structures of proflavin and acridine orange, which intercalate with DNA and cause frameshift mutations.

the mutations that have led to the devastating sex-linked disorder **muscular dystrophy.**

The ABO system is based on a series of antigenic determinants found on erythrocytes and other cells, particularly epithelial types. As previously discussed, three alleles of a single gene exist, the product of which is designed to modify the H substance. The modification involves glycosyltransferase activity, converting the H substance to either the A or B antigen, as a result of the product of the I^Aor I^B allele, respectively, or failing to modify the H substance, as a result of the I^O allele.

Using recombinant DNA technology, the responsible gene has been examined in 14 cases of varying ABO status. Four consistent nucleotide substitutions were found when the DNAs of the I^A and I^B alleles were compared. It is assumed that the changes in the amino acid sequence of the glycosyltransferase gene product resulting from these substitutions leads to the different modifications of the H substance.

The situation with the I^O allele is unique and interesting. Individuals homozygous for this allele are type O, lack glycosyltransferase activity, and fail to modify the H substance. Analysis of the DNA of this allele shows one consistent change compared to that of the other alleles: the deletion of a single nucleotide early in the coding sequence, causing a frameshift mutation. A complete messenger RNA is transcribed, but upon its translation, the frame of reading shifts at the point of deletion and continues out of frame for about 100 nucleotides before a "stop" codon is encountered. At this point, premature termination of the resulting polypeptide chain occurs, producing a nonfunctional product.

These results provide a direct molecular explanation of the ABO allele system and the basis for the biosynthesis of the corresponding antigens. The molecular basis for the antigenic phenotypes is clearly the result of structural alterations, or mutations, of the nucleotide sequence of the gene encoding the glycosyltransferase enzyme.

The second case of mutational analysis involves the severe disorder muscular dystrophy. It is characterized by progressive muscle degeneration, or myopathy, resulting in the death of affected individuals by early adulthood. Because the condition is recessive and sex linked, and since affected males do not reproduce, females are rarely affected by the disorder. The incidence of 1/3500 live male births makes muscular dystrophy one of the most common life-shortening hereditary disorders known. Two forms exist: **Duchenne muscular dystrophy (DMD)** is more common and more severe than the allelic form called **Becker muscular dystrophy (BMD).**

The region containing the gene has been analyzed extensively and consists of over two million base pairs. In unaffected individuals, transcription results in a messenger RNA containing about 14,000 bases (14kb) that is translated into the protein **dystrophin** consisting of 3685 amino acids. This protein can be detected in most cases of the less severe BMD, but is rarely found in DMD. This has led to the hypothesis that most mutations causing BMD do not alter the reading frame, but that most DMD mutations change the reading frame early in the gene, resulting in premature termination of dystrophin translation. This hypothesis is consistent with the observed differences in severity of these two forms.

In an extensive analysis of the DNA of 194 patients (160 DMD and 34 BMD), J. T. Den Dunnen and associates found that 128 of these mutations (65 percent) consisted of substantial deletions or duplications. Of 115 deletions, 17 occurred in BMD, and of 13 duplications, 1 was in BMD, with the remainder being found in DMD cases. In most cases, the above results were consistent with the "reading frame" hypothesis. With few exceptions, DMD mutations changed the frame of reading of exon areas. BMD mutations usually did not alter the reading frame.

Perhaps the most noteworthy finding is the high percentage of deleterious mutations studied that represent the deletion or duplication of nucleotides within the gene. This observation reflects the fact that a mutation caused by a random single nucleotide substitution within a gene is more likely to be tolerated without the devastating effect of muscular dystrophy than the addition or loss of numerous nucleotides that may alter the frame of reading. There are three reasons for this:

1 A nucleotide substitution may not change the encoded amino acid since the code is degenerate.

2 If an amino acid substitution does result, the change may not be present at a location within the protein that is critical to its function.

3 Even if the altered amino acid is present at a critical region, it may still have little or no effect on the function of the protein. For example, an amino acid might be changed to another with nearly identical chemical properties or to one with very similar recognition properties, such as shape.

As a result, single-base substitutions may have little or no effect on protein function, or may simply reduce the efficiency but not eliminate the functional capacity of the gene product. It is clear that we cannot look at mutations with the oversimplified expectation that most of them are single-base substitutions. As more of them are analyzed directly, our picture of mutation will become increasingly clear.

DETECTION OF MUTAGENICITY: THE AMES TEST

There is particular concern about the possible mutagenic properties of any chemical that enters the human body, whether through the skin, digestive tract, or respiratory tract. For example, great attention has been given to residual materials of air and water pollution, food preservatives and additives, artificial sweeteners, herbicides, pesticides, and pharmaceutical products. While mutagenicity may be tested in various organisms, including *Drosophila,* mice, and cultured mammalian cells, the most common test involves bacteria and was devised by Bruce Ames.

The **Ames test** utilizes four tester strains of the bacterium *Salmonella typhimurium* that were selected for sensitivity and specificity for mutagenesis. One strain is used to detect base-pair substitutions, and the other three detect various frameshift mutations. Each mutant strain requires histidine for growth (*his*$^-$). The assay measures the frequency of reverse mutation, which yields wild-type (*his*$^+$) bacteria. Greater sensitivity to mutagens occurs because these strains bear other mutations that eliminate the DNA excision repair system (discussed later in this chapter) and the lipopolysaccharide barrier that coats and protects the surface of the bacteria.

It is very interesting to note that many substances that enter the human body are relatively innocuous until activated metabolically to a more chemically reactive electrophilic product. This usually occurs in the liver. Thus, the Ames test, which is performed *in vitro,* includes a step in which the test compound is incubated in the presence of a mammalian liver extract. Or, test compounds are actually injected into the mouse, which is later sacrificed and the liver removed. Extracts are then tested.

In the initial use of Ames testing in the 1970s, a large number of known carcinogens were examined. Over 80 percent of these were shown to be strong mutagens! This is not surprising, since transformation of cells to the malignant state undoubtedly occurs as a result of some alteration of DNA. Although a positive response as a mutagen does not prove the carcinogenic nature of a test compound, the Ames test is useful as a preliminary screening device. It is used extensively in conjunction with the industrial and pharmaceutical development of chemical compounds.

RADIATION, DNA DAMAGE, AND DNA REPAIR

In addition to chemicals, other components of the natural and man-made environment are capable of inducing mutations. The following sections consider **ultraviolet radiation (UV light)** and **high-energy radiation** as two such components. Both are part of the electromagnetic spectrum and have shorter wavelengths than visible light (Figure 11.10). X rays and gamma rays are characterized by exceedingly short wavelengths. Since the inherent energy associated with components of this spectrum is inversely proportional to wavelength, both X rays and gamma radiation are considered together as high-energy radiation.

As we shall see, the study of the mutagenicity of UV light paved the way for the discovery of DNA damage and subsequently of how natural repair of many forms of DNA damage occurs. The ability to perform such repair is essential to the survival of all organisms, including humans. Our understanding of UV-induced DNA damage and its repair has become more important as our concern grows over the potential loss of the protective ozone layer.

Since we still live in the nuclear age, concern about high-energy radiation is also particularly keen. We shall consider not only the mutagenic effects of high-energy radiation but also some of the issues involving radiation safety. The relevance is obvious as we approach the end of the twentieth century.

UV Radiation, Thymine Dimers, and Photoreactivation Repair

In Chapters 7 and 8, we emphasized the fact that purines and pyrimidines absorb **ultraviolet light (UV)** most intensely at a wavelength around 260 nm. This property has been used extensively in the detection and analysis of nucleic acids. In 1934, as a result of studies involving *Drosophila* eggs, it was discovered that ultraviolet light is mutagenic. By 1960, several studies concerning the *in vitro* effect of UV light on the components of nucleic acids had been completed with the following conclusions. The major effect of UV light is on pyrimidines, where dimers are formed, particularly between two thymine residues (Figure 11.11). While cytosine-cytosine and thymine-cytosine dimers may also be formed, they are less prevalent. It is believed that the dimers distort the DNA conformation and inhibit normal replication, which seems to be responsible for the killing effects of UV light on microorganisms.

Other studies provide more complete information on the relationship of UV light to genetic phenomena. These studies demonstrate that an intricate set of repair processes function to counteract the lesions produced in DNA as a result of UV irradiation. It is during the repair process that most errors leading to UV-induced mutations occur. In the following discussion, we will examine

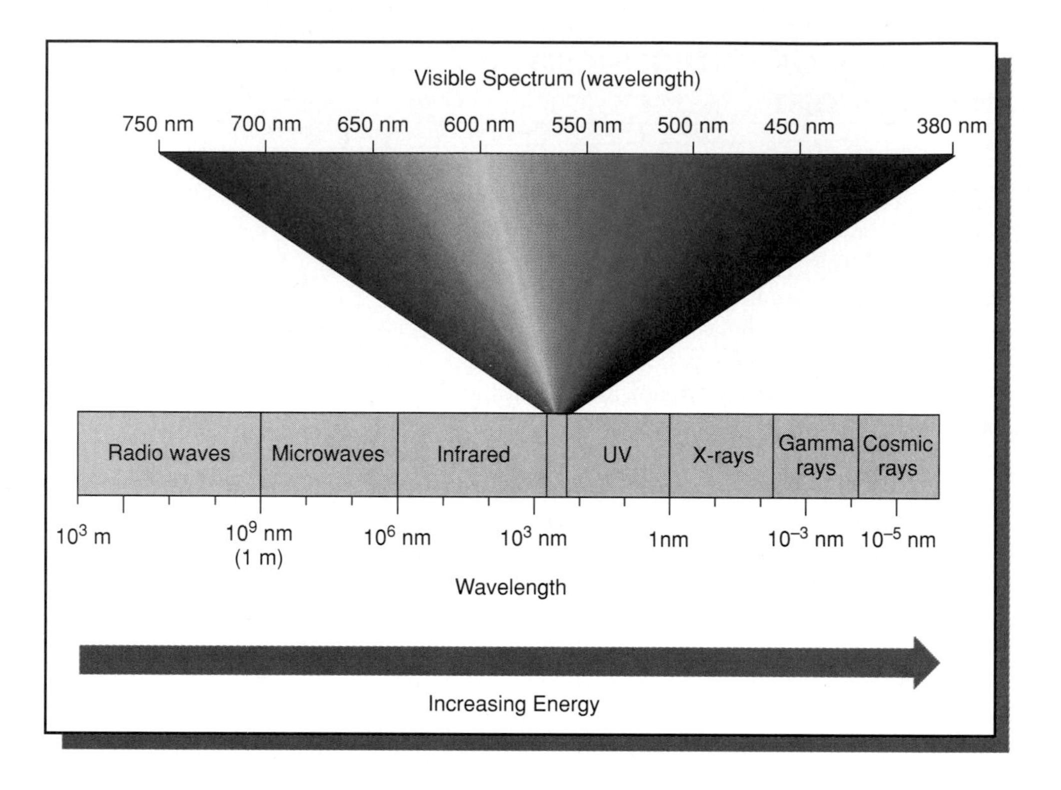

FIGURE 11.10
The components of the electromagnetic spectrum and their associated wavelengths.

DNA repair systems, as initially elucidated in the bacterium *E. coli.*

The first relevant discovery concerning UV repair in bacteria was made in 1949 when Albert Kelner observed the phenomenon of **photoreactivation repair.** He showed that the UV-induced damage to *E. coli* DNA could be partially reversed if, following irradiation, the cells were exposed briefly to light in the blue range of the visible spectrum. The photoreactivation repair process has subsequently been shown to be temperature dependent, thereby suggesting that the light-incuded mechanism involves an enzymatically controlled chemical reaction. Visible light appears to induce the repair process of the DNA damaged by UV light.

Further studies of photoreactivation revealed that the process is due to a protein called the **photoreactivation enzyme (PRE).** This molecule may be isolated from extracts of *E. coli* cells. The enzyme's mode of action is to cleave the bonds between thymine dimers, thus reversing the effect of UV light on DNA [Figure 11.12(a)]. While the enzyme will associate with a dimer in the dark, it must absorb a photon of light to cleave the dimer.

Excision Repair of DNA

Investigations in the early 1960s suggested that a repair system or systems that do not require light also exist in

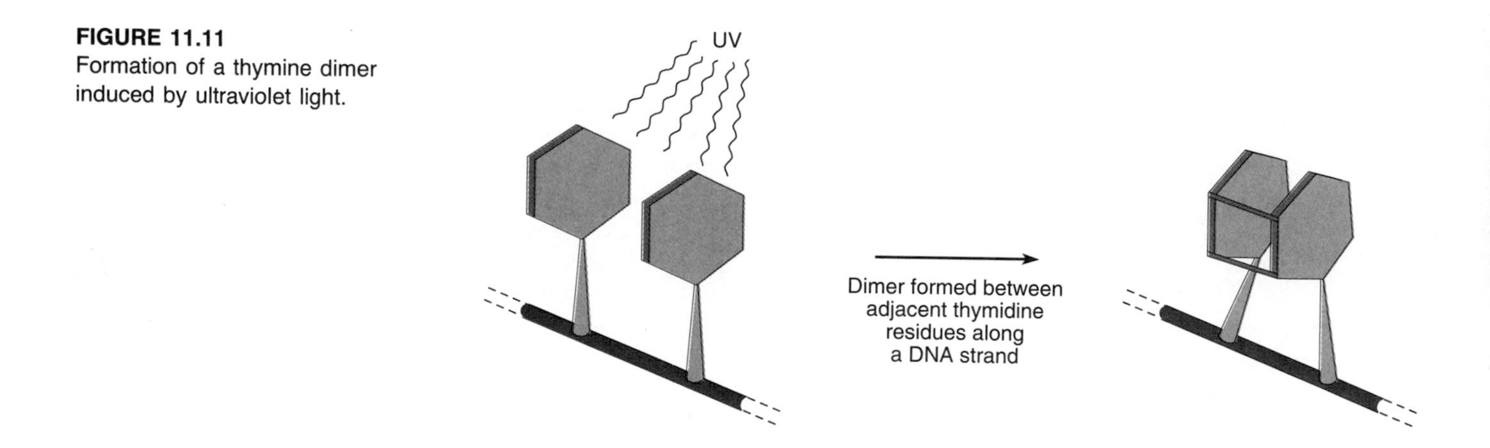

FIGURE 11.11
Formation of a thymine dimer induced by ultraviolet light.

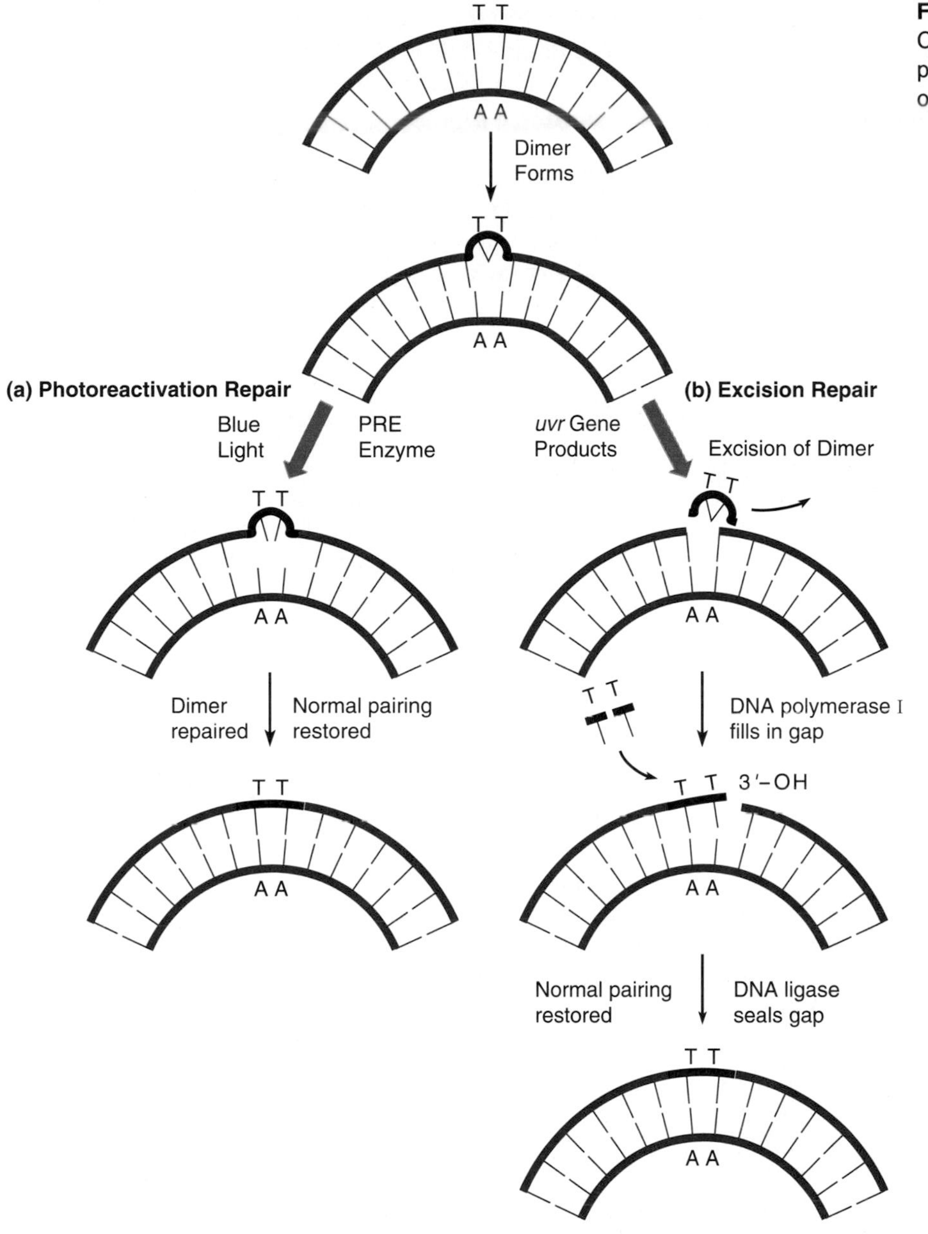

FIGURE 11.12
Contrasting diagrams of (a) photoreactivation and (b) excision repair of UV-induced thymine dimers.

E. coli. Paul Howard-Flanders and coworkers isolated several independent mutants demonstrating increased sensitivity to ultraviolet radiation. One group was designated *uvr* (ultraviolet repair) and included the *uvrA*, *uvrB*, and *uvrC* mutations. These genes and their protein products were subsequently shown to be involved in a process called **excision repair.** During this process, three steps have been shown to occur, as illustrated in Figure 11.12(b):

1 The distortion of the strand caused by the UV-induced dimer is recognized and enzymatically clipped out by a nuclease that cleaves the phosphodiester bonds. This "excision," which may include several nuceotides adjacent to the dimer as well, leaves a gap in the helix. The *uvr* gene products operate at this step.

2 **DNA polymerase I** fills this gap by inserting d-ribonucleotides complementary to those on the intact strand. The enzyme adds these bases to the 3′-OH end of the clipped DNA.

3 The joining enzyme **DNA ligase** seals the final "nick" that remains at the 3′-OH end of the last base inserted, closing the gap.

DNA polymerase I, the enzyme discovered by Kornberg, was once assumed to be the universal DNA-replication enzyme (see Chapter 8). The discovery of the *polA1* mutation demonstrated that this is not the case. *E. coli* cells carrying the *pol A1* mutation lack functional polymerase I. However, replication of DNA takes place normally. Cells with this mutation are unusually sensitive to UV light. Apparently, such cells are unable to fill the gap created by the excision of the thymine dimers. This finding demonstrates the importance of excision repair mechanisms in counteracting the effects of UV light.

Proofreading and Mismatch Repair

Since we have chosen to introduce DNA repair mechanisms during our consideration of mutations, it is logical to complete our discussion of this topic here. Spontaneous errors that occur during replication are introduced at a rate of approximately one in 100,000 (10^{-5}). Each error, involving a **mismatched base pair,** is immediately subject to proofreading associated with DNA polymerase III, as discussed in Chapter 8. Proofreading is thought to increase fidelity by two orders of magnitude, leaving only one in 10,000,000 (10^{-7}) mismatches.

Once proofreading has occurred, still another mechanism, called **mismatch repair,** may be activated. Proposed over 20 years ago by Robin Holliday, the molecular basis of this process is now well established. Like other DNA lesions, (1) the alteration or mismatch must be recognized, (2) the incorrect nucleotide must be removed, and (3) replacement with the correct base must occur. Once recognition of a mismatch occurs in *E. coli,* methylation of the parental strand results, leaving the newly synthesized strand with the incorrect base unmethylated. The repair enzyme then preferentially binds to the unmethylated strand, which is excised, and the correct complement is inserted! The products of a series of *E. coli* genes *mutH*, *L*, *S*, and *U* are involved in the discrimination step. Mutations in each of these genes result in strains deficient in mismatch repair. Such a repair mechanism undoubtedly also occurs in the DNA of higher organisms.

Recombinational Repair

The final mode of repair to be discussed was discovered in an excision-defective strain of *E. coli.* Called **recombinational repair,** this system is activated when damaged DNA has escaped repair and the resultant distortion subsequently disrupts the process of replication. The cells that show this phenomenon are dependent on the product of the **rec A** gene which is involved in several types of recombination in *E. coli.*

When such DNA is being replicated, DNA polymerase skips over the distortion, creating a gap on one of the newly synthesized strands. To counteract this, the *rec A* protein directs an exchange process whereby this gap is filled as a result of the insertion of the homologous segment present on the intact strand. The resulting gap now present on the "donor" strand is filled by repair synthesis as replication proceeds.

This mode of repair is a "rescue operation" that allows replication to create accurate and complete DNA copies from a damaged template. Many different gene products are involved. The activation of this repair system is called the **SOS response,** because it represents the final effort to repair damaged DNA.

UV Light and Human Skin Cancer: Xeroderma Pigmentosum

Xeroderma pigmentosum (XP), a rare autosomal recessive disorder in humans, predisposes individuals to epidermal pigment abnormalities. Exposure to sunlight results in malignant growth of the skin.

Figure 11.13 contrasts two XP individuals, one of whom has been detected early and protected from sunlight. The condition is very severe and may be lethal. Since sunlight contains UV radiation, a causal relationship has been predicted between thymine dimer production and XP. It has also been of great interest to determine which of the three forms of repair processes counteracting the effects of UV-induced damage to DNA (if any) are operating in humans. Furthermore, it was suspected that XP individuals might lack one or more repair systems, which may cause them to be susceptible to UV-induced skin damage.

The various modes of repair of UV-induced lesions have been investigated in human fibroblast cultures derived from XP and normal individuals. Fibroblasts are undifferentiated connective tissue cells. The results are varied and suggest that the XP phenotype may be caused by more than one mutant gene.

In 1968, James Cleaver showed that cells from XP patients were deficient in the **unscheduled DNA synthesis** elicited in normal cells by UV light. This assay is thought to represent activity of the excision repair system, suggesting that XP cells are deficient in this form of repair. In 1974, the presence of a photoreactivation enzyme **(PRE)** was established in normal human cells. Betsy Sutherland identified the enzyme first in leukocytes and subsequently in fibroblast cells. Sutherland demonstrated further that some XP cultures contain a lower PRE activity than control cultures. The activity in various XP cell strains range from 0 to 50 percent in cultures established from different patients.

Cultured cells from any two XP variants may be induced to fuse together, forming a **heterokaryon** where the two nuclei share a common cytoplasm. This process is one aspect of **somatic cell genetics.** When fusion is performed with cells of different XP variants, excision repair, as assayed by unscheduled DNA synthesis, is

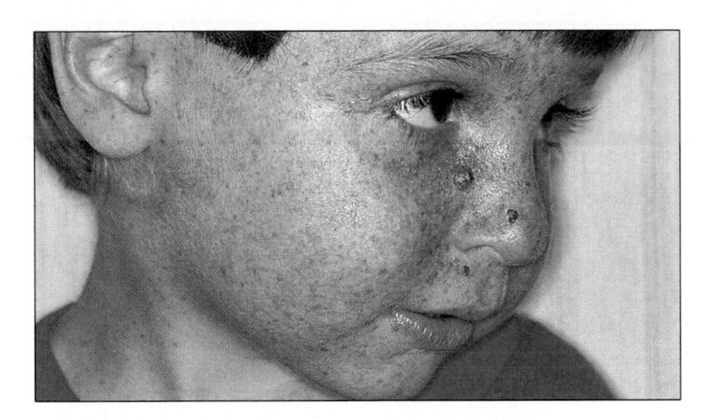

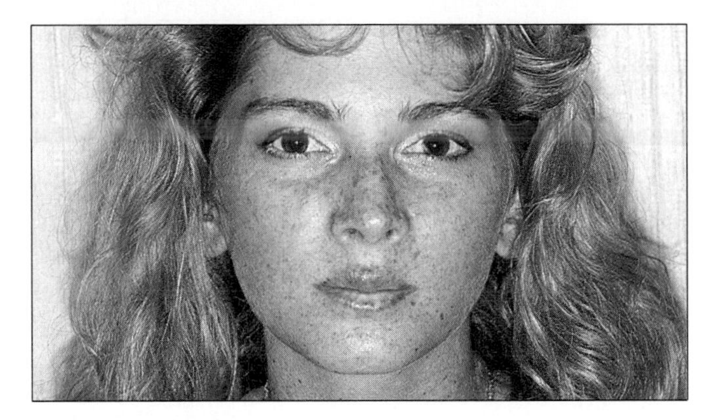

FIGURE 11.13
Xeroderma pigmentosum. The four-year-old boy on the left shows marked skin lesions induced by sunlight. Mottled redness (erythema) and irregular pigment changes that are a response to cellular injury are apparent. Two nodular cancers are present on his nose. The eighteen-year-old girl on the right has been carefully protected from sunlight since the diagnosis of xeroderma pigmentosum made in infancy. Several cancers have been removed and she works as a successful model.

sometimes reestablished. When this occurs, the two variants are said to undergo **complementation.** Alone, neither cell type demonstrates excision repair, but together in a heterokaryon the process occurs. In genetic terms, this is strong evidence that different genes are involved in the defect of each variant. Complementation occurs because the heterokaryon has at least one normal copy of each gene. In many studies, variants have been divided into eight complementation groups, which suggests that as many as eight different genes may be involved in causing XP.

These findings establish that normal individuals are probably susceptible to UV-induced damage of DNA by exposure to sunlight. However, this damage activates the repair systems that counteract it. We can expect that future work will clarify the precise role and mechanism of repair of UV-induced damage in humans.

High-Energy Radiation

Within the electromagnetic spectrum, energy varies inversely with wavelength. We earlier compared the relative wavelengths of the various portions of the electromagnetic spectrum (see Figure 11.10). **X rays, gamma rays,** and **cosmic rays** have even shorter wavelengths than ultraviolet light and are therefore more energetic. As a result, they are strong enough to penetrate deeply into tissues, causing ionization of the molecules encountered along the way. These sources of **ionizing radiation** could be predicted to be mutagenic. Indeed, Herman Muller detected such effects in 1927 while working with *Drosophila*. Since that time, the effects of ionizing radiation, particularly X rays, have been studied intensely.

As X rays penetrate cells, electrons are ejected from the atoms of molecules encountered by the radiation. Thus, stable molecules and atoms are transformed into free radicals and reactive ions. Along the path of a high-energy ray, a trail of ions is left that can initiate a variety of chemical reactions. These reactions can affect directly or indirectly the genetic material, altering the purines and pyrimidines in DNA and resulting in point mutations. Such ionizing radiation is also capable of breaking chromosomes, resulting in a variety of aberrations.

Figure 11.14 shows a plot of induced sex-linked recessive lethal mutations versus the dose of X rays adminis-

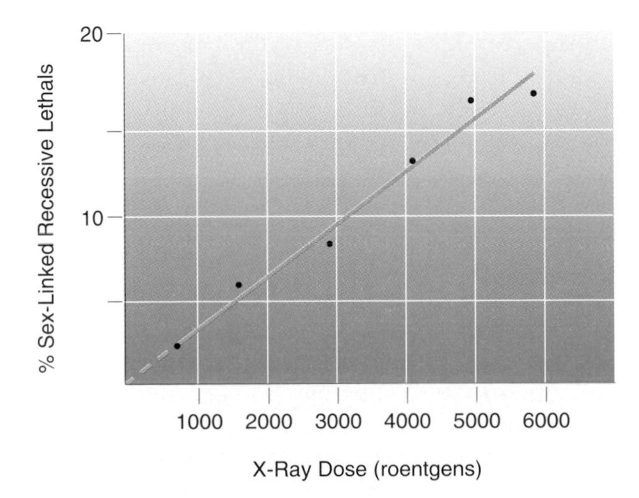

FIGURE 11.14
Plot of sex-linked recessive mutations induced by increasing doses of X-rays. If extrapolated, the graph intersects the zero axis (dashed line).

tered. The graph shows a straight line that, if extrapolated, intersects near the zero axis. A linear relationship is evident between X-ray doses and the induction of mutation. For each doubling of dose, twice as many mutations are induced. Because the line intersects near the zero axis, this graph suggests that even very small doses of irradiation are mutagenic.

These observations may be interpreted in the form of the **target theory,** first proposed in 1924 by J. A. Crowther and F. Dessauer. The theory proposes that there are one or more sites, or targets, within cells and that a single event of irradiation at one site will bring about a damaging effect, or mutation. In a simple form, the target theory says that one "hit" of irradiation will cause one "event" or mutation, suggesting that the X rays interact directly with the genetic material.

Two other observations concerning irradiation effects are of particular interest. First, in some organisms studied, the **intensity of the dose** (dose-rate) administered seems to make little difference in mutagenic effect. That is, a 100 roentgen exposure (a **roentgen** is a measure of energy dose), whether occurring in a single acute dose or cumulatively in many smaller chronic doses, seems to produce the same mutagenic effect. *Drosophila,* for example, shows this response. In mammals such as mice and humans, however, this is not the case. Repair of the damage seems to occur during the intervals between irradiation. Thus, several smaller doses are not as potent as a single large dose.

The second observation is that certain portions of the cell cycle are more susceptible to irradiation effects. As mentioned previously, in addition to a mutagenic effect, X rays can also break chromosomes, resulting in terminal or intercalary deletions, translocations, and general chromosome fragmentation. Damage occurs most readily when chromosomes are greatly condensed in mitosis. This property constitutes one of the reasons why radiation is used to treat human malignancy. Since tumorous cells are more often undergoing division than their nonmalignant counterparts, they are more susceptible to the immobilizing effect of radiation.

RADIATION SAFETY

Technological advances during the past 50 years have led to widespread applications of radiation in our society. Largely due to the pioneering efforts of Muller, the potential dangers of radiation to humans are known, and radiation safety has received increasing attention. In addition to natural sources of radiation, therapeutic and diagnostic use, industrial application, and atomic and nuclear fallout represent the sources of major concern.

It is now known that a radiation dose of 50 to 150 roentgens (r) is lethal to 50 percent of human cells in culture. Whole-body irradiation of 400 r is usually fatal to individuals, and 100 r is sufficient to cause radiation sickness. Sublethal doses of radiation may also shorten the human life span significantly and increase the occurrence of various forms of cancer, particularly leukemia.

Since there seems to be no lower limit for the amount of radiation that will induce mutation, and since the effects of ionizing radiation can be cumulative, precautionary measures are essential to the human population. In 1970, a National Academy of Sciences panel was commissioned to investigate radiation safety standards. At that time, the federal guidelines stated that the general population should not receive more than 170 millirems (mrem) of artificial radiation per year, exclusive of medical sources. A **rem** is a "roentgen equivalent in man," and a millirem is 1/1000 of a rem. The committee's report included the following statement:

> The present guides of 170 millirems per year grew out of an effort to balance societal needs against genetic risks. It appears that these needs can be met with far lower average exposures and lower genetic and somatic risk than permitted by the current radiation protection guide. To this extent, the current guide is unnecessarily high.

The committee strongly admonished the medical profession and urged it to improve the efficiency of X-ray equipment; increase shielding, particularly of the reproductive organs; and avoid unnecessary mass screening for many diseases. Since then, the result has been greater adherence to shielding and the development and use of more sensitive X-ray film, thus diminishing the radiation dose during X rays. X-ray generating equipment also creates a smaller, more focused radiation path.

Table 11.2 summarizes the report's estimate of annual exposure from various sources of radiation.

TABLE 11.2
Estimates of annual whole-body exposure to radiation from various sources in the United States (1970).

Source	Average Dose Rate (mrem/yr)
Natural	102
Global fallout	4
Nuclear power	0.003
Diagnostic	72
Radiopharmaceuticals	1
Occupational	0.8
Miscellaneous	2
TOTAL	182

In 1993, there appears to be no less reason to be concerned about low radiation doses. One way to assess risk is to estimate the dose of radiation that doubles the spontaneous mutation rate—the **doubling dose.** In its most recent report (1988), the United Nations Scientific Committee on the Effects of Atomic Radiation (UNSCEAR) estimates 100 mrem at low dose rates as the doubling dose. At high dose rates, 40 mrem is estimated.

Even though the spontaneous mutation rate is very low, the above information is sufficient to cause concern about the well-being of the human species. Many scientists have projected that the human population will continue to accumulate deleterious mutations as a result of exposure to mutagens such as radiation. The concept of **genetic burden** refers to deleterious alleles that are already present in the human gene pool. It is estimated that the average individual carries two to ten recessive lethal alleles in the heterozygous condition. Before absolute conclusions concerning the dangers of radiation to the genetic well-being of future human populations can be drawn, more direct evidence is needed. Tragically, the atomic bomb attacks on Japan in 1945 have allowed us to investigate this question.

Immediately following World War II, the **Atomic Bomb Casualty Commission (ABCC)** was established. Its large-scale study began in 1946 as a joint venture between the U.S. National Research Council and the Japanese National Institute of Health. Conducted at the Radiation Effects Research Foundation in Hiroshima, this scientific project is still active in the 1990s. The overall study considered various parameters of radiation effects on somatic and germinal tissue.

The results of massive radiation exposure to the human population were devastating. Between 100,000 and 250,000 fatalities were estimated. Somatic effects on survivors were severe. Sublethal radiation doses led to chromosome aberrations, brain damage and retardation, decreased lifespans, and an elevation of various cancers, particularly leukemia.

In spite of these dire consequences, one aspect of the study was more encouraging and came from investigations to assess the impact of radiation exposure on germinal tissue and thus on future generations. Scientists studied children conceived after August, 1945, who had at least one parent exposed in Hiroshima or Nagasaki. From 1948 to 1953, over 70,000 such cases were located. All newborns were examined, and 40 percent of the children were reexamined when they were 8 to 10 months old.

The results of the study through 1953 revealed no statistically significant effect of increasing levels of radiation exposure. Increases in congenital malformations, stillbirths, altered birth weights, or abnormal physical development at age 8 to 10 months could not be detected. In an ongoing investigation, members of this group as well as other children born after 1953 to exposed parents are being examined. The focus of this study has been on mortality, chromosome abnormalities, and alterations in specific proteins. The findings parallel the results of the examination of newborns. No statistically significant increases have been observed as a result of radiation exposure. While the development of cancer has been the major delayed effect on those individuals exposed directly, their children, thus far, fail to exhibit such an effect. All indicators suggest that humans are less sensitive to the genetic effects of ionizing radiation than expected. We hope that this conclusion is upheld as this important study progresses.

SITE-DIRECTED MUTAGENESIS

This section introduces a useful experimental technique that allows tailor-made changes within genes: **site-directed mutagenesis.** The technique relies on a number of manipulations involving recombinant DNA technology, which will be introduced in Chapter 15. However, the underlying principles are based on previous information. This technique is of great interest here because it illustrates the application of basic knowledge of a topic such as mutation to current research investigations.

The goal of the technique is to specifically alter one or more nucleotides within a gene in order to change a specific triplet codon. Upon transcription and translation, this change will cause the insertion of a "mutant" amino acid in the protein encoded by the original gene. Such designed mutations are particularly useful in studying the effects of specific mutations on protein function.

The first step is to determine the nucleotide sequence of the gene being studied. This can be accomplished by DNA sequencing techniques, or it can be predicted if the amino acid sequence of the protein is known by utilizing our knowledge of the genetic code.

The next step is to isolate the DNA of known sequence and obtain from it one of the two complementary strands. A decision is then made as to which nucleotides are to be changed. Then, a small piece of DNA complementary to that region is chemically synthesized. It is complementary at all points, except in the triplet sequence (or sequences) that is to be altered. This triplet encodes the amino acid that will change in the protein.

As illustrated in Figure 11.15, this short piece of DNA hybridizes with the original parent strand, forming a partial duplex because of its complementarity along most of its length. If DNA polymerase and DNA ligase are then added to this hybrid complex, the short sequence is extended so that a duplex of the entire gene is formed. The

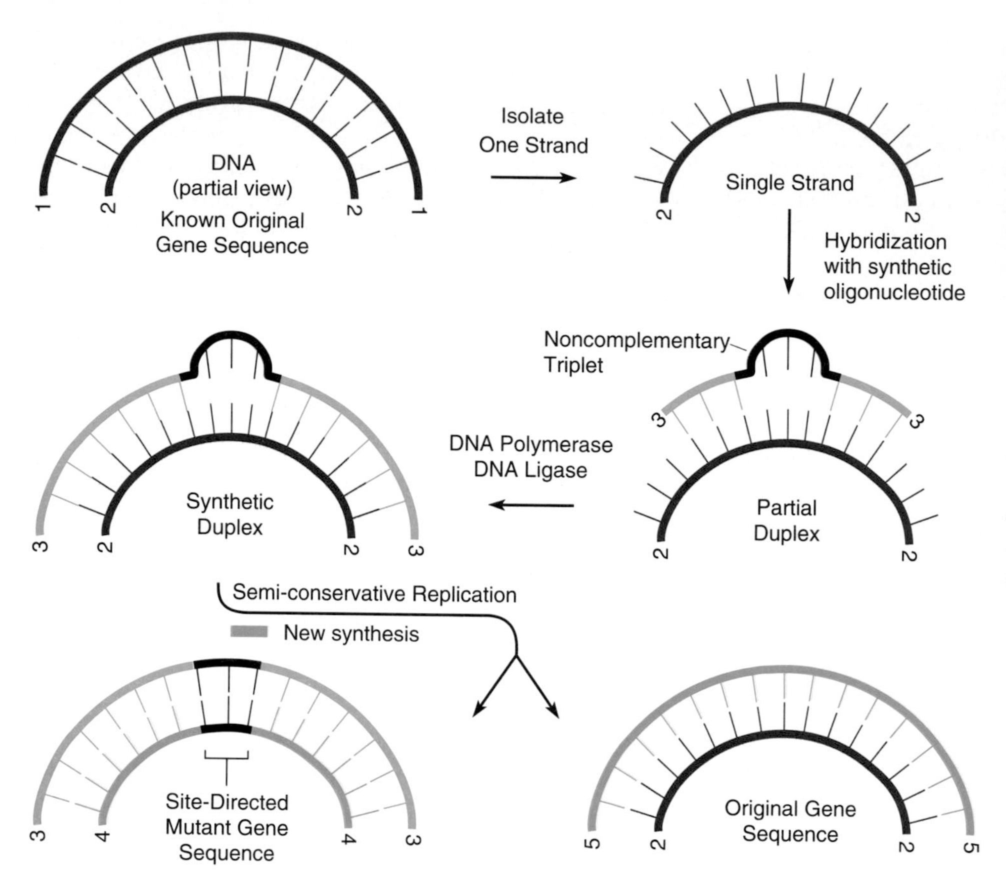

FIGURE 11.15
Site-directed mutagenesis, achieved by first obtaining a single strand of DNA from a gene of interest. This is hybridized with a synthetic oligonucleotide containing a triplet altered so as to encode an amino acid of choice. The partial duplex is completed under the direction of DNA polymerase and DNA ligase. Following semiconservative replication, a different complementary base pair is present in one of the new duplexes. Upon transcription and translation, a mutant protein, designed in the laboratory, will be produced.

two strands are perfectly complementary except at the point of alteration.

When this DNA undergoes semiconservative replication, two types of duplexes are formed: one is like the original, unaltered gene and the other contains the newly designed sequence. Using recombinant DNA technology, it is possible not only to complete the above manipulations, but also to allow the altered gene to be expressed so that large amounts of the desired protein are available for study. This approach is useful in investigating the role of specific amino acids during protein function.

TRANSPOSABLE GENETIC ELEMENTS

We conclude this chapter by introducing a unique source of genetic variation: **Transposable elements.** In bacteria, these were first studied during the early 1970s when several mutations in *E. coli* were shown to be the result of the insertion into existing genes of short segments of DNA, about 1000 base pairs in length. Two observations are particularly relevant. First, such insertions into genes often disrupt their expression and interrupt the function of their products, thus altering the respective phenotypes. Therefore, transposable elements may be considered as a major class of mutagenic agents. Second, these sequences demonstrate mobility, maintaining the ability to move in and out of the bacterial chromosome. As a result, in bacteria, they have been called **insertion sequences.**

We shall return to a discussion of insertion sequences in Chapter 14 in an extended discussion of bacterial genetics. Here, we shall address the issue of similar DNA sequences in eukaryotes. They are present in all organisms studied, often in abundance. The genetic impact of one specific example was investigated by Barbara McClintock over 40 years ago during her study of maize.

The Ac-Ds System in Maize

In the late 1940s, McClintock studied mottling in corn kernels and proposed that the observed phenotypic variation (Figure 11.16) was the result of chromosome breakage due to an element she called **Ds (Dissociation).** She concluded that the *Ds* element induced breakage at its point of insertion in the chromosome, but that it did not have a constant location. Instead it appeared to be mobile within the genome and to induce breakage at different sites, depending on its location. The expression of genes near the breakpoints was altered, effecting the kernel phenotype. She showed further that mobility of *Ds* was dependent on a second element called **Ac (Activator),** which is also mobile.

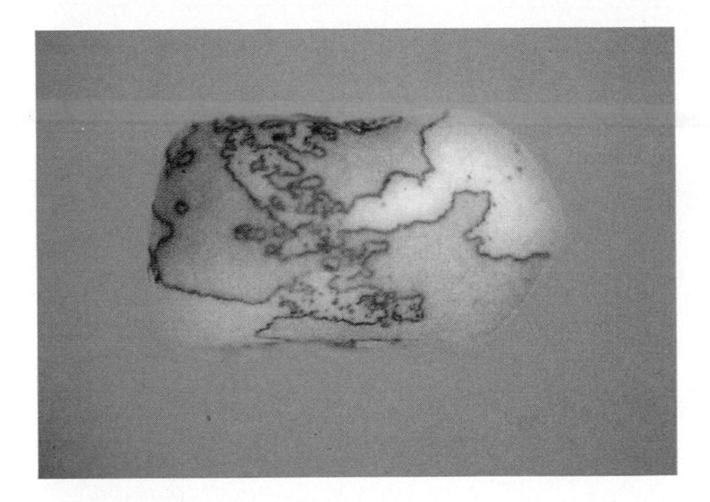

FIGURE 11.16
Corn kernel showing spots of colored aleurone produced by genetic transposition involving the *Ac/Ds* system.

While the validity of McClintock's proposed mobile elements was questioned following her initial observations, molecular analysis has since verified their presence and clarified their mode of action. For her work, Barbara McClintock was awarded the Nobel Prize in 1983 (see the accompanying essay).

Transposons in *Drosophila*

Transposable elements, often called **transposons,** have been discovered in other eukaryotic organisms, notably in yeast, *Drosophila,* and primates, including humans. In 1975, David Hogness and his colleagues David Finnigan, Gerald Rubin, and Michael Young identified a class of genes in *Drosophila melanogaster,* which they designated as **copia.** These genes transcribe "copious" amounts of RNA (thus, their name). Present up to 30 times in the genome of cells, *copia* genes are nearly identical in nucleotide sequence. Mapping studies show that they are transposable to different chromosomal locations and are dispersed throughout the genome.

Copia appears to be only one of approximately 30 families of transposable elements in *Drosophila,* each of which is present 20 to 50 times in the genome. Together, the many families constitute about 5 percent of the *Drosophila* genome and over half of the middle repetitive DNA of this organism. One estimate projects that 50 percent of all visible mutations in *Drosophila* are the result of the insertion of transposons into otherwise wild type genes!

Despite the variability in DNA sequence between the members of different families, they share a common structural organization thought to be related to the insertion and excision processes of transposition. Each *copia* gene consists of approximately 5000 base pairs of DNA, including a long family-specific **direct terminal repeat (DTR)** sequence of 276 base pairs at each end. Within each repeat is a short **inverted terminal repeat (ITR)** of 17 base pairs. These features are illustrated in Figure 11.17. The DTR sequences are found in other transposons in other organisms but are not universal. However, the shorter ITR sequences are considered universal.

Another set of transposons, the P and I elements, are associated with a phenomenon in *Drosophila* called **hybrid dysgenesis.** When P or I insertions are contributed paternally to an offspring whose mother lacks them, increased levels of sterility, male recombination, chromosomal rearrangements, and mutations may result. The offspring is said to be "dysgenic" and thus frequently fails to reproduce. The transposability of these elements, which induce most of these problems, is under the control of the maternal cytoplasm.

Transposons in Humans

Another class of mobile eukaryotic genetic units is the ***Alu* family.** Found in mammals, including humans, the *Alu* family consists of large numbers of short repetitive sequences, originally detected using reassociation kinetics. In humans, there are about 300,000 copies found interspersed throughout the genome. These sequences represent much of the repetitive DNA in humans and constitute 3 percent of the total genome. The family has received its name because a high percentage of these sequences are cleaved by the restriction endonuclease *Alu I.*

The significant role of the *Alu* family in genetic processes is supported by the following observations:

1 These sequences are found in all primate and rodent DNA.

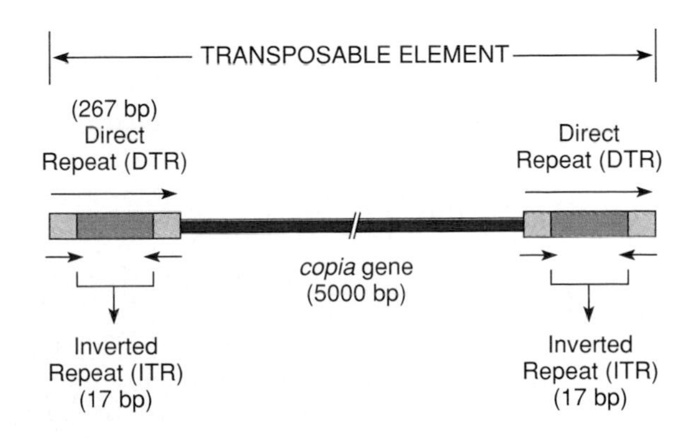

FIGURE 11.17
Structural organization of a *copia* transposable element in *Drosophila melanogaster.*

BARBARA McCLINTOCK
PIONEER GENETICIST

Barbara McClintock was the 1983 Nobel laureate in Medicine and thereby became only the third woman to win an unshared Nobel Prize in science. This award was the culmination of more than fifty years of professional commitment by a scientist of remarkable intellect and dedication.

Born in 1902, Barbara McClintock obtained her B.Sc (1923) and Ph.D. (1927) from Cornell University. Recognition came quickly as a result of her genetic and cytological studies of chromosomes in corn (*Zea mays*). By the early 1930's she was established as a leading cytogeneticist. In 1931, she and Harriet Creighton published a land-mark paper which proved that genetic recombination during meiosis was accompanied by physical exchange of material between chromosomes.

In 1941 she moved to Cold Spring Harbor Laboratory on Long Island, New York, where she worked and lived until her death in 1992. At Cold Spring Harbor she investigated pigment "variegation", the irregular color patches found in individual corn kernels. These patches are the result of somatic mutations occurring at unexpectedly high frequencies and are evidence for particularly unstable genes. In some of the genetically unstable plants, McClintock noted that breakage of a specific chromosome occurred repeatedly.

Painstaking application of classical genetic and cytogenetic techniques established inheritance patterns for the factors responsible and led to an unorthodox conclusion: McClintock proposed that a mobile genetic factor (which she called *Ds* for "dissociation") was responsible. The breaks, she concluded, occurred at the original *Ds* site in a chromosome. But *Ds*, it seemed, could move to other sites, interpose itself into some other gene and thus generate a mutation at the new site. McClintock also deduced that *Ds* was not always actively mobile. A second factor, which she called *Ac* (activator) was needed to make it move. This element, it turned out, was also mobile and can cause mutations.

Deciphering the system not only required remarkable intellectual powers, but considerable academic courage. McClintock reached her conclusions at a time when most researchers believed in the stability of the gene. In her own words, her ideas were met with a mixture of "puzzlement and hostility." As difficult as it was to reach her conclusion, it must have been even harder to sustain a personal belief in it while encountering scientific incomprehension. Although her ideas seemed unorthodox, many of her contemporaries regarded her with enormous respect, for her intellect and dedication were evident to anyone who met her.

In the late 1960's molecular evidence for the existence of mobile ("transposable") genetic elements in lower organisms began to appear. Since then, the occurrence of transposition in many species, including man, has been well documented. In the light of molecular analysis, McClintock's work was "rediscovered". The Nobel prize rewarded her for an astounding feat of deduction, made at a time when the nature of even the conventional gene was only dimly perceived.

Although McClintock's description of transposition is now accepted, other facets of her ideas have yet to be fully assessed. Many biologists believe that transposable elements play no useful role in the organism in which they are found, seeing them as basically parasitic, surviving simply because of an intrinsic ability to replicate.

McClintock, however, suggested that transposition will prove important to normal organisms. It might, she claimed, be a mechanism for controlling cell differentiation during development. Recently, investigators have turned attentions to the evidence on which she based this hypothesis. She noted that the activity of certain elements changed during the development of the plant, suggesting that transposition is a regulated function. In the late 1970's she suggested another possibility, that transposable elements may be important in restructuring the genome in response to stress and have important evolutionary significance.

The jury is still out on these radical ideas; it is by no means certain that new investigations will provide support for McClintock's views. However, the acuity of her judgement on transposition suggests that we should not discard them lightly. Whatever transpires, Barbara McClintock's story provides a marvelous parable of the workings of one side of science; it is a needed reminder that many things of scientific importance make no headlines in the next day's newspapers.

2 There is conserved within the sequences in mammals a specific 40-base-pair unit.

3 These sequences are reflected in nuclear transcripts generated by RNA polymerase II and III.

These sequences are considered transposable based on several lines of evidence. Most important is the fact that they contain a 300-bp sequence flanked on either side by direct repeat sequences consisting of 7 to 20 bp. This observation is similar to that of bacterial insertion sequences. These flanking regions are related to the insertion process during transposition. The second line of evidence involves the observation that clustered regions of *Alu* sequences vary in the DNA of normal and diseased individuals and in different tissues of the same individual. Additionally, the sequences have been found extrachromosomally.

The potential mobility and mutagenic effect of these and other elements have far-reaching effects. A recent example involves a situation where a transposon has been "caught in the act." The case involves a male child with **hemophilia,** as investigated by Haig Kazazian and his colleagues. One cause of hemophilia is a defect in blood-clotting factor VIII, the product of an X-linked gene. Kazazian found two cases where, sitting within this gene, was a transposable element called a **LINE** (a *long, interspersed nucleotide sequence*).

It is estimated that there are 100,000 LINEs of various sequences scattered throughout the human genome. They are much longer than Alu sequences. In the factor VIII case, there has been great interest in determining if one of the mother's X chromosomes also contains this specific LINE. If so, the unaffected mother is heterozygous and passes the LINE-containing chromosome to her son. The startling finding is that the LINE sequence is not present on either of her *X chromosomes,* but *was* detected on *chromosome 22* of both parents. This suggests that this mobile element may have "moved" from one chromosome to another in the gamete-forming cells of one of the parents, prior to being transmitted to the son.

Many questions remain concerning this and other transposable elements. What is their origin? Were they once some sort of retrovirus (see Chapter 14)? Exactly how do they move, and what has been their role during evolution? These and other questions will intrigue researchers for many years to come.

CHAPTER SUMMARY

1 The phenomenon of mutation not only provides the basis for most of the inherent variation present in living organisms, but also serves as the working tool of the geneticist in studying and understanding the nature of genetic processes.

2 Mutations are distinguished by the tissues affected. Somatic mutations are those that may affect the individual, but are not heritable. Mutations arising in gametes may produce new alleles that can be passed on to offspring and can enter the gene pool.

3 Another classification of mutations relies on their effect. Morphological mutations, for example, may be detected visibly. Other types include biochemical, lethal, conditional, and regulatory mutations, groups that are not mutually exclusive.

4 Spontaneous mutations may arise naturally as a result of rare chemical rearrangements of atoms, or tautomeric shifts, and as the result of errors occurring during DNA replication. Although spontaneous mutations are very rare, their rate of occurrence may be increased experimentally by a variety of mutagenic agents.

5 Mutagenic agents such as base analogues and alkylating agents cause chemical changes in nucleotides that alter their base-pairing affinities. As a result, base substitution mutations arise following DNA replication.

6 Frameshift mutations arise when the addition or deletion of one or more nucleotides (but not multiples of three) occurs.

7 Direct analysis of DNA from individuals with specific ABO blood types and muscular dystrophy has been informative. Complete loss of function, as is the

case in blood type O and the Duchenne form of muscular dystrophy, has occurred when deletions or duplications of nucleotides have shifted the reading frame, eventually causing premature termination of translation of the resulting messenger RNA.

8 Ultraviolet light and high-energy radiation from gamma, cosmic, or X ray sources are also potent mutagenic agents. UV light induces the formation of pyrimidine dimers in DNA, while high-energy radiation causes the ionization of molecules in its path and is more penetrating that UV.

9 Various systems of genetically controlled DNA repair have been discovered, including photoreactivation and excision, mismatch, and recombinational repair. Loss of repair function by mutation in humans results in the severe disorder xeroderma pigmentosum.

10 Site-directed mutagenesis is a technique allowing researchers to create specific alterations in the nucleotide sequence of the DNA of genes.

11 Insertion sequences in bacteria and other transposable elements in eukaryotes have a profound effect on genetic expression, thus serving as a distinct category of mutagenic agents. A recent example involves a mutation causing hemophilia in humans.

KEY TERMS

ABO antigens
acridine dyes
Activator (Ac) element
alkylating agents
Alu family
Ames test
attached-X procedure
auxotroph
base analogue
base substitution
behavior mutation
biochemical mutation
complementation
conditional mutation
direct terminal repeat (DTR)
Dissociation (Ds) element
dominant mutation
doubling dose
dystrophin
ethylmethane sulfonate (EMS)
excision repair
5-bromouracil (5 BU)
frameshift mutation
gene mutation
genetic burden
hemophilia
heterokaryon
high-energy radiation
hybrid dysgenesis
induced mutation
insertion sequences
inverted terminal repeat (ITR)
ionizing radiation
lethal mutation
LINE
mismatch repair
muscular dystrophy
nutritional mutation
opaque-2 mutant strain
photoreactivation repair
proflavin
prototroph
recombinational repair
regulatory mutation
rem
reverse mutation
roentgen
sex-linked mutation
site-directed mutagenesis
somatic cell genetics
SOS response
spontaneous mutation
target theory
tautomeric shift
temperature-sensitive mutation
transition mutation
transposable elements
transversion mutation
2-amino purine (2-AP)
ultraviolet radiation (UV)
X rays
xeroderma pigmentosum (XP)

1 How could you isolate a mutant strain of bacterial cells that is resistant to penicillin, an antibiotic that inhibits cell wall synthesis?

SOLUTION: Grow a culture of bacterial cells in liquid medium and plate the cells on agar medium to which penicillin has been added. Only penicillin-resistant cells will reproduce and form colonies. Each colony will, in all likelihood, represent a cloned group of cells with the identical mutation. Isolate members of each colony. To enhance the chance of such a mutation arising, you might want to add a mutagen to the liquid culture.

2 The base analogue 2-amino purine (2-AP) substitutes for adenine during DNA replication, but it may base pair with cytosine. The base analogue 5-bromouracil (5-BU) substitutes for thymidine, but it may base pair with guanine. Follow the double-stranded trinucleotide sequence shown below through three rounds of replication, assuming that in the first round, both analogues are present and become incorporated wherever possible. In the second and third round of replication, they are removed. What final sequences occur?

SOLUTION: The solution to this problem is illustrated below.

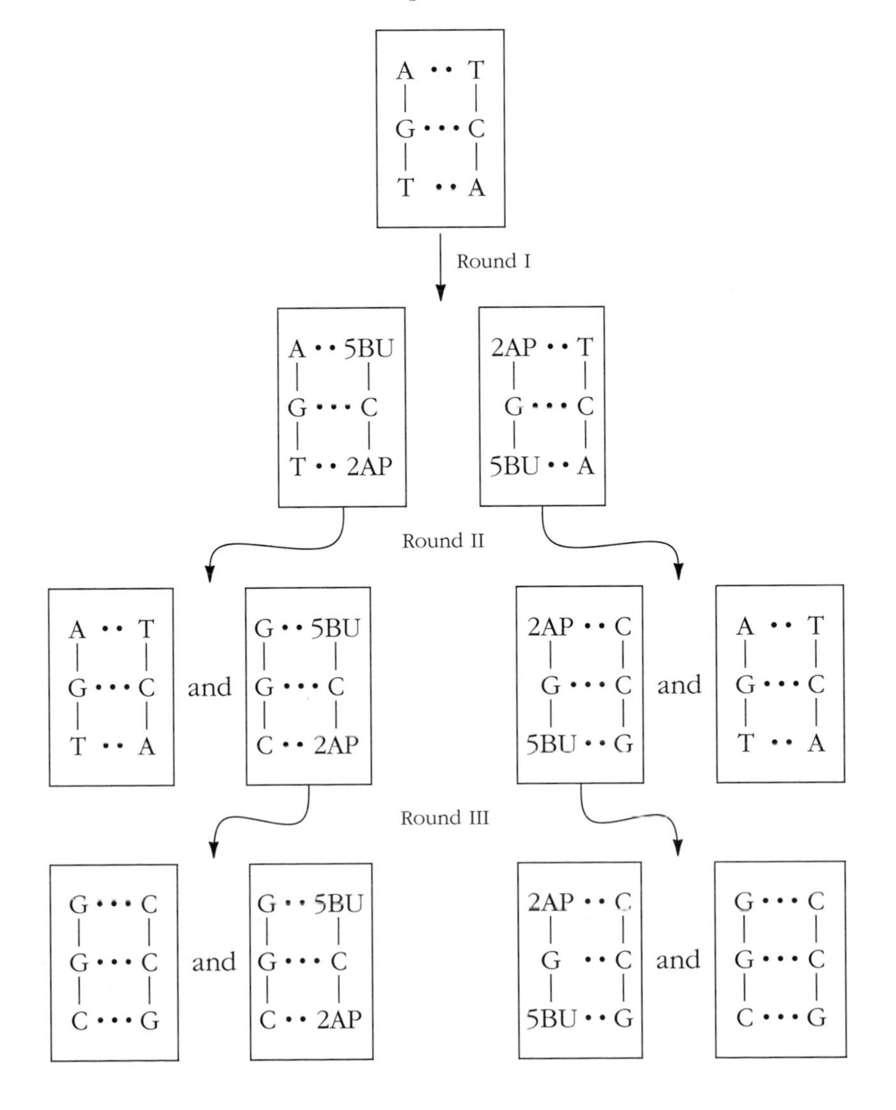

PROBLEMS AND DISCUSSION QUESTIONS

1 What is the difference between a chromosomal mutation and a gene mutation? Between a somatic and a gametic mutation?

2 Why do you suppose that a random mutation is more likely to be deleterious than beneficial?

3 Most mutations in a diploid organism are recessive. Why?

4 In *Drosophila,* induced mutations on chromosome 2 that are recessive lethals may be detected using a second chromosomal stock, *Curly, Lobe/Plum (Cy L/Pm).* These alleles are all dominant and lethal in the homozygous condition. Detection is performed by crossing *Cy L/Pm* females to wild-type males that have been subjected to a mutagen. Three generations are required. In the F_1, *Cy L* males are selected and individually backcrossed to *Cy L/Pm* females. In the F_2 of each cross, flies expressing *Cy* and *L* are mated to produce a series of F_3 generations. Diagram these crosses and predict how an F_3 culture will vary if a recessive lethal was induced in the original test male compared to the case where no lethal mutation resulted. In which F_3 flies would a recessive morphological mutation be expressed?

5 Describe tautomerism and the way in which this chemical event may lead to mutation.

6 Acridine dyes induce frameshift mutations. Is such a mutation likely to be more detrimental than a point mutation where a single pyrimidine or purine has been substituted? If so, why?

7 Contrast the various types of DNA repair mechanisms known to counteract the effects of UV light and other DNA damage. What is the role of visible light in photoreactivation?

8 Mammography is an accurate screening technique for the early detection of breast cancer in humans. Because this technique uses X rays diagnostically, it has been highly controversial. Can you explain why?

9 Presented below are theoretical findings from studies of heterokaryons formed from human xeroderma pigmentosum cell strains:

	XP1	*XP2*	*XP3*	*XP4*	*XP5*	*XP6*	*XP7*
XP1	0						
XP2	0	0					
XP3	0	0	0				
XP4	+	+	+	0			
XP5	+	+	+	+	0		
XP6	+	+	+	+	0	0	
XP7	+	+	+	+	0	0	0

NOTE: + = complementing; 0 = noncomplementing.

These data represent the occurrence of unscheduled DNA synthesis in the fused heterokaryon when neither of the strains alone showed synthesis. What does unscheduled DNA synthesis represent? Which strains fall into the same complementation groups? How many groups are revealed based on these limited data? How do we interpret the presence of these complementation groups?

10 If the human genome contains 100,000 genes, and the mutation rate at each of these loci is 5×10^{-5} per gamete formed, what is the average number of new mutations that exist in each individual? If the current population is 4.3

billion people, how many newly arisen mutations exist in the current populace?

11 Speculate as to how insertion sequences and other transposable elements disrupt genetic expression when inserted into wild type genes.

SELECTED READINGS

AMES, B. N., McCANN, J., and YAMASAKI, E. 1975. Method for detecting carcinogens and mutagens with the *Salmonella*/mammalian microsome mutagenicity test. *Mut. Res:*31:347–64.

AUERBACH, C. 1978. Forty years of mutation research: A pilgrim's progress. *Heredity* 40:177–87.

BEADLE, G. W., and TATUM, E. L. 1945. Neurospora II. Methods of producing and detecting mutations concerned with nutritional requirements. *Amer. J. Bot.* 32:678–86.

BERG, D., and HOWE, M., eds. 1989. *Mobile DNA.* Washington, D.C.: *Amer. Soc. Microbiol.*

CARTER, P. 1986. Site-directed mutagenesis. *Biochem. J.* 237:1–7.

CLEAVER, J. E. 1968. Defective repair replication of DNA in xeroderma pigmentosum. *Nature* 218:652–56.

CLEAVER, J. E., and KARENTZ, D. 1986. DNA repair in man: Regulation by a multiple gene family and its association with human disease. *Bioessays* 6:122–27.

COHEN, S. N., and SHAPIRO, J. A. 1980. Transposable genetic elements. *Scient. Amer.* (Feb.) 242:40–49.

DEERING, R. A. 1962. Ultraviolet radiation and nucleic acids. *Scient. Amer.* (Dec.) 207:135–44.

DEN DUNNEN, J. T., et al. 1989. Topography of the Duchenne muscular dystrophy (DMD) gene. *Am. J. Hum. Genet.* 45:835–47.

DEVORET, R. 1979. Bacterial tests for potential carcinogens. *Scient. Amer.* (Aug.) 241:40–49.

HANAWALT, P. C., and HAYNES, R. H. 1967. The repair of DNA. *Scient. Amer.* (Feb.) 216:36–43.

HASELTINE, W. A., 1983. Ultraviolet light repair and mutagenesis revisited. *Cell* 33:13–17.

HOWARD-FLANDERS, P. 1981. Inducible repair of DNA. *Scient. Amer.* (Nov.) 245:72–80.

KELNER, A. 1951. Revival by light. *Scient. Amer.* (May) 184:22–25.

KNUDSON, A. G. 1979. Our load of mutations and its burden of disease. *Am. J. Hum. Genet.* 31:401–13.

KRAEMER, F. H., et al. 1975. Genetic heterogeneity in xeroderma pigmentosum: Complementation groups and their relationship to DNA repair rates. *Proc. Natl. Acad. Sci.* 72:59–63.

LITTLE, J. W., and MOUNT, D. W. 1982. The SOS regulatory system of *E. coli. Cell* 29:11–22.

MASSIE, R. 1967. *Nicholas and Alexandra.* New York: Atheneum.

MASSIE, R., and MASSIE, S. 1975. *Journey.* New York: Knopf.

McCANN, J., CHOI, E., YAMASAKI, E., and AMES, B. 1975. Detection of carcinogens as mutagens in the *Salmonella*/microsome test: Assay of 300 chemicals. *Proc. Natl. Acad. Sci.* 72:5135–39.

McKUSICK, V. A. 1965. The royal hemophilia. *Scient. Amer.* (Aug.) 213:88–95.

MULLER, H. J. 1927. Artificial transmutation of the gene. *Science* 66:84–87.

——— 1955. Radiation and human mutation. *Scient. Amer.* (Nov.) 193:58–68.

OSSANA, N., PETERSON, K. R., and MOUNT, D. W. 1986. Genetics of DNA repair in bacteria. *Trends Genet.* 2:55–58.

RADMAN, M., and WAGNER, R. 1988. The high fidelity of DNA duplication. *Scient. Amer.* (August) 259(2):40–46.

SCHULL, W. J., OTAKE, M., and NEEL, J.V. 1981. Genetic effects of the atomic bombs: A reappraisal. *Science* 213:1220–27.

SHORTLE, D., DiMARIO, D., and NATHANS, D. 1981. Directed mutagenesis. *Ann. Rev. Genet.* 15:265–94.

SIGURBJORHSSON, B. 1971. Induced mutations in plants. *Scient. Amer.* (Jan.) 224:86–95.

VOGEL, F. 1992. Risk calculations for hereditary effects of ionizing radiation in humans. *Hum. Genet.* 89:127–46.

WILLS, C. 1970. Genetic load. *Scient. Amer.* (March) 222:98–107.

YAMAMOTO, F., et al. 1990. Molecular genetic basis of the histo-blood group ABO system. *Nature* 345:229–33.

YOSHIMOTO, Y. 1990. Cancer risk among children of atomic bomb survivors. *JAMA* 264:596–600.

Organization of Genes and Chromosomes

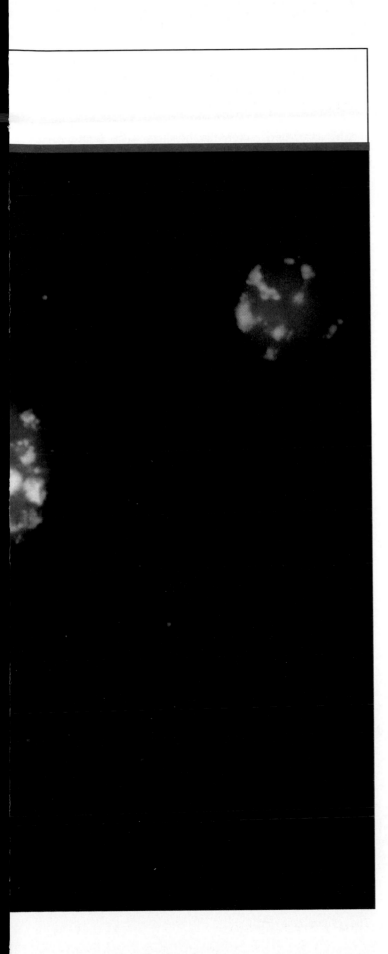

Chapter Concepts

DNA is organized in a manner consistent with the complexity of the host structure with which it is associated. Viruses, bacteria, mitochondria, and chloroplasts contain a shorter, often circular, DNA molecule relatively free of proteins. Eukaryotic cells contain greater amounts of DNA organized into nucleosomes and present as chromatin fibers. This increase in complexity is related to the larger amount of genetic information present as well as the greater complexity associated with their genetic function. Studies designed to elucidate the organization of genes reveal diverse forms of organization, ranging from single copies of genes to large numbers of tandem repeats of families of related genes. A most striking feature is the degree to which noncoding DNA is associated with the organization of genes throughout the eukaryotic genome.

Once it was understood that DNA is the genetic material, it was very important to determine the way in which DNA is organized into chromosomes and how genes are organized within these genetic structures. There has been much interest in these topics because the determination of the spatial arrangement of the genetic material and associated molecules will undoubtedly provide valuable insights into other aspects of genetics. For example, how the genetic information is stored, expressed, and regulated must be related to the organization of the genetic molecule, DNA. In eukaryotes, how the chromatin fibers characteristic of interphase are condensed into chromosome structures visible during mitosis and meiosis is also of great interest.

In this chapter, we first provide a survey of the various forms of chromosome organization, including examples from viruses, bacteria, and various eukaryotes. The genetic material has been studied using numerous approaches, including molecular analysis and direct visualization by light and electron microscopy.

Then, we shall narrow our focus to the level of the gene. While genes in phage and bacteria are relatively basic, studies of eukaryotes have revealed an increased level of complexity. Phage and bacterial genes consist of a linear series of nucleotides, most of which encode genetic information. In comparison, a large portion of many eukaryotic genes consists of noncoding intron and flanking sequences that do not become part of the final mRNA transcript. Further, we now know that a large portion of the total DNA comprising the haploid genome consists of noncoding areas between genes, some of which are occupied by repetitive DNA sequences. Beyond this information, we are learning much more about the organization of genes within the genome. We will review some of these topics and expand upon them with particular emphasis on the organization of multigene families. These topics represent some of the most exciting and unexpected discoveries made in the field of genetics.

VIRAL AND BACTERIAL CHROMOSOMES

In comparison with eukaryotes, the chromosomes of viruses and bacteria are much less complicated. They usually consist of a single nucleic acid molecule, largely devoid of associated proteins. Contained within the single chromosome of viruses and bacteria is much less genetic information than in the multiple chromosomes comprising the genome of higher forms. These characteristics have greatly simplified analysis, which provides a fairly comprehensive view of viral and bacterial chromosomes.

The chromosomes of viruses consist of a nucleic acid molecule—either DNA or RNA—which is single- or double stranded. They may exist as circular structures (closed loops), or they may take the form of linear molecules. The single-stranded DNA of the ϕX174 bacteriophage and the double-stranded DNA of the polyoma virus are ring-shaped molecules within the protein coat of the mature virus and within the host cell. The **bacteriophage lambda (λ),** on the other hand, possesses a linear double-stranded DNA molecule prior to infection, which closes to form a ring upon infection of the host cell. Still other viruses, such as the **T-even series of bacteriophages,** have linear, double-stranded chromosomes of DNA, which do not form circles inside the bacterial host. Thus, circularity is not an absolute requirement for replication in some viruses.

Viral nucleic acid molecules have been visualized with the electron microscope. Figure 12.1 shows a mature bacteriophage lambda with its double-stranded DNA molecule in the circular configuration. One constant feature shared by viruses, bacteria, and eukaryotic cells is the ability to package an exceedingly long DNA molecule into a relatively small volume. In λ, the DNA is 17 μm long and must fit into the phage head, which is less than 0.1 μm on any side. As you examine Figure 12.1, note that the phage is magnified six times more than the DNA molecule.

In Table 12.1, the measured lengths of the genetic molecules are compared with the size of the head in several viruses. In each case, a similar packaging feat must be accomplished. The dimensions given for phage T2 may be compared with the micrograph of both the DNA and viral particle shown in Figure 12.2. Seldom does the space available in the head of a virus exceed the chromosome volume by more than a factor of two. In many cases,

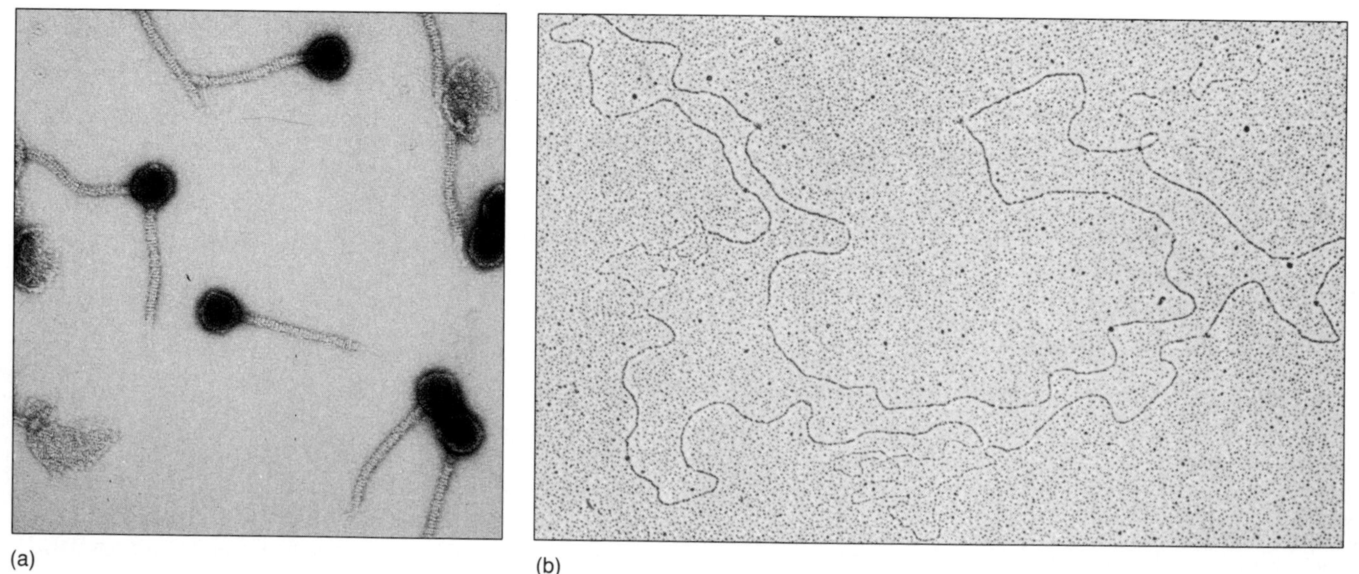

FIGURE 12.1
(a) Electron micrograph of phage λ and (b) the DNA isolated from it. The chromosome is 17 μm long.

TABLE 12.1
The genetic material of representative viruses and bacteria.

Organism	Nucleic Acid			Overall Size of Viral Head or Bacteria (μm)
	Type	SS or DS	Length (μm)	
VIRUSES				
ϕX174	DNA	SS	2.0	0.025 × 0.025
Tobacco mosaic virus	RNA	SS	3.3	0.30 × 0.02
Lambda phage	DNA	DS	17.0	0.07 × 0.07
T2 phage	DNA	DS	52.0	0.07 × 0.10
BACTERIA				
Hemophilus influenzae	DNA	DS	832.0	1.00 × 0.30
Escherichia coli	DNA	DS	1200.0	2.00 × 0.50

SS = single standard; DS = double stranded.

almost all space is filled, indicating nearly perfect packing. Once packed within the head, the genetic material is functionally inert until released into a host cell.

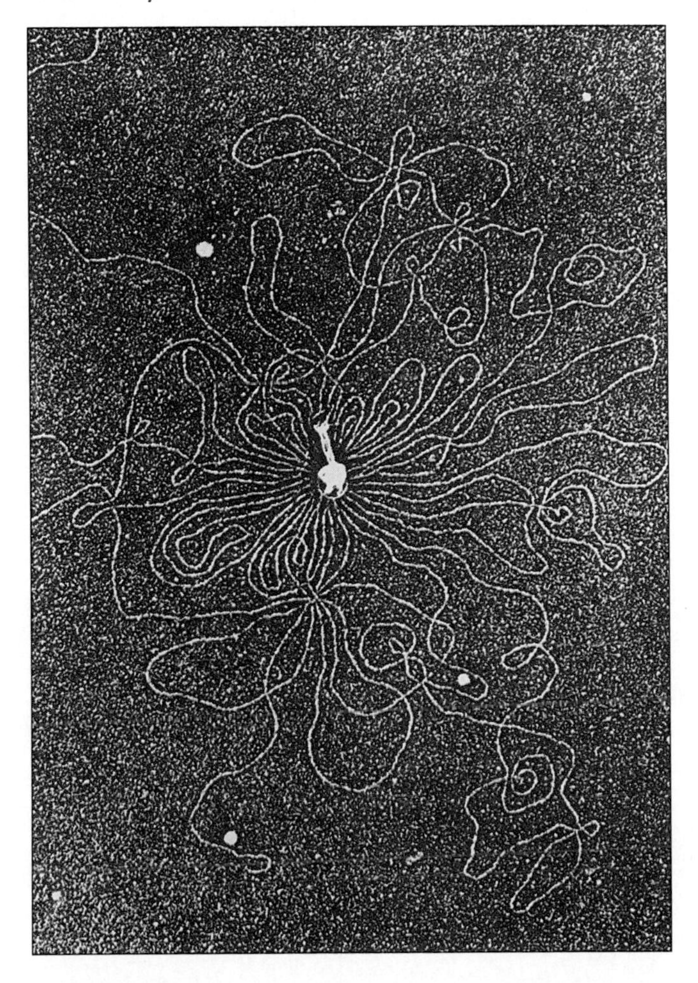

FIGURE 12.2
Electron micrograph of bacteriophage T2, which has had its DNA released by osmotic shock. The chromosome is 52 μm long.

Bacterial chromosomes are also relatively simple in form compared with those of eukaryotic cells. They always consist of a double-stranded DNA molecule, compacted into a structure sometimes referred to as the **nucleoid.** *Escherichia coli,* the most extensively studied bacterium, has a large, circular chromosome, measuring approximately 1200 μm (1.2 mm) in length. When the cell is gently lysed and the chromosome released, it can be visualized under the electron microscope (Figure 12.3).

This DNA is found to be associated with several types of **DNA-binding proteins,** including those called **HU** and **H.** They are small but abundant in the cell and contain a high percentage of positively charged amino acids that can bond ionically to the negative charges of the phosphate groups in DNA. As we will soon see, these proteins resemble structurally similar molecules called **histones** that are found associated with eukaryotic DNA. Unlike the tightly packed chromosome of a virus, the bacterial chromosome is not functionally inert. In spite of the compacted condition of the bacterial chromosome, replication and transcription readily occur.

MITOCHONDRIAL AND CHLOROPLAST DNA

Numerous observations have demonstrated that both **mitochondria** and **chloroplasts** contain their own genetic information. Most convincing was the discovery of mutations in yeast, other fungi, and plants that were shown to alter the function of these organelles. Furthermore, transmission of these mutations did not demonstrate the biparental inheritance patterns characteristic of nuclear genes. Instead, a **maternal mode of inheritance** was observed, with traits being passed to offspring only from their mother. Since the origin of both mitochondria and

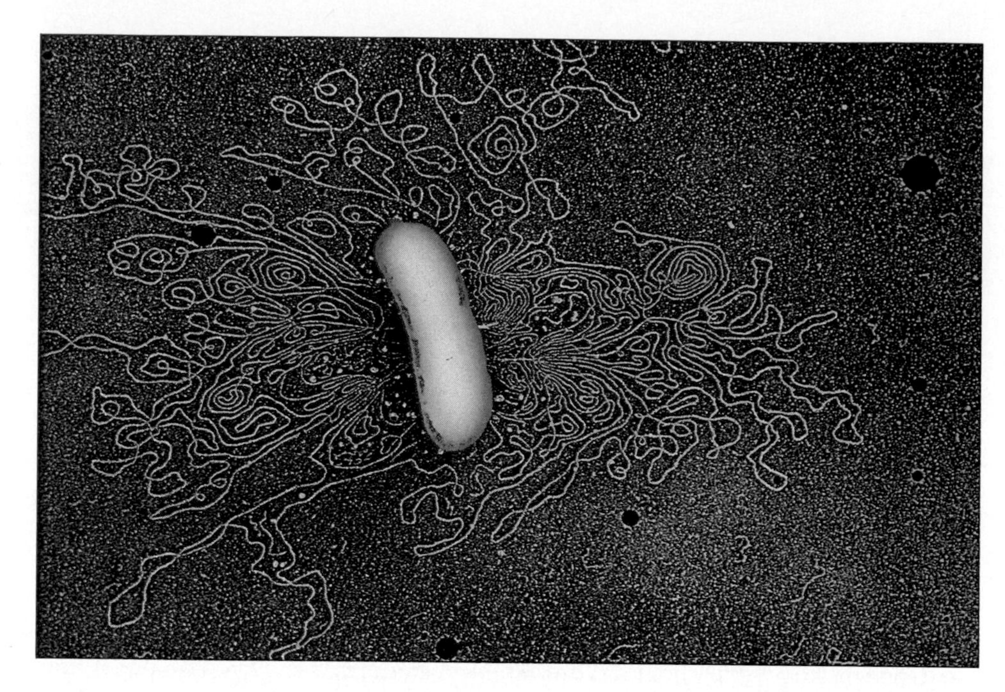

FIGURE 12.3
Electron micrograph of the bacterium *Escherichia coli,* which has had its DNA released by osmotic shock. The chromosome is 1200 μm long.

chloroplasts is also maternal, these observations suggest that these organelles house their own DNA, separate from the nucleus, that influences their function.

Thus, geneticists set out to look for more direct evidence of DNA in these organelles. Electron microscopists not only documented the presence of DNA in both organelles, but they also saw DNA in a form quite unlike that seen in the chromosomes of eukaryotic cells. Instead, this DNA looked remarkably similar to that seen in viruses and bacteria! As we shall see, this similarity is thought to be related to the idea that both mitochondria and chloroplasts were once primitive, free-living, bacterialike organisms.

Molecular Organization and Function of Mitochondrial DNA

Extensive information is now available about the molecular aspects of **mitochondrial DNA (mtDNA)** and related gene function. In most eukaryotes, mtDNA (Figure 12.4a) is a circular duplex that replicates semiconservatively and is free of the chromosomal proteins characteristic of eukaryotic DNA. In size, mtDNA differs among organisms. This can be seen by examining the information presented in Table 12.2. In a variety of animals, mtDNA consists of about 16 to 18 kb. In vertebrates, there are 5 to 10 molecules per organelle. There is a considerably greater amount of DNA present in plant mitochondria, where 100 kb is not unusual.

Several general statements can now be made concerning mtDNA. There appear to be few or no gene repetitions, and replication is dependent upon enzymes encoded by nuclear DNA. Genes have been identified that code for the ribosomal RNAs, over 20 tRNAs, and numerous products essential to the cellular respiratory functions of the organelles.

The protein-synthesizing apparatus as well as the molecular components for cellular respiration are jointly derived from nuclear and mitochondrial DNA. Ribosomes found in the organelle are different from those present in the cytoplasm. The sedimentation coefficient of these particles varies among different species. The data in Table 12.3 show that mitochondrial ribosomes vary considerably in their coefficients (55*S* to 60*S* in vertebrates, to 70*S* in some algae and fungi, 80*S* in certain protozoans and fungi).

Many nuclear-coded gene products are essential to biological activity within mitochondria: DNA and RNA polymerases, initiation and elongation factors essential for translation, ribosomal proteins, aminoacyl tRNA synthetases, and some tRNA species. As in chloroplasts, these imported components are generally regarded as distinct from their cytoplasmic counterparts, even though both sets are coded by nuclear genes. For example, the synthe-

TABLE 12.2
The size of mtDNA in different organisms.

Organism	Size in Kilobases
Human	16.6
Mouse	16.2
Xenopus (frog)	18.4
Drosophila (fruit fly)	18.4
Saccharomyces (yeast)	84.0
Pisum sativum (pea)	110.0

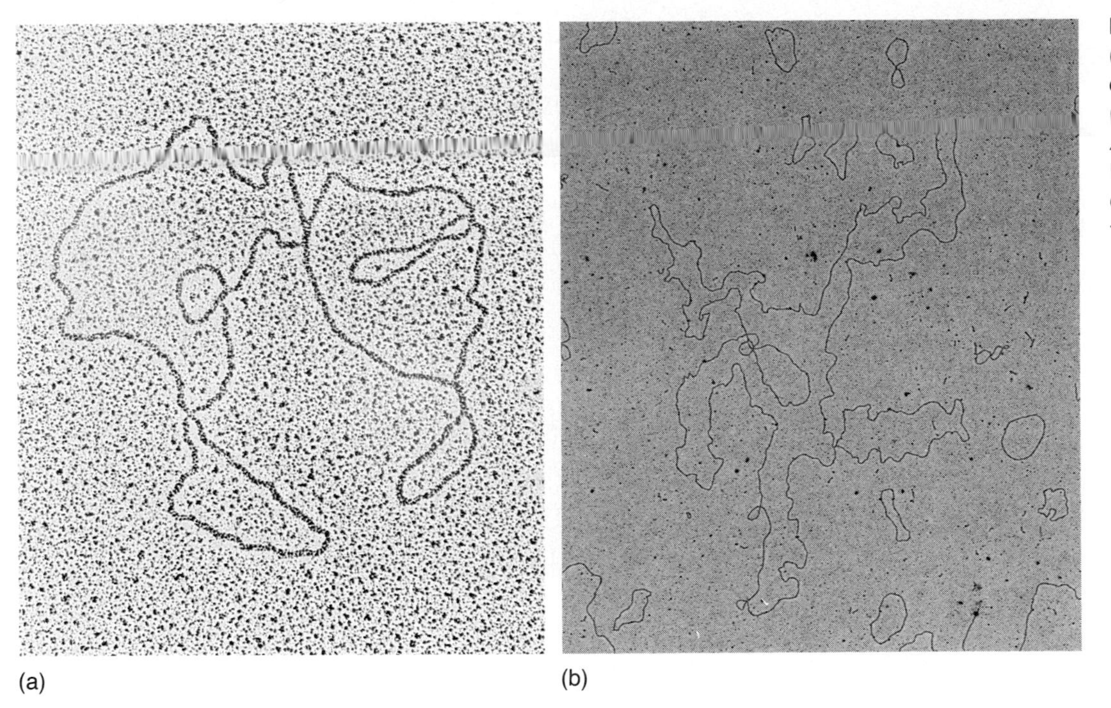

FIGURE 12.4
(a) Electron micrograph of mitochondrial DNA (mtDNA) derived from *Xenopus laevis.* (b) Electron micrograph of chloroplast DNA derived from corn.

tases essential to charging tRNA molecules (a process essential to translation) show a distinct affinity for the mitochondrial tRNA species as compared with the cytoplasmic tRNAs. Similar affinity has been shown for the initiation and elongation factors. Furthermore, while bacterial and nuclear RNA polymerases are known to be composed of numerous subunits, the mitochondrial variety consists of only one polypeptide chain. This polymerase is generally susceptible to antibiotics that inhibit bacterial RNA synthesis but not to eukaryotic inhibitors. The relative contributions of nuclear and mitochondrial gene products are illustrated in Figure 12.5.

The above information, along with that about to be presented concerning chloroplast DNA function serve as the underlying basis for understanding modes of inheritance that are "extrachromosomal." Chapter 13 is devoted to this broad topic.

TABLE 12.3
Sedimentation coefficients of mitochondrial ribosomes.

Kingdom	Examples	Sedimentation Coefficient (*S*)
Animals	Vertebrates	55–60
	Insects	60–71
Protists	*Euglena*	71
	Tetrahymena	80
Fungi	*Neurospora*	73–80
	Saccharomyces	72–80
Plants	Maize	77

Molecular Organization and Function of Chloroplast DNA

There is now available a substantial amount of molecular information about **chloroplast DNA (cpDNA)** and about the genetic function of this organelle. The chloroplast, like the mitochondrion, contains an autonomous genetic system distinct from that found in the nucleus and cytoplasm. This system includes DNA as a source of genetic information and a complete protein-synthesizing apparatus. However, the molecular components of the translation apparatus are jointly derived from both nuclear and chloroplast genetic information. As seen in Figure 12.4(b), chloroplast DNA is much larger than mitochondrial DNA, but it is nevertheless very similar to that found in prokaryotic cells.

DNA isolated from chloroplasts is found to be circular, double stranded, replicated semiconservatively, and free of the associated proteins characteristic of eukaryotic DNA. Compared with nuclear DNA of the same organism, it invariably shows a different buoyant density and base composition. Furthermore, cpDNA lacks 5-methylcytosine, a modified base present in nuclear DNA of many plants.

In *Chlamydomonas,* there are about 75 copies of the chloroplast DNA molecule per organelle. Each copy consists of a length of DNA that contains 195,000 bases (195 kb), almost twice the size of a T-even bacteriophage genome. In higher plants such as the sweet pea, multiple copies of the DNA molecule are present in each organelle, but the molecule is considerably smaller than that in *Chlamydomonas,* consisting of 134 kb.

Some chloroplast gene products function to synthesize

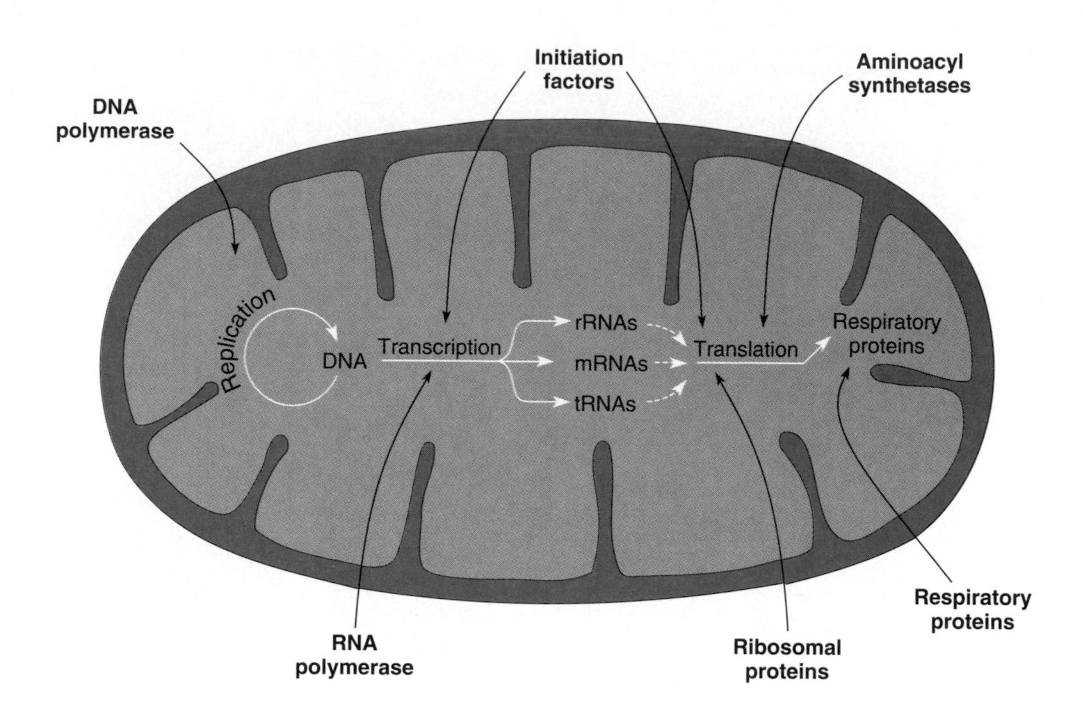

FIGURE 12.5
A comparison of the origin of gene products that are essential to mitochondrial function. Those shown entering the organelle are derived from the cytoplasm and encoded by the nucleus.

proteins. In a variety of higher plants (beans, lettuce, spinach, maize, and oats), two sets of the genes coding for the ribosomal RNAs—5*S*, 16*S*, and 23*S* rRNA are present. Additionally, chloroplast DNA codes for at least 25 tRNA species and a number of ribosomal proteins specific to the chloroplast ribosomes. These ribosomes have a sedimentation coefficient slightly less than 70*S*, similar to that of bacteria.

Chloroplast ribosomes are sensitive to the same protein-synthesis-inhibiting antibiotics as bacterial ribosomes: chloramphenicol, erythromycin, streptomycin, and spectinomycin. Even though ribosomal proteins are derived from nuclear and chloroplast DNA, most, if not all, are distinct from the equivalent proteins of cytoplasmic ribosomes.

Still other chloroplast genes have been identified that are specific to photosynthetic function. Mutations in these genes may have the effect of inactivating photosynthesis in those chloroplasts bearing such a mutation. One of the major chloroplast gene products is the large subunit of the photosynthetic enzyme **ribulose-1-5-biphosphate carboxylase (RuBP).*** Interestingly, the small subunit of this enzyme is encoded by a nuclear gene, while the large subunit, like all other chloroplast-derived gene products, is synthesized and remains within the organelle.

Based on the observation that mitochondrial and chloroplast DNA and their genetic apparatus are similar to their counterparts in bacteria, a theory of the origin of these organelles has been proposed. Championed by Lynn Margulis and others, this hypothesis, called the **endosymbiont theory,** states that mitochondria and chloroplasts may have originated as distinct bacterialike particles that became incorporated into primitive eukaryotic cells. In the evolution of this symbiotic relationship, the particles lost their ability to function independently and underwent an extensive evolution as the eukaryotic host cell became dependent on them. While there are many unanswered questions, the basic tenets of this theory are difficult to dispute.

ORGANIZATION OF DNA IN CHROMATIN

The structure and organization of the genetic material in **eukaryotic cells** is much more intricate than in viruses or bacteria. This complexity is due to the greater amount of DNA per chromosome and the presence of large numbers of proteins associated with DNA in eukaryotes. For example, while DNA in the *E. coli* chromosome is 1200 μm long, the DNA in human chromosomes ranges from 14,000 to 73,000 μm in length. In a single human nucleus, all 46 chromosomes contain sufficient DNA to extend almost 2 meters! This genetic material, along with its associated proteins, is contained within a nucleus that usually measures about 5 μm in diameter.

For many reasons this intricacy is to be expected. It parallels the structural and biochemical diversity of the many types of eukaryotic cells present in a multicellular organism. Different cells assume specific functions by division of labor. It is assumed that any functional specialization is based upon highly specific biochemical activity under the direction of specific genetic information. A highly ordered regulatory system governing the readout

*The enzyme is also named ribulose-1-5-diphosphate carboxylase.

of this information must exist if dissimilar cells performing different functions are present. Such a system must in some way be imposed on or related to the molecular structure of the genetic material.

While bacteria can reproduce themselves, they never exhibit a detailed process similar to mitosis. As described in Chapter 2, eukaryotic cells exhibit a highly organized cell cycle. During interphase, the genetic material is dispersed finely throughout the nucleus as chromatin. When mitosis begins, the chromatin condenses greatly, and during prophase it is compressed into recognizable chromosomes. This condensation represents a contraction in length of some 10,000 times for each chromatin fiber and is the basis of the **folded fiber model** of chromosome structure. This highly regular condensation-uncoiling cycle poses special organizational problems in eukaryotic genetic material.

Early studies of the structure of eukaryotic genetic material concentrated on intact chromosomes, preferably large ones, because of the limitation of light microscopy. Discussion of two such structures—polytene and lampbrush chromosomes—was presented in Chapter 2. Subsequently, the development of techniques for biochemical analysis, as well as the examination of relatively intact eukaryotic chromatin and mitotic figures under the electron microscope, have greatly enhanced our understanding of chromosome structure.

As established earlier, the genetic material of viruses and bacteria consists of essentially naked strands of DNA or RNA. In eukaryotes, the structure of the chromosome is much more complex. A substantial amount of protein is associated with the chromosomal DNA in all phases of the eukaryotic cell cycle. Such material is referred to as **chromatin** during interphase, when it is uncoiled. The associated proteins are divided into basic, positively charged **histones** and less positively charged **nonhistones.**

Nucleosome Structure

The general model for chromatin structure is based on the assumption that chromatin fibers, composed of DNA and protein, undergo extensive coiling and folding as they are condensed within the cell nucleus. Of the proteins associated with DNA, the histones clearly play the most essential structural role. Histones contain large amounts of the positively charged amino acids lysine and arginine, making it possible for them to bond electrostatically to the negatively charged phosphate groups of nucleotides. Recall that a similar interaction has been proposed for several bacterial proteins. There are five main types of histones (Table 12.4).

X ray diffraction studies confirm that histones play an important role in chromatin structure. Chromatin produces regularly spaced diffraction rings, suggesting that repeating structural units occur along the chromatin axis.

TABLE 12.4
Categories and properties of histone proteins.

Histone Type	Lysine-Arginine Content	Molecular Weight (daltons)
H1	Lysine-rich	21,000
H2A	Slightly lysine-rich	14,500
H2B	Slightly lysine-rich	13,700
H3	Arginine-rich	15,300
H4	Arginine-rich	11,300

If the histone molecules are chemically removed from chromatin, the regularity of this diffraction pattern is disrupted.

The basic model for chromatin structure was worked out in the mid-1970s. Several observations are relevant to the development of this model:

1. Digestion of chromatin by certain endonucleases, such as micrococcal nuclease, yields DNA fragments that are approximately 200 base pairs in length, or multiples thereof. This demonstrates that enzymatic digestion is not random, for if it were, we would expect a wide range of fragment sizes. Thus, chromatin consists of some type of repeating unit, each of which is protected from enzymatic cleavage, except where any two units are joined. It is the area between all units that is attacked and cleaved by the nuclease.

2. Electron microscopic observations of chromatin have revealed that chromatin fibers are composed of linear arrays of spherical particles (Figure 12.6). Discovered by Ada and Donald Olins, the particles occur regularly along the axis of a chromatin strand and resemble beads on a string. These particles are now referred to as **ν-bodies** or **nucleosomes** (ν is the Greek letter nu). These findings conform nicely to the earlier proposal, which suggests the existence of repeating units.

3. Results of the study of precise interactions of histone molecules and DNA in the nucleosomes constituting chromatin show that histones H2A, H2B, H3, and H4 occur as two types of tetramers: $(H2A)_2 \bullet (H2B)_2$, and $(H3)_2 \bullet (H4)_2$. This suggests that each repeating nucleosome unit consists of one of each tetramer. This octamer interacts with about 200 base pairs of DNA. These data correlate well with previous observations and provide the basis for a model that explains the interaction of histone and DNA in chromatin.

4. When nuclease digestion time is extended, DNA is removed from both the entering and exiting strands, creating a **nucleosome core particle** consisting of 146 base pairs. This number is consistent in all organ-

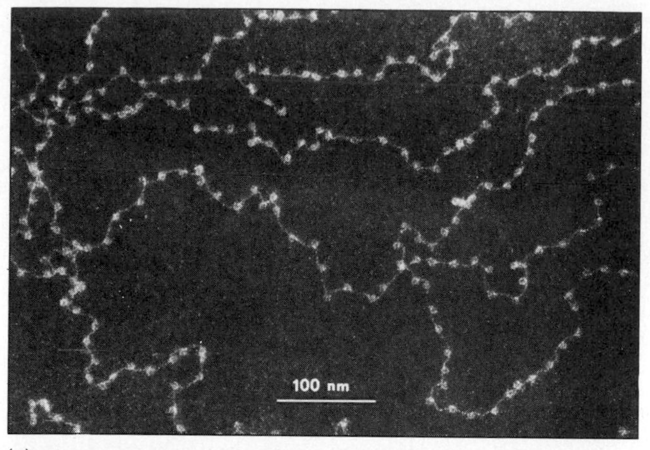

(a)

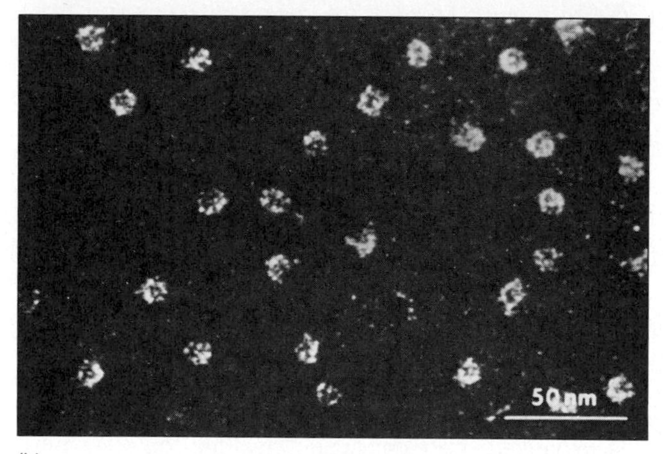

(b)

FIGURE 12.6
(a) Dark-field electron micrograph of nucleosomes present in chromatin derived from a chicken erythrocyte nucleus. (b) Dark-field electron micrograph of nucleosomes produced by micrococcal nuclease digestion.

isms studied. The DNA lost in this prolonged digestion is responsible for linking nucleosomes together. This **linker DNA** is associated with histone H1.

5 On the basis of the above information as well as X ray and neutron scattering analysis of crystallized core particles by John T. Finch, Aaron Klug, and others, a detailed model of the nucleosome was put forward (Figure 12.7a). In this model, the 146 base-pair DNA core exists as a secondary helix surrounding the octamer of histones. The coiled DNA does not quite complete two full turns. The entire nucleosome is ellipsoidal in shape, measuring about 100Å in its greatest dimension.

The extensive investigation of nucleosomes now provides the basis for predicting how the packaging of chromatin within the nucleus occurs. In its most extended state under the electron microscope, the chromatin fiber is about 100 Å in diameter, a size consistent with the longer dimension of the ellipsoidal nucleosome. Thus, it is believed that chromatin fibers consist of long strings of repeating nucleosomes (Figure 12.7b).

The 100-Å chromatin fiber is apparently further folded into a structure 300 Å in diameter, as viewed under the electron microscope. This transition has been studied carefully and can be accounted for in our nucleosome model (Figure 12.7c). The 300-Å fiber appears to consist of 5 or 6 nucleosomes coiled closely together.

In the overall transition from a fully extended chromatin fiber to the extremely condensed status of the mitotic chromosome, a packing ratio (the ratio of DNA length to the length of the structure containing it) of about 500 must be achieved. Our model accounts for a ratio of about 50. Obviously, the larger fiber can be further bent, coiled, and packed as even greater condensation occurs during the formation of a mitotic chromosome.

Heterochromatin

All recent evidence supports the concept that the DNA of each eukaryotic chromosome consists of one continuous double-helical fiber along its entire length. A continuous fiber is the basis of the **unineme model of DNA** within a chromosome. This finding, along with our knowledge of nucleosomes, might lead you to suspect that the chromosome would demonstrate structural uniformity along its length. However, in the early part of this century, it was observed that some parts of the chromosome remain condensed and stain deeply during interphase, but most do not. In 1928, the terms **heterochromatin** and **euchromatin** were coined to describe the parts of chromosomes that remain condensed and those that are uncoiled, respectively.

Subsequent investigation has revealed a number of characteristics of heterochromatin that distinguish it from euchromatin. Heterochromatic areas are genetically inactive either because they lack genes or contain genes that are repressed. Also, heterochromatin replicates later during the S phase of the cell cycle than does euchromatin. The discovery of heterochromatin provided the first clues that parts of eukaryotic chromosomes are not related to storage of genetic information. Instead, some chromosome regions are clearly involved in the maintenance of the chromosome's structural integrity.

Heterochromatin is characteristic of the genetic material of eukaryotes. Early cytological studies showed that areas of the centromeres are composed of heterochromatin. The ends of chromosomes, called telomeres, are also heterochromatic. In some cases, whole chromosomes are heterochromatic. Such is the case with the mammalian Y chromosome, which for the most part is genetically inert. And, as we discussed in Chapter 6, the inactivated X chromosome in mammalian females is condensed into an inert Barr body. In some species, such as Mealy bugs, all chromosomes of one entire haploid set are heterochromatic.

When certain heterochromatic areas from one chro-

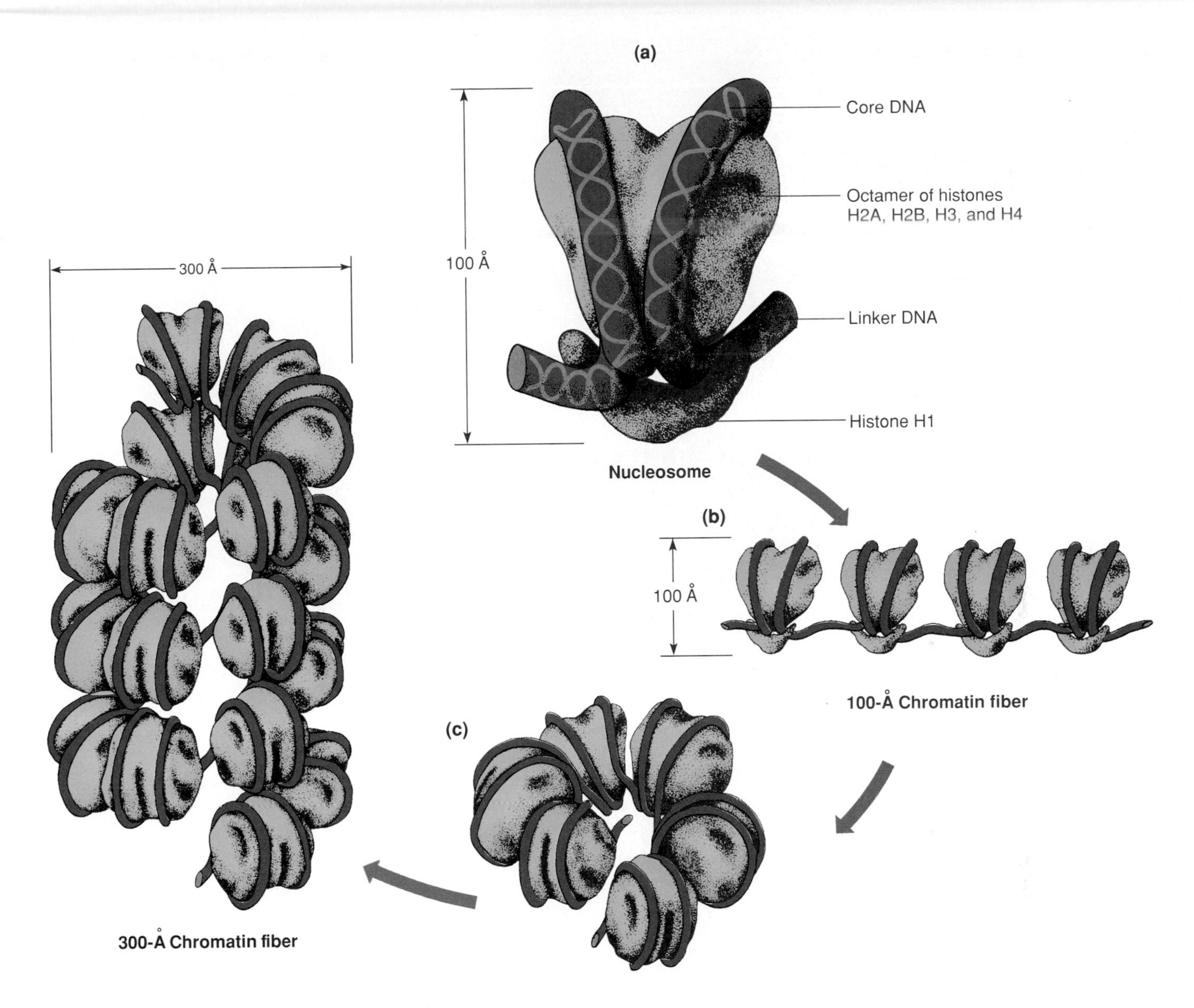

FIGURE 12.7
(a) General model of the association of histones and DNA in the nucleosome. (b) and (c) Schematic illustration of the way in which the chromatin fiber may be coiled into a more condensed structure.

mosome are translocated to a new site on the same or another nonhomologous chromosome, genetically active areas sometimes become genetically inert if they now lie adjacent to the translocated heterochromatin. This influence on existing euchromatin is one example of what is more generally referred to as a **position effect.** That is, the position of a gene or groups of genes relative to all other genetic material may affect their expression. We first introduced this term in Chapter 6.

Chromosome Banding

Until about 1970, mitotic chromosomes viewed under the light microscope could be distinguished only by their relative sizes and the positions of their centromeres. Even in organisms with a low haploid number, two or more chromosomes are often indistinguishable from one another. However, around 1970, differential staining along the longitudinal axis of mitotic chromosomes was made possible by new cytological procedures. These methods are now called **chromosome banding techniques** because the staining patterns resemble the bands of polytene chromosomes.

One of the first chromosome banding techniques was devised by Mary Lou Pardue and Joe Gall. They found that if chromosome preparations were heat denatured and then treated with Giemsa stain, a unique staining pattern emerged. The centromeric regions of mitotic chromosomes preferentially took up the stain! Thus, this cytological technique stained a specific area of the chromosome composed of heterochromatin. A diagram of the mouse karyotype treated in this way is shown in Figure 12.8. The

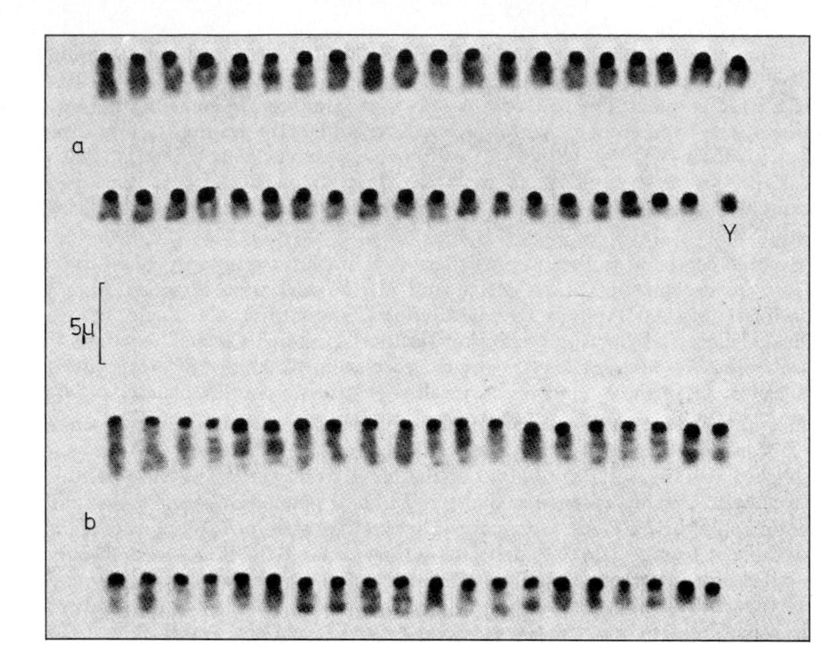

FIGURE 12.8
Karyotypes of a male (a) and female (b) mouse where chromosome preparations were processed to demonstrate C-bands, where only the centromeres stain.

staining pattern is referred to as **C-banding.**

Still other chromosome banding techniques were developed about the same time. A group of Swedish researchers, led by Tobjorn Caspersson, used a technique that provided even greater staining differentiation of metaphase chromosomes. They used fluorescent dyes that bind to nucleoprotein complexes and produce unique banding patterns. When the chromosomes are treated with the fluorochrome quinacrine mustard and viewed under a fluorescent microscope, precise patterns of differential brightness are seen. Each of the 23 human chromosome pairs can be distinguished by this technique. The bands produced by this method are called **Q-bands.**

Another banding technique, produces a staining pattern nearly identical to the Q-bands. This method, producing **G-bands** (Figure 12.9), involves the digestion of the mitotic chromosomes in the cytological preparations with the proteolytic enzyme trypsin followed by Giemsa staining. Another technique results in the reverse G-band staining pattern, called an **R-band** pattern. In 1971, a meeting was held in Paris to establish the nomenclature for these various patterns (Figure 12.10). Still other banding techniques are now available.

Intense efforts are currently underway to elucidate the molecular mechanisms involved in producing these banding patterns. The variety of staining reactions under different conditions reflects the heterogeneity and complexity of chromosome composition.

Centromeres and Telomeres

Linear chromosomes of most eukaryotes contain two regions that have essential structural roles. The first is the **centromere,** the region of the chromosome that attaches to one or more spindle fibers and that is pulled to one of the poles during the anaphase stages of mitosis and meiosis.

Separation of chromatids is essential to the fidelity of chromosome distribution during mitosis and meiosis.

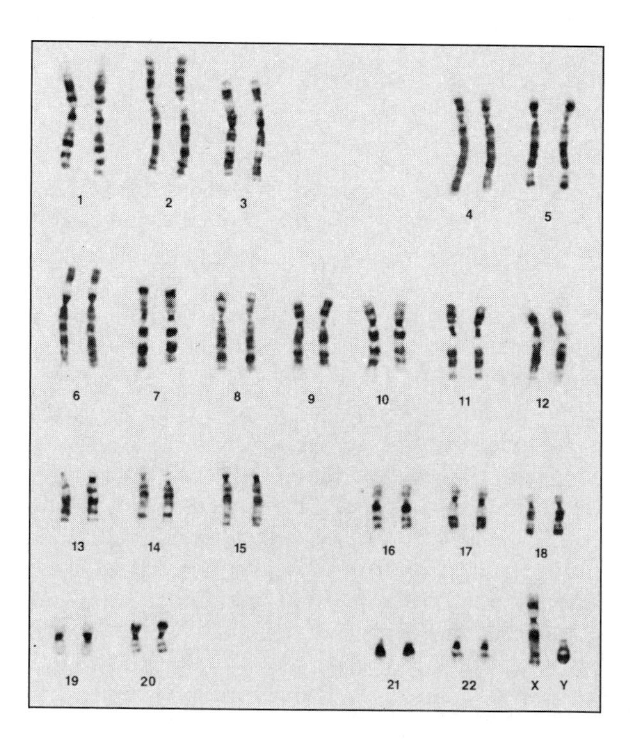

FIGURE 12.9
G-banded karyotype of a normal human male showing approximately 400 bands per haploid set of chromosomes. Chromosomes were derived from cells in metaphase.

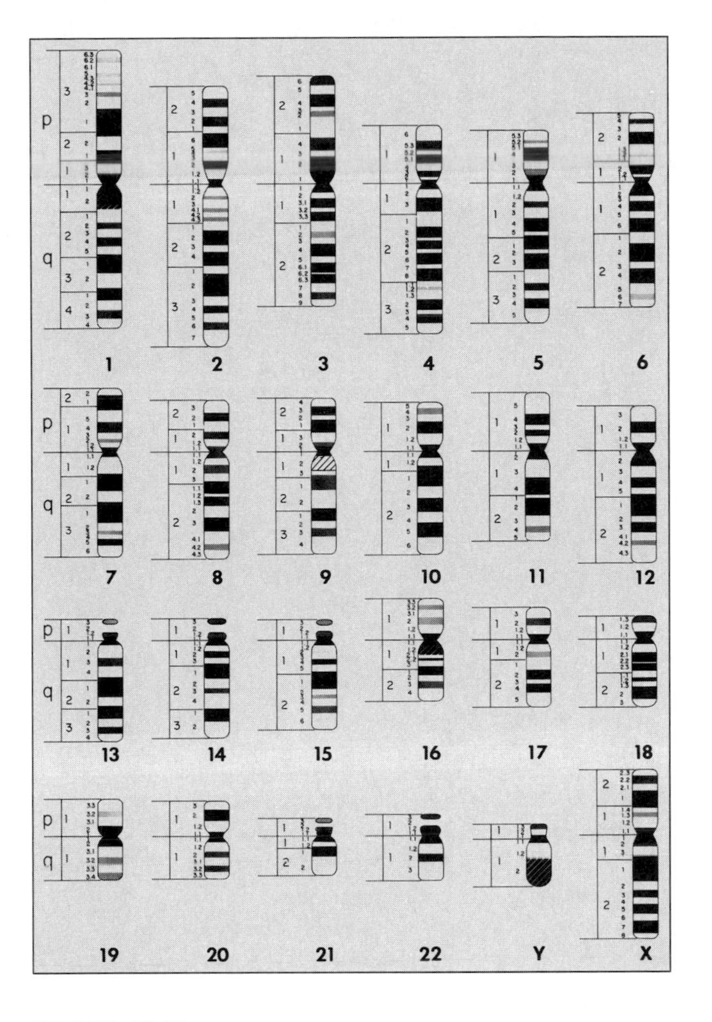

FIGURE 12.10
Schematic representation and nomenclature of human chromosomes at the 400-band stage using G-banding. Different shades represent varying intensities of bands.

Most estimates of infidelity during the transmission of chromosomes are exceedingly low: 10^{-5} to 10^{-6}, or 1 error per 100,000 to 1,000,000 cell divisions. As a result, it has been generally assumed that analysis of the DNA sequence of centromeric regions will provide insights into the rather remarkable features of this chromosomal region. This DNA region is designated the **CEN** in eukaryotes.

The CEN regions of yeast chromosomes provide an excellent model system and have now been isolated and sequenced. Since each centromere really serves an identical function, it is not surprising that all CENs are remarkably similar. The region consists of about 225 base pairs, which can be divided into three regions (Figure 12.11). The first and third (I and III) are relatively short and highly conserved, consisting of only 8 and 25 base pairs, respectively. Region II, which is larger (80–85 base pairs) and extremely A-T rich, varies in sequence between different chromosomes. Up to 95 percent of the sequences are adenine or thymine, making it highly unlikely that this region might be transcribed into a meaningful RNA molecule.

Mutational analysis suggests that Regions I and II are less critical to centromere function than Region III. Genetic alteration of the former regions are often tolerated, but those in Region III are usually disruptive to the function of the centromere. The DNA sequence of this region may be essential to binding to the proteins constituting the **kinetochore,** that portion of the centromere that binds to the spindle fibers.

The second important structure is that portion of the linear chromosome constituting each end, called the **telomere.** Its major function is to provide stability, aiding in the maintenance of individual integrity of the chromosome. This role is accomplished by virtue of rendering chromosome ends generally inert in interactions with other chromosome ends. In contrast to broken chromosomes, whose ends are "sticky" and readily rejoin other such ends, telomere regions do not fuse with one another or with broken ends. Thus, it has been thought that some aspect of the DNA sequence of chromosome ends must be unique compared with most other chromosome regions.

As with centromeres, the problem has been approached by investigating the smaller chromosomes of less-advanced eukaryotes, such as protozoans and yeast. The idea that all telomeres in a given species might share a common nucleotide sequence has now been borne out.

Two types of sequences have been discovered. The first type, called simply **telomeric DNA sequences,** consists of short tandem repeats. It is this group that contributes to the stability and integrity of the chromosome. In the ciliate *Tetrahymena,* over fifty tandem repeats of the hexanucleotide sequence TTGGGG occur. In another ciliate *Stylonychia,* TTTTGGGG is repeated many times. In the slime mold *Physarum,* the tandem repeat TAGGG occurs.

While it is not yet clear how such sequences might relate to the creation of inert chromosome ends, it is now

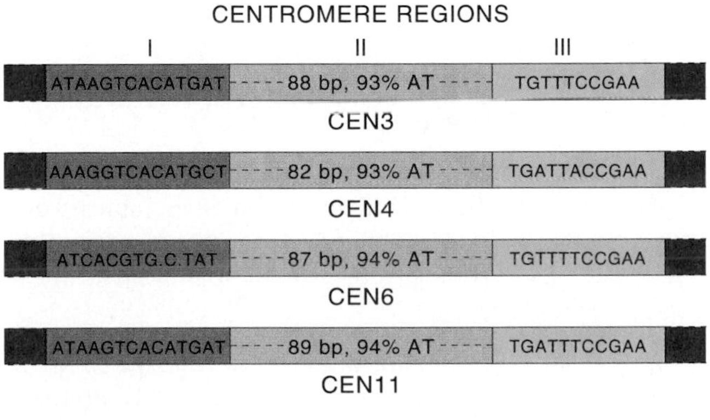

CENTROMERE REGIONS
I II III
ATAAGTCACATGAT ----88 bp, 93% AT---- TGTTTCCGAA
CEN3
AAAGGTCACATGCT ----82 bp, 93% AT---- TGATTACCGAA
CEN4
ATCACGTG.C.TAT ----87 bp, 94% AT---- TGTTTTCCGAA
CEN6
ATAAGTCACATGAT ----89 bp, 94% AT---- TGATTTCCGAA
CEN11

FIGURE 12.11
Nucleotide sequence information derived from DNA of the three major centromere regions of chromosomes 3, 4, 6, and 11 of yeast.

clear that the repeated sequences are related to the process of DNA replication. At chromosome ends, replication of the lagging strand poses difficulty for initiation of synthesis. A unique enzyme, **telomerase,** overcomes this problem.

The analysis of telomeric DNA sequences has shown them to be highly conserved throughout evolution, reflecting the critical role they play in maintaining the integrity of chromosomes.

ORGANIZATION OF THE EUKARYOTIC GENOME

Having established the general view of the chromatin fiber, we now turn to a consideration of the gene structure as well as the organization of the **genome** of eukaryotic organisms. The genome includes the entire haploid set of chromosomes constituting the genetic material of an organism.

Prior to 1970, there was little evidence at the molecular level involving eukaryotic gene structure and organization. Nevertheless, many geneticists believed that when the eukaryotic gene could be examined directly, it might well be quite different from that of viruses and bacteria. Multicellular eukaryotes, it was reasoned, have several unique requirements distinguishing them from phages and bacteria. Primarily, cell differentiation, tissue organization, and coordinated development and function depend on genetic expression and its regulation. Do these requirements also depend on different modes of gene structure and organization?

When the technologies of recombinant DNA and rapid nucleotide sequencing were developed (Chapter 15), the answer to this question began to emerge. The eukaryotic gene is far more complex than we could ever have imagined. In this section, we will review many findings related to gene structure and the organization of the eukaryotic genome.

Eukaryotic Genomes and the *C* Value Paradox

The amount of DNA contained in the haploid genome of a species is called the ***C* value.** When such values were determined for a large variety of eukaryotic organisms (Figure 12.12) several trends were apparent. The most notable trends are that:

1 Eukaryotes contain substantially more DNA in their genomes than viruses or prokaryotes and exhibit a wide variation among different groups.

2 Evolutionary divergence has been accompanied by increased amounts of DNA. In many major phylogenetic groups, more evolutionarily advanced forms generally contain more DNA than most primitive forms.

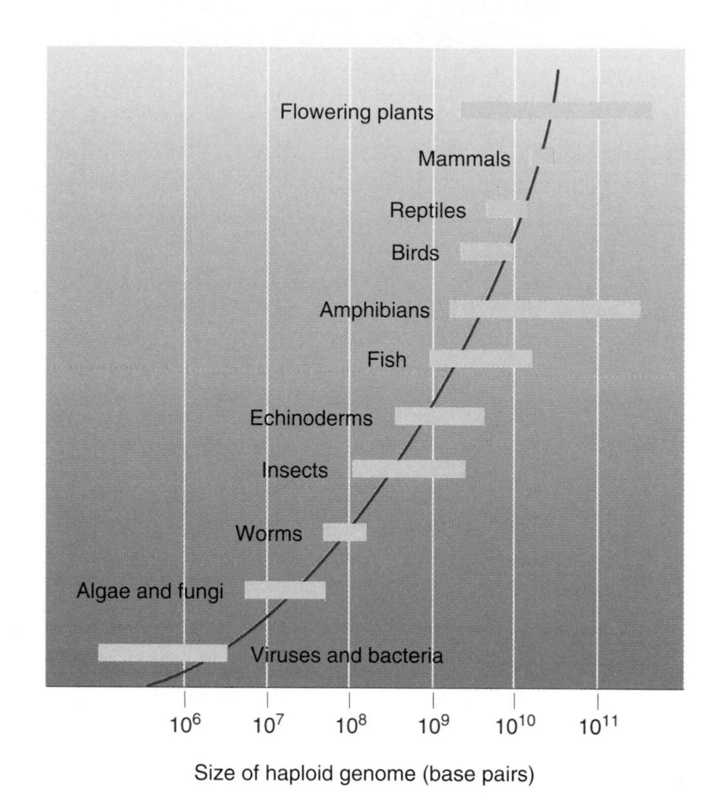

FIGURE 12.12
The range of DNA content of the haploid genome of many representative groups of organisms.

It might be argued that increasing C values are simply the result of a greater need for increased amounts and varieties of gene products in more complex organisms. However, several observations make this explanation unacceptable. First, the increases are very dramatic. While viruses and bacteria contain 10^4 to 10^6 nucleotide pairs in their single chromosomes, eukaryotes contain 10^7 to 10^{11} nucleotide pairs in each set of chromosomes. Does a eukaryote require 10^5 (100,000) more genes, as these figures might suggest? Second, closely related organisms with the same degree of complexity in body form and tissue and organ types often vary tenfold or more in DNA content. Third, amphibians and flowering plants, which vary as much as 100-fold within their taxonomic classes, often contain much more DNA than other more recently evolved eukaryotes. In spite of this observation, there is no correlation between increased genome size and the morphological complexity of these groups.

Therefore, it is doubtful that the development of greater complexity during evolution can account for the amount of DNA found in eukaryotic genomes. This conclusion is the basis of what has been called the ***C* value paradox.** Excess DNA is present that does not seem to be essential to the development or evolutionary divergence of eukaryotes. Is such DNA vital to the organisms carrying it in their genomes, or is it simply excess DNA that has somehow accumulated during evolution and has no function?

Recent studies of gene structure are beginning to unravel these mysteries. Some insights have been gained from findings presented earlier in this text. For example, we have already established that noncoding sequences are indeed present in the genomes of higher organisms in the form of **repetitive DNA** (see Chapter 7). Can this category account for the excess DNA?

To answer this question, consider the human genome. Only about 5 percent of human DNA is highly repetitive. Some of this is in the form of satellite DNA located at heterochromatic regions associated with the centromere and telomeres. Another major source of repetitive DNA is the **Alu family** characteristic of mammals, which we discussed in the previous chapter. In humans, there are about 300,000 copies of a repeated sequence of about 300 base pairs. These are interspersed throughout the genome and constitute roughly 3 percent of the total DNA.

Middle repetitive DNA (see Chapter 7) comprises another 15 to 30 percent of the human genome. This fraction includes duplicated genes as well as nontranscribable DNA sequences interspersed between unique or single-copy genes. This leaves over 60 percent of the genome still unaccounted for.

Such observations are not uncommon. While the proportion of the genome consisting of repetitive DNA varies between organisms, one feature seems to be shared: **Only a small part of the genome codes for proteins.** For example, sea urchins contain an estimated 20,000 to 30,000 genes coding for proteins, which occupy less than 10 percent of the genome. In *Drosophila,* only 5 to 10 percent of the genome is occupied by genes coding for proteins. In humans, only 1 to 2 percent of the genome encodes proteins! We must conclude that while there is sufficient DNA to code for over one million genes in many eukaryotic organisms, the majority of DNA does not serve this function.

Eukaryotic Gene Structure

We turn now to a brief review of the structure of the eukaryotic gene. This topic has been previously introduced in Chapter 9. As this discussion ensues, you should refer to Figure 12.13, which presents a model of the eukaryotic gene.

Important insights have come from the study of RNA. The encoded information is present in the ribonucleotide sequence of mRNA. This information originates in DNA, where complementary sequences of deoxyribonucleotides exist. In bacteria, the relationship between DNA and RNA appears to be quite direct. The DNA base sequence is transcribed into an mRNA sequence, which is then translated into an amino acid sequence according to the genetic code. This is the basis for the concept of **colinearity.**

However, in eukaryotes the situation is very different from that in bacteria. It has been found that many internal base sequences of genes, called **introns,** never appear in the mature mRNAs that are translated. The remaining areas of each gene, which are ultimately translated into the amino acid sequence of the encoded protein, are called **exons.** While the entire gene is transcribed into a large precursor mRNA, called **heterogeneous RNA (hnRNA),** the introns are excised and the exons spliced back together prior to translation.

Given this information about internal gene sequences, we can ask whether the nucleotide sequences found in introns solve the *C* value paradox? That is, do introns account for the remainder of the noncoding sequences beyond those of repetitive DNA? The answer is emphatically no! As shown in Table 12.5, the presence of introns may increase by four or five times our estimate of the nucleotide length occupied by single-copy genes. However, since this estimate is seldom above 10 percent of

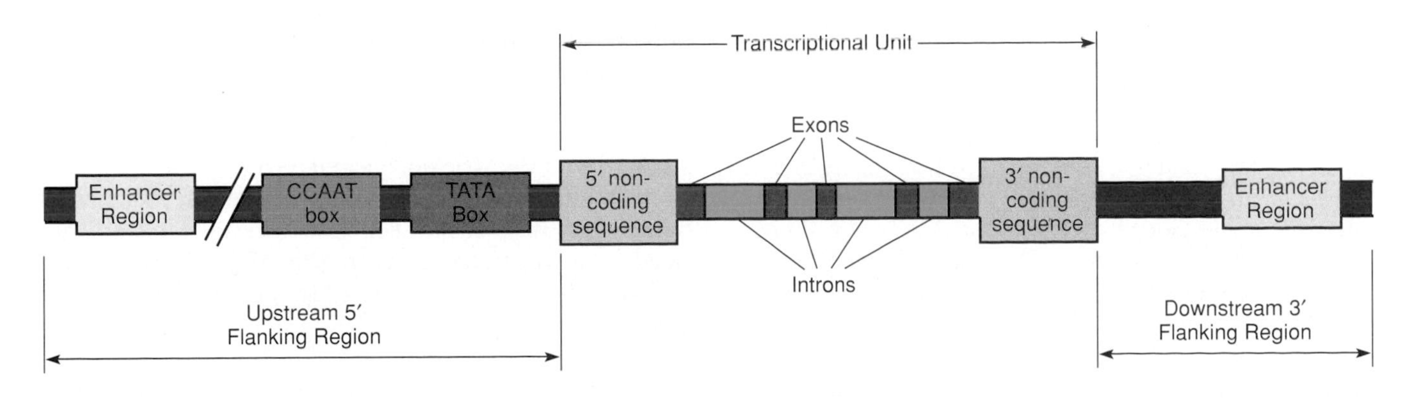

FIGURE 12.13
Modern concept of the eukaryotic gene.

TABLE 12.5
A comparison of mRNAs and the initial transcripts from which they are derived.

Gene	mRNA Size	Minimum Transcript Length	Ratio
Rabbit β-globin	589	1295	2.2
Mouse β^{maj}-globin	620	1382	2.2
Mouse β^{min}-globin	575	1275	2.2
Mouse α-globin	585	850	1.5
Rat insulin I	443	562	1.3
Rat insulin II	443	1061	2.4
Chick ovalbumin	1859	7500	4.0
Chick ovomucoid	883	5600	6.3
Chick lysozyme	620	3700	6.0

the genome and usually much less, a great deal of noncoding DNA is still unaccounted for.

Aside from introns, we shall review briefly the presence and nature of the noncoding flanking regions to complete our discussion of the eukaryotic gene. Recall from Chapter 9 that the 5′ region upstream from the coding sequence is critical to efficient transcription. Closest to the coding sequence is the **promoter** with its **TATA box,** where RNA polymerase binds prior to initiation of transcription. Farther upstream is the **CAAT box** and in numerous cases, an **enhancer** region is also present. These regions play a role in modulating transcription. Enhancers may also be found within the coding sequence or as part of the 3′ downstream region, beyond the coding sequence.

Since flanking regions are critical to specific gene function, they are considered part of each gene. Their discovery has extended our knowledge of gene structure considerably.

Exon Shuffling and Protein Domains

At first glance, the discovery of introns and exons is most puzzling. Why, during the evolution of eukaryotes, would their genes acquire noncoding introns between the coding portions of genes whose presence requires precision splicing following transcription and whose replication requires a substantial expenditure of energy? While we do not yet have an absolute answer, important clues are emerging primarily as a result of our ability to determine rapidly the nucleotide sequence of many genes. With the knowledge of the sequence of the exons of genes, the amino acid sequences of the corresponding proteins can be predicted accurately. When composite data bases are established and compared by computer analysis, the uniqueness of genes can be examined. As a result of this approach, some very interesting findings have been forthcoming, as first recognized and proposed by Walter Gilbert in 1977. He suggested that genes in higher organisms may consist of collections of exons present in ancestral genes that are brought together through recombination during the course of evolution. Referred to as **exon shuffling,** he proposed that exons are modular in the sense that each might encode a functional domain in the structure of a protein. For example, a particular amino acid sequence encoded by a single exon might always create a specific type of fold in a protein. This **protein domain** may impart a unique but characteristic type of molecular interaction to all molecules containing such a fold. Serving as the basis of useful parts of a protein, many similar domains might be functional in a variety of proteins. In Gilbert's proposal, during evolution, many exons could be mixed and matched to form unique genes in eukaryotes.

Several observations lend support to this proposal. Most exons are fairly small, averaging about 100 to 150 nucleotides and encoding 30 to 50 amino acids. This size is consistent with the production of functional domains in proteins. Second, recombinational events that lead to exon shuffling would be expected to occur within areas of genes represented by introns. Since introns are free to accumulate mutations without harm to the organism, recombinational events would tend to further randomize their nucleotide sequences. Over extended evolutionary periods, diverse sequences would tend to accumulate. This is, in fact, what is observed. Introns range from 50 to 20,000 bases and exhibit fairly random base sequences.

Since 1977, a vast research effort has been directed toward the analysis of split genes. In 1985, more direct evidence in favor of Gilbert's proposal of exon modules was presented. Analysis was made of the human gene encoding the membrane receptor for low density lipoproteins (LDL). This **LDL receptor protein** is essential to the transport of plasma cholesterol into the cell. It mediates **endocytosis** and is expected to have numerous functional domains. These include the capability of this protein to bind specifically to the LDL substrates and to interact with other proteins at different levels of the membrane during transport across it. In addition, this receptor molecule is modified posttranslationally by the addition of a carbohydrate; a domain must exist that links to this carbohydrate.

Detailed analysis of the gene encoding this protein supports the concept of exon modules and their shuffling during evolution. The gene is quite large—45,000 nucleotides—and contains 18 exons. These represent only slightly less than 2600 nucleotides. These exons are related to the functional domains of the protein *and* appear to have been recruited from other genes during evolution.

Figure 12.14 illustrates these relationships. The first exon encodes a signal sequence that is removed from the protein before the LDL receptor becomes part of the membrane. The next five exons represent the domain

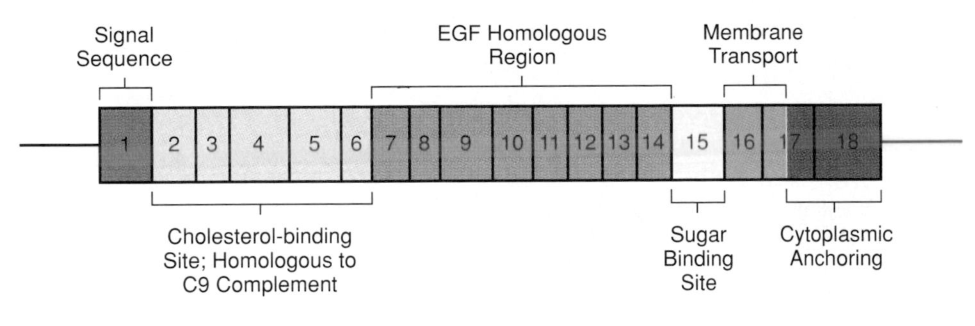

FIGURE 12.14
A comparison of the 18 exons making up the LDL receptor protein gene and their organization into five functional domains and one signal sequence.

specifying the binding site for cholesterol. This domain is made up of a 40-amino-acid sequence repeated seven times. The next domain consists of a sequence of 400 amino acids bearing a striking homology to the mouse peptide hormone **epidermal growth factor (EGF).** This region is encoded by eight exons and contains three repetitive sequences of 40 amino acids. A similar sequence is also found in three blood-clotting proteins. The fifteenth exon specifies the domain for the posttranslational addition of the carbohydrate, while the remaining two exons specify regions of the protein that are part of the membrane, anchoring the receptor to specific sites called coated pits on the cell surface.

These observations are fairly compelling in favor of the theory of exon shuffling during evolution. Certainly, there is no disagreement concerning the concept of protein domains being responsible for specific molecular interactions.

What does remain controversial and evocative in the exon shuffling theory is the question of when introns first appeared on the evolutionary scene. In 1978, W. Ford Doolittle proposed that these intervening sequences were part of the genome of the most primitive ancestors of modern-day eukaryotes. In support of this idea, Gilbert has argued that if the same introns are found in identical positions within genes shared by totally unrelated eukaryotes (such as humans, chickens, and corn), they must have also been present in primitive ancestral genomes.

If Doolittle's proposal is correct, why are introns absent in today's prokaryotes? Gilbert argues that they were present at one point during evolution, but as the genome of these primitive organisms evolved, they were lost. This occurred as a result of strong selection pressure to streamline their chromosomes to minimize energy expenditure supporting replication and gene expression. Further, streamlining leads to more error-free mRNA production. However, supporters of the opposing "intron-late" school, including Jeffrey Palmer, argue against this speculation and believe that much later during evolution, introns became a part of a single group of eukaryotes that are ancestral to modern-day members but not prokaryotes. While it may be difficult to resolve, it seems certain that this most interesting controversy will persist for many years to come.

Multigene Families: The α- and β-Globin Genes

The study of still another level of eukaryotic gene organization has contributed to our understanding of the distribution of genetic units within the genome. This level consists of the organization of **multigene families,** where genes identical or closely related in DNA sequence and function are found together at a single location along a chromosome. The examination of multigene families further provides valuable insights into the *C* value paradox.

We will first examine cases where groups of genes encode very similar, but not identical, polypeptide chains that become part of proteins that are very closely related in function. The globin gene families, responsible for encoding the various polypeptides that are part of hemoglobin molecules, are examples that provide many insights into the organization of the genome. Then, we will consider the case of identical, tandemly repeated genes, as represented by ribosomal RNA genes.

The human **alpha- and beta-globin gene families** are two of the most extensive and best characterized groups. The alpha family resides on the short arm of chromosome 16 and contains five genes, while the beta family resides on the short arm of chromosome 11 and contains six genes. Members of both families show homology to one another, but not nearly to the extent members within the same family exhibit homology to one another. Members of both families encode globin polypeptides that combine into a single tetrameric molecule, which interacts with a heme group that can reversibly bind to oxygen. And, within each group of genes, members are coordinately turned on and off during the embryonic, fetal, and adult stages of development in the precise order in which they occur along the chromosome within each family.

The alpha family [Figure 12.15(a)] spans more than

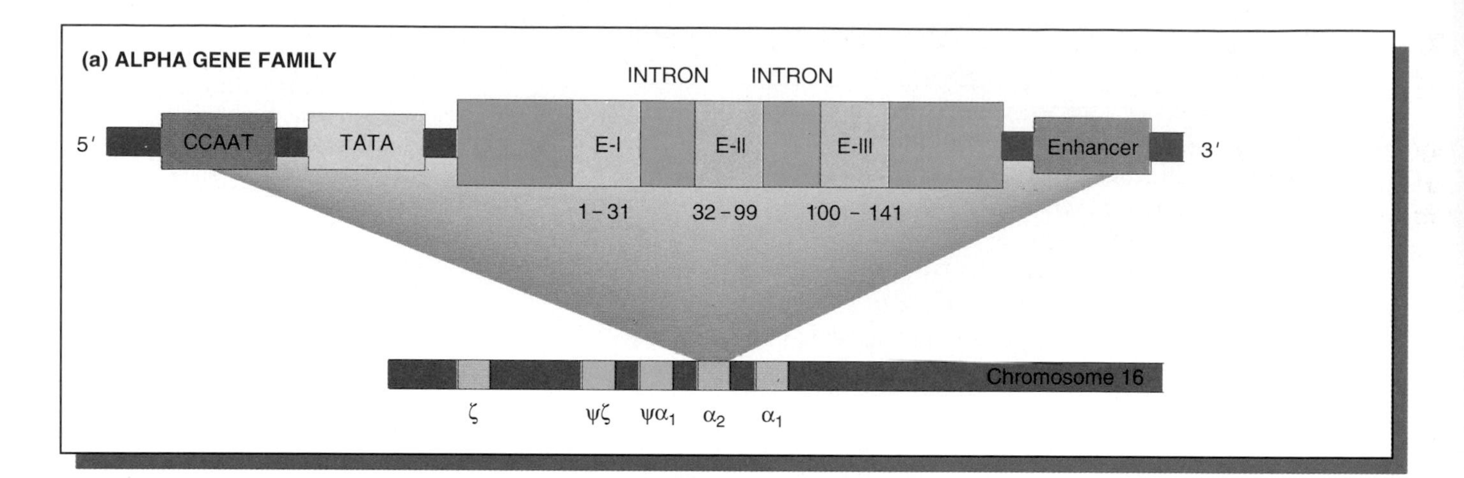

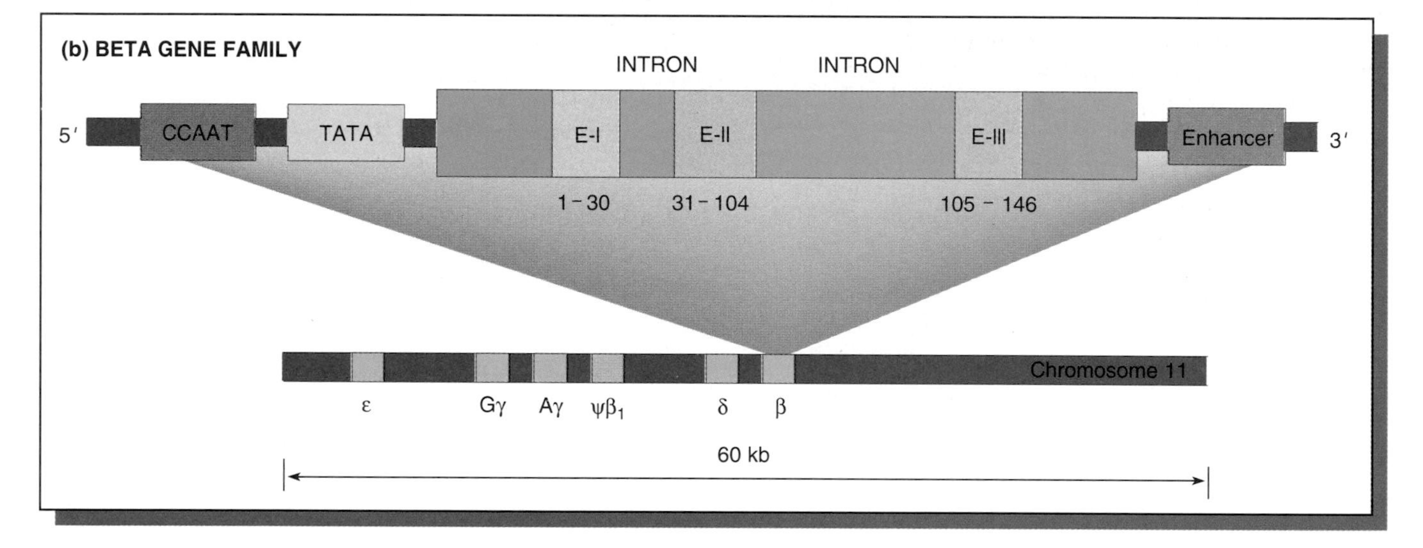

FIGURE 12.15
The comparative arrangement of the human alpha and beta gene families, depicted in parts (a) and (b), respectively. The internal organization of the functional genes from each family are also shown. All genes contain three exons (E–I, II, and III) and two introns, as well as 5′ and 3′ nocoding regions. The arabic numbers designate the amino acid positions encoded by each exon.

30,000 nucleotide pairs (30 kb) and contains five genes: the zeta gene (ζ), expressed only in the embryonic stage, two nonfunctional **pseudogenes,** and two copies of the alpha (α) gene, expressed during the fetal and adult stages. Pseudogenes, found in both families, are similar in sequence to the other genes in their family, but contain significant nucleotide substitutions, deletions, and duplications that prevent their transcription. The first pseudogene in the alpha family shows greater homology to the zeta gene, while the second shows greater homology to the alpha gene. Pseudogenes are designated by the prefix ψ (the greek letter psi), followed by the symbol of the gene they most resemble. Thus, the designation $\psi\alpha1$ indicates a pseudogene of the first alpha gene.

While we will do a more exact analysis of the beta family, an examination of the organization of the members of the alpha family as well as their intragenic distribution of introns and exons [Figure 12.15(a)], reveals several interesting features. First, only a small portion of the chromosomal region housing the entire family is occupied by the three functional genes. The region consists almost entirely of intergenic regions. Second, the functional genes in this family contain two introns at precisely the same positions within the genes. The nucleotide sequences within corresponding exons are nearly identical when the zeta and alpha genes are compared. Both encode polypeptide chains that are 141 amino acids long. However, the intron sequences are extremely divergent, even though they are about the same size. Significantly, a high percentage of the nucleotide sequence within each gene is contained in these noncoding introns. The above information bears on both the structural organization of a gene family as well as on our earlier discussion of the *C* value paradox.

The beta-globin family in humans is even more extensive, containing six genes and occupying a 60-kb sequence of DNA [Figure 12.15(b)]. As with the alpha family, the sequence of genes along the chromosome parallels the order of expression during development. A pseudogene is also present.

Of the six genes, three are expressed prior to birth: the epsilon (ϵ) gene is embryonic, while two nearly identical gamma genes ($^{G}\gamma$ and $^{A}\gamma$) are fetal. The two gamma gene products vary by only a single amino acid. The two remaining functional genes, delta (δ) and beta (β), are expressed following birth. Finally, a single pseudogene $\psi\beta1$ is present within the family.

All five functional genes encode products that are 146 amino acids long and contain two similarly sized introns at exactly the same positions. The second intron is significantly larger than its counterpart in the functional alpha genes. These introns and their locations within the gene are compared in Figure 12.15.

The flanking regions of each gene have also been studied. In all functional genes, both a TATA box (29 or 30 base pairs upstream) and a CCAAT region (70 to 78 base pairs upstream from the initial codon) have been observed. Downstream, globin enhancers have been identified.

Several observations of the β-globin genes pertain to the *C* value paradox. Only about 5 percent of the 60-kb region consists of coding sequences. The remaining 95 percent includes the introns, flanking regions, and spacer DNA found between genes. Of this percentage, only about 11 percent consists of introns. The remainder of the DNA serves no known function and consists of sequences not found elsewhere in the genome. Because it represents the majority of DNA in the cluster, we must conclude that it is either excess nonfunctional DNA *or* DNA whose function has yet to be discovered.

Tandem Repeat Families: rRNA Genes

We conclude our discussion of the organization of multigene families by considering the example of **tandem repeats** of the gene family encoding rRNA molecules present in ribosomes. This gene family encodes separate gene products. In this case, three molecules are encoded in the following order along the chromosome: 18*S*, 5.8*S*, and 28*S* rRNAs (Figure 12.16). The rRNA family contains substantial regions between each gene that are part of the initial transcript, but that are not found in the final rRNA gene products. Thus, the initial transcript must be processed in a manner analogous to the removal of introns from the primary transcript of most mRNAs. In addition, there is a nontranscribed spacer region between each gene family unit.

The basic repeating unit that is transcribed varies in length in different organisms from almost 7 kb in yeast to 13 kb in humans (Figure 12.16). These spacer regions between genes are clearly visible in the insert in Figure 12.16, where transcription of tandem repeats has been visualized under the electron microscope. Together, each gene family unit and its accompanying spacer regions occupy a substantial portion of DNA. In the human rRNA gene family, a length of about 43 kb of DNA is utilized in each tandem repetition. This amount of DNA is as large as the smaller globin gene cluster.

Multiple copies of a single transcriptional unit encoding the rRNAs exist in bacteria. In *E. coli,* seven dispersed copies are present, presumably to accommodate the need for the rapid production of the approximately 15,000 ribosomes found in each cell. It is not surprising, therefore, to find many more than seven copies in the larger, more complex eukaryotic cells. Yeast contains over 100, while *Xenopus, Drosophila,* and humans have up to 400 copies per genome. A single eukaryotic cell

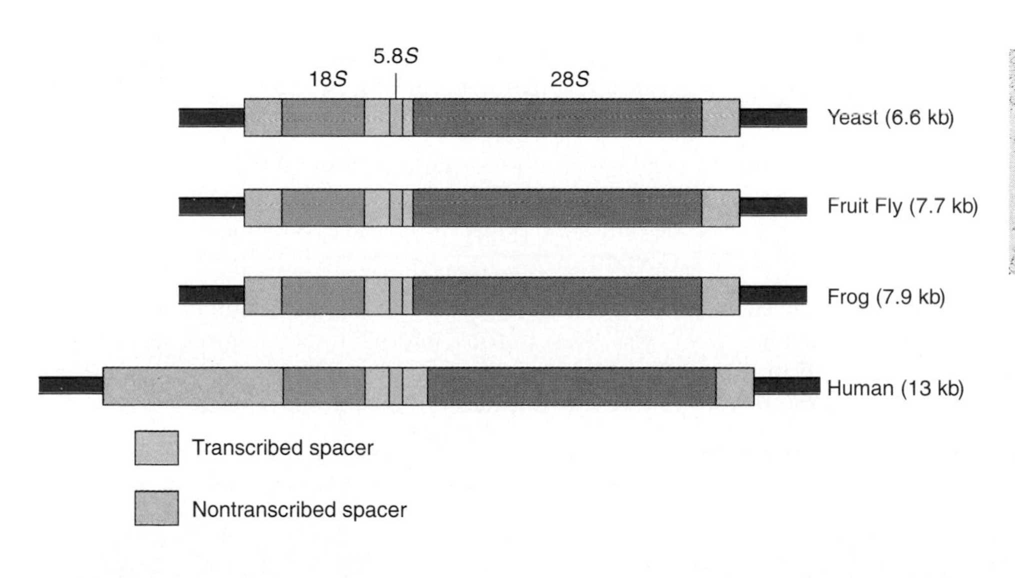

FIGURE 12.16
Comparison of the repeating rRNA gene unit in various eukaryotic organisms. The electron micrograph visualizes simultaneous transcription of several members of the repeating rRNA unit of the newt.

may have a million or more ribosomes if it is actively producing proteins.

If all of these genes were present in a single cluster in an organism such as humans, a substantial part of one chromosome would be occupied by them (400×45 kb = 18,000 kb or 18 million base pairs). This amount of DNA is almost five times greater than the entire *E. coli* chromosome (4×10^6 base pairs). As it turns out, they are present in clusters dispersed throughout the genome. In humans, for example, the clusters are found near the ends of chromosomes 13, 14, 15, 21, and 22. In *Drosophila,* they are present on both the X and Y chromosomes. Within the nucleus, these various gene regions are collectively referred to as the **nucleolus organizer regions (NORs),** because of their association with the production of ribosomes in the nucleolus.

This information extends our knowledge and overall picture of how some genes are organized within the genome. It also provides insights into the vast amount of DNA that may be utilized in order to provide cells with one set of gene products. Examination of the same gene family in different organisms also provides information valuable to the study of evolution at the molecular level.

CHAPTER SUMMARY

1 Knowledge of the organization of the molecular components forming chromosomes is essential to the understanding of the function of the genetic material. Largely devoid of associated proteins, bacteriophage and bacterial chromosomes are naked DNA molecules present in a form equivalent to the Watson–Crick model.

2 Mitochondria and chloroplasts contain DNA that encodes products essential to their biological function. This DNA is remarkably similar in form and appearance to bacterial and phage DNA, lending support to the endosymbiont theory which suggests that these organelles were once free-living organisms.

3 The eukaryotic chromatin fiber is a nucleoprotein organized into repeating units called nucleosomes. Composed of about 200 base pairs of DNA and an octamer of four types of histones, the nucleosome is important in facilitating the conversion of the extensive chromatin fiber characteristic of interphase into the highly condensed chromosome seen in mitosis.

4 The structural heterogeneity of the chromosome axis has been established as a result of both biochemical and cytological investigation. Heterochromatin, prematurely condensed in interphase, is genetically inert. The centromeric and telomeric regions, the Y chromosome, and the Barr body are examples.

5 DNA analysis has revealed highly conserved, unique nucleotide sequences in both the heterochromatic centromere and telomere regions. These sequences distinguish these structures from other parts of the chromosome and provide the basis for the critical features imparted to the chromosome by these regions.

6 The *C* value paradox, made apparent during the study of the organization of eukaryotic DNA, suggests that eukaryotic organisms contain much more DNA than necessary to encode only gene products essential to the development and normal functions of an organism.

7 Detailed analysis of eukaryotic gene structure has revealed numerous categories of noncoding DNA sequences. Introns, internal noncoding sequences, are represented in the initial mRNA molecules or heterogeneous RNA, but never appear in the mature mRNAs from which proteins are synthesized.

8 The flanking regions of genes, particularly those giving rise to the 5′ initiation point of mRNA, have also been analyzed. Three upstream regions that appear to be essential to efficient transcription are the TATA box, the CAAT box, and the enhancer element.

9 Multigene families are composed of structurally related genes, usually clustered together, encoding functionally similar products. The human alpha- and beta-globin families are two examples.

10 Ribosomal RNA genes provide an example of tandem repeat families, where the same gene is duplicated many times.

KEY TERMS

alpha- and beta-globin gene families
Alu family
bacteriophage lambda
CAAT box
C value paradox
centromere
chloroplast
chloroplast DNA (cpDNA)
chromatin
chromosome banding
colinearity
DNA-binding proteins
endosymbiont theory
enhancer
cuchromatin
exon
exon shuffling
folded fiber chromosome model
genome
heterochromatin
heterogeneous RNA (hnRNA)
histones
intron
kinetochore
maternal mode of inheritance
mitochondrial DNA (mtDNA)
mitochondrion
multigene families
nu (ν) bodies
nucleoid
nucleolus organizer region (NOR)
nucleosome
position effect
promoter
protein domain
pseudogenes
repetitive DNA
tandem repeats
TATA box
telomerase
telomere
telomeric DNA sequences
T-even series of bacteriophages
unineme model of DNA

A previously undiscovered single-celled organism was found living at a great depth on the ocean floor. Its nucleus contained only a single linear chromosome containing 7×10^6 nucleotide pairs of DNA coalesced with three types of histonelike proteins.

1 A short micrococcal nuclease digestion yielded DNA fractions consisting of 700, 1400, and 2100 base pairs. Predict what these fractions represent. What conclusions can be drawn?

SOLUTION: The chromatin fiber may consist of a variation of nucleosomes containing 700 base pairs of DNA. The 1400 and 2100 bp fractions represent two and three nucleosomes, respectively, linked together. Enzymatic digestion has been incomplete in the latter two fractions.

2 Analysis of individual nucleosomes revealed that each unit contained one copy of each protein, and that the short linker DNA contained no protein bound to it. If the entire chromosome consists of nucleosomes, how many are there, and how many total proteins are needed to form them?

SOLUTION: Since the chromosome contains 7×10^6 base pairs of DNA, the number of nucleosomes, each containing 7×10^2 base pairs, is equal to

$$\frac{7 \times 10^6}{7 \times 10^2} = 10^4 \text{ nucleosomes}$$

The chromosome thus contains 10^4 copies of each of the three proteins, for a total of 3×10^4 molecules.

3 Analysis then revealed DNA to be a double helix similar to the Watson–Crick model, but containing 20 base pairs per complete turn of the right-handed helix. The physical size of the nucleosome was exactly double the volume occupied by that found in all other known eukaryotes, by virtue of increasing the distance along the fiber axis by a factor of two. Compare the degree of compaction of this organism's nucleosome to that found in other eukaryotes.

SOLUTION: The unique organism compacts a length of DNA consisting of 35 complete turns of the helix (700 base pairs per nucleosome/20 base pairs per turn) into each nucleosome. The normal eukaryote compacts a length of DNA consisting of 20 complete turns of the helix (200 base pairs per nucleosome/10 base pairs per turn) into a nucleosome one-half the volume of that in the unique organism. The degree of compaction is therefore less in the unique organism.

PROBLEMS AND DISCUSSION QUESTIONS

1 Compare and contrast the chemical nature, size, and form assumed by the genetic material of viruses and bacteria.

2 Contrast the DNA associated with mitochondria and chloroplasts.

3 Describe the sequence of research findings leading to the model of chromatin structure. What is the molecular composition and arrangement of the nucleosome?

4 Provide a comprehensive definition of heterochromatin and list as many examples as you can.

5 Mammals contain a diploid genome consisting of at least 10^9 base pairs. If this amount of DNA is present as chromatin fibers where each group of 200 base pairs of DNA is combined with 9 histones into a nucleosome, and each group of 5 nucleosomes is further condensed into a structure called a solenoid, achieving a final packing ratio of 50, determine:
(a) the total number of nucleosomes in all fibers.
(b) the total number of solenoids in all fibers.
(c) the total number of histone molecules combined with DNA in the diploid genome.
(d) the combined length of all fibers.

6 Assume that a viral DNA molecule is in the form of a 50-μm-long, circular rod of a uniform 20-Å diameter. If this molecule is contained within a viral head that is a sphere with a diameter of 0.08 μm, will the DNA molecule fit into the viral head, assuming complete flexibility of the molecule? Justify your answer mathematically.

7 Describe what is meant by exon shuffling.

8 The β-globin gene family consists of 60 kb of DNA yet only 5 percent of the DNA encodes β-globin gene products. Account for as much of the remaining 95 percent of the DNA as you can.

9 Describe the rRNA gene family in eukaryotes. What accounts for the major difference in the size of tandem repeats in different organisms?

10 "In the 1990s, the *C* value paradox is not so paradoxical." Agree or disagree with this statement and support your opinion.

SELECTED READINGS

BLACKBURN, E. H. 1990. Telomeres: Structure and synthesis. *J. Biol. Chem.* 265:5919–21.

BLACKBURN, E. H., and SZOSTAK, J. W. 1984. The molecular structure of centromeres and telomeres. *Ann. Rev. Biochem.* 53:163–94.

BLOOM, K., HILL, A., and YEH, E. 1986. Structural analysis of a yeast centromere. *BioEssays* 4:100–05.

CARBON, J. 1984. Yeast centromeres: Structure and function. *Cell* 37:352–53.

CHEN, T. R., and RUDDLE, F. H. 1971. Karyotype analysis utilizing differential stained constitutive heterochromatin of human and murine chromosomes. *Chromosoma* 34:51–72.

CRICK, F. 1979. Split genes and RNA splicing. *Science* 204:264–71.

FRITSCHE, E. F., LAWN, R. M., and MANIATIS, T. 1980. Molecular cloning and characterization of the human β-like globin gene cluster. *Cell* 19:959–72.

GREEN, B. R., and BURTON, H. 1970. *Acetabularia* chloroplast DNA: Electron microscopic visualization. *Science* 168:981–82.

HEWISH, D. R., and BURGOYNE, L. 1973. Chromatin sub-structure. The digestion of chromatin DNA at regularly spaced sites by a nuclear deoxyribonuclease. *Biochem. Biophys. Res. Comm.* 52:504–10.

HSU, T. C. 1973. Longitudinal differentiation of chromosomes. *Ann. Rev. Genet.* 7:153–77.

KORNBERG, R. D. 1975. Chromatin structure: A repeating unit of histones and DNA. *Science* 184:868–71.

KORNBERG, R. D., and KLUG, A. 1981. The nucleosome. *Scient. Amer.* (Feb) 244:52–64.

MANIATIS, T., et al. 1980. The molecular genetics of human hemoglobins. *Ann. Rev. Genet.* 14:145–78.

OLINS, A. L., and OLINS, D. E. 1974. Spheroid chromatin units (ν bodies). *Science* 183:330–32.

———, 1978. Nucleosomes: The structural quantum in chromosomes. *Amer. Scient.* 66:704–11.

PALMER, J. D., and LOGSDON, J. 1991. The recent origins of introns. *Curr. Opin. Genet. Dev.* 1:470.

SCHMID, C. W., and JELINEK, W. R. 1982. The Alu family of dispersed repetitive sequences. *Science* 216:1065–70.

SUDHOF, T. C., et al. 1985. The LDL receptor gene: A mosaic of exons shared with different proteins. *Science* 228:815–22.

van HOLDE, K. E. 1989. *Chromatin.* New York: Springer-Verlag.

VERMA, R. S., ed. 1988. *Heterochromatin: Molecular and structural aspects.* Cambridge, England: Cambridge University Press.

YUNIS, J. J. 1981. Chromosomes and cancer: New nomenclature and future directions. *Hum. Pathol.* 12:494–503.

Extrachromosomal Inheritance

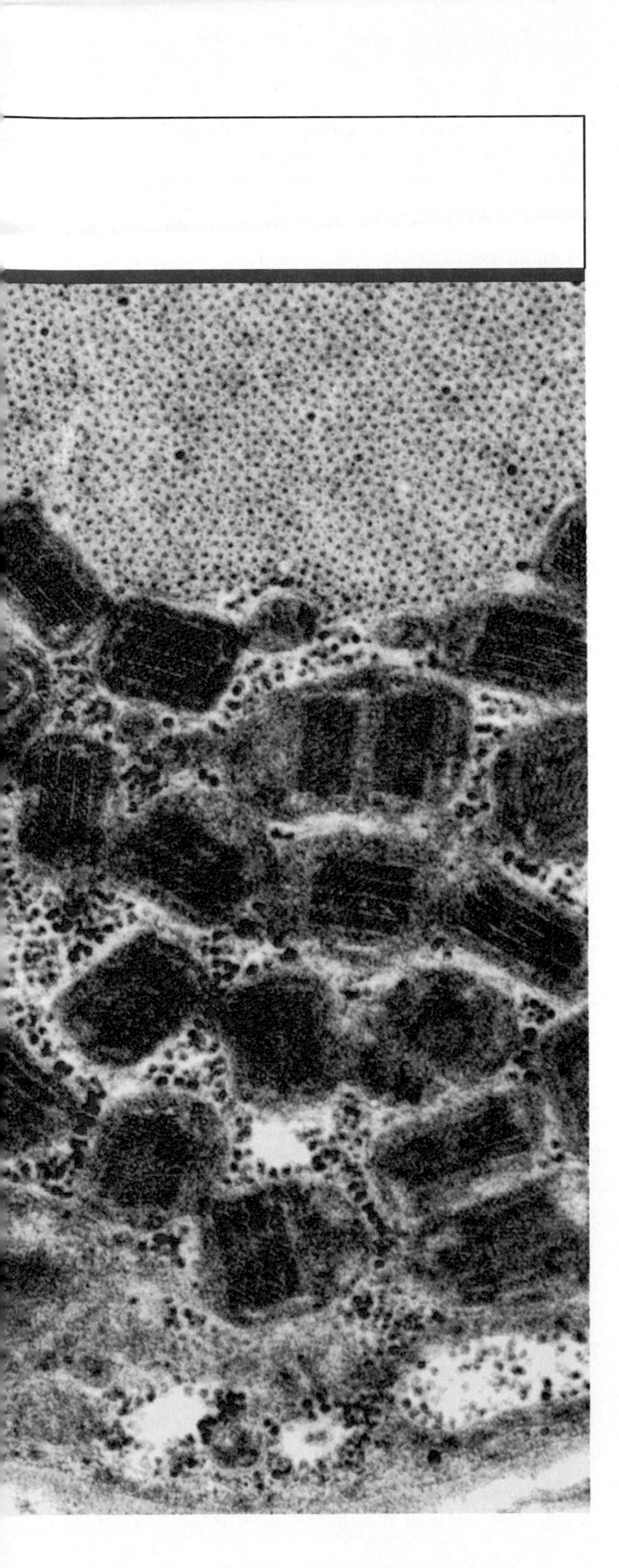

Chapter Concepts

Some traits in eukaryotes do not adhere to expected patterns associated with biparental inheritance. In most cases, these traits are controlled by genes located outside the nucleus. Transmission is usually through the female parent, whose gametes either contain maternal gene products that influence development, or are the sole source of chloroplasts and mitochondria that affect the offspring's phenotype. Plasmids in bacteria are still another example of extrachromosomal units.

Throughout the history of genetics, occasional reports have challenged the basic tenets of transmission genetics—the production of the phenotype through the transmission of genes located on chromosomes of both parents. Instead, genetic results have been observed that do not reflect either Mendelian or neo-Mendelian principles. These reports have indicated an apparent extranuclear or extrachromosomal influence on the phenotype. Such observations were often regarded with skepticism. However, with the increasing knowledge of molecular genetics and the discovery of DNA in mitochondria and chloroplasts, **extrachromosomal inheritance** is now recognized as an important aspect of genetics.

There are many diverse examples of these unusual modes of inheritance. In this chapter we will focus on three general types of extrachromosomal genetic phenomena: (1) maternal influence resulting from the effect of stored products of nuclear genes of the female parent during early development; (2) organelle heredity resulting from the expression of DNA contained in mitochondria and chloroplasts; and (3) infectious heredity resulting from the symbiotic or parasitic association of microorganisms with eukaryotic cells. Each has the effect of producing inheritance patterns that vary from those predicted by the concepts of Mendelian and neo-Mendelian genetics.

MATERNAL EFFECT

Maternal effect, also referred to as maternal influence, implies that an offspring's phenotype for a particular trait is strongly influenced by the nuclear genotype of the maternal parent. This is in contrast to most cases, where inheritance of traits is biparental. In cases of maternal effect, the genetic information of the female gamete is transcribed, and these genetic products (either proteins or yet untranslated mRNAs) are present in the egg cytoplasm. Following fertilization, these products influence patterns or traits established during early development. Two examples will illustrate the influence of the maternal genome on particular traits.

Ephestia Pigmentation

In the Mediterranean meal moth, *Ephestia kuehniella,* the wild-type larva has a pigmented skin and brown eyes as a result of the dominant gene *A*. The pigment is derived from a precursor molecule, kynurenine, which is in turn a derivative of the amino acid tryptophan. A mutation, *a*, results in red eyes and little pigmentation in larvae when homozygous. As illustrated in Figure 13.1, different results are obtained in the cross *Aa* × *aa*, depending on which parent carries the dominant gene. When the male is the heterozygous parent, a 1:1 brown/red-eyed ratio is observed in larvae as predicted by Mendelian segregation. However, when the female is heterozygous for the *A* gene, all larvae are pigmented and have brown eyes. As these larvae develop into adults, one-half of them gradually develop red eyes, reestablishing the 1:1 ratio.

One explanation of these results is that the *Aa* oocytes synthesize kynurenine or an enzyme necessary for its synthesis and accumulate it in the ooplasm prior to the completion of meiosis. Even in *aa* progeny (whose mothers were *Aa*), this pigment is distributed in the cytoplasm of the cells of the developing larvae—thus, they develop pigmentation and brown eyes. Eventually, the pigment is diluted among many cells and used up, resulting in the conversion to red eyes as adults. The *Ephestia* example demonstrates the maternal effect in which a cytoplasmically stored nuclear gene product influences the larval phenotype and at least temporarily overrides the genotype of the progeny.

Limnaea Coiling

Shell coiling in the snail *Limnaea peregra* represents a permanent rather than a transitory maternal effect. Some strains of this snail have left-handed or sinistrally coiled shells (*dd*), while others have right-handed or dextrally coiled shells (*DD* or *Dd*). These snails are hermaphroditic and may undergo either cross- or self-fertilization, thus providing a variety of types of matings.

As shown in Figure 13.2, the coiling pattern of the progeny snails is determined by **the genotype of the parent producing the egg, regardless of the phenotype of that parent.** Investigation of the developmental events in these snails reveals that the orientation of the spindle in the first cleavage division after fertilization apparently determines the direction of coiling. Spindle orientation is thought to be controlled by maternal genes acting on the developing eggs in the ovary. The orientation of the spindle, in turn, influences cell divisions following fertilization and quickly establishes the permanent adult coiling pattern. Therefore, the progeny phenotypes are determined by the maternal genotype.

The dextral allele (*D*) produces an active gene product that causes right-handed coiling. If ooplasm from dextral eggs is injected into uncleaved sinistral eggs, they cleave in a dextral pattern. However, in the converse experi-

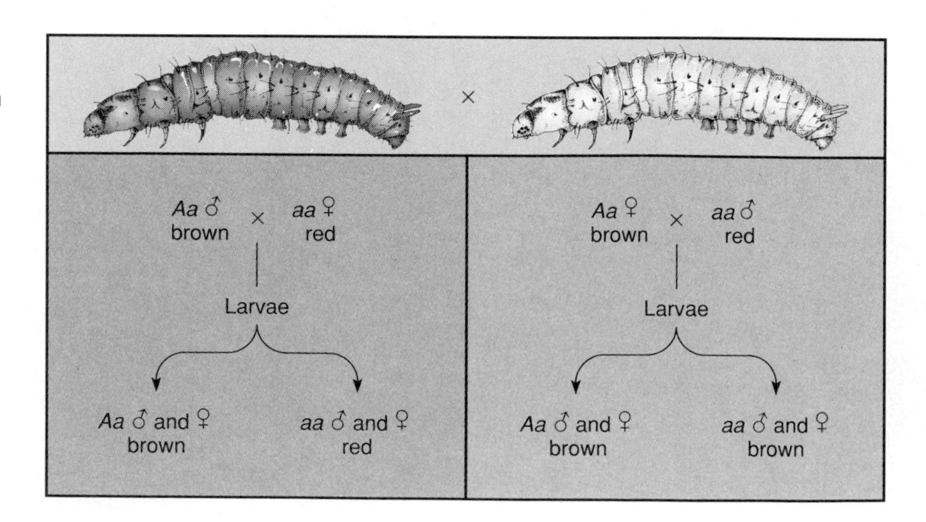

FIGURE 13.1
Illustration of maternal influence in the inheritance of eye pigment in the meal moth *Ephestia kuehniella.* Multiple light receptor structures (eyes) are present on each side of the anterior portion of larvae.

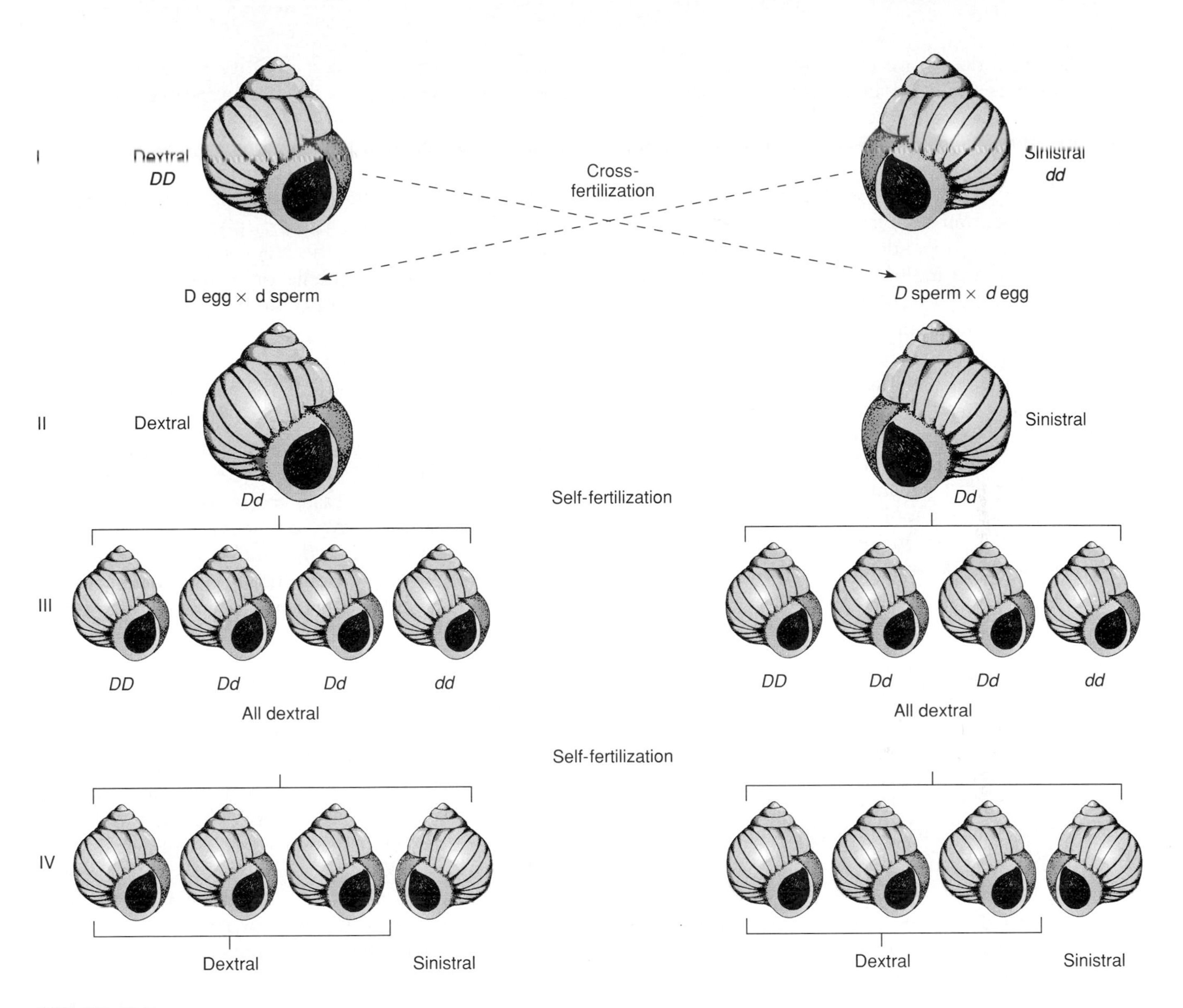

FIGURE 13.2
Inheritance of coiling in the snail *Limnaea peregra.* Coiling is either dextral (right handed) or sinistral (left handed). A maternal effect is evident in generations II and III, where the genotype of the maternal parent controls the phenotype of the offspring rather than the offspring's own genotype.

ment, sinistral ooplasm has no effect when injected into dextral eggs. Apparently, the sinistral allele is the result of a classical recessive mutation inactivating the gene product.

ORGANELLE HEREDITY: MATERNAL INHERITANCE

In this section we will examine examples of inheritance patterns of phenotypes related to chloroplast and mitochondrial function. Prior to the discovery of DNA in these organelles and extensive characterization of genetic processes occurring within them, these patterns were grouped under the category of **cytoplasmic inheritance.** That is, certain mutant phenotypes seemed to be inherited through the cytoplasm rather than through the genetic information of chromosomes. Transmission was most often from the maternal parent through the ooplasm, and thus the results of reciprocal crosses varied. Such patterns are now more appropriately considered examples of **maternal inheritance.**

While such results are similar to the examples of maternal effect presented in the preceding section, a major difference exists. Maternal effects (due to the presence of nuclear gene products) are transitory in the sense that they are not necessarily heritable. With maternal inheritance, however, the phenotype is stable and is continually passed to future generations.

Analysis of the hereditary transmission of mutant alleles of chloroplast and mitochondrial DNA has been difficult owing to a number of factors. First, the function of these organelles is dependent upon gene products of both nuclear and organelle DNA. Second, the transmission of mitochondrial or chloroplast DNA to progeny is necessarily through the transmission of the organelle itself. Sometimes, it is difficult to determine the origin (maternal vs. paternal) of the organelle contributed to the zygote. Third, the number of organelles contributed to each progeny often exceeds one. If many chloroplasts and/or mitochondria are contributed and only one or a few of them contain a mutant gene, the corresponding mutant phenotype may not be revealed. Taken together, these complexities have made analysis much more involved than for Mendelian characters.

In this section, for both chloroplasts and mitochondria, we shall discuss examples of inheritance patterns seemingly related to the respective organelle. We have previously provided a fairly detailed analysis of the molecular genetics of these organelles in the previous chapter. You should review and relate this information to the discussion that follows.

Chloroplasts: Variegation in Four O'Clock Plants

In 1908, Carl Correns (one of the rediscoverers of Mendel's work) provided the earliest example of inheritance linked to chloroplast transmission. Correns discovered a variety of the four o'clock plant, *Mirabilis jalapa,* which had branches with either white, green, or variegated leaves. As shown in Table 13.1, inheritance in all possible combinations of crosses is strictly determined by the *phenotype* of the ovule source. For example, if the seeds (representing the progeny) were derived from ovules on branches with green leaves, all progeny plants bore only green leaves regardless of the phenotype of the source of pollen.

Correns concluded that inheritance was through the cytoplasm of the maternal parent because the pollen, which contributes little or no cytoplasm to the zygote, had no influence on the progeny phenotypes. Since the leaf coloration involves the chloroplast, either genetic information in that organelle or in the cytoplasm and influencing the chloroplast could be responsible for this inheritance pattern.

Iojap in Maize

A phenotype similar to *Mirabilis* but with a different pattern of inheritance has been analyzed in maize by Marcus M. Rhoades. In this case, green, colorless, or green-and-colorless striped leaves are under the influence not only of the cytoplasm, but, in addition, of a nuclear gene located on the seventh linkage group. This locus is called *iojap,* after the recessive mutation *ij* located there. The wild-type allele is designated *Ij*. Plants homozygous for the mutation *ij/ij* may have green-and-white striped leaves. However, when reciprocal crosses are made between plants with striped leaves and green leaves, the results are seen to vary, depending upon which parent is mutant (Figure 13.3). If the female is striped (*ij/ij*) and the male is green (*Ij/Ij*), plants with colorless, striped, and green leaves are observed as progeny. If the male parent is striped and the female is green, only green plants are produced! In both types of cross, all offspring have identical genotypes (*Ij/ij*). We may conclude that although a nuclear gene is somehow involved, the inheritance pattern is influenced maternally.

We can understand this pattern better by examining the offspring resulting from self-fertilization of these heterozygous plants (Figure 13.3). The striped plant gives rise to progeny with colorless, striped, and green leaves, regardless of their genotype. The green plants give rise to both green and striped progeny in a 3:1 ratio. The results

TABLE 13.1
Crosses between flowers from various branches of variegated four o'clock plants.

	Source of Ovule		
Source of Pollen	**White Branch**	**Green Branch**	**Variegated Branch**
White branch	White	Green	White, green, or variegated
Green branch	White	Green	White, green, or variegated
Variegated branch	White	Green	White, green, or variegated

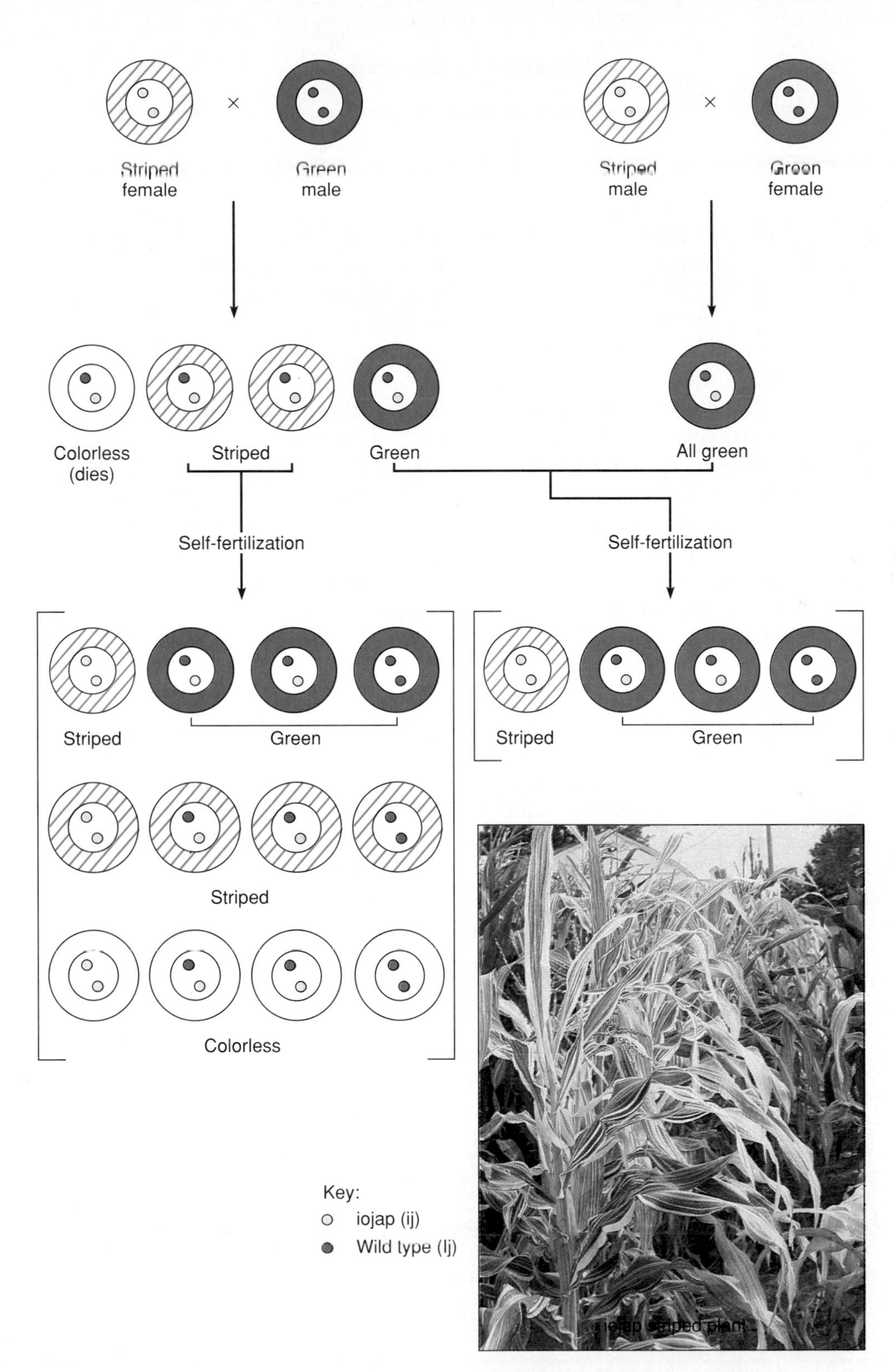

FIGURE 13.3
Maternal inheritance of striping in maize. Regardless of the genotype of the *iojap* genes, offspring reflect the maternal phenotypes in their appearance. If the maternal parent is striped (containing green *and* colorless areas), progeny are either colorless, green, or striped. If the maternal parent has green leaves, so do all progeny plants. Since color is due to chloroplasts in the leaves, inheritance is controlled by these organelles as they are passed through the maternal cytoplasm.

THE GENETIC STUDY OF HUMAN ORIGINS

Historically, we have relied on fossil remains and artifacts to map human evolutionary history. Recent advances in genetic technology have allowed us to look at evolution of genes and proteins at the molecular level. As a result, we are obtaining much clearer insights into the biological origins of humankind.

Basically, evolution is the result of genetic changes in a population occurring over time. At the DNA level this is seen as changes in the nucleotide sequence. As evolution proceeds, mutational changes accumulate such that the sequence may become quite different from the ancestral sequence. By comparing equivalent sequences from related organisms, we can reconstruct the evolutionary history of a group of populations or species. The number of changes in the DNA sequence can thus be regarded as a clock, measuring evolutionary time. In appropriate instances genealogical trees can be constructed to illustrate ancestral relationships. The root of the tree is the ancestral type and the evolutionary derivatives are depicted as branches and sub-branches. Complex statistical computer programs are used to define the most parsimonious tree, that is, the one that fits the data best. In such studies, it is assumed that long-established populations will contain more variation than newly emerged populations.

Conclusions from molecular genetic analysis are not always the same as those derived at by the study of fossils. A controversial paper published in 1987 by R. Cann and her associates sparked a debate about the origins of human beings. They carried out a molecular analysis of mitochondrial (mt) DNA from 147 individuals representing five very different populations. Africans, Caucasians, Asians, native Australians, and New Guineans. Mt DNA is a small circular DNA molecule. It was chosen for analysis because, unlike nuclear DNA, it is inherited solely from the mother; consequently, only female ancestors need be considered. The fact that mitochondrial DNA cannot be altered by recombination with paternal DNA also significantly simplifies the analysis. Another advantage is an apparent high mutation rate, which improves resolution of ancestry over a relatively short evolutionary period. Cann found 133 distinct types of mt DNA in the 147 samples. Parsimony analysis gave a genealogical tree with two primary branches, one composed entirely of Africans, the other with some Africans and all the other groups. However, their main and most controversial conclusion was that all the variation seen could be traced back to a single female who lived in Africa about 200,000 years ago. This hypothetical individual became popularly known as Eve, whose descendants presumably left Africa and gave rise to all other human populations.

Cann's conclusions are in sharp contrast to a theory of multiregional evolution, which also proposes that humans originated in Africa but migrated to other areas at least a million years ago, undergoing parallel evolution within widespread populations. The interpretation of the DNA data suggests that descendants of the migrant African population totally replaced all the others since that time. However, there is no archaeological evidence supporting this concept.

The debate is not over. Arguments now have been advanced concerning the validity of the genetic interpretation of Cann's data, which used a particular computer program to generate the most parsimonious tree. It has since been shown that there are many different trees that could have been generated, many of which did not have an African origin. In addition, computer analysis may be influenced by the order of entry of the data. The flaws in Cann's analysis are now generally acknowledged, and it is conceded that the analysis does not confirm or rule out her theory.

We may have learned from this study more about the fallibility of science than about the human origins. Nonetheless, DNA remains a rich source of clues about evolutionary history, once the sole province of archaeologists and palaeontologists.

of these self-fertilizations substantiate that the mutant phenotype is controlled solely through the female cytoplasm, regardless of the plant's nuclear genotype.

The role of the nucleus in the origin of the mutant phenotype is unclear. However, the colorless areas of the leaf are due to the lack of the green chlorophyll pigment in chloroplasts. Once acquired, colorless chloroplasts are transmitted through the egg cytoplasm, establishing the phenotypes of leaves of progeny plants.

Chlamydomonas Mutations

The unicellular green alga *Chlamydomonas reinhardi* has provided an excellent system for the investigation of maternal inheritance. The organism is eukaryotic and contains a single large chloroplast as well as numerous mitochondria. Matings are followed by meiosis and the various stages of the life cycle are easily studied in culture in the laboratory. The first cytoplasmic mutant, *streptomycin resistance* (*sr*), was reported in 1954 by Ruth Sager. Although *Chlamydomonas'* two mating types —mt^+ and mt^-—appear to make equal cytoplasmic contributions to the zygote, Sager determined that the *sr* phenotype is transmitted only through the mt^+ parent.

Since this discovery, a number of other *Chlamydomonas* mutations (including an acetate requirement as well as resistance to or dependence on a variety of bacterial antibiotics) have been discovered that show a similar maternal inheritance pattern. These mutations have been linked to the transmission of the chloroplast, and their study has extended our knowledge of chloroplast inheritance.

Following fertilization, the single chloroplasts of the two mating types fuse. After the resulting zygote has undergone meiosis, it is apparent that the genetic information of the chloroplasts of progeny cells is derived only from the mt^+ parent. The mt^- chloroplast has been destroyed!

Further, studies suggest that these chloroplast mutations, representing numerous gene sites, form a single circular linkage group—the first such extrachromosomal unit to be established in a eukaryotic organism. It is also apparent that the linkage groups from the two mating types are capable of undergoing recombination following zygote formation. All available evidence supports the hypothesis that this linkage group consists of DNA residing in the chloroplast.

Mitochondria: *poky* in *Neurospora*

Mitochondria, like chloroplasts, play a critical role in cellular bioenergetics and contain a distinctive genetic system. Mutants affecting mitochondrial function have been discovered and studied. As with chloroplast mutants, these are transmitted through the cytoplasm and result in non-Mendelian inheritance patterns. Additionally, the mitochondrial genetic system has now been extensively characterized.

In 1952, Mary B. and Hershel K. Mitchell discovered a slow-growing mutant strain of the mold *Neurospora crassa* and called it *poky*. (It is also designated *mi-1* for maternal inheritance.) Studies have shown slow growth to be associated with impaired mitochondrial function specifically related to certain cytochromes essential to electron transport. The results of genetic crosses between wild-type and *poky* strains suggest that the trait is maternally inherited. If the female parent is *poky* and the male parent is wild type, all progeny colonies are *poky*. The reciprocal cross produces normal colonies.

Studies with *poky* mutants illustrate the use of **heterokaryon** formation during the investigation of maternal inheritance in fungi. Occasionally, hyphae from separate mycelia fuse with one another, giving rise to structures containing two or more nuclei in a common cytoplasm. If the hyphae contain nuclei of different genotypes, the structure is called a heterokaryon. The cytoplasm will contain mitochondria derived from both initial mycelia. A heterokaryon may give rise to haploid spores, or **conidia,** that produce new mycelia. The phenotypes of these structures may be determined.

Heterokaryons produced by the fusion of *poky* and wild-type hyphae initially show normal rates of growth and respiration. However, mycelia produced through conidia formation become progressively more abnormal until they show the *poky* phenotype. This occurs in spite of the presumed presence of both wild-type and *poky* mitochondria in the cytoplasm of the hyphae.

To explain the initial growth and respiration pattern, it is assumed that the wild-type mitochondria support the respiratory needs of the hyphae. The subsequent expression of the *poky* phenotype suggests that the presence of the *poky* mitochondria may somehow prevent or depress the function of these wild-type mitochondria. Perhaps the *poky* mitochondria replicate more rapidly and "wash out" or dilute wild-type mitochondria numerically. Another possibility is that *poky* mitochondria produce a substance that inactivates the wild-type organelle or interferes with the replication of its DNA (mtDNA). As a result of this type of interaction, mutations such as *poky* are labeled **suppressive.** This general phenomenon is characteristic of many other suspected mitochondrial mutations of *Neurospora* and yeast.

Petite in *Saccharomyces*

Another extensive study of mitochondrial mutations has been performed with the yeast *Saccharomyces cerevisiae*. The first such mutation, *petite,* was described by Boris

Ephrussi and his coworkers in 1956. The mutant is so named because of the small size of the yeast colonies. Many independent *petite* mutations have since been discovered and studied. They all have a common characteristic: deficiency in cellular respiration involving abnormal electron transport. Fortunately, this organism is a facultative anaerobe and can grow by fermenting glucose through glycolysis. Thus, although colonies are small, the organism may survive the loss of mitochondrial function by generating energy anaerobically.

The complex genetics of *petite* mutations is diagramed in Figure 13.4. A small proportion of these mutants exhibit Mendelian inheritance and are called **segregational petites,** indicating that they are the result of nuclear mutations. The remainder demonstrates cytoplasmic transmission, producing one of two effects in matings. The **neutral petites,** when crossed to wild type, produce ascospores that give rise only to wild-type or normal colonies. The same pattern continues if progeny of this cross are back-crossed to *neutral petites.* This is because the majority of neutrals lack mtDNA completely or have lost a substantial portion of it. Thus, the wild-type gamete is the effective source of normal mitochondria capable of reproduction. Neutral petites have now been shown to lack most, if not all, mitochondrial DNA (mtDNA).

A third type, the **suppressive petites,** behaves similarly to *poky* in *Neurospora.* Crosses between mutant and wild type give rise to mutant diploid zygotes, which, upon undergoing meiosis, immediately yield all mutant

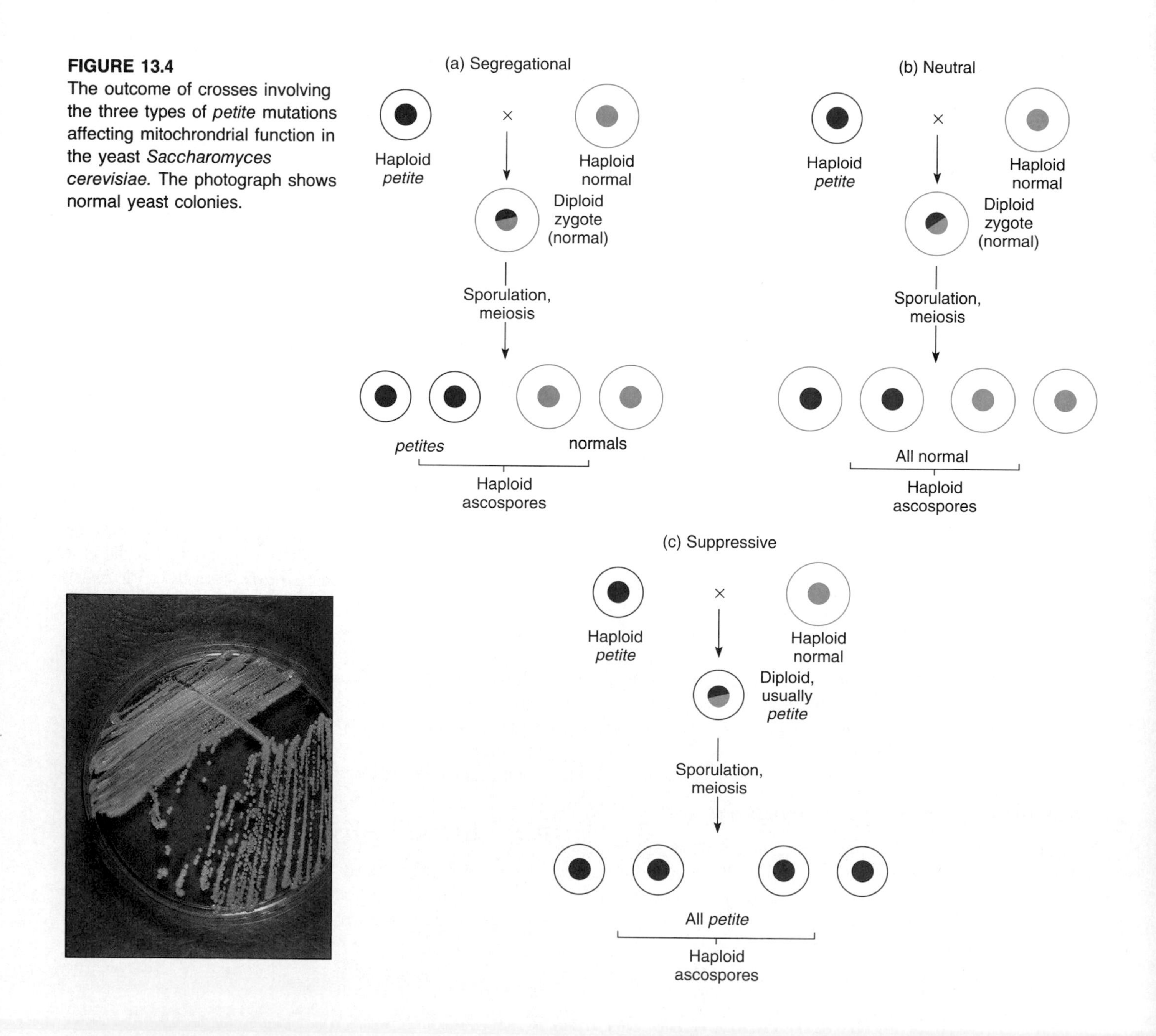

FIGURE 13.4
The outcome of crosses involving the three types of *petite* mutations affecting mitochrondrial function in the yeast *Saccharomyces cerevisiae.* The photograph shows normal yeast colonies.

cells. Under these conditions, the *petite* mutation behaves "dominantly" and seems to suppress the function of the wild-type mitochondria. *Suppressive petites* also represent deletions of mtDNA, but not nearly to the extent of the *neutral petites.* Buoyant density studies show a lower G-C content in suppressive mtDNA compared with normal mtDNA.

Suppressiveness remains unexplained. Two major hypotheses have been advanced. One explanation suggests that the mutant (or deleted) mtDNA replicates more rapidly, and thus mutant mitochondria "take over" or dominate the phenotype by numbers alone. The second explanation suggests that recombination occurs between the mutant and wild-type mtDNA, introducing errors into or disrupting the normal mtDNA. It is not yet clear which, if either, of these explanations is correct.

Mitochondrial DNA and Human Diseases

As you may recall from Chapter 12, the DNA found in mitochondria **(mtDNA)** has been extensively characterized in a variety of organisms, including humans. Human mtDNA, which is circular, contains 16,569 base pairs and is strictly inherited maternally. The mitochondrial gene products encoded include:

13 proteins, required for oxidative respiration
22 tRNAs, required for translation
2 rRNAs, required for translation

Unlike the nuclear genome, mitochondria contain little extraneous DNA that fails to code for gene products. Thus, mtDNA would seem particularly vulnerable to mutations, since most genetic alterations can potentially result in a disruption of either translation or oxidative respiration within the organelle.

On the other hand, a zygote receives a large number of organelles through the egg, so if only one of them contains a mutation, its impact is diluted because there will be many more mitochondria that will function normally. During early development, cell division disperses the initial population of mitochondria present in the zygote, and in newly formed cells, these organelles reproduce autonomously. Therefore, adults will exhibit cells with a variable mixture of normal and abnormal organelles, should a deleterious mutation arise or already be present in the initial population of organelles. This is a condition called **heteroplasmy.**

In order for a human disorder to be attributable to genetically altered mitochondria, several criteria must be met:

1. Inheritance must exhibit a maternal rather than a Mendelian pattern.
2. The disorder must reflect a deficiency in the bioenergetic function of the organelle.
3. A specific genetic mutation in one of the mitochondrial genes must be documented.

Thus far, several such cases are known that demonstrate these characteristics.

Myoclonic epilepsy and ragged red fiber disease (MERRF) demonstrates a pedigree consistent with maternal inheritance. Individuals with this rare disorder express deafness and dementia in addition to seizures. Both muscle fibers and mitochondria are abnormal in appearance. Such aberrant mitochondria are evident in Figure 13.5. Upon analysis of mtDNA, a single nucleotide

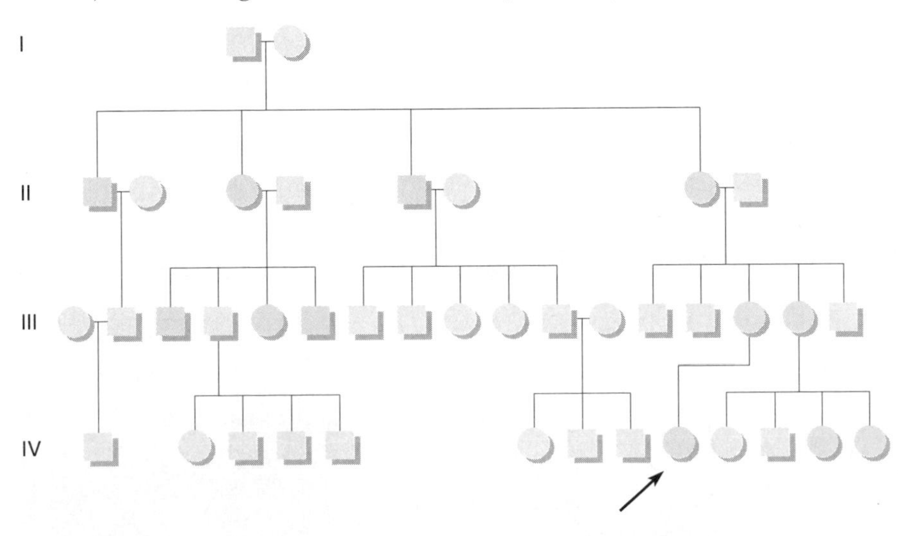

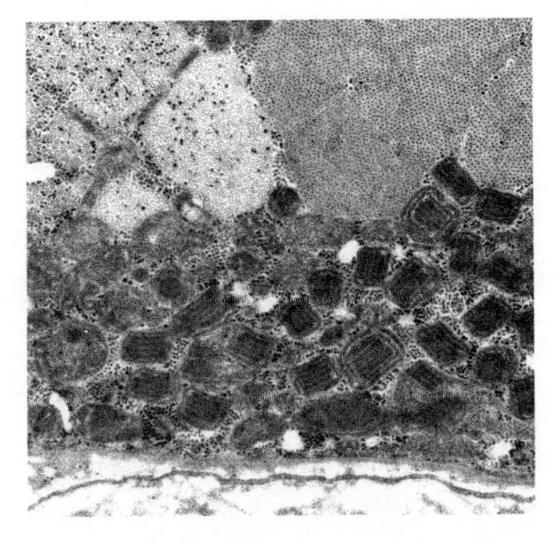

FIGURE 13.5
A partial pedigree illustrating maternal inheritance of myoclonic epilepsy. Some of the individuals were not available for analysis, including the great-grandparents (I) of the proband (arrow). The photograph demonstrates abnormal mitochondria derived from cells of an afflicted individual. Note the highly abnormal cristae.

substitution has been found in the transfer RNA specific for lysine ($tRNA^{lys}$). This change apparently interferes with translation within the organelle. Presumably, the deficiency in efficient mitochondrial function is related to the various manifestations of the disorder.

A second disorder, **Leber's hereditary optic neuropathy (LHON)** also exhibits maternal inheritance as well as mtDNA lesions. In this case, a missense mutation has been identified in the gene encoding subunit 4 of NADH:ubiquinone reductase (ND4). Some patients display still other nucleotide changes. In a third disorder, **Kearns-Sayre syndrome,** patients develop encephalomyopathy and bear large deletions at various positions within their mtDNA. It is felt that these heterogeneous deletions in different patients lead to a common phenotype due to the disruption of the translational machinery within these organelles.

The study of hereditary, mitochondrial-based disorders provides insights into the importance of this organelle during normal development as well as the relationship between mitochondrial function and neuromuscular disorders.

INFECTIOUS HEREDITY

There are numerous examples of cytoplasmically transmitted phenotypes in eukaryotes that are due to an invading microorganism or particle. The foreign invader coexists in a symbiotic relationship, is usually passed through the maternal ooplasm to progeny cells or organisms, and confers a specific phenotype that may be studied. We shall consider several examples illustrating this phenomenon.

Kappa in *Paramecium*

First described by Tracy Sonneborn, certain strains of *Paramecium aurelia* are called **Killers** because they release a cytoplasmic substance called **paramecin** that is toxic and sometimes lethal to sensitive strains. This substance is produced by particles called **kappa** that replicate in the Killer cytoplasm, contain DNA and protein, and depend for their maintenance on a dominant nuclear gene *K*. One cell may contain 100 to 200 such particles.

Paramecia are diploid protozoans that can undergo

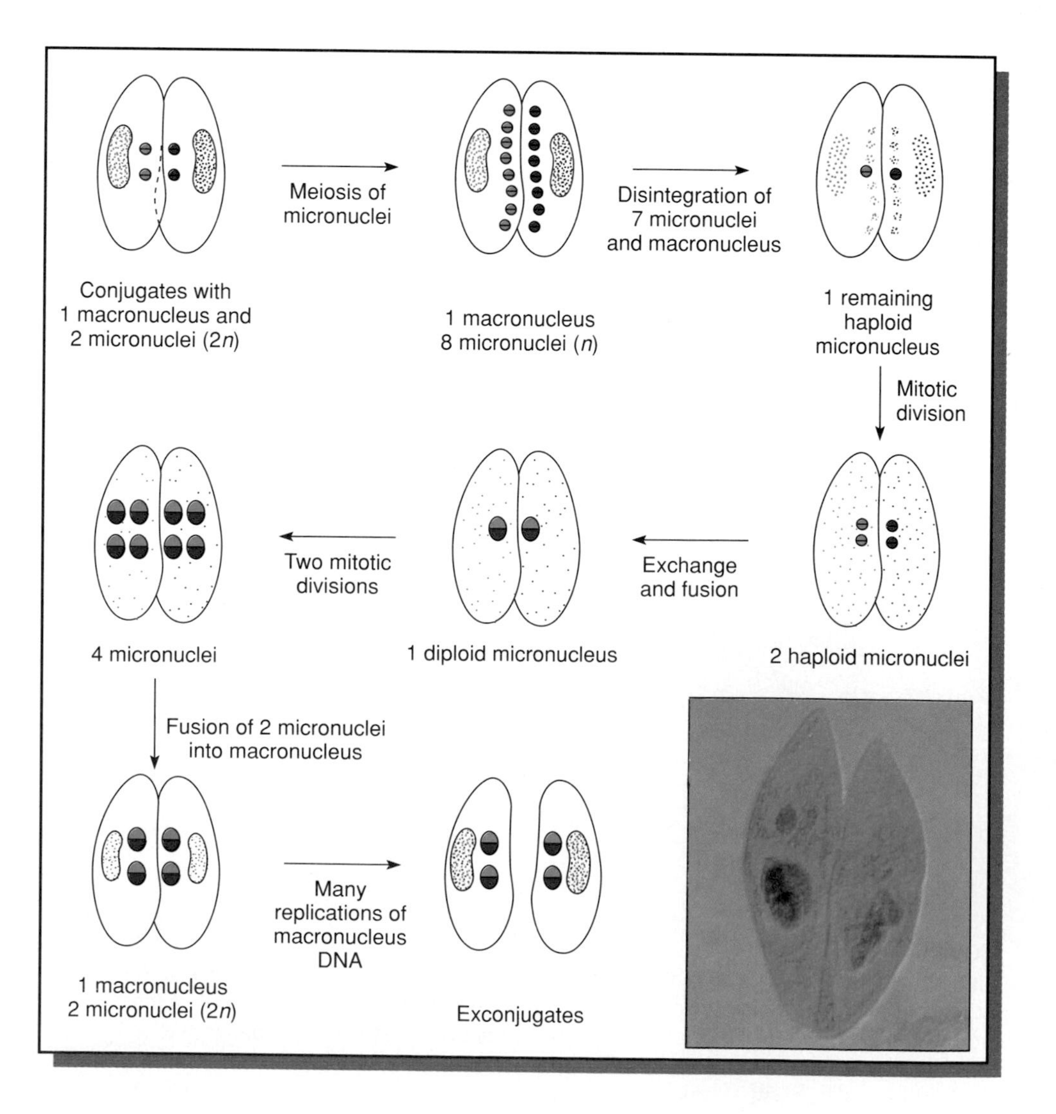

FIGURE 13.6 Genetic events occurring during conjugation in *Paramecium.* The photographic insert shows a pair of organisms undergoing conjugation.

sexual exchange of genetic information through the process of **conjugation.** In some instances, cytoplasmic exchange also occurs. Thus, there is a variety of ways in which the *K* gene and kappa can be transmitted.

The genetic events occurring during conjugation in *Paramecium aurelia* are shown in Figure 13.6. Paramecia contain two diploid micronuclei. Early in conjugation, both micronuclei in each mating pair undergo meiosis, resulting in eight haploid nuclei. However, seven of these degenerate, and the remaining one undergoes a single mitotic division. Each cell then donates one of the two haploid nuclei to the other, recreating the diploid condition in both cells. As a result, exconjugates are of identical genotypes.

In a similar process involving only a single cell, **autogamy** occurs. Following meiosis of both micronuclei, seven products degenerate and one survives. This nucleus divides, and the resulting nuclei fuse to recreate the diploid condition. If the original cell was heterozygous, autogamy results in homozygosity because the newly formed diploid nucleus was derived solely from a single haploid meiotic product. In a population of cells that were originally heterozygous, half of the new cells express one allele and half express the other allele.

Figure 13.7 illustrates the results of crosses between *KK* and *kk* cells, without and with cytoplasmic exchange. When no cytoplasmic exchange occurs, even though the resultant cells may be *Kk* (or *KK* following autogamy), they remain sensitive if no kappa particles are transmitted. When exchange occurs, the cells become Killers provided the kappa particles are supported by at least one dominant *K* allele.

Kappa particles are bacterialike and may contain temperate bacteriophages. One theory holds that these viruses of kappa may become vegetative; during this multiplication, they produce the toxic products that are released and kill sensitive strains.

Infective Particles in *Drosophila*

Two examples of similar phenomena are known in *Drosophila:* **CO_2 sensitivity** and **sex-ratio.** In the former, flies that would normally recover from carbon dioxide anesthetization instead become permanently paralyzed and are killed by CO_2. Sensitive mothers pass this trait to all offspring. Furthermore, extracts of sensitive flies induce the trait when injected into resistant flies. Phillip L'Heritier has postulated that sensitivity is due to the presence of a virus, **sigma.** The particle has been visualized and is smaller than kappa. Attempts to transfer the virus to other insects have been unsuccessful, demonstrating that specific nuclear genes support the presence of sigma in *Drosophila.*

A second example of infective particles comes from the study of *Drosophila bifasciata.* A small number of these flies were found to produce predominantly female offspring if reared at 21°C or lower. This condition, designated sex-ratio, was shown to be transmitted to daughters but not to the low percentage of males produced. This phenomenon was subsequently investigated in *Drosophila willistoni.* In these flies, the injection of ooplasm from sex-ratio females into normal females induced the condition. This observation suggests that an extrachromosomal element is responsible for the sex-ratio pheno-

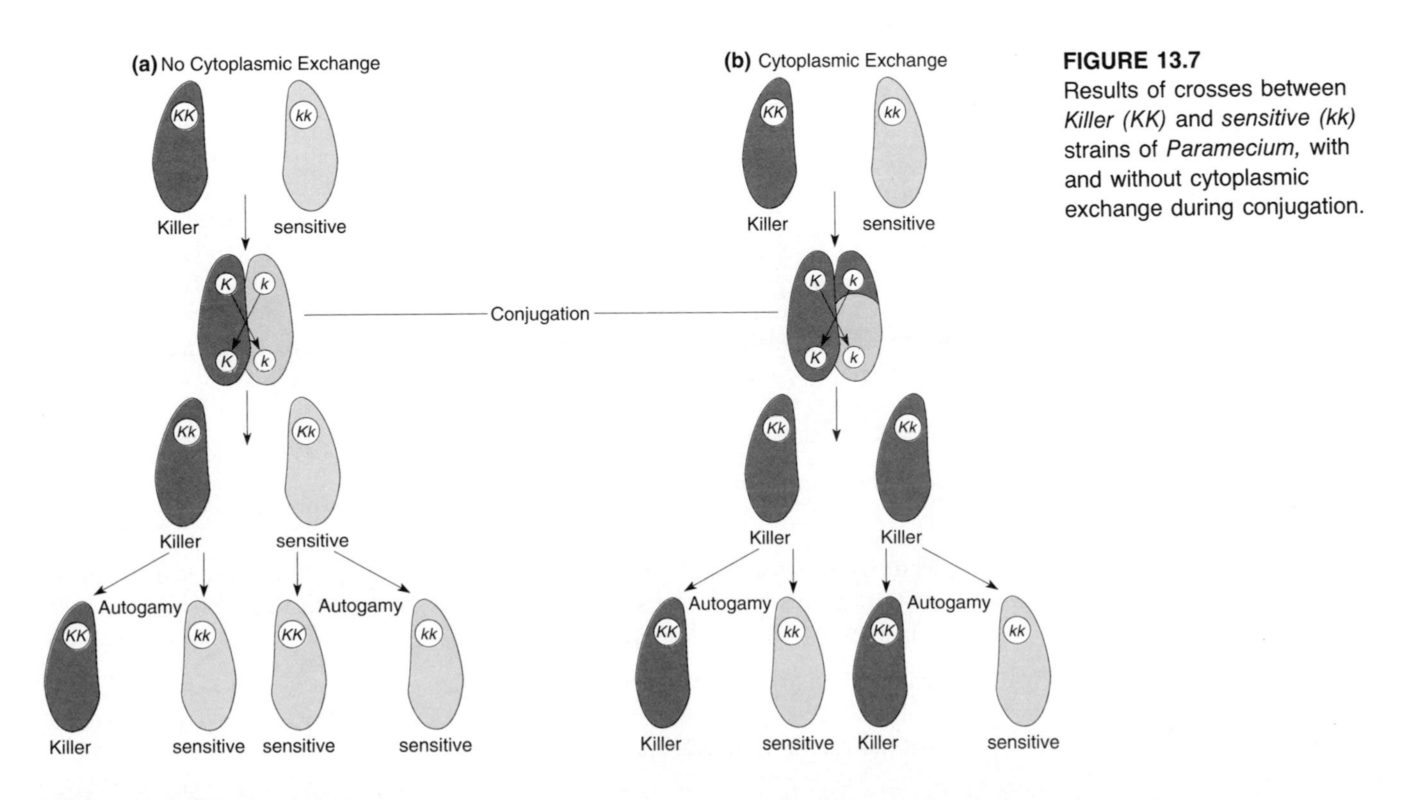

FIGURE 13.7
Results of crosses between *Killer (KK)* and *sensitive (kk)* strains of *Paramecium,* with and without cytoplasmic exchange during conjugation.

type. The agent has now been isolated and shown to be a protozoan. While the protozoan has been found in both males and females, it is lethal primarily to developing male larvae. There is now some evidence that a virus harbored by the protozoan may be responsible for producing a male-lethal toxin.

PLASMIDS IN BACTERIA

We conclude this chapter with a brief introduction to extrachromosomal genetic elements found in bacteria. Referred to generally as **plasmids,** these are small circular DNA units that are distinct from the larger bacterial chromosome. In some cases, plasmids *also* demonstrate the ability to integrate into the host chromosome, in which case they are more appropriately called **episomes.** While in the cytoplasm, plasmids replicate along with the chromosomes and are distributed to daughter cells following division.

While plasmids will constitute an important topic in the next chapter, we will provide a short survey of the various types since inheritance patterns involving their stored genetic information varies from that of chromosomal genes. They thus qualify as examples of extrachromosomal inheritance.

F Factors

The plasmid that confers the ability for a bacterium to undergo conjugation is called the **F factor** (F for fertility). In *E. coli,* this is about 10^5 base pairs in size and contains genes encoding molecules essential to mating. Cells containing the F factor (called $\mathbf{F^+}$) can produce a conjugation tube that will fuse with cells lacking the F factor (called $\mathbf{F^-}$). During conjugation, the F factor is replicated with one copy being passed to the recipient cell. Thus, following conjugation, the recipient cell becomes F^+:

$$F^+ \times F^- \longrightarrow \text{all } F^+ \text{ cells}$$

As we shall see in Chapter 14, under certain conditions, F^+ cells may also transfer part of their chromosome to the recipient cell, altering its genotype. This occurs when the F factor integrates into the host chromosome. When this occurs, the cell remains as a special type of F^+ cell (called **Hfr**). The ability to integrate into the host chromosome distinguishes the F factor from the subsequent examples of plasmids, which we will discuss below.

R Factors

R factors are plasmids that, like the F factor, provide the ability to induce conjugation. But R factors are specifically characterized by the presence of additional genes that confer resistance to antibiotics such as sulfonamides, streptomycin, and tetracycline. The portion of the R factor plasmid that confers the ability to be transferred to a recipient cell is called the **resistance transfer factor (RTF).**

The R plasmid, which may be present in one to three copies per cell, is significant because it provides the means for rapid conferral of multiple drug resistance to cells previously sensitive to antibiotics. This phenomenon came to the attention of researchers in the late 1950s in Japan when multiply-resistant strains of *Shigella* (which causes dysentery) were found in the gut of humans. *Shigella* had acquired this resistance following conjugation with *E. coli,* where the R factor originated. In such cases, antibiotic treatment kills off the sensitive cells, decreasing the competition resistant cells encounter. The resistant cells then become the dominant "flora" of the gut.

Col Factors

A third type of plasmid carries genes encoding **colicins,** proteins that kill susceptible bacterial cells lacking the **Col factor.** There are numerous varieties of this factor, which is small (10–20,000 base pairs) but present in 10–20 copies per cell. The Col factor includes a gene that confers immunity to the colicins that kill cells lacking the plasmid. The Col factor, unlike the RTF portion characteristic of R factors, is not normally transferred to other cells.

We shall return to a more detailed consideration of plasmids in the following chapter. For now, we include them in the rather broad category of examples of extrachromosomal inheritance.

CHAPTER SUMMARY

1 Patterns of inheritance sometimes vary from that expected of nuclear genes. In such instances, phenotypes most often appear to result from genetic information transmitted through the egg.

2 Maternal-effect patterns result when nuclear gene products controlled by the maternal genotype of the egg influence early development. *Ephestia* pigmentation and coiling in snails are examples.

3 Maternal inheritance patterns result from the genotypes of chloroplast and mitochondrial DNA as these organelles are transmitted through the egg. These are cases of true extrachromosomal inheritance.

4 Chloroplast mutations affect the photosynthetic capabilities of plants, while mitochondrial mutations affect cells highly dependent on ATP generated through cellular respiration. The resulting mutants display phenotypes related to the loss of function of these organelles.

5 Another form of extrachromosomal inheritance is due to the transmission of infectious microorganisms, which establish symbiotic relationships with their host cells. Kappa particles and CO_2-sensitivity and sex-ratio determinants are examples.

6 Plasmids are extrachromosomal genetic units that are characteristic of many bacterial cells, conferring various capabilities on cells bearing them.

KEY TERMS

autogamy
CO_2 sensitivity
Col factors
conjugation
cytoplasmic inheritance
episomes
F factor
heterokaryon
heteroplasmy
iojap locus (*ij*)
kappa
Kearns-Sayre syndrome
Killer strains
Leber's hereditary optic neuropathy (LHON)
maternal effect
maternal inheritance
mtDNA
myoclonic epilepsy and ragged red fiber disease (MERRF)
paramecin
petite mutation
plasmids
poky mutation
resistance transfer factor (RTF)
segregational petites
sex-ratio
suppressive mutation
suppressive petites

1 Analyze the following theoretical pedigree and determine the most consistent interpretation of how the trait is inherited. Given your interpretation, are there any inconsistencies in the various individuals?

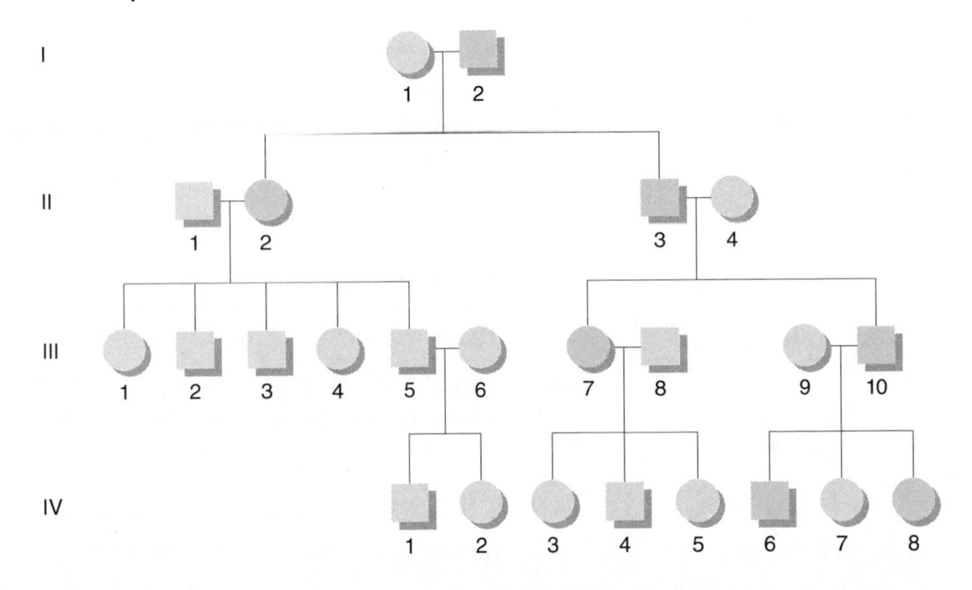

SOLUTION: The trait is passed from all male parents to all but one offspring and *never* passed maternally. Individual IV-7 (a female) is the only exception.

2 Can the above explanation be attributed to a gene on the Y chromosome? Defend your answer.

ANSWER: No, since male parents pass the trait to their daughters as well as to their sons.

3 Is the above case an example of a paternal effect or of paternal inheritance?

ANSWER: It has all the earmarks of paternal inheritance since males pass the trait to all of their offspring. To assess whether the trait is due to a paternal effect (resulting from a nuclear gene in the male gamete), analysis of further matings would be needed.

PROBLEMS AND DISCUSSION QUESTIONS

1 What genetic criteria distinguish a case of extrachromosomal inheritance from a case of Mendelian autosomal inheritance? from a case of sex-linked inheritance?

2 In *Limnaea,* what results would be expected in a cross between a *Dd* dextrally coiled and a *Dd* sinistrally coiled snail, assuming cross-fertilization occurs as shown in Figure 13.2? What results would occur if the *Dd* dextral produced only eggs and the *Dd* sinistral produced only sperm?

3 Streptomycin resistance in *Chlamydomonas* may result from a mutation in a chloroplast gene or in a nuclear gene. What phenotypic results would occur in a cross between a member of an mt^+ strain resistant in both genes and a member of an mt^- strain sensitive to the antibiotic? What results would occur in the reciprocal cross?

4 A plant may have green, white, or green and white (variegated) leaves on its branches owing to a mutation in the chloroplast (which produces the white leaves). Predict the results of the following crosses.

	Ovule Source		Pollen Source
(a)	Green branch	×	White branch
(b)	White branch	×	Green branch
(c)	Variegated branch	×	Green branch
(d)	Green branch	×	Variegated branch

5 In diploid yeast strains, sporulation and subsequent meiosis can produce haploid ascospores. These may fuse to reestablish diploid cells. When ascospores from a *segregational petite* strain fuse with those of a normal wild-type strain, the diploid zygotes are all normal. However, following meiosis, ascospores are 1/2 petite and 1/2 normal. Is the *segregational petite* phenotype inherited as a dominant or recessive gene?

6 Predict the results of a cross between ascospores from a *segregational petite* strain and a *neutral petite* strain. Indicate the phenotype of the zygote and the ascospores it may subsequently produce.

7 Described below are the results of three crosses between strains of *Paramecium.* Determine the genotypes of the parental strains.

(a) Killer × sensitive → 1/2 Killer: 1/2 sensitive
(b) Killer × sensitive → all Killer
(c) Killer × sensitive → 3/4 Killer: 1/4 sensitive

8 *Chlamydomonas,* a eukaryotic green alga, is sensitive to the antibiotic erythromycin that inhibits protein synthesis in prokaryotes.
  (a) Explain why.
  (b) There are two mating types in this alga, mt^+ and mt^-. If an mt^+ cell sensitive to the antibiotic is crossed with an mt^- cell that is resistant, all progeny cells are sensitive. The reciprocal cross (mt^+ resistant and mt^- sensitive) yields all resistant progeny cells. Assuming that the mutation for resistance is in cpDNA, what can be concluded?

9 In *Limnaea,* a cross where the snail contributing the eggs was dextral but of unknown genotype mated with another snail of unknown genotype and phenotype. All F_1 offspring exhibited dextral coiling. Ten of the F_1 snails were allowed to undergo self-fertilization. One-half produced only dextrally coiled offspring, while the other half produced only sinistrally coiled offspring. What were the genotypes of the original parents?

10 In *Drosophila subobscura,* the presence of a recessive gene called *grandchildless* (*gs*) causes the offspring of homozygous females, but not homozygous males, to be sterile. Can you offer an explanation as to why females but not males are affected by the mutant gene?

SELECTED READINGS

BOGORAD, L. 1981. Chloroplasts. *J. Cell Biol.* 91:256s–70s.
COHEN, S. 1973. Mitochondria and chloroplasts revisited. *Amer. Sci.* 61:437–45.
FREEMAN, G., and LUNDELIUS, J. W. 1982. The developmental genetics of dextrality and sinistrality in the gastropod *Lymnaea peregra. Wilhelm Roux Arch.* 191:69–83.
GILLHAM, N. W. 1978. *Organelle hereditary.* New York: Raven Press.
GOODENOUGH, U., and LEVINE, R. P. 1970. The genetic activity of mitochondria and chloroplasts. *Scient. Amer.* (Nov.) 223:22–29.
GRIVELL, L. A. 1983. Mitochondrial DNA. *Scient. Amer.* (March) 248:78–89.
LANDER, E. S., et al. 1990. Mitochondrial diseases: Gene mapping and gene therapy. *Cell* 61:925–26.
LEVINE, R. P., and GOODENOUGH, U. 1970. The genetics of photosynthesis and of the chloroplast in *Chlamydomonas reinhardi. Ann. Rev. Genet.* 4:397–408.
MARGULIS, L. 1970. *Origin of eukaryotic cells.* New Haven, Conn.: Yale University Press.
MITCHELL, M. B., and MITCHELL, H. K. 1952. A case of maternal inheritance in *Neurospora crassa. Proc. Natl. Acad. Sci.* 38:442–49.
NOVICK, R. P. 1980. Plasmids. *Scient. Amer.* (Dec.) 243:102–27.
PREER, J. R. 1971. Extrachromosomal inheritance: Hereditary symbionts, mitochondria, chloroplasts. *Ann. Rev. Genet.* 5:361–406.
ROSING, H. S., et al. 1985. Maternally inherited mitochondrial myopathy and myoclonic epilepsy. *Ann. Neurol.* 17:228–37.
SAGER, R. 1965. Genes outside the chromosomes. *Scient. Amer.* (Jan.) 212:70–79.
———. 1985. Chloroplast genetics. *BioEssays* 3:180–84.
SCHWARTZ, R. M., and DAYHOFF, M. O. 1978. Origins of prokaryotes, eukaryotes, mitochondria and chloroplasts. *Science* 199:395–403.
SLONIMSKI, P. 1982. *Mitochondrial genes.* Cold Spring Harbor, N. Y.: Cold Spring Harbor Laboratory.
SONNEBORN, T. M. 1959. Kappa and related particles in *Paramecium. Adv. in Virus Res.* 6:229–356.
STRATHERN, J. N., et al., eds. 1982. *The molecular biology of the yeast* Saccharomyces: *Life cycle and inheritance.* Cold Spring Harbor, N.Y.: Cold Spring Harbor Laboratory.
STURTEVANT, A. H. 1923. Inheritance of the direction of coiling in *Limnaea. Science* 58:269–70.
TZAGOLOFF, A. 1982. *Mitochondria.* New York: Plenum Press.
WALLACE, D. C. 1992. Mitochondrial genetics: A paradigm for aging and degenerative diseases. *Science* 256:628–32.

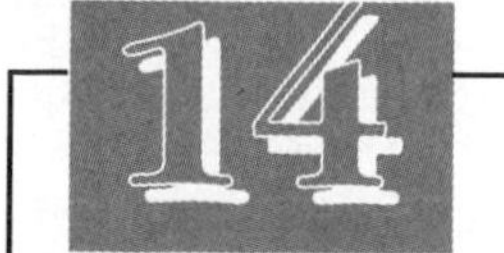

14 Genetics of Bacteria and Viruses

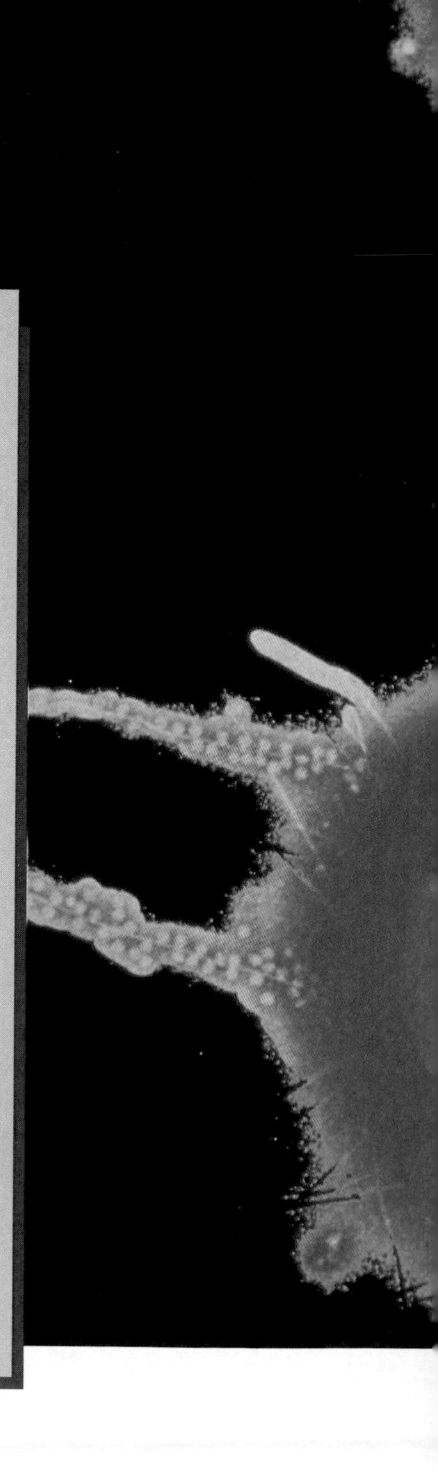

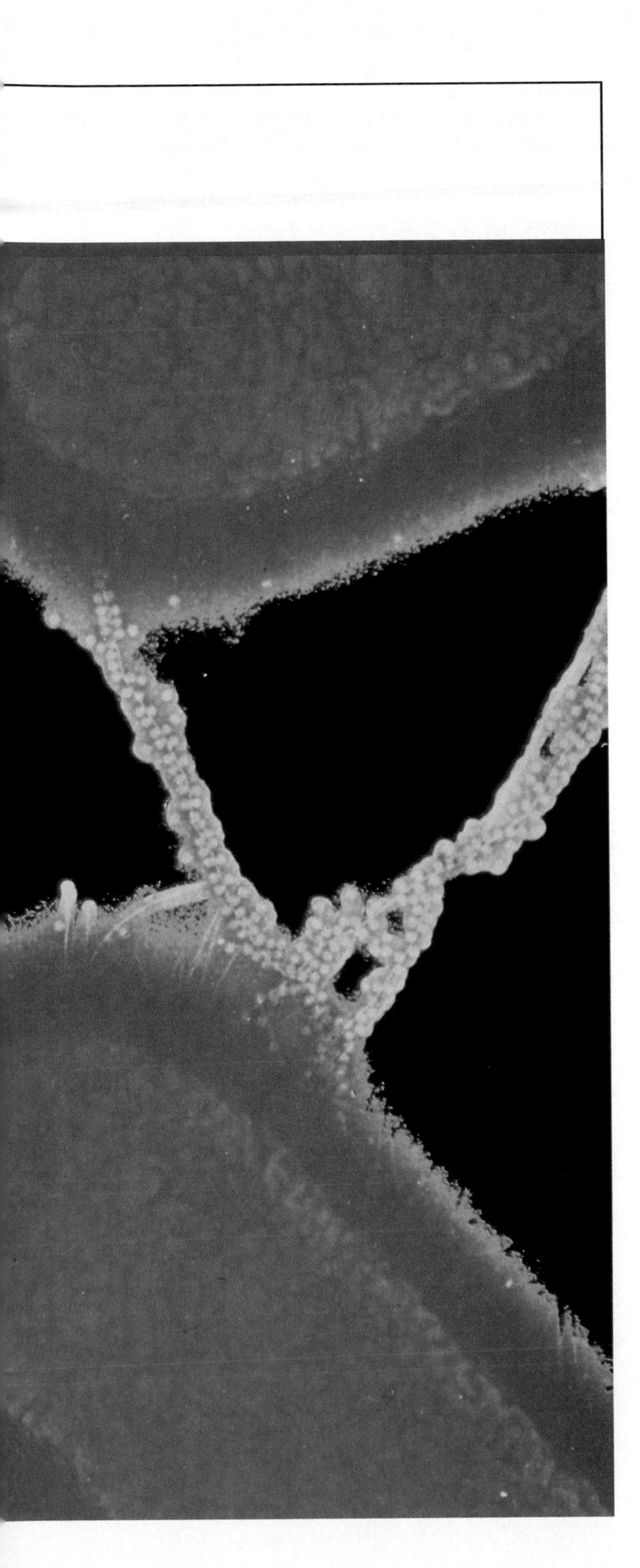

Chapter Concepts

Bacteria and viruses that have bacteria as their hosts (bacteriophages) have been the subject of extensive genetic analysis and demonstrate mechanisms by which genetic exchange occurs. These processes serve as the basis for genetic mapping. Bacteria may contain extrachromosomal DNA in the form of plasmids, which, in addition to bacteriophage DNA, can be integrated into the bacterial chromosome.

The use of bacteria and bacteriophages has been essential to the accumulation of knowledge in many areas of genetic study. For example, much of what is known about molecular genetics, recombinational phenomena, and gene structure has been initially derived from experimental work with these organisms. Their successful use in genetic studies is due to numerous factors. Both bacteria and their viruses have extremely short reproductive cycles. Literally hundreds of generations, giving rise to billions of organisms, can be produced in short periods of time. Importantly, they can be studied in pure cultures. That is, a single species or mutant strain of bacteria or one type of virus can be investigated independently of other similar organisms. Pure culture techniques for the study of microorganisms were developed in the latter part of the nineteenth century.

In this chapter, we will review a number of the historical developments leading to the extensive use of bacteria and bacteriophages in genetics and examine in depth the processes by which they undergo genetic recombination.

BACTERIAL MUTATION AND GROWTH

Although it was known well before 1943 that pure cultures of bacteria could give rise to small numbers of cells exhibiting heritable variation, particularly with respect to survival under different environmental conditions, the source of the variation was hotly debated. The majority of bacteriologists believed that environmental factors induced changes in certain bacteria that led to their survival or adaptation to the new conditions. For example, strains of *E. coli* are known to be *sensitive* to infection by the bacteriophage T1. Infection by the bacteriophage leads to the reproduction of the virus at the expense of the bacterial cell, which is lysed or destroyed (see Figure 7.4). If a plate of *E. coli* is homogeneously sprayed with T1, almost all cells are lysed. Rare *E. coli* cells, however, survive infection and are not lysed. If these cells are isolated and established in pure culture, all descendants are *resistant* to T1 infection. The **adaptation hypothesis,** put forth to explain this type of observation, implied that the interaction of the phage and bacterium is essential to the acquisition of immunity. In other words, the phage had "induced" resistance in the bacteria.

The occurrence of **spontaneous mutations** provided an alternative model to explain the origin of T1 resistance in *E. coli.* In 1943, Salvador Luria and Max Delbruck presented the first convincing evidence that bacteria, like eukaryotic organisms, are capable of spontaneous mutation. This experiment (called the **fluctuation test**) marked the initiation of modern bacterial genetic study. With only minor debate (see the essay on pp. 323) spontaneous mutation is considered to be the source of genetic variation in bacteria.

Mutant cells that arise spontaneously in an otherwise pure culture can be isolated and established independently from the parent strain by using efficient selection techniques. As a result, mutations for almost any desired characteristic can now be induced and isolated. Since bacteria and viruses are haploid, all mutations are expressed directly in the descendants of mutant cells, adding to the ease with which these microorganisms can be studied.

Bacteria are grown in either a liquid culture medium or in a Petri dish on a semisolid agar surface. If the nutrient components of the growth medium are very simple and consist only of an organic carbon source (such as a glucose or lactose) and a variety of ions, including Na^+, K^+, Mg^{++}, Ca^{++}, and NH_3^+ present as inorganic salts, it is called **minimal medium.** In order to grow on such a medium, a bacterium must be able to synthesize all essential organic compounds (e.g., amino acids, purines, pyrimidines, sugars, vitamins, fatty acids). A bacterium that can accomplish this remarkable biosynthetic feat—one that we ourselves cannot duplicate—is termed a **prototroph.** It is said to be wild type for all growth requirements. On the other hand, if a bacterium loses, through mutation, the ability to synthesize one or more organic components, it is said to be an **auxotroph.** For example, if it loses the ability to make histidine, then this amino acid must be added as a supplement to the minimal medium in order for growth to occur. The resulting bacterium is designated as a his^- auxotroph, as opposed to its prototrophic his^+ counterpart.

In order to study mutant bacteria in a quantitative fashion, an inoculum of bacterium is placed in liquid culture medium. A characteristic growth pattern is exhibited, as illustrated in Figure 14.1. Initially, during the **lag phase,** growth is slow. Then, a period of rapid growth ensues called the **log phase.** During this phase, cells divide many times with a fixed time interval between cell divisions, resulting in logarithmic growth. When a cell density of about 10^9 cells per ml is reached, nutrients and oxygen become limiting and cells enter the **stationary phase.** Since the doubling time during the log phase may be as short as 20 minutes, an initial inoculum of a few thousand cells can easily achieve a maximum cell density in an overnight culture.

Cells grown in liquid medium may be quantitated by transferring them to semisolid medium in a Petri dish. Following incubation and many divisions, each cell gives rise to a visible colony on the surface of the medium. If the number of colonies is too great to count, then serial dilutions of the original liquid culture can be made and plated, until the colony number is reduced to the point where it can be counted (Figure 14.2). This technique

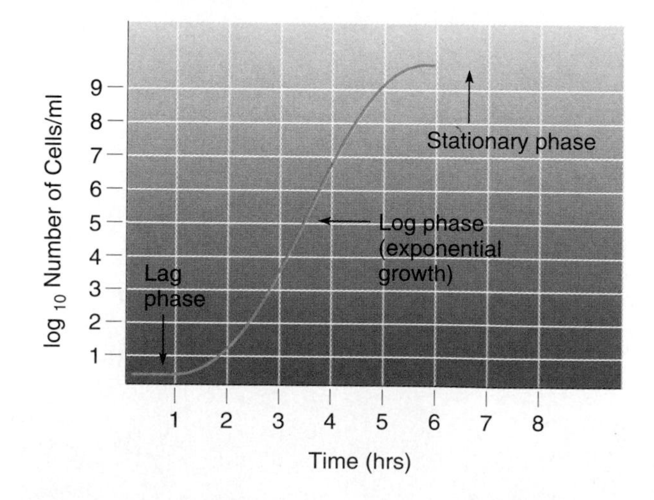

FIGURE 14.1
A typical bacterial population growth curve illustrating the initial lag phase; the log phase, where exponential growth occurs; and the stationary phase that results as nutrients are exhausted.

ADAPTIVE MUTATION IN BACTERIA

While spontaneous mutation in viruses, bacteria, and higher organisms is no longer disputed, the possibility that organisms such as bacteria might also be capable of producing adaptive mutations occurring in response to specific environmental pressures has long intrigued geneticists. Two independent investigations, published in 1988 by John Cairns and Barry Hall and their colleagues, provided preliminary evidence that this might be the case. Further research has continued to support these initial findings.

Cairns devised an experimental protocol that was designed to detect such adaptive mutations arising in response to factors in the environment in which bacteria are cultured. The results of this work suggest that some bacteria may actually select mutations that are "adaptive" to the environment, a notion with Larmarckian overtones. Cairns' work involved a mutant strain of bacteria that cannot use lactose as a carbon source (lac^-). This gene was followed by plating lac^- cells on Petri plates containing minimal medium to which lactose has been added. The lac^- cells, present in the vast majority, will survive on the plates; but since they lack a carbon source that they can metabolize, they cannot proliferate and form colonies. On the other hand, cells that have mutated to lac^+, may be detected because they proliferate and form colonies.

Using a sophisticated mathematical analysis of the appearance of mutations over time, Cairns attempted to identify spontaneous versus adaptive mutations. Cairns' analysis revealed both, and he hypothesized that adaptive mutations arose as a response to the presence of lactose on the Petri plates.

But how was Cairns to prove that, in fact, these mutations occurred on the plates in response to lactose? To answer this question, Cairns asked whether another mutation, unrelated to the metabolism of lactose, had also occurred on the plates. He chose to assay for the production of another mutation val^R, which confers resistance to high concentrations of the amino acid valine. The assay was accomplished easily by overlaying the plates with agar containing valine and glucose. Only val^R mutants will grow. Cairns reasoned that if the lac^+ mutations were not necessarily in response to the lactose, then val^R mutations should arise with a similar distribution over time as observed earlier for the newly arisen lac^+ mutations. The result was that plates accumulating lac^+ mutants were not, at the same time, accumulating val^R mutations. He concluded that the increased frequency of mutations affecting lactose metabolism was the result of the presence of lactose in the medium.

Cairns' work suggests that the bacterial cell's genetic machinery can profit from and respond to its environment by producing adaptive mutations. This conclusion is contrary to the current thinking that all mutations are spontaneous events, many of which occur as random errors during DNA replication. Cairns' observations are not isolated examples.

Barry Hall's work demonstrated a similar adaptive response by *E. coli* to growth on salicin. In this case, interestingly, a two-step genetic change was required. The first occurred in response to salicin even though that mutation offered no selective advantage. The second, the deletion of an insertion sequence (IS), must occur sequentially. They do so at a much higher frequency than could be predicted, provided salicin is present in the growth medium. The observed frequency is several orders of magnitude higher!

What possible explanation can account for observations such as these? One suggestion is that under stressful nutritional conditions (starvation), bacteria may be capable of activating mechanisms that create a hypermutable state in genes that will enhance their survival. In the context of a textbook such as this, an important idea to be realized from this discussion is that knowledge may be derived often from observations that initially cannot be explained. This and the ensuing debates and surrounding controversy constantly maintain intrigue and interest in science!

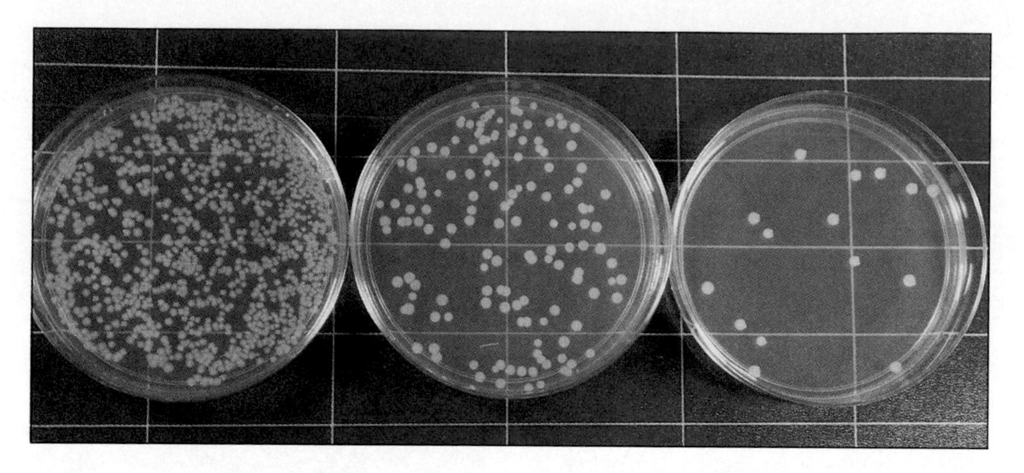

FIGURE 14.2
Results of the serial dilution technique and subsequent culture of bacteria. Each dilution varies by a factor of 10. Each colony was derived from a single bacterial cell.

allows one to calculate the number of bacteria present in the original culture. Such calculations are useful in a variety of studies.

GENETIC RECOMBINATION IN BACTERIA: CONJUGATION

The development of techniques that allowed the identification and study of bacterial mutations led to detailed investigations of the arrangement of genes on the bacterial chromosome. In 1946, Joshua Lederberg and Edward Tatum initiated these studies. They showed that bacteria undergo **conjugation,** a parasexual process in which the genetic information from one bacterium is transferred to and recombined with that of another bacterium. Like meiotic crossing over in eukaryotes, genetic recombination in bacteria led to methodology for chromosome mapping.

Lederberg and Tatum's initial experiments were performed with two multiple auxotroph strains of *E. coli* K12. Strain A required methionine and biotin in order to grow, while strain B required threonine, leucine, and thiamine (Figure 14.3). Neither strain would grow on minimal medium. The two strains were first grown separately in supplemented media, and then cells from both were mixed and grown together for several more generations. They were then plated on minimal medium. Any bacterial cells that grew on minimal medium would be prototrophs. It was highly improbable that any of the cells that contained two or three mutant genes would undergo spontaneous mutation simultaneously at two or three locations. Therefore, any prototrophs recovered must have arisen as a result of some form of genetic exchange and recombination.

In this experiment, prototrophs were recovered at a rate of $1/10^7$ (10^{-7}) cells plated. The controls for this experiment involved separate plating of cells from strains A and B on minimal medium. No prototrophs were recovered. Lederberg and Tatum therefore concluded that genetic exchange had occurred!

F^+ and F^- Bacteria

There soon followed numerous experiments designed to elucidate the genetic basis of conjugation. It became evident quickly that different strains of bacteria were involved in a unidirectional transfer of genetic material. Cells of one strain could serve as donors of parts of their chromosomes and are designated as **F^+ cells** (F for "fertility"). Recipient bacteria, which undergo genetic recombination by receiving donor DNA and exchanging it with part of their own, are designated as **F^- cells.**

It was established subsequently that cell contact is essential to chromosome transfer. Support for this concept was provided by Bernard Davis, who designed a U-tube in which to grow F^+ and F^- cells (Figure 14.4). At the base of the tube was a sintered glass filter with a pore size that allowed the passage of the liquid medium, but was too small to allow the passage of bacteria. F^+ cells were placed on one side and F^- cells on the other side of the filter. The medium was moved back and forth across the filter so that the cells essentially shared a common medium during bacterial incubation. Samples from both sides of the tube were then plated on minimal medium, but no prototrophs were found. Davis concluded that **physical contact is essential to genetic recombination.** This physical interaction is the initial stage of the process of conjugation and is mediated through a structure called the **F sex pilus.** Bacteria often have many pili, which are microscopic extensions of the cell. After contact has been initiated between mating pairs (Figure 14.5), transfer of DNA begins.

Evidence was provided subsequently that F^+ cells contained a **fertility factor,** conferring their ability to do-

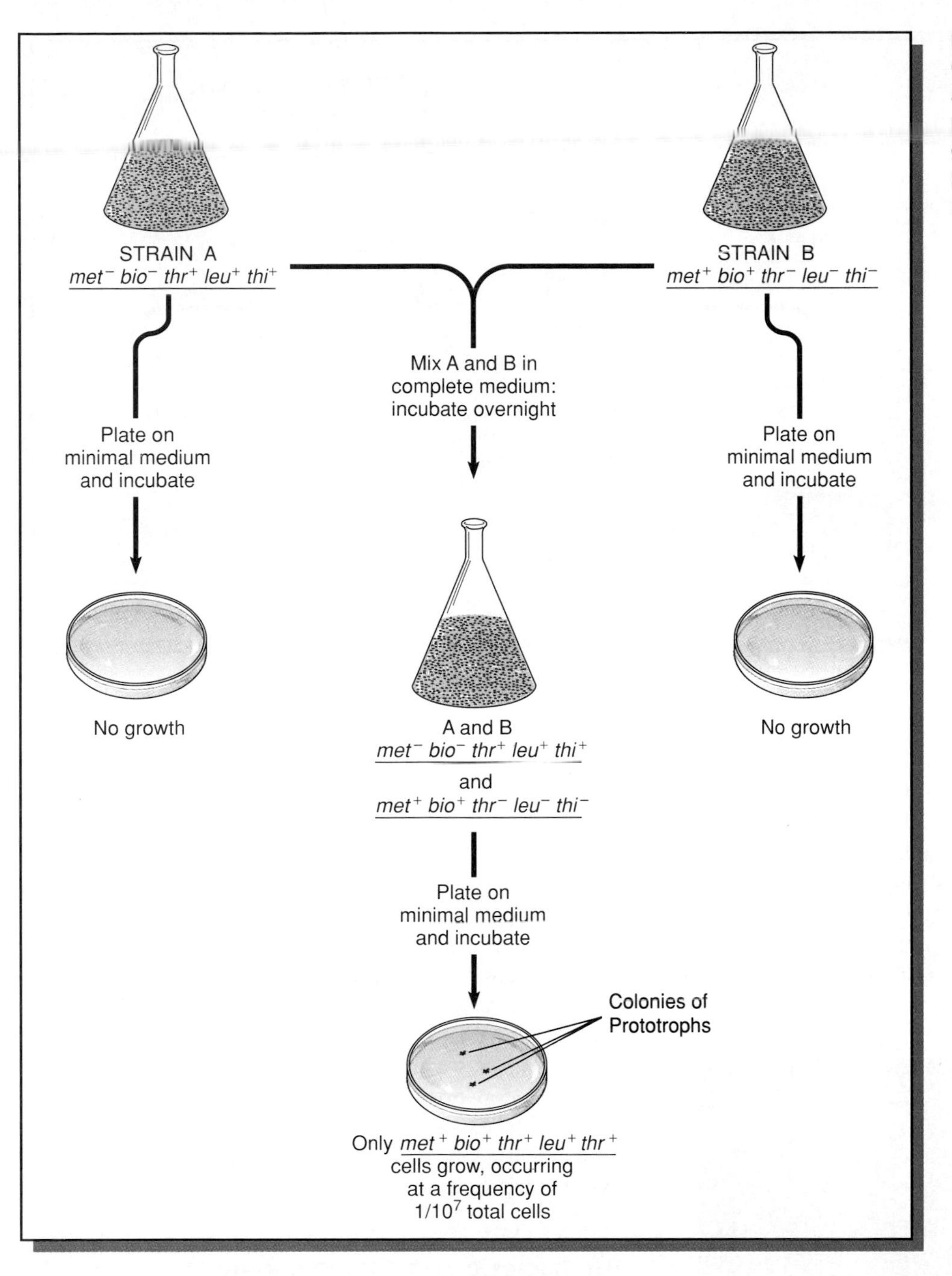

FIGURE 14.3
Genetic exchange between two auxotrophic strains resulting in prototrophs. Neither auxotroph will grow on minimal medium; but prototrophs will, allowing their recovery.

nate part of their chromosome during conjugation. In experiments by the Lederbergs and by William Hayes and Luca Cavalli-Sforza, it was shown that certain conditions could eliminate donor ability in otherwise fertile cells. However, if these cells were then grown with fertile donor cells, fertility was reestablished. The conclusion was drawn that an F factor exists that cells can lose and regain.

This conclusion was further supported by the observation that following conjugation and genetic recombination, recipient cells always become F^+. Thus, in addition to the *rare* case of gene transfer, the F factor is passed to *all* recipient cells. On this basis, the initial crosses of the Lederberg and Tatum (Figure 14.3) may be designated:

STRAIN A		STRAIN B
F^+	×	F^-
DONOR		RECIPIENT

Confirmation of these conclusions is based on the isolation of the F factor. It has been shown to consist of a circular, double-stranded DNA molecule that is distinct

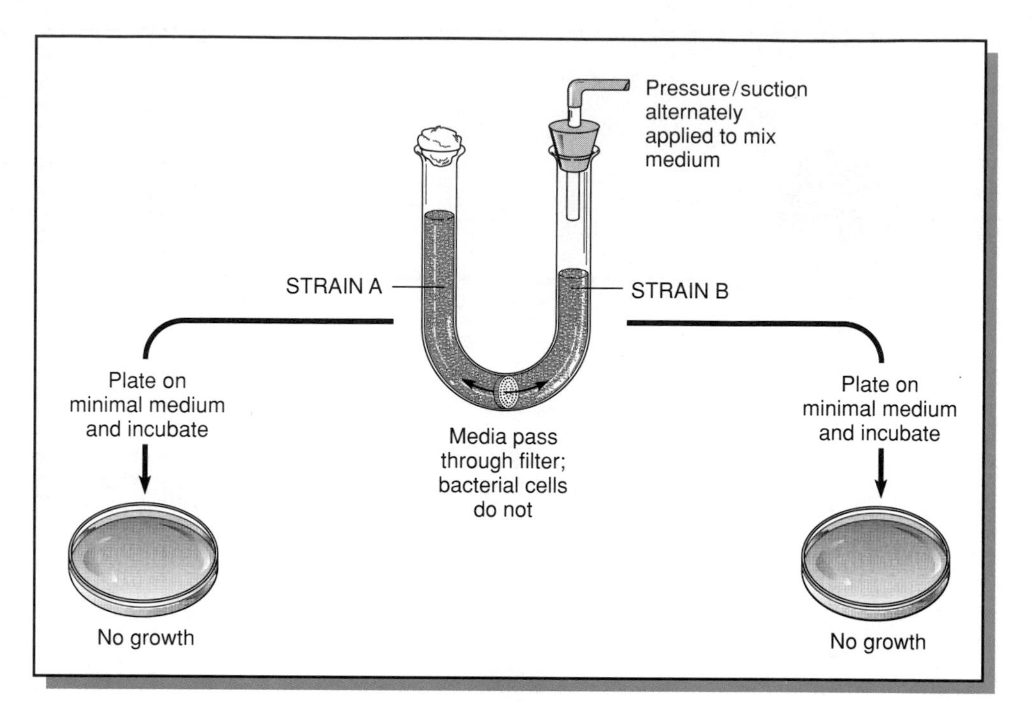

Pressure/suction alternately applied to mix medium
STRAIN A
STRAIN B
Plate on minimal medium and incubate
Plate on minimal medium and incubate
Media pass through filter; bacterial cells do not
No growth
No growth

FIGURE 14.4
When strain A and B auxotrophs are grown in a common medium but are separated by a filter, no genetic exchange occurs and no prototrophs are produced. This apparatus is called a Davis U-tube.

from the bacterial chromosome and contains about 100,000 nucleotide base pairs (100 kb). This amount of DNA is equivalent to about 2 percent of that making up the bacterial chromosome. Contained in the F factor DNA, among others, are 19 genes involved in the transfer of genetic information (*tra* genes) including those essential to the formation of the sex pilus.

It is believed that the transfer of the F factor during conjugation involves strand separation of the F factor DNA and the movement of one of the two strands into the recipient. Both parental strands, one remaining in the donor cell, complete semiconservative replication, resulting in two F^+ cells. This process is diagrammed in Figure 14.6.

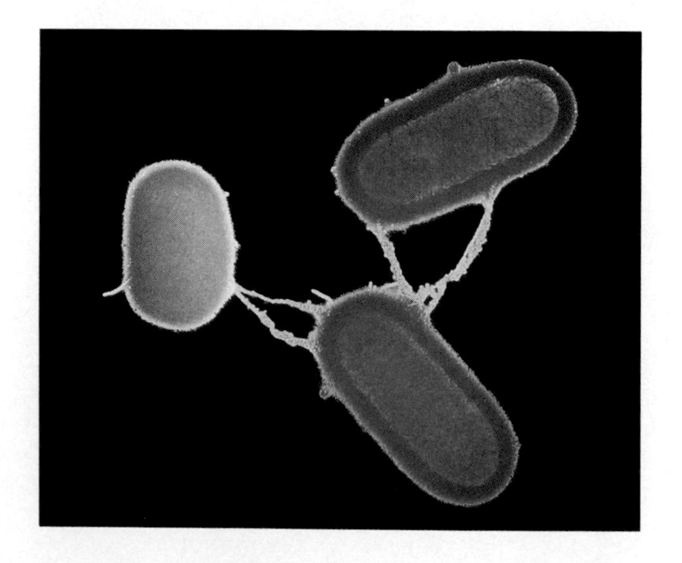

FIGURE 14.5
An electron micrograph of conjugation between an F^+ *E. coli* cell and two F^- cells. The sex pili are visible.

To summarize, *E. coli* cells may or may not contain the F factor. When this factor is present, the cell is able to form a sex pilus and potentially serve as a donor of genetic information. During conjugation, a copy of the F factor is always transferred from the F^+ cell to the F^- recipient, converting it to the F^+ state. The question remains as to exactly how a very low number of F^- cells undergo genetic recombination. As we shall see, the answer awaited further experimentation.

Hfr Bacteria and Chromosome Mapping

Subsequent discoveries were to not only clarify how genetic recombination occurs, but to define a mechanism by which the *E. coli* chromosome could be mapped. We shall first address chromosome mapping.

In 1950, Cavalli-Sforza treated an F^+ strain of *E. coli* K12 with nitrogen mustard, a potent mutagen. From these treated cells he recovered a strain of donor bacteria that underwent recombination at a rate of $1/10^4$ (10^{-4}), 1000 times more frequently than the original F^+ strains. In 1953, Hayes isolated another strain demonstrating a similar elevated frequency. Both strains were designated **Hfr,** or **high-frequency recombination.** Since Hfr

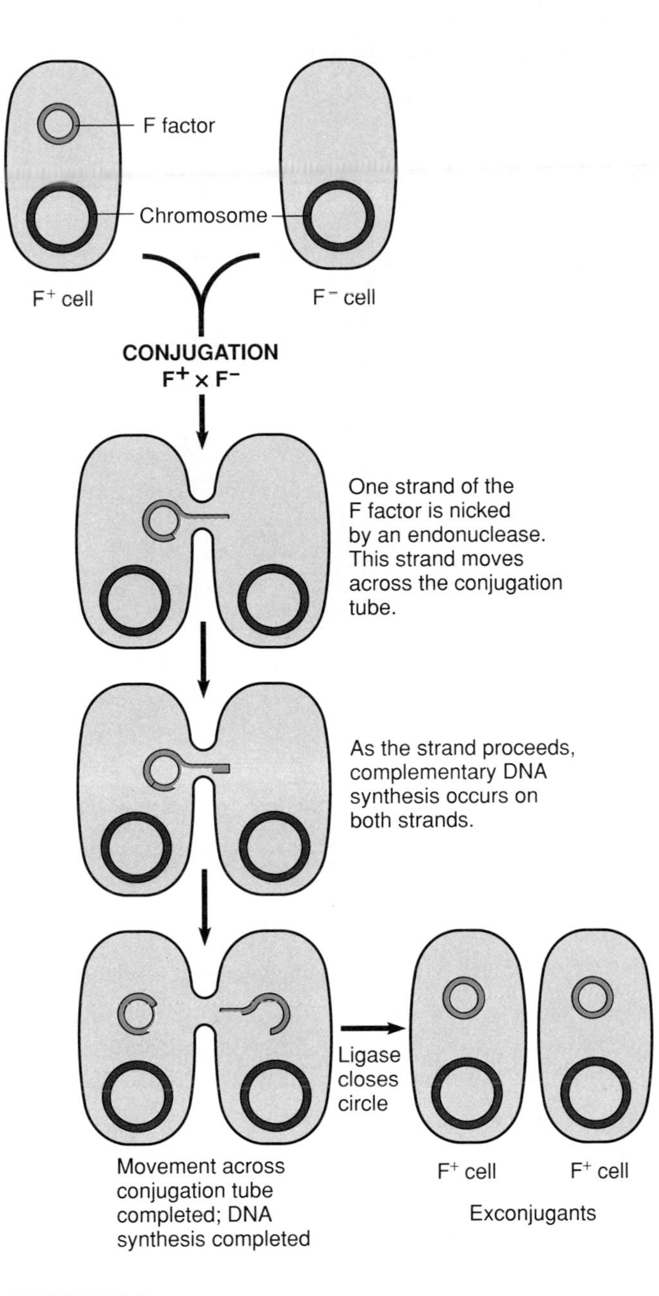

FIGURE 14.6
An $F^+ \times F^-$ mating demonstrating how the recipient F^- cell is converted to F^+. During conjugation, the DNA of the F factor is replicated with one new copy entering the recipient cell, converting it to F^+.

cells behave as donors, they are a special class of F^+ cells.

Another important difference was noted between Hfr strains and the original F^+ strains. If the donor is an Hfr strain, recipient cells, while sometimes displaying genetic recombination, never become Hfr; that is, they remain F^-. In comparison, then,

$$F^+ \times F^- \rightarrow F^+ \text{ (low rate of recombination)}$$

$$\text{Hfr} \times F^- \rightarrow F^- \text{ (higher rate of recombination)}$$

Perhaps the most significant characteristic of Hfr strains is the nature of recombination. In any given strain, certain genes are more frequently recombined than others, and some not at all. This nonrandom pattern was shown to vary from Hfr strain to Hfr strain. While these results were puzzling, Hayes interpreted them to mean that some physiological alteration of the F factor had occurred, resulting in the production of Hfr strains of *E. coli.*

In the mid-1950s, experimentation by Ellie Wollman and Francois Jacob helped explain why cells are Hfr compared to F^+ and showed how these strains allow genetic mapping of the *E. coli* chromosome. In their experiments, Hfr and F^- strains with suitable marker genes were mixed and recombination of specific genes assayed at different times. To accomplish this, a culture containing a mixture of an Hfr and an F^- strain was first incubated and samples removed at various intervals and placed in a blender. The shear forces created in the blender separated conjugating bacteria so that the transfer of the chromosome was terminated effectively. The cells were then assayed for genetic recombination.

This process, called the **interrupted mating technique,** demonstrated that specific genes of a given Hfr strain were transferred and recombined sooner than others. Figure 14.7 illustrates this point. During the first 8 minutes after the two strains were initially mixed, no genetic recombination could be detected. At about 10 minutes, recombination of the *azi* gene could be detected, but no transfer of the ton^s, lac^+, or gal^+ genes was noted. By 15 minutes, 70 percent of the recombinants were azi^+; 30 percent were now also ton^s; but none was lac^+ or gal^+. By 20 minutes, the lac^+ gene was found among the recombinants; and by 30 minutes, gal^+ was also being transferred. Therefore, Wollman and Jacob had demonstrated an oriented transfer of genes that was correlated with the length of time conjugation was allowed to proceed.

It appeared that the chromosome of the Hfr bacterium was transferred linearly and that the sequence and distance between the genes, as measured in minutes, could be predicted from such experiments (Figure 14.8). This information served as the basis for the first genetic map of the *E. coli* chromosome. Apparently, conjugation does not usually last long enough to allow the entire chromosome to pass across to the recipient cell.

Wollman and Jacob then repeated the same type of experimentation with other Hfr strains, obtaining similar results with one important difference. While genes were always transferred linearly with time, as in their original experiment, which genes entered first and which followed later seemed to vary from Hfr strain to Hfr strain [Figure 14.9(a)]. When they reexamined the rate of entry of genes, and thus the different genetic maps for each

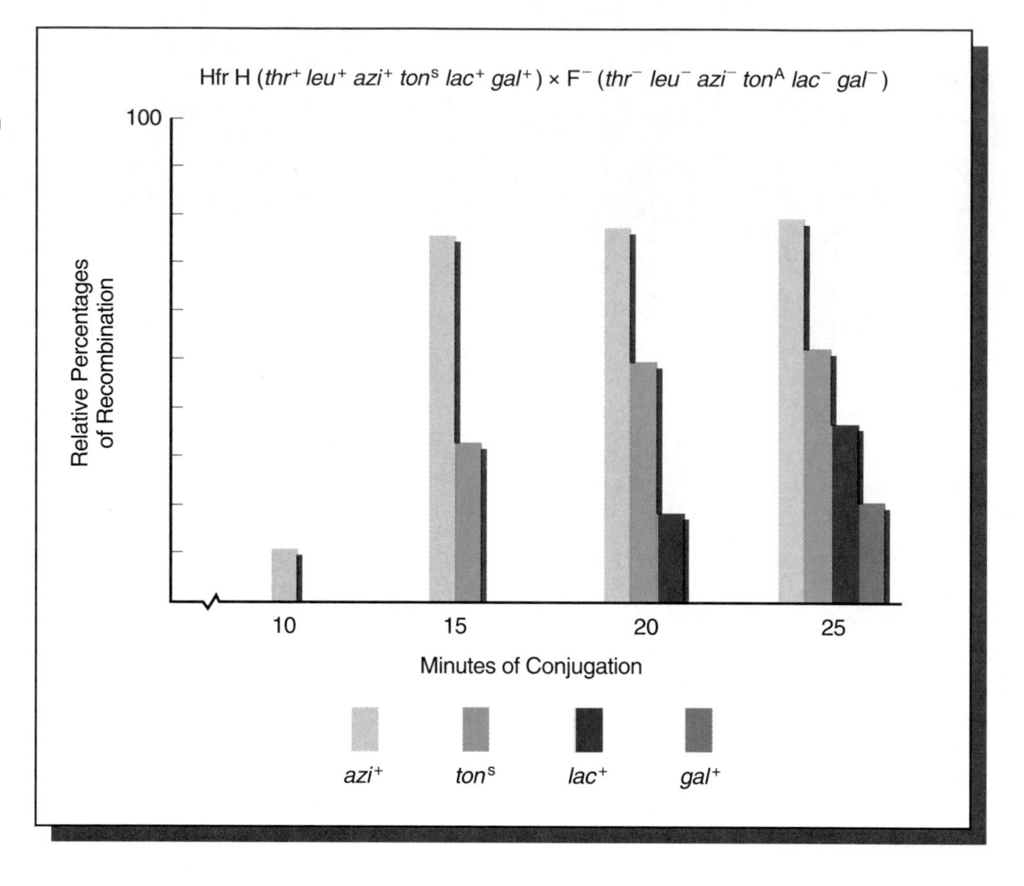

FIGURE 14.7
The progressive transfer during conjugation of various genes from a specific Hfr strain of *E. coli* to an F^- strain. Certain genes (*azi* and *ton*) are transferred sooner than others and recombine more frequently. Others (*lac* and *gal*) take longer to be transferred and recombine with a lower frequency.

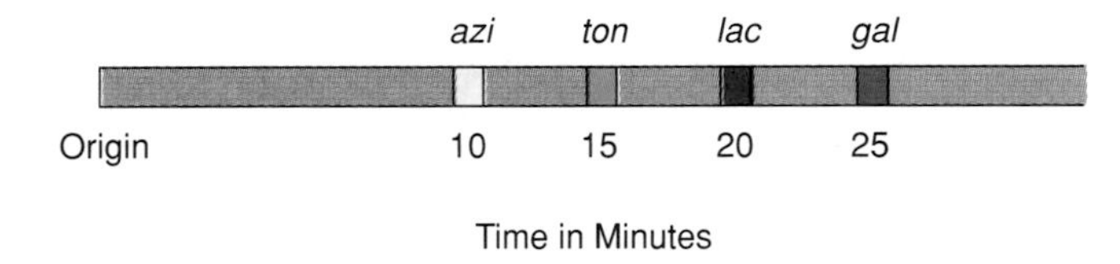

FIGURE 14.8
A time map of the genes studied in the experiment depicted in Figure 14.7.

strain, a definite pattern emerged. The major differences between all strains were simply the point of the origin (*O*) and the direction in which entry proceeded from that point [Figure 14.9(b)].

In order to explain these results, Wollman and Jacob postulated that the *E. coli* chromosome is circular. If the point of origin (*O*) varied from strain to strain, a different sequence of genes would be transferred in each case. But what determines *O*? They proposed that in various Hfr strains, the F factor integrates into the chromosome at different points. Its position determines the *O* site. One such case is shown in Figure 14.10. During conjugation between an Hfr and F^- cell, the position of the F factor determines the initial point of transfer. Those genes adjacent to *O* are transferred first. The F factor becomes the last part to be transferred. Apparently, conjugation rarely, if ever, lasts long enough to allow the entire chromosome to pass across the conjugation tube. This proposal explains why recipient cells, when mated with Hfr cells, remain F^-.

The use of the interrupted mating technique with different Hfr strains has provided the basis for mapping the entire *E. coli* chromosome. Mapped in time units, strain K12 (or *E. coli* K12) is 90 minutes long. Over 900 genes have now been placed on the map. In most instances, only a single copy of each gene exists.

Recombination in $F^+ \times F^-$ Matings: A Reexamination

The above model has helped geneticists to understand better how genetic recombination occurs during the $F^+ \times F^-$ matings. Recall that recombination occurs much less frequently in them than in Hfr $\times$ F^- matings, and that random gene transfer is involved. The current belief is that when F^+ and F^- cells are mixed, conjugation occurs readily and that each F^- cell involved in conjugation receives a copy of the F factor, but that no genetic recom-

(a)

Hfr Strain	Order of Transfer →							
H	*thr* –	*leu* –	*azi* –	*ton* –	*pro* –	*lac* –	*gal* –	*thi*
1	*leu* –	*thr* –	*thi* –	*gal* –	*lac* –	*pro* –	*ton* –	*azi*
2	*pro* –	*ton* –	*azi* –	*leu* –	*thr* –	*thi* –	*gal* –	*lac*
7	*ton* –	*azi* –	*leu* –	*thr* –	*thi* –	*gal* –	*lac* –	*pro*

(b)

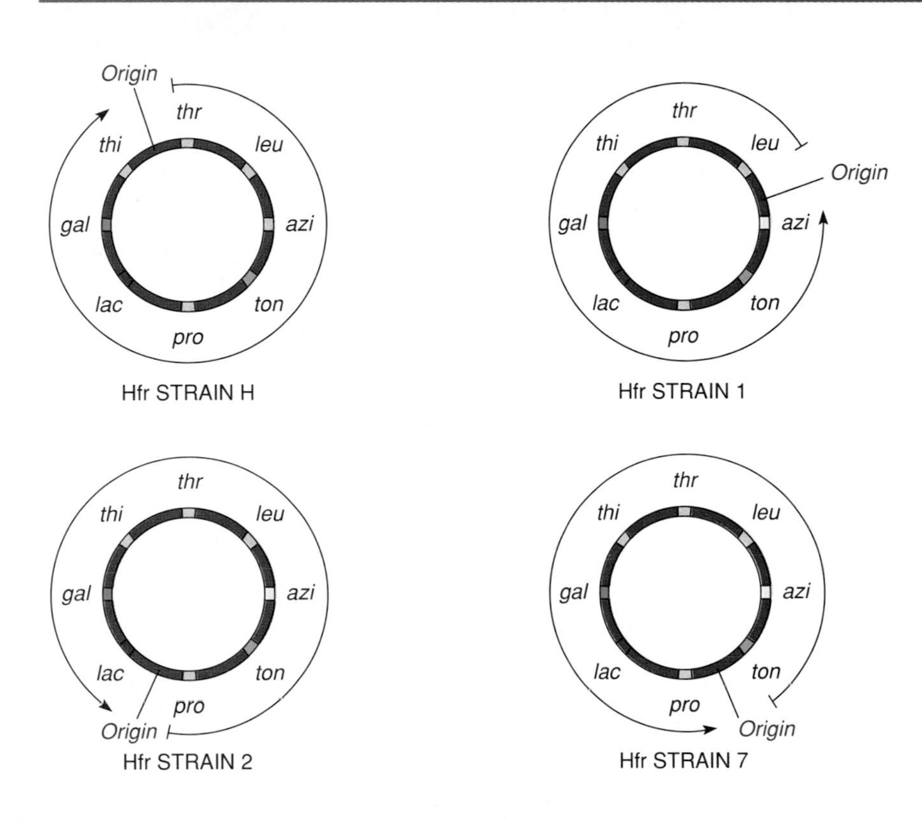

FIGURE 14.9
The order of gene transfer in four Hfr strains, suggesting that the *E. coli* chromosome is circular. The position of the origin of transfer is identified in each strain. Note that transfer can proceed in either direction, depending on the strain. The origin is determined by the point of integration into the chromosome of the F factor, and the direction of transfer is determined by the orientation of the F factor as it integrates.

bination occurs. However, in a very low frequency in a population of F^+ cells, the F factor integrates spontaneously at a random point into the bacterial chromosome. This integration converts each such F^+ cell to the Hfr state. Therefore, in $F^+ \times F^-$ crosses, the very low frequency of genetic recombination (10^{-7}) is attributed to the rare, newly formed Hfr cells, which then also undergo conjugation with F^- cells. Since the point of integration is random, a nonspecific gene transfer ensues, leading to the low-frequency, random genetic recombination observed in the $F^+ \times F^-$ mating.

The F′ State and Merozygotes

In 1959, during experiments with Hfr strains of *E. coli,* Edward Adelberg discovered that the F factor could lose its integrated status, causing reversion to the F^+ state (Figure 14.11). When this occurs, the F factor frequently carries several adjacent bacterial genes along with it. He labeled this condition **F′** to distinguish it from F^+ and Hfr. F′ is thus a special case of F^+.

The presence of bacterial genes within a cytoplasmic F factor creates an interesting situation. An F′ bacterium

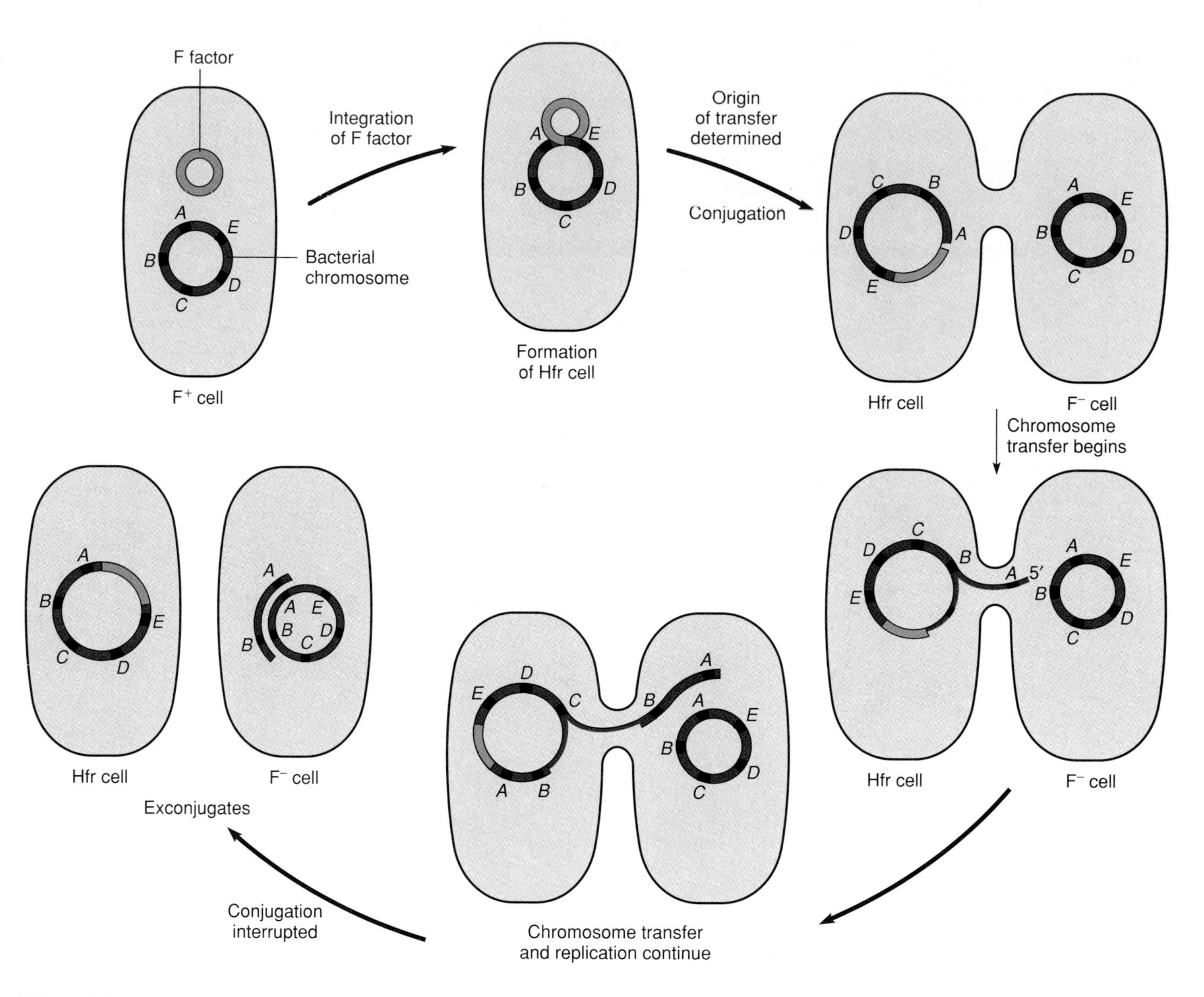

FIGURE 14.10
Conversion of F$^+$ to an Hfr state occurs by the integration of the F factor into the bacterial chromosome. During conjugation, the F factor is nicked by an enzyme, creating the origin of transfer of the chromosome. The major portion of the F factor is now on the end of the chromosome opposite the origin and is the last part to be transferred. Conjugation is usually interrupted prior to complete transfer. Above, only the A and B genes are transferred.

behaves as an F$^+$ cell, initiating conjugation with F$^-$ cells. When this occurs, the F factor, along with the chromosomal genes, is always transferred to the F$^-$ cell. As a result, whatever chromosomal genes are part of the F factor are now present in duplicate in the recipient cell because they are also present in the recipient cell's chromosome. This creates a partially diploid cell called a **merozygote.** Pure cultures of F′ merozygotes may be established. They have been extremely useful in the study of bacterial genetics, particularly in genetic regulation.

THE *rec* PROTEINS AND RECOMBINATION

Once part of the chromosome of a bacterial donor has entered a recipient cell, just how does actual genetic recombination with the recipient chromosome occur? The first clue in answering this question was the discovery of a series of mutations that diminished the rate of genetic exchange, called *recA, recB, recC,* and *recD.* Two protein products result from these genes: *recA* and *recBCD*. They

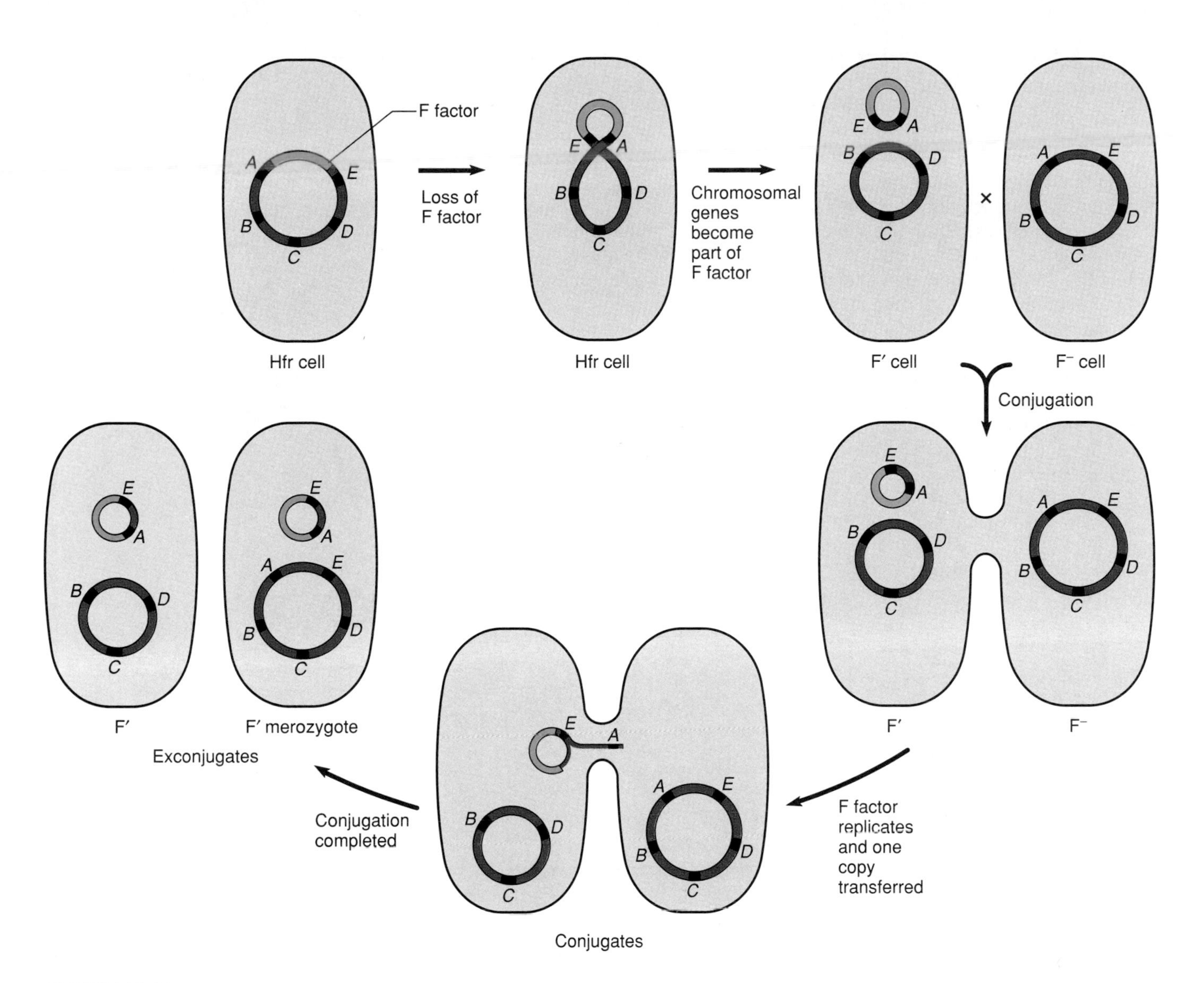

FIGURE 14.11
Conversion of an Hfr bacterium to F′ and its subsequent mating with an F^- cell. The conversion occurs when the F factor loses its integrated status and carries with it one or more chromosomal genes (A and E). Following conjugation with such an F′ cell, an F^- recipient becomes partially diploid and is called a merozygote. It also becomes F′.

have been isolated and studied and their roles elucidated.

The *recA* protein binds to the invading DNA molecule and promotes its invasion into the homologous region of the host chromosome. Most often, only one strand of DNA invades the recipient cell. The *recBCD* protein complex displays helicase activity and plays a role in opening up the double helix of the host DNA, thus facilitating recombination. It is also important during renaturation of the hybrid strands. Thus, together, these proteins have the effect of correctly positioning the donor DNA and facilitating genetic exchange (Figure 14.12).

PLASMIDS

In the above sections, we introduced and discussed the F factor extensively. When existing autonomously in the bacterial cytoplasm, this unit takes the form of a double-stranded closed circle of DNA. These characteristics place the F factor in the more general category of the genetic structure called a **plasmid.** These structures contain one or more genes and often, quite a few. Their replication depends on the host cell's enzymes and they are distributed to daughter cells along with the host chromosome during cell division.

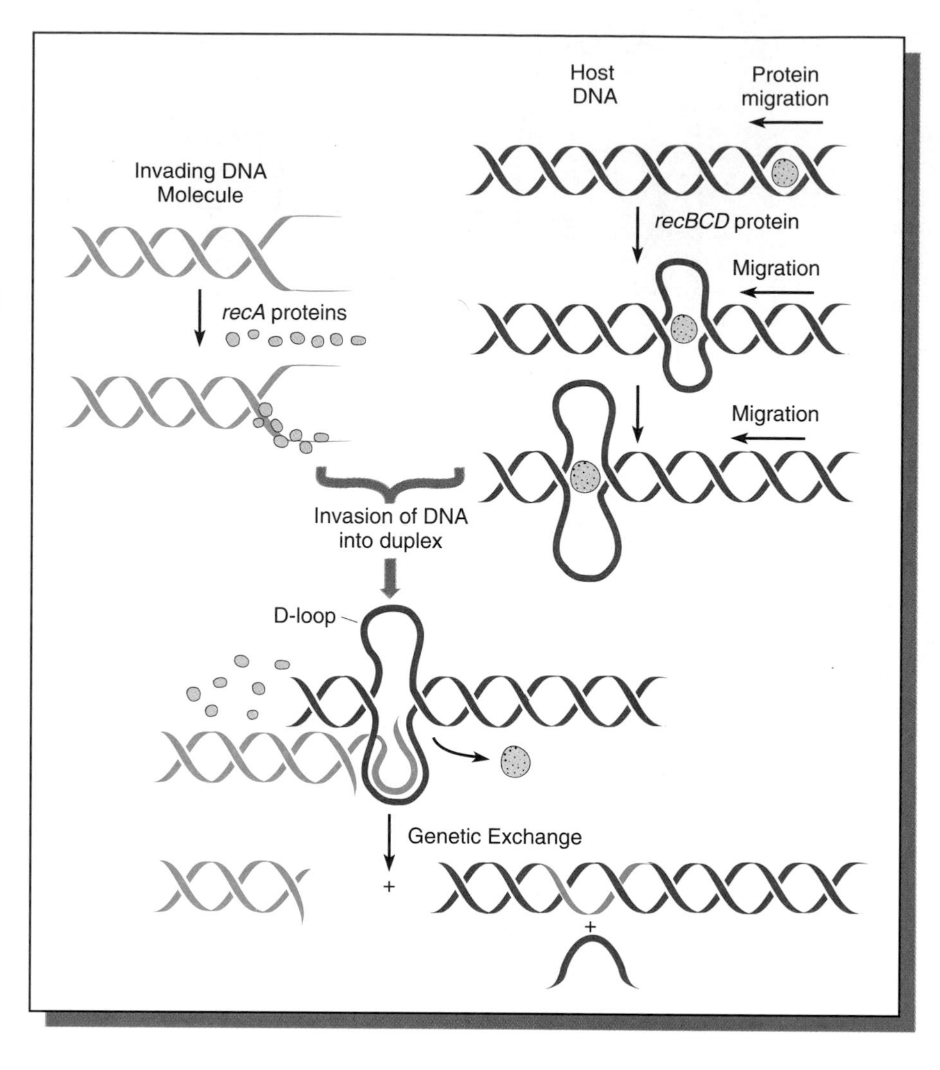

FIGURE 14.12
A theoretical model illustrating the possible role of the *recA* and *recBCD* proteins in the process of genetic recombination between bacterial DNA molecules. The *recA* protein has an affinity to bind to single-stranded regions of DNA and to facilitate their invasion of double-stranded DNA. The *recBCD* protein has an affinity for double-stranded DNA. This protein migrates along the helix, causing the complementary strands to dissociate. Once the invasive region finds its homologous DNA sequence, the *recA* proteins are shed, and a hybrid complement is formed. The displaced strand forms a D-loop, and genetic exchange occurs.

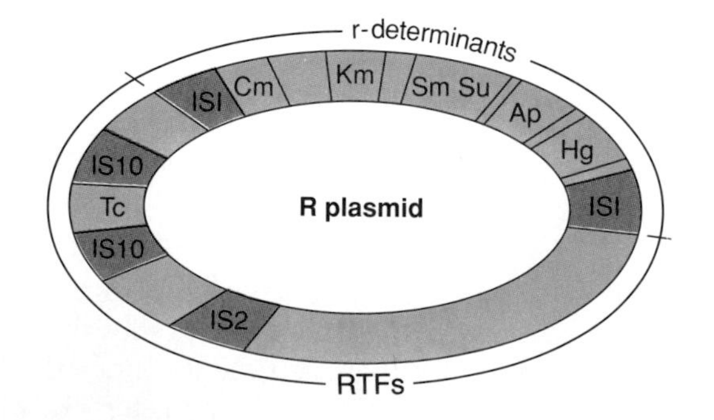

FIGURE 14.13
R plasmid containing resistance transfer factors (RTFs) and multiple r-determinants (Tc—tetracycline; Km—kanamycin; Sm—streptomycin; Su—sulfonamide; Ap—ampicillin; Hg—mercury).

Plasmids are generally classified according to the genetic information specified by their DNA. The F factor confers fertility and contains genes essential for sex pilus formation. Other examples include the **R** and the **Col plasmids.** Recall from Chapter 13, that these plasmids confer multiple resistance to antibiotics and the ability to release toxic colicins, respectively. An R plasmid is illustrated in Figure 14.13.

Interest in plasmids has increased dramatically because of their role in the genetic technology referred to as **recombinant DNA research.** Specific genes from any source, e.g., a human cell, may be inserted into a plasmid, which may be inserted into a bacterial cell. As the cell replicates its DNA and undergoes division, the foreign gene is treated like one of the cell's own. This is the major topic of the next chapter.

INSERTION SEQUENCES AND TRANSPOSONS IN BACTERIA

In the early 1970s, a number of independent workers, including Peter Starlinger and James Shapiro, discovered a unique class of mutations in *E. coli.* These mutations affected several genes in different strains of bacteria. For example, transcription of a cluster of genes controlling galactose metabolism in *E. coli* was repressed or shut off as a result of one of these mutations. This phenotypic effect was heritable, but was found not to be caused by a nucleotide change characteristic of conventional mutations. Instead, it was shown that a short, specific DNA segment had been inserted into the bacterial chromosome at the beginning of the galactose gene cluster. When this segment was spontaneously excised from the bacterial chromosome, wild-type function was restored!

These and other DNA segments have now been characterized. They are relatively short, not exceeding 2000 base pairs (2 kb or 2 kilobases), and are now called **insertion sequences (IS).** The genetic information contained within an IS unit appears to be involved only in the insertion process itself.

Analysis of the DNA sequences of most IS units has revealed a most interesting feature. At each end of any given double-stranded unit, the nucleotide sequence consists of a perfect **inverted repeat** of the other end (Figure 14.14). Note that an identical set of five base pairs runs from each end toward the middle of the sequence they surround. While this figure shows this terminal repeating unit to consist of only a few nucleotides, many more are actually involved. For example, in *E. coli,* the IS1 termini contain 23 nucleotide pairs, IS2 41 pairs, and IS4 18 pairs. It seems likely that these terminal sequences are involved closely in the mechanism of insertion of IS units into DNA. This contention is supported by the fact that insertion of IS units is not random but instead is more likely to occur at certain DNA regions. This suggests that the termini may be involved in recognition sites during the insertion process.

In the *E. coli* chromosome, variable numbers of IS units are found. They are found also in plasmids (Figure 14.15). Their presence does not always result in mutation. Their major role appears to involve the integration and excision of other DNA sequences. For example, the integration of the F factor into the chromosome is thought to be dependent on IS units.

In such a capacity, IS units play a significant role in the formation and movement of the larger **transposon (Tn) elements.** Transposons in bacteria consist of IS units that are covalently joined to other genes whose functions

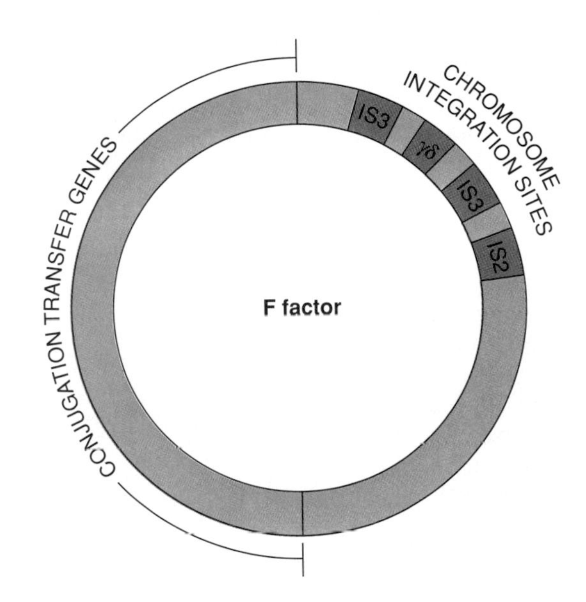

FIGURE 14.15
Diagram of the F factor in *E. coli* showing four insertion sequences that serve as integration sites into the bacterial chromosome.

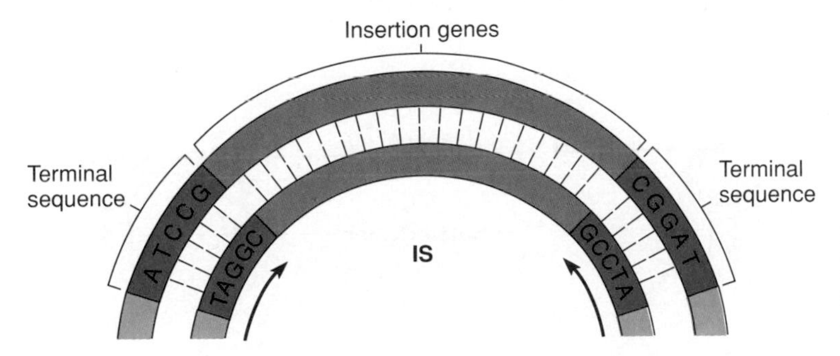

FIGURE 14.14
Diagrammatic representation of an insertion sequence (IS). Inverted repeat sequences are at each end (see arrows).

are unrelated to the insertion process. Like IS units, Tn elements can insert into bacterial and viral chromosomes as well as plasmids. Thus, Tn elements provide a mechanism for movement of genetic information from place to place both within and between organisms.

Transposons were first discovered to move between bacterial DNA molecules as a result of observations of antibiotic-resistant bacteria. In the mid-1960s, Susumu Mitsuhashi suggested that an R plasmid containing a gene responsible for resistance to antibiotic chloramphenicol had been transferred to the bacterial chromosome. Since then, it has been demonstrated that antibiotic-resistant genes can move from plasmid to plasmid as well as between plasmids and chromosomes. Electron microscopic studies have confirmed that the two terminal inverted repeat sequences are indeed present within plasmids harboring transposons. As shown in Figure 14.16, when the strands of DNA from such a plasmid are separated and allowed to reanneal separately, the inverted repeat units are complementary, and a heteroduplex is formed as expected. All areas other than the repeat units remain single stranded and form loops on either end of the double-stranded stem.

Transposons have become the focus of increased interest, particularly because they have been found not to be unique to bacteria. Bacteriophages that demonstrate the ability to insert their genetic material into the host chromosome behave in a similar fashion. (See the discussion of this topic later in this chapter.) The **bacteriophage mu,** consisting of over 35,000 nucleotides, can insert its DNA anywhere in the *E. coli* chromosome! And like IS units, insertion into a gene results in mutant behavior at that locus. Transposons have also been discovered in higher organisms, including yeast, corn, and *Drosophila*.

BACTERIAL TRANSFORMATION

In Chapter 7 we described how the study of transformation provided direct evidence that DNA is the genetic material. Transformation, like conjugation, provides a

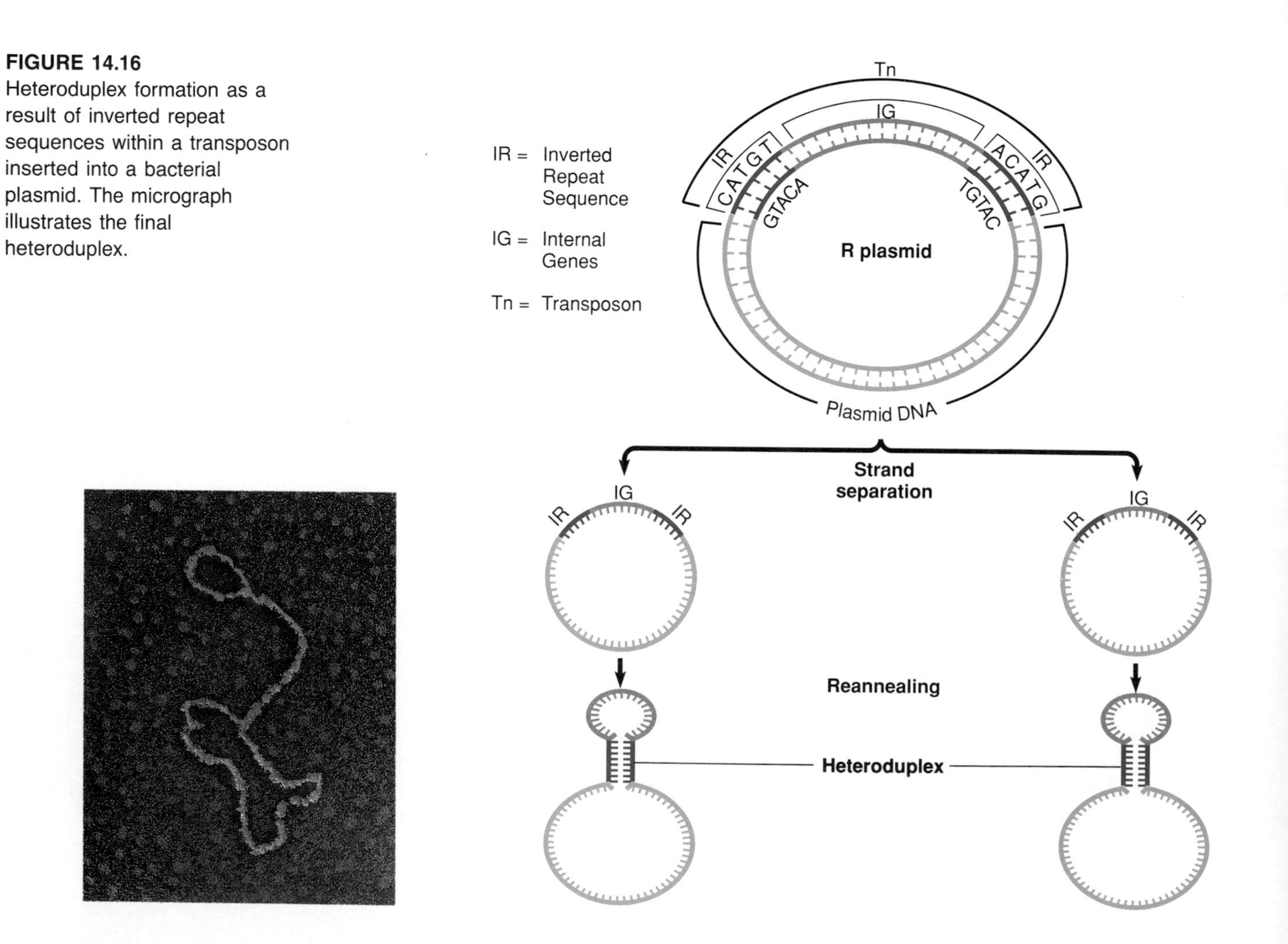

FIGURE 14.16
Heteroduplex formation as a result of inverted repeat sequences within a transposon inserted into a bacterial plasmid. The micrograph illustrates the final heteroduplex.

mechanism for the recombination of genetic information in some bacteria. Transformation can also be used to map bacterial genes, although in a more limited way.

This process consists of numerous steps that can be divided into two main categories: (1) entry of exogenous DNA into a recipient cell, and (2) recombination of the donor DNA with its homologous region in the recipient chromosome. In a population of cells, only those in a particular physiological state, referred to as **competence,** take up DNA. Entry apparently occurs at a limited number of receptor sites on the surface of the bacterial cell. The efficient length of transforming DNA is about 10,000 to 20,000 base pairs, an amount equal to about 1/200 of the *E. coli* chromosome. Passage across the cell wall and membrane is an active process requiring energy and specific transport molecules. This concept is supported by the fact that substances that inhibit energy production or protein synthesis in the recipient cell also inhibit the transformation process.

During the process of entry, one of the two strands of the double helix is digested by nucleases, leaving only a single strand to participate in transformation. This DNA segment aligns with its complementary region of the bacterial chromosome. In a process involving several enzymes, the segment replaces its counterpart in the chromosome, which is excised and degraded (Figure 14.17).

In order for recombination to be detected, the transforming DNA must be derived from a different strain of bacteria, bearing some genetic variation, such as a mutation. Once it is integrated into the chromosome, that region contains one host strand (originally present) and one mutant strand. Since these strands are from different

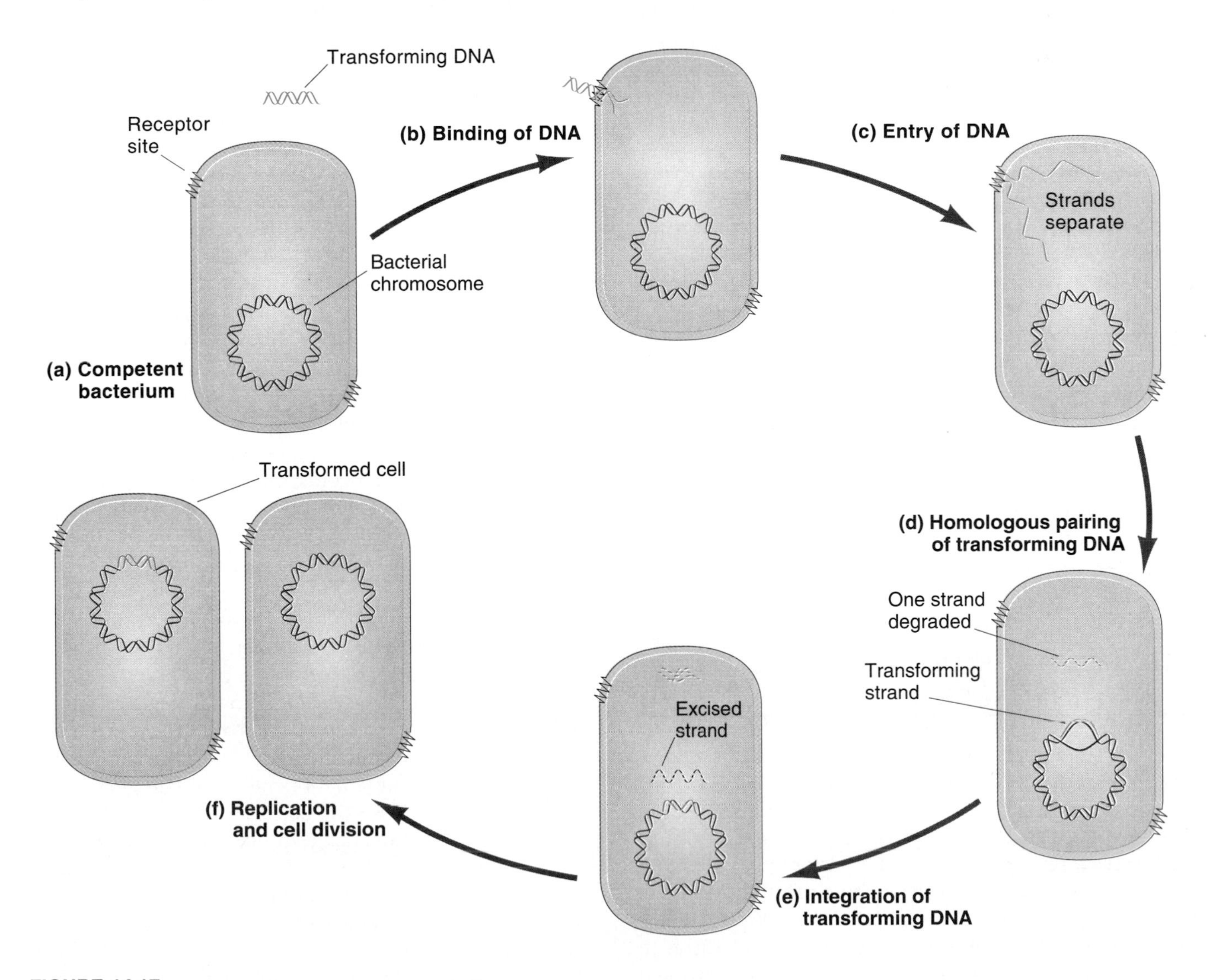

FIGURE 14.17
Proposed steps leading to transformation of a bacterial cell by exogenous DNA. Only one of the two strands of the entering DNA is involved in the transformation event, which is completed following cell division.

sources, this helical region is referred to as a **heteroduplex.** Following one round of semiconservative replication, one chromosome is identical to that of the original recipient, and the other contains the mutant gene. Following cell division, one nonmutant and one mutant cell are produced.

Transformation and Linked Genes

Because transforming DNA is usually about 10,000 to 20,000 base pairs long, it contains a sufficient number of nucleotides to encode several genes. Genes that are adjacent or very close to one another on the bacterial chromosome may be carried on a single piece of DNA of this size. Because of this fact, a single event may result in the **cotransformation** of several genes simultaneously. Genes that are close enough to each other to be cotransformed are said to be **linked.** Note that here *linkage* refers to the proximity of genes, in contrast to the use of this term in eukaryotes to indicate all genes on a single chromosome.

If two genes are not linked, simultaneous transformation can occur only as a result of two independent events involving two distinct segments of DNA. As in double crossing over in eukaryotes, the probability of two independent events occurring simultaneously is equal to the product of the individual probabilities. Thus, the frequency of two unlinked genes being transformed simultaneously is much lower than if they are linked.

Subsequent studies have shown that a variety of bacteria readily undergo transformation (e.g., *Bacillus subtilis, Shigella paradysenteriae,* and, to a lesser extent, *E. coli*). Under certain conditions, studies have shown that relative distances between linked genes can be determined from the transformation data provided. While analysis is more complex, such data are interpreted in a manner analogous to chromosome mapping in eukaryotes.

THE GENETIC STUDY OF BACTERIOPHAGES

There is still a third mode involving genetic transfer between bacteria, called **transduction.** We shall delay our discussion temporarily until we have considered the genetics of bacteriophages and other viruses, since transduction is a process mediated by bacteriophages. As we shall see, bacterial viruses are capable of either lysing the bacterial host or inserting their DNA into the host chromosome where it may remain genetically quiescent. In some cases, it is when this integrated DNA is extricated from the host chromosome that transduction becomes possible.

Recombination has also been observed in bacterial viruses, but only when genetically distinguishable strains have simultaneously infected the same host cell. In this section, we will first review briefly the life cycle of a typical bacteriophage, one from the so-called T-even series. Then, we will discuss how these phages are studied during their infection of bacteria. Finally, we will contrast the two possible modes of behavior once the phage DNA is injected into the bacterial host. This information will serve as background for our discussion of transduction.

The T4 Phage Life Cycle

In Chapter 7, we discussed the Hershey–Chase experiment, which utilized bacteriophage T2 in providing support for the concept that DNA is the genetic material. Phage T4 is very similar and typical of the family of T-even phages. It contains double-stranded DNA in a quantity sufficient to encode more than 150 average-sized genes. The genetic material is enclosed by an icosahedral protein coat (a polyhedron with 20 faces) making up the head of the virus. This is connected to a tail that contains a collar and a contractile sheath that surrounds a central core. Six tail fibers protrude from the tail; their tips contain binding sites that specifically recognize unique areas of the external surface of *E. coli* cell wall.

Binding of tail fibers to these unique areas of the cell wall is the first step of infection. Then, an ATP-driven contraction of the tail sheath causes the central core to penetrate the cell wall. The DNA in the head is extruded where it then moves across the cell membrane into the bacterial cytoplasm. Within minutes, all *bacterial* DNA, RNA, and protein synthesis is inhibited, and synthesis of viral macromolecules begins. This initiates the **latent** or **eclipse period,** in which the production of viral components occurs prior to the assembly of virus particles.

RNA synthesis is initiated by the transcription of viral genes by the host cell RNA polymerase. A plethora of early and late gene products are subsequently synthesized using the host cell ribosomes. For example, about 20 different proteins are part of the protein capsid of the head. Many others are structural components of the tail and its fibers. Additionally, numerous enzymes participate in assembly of mature bacteriophages, but are not themselves structural components. One late gene product, **lysozyme,** digests the bacterial cell wall, leading to the release of mature phages once they are assembled.

Prior to the synthesis of most viral gene products, semiconservative DNA replication begins, leading to a pool of viral DNA molecules that are available to be packaged into phage heads. It is at this time that recombination may occur between DNA molecules.

Assembly of mature viruses is a complex process that has been well studied by William Wood, Robert Edgar, and others. Three major independent pathways occur, leading to: (1) DNA packaged into heads; (2) construction of tails; and (3) synthesis and assembly of tail fibers. A mutation that interrupts any one of the three pathways does not inhibit the other two. As shown in Figure 14.18, the head is assembled and DNA is packaged into it. This complex then combines with the tail, and only then are the tail fibers added. Total construction is a combination of self-assembly and enzyme-directed processes. When approximately 200 viruses are assembled, the bacterial cell is ruptured by the action of lysozyme and the mature phages are released. If this occurs on a lawn of bacteria,

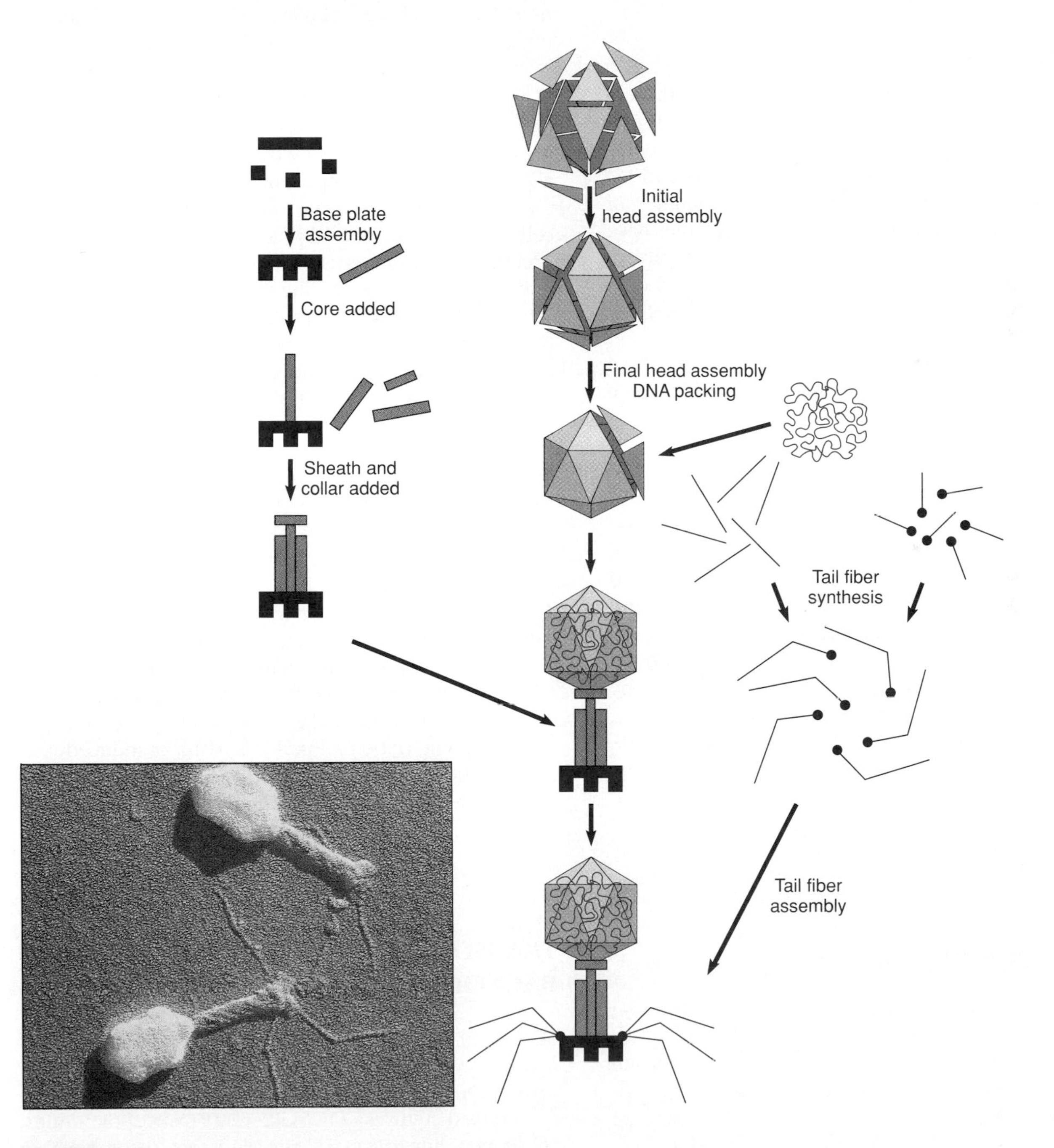

FIGURE 14.18
The assembly of bacteriophage T4 following infection of *E. coli*. Three separate pathways lead to the formation of an icosahedral head filled with DNA, a tail containing a central core, and tail fibers. The tail is added to the head, and then tail fibers are added. Each step of each pathway is influenced by one or more phage genes. The electron micrograph shows two mature phages.

the 200 phages will infect other bacterial cells and the process will be repeated over and over again. On the lawn, infection leads to a clear area where many bacteria have been lysed. Each such area is called a **plaque** and represents a clone of the single infecting T4 bacteriophage.

The Plaque Assay

The experimental study of bacteriophages and other viruses has played a critical role in our understanding of molecular genetics. During infection of bacteria, enormous quantities of bacteriophages may be obtained for investigation. Often, over 10^{10} viruses per milliliter of culture medium are produced. Many genetic studies have relied on the ability to quantitate the number of phages produced following infection under specific culture conditions. The technique that works nicely is called the **plaque assay.**

This assay is illustrated in Figure 14.19, where actual plaque morphologies are also shown. A serial dilution of the original virally infected bacterial culture is first performed. Then, a 0.1-ml aliquot from one or more dilutions is added to a small volume of melted nutrient agar (about 3 ml) to which a few drops of a healthy bacterial culture have been mixed. The solution is then poured evenly over a base of solid nutrient agar in a Petri dish and allowed to solidify prior to incubation. As described in the above section and illustrated in Figure 14.19, viral plaques occur at each place where a single virus has initially infected one bacterium in the lawn that has grown up during incubation. If the dilution factor is too low, the plaques are plentiful and will fuse, lysing the entire lawn. This has occurred in the 10^{-3} dilution in Figure 14.19. On the other hand, if the dilution factor is increased, plaques can be counted and a density of viruses in the initial culture can be estimated. The calculation is the same type as that used for determining bacterial density by counting colonies following serial dilution of an initial culture:

$$(\text{plaque number/ml}) \times (\text{dilution factor})$$

Using the results shown in Figure 14.19, it is observed that there are 23 phage plaques derived from the 0.1-ml aliquot of the 10^{-5} dilution. Therefore, we can estimate that there are 230 phage per ml at this dilution. The **initial phage density** in the undiluted sample, where 23 plaques are observed from 0.1 ml, is calculated as:

$$(230/\text{ml}) \times (10^5) = 230 \times 10^5/\text{ml}$$

Since this figure is derived from the 10^{-5} dilution, we can estimate that there will be only 0.23 phage per 0.1 ml in the 10^{-7} dilution. As a result, when 0.1 ml from this tube is assayed, it is not unexpected that no phage particles will be present. This possibility is borne out in Figure 14.19, where an intact lawn of bacteria exists. The dilution factor is simply too great.

The use of the plaque assay has been invaluable in mutational and recombinational studies of bacteriophages.

Lysis and Lysogeny

The relationship between virus and bacterium does not always result in viral reproduction and lysis. As early as the 1920s it was known that a virus could enter a bacterial cell and establish a symbiotic relationship with it. The precise molecular basis of this symbiosis is now well understood. Upon entry, the viral DNA, instead of replicating in the bacterial cytoplasm, is integrated into the bacterial chromosome, a step that characterizes the **lysogenic pathway.** Subsequently, each time the bacterial chromosome is replicated, the viral DNA is also replicated and passed to daughter bacterial cells following division. No new viruses are produced and no lysis of the bacterial cell occurs. However, under certain stimuli, such as chemical or ultraviolet light treatment, the viral DNA may lose its integrated status and initiate replication, phage reproduction, and lysis of the bacterium.

Several terms are used to describe this relationship. The viral DNA, integrated in the bacterial chromosome, is called a **prophage.** Viruses that can either lyse the cell or behave as a prophage are called **temperate.** Those that can only lyse the cell are referred to as **virulent.** A bacterium harboring a prophage is said to be **lysogenic;** that is, it is capable of being lysed as a result of induced viral reproduction. The viral DNA, which can replicate either in the bacterial cytoplasm or as part of the bacterial chromosome, may be classified as an **episome.**

TRANSDUCTION: VIRUS-MEDIATED BACTERIAL DNA TRANSFER

In 1952, Joshua Lederberg and Norton Zinder were investigating possible recombination in the bacterium *Salmonella typhimurium.* Although they recovered prototrophs from mixed cultures of two different auxotrophic strains, subsequent investigations revealed that recombination was occurring in a manner different from that attributable to the presence of an F factor, as in *E. coli.* What they were to discover was still a third mode of bacterial recombination, one mediated by bacteriophages and now called **transduction.**

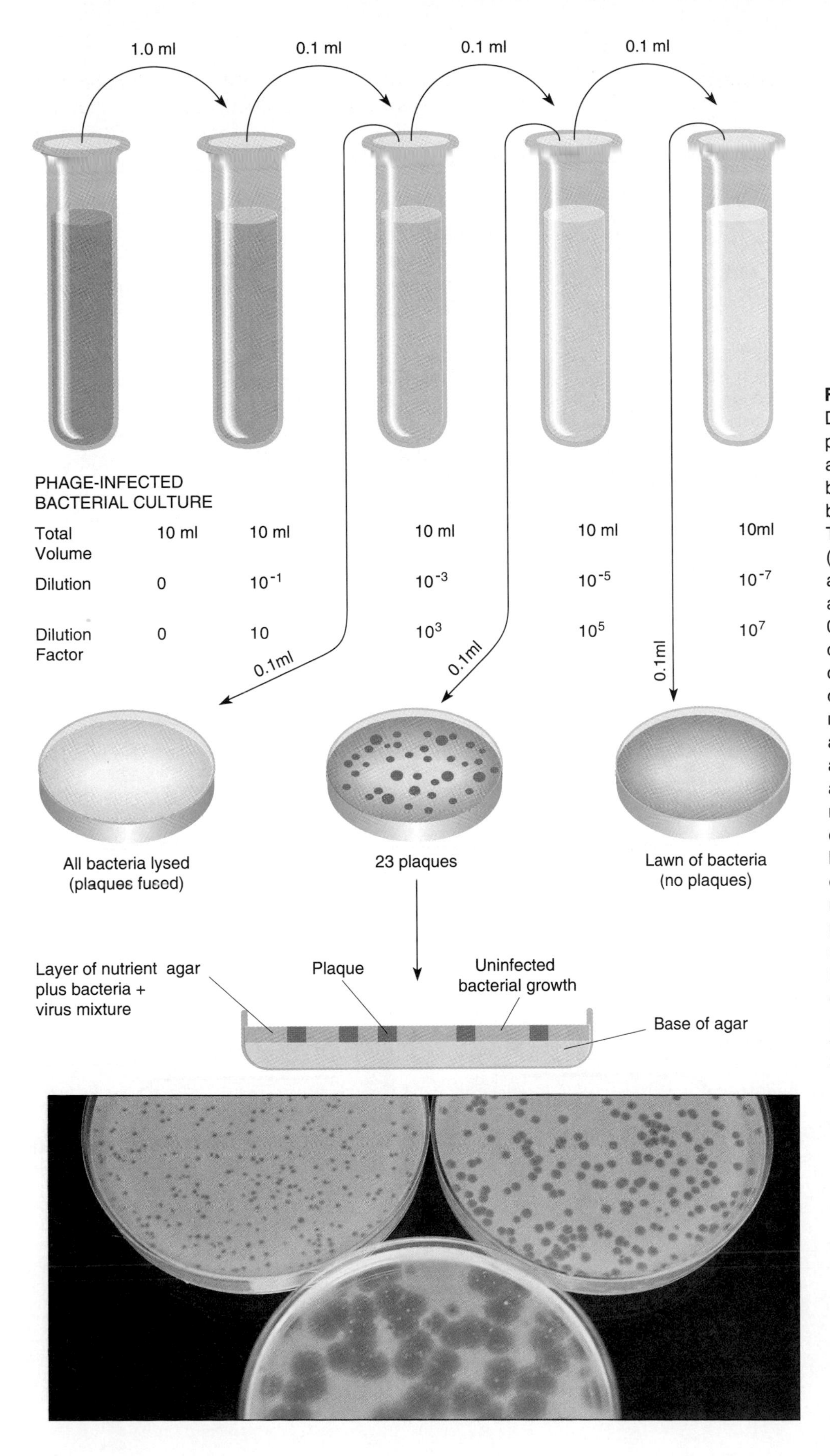

FIGURE 14.19
Diagrammatic illustration of the plaque assay for bacteriophage analysis. Serial dilutions of a bacterial culture infected with bacteriophage are first made. Then, three of the dilutions (10^{-3}, 10^{-5}, and 10^{-7}) are analyzed using the plaque assay technique. In each case, 0.1 ml of the bacterial–viral culture is mixed with a few drops of a healthy bacterial culture in nutrient agar. This mixture is spread over and allowed to solidify on a base of agar. Each clear area that arises is a plaque and represents the initial infection of one bacterial cell by one bacteriophage. Each generation of viral reproduction involves more bacteria, creating a visible plaque. In the 10^{-3} dilution, so many phage are present that all bacteria are lysed. In the 10^{-5} dilution, 23 plaques are produced. In the 10^{-7} dilution, the dilution factor is so great that no phage are present in the 0.1-ml sample, and thus no plaques form. From the 0.1-ml aliquot of the 10^{-5} dilution, the original bacteriophage density can be calculated as $23 \times 10 \times 10^5$ phage/ml (23×10^6 or 230×10^5), as described in the text. The photo illustrates plaque morphology of phage T2 (upper left), T4 (upper right), and lambda (bottom), all grown on *E. coli.* Note that plaques appear as clear areas in the drawing but as purple areas in the color-enhanced photograph.

The Lederberg–Zinder Experiment

Lederberg and Zinder mixed the *Salmonella* auxotrophic strains LA-22 and LA-2 together and recovered prototroph cells when the mixture was plated on minimal medium. LA-22 was unable to synthesize the amino acids phenylalanine and tryptophan (phe^- trp^-), and LA-2 could not synthesize the amino acids methionine and histidine (met^- his^-). Prototrophs (phe^+ trp^+ met^+ his^+) were recovered at a rate of about $1/10^5$ (10^{-5}) cells.

Although these observations at first appeared to suggest that the type of recombination involved was the kind observed earlier in *E. coli,* experiments using the Davis U-tube soon showed otherwise (Figure 14.20). When the two auxotrophic strains were separated by a glass-sintered filter, thus preventing cell contact but allowing growth to occur in a common medium, a startling observation was made. When samples were removed from both sides of the filter and plated independently on minimal medium, prototrophs were recovered, but only from one side of the tube. Recall that if conjugation were responsible, the conditions in the Davis U-tube would have prevented recombination.

Prototrophs were recovered only when cells from the side of the tube containing LA-22 bacteria were plated. The presence of LA-2 cells on the other side of the tube was essential for recombination, since LA-2 cells were the source of the new genetic information. Because the genetic information responsible for recombination had somehow passed across the filter, but in an unknown form, it was initially designated simply as a **filterable agent (FA).**

Three subsequent observations made it clear that this recombination was quite distinct from any other form of recombination:

1 The FA would not pass across filters with a pore diameter of less than 100 nm, a size that normally allows passage of small DNA molecules.

2 Testing the FA in the presence of DNase, which will enzymatically digest DNA, showed that the FA was not destroyed. These first two observations demonstrate that FA is not naked DNA.

3 The third observation was particularly important. It was observed that FA is produced by the LA-2 cells only when they are grown in association with LA-22 cells. If LA-2 cells are grown independently and that culture medium is then added to LA-22 cells, recombination is not observed. Therefore, LA-22 cells play some role in the production of FA by LA-2 cells and do so only when sharing common growth medium. This was a key observation in determining the basis of genetic exchange.

These observations are explained by the existence of a

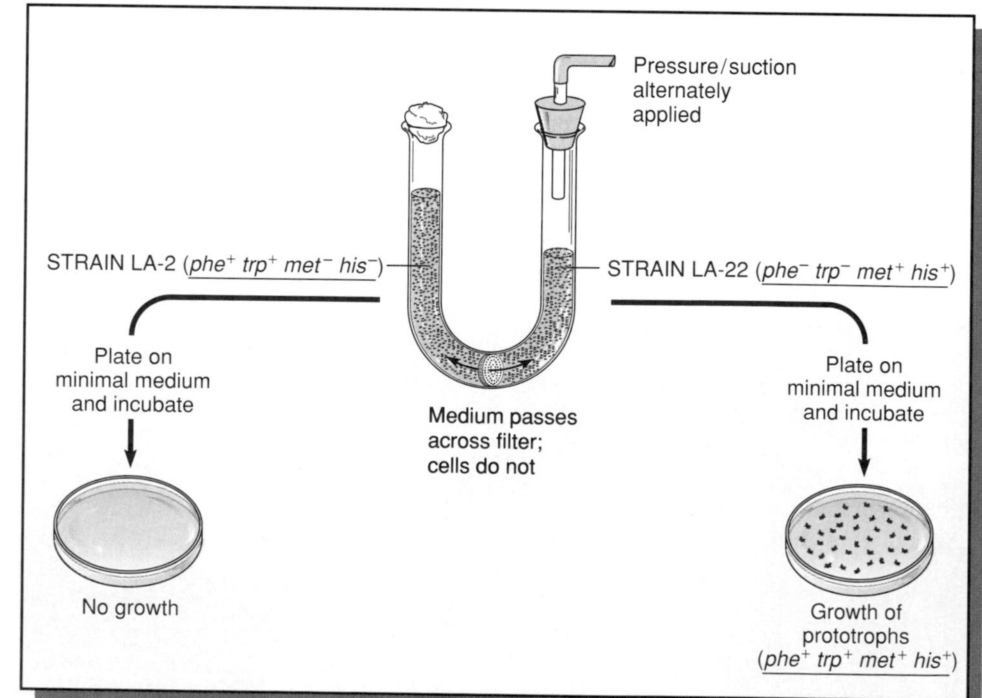

FIGURE 14.20
The Lederberg–Zinder experiment using *Salmonella.* After mixing two auxotrophic strains in a Davis U-tube, Lederberg and Zinder recovered prototrophs from the side containing LA-22 but not from the side containing LA-2. These initial observations led to the discovery of the phenomenon called transduction.

prophage (P22) present in the LA-22 *Salmonella* cells. Rarely, P22 prophages enter the vegetative or lytic phase, reproduce, and lyse some of the LA-22 cells. This phage, being much smaller than a bacterium, is then able to cross the filter and lyse some of the LA-2 cells, because this strain is not immune to attack. In the process of lysis, the P22 phages produced in LA-2 often acquire a region of the LA-2 chromosome along with their own genetic material. If this region contains the phe^+ and trp^+ genes present in LA-2, and if the phages subsequently pass back across the filter and reinfect LA-22 cells, prototrophs are produced. The exact nature of how this occurs was not immediately clear.

The Nature of Transduction

Further studies have revealed the existence of transducing phages in other species of bacteria. For example, *E. coli, Bacillus subtilis,* and *Pseudomonas aeruginosa* can be transduced by the phages P1, SP10, and F116, respectively. The precise mode of transfer of DNA during transduction has also been established. The process most often begins when a prophage enters the lytic cycle and progeny viruses subsequently infect and lyse other bacteria. During infection, the bacterial DNA is degraded into small fragments and the viral DNA is replicated. As packaging of the viral chromosomes in the protein head of the phage occurs, errors are sometimes made. Rarely, instead of viral DNA, segments of bacterial DNA are packaged.

Even though the initial discovery of transduction involved lysogenic bacteria, the same process can occur during the normal lytic cycle. Following infection, the bacterial chromosome is degraded into small pieces. If, during bacteriophage assembly, a small piece of bacterial DNA is packaged along with the viral chromosome, subsequent transduction is possible.

Sometimes, only bacterial DNA is packaged! Regions as large as 1 percent of the bacterial chromosome may become enclosed randomly in the viral head. Following lysis, these aberrant phages, lacking their own genetic material, are released in the culture medium. Because the ability to infect is a property of the protein coat, they can initiate infection of other unlysed bacteria. When this occurs, bacterial rather than viral DNA is injected into the bacterium and can either remain in the cytoplasm or recombine with its homologous region of the bacterial chromosome. If the bacterial DNA remains in the cytoplasm, it does not replicate but may remain in one of the progeny cells following each division. When this happens, only a single cell, partially diploid for the transduced genes, is produced—a phenomenon called **abortive transduction.** If the bacterial DNA recombines with its homologous region of the bacterial chromosome, the transduced genes are replicated as part of the chromosome and passed to all daughter cells. This process is called **complete transduction.** Both abortive and complete transduction are subclasses of the broader category called **generalized transduction.** As described above, transduction is characterized by the random nature of DNA fragments and genes transduced. Each fragment has a finite but small chance of being packaged in the phage head. Most cases of generalized transduction are of the abortive type; some data suggest that complete transduction occurs 10 to 20 times less frequently. This finding may be related to the fact that double-stranded DNA is involved. In comparison with single-stranded DNA, which is integrated during transformation, it may be much more difficult for double-stranded DNA to become integrated.

MUTATION AND RECOMBINATION IN VIRUSES

Much of what is known concerning viral genetics has been derived from studies of bacteriophages. Phage mutations often affect the morphology of the plaques formed following lysis of the bacterial cells. For example, in 1946 Alfred Hershey observed unusual T2 plaques on plates of *E. coli* strain B. While the normal T2 plaques are small and have a clear center surrounded by a turbid, diffuse halo, the unusual plaques are larger and possess a sharp outer perimeter. When the viruses are isolated from these plaques and replated on *E. coli* B cells, an identical plaque appearance is noted. Thus, the plaque phenotype is an inherited trait. Hershey named the mutant *rapid lysis* (r) because the plaques are larger, apparently resulting from a more rapid or more efficient life cycle of the phage (Figure 14.21). It is now known that wild-type phages undergo an inhibition of reproduction once a particular-sized plaque has been formed. The r mutant T2 phages are able to overcome this inhibition, producing larger plaques with a sharp perimeter.

Another bacteriophage mutation, *host range* (h), was discovered by Luria. This mutation extends the range of bacterial hosts that the phage can infect. Although wild-type T2 phages can infect *E. coli* B, they cannot normally attach or be absorbed to the surface of *E. coli* B-2. The h mutation, however, provides the basis for adsorption and subsequent infection.

Table 14.1 lists other representative types of mutations that have been isolated and studied in the T-even series of bacteriophages (T2, T4, T6, etc.). These mutations are important to the study of genetic phenomena in bacterio-

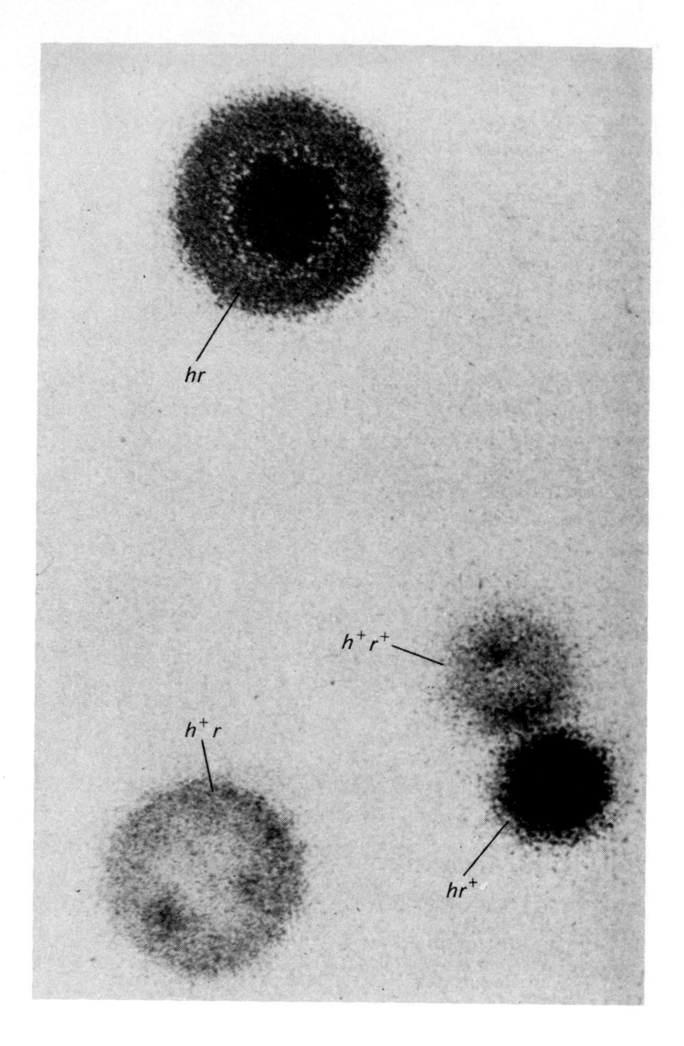

FIGURE 14.21
Actual plaque phenotypes observed following simultaneous infection of *E. coli* by two strains of phage T2, h^+r and hr^+. In addition to the parental genotypes, recombinant plaques hr and h^+r^+ were also recovered.

phages. Conditional, temperature-sensitive mutations have been particularly valuable in the study of essential genes where mutations eliminate the ability of the phage to reproduce.

Around 1947, several research teams demonstrated that recombination occurs between bacteriophages. This discovery of genetic exchange was made during experiments in which two mutant strains were allowed to simultaneously infect the same bacterial culture. These **mixed infection experiments** were designed so that the number of viral particles sufficiently exceeded the number of bacterial cells in order to ensure simultaneous infection of most cells by both viral strains.

For example, in one study using the T2/*E. coli* system, the viruses were of either the h^+r or hr^+ genotype. If no recombination occurred, these two parental genotypes (wild-type host range restriction, rapid lysis; and extended host range, normal lysis) would be the only expected phage progeny. However, the recombinant h^+r^+ and hr were detected in addition to the parental genotypes (Figure 14.21). The percentage of recombinant plaques divided by the total number of plaques reflects the relative distance between the genes:

$$\frac{(h^+r^+) + (hr)}{\text{total plaques}} \times 100 = \text{recombinational frequency}$$

Sample data for the h and r loci are shown in Table 14.2.

Similar recombinational studies have been performed with large numbers of mutant genes in a variety of bacteriophages. Data are analyzed in much the same way as they are in eukaryotic mapping experiments. Two- and three-point mapping crosses are possible, and the percentage of recombinants in the total number of phage progeny is calculated. This value is proportional to the

TABLE 14.1
Some mutant types of T-even phages.

Name	Description
minute	Small plaques
turbid	Turbid plaques on *E. coli* B
star	Irregular plaques
uv-sensitive	Alters UV sensitivity
acriflavin resistant	Forms plaques on acriflavin agar
osmotic shock	Withstands distilled water
lysozyme	Does not produce lysozyme
amber	Grows in *E. coli* K12, but not B
temperature sensitive	Grows at 25°C, but not at 42°C

TABLE 14.2
Results of a cross involving the *h* and *r* genes in phage T2.

Genotype	Plaques	Designation
hr^+	42	Parental Progeny 76%
h^+r	34	
h^+r^+	12	Recombinants 24%
hr	12	

SOURCE: Data derived from Hershey and Rotman, 1949.

relative distance between two genes along the chromosome.

An interesting observation in phage crosses is **negative interference.** Recall that in eukaryotic mapping crosses, positive interference is the rule. Fewer than expected double-crossover events are observed. In phage crosses, often just the reverse occurs. In three-point analysis, a greater than expected frequency of double exchanges is observed. Negative interference is explained on the basis of the dynamics of the conditions leading to recombination within the bacterial cell.

How phage recombination occurs has been investigated. The available evidence supports the concept that recombination between phage chromosomes involves a breakage and reunion process similar to that of eukaryotic crossing over. The process is facilitated by nucleases that nick and reseal DNA strands. A fairly clear picture of the dynamics of viral recombination has emerged.

Following the early phase of infection, the chromosomes of each phage begin replication. As this stage progresses, a pool of chromosomes accumulates in the bacterial cytoplasm. If double infection by phages of two genotypes has occurred, then the pool of chromosomes initially consists of the two parental types. Genetic exchange between these two types will occur, producing recombinant chromosomes.

In the case of the h^+r and hr^+ example just discussed, h^+r^+ and hr chromosomes are produced. Each of these recombinant chromosomes may undergo replication and is also free to undergo new exchange events with the other and with the parental types. Furthermore, recombination is not restricted to exchange between two chromosomes—three or more may be involved. As phage development progresses, chromosomes are removed randomly from the pool and packaged into the phage head, forming mature phage particles. Thus, parental and recombinant genotypes are produced.

Recombination events in viruses are not restricted to exchanges between genes. With powerful selection systems, it is possible to detect intragenic recombination, where exchanges occur within a single gene.

OTHER STRATEGIES FOR VIRAL REPRODUCTION

In contrast to the T-even phages, other bacterial viruses have evolved many different strategies in order to reproduce. For example, during the infection of *E. coli,* the smaller **bacteriophage ϕX174** contains only a circular single strand of DNA, called the plus (+) strand. It infects the host and immediately serves as the template for the synthesis of the complementary negative (−) strand. This results in a circular double-stranded DNA molecule called the **replicative form (RF).** The RF is itself replicated semiconservatively, producing about 50 progeny molecules. These then serve as templates for the production of only (+) strands. As hundreds of these DNA strands are produced, they are packaged into protein capsids synthesized under the direction of the viral genetic information.

While DNA viruses can pirate the use of the host DNA polymerase in order to replicate their nucleic acid, no comparable enzyme exists when RNA-containing viruses invade their host cells. Some RNA viruses require an RNA-dependent RNA polymerase, which is sometimes more simply called **RNA replicase.** For example, when the small **bacteriophage Qβ** infects *E. coli* as its host cell, the infective single-stranded RNA molecule first serves as a messenger RNA (mRNA), encoding three proteins. Two are components of the coat protein, while one is a subunit of RNA replicase. This subunit combines with three other subunits that are bacterial in origin. The resultant enzyme is specific for the replication of the phage RNA (as compared with the bacterial RNAs that are present). First, complementary RNA virus strands (−) are created. These serve as the template for the synthesis of many (+) strands. Ultimately, the (+) strands are encapsulated by coat proteins as mature viruses are formed. A similar mode of replication occurs in the eukaryotic, RNA-containing **poliovirus.**

Reverse Transcriptase

The final type of viral reproduction to be discussed occurs following the infection of animal cells rather than bacteria and involves an extraordinary finding made in the early 1960s. A long series of investigations by Howard Temin and David Baltimore led to the discovery of **reverse transcriptase,** an enzyme capable of synthesizing DNA on an RNA template. This enzyme is also called **RNA-directed DNA polymerase.** Temin proposed that the RNA that is the genetic material of certain RNA-containing animal tumor viruses is transcribed reversibly into DNA, which then serves as the template for replication and transcription of RNA during viral infection. These viruses are called **oncogenic** because they induce malignant growth.

Temin's proposed reverse transcriptase was largely ignored because it contradicted the previously accepted notion that genetic information flows only from DNA to RNA. In addition, other RNA-containing viruses (polio virus and double- and single-stranded RNA bacteriophages), whose RNA is replicated by the enzyme RNA replicase, did not demonstrate reverse transcriptase activity. Thus, molecular biologists questioned why the tumor viruses should require such an unusual enzyme.

Working with the RNA-containing oncogenic **Rous sarcoma virus (RSV),** which produces cancer in chickens, Temin added the antibiotic actinomycin D to cultures of cells that had been infected. He found that the antibiotic inhibits viral multiplication. Actinomycin D is known to specifically inhibit DNA-directed RNA synthesis and not to affect RNA-directed RNA synthesis. Temin also found that inhibitors of DNA synthesis such as 5-fluorodeoxyuridine also block infection. Thus, he reasoned that DNA must therefore serve as an essential intermediate molecule during infection.

Temin subsequently demonstrated directly that DNA is synthesized by RSV during infection. He announced these results in 1970 and found that other workers, using a mouse leukemia virus, had arrived at the same conclusions. These reports stimulated a tremendous amount of work on the enzyme and its role in tumor production. Reverse transcriptase was subsequently isolated independently by Temin and Baltimore. It has since been found in all RNA oncogenic viruses. Because they "reverse" the flow of genetic information they are called **retroviruses.**

The enzyme uses the infecting (+) RNA strand as a template, synthesizing DNA in the 5′-to-3′ direction. The enzyme then uses this DNA strand as a template and synthesizes the complementary strand, creating a double-stranded DNA molecule. This is capable of integrating into the genome of the host cell, forming what is called a **provirus.** When the host genome is transcribed into RNA, so is the proviral DNA. This produces (+) RNA strands, some of which are translated into viral proteins. Packaging of other newly formed RNA molecules into the viral capsid occurs to form mature viruses. Such a host cell is said to be **transformed.** A similar life cycle occurs in all retroviruses.

Acquired Immunodeficiency Syndrome (AIDS)

Interest in retroviruses has heightened considerably with the discovery of the **human immunodeficiency virus (HIV),** considered to be the causative agent in **acquired immunodeficiency syndrome (AIDS).** HIV is a member of a subgroup of retroviruses transmitted by sexual contact or the exchange of body fluids. The genetic material consists of single-stranded RNA containing about 9000 ribonucleotides, surrounded by a protein capsid. Genetically programmed surface glycoproteins that are associated with this capsid are essential to the infection process. One set of these, called gp120 (a glycoprotein of 120,000 molecular weight), interacts with a surface receptor on the host cell called CD4, which is found mainly on a subset of T lymphocytes, the T helper cells.

Viral infection and reproduction is dependent on DNA produced under the direction of reverse transcriptase. Following assembly, the viruses bud through the cell

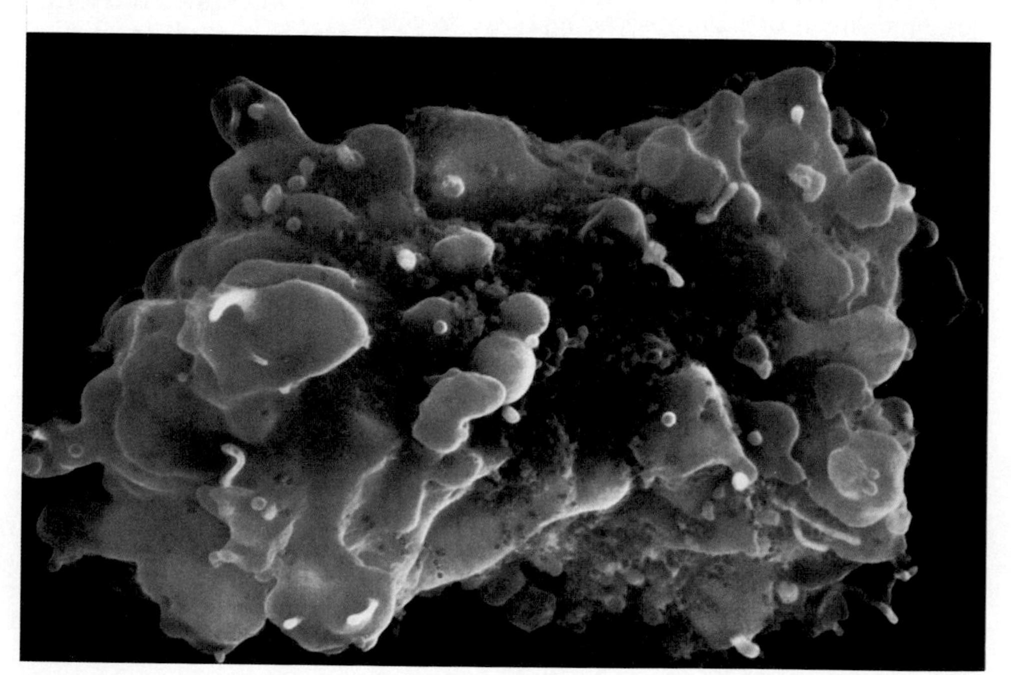

FIGURE 14.22
A scanning electron micrograph of a T lymphocyte infected by HIV viruses. The projections from the surface are microvilli. The virus particles are present on the cell surface and are seen as small, blue, rounded dots.

membrane of the infected cell (Figure 14.22). In addition to the production of viral particles, an excess amount of gp120 is produced that is inserted into the surface of the infected cell. Its presence facilitates fusion with other uninfected T helper cells, leading to the death of all involved cells. The depletion of the T lymphocyte helper cell population has a catastrophic effect on the immune system, eventually immobilizing it almost completely.

Infected individuals are, as a result, susceptible to so-called opportunistic infections that healthy immune systems fend off fairly easily. Such diseases, including *Pneumocystitis carini* pneumonia and Kaposi's sarcoma, are often the cause of death in AIDS patients.

HIV has an unusually high error rate during its replication, in comparison to many other viruses. This characteristic leads to the production of virus particle populations with substantial genetic diversity, particularly in regard to the chemical specificity of the envelope protein gp120. As a result, the development of a vaccine against this devastating virus has been extremely difficult.

CHAPTER SUMMARY

1 Genetic analysis of bacteria and viruses has been pursued successfully since the 1940s. The ease of obtaining large quantities of pure cultures of bacteria and mutant strains of viruses as experimental material, and the efficient selection and isolation of naturally occurring and induced mutations have made these the organisms of choice in numerous types of genetic studies.

2 Although the concept of spontaneous mutation is supported experimentally and is widely upheld, recent evidence suggests that some mutations that are adaptive for the survival of the organism may arise in response to selective pressure from the environment.

3 Exchange of genetic information between bacteria involves three different modes of recombination: conjugation, transformation, and transduction.

4 Conjugation is initiated by a bacterium housing a plasmid called the F factor. If the F factor is in the cytoplasm (F^+), following its replication one copy is transferred to the recipient cell, converting it to the F^+ status.

5 If the F factor is integrated into the donor cell chromosome (Hfr), recombination is initiated with the recipient cell, with genetic information flowing unidirectionally. Time mapping of the bacterial chromosome is based on the orientation and position of the F factor in the donor chromosome.

6 The products of a group designated as the *rec* genes have been found to be directly involved in the process of recombination between the invading DNA and the recipient bacterial chromosome.

7 Insertion sequences, unique DNA sequences present in both plasmids and the bacterial chromosome, facilitate movement of genetic information such as the insertion and excision of the F factor into and out of the chromosome. These sequences may combine with genes, producing transposons, which also move from one genetic molecule or position to another.

8 The phenomenon of transformation, which does not require cell contact, involves the entry of single-stranded exogenous DNA into the host chromosome of a recipient bacterial cell. Linkage mapping of closely aligned genes may be performed using this process.

9 Transduction, or virus-mediated bacterial DNA transfer, requires the formation of a symbiotic relationship between bacteria and bacteriophages that infect them. The viral genetic material may become integrated into the bacterial chromosome as a prophage and lysogenize the host cell.

10 Phages are studied using the plaque assay. The discovery of mutations in phages, such as those causing variation in plaque morphology and the alteration of the host range of infectivity, have allowed the analysis of recombination between viruses.

11 Retroviruses are RNA oncogenic viruses that depend on reverse transcriptase activity for infection. This enzyme reverses the flow of genetic information from RNA to DNA, which most often integrates into the genome of the host chromosome. HIV, a subgroup of the retroviruses, debilitates the immune system by infecting T lymphocyte helper cells, consequently increasing the patients' susceptibility to diseases characteristic of AIDS.

KEY TERMS

acquired immunodeficiency syndrome (AIDS)
adaptation hypothesis
auxotroph
bacteriophage
competence
conjugation
cotransformation
episome
F^- cells
F^+ cells
F′ cells
F sex pilus
fluctuation test
heteroduplex
Hfr (high-frequency recombination)
Human immunodeficiency virus (HIV)
insertion sequence
inverted repeat sequence
lysogenic pathway
merozygote
minimal medium
oncogenic virus
plaque
plaque assay
plasmid
prophage
prototroph
recombinant DNA research
replicative form (RF)
retrovirus
reverse transcriptase
RNA replicase
spontaneous mutation
temperate phage
transduction
transposon (Tn)
virulent phage

1 Time mapping was performed in a cross involving the genes *his, leu, mal,* and *xyl.* After 25 minutes, mating was interrupted with the following results in recipient cells.

90% were *xyl*
80% were *mal*
20% were *his*
none were *leu*

What are the positions of these genes relative to the origin (*O*) of the F factor and to one another?

SOLUTION: Since the *xyl* gene was transferred most frequently, it is closest to *O* (*very* close). The *mal* gene is next and reasonably close to *xyl,* followed by the *his* gene. The *leu* gene is well beyond these three, since no recombinants are recovered that include it.

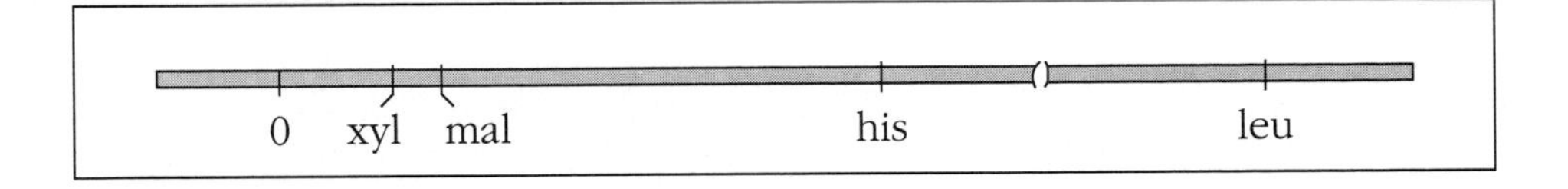

2 In four Hfr strains of bacteria, all derived from an original F^+ culture grown over several months, a group of hypothetical genes were studied and shown to be transferred in the following orders:

Hfr Strain	Order of Transfer
1	E R I U M B
2	U M B A C T
3	C T E R I U
4	R E T C A B

Assuming that B is the first gene along the chromosome, determine the sequence of all genes shown. One strain creates an apparent dilemma. Which one is it? Explain why the dilemma is only apparent and not real.

SOLUTION: The solution is arrived at by overlapping the genes in each strain in the sequence in which they were transferred:

	2	U M **B A C** T
Strain:	3	C **T E R** I U
	1	E R **I U M** B

Starting with B, the sequence of genes is: BACTERIUM!

Strain 4 creates an apparent dilemma, which is resolved by realizing that the F factor is integrated in the opposite orientation; thus the genes enter in the opposite sequence; starting with gene R.

MUIRETCAB (←, starting at R)

PROBLEMS AND DISCUSSION QUESTIONS

1 Distinguish between the three modes of recombination in bacteria.

2 With respect to F^+ and F^- bacterial matings, answer the following questions:
(a) How was it established that physical contact was necessary?
(b) How was it established that chromosome transfer was unidirectional?
(c) What is the basis of a bacterium being F^+ or Hfr?

3 List all major differences between (a) the $F^+ \times F^-$ and the Hfr $\times$ F^- bacterial crosses and (b) F^+, F^-, Hfr, and F' bacteria.

4 Describe the basis for chromosome mapping in the Hfr $\times$ F^- crosses.

5 When the interrupted mating technique was used with five different strains of Hfr bacteria, the following order of gene entry and recombination was observed. On the basis of these data, draw a map of the bacterial chromosome. Do the data support the concept of circularity?

Hfr Strain	Order
1	T C H R O
2	H R O M B
3	M O R H C
4	M B A K T
5	C T K A B

6 Why are the recombinants produced from an Hfr × F^- cross never F^+?

7 Describe the origin of F′ bacteria and merozygotes.

8 Explain the observations that led Zinder and Lederberg to conclude that the prototrophs recovered in their transduction experiments were not the result of F-mediated conjugation.

9 Define plaque, lysogeny, and prophage.

10 If a single bacteriophage infects one *E. coli* cell present on a lawn of bacteria, and upon lysis yields 200 viable viruses, how many phage will exist in a single plaque if only three more lytic cycles occur?

11 A culture of an auxotrophic *leu*$^-$ strain of bacteria was irradiated and incubated until it reached the stationary phase. A control culture, which was not irradiated, was also studied. These cultures were then serially diluted, and 0.1 ml of various dilutions was plated on minimal medium plus leucine and on minimal medium. The results, shown below, were used to determine the spontaneous and X-ray-induced mutation rate of *leu*$^-$ to *leu*$^+$.

Culture Condition	Culture Medium	Dilution	Number of Colonies
Irradiated	(1) Minimal medium plus leucine	10^{-9}	24
	(2) Minimal medium	10^{-2}	12
Control	(3) Minimal medium plus leucine	10^{-9}	12
	(4) Minimal medium	10^{-1}	3

(a) Describe what is represented by each value obtained. Which values should be approximately equal? Are they?

(b) Determine the induced and spontaneous mutation rate leading to prototrophic growth (*leu*$^-$ to *leu*$^+$).

12 Distinguish between insertion sequences and transposons in bacteria.

13 Describe the role of reverse transcriptase during the infection of an animal cell by a retrovirus.

SELECTED READINGS

ADELBERG, E. A. 1960. *Papers on bacterial genetics.* Boston: Little, Brown.

BIRGE, E. A. 1988. *Bacterial and bacteriophage genetics—An introduction.* New York: Springer-Verlag.

BRODA, P. 1979. *Plasmids.* New York: W.H. Freeman.

BUKHARI, A. I., SHAPIRO, J. A., and ADHYA, S. L., eds. 1977. DNA *insertion elements, plasmids, and episomes.* Cold Spring Harbor, NY: Cold Spring Harbor Laboratory.

CAIRNS, J., OVERBAUGH, J., and MILLER, S. 1988. The origin of mutants. *Nature* 335:142–45.

CAIRNS, J., STENT, G. S., and WATSON, J. D., eds. 1966. *Phage and the origins of molecular biology.* Cold Spring Harbor, NY: Cold Spring Harbor Laboratory.

CAMPBELL, A. M. 1976. How viruses insert their DNA into the DNA of the host cell. *Scient. Amer.* (Dec.) 235:102–13.

CAVALLI-SFORZA, L. L., and LEDERBERG, J. 1956. Isolation of preadaptive mutants in bacteria by sib selection. *Genetics* 41:367–81.

COHEN, S. N., and SHAPIRO, J. A. 1980. Transposable genetic elements. *Scient. Amer.* (Feb.) 242:40–49.

HALL, B. G. 1990. Spontaneous point mutations that occur more often when advantageous than when neutral. *Genetics* 126:5–16.

HAYES, W. 1968. *The genetics of bacteria and their viruses.* 2nd ed. New York: Wiley.

HERSHEY, A. D., and ROTMAN, R. 1949. Genetic recombination between host range and plaque-type mutants of bacteriophage in single cells. *Genetics* 34:44–71.

JACOB, F., and WOLLMAN, E. L. 1961. Viruses and genes. *Scient. Amer.* (June) 204:92–106.

LEDERBERG, J. 1986. Forty years of genetic recombination in bacteria: A fortieth anniversary reminiscence. *Nature* 324:627–28.

LURIA, S. E., and DELBRUCK, M. 1943. Mutations of bacteria from virus sensitivity to virus resistance. *Genetics* 28:491–511.

LWOFF, A. 1953. Lysogeny. *Bacteriol. Rev.* 17:269–337.

MORSE, M. L., LEDERBERG, E. M., and LEDERBERG, J. 1956. Transduction in *Escherichia coli* K 12. *Genetics* 41:141–56.

NOVICK, R. P. 1980. Plasmids. *Scient. Amer.* (Dec.) 243:102–27.

SMITH-KEARY, P. F. 1989. *Molecular genetics of Escherichia coli.* New York: Guilford Press.

STAHL, F. W. 1987. Genetic recombination. *Scient. Amer.* (Nov.) 256:91–101.

STARLINGER, P. 1980. IS elements and transposons. *Plasmid* 3:241–59.

STENT, G. S. 1963. *Molecular biology of bacterial viruses.* New York: W.H. Freeman.

———, 1966. *Papers on bacterial viruses.* 2nd ed. Boston: Little, Brown.

TEMIN, H. 1972. RNA-directed DNA synthesis. *Scient. Amer.* (Jan.) 24–33.

VARMIS, H. 1987. Reverse transcription. *Scient. Amer.* (Sept.) 257:56–64.

VISCONTI, N., and DELBRUCK, M. 1953. The mechanism of genetic recombination in phage. *Genetics* 38:5–33.

WOLLMAN, E. L., JACOB, F., and HAYES, W. 1956. Conjugation and genetic recombination in *Escherichia coli* K 12. *Cold Spr. Harb. Symp.* 21:141–62.

ZINDER, N. D. 1958. Transduction in bacteria. *Scient. Amer.* (Nov.) 199:38.

15 Recombinant DNA and Genetic Technology

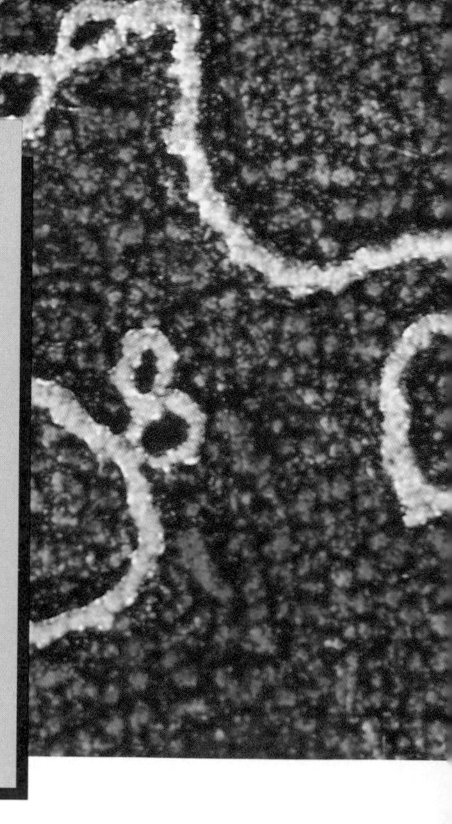

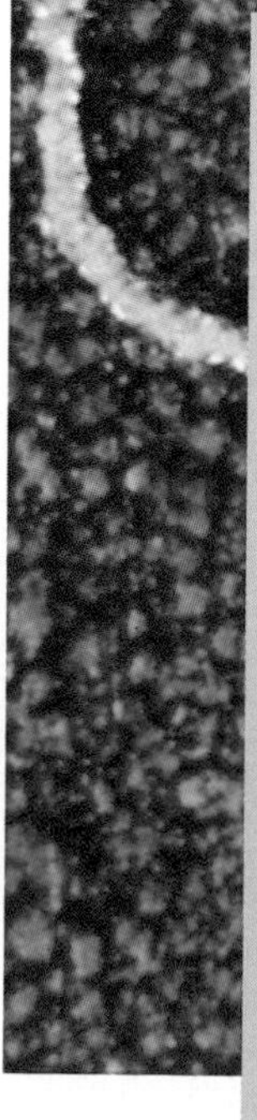

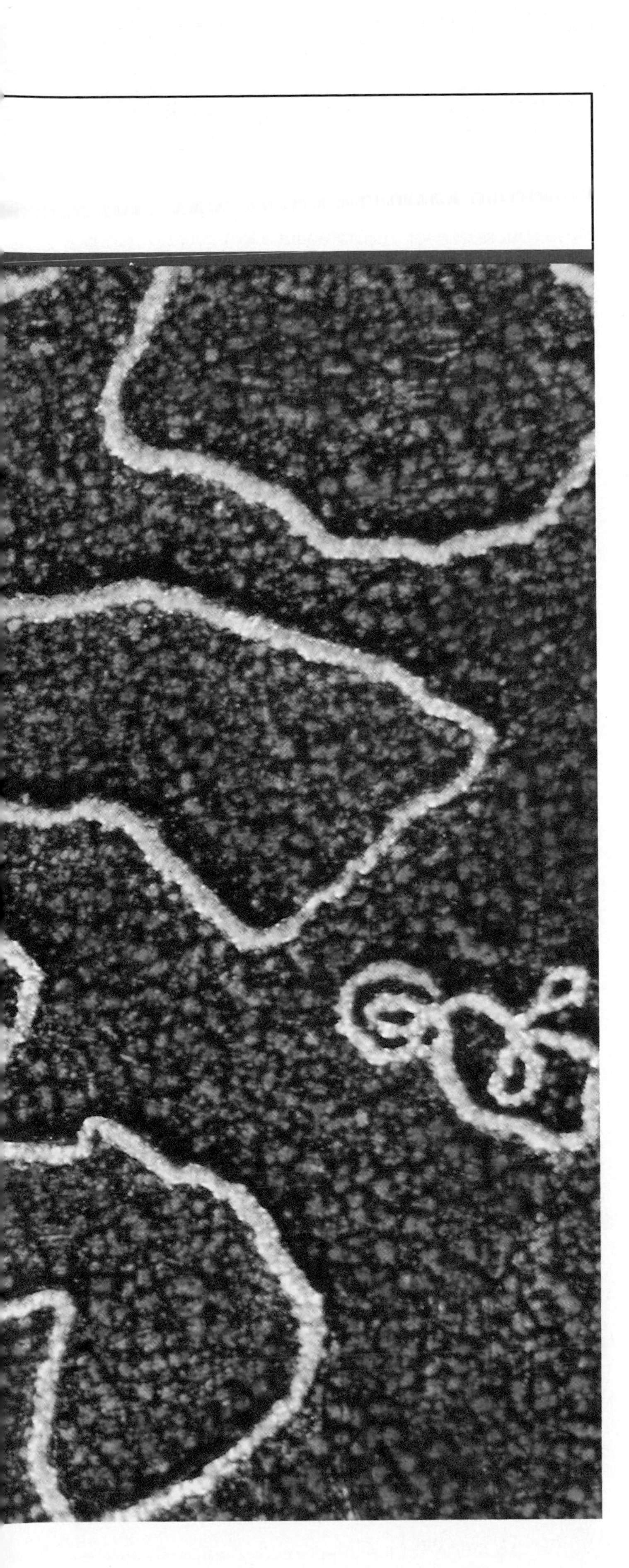

Chapter Concepts

Recombinant DNA technology depends in part on the ability to cleave and rejoin DNA segments at specific base sequences. Using this methodology, individual DNA segments can be transferred to viruses, bacteria, or yeast and amplified, isolated, and identified. The use of this technology has brought about significant advances in gene mapping, disease diagnosis, the commercial production of human gene products, and the trans-specific transfer of genes in plants and animals.

We are now in the beginning stages of perhaps the most profound transition in the history of genetics—the development and application of **recombinant DNA technology,** also known as **gene splicing** or **genetic engineering.** This technology is being used to generate new knowledge and to develop new commercial products such as vaccines and drugs. It has also raised fears about epidemics or widespread ecological changes that might result from the release of genetically engineered organisms into the environment. In this chapter, we will review some of the methods used in recombinant DNA technology to isolate, replicate, and analyze genes, and we will discuss some of the applications of this technology to agriculture, medicine, and industry.

RECOMBINANT DNA TECHNOLOGY: AN OVERVIEW

The term **recombinant DNA** refers to the creation of a new association between DNA molecules or segments of DNA molecules that are not found together naturally. Although genetic mechanisms such as crossing over technically produce recombinant DNA, the term is generally reserved for DNA molecules produced by joining segments derived from different biological sources.

Recombinant DNA technology uses techniques derived from the biochemistry of nucleic acids coupled with genetic methodology originally developed for the study of bacteria and viruses. The basic procedures involve a series of steps:

1 DNA fragments are generated by using enzymes that recognize and cut DNA molecules at specific nucleotide sequences.

2 These segments are joined to other DNA molecules that serve as **vectors.** Vectors can replicate autonomously and thus facilitate the manipulation and identification of the newly created recombinant DNA molecule.

3 The vector, carrying an inserted DNA segment, is transferred to a host cell. Within this cell, the recombinant DNA molecule is replicated, producing copies of the inserted DNA segment known as **clones.**

4 The cloned DNA segments can be recovered from the host cell, purified, and analyzed.

5 Potentially, the cloned DNA can be transcribed, its mRNA translated, and the gene product isolated and studied.

Restriction Enzymes

The cornerstone of recombinant DNA technology is a class of enzymes called **restriction endonucleases.** These enzymes, isolated from bacteria, received their name because they restrict or prevent viral infection by degrading the invading nucleic acid. These restriction enzymes recognize a specific nucleotide sequence and produce a double-stranded break within that sequence. In most cases, the nucleotide sequence in the recognition site is **palindromic** (that is, it reads the same in both directions).

Some restriction enzymes produce DNA fragments with single-stranded complementary tails. For example, one of the first such enzymes discovered was from *E. coli* and was designated ***Eco*RI.** Its palindromic recognition site and points of cleavage are shown in the following diagram:

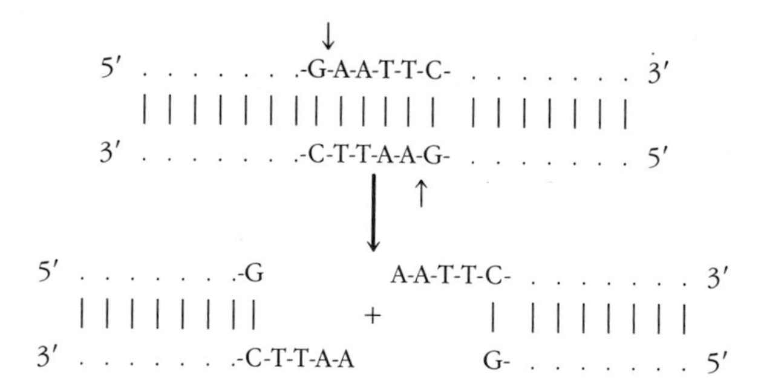

The complementary tails are said to be "sticky" be-

TABLE 15.1
Some restriction enzymes cleavage pattern, and source

Enzyme	Recognition, Clevage Sequence	Source	Enzyme	Recognition, Clevage Sequence	Source
*Eco*RI	↓ GAATTC CTTAAG ↑	*E. coli*	*Alu*I	↓ AGCT TCGA ↑	*Arthrobacter luteus*
*Hind*III	↓ AAGCTT TTCGAA ↑	*Hemophilus influenza*	*Hae*III	↓ GGCC CCGG ↑	*Hemophilus aegypticus*
*Bam*HI	↓ GGATCC CCTAGG ↑	*Bacillus amyloliquefaciens*	*Bal*I	↓ TGGCCA ACCGGT ↑	*Brevibacterium albidum*
*Taq*I	↓ TCGA AGCT ↑	*Thermus aquaticus*	*Sau*3A	↓ GATC CTAG ↑	*Staphylococcus aureus*

cause under hybridization conditions, they can reanneal with each other. If DNAs from different sources share the same palindromic recognition sites, then both will contain complementary single-stranded tails when treated with a restriction endonuclease. If placed together under proper conditions, the DNA fragments from these two sources can form recombinant molecules by annealing of their sticky ends. The enzyme DNA ligase is used *in vitro* to chemically join these fragments (Figure 15.1). Table 15.1 lists some common restriction enzymes and their recognition sequences.

Vectors

By using a vector or cloning vehicle, a DNA segment can gain entry into a host cell and be replicated or cloned. Vectors are, in essence, carrier DNA molecules. There are a number of vectors currently in use, including **plasmids, bacteriophage,** and **cosmids.**

Plasmids are naturally occurring, extrachromosomal, double-stranded DNA molecules that replicate autonomously within bacterial cells (Figure 15.2). Some plasmids can also be amplified within cells, increasing their copy number from 10–20 to almost 1000, allowing more copies of cloned DNA to be produced. Many plasmids have been modified or engineered to contain a limited number of restriction sites and specific antibiotic resistance genes. The plasmid pBR322, shown in Figure 15.3, carries cleavage sites for several restriction enzymes and genes for resistance to ampicillin and tetracycline. A DNA segment inserted into the tetracycline gene will inactivate this gene, but not affect the gene for ampicillin resistance. Bacterial cells that acquire the recombinant plasmid can then be selected because they are sensitive to tetracycline, but resistant to ampicillin.

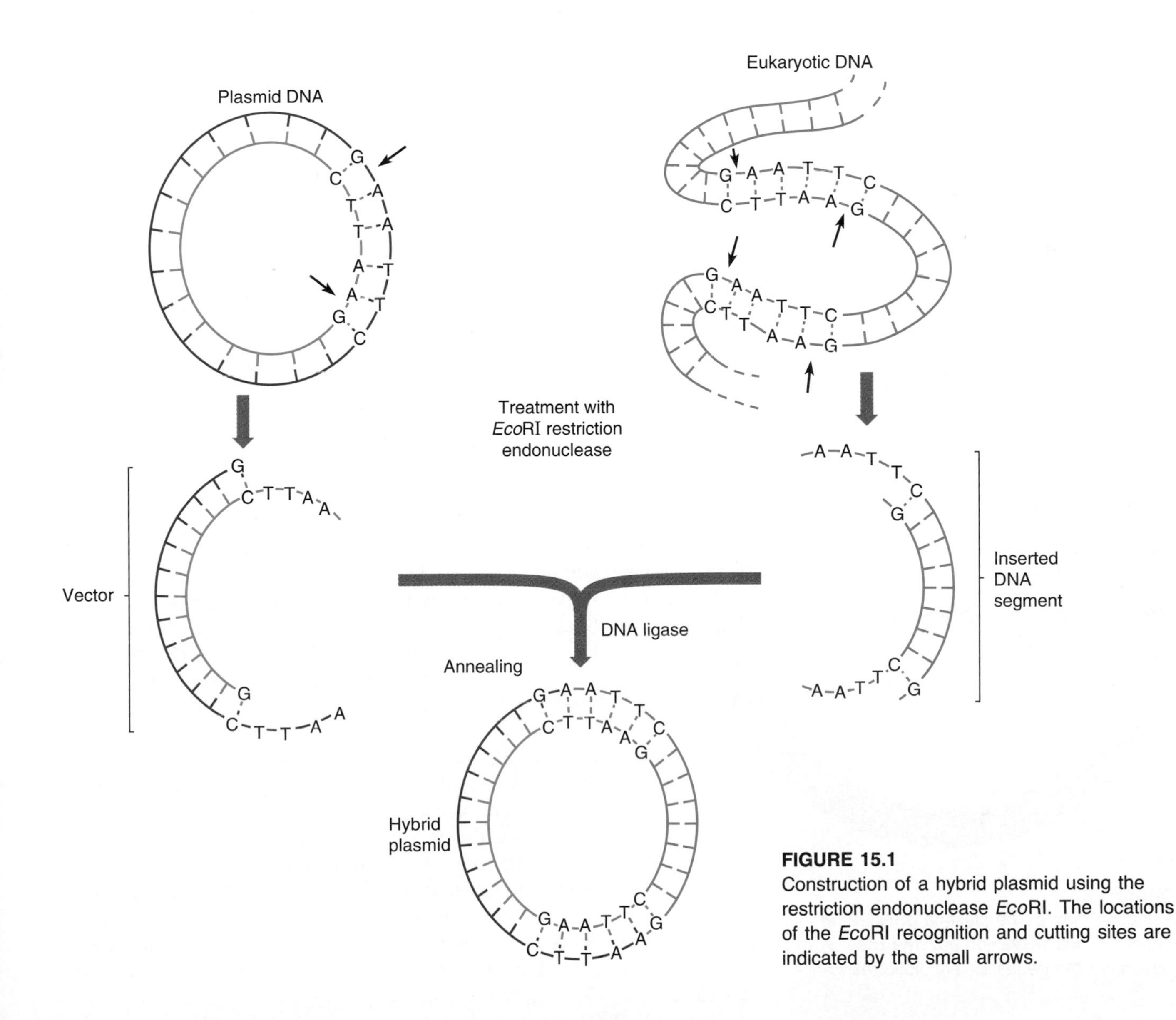

FIGURE 15.1
Construction of a hybrid plasmid using the restriction endonuclease *Eco*RI. The locations of the *Eco*RI recognition and cutting sites are indicated by the small arrows.

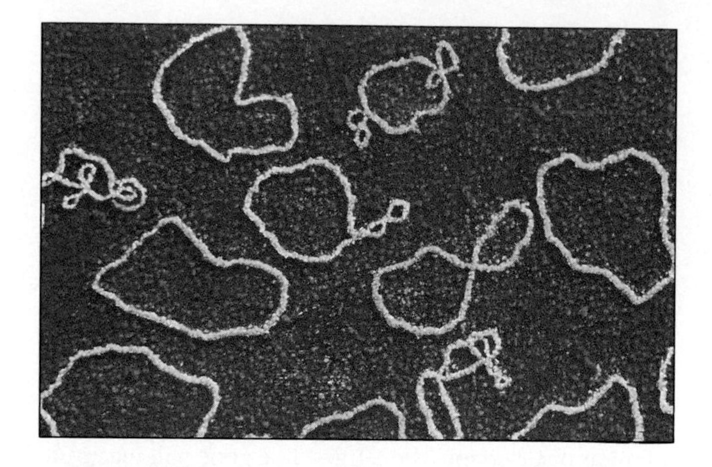

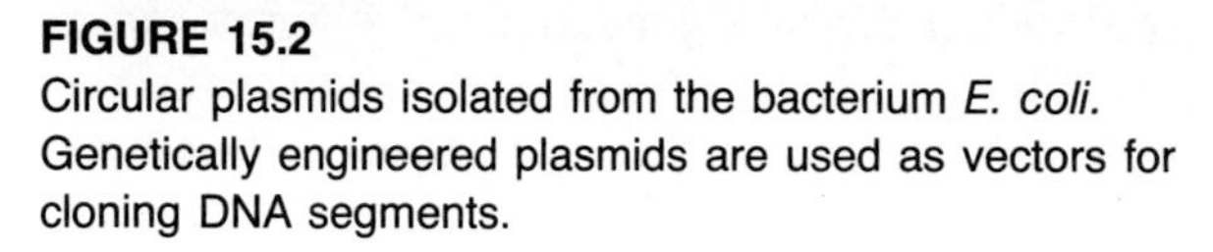

FIGURE 15.2
Circular plasmids isolated from the bacterium *E. coli.* Genetically engineered plasmids are used as vectors for cloning DNA segments.

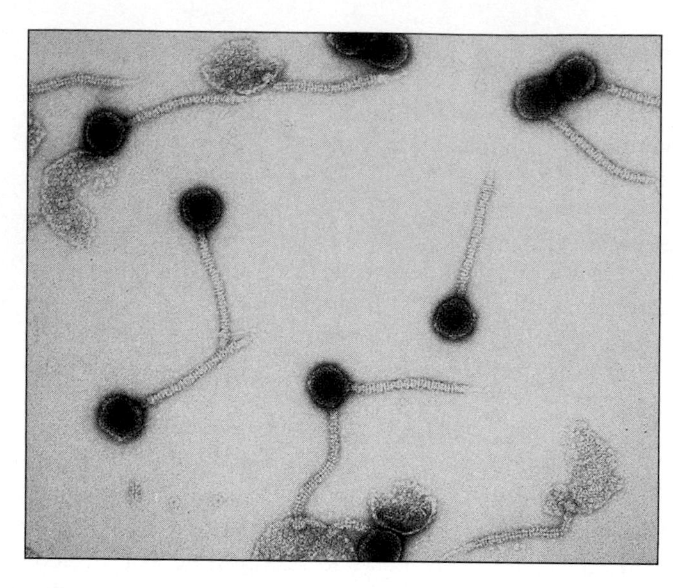

FIGURE 15.4
An electron micrograph showing a cluster of the bacteriophage lambda, widely used as a vector in recombinant DNA work.

A bacteriophage widely used in recombinant DNA work is the phage lambda (Figure 15.4). Its genes have all been identified and mapped, and the DNA sequence of the entire genome is known. The middle third of the lambda chromosome contains a cluster of genes necessary for the lysogenic phase of its life cycle, but not for the lytic phase. As a result, this region can be replaced with foreign DNA without affecting the ability of the phage to infect cells and form plaques (Figure 15.5). Over one hundred vectors based on lambda phage have been derived by removing various portions of the central gene cluster, which is unnecessary for production of viral particles. DNA to be cloned can be inserted in place of this region. Lambda vectors containing inserted DNA can be introduced into bacterial host cells in a process called transfection, in which the host cells have been made permeable by a chemical treatment.

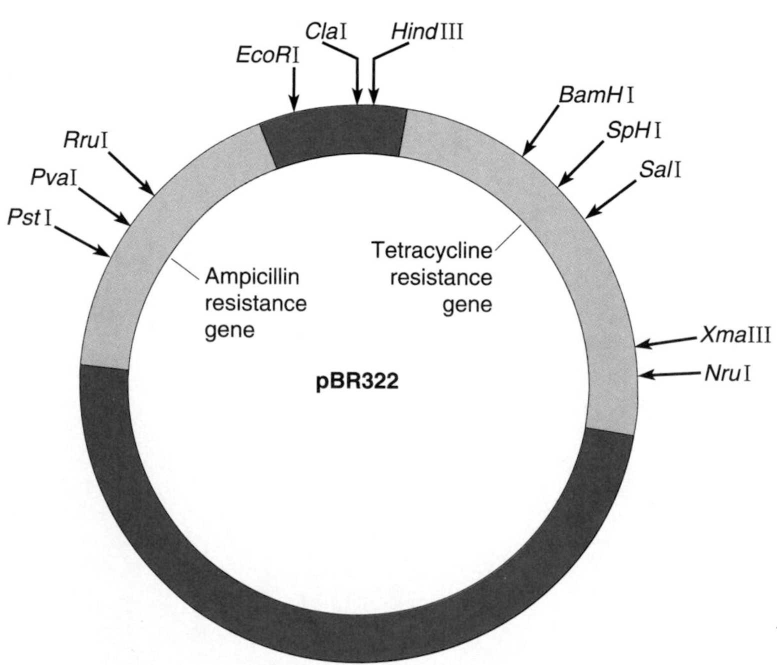

FIGURE 15.3
Restriction map of the plasmid pBR322; arrows show the locations of the restriction enzyme sites. Also shown are the locations of the genes for antibiotic resistance.

Cosmids are artificial vectors constructed from the *cos* sequence of phage lambda and plasmid (the *mid* in cosmid) sequences for antibiotic resistance and replication (Figure 15.6). Cosmid DNA containing DNA inserts is packaged inside lambda protein heads. These constructs will infect bacterial cells and replicate as plasmids within host cells. Cosmids allow the cloning of large segments of DNA, up to 50 kb in length. Phage vectors, on the other hand, can accommodate DNA inserts of about 20 kb, while plasmids are usually limited to inserts of 10 kb of foreign DNA.

A variety of cell types can be used as hosts for replication of DNA segments in vectors. The most commonly used host is a laboratory strain of *E. coli* known as K12. *E. coli* strains are genetically well characterized, and can serve as hosts for a wide range of plasmid, phage, and cosmid vectors. The bacteria are plated on a solid medium and form discrete colonies (Figure 15.7). Since each colony consists of members derived from a single ancestor, all members of the colony are genetically identical, or **clones.** If these cells contain a recombinant plas-

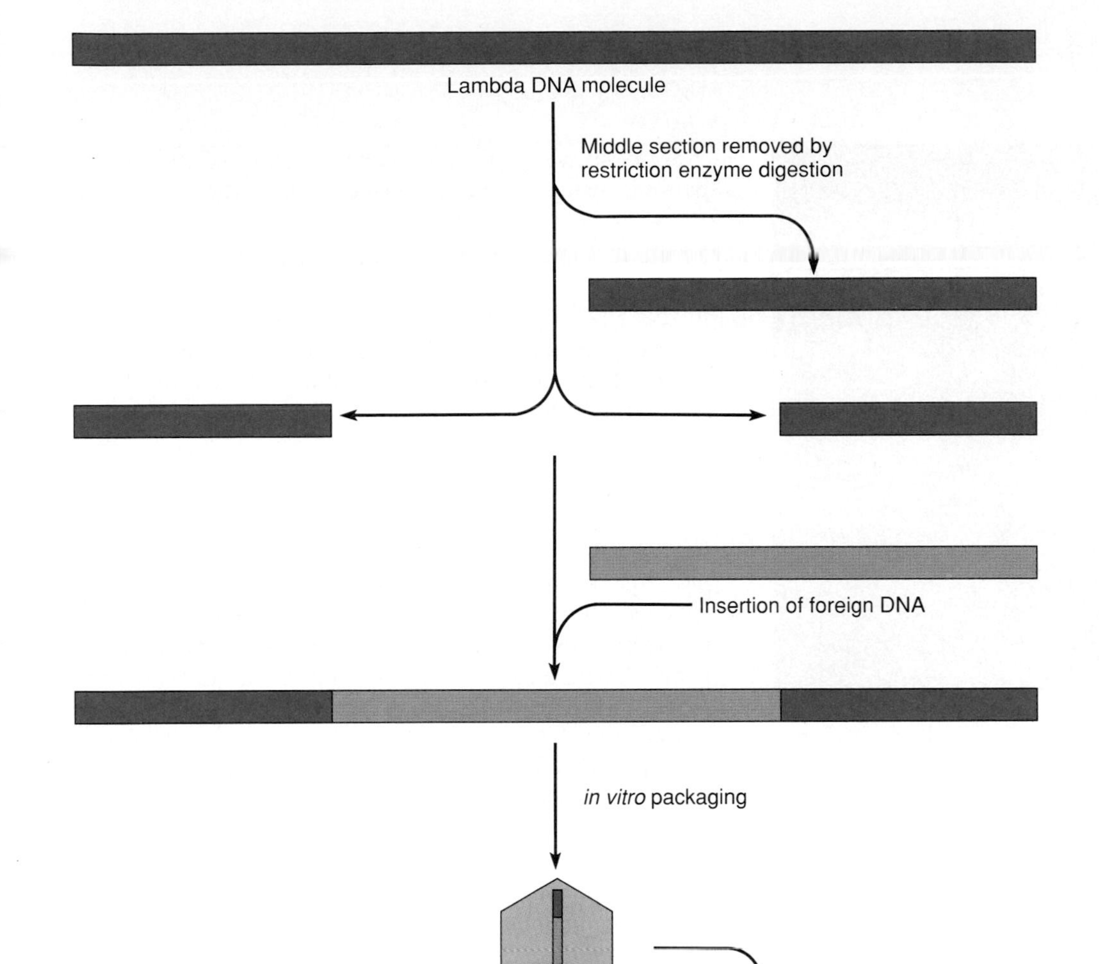

FIGURE 15.5
A diagrammatic representation of the lambda chromosome, showing how the middle section can be removed and replaced with foreign DNA. The recombinant chromosome is then packaged into phage proteins to form a recombinant virus. This virus is able to infect bacterial cells and replicate its chromosome, including the foreign DNA insert.

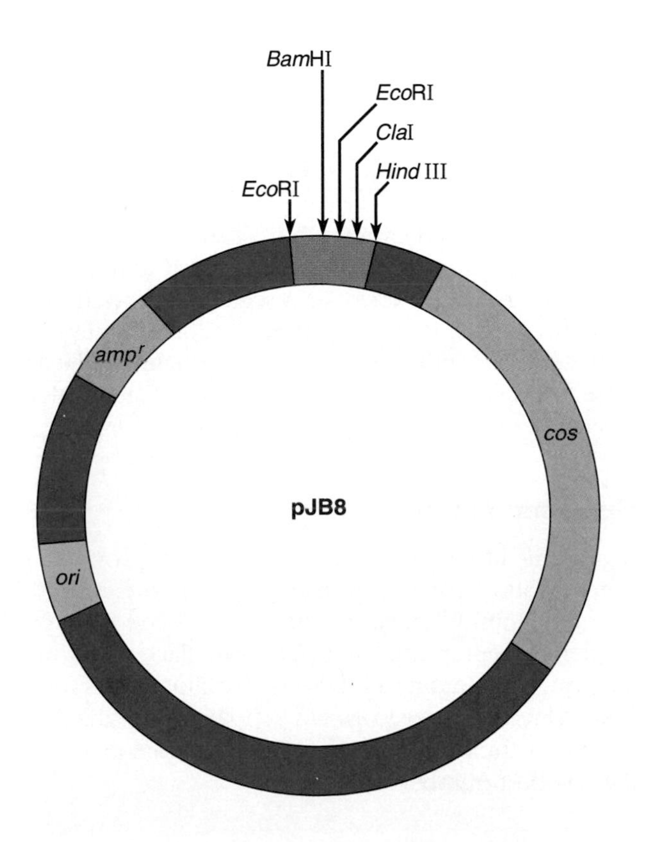

FIGURE 15.6
The artifically constructed cosmid pJB8 contains a *cos* sequence from phage lambda, plasmid regions containing antibiotic resistance sequences, and an origin of replication. Also included is a region containing restriction sites for cloning (arrowheads). The vector can accept foreign DNA segments up to 46 kb in length. The *cos* site allows cosmids with large inserts to be packaged inside lambda protein coats.

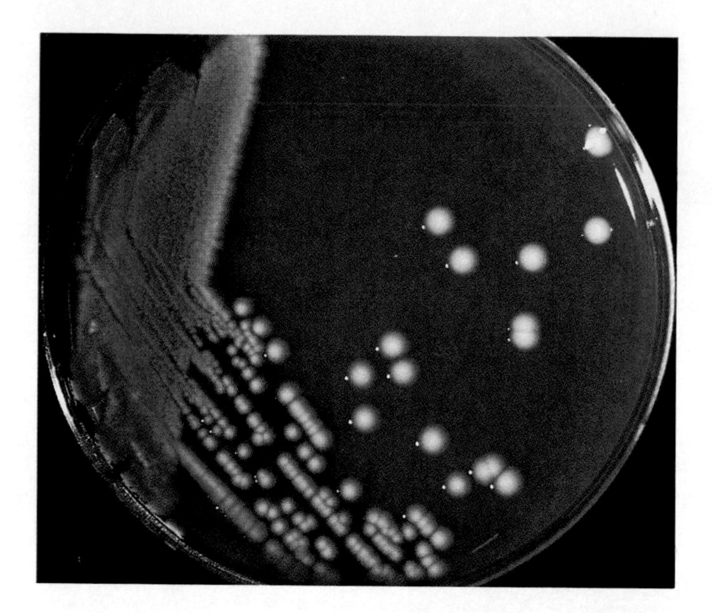

FIGURE 15.7
A petri plate containing bacterial colonies. Each colony is composed of millions of cells, all derived from a single ancestor; each colony, therefore, can be regarded as a clone.

mid, many copies of the foreign DNA have been cloned. Similarly, phage containing foreign DNA can be grown in *E. coli,* and the resulting plaques each represent a cloned descendant of a single ancestral phage and are regarded as clones. Cosmids behave as phage and infect bacterial cells, but then replicate as plasmids inside the host cell, and can be identified and recovered in the same way as plasmids.

LIBRARY CONSTRUCTION

Many separate clones must be constructed in order to include even a small portion of the total genome of an organism. A set of cloned DNA segments derived from a single individual is called a **library.** Cloned libraries can represent the entire genome of an individual, the DNA from a single chromosome, or the set of genes that is actively transcribed in a single cell type.

Genomic Libraries

A genomic library is usually constructed to provide a reference source for clones from a specific organism. Ideally, a genomic library contains at least one copy of all sequences represented in the genome. Each vector molecule can contain only a relatively few kilobases of foreign DNA. One task in preparing a genomic library is the selection of the proper vector to contain all sequences in the smallest number of clones.

Suppose we wish to prepare a library of the human genome using a lambda phage vector. The genome is 3.0×10^6 kb, and the average size of cloned inserts in the vector is 17 kb; therefore, a library of about 8.1×10^5 phage would be required to have a 99 percent probability that any human gene is present in at least one copy. If we had selected pBR322 as the vector, with an average insert of about 5 kb, several million clones would be needed to contain the library.

Chromosome-Specific Libraries

A library made from a subgenomic fraction such as a single chromosome can be of great value in the study of specific genes and the study of chromosome organization. Libraries derived from individual human chromosomes have been prepared using a technique known as **flow cytometry.** In this procedure, chromosomes from mitotic cells are stained with two fluorescent dyes, one that binds to A-T pairs, the other to G-C pairs. The stained chromosomes flow past a laser beam that stimulates them to fluoresce, and a photometer identifies the chromosomes by differences in dye binding (Figure 15.8). Using this technique, the National Laboratories at Los Alamos and Lawrence-Berkley have prepared cloned libraries for each of the human chromosomes and made them available for distribution to research scientists all over the world.

Recently, using a modification of electrophoresis known as **pulse field gel electrophoresis,** yeast chro-

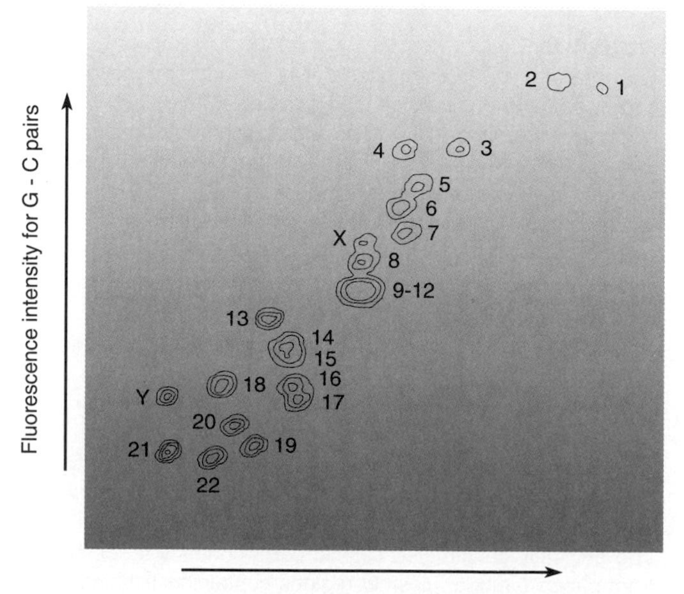

FIGURE 15.8
A sample of human chromosomes after separation by flow sorting.

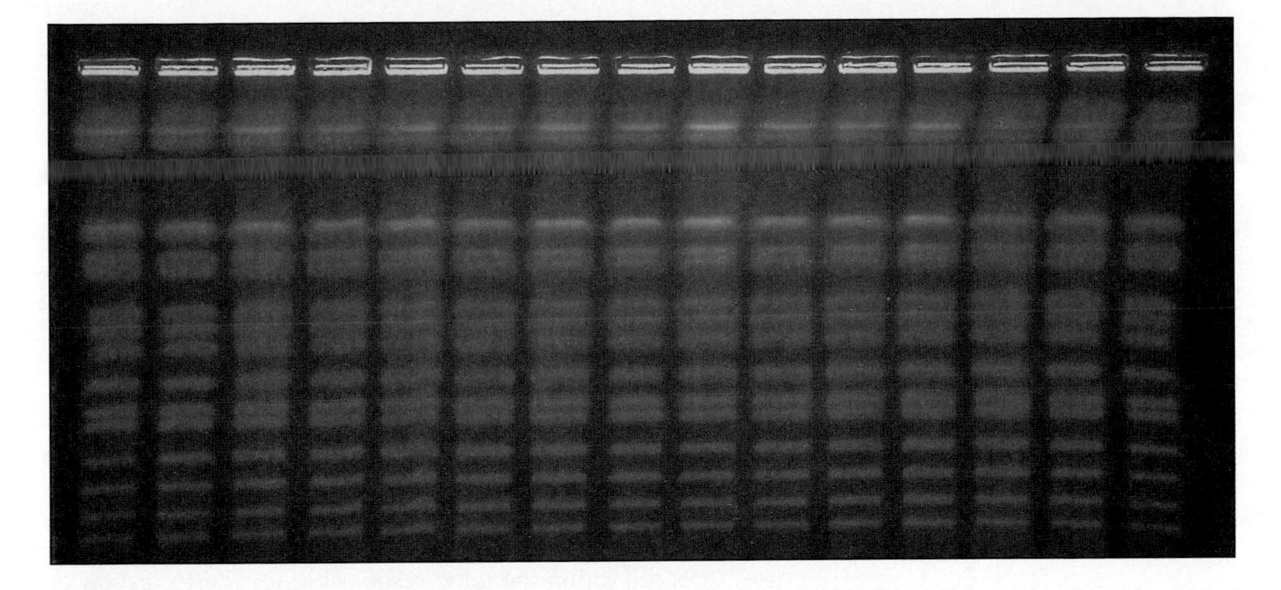

FIGURE 15.9
Intact yeast chromosomes separated using a method of electrophoresis employing contour-clamped homogeneous electric field (CHEF). In each lane, 15 of the 16 yeast chromosomes are visible, separated by size, with the largest chromosomes at the top.

mosomes have been separated from each other and used to construct chromosome-specific libraries (Figure 15.9). Libraries constructed using these techniques can be used to gain access to genetic loci where other, more widely established techniques fail. In addition, chromosome-specific libraries provide a means for studying the molecular organization and even the nucleotide sequence in a defined region of the genome.

cDNA Libraries

A library representing the genes active in a specific cell type at a specific time can be constructed by first synthesizing **cDNA (complementary DNA) molecules.** If we begin with an mRNA molecule containing a 3′ poly A tail, poly dT can be used as a primer to pair with the poly A residues (Figure 15.10). This poly dT serves as the start-

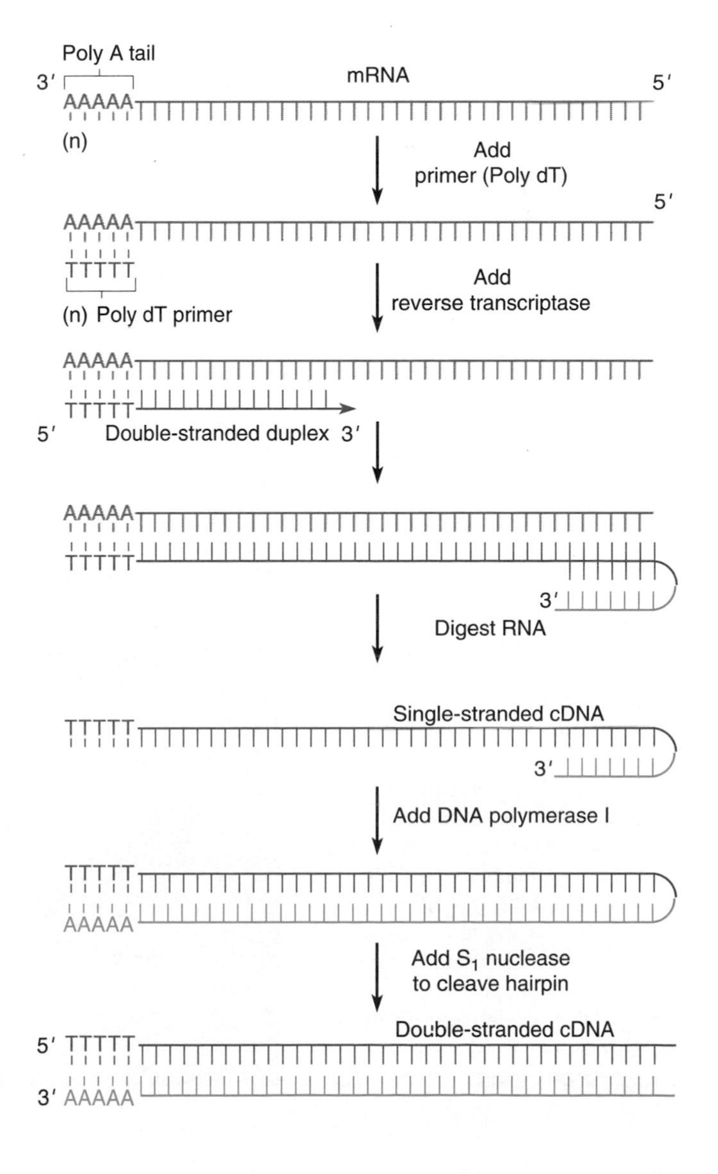

FIGURE 15.10
The production of cDNA from mRNA. Since many eukaryotic mRNAs have a polyadenylated tail of variable length (A_n) at their 3′ end, a short oligonucleotide sequence composed of thymidine nucleotides can be annealed to this tail. This oligonucleotide acts as a primer for the enzyme reverse transcriptase, which uses the mRNA as a template to synthesize a complementary DNA strand. A characteristic hairpin loop is often formed as synthesis is terminated on the template. The mRNA can be removed by alkaline treatment of the complex, and DNA polymerase is used to synthesize the second DNA strand. S_1 nuclease is used to open the hairpin loop, and the result is a double-stranded cDNA molecule that can be cloned into a suitable vector or used as a probe in library screening.

ing point for the synthesis of a DNA strand using the enzyme **reverse transcriptase.** The result is an RNA-DNA double-stranded duplex. The RNA strand can be removed by treatment with alkali or the enzyme **ribonuclease H.** The single strand of DNA is then used as a template to synthesize a complementary strand of DNA using the enzyme **DNA polymerase I.** The 3′ end of the single strand of DNA often loops back upon itself; it can serve as the primer for the synthesis of the second strand. The result is a DNA duplex with the strands joined together at one end. The hairpin loop can be opened using the enzyme **S_1 nuclease.** The result is a double-stranded DNA molecule that can be cloned by attaching sequences to both ends to allow it to be inserted into a plasmid or phage vector.

SELECTION OF RECOMBINANT CLONES

As we calculated above, a library can contain several thousand or several hundred thousand clones. The problem is to identify and select only the clone or clones that contain a DNA sequence or gene of interest. Several techniques can be employed to accomplish this task, and the choice often depends on circumstances and available information. The methods described as follows utilize somewhat different approaches to this problem.

Colony and Plaque Hybridization

Once DNA segments have been produced, inserted into a vector, and transferred into a host cell, the next task is to ensure that the host cells contain vectors that can be used as a viable source of recombinant DNA clones. In the case of plasmids such as pBR322, this can be accomplished by transfer of bacterial cells from a given colony to plates containing ampicillin or tetracycline. If, for example, foreign DNA is inserted into a site within the tetracycline resistance gene (Figure 15.3), the gene will be inactivated. Colonies formed from the cell carrying this vector will not grow in the presence of tetracycline, but will grow on plates containing ampicillin. If the cell had incorporated a vector with no insert, it would grow on both tetracycline and ampicillin.

Other plasmid and phage vectors have been engineered to produce blue plaques and colonies when plated on medium containing a compound called **Xgal.** The restriction sites for the insertion of foreign DNA occur within the gene responsible for the formation of blue plaques and colonies, and consequently, phage carrying insertions produce clear plaques and plasmids carrying insertions produce blue colonies (Figure 15.11). These methods represent the initial screening steps in the isolation of suitable clones.

Probes to Identify Specific Clones

Most procedures to select specific clones from a library employ a radioactive polynucleotide that contains a base sequence complementary to all or part of the gene of interest. This radioactively-labeled nucleic acid is commonly referred to as a **probe.** Probes can be derived

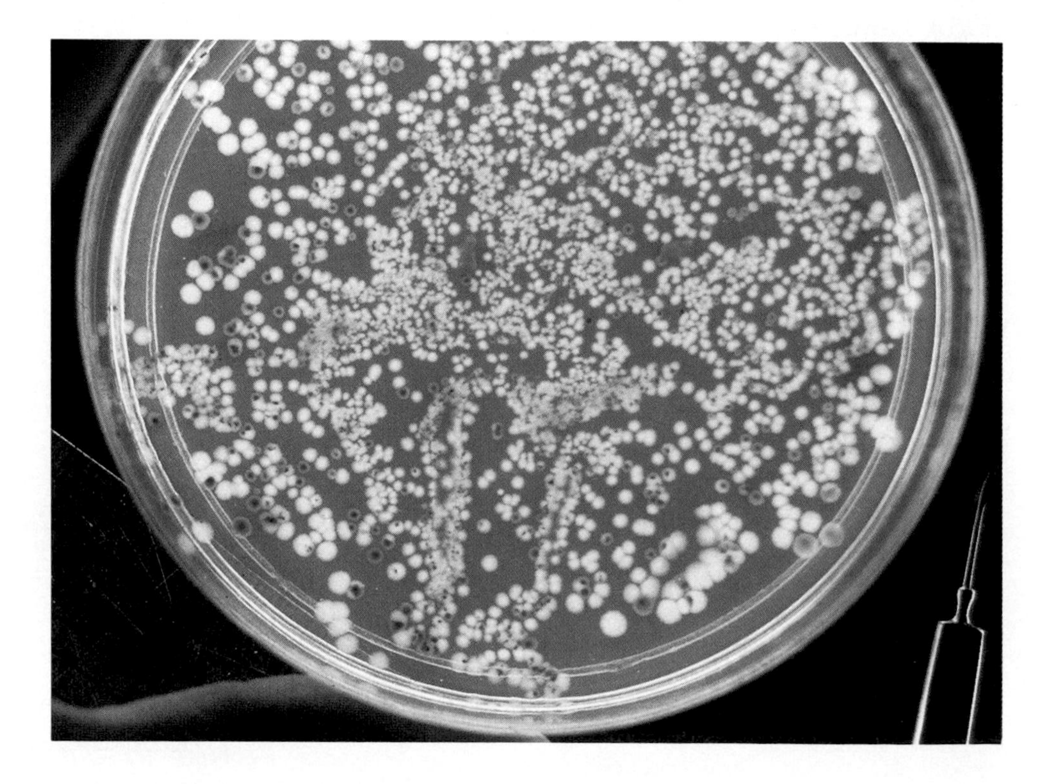

FIGURE 15.11
A petri plate showing colonies derived from a cloning experiment. The medium on the plate contains the compound Xgal. The vector has been engineered so that foreign DNA inserts disrupt the gene responsible for the formation of blue colonies. As a result, blue colonies do not carry any cloned foreign DNA, while clear colonies contain cloned vectors carrying foreign DNA inserts.

from a variety of sources; even related genes isolated from other species can serve as probes if enough of the DNA sequence has been conserved. For example, extrachromosomal copies of the ribosomal genes of the clawed frog *Xenopus laevis* can be isolated by centrifugation and cloned into plasmid vectors. Because ribosomal gene sequences have been conserved during evolution, the *Xenopus* clones can be used as probes to identify the human ribosomal genes from a cloned library.

If the gene to be selected from a genomic library is transcriptionally active in certain cell types, a cDNA probe can be prepared. This technique is particularly useful when purified mRNA molecules can be obtained. For example, hemoglobin mRNA is the predominant mRNA produced at a certain stage of red blood cell development. Purified mRNA can be copied by reverse transcriptase into a cDNA molecule that can be used as a probe to recover the structural gene for hemoglobin from a cloned genomic library.

PCR Analysis

Recombinant DNA techniques were developed in the early 1970s, and in the subsequent years, they revolutionized the way geneticists and molecular biologists conduct research. In 1986, another technique, called the **polymerase chain reaction (PCR),** was developed, and has extended the power of recombinant DNA research, and has even replaced some of the earlier methods. PCR analysis has found applications in a wide range of disciplines, including molecular biology, human genetics, evolution, development, and even forensics.

One of the prerequisites for many recombinant DNA techniques is the availability of large quantities of a specific DNA segment. Often such quantities are obtained through tedious and labor-intensive efforts in cloning and recloning. PCR allows a more direct amplification of specific DNA segments and can be used on fragments of DNA that are initially present in infinitesimally small quantities. The PCR method is based on the amplification of a DNA segment using DNA polymerase and oligonucleotide primers that hybridize to opposite strands of the sequence to be amplified (Figure 15.12). There are three basic steps in the PCR reaction, and the amount of amplified DNA produced is limited only by the number of cycles through which this series of steps is repeated.

The first step is the denaturation of the DNA to be amplified. The DNA to be amplified does not have to be purified and can come from any number of sources, including genomic DNA, forensic samples such as dried blood or semen, dried samples stored as part of medical

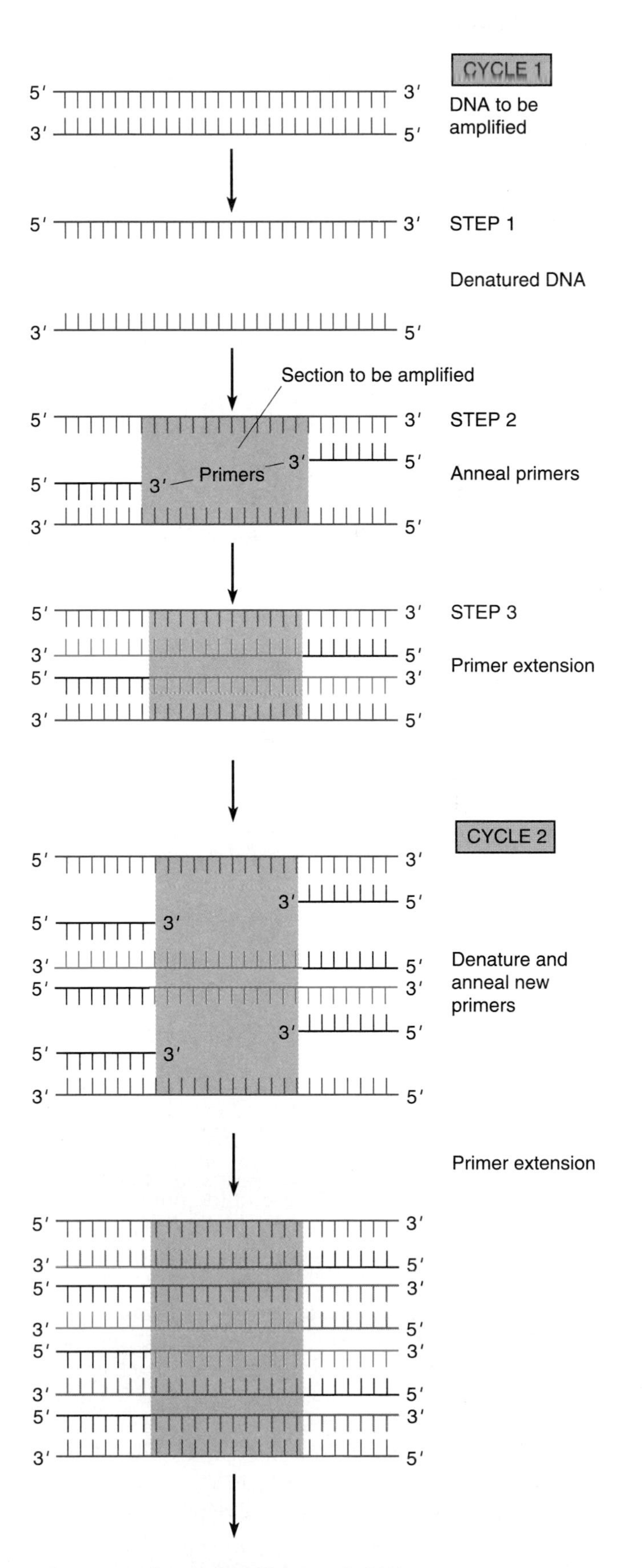

FIGURE 15.12
PCR amplification.

records, single hairs, mummified remains, and fossils. The double-stranded DNA is denatured by heating until it dissociates into single strands.

The second step is the annealing of primer sequences to the single strands of DNA. These primers are synthetic oligonucleotides that anneal to sequences flanking the segment to be amplified. Two different primers are most often used in PCR. Each primer has a sequence complementary to only one of the two strands of DNA. The primers align themselves with their 3′ ends facing each other since they anneal to opposite strands (Figure 15.12).

The third step in the PCR is the actual amplification event, carried out by extension of the primers using the single-stranded DNA as a template (Figure 15.12). This is accomplished by adding DNA polymerase to the reaction mixture. The polymerase extends the oligonucleotide primers in the 5′-to-3′ direction, using the single-stranded DNA bound to the primer as a template. The product is a double-stranded DNA molecule with the primers incorporated into the final product (Figure 15.12).

Each set of three steps—denaturation of the double-stranded product, annealing of primers, and extension by polymerase—is referred to as a cycle. The cycle can be repeated by carrying out each of the steps again. Twenty-five cycles of amplification results in over a one million-fold increase in the amount of target DNA. The process has been automated through the development of instruments that can be programmed to carry out a predetermined number of cycles, yielding large amounts of amplified DNA segments that can be used in other procedures such as cloning, sequencing, diagnosis of infectious disease, and genetic screening.

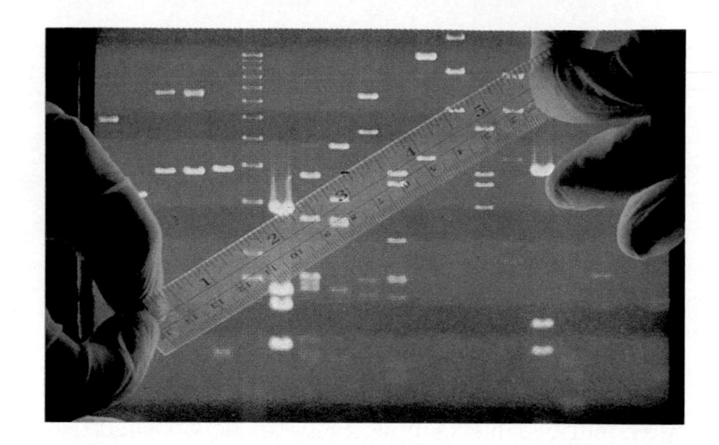

FIGURE 15.13
An agarose gel showing separated DNA fragments. Smaller fragments migrate faster and farther than larger fragments, resulting in the distribution shown.

CHARACTERIZATION OF CLONED SEQUENCES

Restriction Mapping

A restriction map is a compilation of the number, order, and distance between restriction enzyme cutting sites along a cloned segment of DNA. The map units are expressed in **base pairs (bp)** or **kilobases (kb).** The fragments generated by cutting the cloned DNA segment with restriction enzymes can be separated by **gel electrophoresis,** a method that separates fragments by size, with the smallest pieces moving farthest. The fragments appear as a series of bands that can be visualized by staining the DNA with ethidium bromide and viewing under ultraviolet illumination (Figure 15.13).

Restriction maps provide an important way of characterizing a DNA segment and can be constructed in the absence of any information about the genetic content or function of the mapped DNA. In conjunction with other techniques, restriction mapping can be used to define the boundaries of a gene and provides a way of dissecting the molecular organization within a gene and its flanking regions (Figure 15.14). Mapping can also serve as a starting point for the isolation of an intact gene from cloned segments of DNA and provides a means for locating mutational sites within genes.

Restriction maps can also be used to construct and refine genetic maps. To a large extent, the accuracy of genetic maps is based on the frequency of recombination between genetic markers and the number of genes or genetic markers used in construction of the map. If there is a large distance between markers and variation in recombination frequency, the genetic map may not correspond to the physical map of the chromosome or chromosome region. In humans, for example, the genome size is large (3.2×10^9 bp), and the number of mapped genes is small (a few thousand), meaning that each map unit is composed of millions of base pairs of DNA. The result is a poor correlation between the genetic and physical maps of chromosomes. Restriction enzyme cutting sites can be used as genetic markers, reducing the distance between sites for map construction, increasing the accuracy of genetic maps, and providing reference points for the correlation of genetic and physical maps.

Restriction sites have played an important role in mapping genes to specific human chromosomes and to defined regions of individual chromosomes. In addition, if a restriction site maps close to a mutant gene, it can be used as a marker in a diagnostic test. This has proven especially useful because the mutant genes underlying

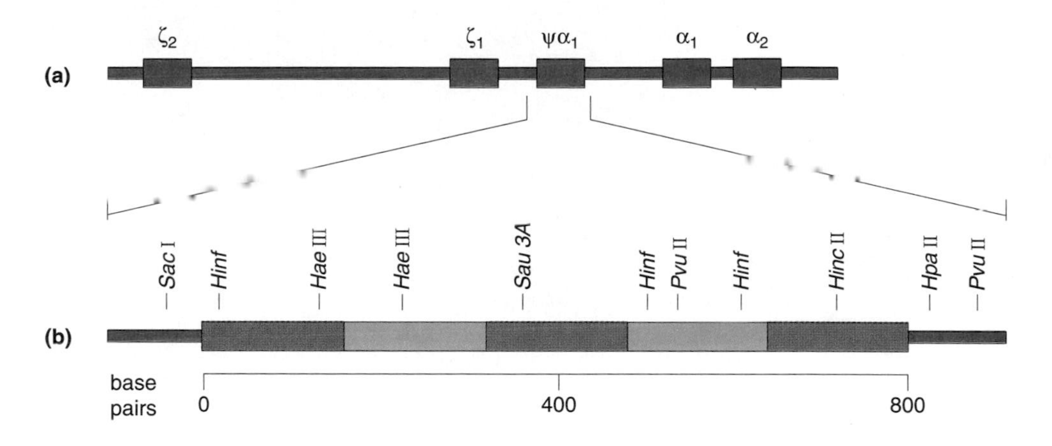

FIGURE 15.14
Restriction map of the alpha globin pseudogene in humans. (a) The alpha gene cluster on human chromosome 16. The cluster contains two copies of the alpha gene (designated $alpha_1$ and $alpha_2$), and three other genes: the embryonically expressed zeta gene and two pseudogenes, pseudo-zeta and pseudo-alpha. Pseudogenes are nonfunctional versions of normal genes. (b) A restriction map of the pseudo-alpha-1 gene, showing the exons, introns and restriction sites within and adjacent to the gene.

many human genetic diseases are poorly characterized at the molecular level, and closely linked restriction sites have been successfully used in the detection of afflicted individuals and heterozygotes at risk for having affected children.

Nucleic Acid Blotting

DNA fragments cloned into vectors can be used in hybridization reactions to characterize the identity of specific genes, to locate coding regions or flanking regulatory regions within cloned sequences, and to study the molecular organization of genomic sequences.

In this procedure, the DNA to be probed is first cut into fragments with one or more restriction enzymes, and the fragments separated by gel electrophoresis (Figure 15.15). The DNA in the gel is denatured into single-stranded fragments by treatment with alkali. These fragments are transferred to a sheet of DNA-binding material, usually nitrocellulose or a nylon derivative. The denatured DNA is then hybridized with a radioactive probe, and the hybridized fragments are visualized by autoradiography (Figure 15.15).

This technique, developed by Edward Southern, has become known as a **Southern blot,** and is a procedure with many applications in genetics. The Southern blot is widely used to identify cloned genes that correspond to particular probes, to identify related genes in genomes of other organisms, and to find related sequences or members of a gene family within a single genome.

A related technique, developed by James Alwine and colleagues, can be used to determine whether a cloned gene is transcriptionally active in a given cell or tissue type. This characterization of gene activity is accomplished by extracting RNA from one or several cell and/or tissue types. The RNA is fractionated by gel electrophoresis and the pattern of RNA bands is transferred to a sheet of RNA-binding membrane. The sheet is then hybridized to a radioactive single-stranded DNA probe, derived from cloned genomic DNA or cDNA. If RNA complementary to the DNA probe is present, it will be detected by autoradiography as a band on the film. Because the original procedure using DNA bound to a filter became known as a Southern blot, this procedure using RNA bound to a filter is called a **Northern blot.** (Following this somewhat perverse logic, another procedure involving proteins bound to a filter is known as a **Western blot.**)

Northern blots provide information about whether RNA transcripts complementary to cloned DNA segments are present in a cell or tissue and also provide two additional bits of information. The size and molecular weight of the transcribed RNA can be calculated, and the amount of transcribed RNA present in the cell or tissue is reflected by the density of the RNA band on the autoradiograph. Thus, Northern blotting can be used to characterize and quantify the transcriptional activity of a cloned DNA segment in different cells and tissues.

DNA Sequencing

In a sense, the ultimate characterization of a cloned DNA segment is the determination of its nucleotide sequence, a technique called DNA sequencing. The ability to se-

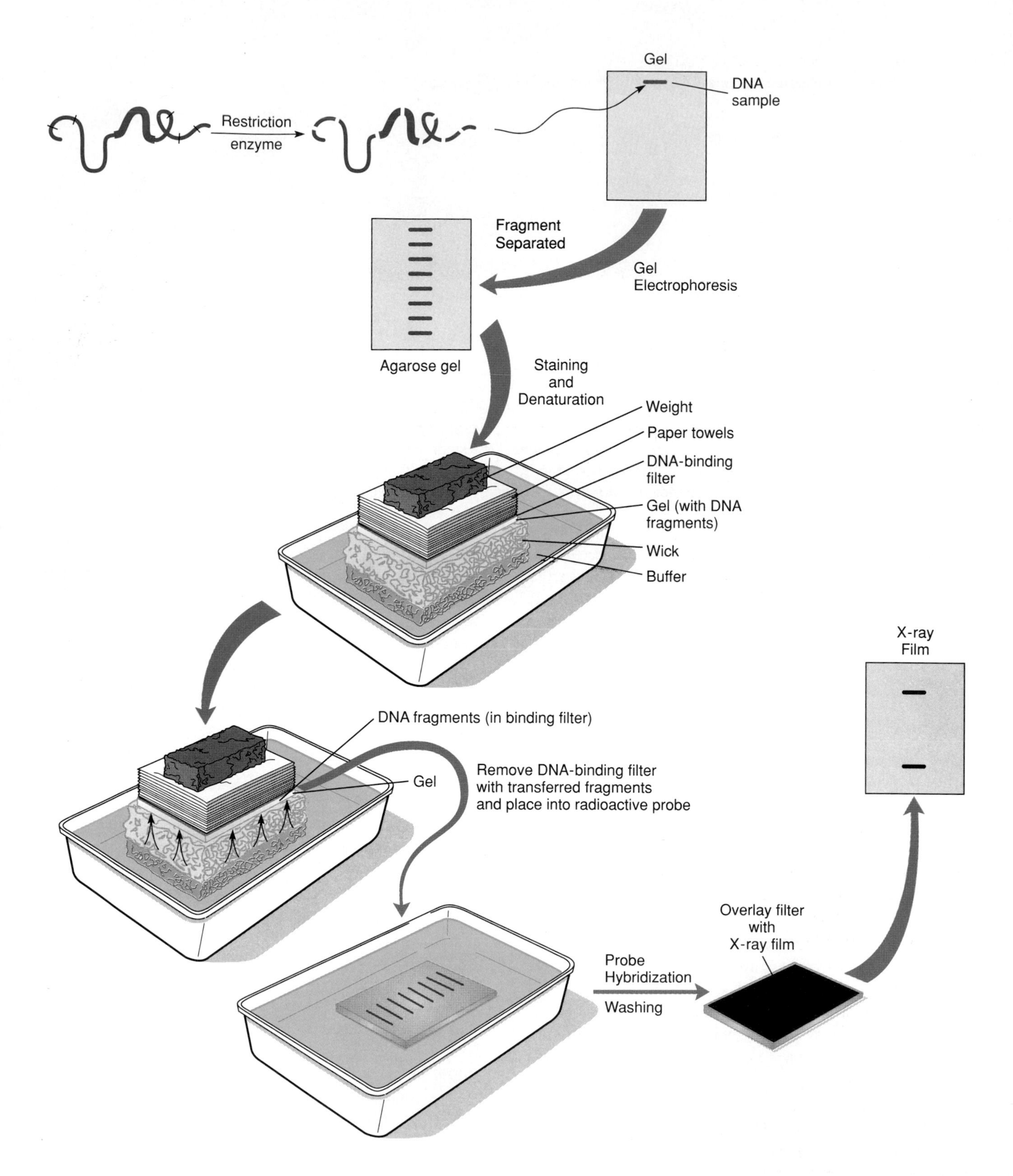
Gel
DNA sample
Restriction enzyme
Fragment Separated
Gel Electrophoresis
Agarose gel
Staining and Denaturation
Weight
Paper towels
DNA-binding filter
Gel (with DNA fragments)
Wick
Buffer
X-ray Film
DNA fragments (in binding filter)
Gel
Remove DNA-binding filter with transferred fragments and place into radioactive probe
Overlay filter with X-ray film
Probe Hybridization
Washing

◀ **FIGURE 15.15**
The Southern blotting technique. DNA is first cut with one or more restriction enzymes and the fragments separated by gel electrophoresis. The pattern of fragments can be visualized and photographed under ultraviolet illumination by staining the gel with ethidium bromide. The gel is then placed on a filter paper wick in contact with a buffer solution and covered with a sheet of a DNA binding filter. Layers of paper towels or blotting paper are placed on top of the DNA binding filter and held in place with a weight. Capillary action draws the buffer through the gel, transferring the pattern of DNA fragments from the gel to the DNA binding filter. The DNA fragments on the binding filter are then denatured and hybridized with a radioactive probe, washed, and overlaid with a piece of X-ray film for autoradiography. The hybridized fragments show up as bands on the X-ray film.

quence cloned DNA has added immensely to our understanding of gene structure and the mechanisms of gene regulation.

The application of DNA sequencing depends on the ability of cloning methods to provide large amounts of specific DNA fragments. The use of special vectors or modifications of the PCR technique is also important in providing single-stranded DNA segments that can be directly sequenced.

DNA sequencing uses a series of four chemical reactions, each in a separate tube, to determine the sequence of the four different bases in a DNA segment. These sequences, differing in length by as little as one base, are separated from each other by gel electrophoresis. The result is a series of bands that form a ladderlike pattern. The sequence can be read directly from the band pattern (Figure 15.16).

DNA sequencing has provided information about the organization of genes and the nature of mutational events that alter both genes and gene products, confirming the genetic conclusion that the sequence of nucleotides determines the sequence of amino acids in proteins. Sequencing has also provided information about the organization of regulatory regions that flank prokaryotic and eukaryotic genes and has been used to infer the structure of proteins. The gene for cystic fibrosis, an autosomal recessive human genetic disorder, was first identified by the fact that it was a specific DNA sequence. The DNA sequence was then used to infer that the gene encodes a protein of 1480 amino acids that functions in chloride ion transport. This discovery serves to illustrate how the use of DNA sequencing plays an important role in genetic analysis.

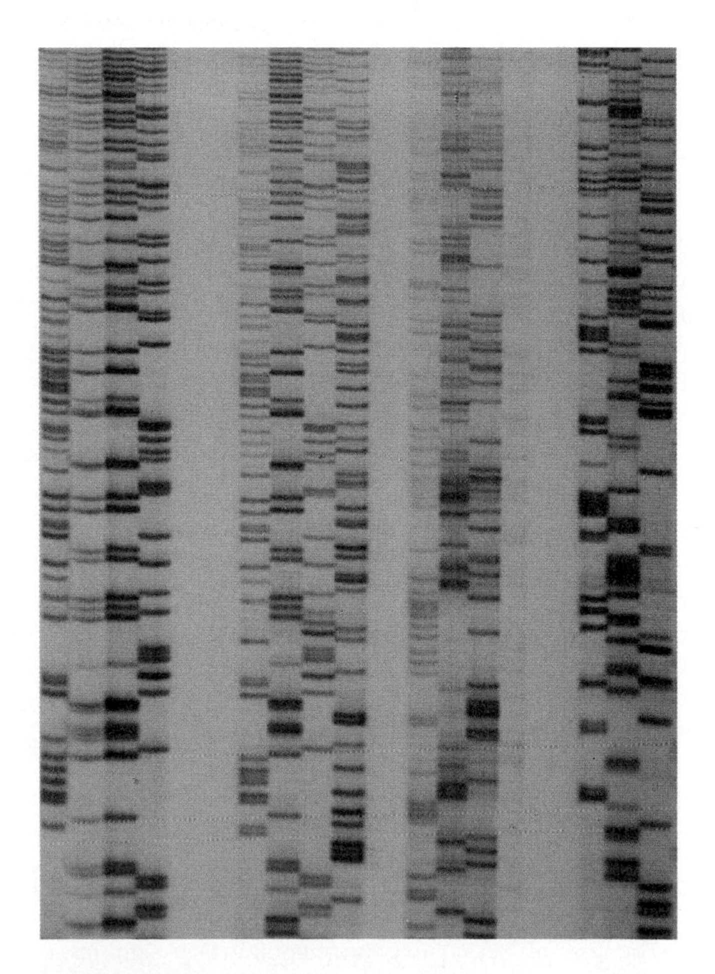

FIGURE 15.16
Photograph of a DNA sequencing gel, showing the separation of bases that are analyzed to reveal the base sequence of the DNA segment.

APPLICATIONS OF RECOMBINANT DNA TECHNOLOGY

Gene Mapping

Although an increasing number of the 50,000 to 100,000 human genes have been localized to chromosomal sites, in the majority of human genetic disorders the primary defect is unknown, and without a marker, the gene cannot be mapped. A defective gene product has been identified in just under 400 single-gene disorders, and the nature of the mutation is known in only about 40 of these. In the majority of cases, the mutant gene product responsible for a genetic disease has not been identified, and as a result, classical genetic methods cannot be used to identify the chromosomal locus of the disorder.

The development of recombinant DNA technology has provided a new method for mapping allelic variants without the need to identify the gene product. This method makes use of the fact that single base changes in the recognition sequence of restriction enzymes can alter the pattern of cuts made in DNA. This gives rise to a detectable variation in DNA fragment length that is inherited in a Mendelian codominant fashion (Figure 15.17). These allelic variants or polymorphisms are called **restriction fragment length polymorphisms** or **RFLPs** and are mapped to specific chromosomes or chromosome regions as they are identified.

The mapping of genetic disorders using RFLPs depends on finding an RFLP that is closely linked to the genetic disorders to be mapped. This is accomplished by studying large, multigenerational families in which both the RFLP and the genetic disorder are inherited. Once the chromosome carrying the genetic disorder has been identified, further analysis using RFLPs along the chromosome can identify the region containing the locus for the disorder. Finally, the search can be narrowed to a small collection of cloned sequences from the region in question. This collection can be screened to identify the gene for the disorder.

Once the DNA sequence of the gene is known, it is then possible to reconstruct the amino acid sequence of the gene product from the nucleotide sequence and deduce its functional properties. This information, in turn, allows the isolation of the gene product and the construction of antibodies to study its pattern of distribution in the cell.

The use of RFLPs in gene mapping is a significant breakthrough in human genetics and has been used to map an increasing number of human genes. Coupled with other techniques, RFLP analysis has revolutionized human gene mapping and promises to produce a human gene map with a level of resolution approaching that of other eukaryotic organisms such as *Drosophila* and the mouse.

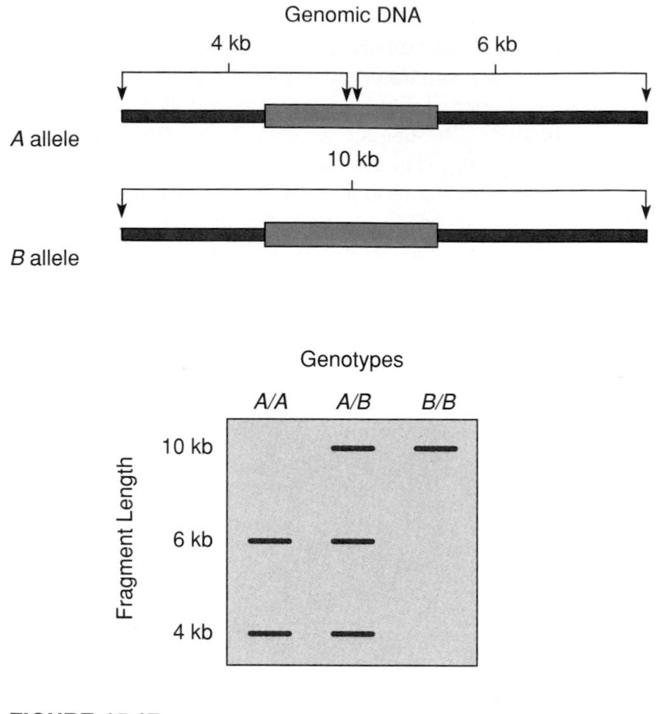

FIGURE 15.17
Restriction fragment length polymorphisms (RFLPs). The *A* and *B* alleles represent DNA segments from homologous chromosomes. The thick region represents a segment that can be detected with a DNA probe. Arrows indicate the pattern of restriction enzyme cutting sites that define the alleles. In allele *A*, three cutting sites generate fragments of 6 and 4 kb. In allele *B*, only two cutting sites are present, generating a fragment 10 kb in length. The absence of the cutting site in *B* could be the result of single base mutation within the enzyme recognition/cutting site. Because the alleles are codominant, three genotypes (*AA*, *AB*, and *BB*) can be generated. The fragment patterns for the three possible genotypes are shown as they would appear on a Southern blot.

Disease Diagnosis

The most widely used methods for prenatal detection of genetic diseases are **amniocentesis** and **chorionic villus sampling (CVS).** In amniocentesis, a needle is used to withdraw amniotic fluid (Figure 15.18). The fluid and the cells it contains can be analyzed for chromosomal or single-gene disorders. In CVS, a catheter is inserted into the uterus and used to retrieve a small tissue sample of the fetal chorion. This tissue is used for cytogenetic and biochemical prenatal diagnosis.

Coupled with these sampling methods, recombinant DNA technology has proven to be a highly sensitive and accurate tool for the detection of genetic disorders. The use of cloned DNA sequences has the advantage of allowing direct examination of the genotype.

One such condition that can be detected prenatally is

THE HUMAN GENOME PROJECT

One fundamental goal in human genetics is the isolation and characterization of the estimated 50,000 to 100,000 genes that control the makeup of our species. Already over 4000 genetic disorders caused by single gene defects are known; an understanding of these genes is central to diagnosis and treatment of affected individuals. Present knowledge is the outcome of several decades spent isolating, mapping and, more recently, sequencing genes one at a time. As technology advanced, the idea of wholesale sequencing was born. By 1987, mapping and sequencing the entire human genome as one big project was being widely debated.

Opponents of the proposal questioned the cost, worrying that funding of such a project would diminish support for other projects. As biology's first "Big Science" project, questions about its fundamental utility were raised, since it is currently believed that more than 90 percent of the human DNA may contain little or no hereditary information.

Although the debate is still not over, the project was supported and, in 1988, the Office of Human Genome Research was established, with Nobel laureate, James Watson, as its first director. This center, together with the Department of Energy, prepared the National Genome Plan which outlined the objectives and goals of the proposal. In October 1990, the Human Genome Project was officially launched.

One of the first goals of the plan is to produce complete physical and linkage maps of the 22 autosomes and the two sex chromosomes by 1995. Linkage maps are created by calculating the amount of genetic recombination occurring between identifiable marker sites along the chromosomes. The units of measurement on linkage maps are recombination frequencies. Physical mapping places the markers on a linkage map into the physical context of DNA. Physical map distances are in units of physical length, generally expressed in terms of the number of nucleotides separating markers. Recently, the completion of detailed physical maps of two human chromosomes has been reported. David Page's group in the United States has mapped the functional sections of the Y chromosome, and Daniel Cohen's group in France has mapped the smallest autosome, chromosome 21. On the basis of these two reports it is now suggested that human chromosomes may be mapped more quickly than first predicted.

The second major goal of the Human Genome Project is to sequence the total genome. It is estimated that it will take 15 years to determine the order of the 3 billion base pairs that make up our chromosomes. Advanced computer technology will be required to cope with the storage, organization, and analysis of such a massive quantity of data, and the development of such technology is also one of the objectives of the project.

The total cost of the Human Genome Project is estimated at 3 billion dollars. This money will be used to finance several large research centers, each responsible for a specific part of the project, and smaller research projects. Several other countries including England, France, Italy, and Japan, have also set up national genome projects. These national programs are now coordinated by the Human Genome Organization (HUGO).

Whatever the immediate scientific value of the project, it will doubtlessly spin-off important technical advances and provide significant information for the health sciences community. For those who support it, this long-range vision of a full analytical understanding of the genetic biology of humans is one of the project's greatest appeals.

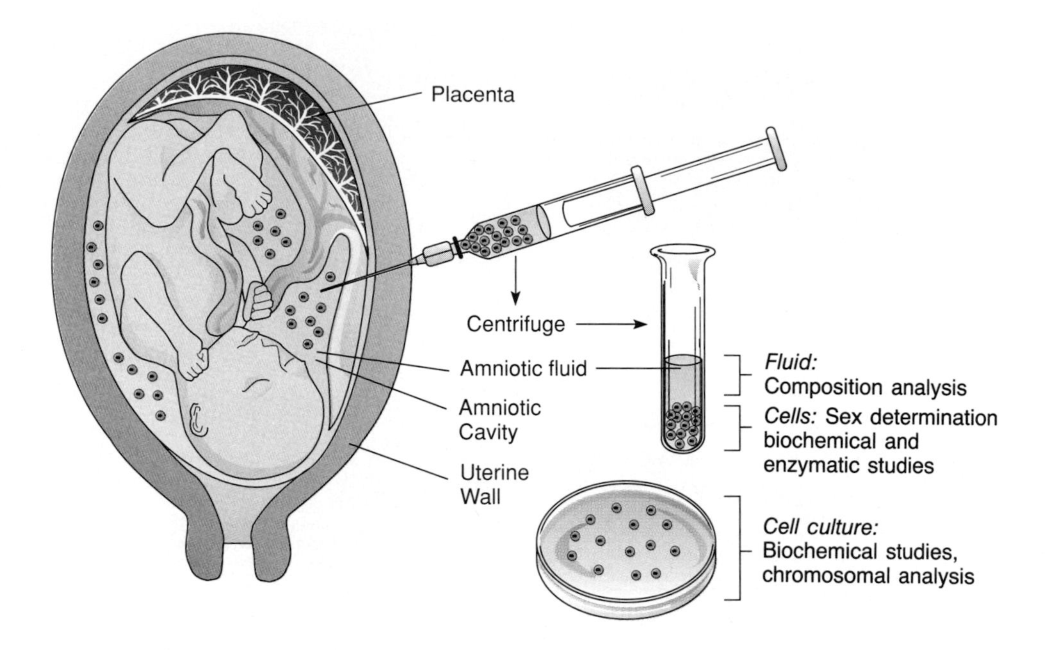

FIGURE 15.18
The technique of amniocentesis. The position of the fetus is first determined by ultrasound, and then a needle is inserted through the abdominal and uterine wall to recover fluid and fetal cells for cytogenetic and/or biochemical analysis.

sickle-cell anemia, an autosomal recessive condition common in those with family origins in areas of West Africa and the Mediterranean basin. Sickle-cell anemia is caused by a single nucleotide substitution in codon 6 (changing the DNA sequence from GAG, encoding glutamine, to GTG, encoding valine). While the mutation does not change the length of the beta globin gene, it does destroy a cutting site for a restriction enzyme. This changes the length of restriction fragments, allowing diagnosis of genotypes by Southern blot analysis (Figure 15.19).

DNA Fingerprints

Restriction fragment length polymorphisms have been used to distinguish normal from mutant copies of single genes and can be used as genetic markers, since they are inherited in a codominant fashion. The phenotype of these genetic markers is DNA fragments of different sizes detected on a Southern blot after the DNA has been cut with a restriction enzyme, and the fragments of different sizes have been separated by gel electrophoresis. When the cloned probe used to detect an RFLP comes from a *minisatellite* region of DNA, the result is a series of multiple bands (Figure 15.20).

Minisatellites are regions of repeated DNA sequences organized as tandem repeats. For example, the base sequence GGAAGGGAAGGGAAGGGAAG consists of four tandem repeats of a GGAAG sequence. Such groups of sequences are widely dispersed in the human genome; the patterns of bands produced when such sequences are cut with restriction enzymes and visualized on gels are

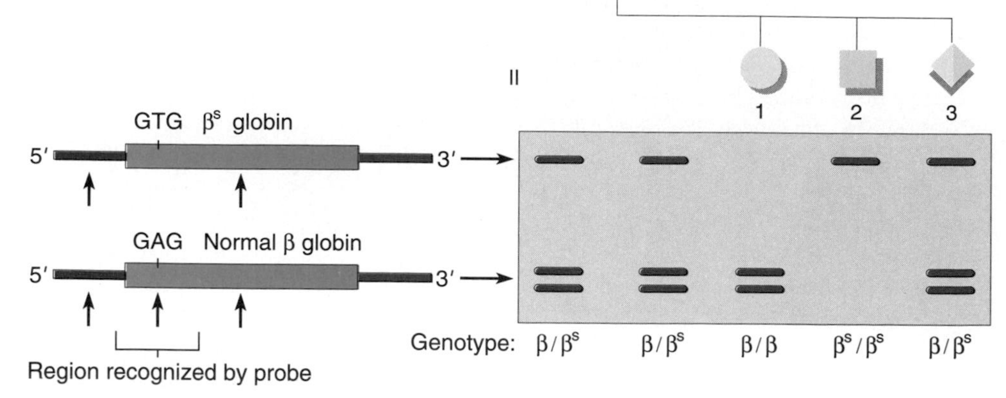

FIGURE 15.19
Southern blot diagnosis of sickle-cell anemia. Arrows below the genes represent restriction enzyme cutting sites. In the mutant (β^S) globin gene, a point mutation (GAG → GTG) has destroyed a restriction enzyme cutting site, resulting in an altered pattern of DNA fragments. In the pedigree, the family has one unaffected homozygous normal daughter (II-1), an affected son (II-2), and an unaffected fetus (II-3).

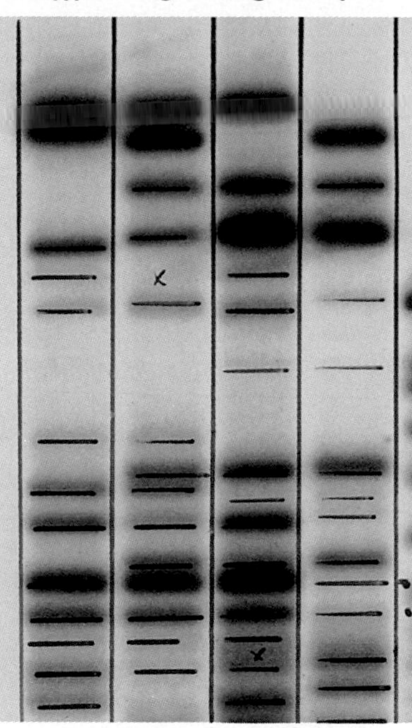

FIGURE 15.20
DNA fingerprinting in a family consisting of mother (M), child (C), child (C) and father (F). The bands represent DNA fragments separated by gel electrophoresis. Note that the pattern of banding in the children is made up of bands from one or the other or both parents.

known as *DNA fingerprints.* These RFLPs are the equivalent of fingerprints because the pattern of bands generated varies from one individual to another, but is always the same for a given individual, no matter what tissue is used as the source of DNA. In addition, there is so much variation from individual to individual in the band pattern that each person's pattern is unique, like the individualized pattern in a fingerprint (Figure 15.20). DNA fingerprint analysis can be performed on very small samples of material (less than 60 microliters (μl) of blood) and on samples that are old (RFLP analysis has been performed on Egyptian mummies over 2400 years old), increasing its usefulness in legal cases.

In the United States, DNA fingerprints have been used as evidence in criminal trials since 1988 and have also been used to settle immigration cases, disputes involving purebred dogs, and in animal conservation studies. The use of DNA fingerprints in criminal trials has been the subject of some dispute, however. The controversy centers on two facets of the method, one technical, and the other scientific. On the technical side, there has been little standardization of the methods used to perform DNA fingerprinting, and the quality control in some commercial laboratories has been called into question. Scientifically, the method depends on showing that the pattern of any individual is unique enough that he or she is the sole source of such a pattern among individuals in a population. Such statements can be made only when information is available about the frequency of specific RFLPs in various populations. If these problems can be resolved in a satisfactory manner, DNA fingerprinting will become a routine part of the evidence in several types of criminal cases.

Gene Therapy

The ability to isolate and clone specific human genes, developed as a research tool, is now being used in medicine to treat inherited disorders by replacing defective genes with copies of normal genes. This process, known as *gene therapy,* is at the heart of a new set of medical treatments for an ever-increasing range of diseases and disorders. Several methods have been developed that allow the delivery of genes into human cells. These include the use of viruses as vectors to deliver genes to cells, chemical-assisted transfer that promotes the transport of genes across cell membranes, and the fusion of cells with artificially constructed vesicles containing cloned DNA sequences. Gene therapy relies on using DNA and sometimes RNA in the same way that drugs are used to treat specific diseases. Several disorders are currently being treated with gene therapy, including severe combined immunodeficiency and familial hypercholesterolemia.

Severe combined immunodeficiency (SCID) is a rare autosomal recessive disorder in which affected individuals have no functioning immune system and usually die from otherwise minor infections (see the 'boy in the bubble' described in Chapter 17). One form of SCID is caused by a defect in a gene that encodes the enzyme **adenosine deaminase (ADA).** A young girl affected with this form of SCID was the first patient to ever receive gene therapy. This treatment, begun in 1990, starts with the isolation of specialized white blood cells called **T cells** (Figure 15.21). These cells are then mixed with a genetically modified virus that carries a normal copy of the human ADA gene. The virus infects the T cells, allowing a functional copy of the ADA gene to be inserted into the T cell's chromosomes (Figure 15.21). The modified T cells are grown in the laboratory, and the patient is treated by periodically re-implanting a billion or so genetically altered T cells into the body. Results on a small number of SCID patients has been encouraging, and most are able to attend school and have been exposed to common childhood diseases, such as chicken pox, which otherwise might be fatal.

Another genetic disorder being treated by gene therapy is familial hypercholesterolemia. This relatively common autosomal dominant condition is characterized by the inability to properly metabolize dietary fats and re-

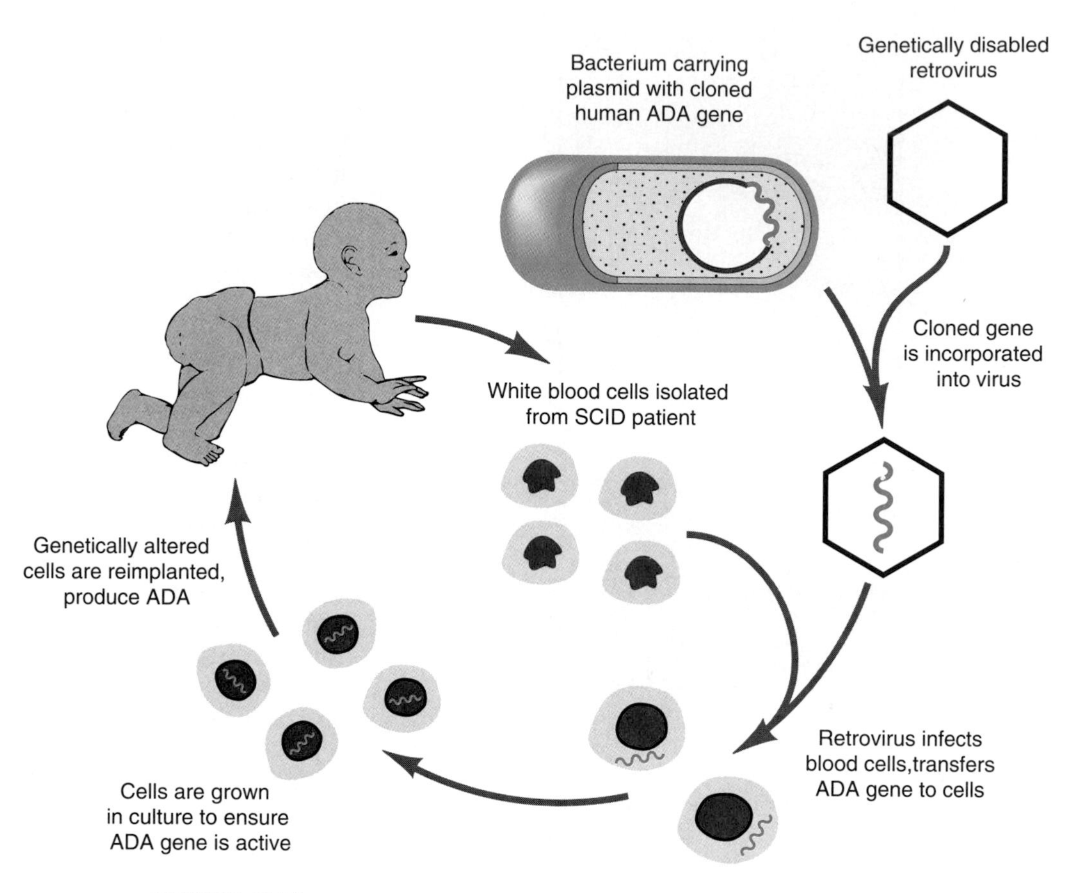
Bacterium carrying
plasmid with cloned
human ADA gene
Genetically disabled
retrovirus
Cloned gene
is incorporated
into virus
White blood cells isolated
from SCID patient
Genetically altered
cells are reimplanted,
produce ADA
Retrovirus infects
blood cells,transfers
ADA gene to cells
Cells are grown
in culture to ensure
ADA gene is active

FIGURE 15.21
Gene therapy for treatment of severe combined immunodeficiency (SCID), a fatal disorder of the immune system, caused by lack of the enzyme adenosine deaminase (ADA). The cloned human ADA gene is transferred into a viral vector, which is used to infect white blood cells removed from the patient. The transferred ADA gene is incorporated into a chromosome, and becomes active. After growth to enhance their numbers, the cells are reimplanted into the patient, where they produce ADA, allowing the development of an immune response.

sults in elevated levels of blood cholesterol, increased deposition of cholesterol in arterial plaques, and premature death from coronary heart disease.

For gene therapy, a section of liver is removed from a person with hypercholesterolemia and separated into individual cells. A genetically modified mouse virus, carrying a human gene for cholesterol metabolism is used to infect the liver cells. The treated cells are then injected into the patient, using a vein that supplies the liver. The injected cells are carried through the vein into the liver where they take up residence and begin to metabolize cholesterol, reducing blood levels to the normal range.

These early trials using gene therapy have been successful, and new applications to treat other diseases are now beginning. Gene therapy is also being used or contemplated as a treatment for skin cancer, breast cancer, brain cancer, and AIDS. In the last few years, at least a dozen biotech companies have been founded specifically to develop products for gene therapy, and a number of such products should reach the marketplace beginning in 1995. In the twenty-first century, gene therapy will be a commonplace method of treatment for a large number of disorders.

Protein Production

Although recombinant DNA techniques were originally developed to facilitate basic research into gene organization and regulation of expression, scientists have not been blind to the commercial possibilities of this technology. As a result, scientists have participated in the formation of biotechnology companies to exploit recombinant DNA technology. The commercial production of human gene products, such as hormones and clotting factors, from sequences cloned into bacterial cells are examples of the commercial use of recombinant DNA. In the last decade, the biotechnology industry has grown into a multibillion dollar segment of the economy.

Commercial applications often represent ingenious solutions to problems of producing eukaryotic gene products in prokaryotic cells, or animal proteins in plant cells. For example, antibodies are proteins involved in the immune response and are composed of two polypeptide chains, an H chain and an L chain. Antibodies are used in many clinical treatments, including cancer therapy. At present, antibodies are commercially produced through a process of cell hybridization that is inefficient, time-consuming, and expensive. To improve the commercial production of antibodies, tobacco plants have been converted into antibody factories. Genes for the H and L chains of an antibody known as IgG were cloned into separate vectors and inserted separately into different plants. Tobacco plants expressing H chains were crossed with plants expressing L chains to produce progeny that expressed complete antibody chains. Not only were the genes expressed, the antibodies amounted to 1.5 percent of the protein in the tobacco leaves. Although this has been accomplished only on an experimental basis, it is easy to see that this method can be adapted to produce commercial quantities of antibodies, such as those that attack tumor cells, from an acre or two of genetically modified plants.

TABLE 15.2
Genetically engineered pharmaceutical products available or in clinical testing.

Gene Product	Condition Being Treated
Atrial natriuretic factor	Heart failure, hypertension
Epidermal growth factor	Burns, skin transplants
Erythropolten	Anemia
Factor VIII	Hemophilia
Gamma interferon	Cancer
Granulocyte colony-stimulating factor	Cancer
Hepatitis B vaccine	Hepatitis
Human growth hormone	Dwarfism
Insulin	Diabetes
Interleukin-2	Cancer
Superoxide dismutase	Transplants
Tissue plasminogen activator	Heart attacks

CHAPTER SUMMARY

1 Recombinant DNA refers to the creation of a new association between DNA molecules or segments not usually found together. The cornerstone of recombinant DNA technology is a class of enzymes called restriction endonucleases that cut DNA at specific recognition sites. The fragments produced are joined with DNA vector segments using the enzyme DNA ligase to form recombinant DNA molecules. Vectors have been constructed from many sources, including bacterial plasmids, viruses, and artificial yeast chromosomes.

2 Recombinant DNA molecules are transferred into a host, and cloned copies are produced during host cell replication. A variety of host cells may be used for replication, including yeast, bacteria, cells of higher plants, and mammalian cells in tissue culture. However, the most common host is *E. coli.* Cloned copies of foreign DNA sequences can be recovered, purified, and analyzed.

3 Cloned libraries have been constructed containing DNA sequences from entire genomes, single chromosomes, or chromosome segments. Researchers use libraries to select probes complementary to all or part of a gene being surveyed.

4 Cloned DNA segments are used in a variety of applications, including gene mapping, disease diagnosis, and forensic analysis. In addition, cloned human genes inserted into bacterial hosts are used for the commercial production of human gene products for therapeutic treatment of a range of disorders, including diabetes and growth defects.

5 The correction of human genetic defects may be possible using retroviruses. Some of the viral genes can be removed from the retroviral genome to create a vector capable of transferring human structural genes into human target tissues, where the human genes become activated.

KEY TERMS

amniocentesis
bacteriophage
base pairs (bp)
chorionic villus sampling (CVS)
cDNA (complementary DNA) molecules
clones
cosmids
DNA polymerase I
EcoRI
flow cytometry
gel electrophoresis
gene splicing
generic engineering
library
Northern blot
palindromic
plasmids
polymerase chain reaction (PCR)
probe
pulse field gel electrophoresis
recombinant DNA technology
restriction endonucleases
restriction fragment length polymorphism (RFLP)
reverse transcriptase
ribonuclease H
S_1 nuclease
severe combined immunodeficiency (SCID)
Southern blot
T cells
vectors
Western blot
Xgal

The recognition site for the restriction enzyme *Sau*3A is GATC (Table 15.1). The recognition site for the enzyme *Bam*HI is GGATCC, where the four internal bases are identical to the *Sau*3A site. This means that the single stranded ends produced by the two enzymes are identical. Suppose you have a cloning vector containing a *Bam*HI site and foreign DNA that you have cut with *Sau*3A.

1 Can this DNA be ligated into the *Bam*HI site of the vector? Why?

ANSWER: DNA cut with *Sau*3A can be ligated into the *Bam*HI site of the vector because the single-stranded ends generated by the two enzymes are identical.

2 Can the DNA segment cloned into this site be cut from the vector with *Sau*3A? With *Bam*HI? What potential problems do you see with the use of *Bam*HI?

ANSWER: The DNA can be cut from the vector with *Sau*3A since the recognition site for this enzyme is maintained. The DNA may be recovered from the vector by *Bam*HI, but there is a potential problem with the use of this enzyme. When the *Sau*3A fragment is ligated into the *Bam*HI site, a base at each end of the *Bam*HI site must be added (arrows). Even though a template is present on the opposite strand, insertion of the wrong base will occur in a fraction of cases, destroying the *Bam*HI site, making it impossible to remove the insert with *Bam*HI.

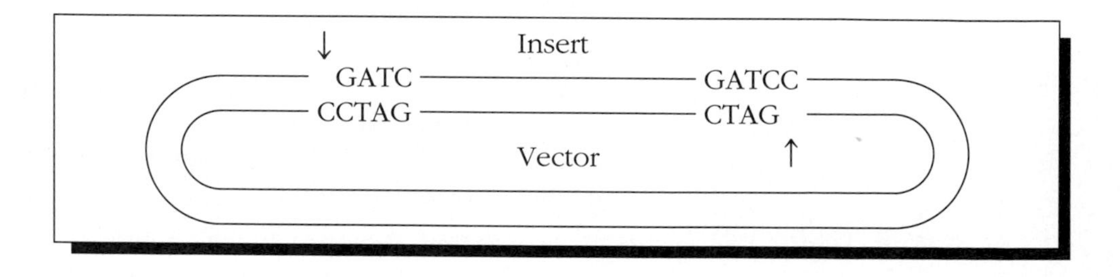

3 Calculate the average size of the fragments generated when cutting DNA with *Sau*3A and with *Bam*HI.

ANSWER: According to the Watson–Crick model of DNA, the sequence of bases in a DNA molecule can be at random. Therefore, if an A is present in a strand of DNA, the next base in that strand can be any base, even another A. Thus, the chance that any given base in DNA can be an A is 1/4, and the chance that the next base is a T is also 1/4. As a result, the chance that the four-base sequence GATC will occur is 4^4, or 1 in every 256 bases. The chance of encountering the six-base sequence GGATCC is 4^6, or 1 in every 4096 bases. Accordingly, cuts with *Sau*3A should generate fragments that average 256 bp in length, while cuts with *Bam*HI should generate fragments that average 4096 base pairs in length. This, of course, is only an approximation and assumes that each base is present about 25 percent of the time. In reality, however, many genomes depart from this ideal and may have only 40 or 42 percent G + C content, instead of the 50 percent ideal.

PROBLEMS AND DISCUSSION QUESTIONS

1 In recombinant DNA studies, what is the role of each of the following: restriction endonucleases, vectors, and host cells?

2 Why is poly dT an effective primer for reverse transcriptase?

3 An ampicillin-resistant, tetracycline-resistant plasmid is cleaved with *Eco*RI, which cuts within the ampicillin gene. The cut plasmid is ligated with *Eco*RI-digested *Drosophila* DNA to prepare a genomic library. The mixture is used to transform *E. coli* K12.

(a) Which antibiotic should be added to the medium to select cells that have incorporated a plasmid?

(b) What antibiotic resistance pattern should be selected to obtain plasmids containing *Drosophila* inserts?

(c) How can you explain the presence of colonies that are resistant to both antibiotics?

4 Clones from the previous question are found to have an average length of 5 kb. Given that the *Drosophila* genome is 1.5×10^5 kb long, how many clones would be necessary to give a 99 percent probability that this library contains all genomic sequences?

5 The human insulin gene contains a number of introns. In spite of the fact that bacterial cells will not excise introns from mRNA, how can a gene like this be cloned into a bacterial cell and produce insulin?

6 Restriction enzymes recognize palindromic sequences in intact DNA molecules and cleave the double-stranded helix at these sites. Inasmuch as the bases are internal in a DNA double helix, how is this recognition accomplished?

7 In a control experiment, a plasmid containing a *Hind*III site within a kanamycin-resistant gene is cut with *Hind*III, religated, and used to transform *E. coli* K12 cells. Kanamycin-resistant colonies are selected, and plasmid DNA from these colonies subjected to electrophoresis. Most of the colonies contain plasmids that produce single bands that migrate at the same rate as the original, intact plasmid. A few colonies, however, produce two bands, one of original size, and one that migrates much higher in the gel. Diagram the origin of this slow band during the religation process.

8 In mice transfected with the rabbit beta-globin gene, the rabbit gene is active in a number of tissues including spleen, brain, and kidney. In addition, some mice suffer from thalassemia, caused by an imbalance in the coordinate production of alpha and beta globins. What problems associated with gene therapy are illustrated by these findings?

9 What facts should you consider in deciding which vector to use in constructing a genomic library of eukaryotic DNA?

10 Although the potential benefits of cloning in higher plants are obvious, the development of this field has lagged behind cloning in bacteria, yeast, and mammalian cells. Can you think of any reason for this?

11 Using DNA sequencing on a cloned DNA segment, you recover the following nucleotide sequence: CAGTATCCTAGGCAT. Does this segment contain a palindromic recognition site for a restriction enzyme? What is the double-stranded sequence of the palindrome? What enzyme would cut at this site (Consult Table 15.1 for a list of restriction enzyme recognition sites)?

12 Table 15.1 lists restriction enzymes that recognize sequences of four bases and six bases. How frequently should four and six base recognition sequences occur in a genome? If the recognition sequence consists of eight bases, how frequently would such sequences be encountered? Under what circumstances would you select such an enzyme for use?

13 You have recovered a cloned DNA segment of interest, and determined that the insert is 1300 bp in length. To characterize this cloned segment, you have isolated the insert and decide to construct a restriction map. Using enzyme I and enzyme II, followed by gel electrophoresis, you determine the number and size of the fragments produced by enzymes I and II alone and in combination as follows:

Enzymes	Restriction Fragment Sizes (bp)
I	350 bp, 950 bp
II	200 bp, 1100 bp
I and II	150 bp, 200 bp, 950 bp

Construct a restriction map from these data, showing the positions of the restriction sites relative to one another, and the distance between them in base pairs.

SELECTED READINGS

ANDERSON, W. F., and DIACUMAKOS, E. G. 1981. Genetic engineering in mammalian cells. *Scient. Amer.* (July) 245:106–21.

ANTONARAKIS, S. 1989. Diagnosis of genetic disorders at the DNA level. *New Engl. J. Med.* 320:153–163.

BARKER, D., et al. 1987. Gene for von Recklinghausen neurofibromatosis is in the pericentromeric region of chromosome 17. *Science* 236:1001–1102.

BERG, P. 1981. Dissections and reconstructions of genes and chromosomes. *Science* 213:296–303.

GRAF, L. H. 1982. Gene transformation. *Amer. Scient.* 70:496–505.

GRIESBACH, R. J., KOIVUNIEMI, P. J., and CARLSON, P. S. 1981. Extending the range of plant genetic manipulation. *Bioscience* 31:754–56.

HOPWOOD, D. A. 1981. Genetic programming of industrial microorganisms. *Scient. Amer.* (September) 245:91–102.

KNORR, D., and SINSKEY, A. J. 1985. Biotechnology in food production and processing. *Science* 229:1224–29.

McKUSICK, V. A. 1988. The new genetics and clinical medicine. *Hospital Practice* 23:177–91.

MOLLIS, K. B. 1990. The unusual origin of the polymerase chain reaction. *Scient. Amer.* (April) 262:56–65.

OLD, R. W., and PRIMROSE, S. B. 1985. *Principles of genetic manipulation: An introduction to genetic engineering.* Palo Alto, CA: Blackwell Scientific.

OSTE, C. 1988. Polymerase chain reaction. *BioTechniques* 6:162–67.

REISS, J., and COOPER, D. N. 1990. Application of the polymerase chain reaction to the diagnosis of human genetic disease. *Human Genetics* 85:1–8.

THOMAS, T. L., and HALL T. C. 1985. Gene transfer and expression in plants: Implications and potential. *BioEssays* 3:149–53.

TORREY, J. G. 1985. The development of plant biotechnology. *Amer. Scient.* 73:354–63.

VERMA, I. 1990. Gene therapy. *Scient. Amer.* (November) 263–68.

WHITE, R. 1985. DNA sequence polymorphisms revitalize linkage approaches in human genetics. *Trends in Genetics.* 1:177–80.

Regulation of Gene Expression

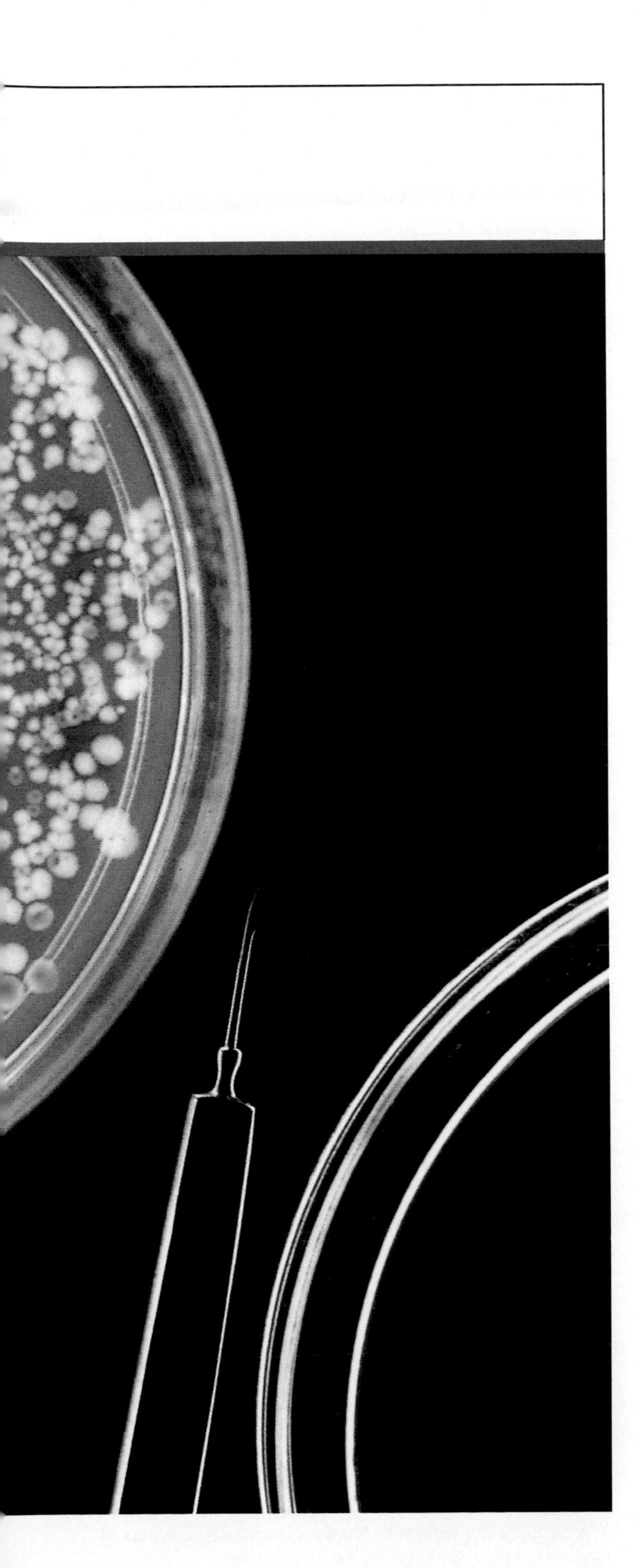

Chapter Concepts

Efficient expression of genetic information in all organisms is dependent on regulatory mechanisms that either activate or repress the transcription of genes. Transcription is modulated by the interaction of various regulatory molecules with short DNA sequences, most often located upstream from affected genes. Genetic regulation in eukaryotes also occurs during post-transcriptional events.

In earlier chapters, we established how DNA is organized into genes, how genes store genetic information, and how this information is expressed. We now consider one of the most fundamental issues in molecular genetics: **How is genetic expression regulated?** The evidence in support of the idea that genes can be turned on and off is very convincing. You need only look at bacterial cells growing on a specific substrate such as lactose. In its absence, enzymes needed to metabolize this sugar are also absent. But in its presence, the genes encoding the enzymes are activated, and the enzymes are synthesized. If all lactose is then used up, these genes shut down!

In most eukaryotic organisms, such as mammals, evidence of genetic regulation is even more apparent. Cells of the pancreas do not make retinal pigment, and retinal cells do not make insulin. In such multicellular eukaryotes, genetic regulation is at the heart of cellular differentiation. Differential gene expression is believed to serve as the underlying basis for phenotypic specialization at the cellular and tissue level in both plants and animals.

In this chapter, we will explore examples of gene regulation in bacteria, viruses, and eukaryotes. Fundamental to this discussion is the realization that all cells in organisms contain a complete set of genetic information characteristic of that species. Thus, regulation is *not* accomplished simply by eliminating unused genetic information. Instead, elaborate mechanisms have evolved that control the expression of genes. Some of these mechanisms are clear to us, while others have yet to be elucidated. It is safe to say that this general topic represents one of the most exciting and significant areas of molecular genetics to which extensive research will be devoted during future years.

GENETIC REGULATION IN PROKARYOTES: AN OVERVIEW

The regulation of genetic expression has been most extensively studied in prokaryotes, particularly in *E. coli.* What has been found is that highly efficient genetic mechanisms have evolved that turn genes on and off, depending on the cell's metabolic need for the respective gene products. Detailed analysis of proteins in *Escherichia coli* has shown, for example, that for the 4000 or so polypeptide chains encoded by the genome, there is a vast range of concentration of gene products. Some proteins may be present in as few as 5 to 10 molecules per cell, whereas others, such as ribosomal proteins and the many proteins involved in the glycolytic pathway, are present in as many as 100,000 copies per cell.

While a basal level of most gene products exists, it is clear that this level can be increased and subsequently decreased in prokaryotes. Bacteria "adapt" to their chemical environment, producing certain enzymes only when specific substrates are present. Such enzymes are referred to as **inducible,** reflecting the role of the substrate in their production. In contrast, other proteins are produced continuously regardless of the chemical makeup of the environment and are called **constitutive.**

Investigations have also revealed cases where the presence of a specific molecule causes inhibition of genetic expression. This is usually true for molecules that are end products of biosynthetic pathways. For example, amino acids can be synthesized by bacterial cells. If an external supply of a specific amino acid is present in the environment or culture medium, it is energetically inefficient to synthesize the enzymes necessary for the production of that amino acid. A mechanism has evolved whereby the amino acid plays a role in repressing transcription of RNA essential to the translation of the appropriate biosynthetic enzymes. In contrast to the inducible system controlling lactose metabolism, the system governing such synthesis is said to be **repressible.**

Instances of regulation, whether inducible or repressible, may be under either **negative** or **positive control.** Under negative control, genetic expression occurs unless it is shut off by some form of regulator. This is in contrast to positive control, where transcription does not occur unless a regulator molecule directly stimulates RNA production. Examples discussed in the ensuing sections will help distinguish between these mechanisms.

LACTOSE METABOLISM IN *E. COLI:* AN INDUCIBLE GENE SYSTEM

The most extensively studied system of gene regulation has involved the metabolism of lactose in *E. coli.* Beginning in 1946 with the studies of Jacques Monod and continuing through the next decade with significant contributions by Joshua Lederberg, François Jacob, and Andre L'woff, genetic and biochemical evidence was amassed. This research provided clear insights into the way in which the genes responsible for lactose metabolism are turned off, or repressed, when lactose is absent, but activated or induced when it is available. In the presence of lactose, the concentration of the enzymes responsible for its metabolism increases rapidly from 5 to 10 molecules to thousands per cell. The enzymes are **inducible,** and lactose serves as the **inducer.**

Paramount to the understanding of genetic regulation in this system was the discovery of two genes that serve strictly in a regulatory capacity. They do not code for enzymes necessary for lactose metabolism. Three other genes are responsible for the production of enzymes involved in lactose metabolism. Together, the five genes function in an integrated fashion and provide a rapid response to the presence or absence of lactose.

lac Structural Genes

Genes coding for the primary structure of the enzymes are called **structural genes.** The so-called *lac z* gene specifies the amino acid sequence of the **β-galactosidase enzyme,** which converts lactose to glucose and galactose (Figure 16.1). This conversion is essential if lac-

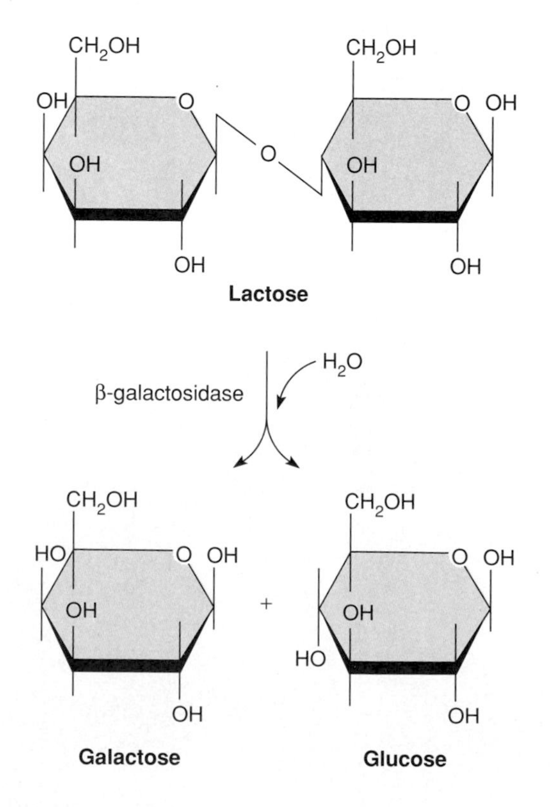

FIGURE 16.1
The catabolic conversion of the disaccharide lactose into its monosaccharide units, galactose and glucose.

tose is to serve as the primary energy source in glycolysis. The second gene, *lac y,* specifies the primary structure of ***β*-galactoside permease,** which facilitates the entry of lactose into the bacterial cell. The third gene, *lac a,* codes for **transacetylase,** an enzyme whose role is unrelated to our discussion.

Studies of the genes coding for these three enzymes relied on the isolation of numerous mutations called lac^- that eliminated the function of one or the other enzyme. Mutant cells fail to produce either active *β*-galactosidase or permease molecules and so are unable to utilize lactose as an energy source. Mapping studies by Lederberg established that all three genes are closely linked to one another in the order *z-y-a.* They are transcribed together, resulting in a single mRNA representative of all three genes. Such a message is called a **polycistronic message** or **mRNA** (Figure 16.2). As a result, when induction occurs, all three genes are activated simultaneously.

The Discovery of Regulatory Mutations

How, then, can lactose activate structural genes and induce the synthesis of the related enzymes? The answer to this question required the study of a second class of mutation, the **constitutive mutants.** In this type of mutant, the enzymes are produced regardless of the presence or absence of lactose. These mutations served as the basis of studies that defined the regulatory scheme for lactose metabolism.

Maps of the first type of constitutive mutation, *lac* i^-, showed that it is located at a site on the DNA close to, but distinct from, the structural genes. As we will soon see, the *i* gene is appropriately called a **repressor gene.** A second set of mutations produced identical effects but was found in a region immediately adjacent to the structural genes. This class is designated *lac* o^c and represents the **operator region.** Because the enzymes are continually produced for both types of constitutive mutation, they clearly represent regulatory units.

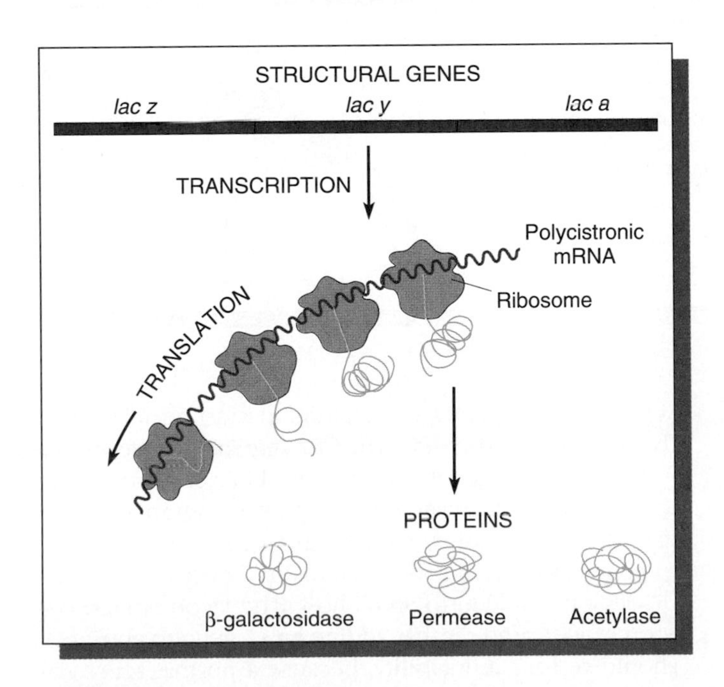

FIGURE 16.2
The structural genes of the *lac* operon in *E. coli.* The three genes are transcribed into a single polycistronic mRNA, which is sequentially translated into the three enzymes encoded by the operon.

The Operon Model: Negative Control

In 1961, Jacob and Monod proposed a scheme of negative control of regulation, which they called the ***lac* operon model** (Figure 16.3). In this model, the operon consists of the structural genes as well as the adjacent region of DNA represented by the *lac* o^c mutation. They proposed that the *lac i* gene regulates the transcription of the structural genes by producing a **repressor molecule.** The repressor was hypothesized to be **allosteric,** reversibly interacting with another smaller molecule and causing a conformational change in three-dimensional shape.

Jacob and Monod suggested that the repressor normally interacts with the DNA sequence of the operator region. When it does so, it inhibits the action of RNA polymerase, effectively repressing the transcription of the structural genes [Figure 16.3(b)]. However, in the presence of lactose, this disaccharide binds to the repressor, causing a conformational change. This change alters the binding site of the repressor capable of interacting with operator DNA [Figure 16.3(c)]. In the absence of the repressor/operator interaction, RNA polymerase transcribes the structural genes, and the enzymes necessary for lactose metabolism are translated. *Since transcription occurs only in the absence of the repressor,* **negative control** is exerted.

The operon model uses these potential molecular interactions to explain the efficient regulation of the structural genes. In the absence of lactose, the enzymes encoded by the genes are not needed and so are repressed. When lactose is present, it indirectly induces the activation of the genes by binding with the repressor. If all lactose is metabolized, none is available to bind to the repressor, which is again free to bind to operator DNA and repress transcription.

Both the i^- and o^c constitutive mutations interfere with these molecular interactions, allowing continuous transcription of the structural genes. In the case of the i^- mutant, the repressor product is altered and cannot bind to the operator region, so the structural genes are always turned on. In the case of the o^c mutant, the nucleotide

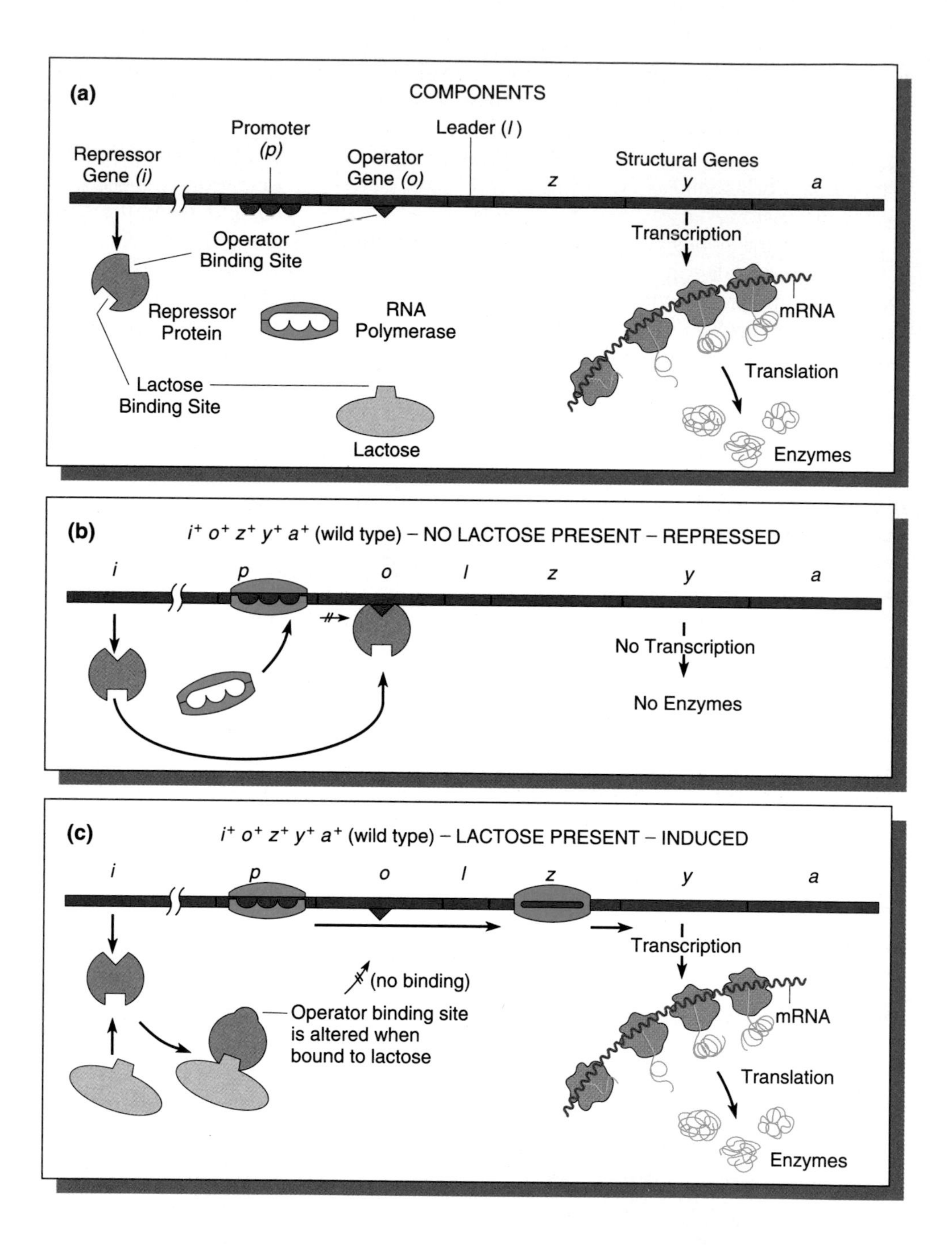

FIGURE 16.3
The components involved in the regulation of the *lac* operon and their interaction under various genotypic conditions, as described in the text.

sequence of the operator DNA is altered and will not bind with a normal repressor molecule. The result is the same: Structural genes are always transcribed. Both types of constitutive mutations are illustrated diagrammatically in Figure 16.3(d) and (e).

Genetic Proof of the Operon Model

The operon model is a good one because there are predictions derived from it that can be tested to determine its validity. The major assumptions to be tested are (1) the *i* gene produces a diffusible cellular product; (2) the *o* region does not; and (3) the *o* region must be adjacent to the structural genes in order to regulate transcription.

The construction of partial diploid bacteria (see Chapter 14) allows an assessment of these assumptions. For example, it is possible to construct genotypes where an i^+ gene has been introduced into an i^- host, or where an o^+ region is added to an o^c host. The Jacob-Monod operon model predicts that adding an i^+ gene to an i^- cell should restore inducibility, because a normal repressor would again be produced. Adding an o^+ region to an o^c cell should have no effect on constitutive enzyme production, since regulation depends on an o^+ region in the

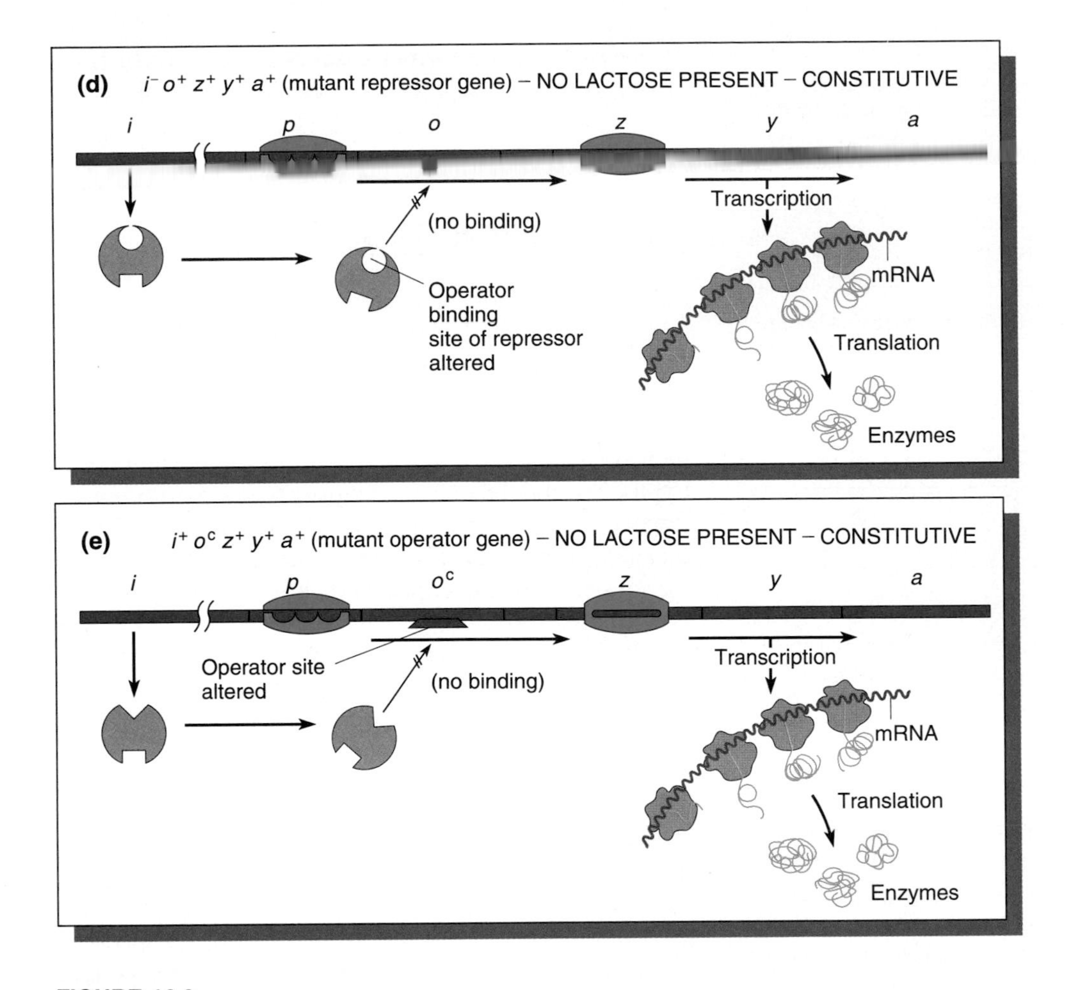

FIGURE 16.3
continued

host DNA, immediately adjacent to the structural genes.

The results of these experiments are shown in Table 16.1, where z represents the structural genes. In both cases just described, the Jacob-Monod model is upheld (part B of Table 16.1). Part C shows the reverse experiments, where either an i^- gene or an o^c region is added to cells of normal inducible genotypes. The model predicts that inducibility will be maintained in these partial diploids, and it is.

Another prediction of the operon model is that certain mutations in the i gene should have the opposite effect of i^-. That is, instead of being constitutive by failing to interact with the operator, mutant repressor molecules should be produced, which cannot interact with the inducer, lactose. As a result, the repressor would always bind to the operator sequence, and the structural genes would be repressed permanently. If this were the case, the presence of an additional i^+ gene would have little or no effect on repression.

In fact, as shown in Table 16.1D, such a mutation, i^s, was discovered. An additional i^+ gene does not effectively relieve repression of gene activity. These observations again support the operon model for gene regulation.

Isolation of the Repressor

The most direct proof of the operon's existence was the isolation and characterization of the repressor molecule predicted by the model. This was a difficult task because there were so few copies of it (ten or so) present among the millions of molecules in each *E. coli* cell. In spite of the enormity of the task, in 1966 Walter Gilbert and Benno Müller-Hill reported its isolation.

The repressor was shown to exhibit numerous characteristics of proteins and to consist of four identical polypeptide subunits. Confirmation of its role was provided by showing that the repressor binds to DNA containing an o^+ gene but not to DNA with an o^c gene. Further, repressor binding activity could not be demonstrated

TABLE 16.1
A comparison of gene activity (+ or −) in the presence or absence of lactose for various *E. coli* genotypes. In Parts B–D, many genotypes are partially diploid, containing an F factor plus attached genes (F′).

Genotype	Presence of β-galactosidase Activity	
	Lactose Present	Lactose Absent
A. $i^+o^+z^+$	+	−
$i^+o^+z^-$	−	−
$i^-o^+z^+$	+	+
$i^+o^cz^+$	+	+
B. $i^-o^+z^+/F'i^+$	+	−
$i^+o^cz^+/F'o^+$	+	+
C. $i^+o^+z^+/F'i^-$	+	−
$i^+o^+z^+/F'o^c$	+	−
D. $i^so^+z^+$	−	−
$i^so^+z^+/F'i^+$	−	−

among the proteins isolated from i^- cells, as predicted.

These observations confirmed the existence of a finely tuned genetic model explaining how a set of inducible genes are activated when needed and repressed when they aren't. Most impressive in their study, for which Jacob, Monod, and L'woff shared the Nobel Prize, was their ability to predict the *molecular basis* of regulation based strictly on *genetic observations.*

THE *ara* REGULATOR PROTEIN: POSITIVE AND NEGATIVE CONTROL

Before turning to a consideration of repressible systems, we want to make brief mention of a rather unique inducible operon in *E. coli* that appears to be under both positive and negative control.

In *E. coli,* the metabolism of the sugar arabinose is under the direction of the enzymatic products of the *ara B, A,* and *D* genes. These genes are regulated by the ***ara* C protein** encoded by the *ara* C gene. In the absence of arabinose, the protein behaves as a repressor, binding to three regulatory sites within the operon. In the presence of arabinose, the *ara* C protein binds to the sugar, and this complex becomes an activator. Attaching to a separate initiator site on the operon stimulates the transcription of the *B, A,* and *D* genes. Thus, the *ara* regulatory protein can alternately behave as a repressor or an activator, depending on the absence or presence of arabinose, respectively. In the former case, negative control is exerted, while in the latter case, positive control occurs.

TRYPTOPHAN METABOLISM IN *E. COLI*: A REPRESSIBLE GENE SYSTEM

Although induction had been known for some time, it was not until 1953 that Monod and his coworkers discovered **enzyme repression.** Wild-type *E. coli* are capable of producing the necessary enzymes essential to the biosynthesis of amino acids as well as other essential biomolecules. Monod focused his studies on the amino acid tryptophan and the enzyme **tryptophan synthetase.** He discovered that if tryptophan is present in sufficient quantity in the growth medium, the enzymes necessary for its synthesis are **repressed.** Energetically, such enzyme repression is highly economical to the cell, because synthesis is unnecessary in the presence of exogenous tryptophan.

Further investigation showed that a series of enzymes encoded by five contiguous genes on the *E. coli* chromosome is involved in tryptophan synthesis. These genes are part of an operon, and in the presence of tryptophan, all genes are coordinately repressed. As a result, none of the enzymes are produced. Because of the great similarity between this repression and the induction of enzymes for lactose metabolism, Jacob and Monod proposed a model of gene regulation analogous to the *lac* system.

To account for repression, they suggested that an inactive repressor is normally made that alone cannot interact with the operator region of DNA within the operon. However, it is allosteric, interacting with tryptophan, if present. The resulting interaction activates the repressor, changing its conformation so that it now binds to the operator, repressing transcription. Thus, when the end product of this anabolic biosynthesis pathway is present, enzymes are not made. Since the regulatory complex inhibits transcription of the operon, this **repressible system,** like the *lac* system, is under **negative control.** Because tryptophan participates in repression, it is referred to as a **corepressor** in this regulatory scheme.

Evidence for and Concerning the *trp* Operon

Support for the concept of a repressible operon was soon forthcoming. Primarily, two distinct categories of constitutive mutations were isolated. The first class, *trp* R^-, maps at a considerable distance from the structural genes. This locus represents genetic information coding for the repressor. Presumably, the R^- mutation either inhibits the interaction of a mutant repressor with tryptophan or inhibits repressor formation entirely. Thus, no repression ever occurs. The presence of an R^+ gene restores repressibility.

The second constitutive mutant behaves like the O^c mutation in the *lac* operon. It is found immediately adjacent to the structural genes. Furthermore, the addition of a wild-type gene into mutant cells does not restore enzyme repression. This is predictable if the mutation represents an operator region that can no longer interact with the repressor-tryptophan complex.

The entire *trp* operon has now been well defined, as shown in Figure 16.4. Five contiguous structural genes (*trp E, D, C, B,* and *A*) are transcribed as a polycistronic message that directs translation of the polypeptide components. These components catalyze the biosynthesis of tryptophan. As in the *lac* operon, a promoter region (*trp P*), representing the binding site for RNA polymerase, and an operator region (*trp O*), which binds the repressor, have been demonstrated. In the absence of binding, transcription is initiated within the overlapping *trp P-trp O* region and proceeds along a **leader sequence.** Transcription is initiated 160 nucleotides prior to the first structural gene (*E*). Within this leader sequence, still another regulatory site has been demonstrated. This component, called an **attenuator,** has been investigated extensively by Charles Yanofsky and his colleagues. As we shall see, this regulatory unit is an integral part of the control mechanism of this operon.

The Attenuator

It has been observed that even in the presence of tryptophan, when repression *should* be complete, transcription may often be initiated. As a result, repression in this system is said to be "weak" in comparison to that which occurs in the *lac* operon. While strong versus weak re-

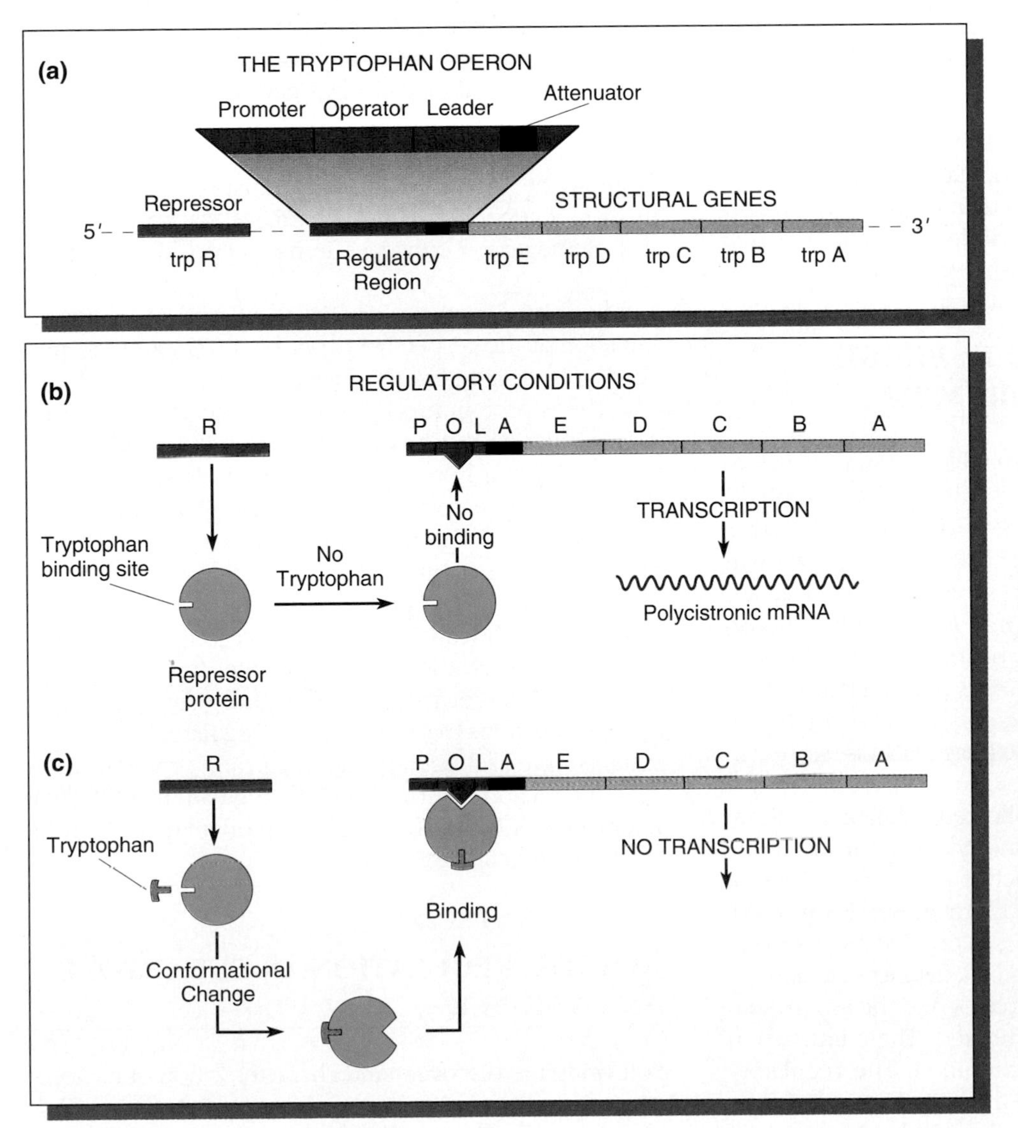

FIGURE 16.4
(a) The components involved in the regulation of the tryptophan operon.
(b) Regulatory conditions involving either activation or (c) repression of the structural genes. In the absence of tryptophan, an inactive repressor is made that cannot bind to the operator (*O*), thus allowing transcription to proceed. In the presence of tryptophan, it binds to the repressor, causing an allosteric transition to occur. This complex binds to the operator region, resulting in repression of the operon.

pression may seem difficult to understand at first, such phenomena are not unexpected. Molecular binding, such as that between the protein repressor and operator DNA, is a dynamic process subject to chemical equilibrium. If the molecular affinity between the two is pronounced, repression at equilibrium will be quite effective. If the affinity is less pronounced, repression will be weak.

The process of **attenuation,** as studied extensively by Yanofsky, provides a second level of regulation of the operon. In this context, attenuate means to reduce in amount. In the presence of tryptophan, mRNA synthesis may be initiated, but is frequently terminated at a point about 140 nucleotides along the transcript. This point is identified in Figure 16.4a. In the absence of tryptophan, the process of attenuation is overcome and transcription of the operon proceeds, leading to the production of enzymes necessary for tryptophan biosynthesis.

The process of attenuation is complex and involves the molecular folding of the RNA transcribed from the leader sequence. When attenuation is occurring, the structure of the leader RNA mimics that found normally at the end of mRNAs, leading to the premature termination of transcription. Yanofsky has described a model of how this process of attenuation occurs and how it is overcome in the absence of tryptophan. Again, like Jacob and Monod's study, the prediction of molecular events was initially based on genetic evidence, including mutations within the leader sequence that abolish attenuation.

GENETIC REGULATION IN PHAGE LAMBDA: LYSOGENY OR LYSIS?

Our understanding of genetic regulation at the transcriptional level has benefited from studies of bacteriophage lambda as well as from studies of operons in bacteria. Lambda DNA contains about 45,000 base pairs, enough bioinformation for about 35 to 40 average-sized genes. After injection of lambda DNA into *E. coli,* either the **lysogenic** or **lytic** pathways may be followed (see Chapter 14). In the lysogenic pathway, the phage DNA is integrated into the host genome and is almost totally repressed; in the lytic pathway, the phage DNA is expressed and viral reproduction ensues.

If the lysogenic pathway is followed, the genes responsible for phage reproduction and lysis are turned off by the **λ repressor protein,** which is produced by one of the virus's own genes, *cI*. If the lytic pathway is followed, the expression of the *cI* gene is repressed by a second protein, **cro,** which is produced by a distinct viral gene of the same name. Both the *cI* repressor and the cro protein have been isolated and characterized. Their interaction with λ DNA has also been determined. The regulatory region of λ DNA is diagrammed in Figure 16.5.

The λ repressor was isolated and characterized by Mark Ptashne in 1967. It is a protein consisting of 236

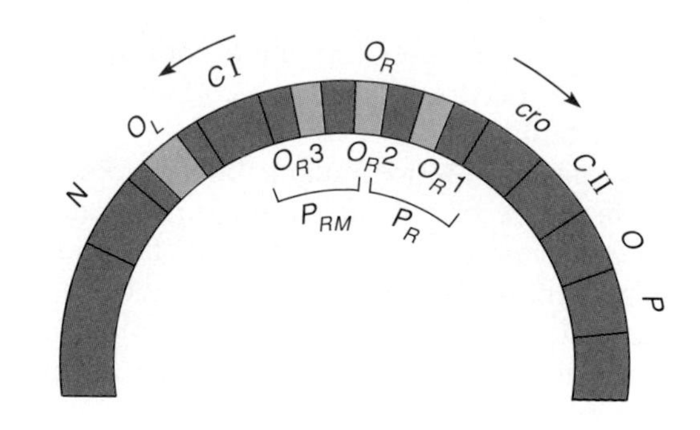

FIGURE 16.5
The regulatory sites controlling lysogeny and lysis in phage λ.

amino acids. The cro protein, isolated more recently, consists of 66 amino acids. When the repressor is produced, it recognizes two different operator regions in the λ DNA, O_L and O_R, found on either side (left or right) of the *cI* gene. When the repressor is bound to these regions, two sets of so-called early genes are repressed, causing the remainder of the λ genes to be turned off as well. In this case, the repressor behaves as a negative control element. It is now clear that during this same binding, the *cI* repressor also stimulates transcription of its own *cI* gene. In this role, it behaves as a positive control element and enhances transcription by a factor of ten.

When the repressor binding to O_L and O_R is absent, initiation at the respective promoters (P_{RM} and P_R) proceeds, resulting in the production of two proteins, N and cro. The N protein functions as an antiterminator, allowing completion of transcription of genes essential to reproduction and lysis. The cro protein, as stated earlier, serves as a repressor of *cI* gene transcription. Mutational analysis supports this model. Mutations in either O_L or O_R prevent repressor binding and abolish the potential for lysogeny. cro^- mutations abolish the potential for lysis.

The research leading to this and accompanying information is remarkable. It illustrates the depth of analysis attainable with regard to what at first may appear as a simple question: How is lysogeny regulated? Important insights have also been gained regarding protein/nucleic acid interactions during genetic regulation. Molecular interactions such as these undoubtedly play major roles in many other genetic phenomena.

GENETIC REGULATION IN EUKARYOTES: AN OVERVIEW

Following the discovery and characterization of bacterial operons, many studies were initiated to locate comparable regulatory systems in eukaryotes. While a similar system has been found in single-celled yeast, no operons

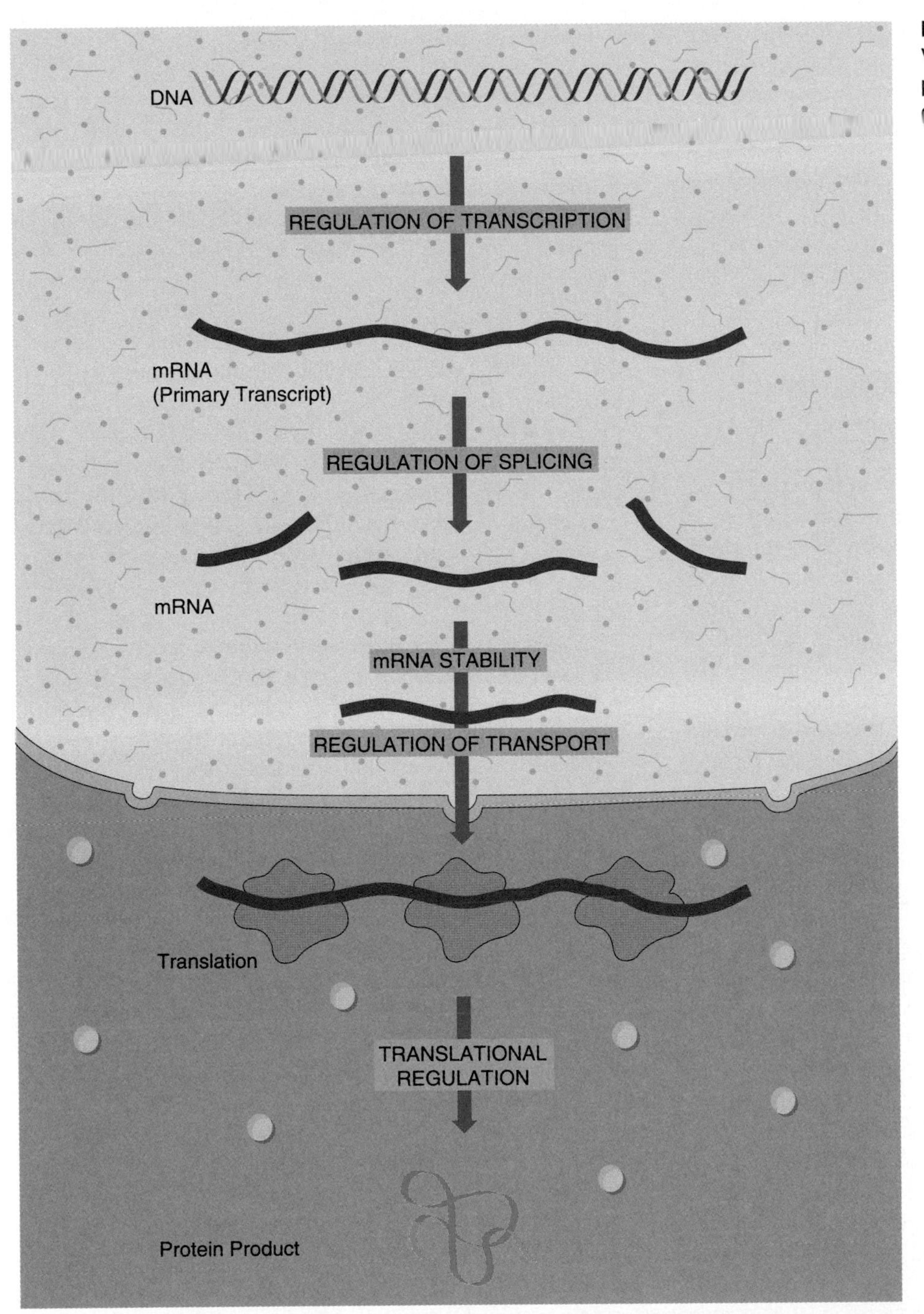

FIGURE 16.6
Various levels of regulation that are possible during the expression of the genetic material.

have yet been discovered in multicellular eukaryotes, in spite of 30 years of investigative research! Careful consideration of the fundamental differences between prokaryotes and eukaryotes provides us with important clues as to why operons do not constitute the basic mechanism for the regulation of eukaryotic genes.

Primarily, eukaryotic genetic systems are extremely complex in comparison to the evolutionarily less advanced prokaryotes. Each eukaryotic cell contains a much greater amount of genetic information, and this DNA is complexed with histones and other proteins to form chromatin, while in prokaryotic organisms, the DNA does not form into chromatin. The genetic information in eukaryotes is carried on many chromosomes, rather than one, and these chromosomes are enclosed within a nuclear membrane. In eukaryotes, the process of transcription is spatially and temporally separated from that of translation. The transcripts of eukaryotic genes are processed, cleaved, and realigned before transport to the cytoplasm, and many transcribed sequences never leave the nucleus. In addition, during the development of multicellular organisms, cellular differentiation and changes in gene expression are often influenced by cellular interactions as well as by external signals such as hormones.

Without belaboring the point, it seems evident that these properties of eukaryotes are highly likely to require more sophisticated regulatory mechanisms than those found in bacteria. Regulation of gene expression may potentially occur at many levels, including transcription, processing, transport, and translation (Figure 16.6). Although the details of regulation in eukaryotes have not yet been described for many genes, enough is known to allow some general features to be presented.

REGULATION OF TRANSCRIPTION IN EUKARYOTES

The first and most apparent level at which gene expression may be controlled is at the point of transcription. As in prokaryotes, control of this process involves the interaction of proteins that bind specifically to DNA sequences. Unlike the operons of bacteria, however, these DNA-binding proteins are usually activators rather than repressors, resulting in positive control. Called **transcription factors,** they may bind to sequences much more distant from the genes they regulate than is the case with operator DNA and repressors characteristic of bacteria. In the following sections we shall review the general topic of **transcriptional control** of eukaryotic genes.

Galactose Metabolism in Yeast

We shall begin by examining the genetic control mechanisms involving the enzymes that participate in the metabolism of galactose in the yeast *Saccharomyces cerevisiae.* As we shall see, there are several similarities between this regulatory model and bacterial operons. However, as with many genes in more advanced eukaryotes, regulation involves a positive system characterized by **transcriptional activation** rather than repression.

The activities of a cluster of three genes, all linked on chromosome 2, are involved in the metabolism of galactose. These genes, *GAL1, GAL7,* and *GAL10,* encode a kinase, a transferase, and an epimerase, respectively. All three enzymes play a role in the conversion of galactose to glucose-6-phosphate. The latter molecule serves as an entry point into glycolysis, thus allowing galactose to serve as an energy source.

In the absence of galactose, these three genes are not transcribed. If galactose is added to the growth medium, immediate transcription of the three genes commences. Constitutive mutations in one or more genes at other sites have been isolated, indicating the existence of genetic regulation of transcription. As a result of further study, two other genes, *GAL4* (on chromosome 16) and *GAL80* (on chromosome 7), have been implicated in such regulation, leading to the model presented below and diagrammed in Figure 16.7.

For the purpose of our discussion, we shall consider *GAL1, 7,* and *10* as a structural unit of genes, since they are all activated together. However, it is important to note their arrangement on chromosome 2. *GAL1* and *10* are closely linked, separated only by a critical 600-base pair sequence called $\mathbf{UAS_G}$ (upstream activating sequence). Their polarities of transcription, noted by the arrows in Figure 16.7(a), are in opposite directions. *GAL7* is not as closely linked and is upstream (in the 5′ direction) from *GAL10.*

Analysis of regulatory mutations has led to and support the following model, depicted in Figure 16.7(b). The product of the *GAL4* gene is a transcription activator that functions by binding to the UAS_G region. If, however, no galactose is present, the *GAL80* gene product binds to the *GAL4* activator, rendering it ineffective in stimulating transcription of the structural genes. If galactose *is* present, this sugar binds to the GAL80 product, preventing it from binding to the *GAL4* activator, and transcription is initiated.

In these cases, we must assume that different combinations of binding induce conformational changes that provide alternate conditions of recognition. This assumption is supported by the findings that different portions of the regulatory molecules play different roles. The *GAL4* activator protein has now been studied extensively by Mark Ptashne and his colleagues. The protein is 881 amino acids long. Three functional regions have been identified. First, the initial 147 amino acids contains sequences referred to as a **zinc finger,** a protein domain that is known to bind to DNA (see the following section). It is this region that physically interacts with UAS_G. While not shown in Figure 16.7, the DNA of UAS_G contains four such binding regions, each 17 base pairs in length. Effective activation requires that all four sites are occupied by a *GAL4* dimer complex.

The second and third regions identified within the *GAL4* protein are responsible for activation of transcription and binding to the *GAL80* product, respectively. Mutations that prevent binding to *GAL80,* lead to constitutive production of the enzymes encoded by the structural genes.

As a result of the detailed analysis of these regions, we now have a much better understanding of the molecular interactions that are possible in eukaryotes between protein activators with DNA as well as with other transcription factors. In the ensuing sections, we will focus on information involving such interactions in other eukaryotes.

Promoters in Eukaryotes

Eukaryotic genes have two major types of regulatory regions. The first consists of nucleotide sequences that serve as the recognition point for RNA polymerase binding. Such regions are called **promoters,** as they are in

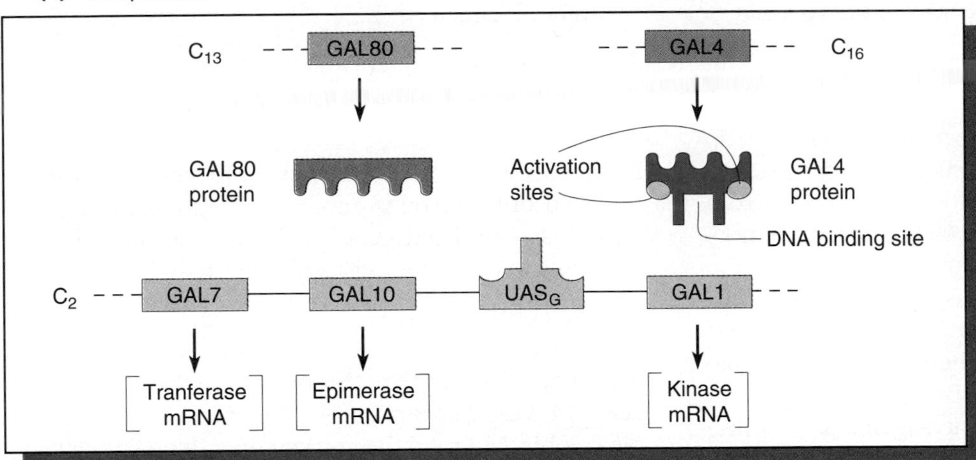

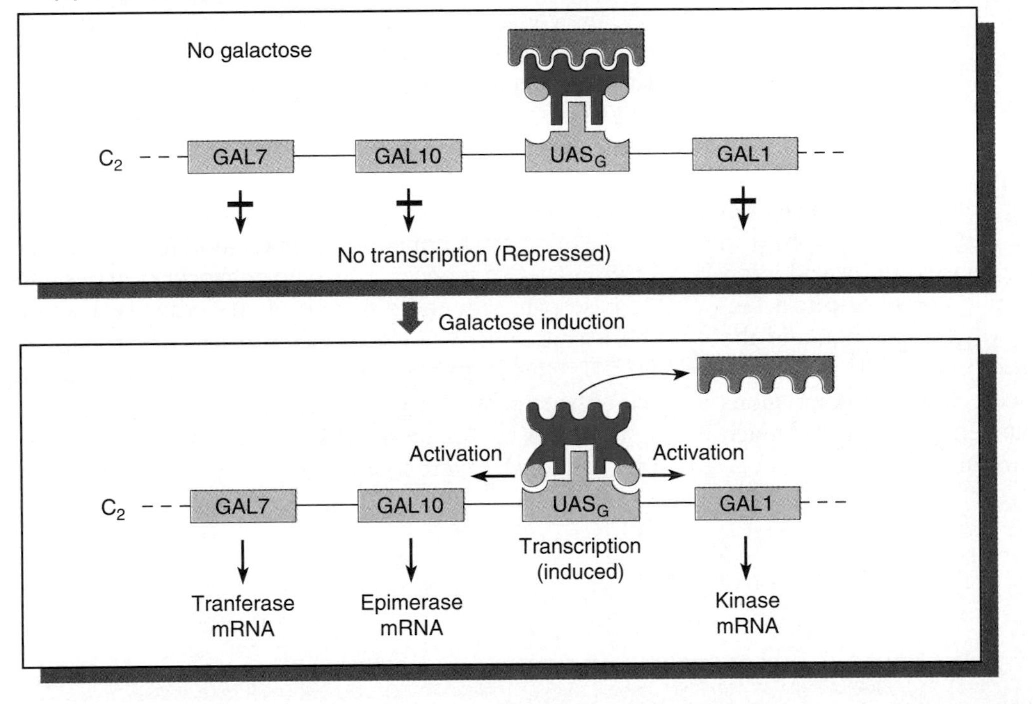

FIGURE 16.7
A representation of the interactions of the products of the *GAL80* and *GAL4* genes during the regulation of the galactose operon in yeast. Part (a) depicts the various genes and their products. Part (b) contrasts the interaction of the *GAL80* and *GAL4* products with the upstream activating sequence (UAS_G) in the absence and in the presence of galactose. Activation of transcription of the operon is under positive control of the *GAL4* protein.

prokaryotes. The promoter represents the region necessary to *initiate* transcription. Promoters might be thought of as establishing the "potential" for transcription to occur.

In spite of having a similar function in both eukaryotes and prokaryotes, major differences exist. Foremost is the fact that, in prokaryotes, regulatory regions are usually very close to the site where transcription is initiated, whereas in eukaryotes, the upstream region is more extensive and may be farther away from the gene that is being regulated. A second major difference is the variation observed in the promoter regions in eukaryotes. This is undoubtedly related to the fact that there are three different forms of **RNA polymerase** that transcribe three different groups of genes. These are classified as type I (ribosomal RNAs), type II (mRNAs and snRNAs), and type III (tRNAs, 5 SRNA, and several other small cellular RNAs). Each category requires different recognition sites within the promoter.

As you might suspect, this general topic has the potential for a very complex discussion. We will simplify our coverage by focussing primarily on the regulatory units associated with type II genes. Promoters that are recognized by RNA polymerase II are composed of short modular sequences of DNA usually located within 100 bp upstream (in the 5′ direction) of the gene. In the promoter of most genes is a sequence called the **TATA box.** Located some 25 to 30 bases upstream from the initial point of transcription (designated as -25 to -30), it consists of an 8-base pair consensus sequence (a sequence conserved in most or all genes studied) composed only of $T = A$ pairs, often flanked on either side by $G \equiv C$-rich regions. Mutations in the TATA box severely restrict transcription, and deletions often alter the initiation point of transcription.

A second modular component of many promoters is called the **CAAT box.** Its consensus sequence is CAAT or CCAAT, it frequently appears in the region −70 to −80 bp from the start site, and it can function in either orientation. Mutational analysis suggests that the CAAT box is critical to the promoter's ability to facilitate transcription. A third modular element of some promoters is called the **GC box,** which has the consensus sequence GGGCGG and is often found at about position −110. This module is found in either orientation and often occurs in multiple copies.

Figure 16.8 illustrates several arrangements of these upstream regulatory units. In addition to the elements we have grouped as part of the promoter, the second major regulatory element, the enhancer, is illustrated, to which we now turn.

Enhancers in Eukaryotes

In addition to the promoter region, transcription of most, if not all, eukaryotic genes is regulated by additional DNA sequences called **enhancers.** They are not involved in the direct binding of RNA polymerase, but instead interact with regulatory molecules called **transcription factors.** Their role is to regulate whether or not, and the rate at which, transcription *actually* occurs. Thus, there is some analogy between enhancers and operator regions in prokaryotes. However, enhancers appear to be much more complex in both structure and function.

Enhancers can be distinguished from promoters by two other characteristics:

1 The position of the enhancer need not be fixed; it can be placed upstream, downstream, or even within the gene it regulates.

2 Enhancers are not restricted to specific genes. If an enhancer is moved to another location in the genome, or if an unrelated gene is placed near an enhancer, transcription of the adjacent gene may be enhanced.

Most eukaryotic genes are under the control of enhancers. In the immunoglobulin heavy chain genes, an enhancer is located in an intron, within the gene it regulates. Downstream enhancers are found in the human beta globin gene and the chicken thymidine kinase gene. In chickens, an enhancer is located between the beta globin and the epsilon globin genes and apparently works in one direction to control the epsilon gene during embryonic development and in the opposite direction to regulate the beta gene during adult life. In yeast, regulatory sequences similar to enhancers, called **upstream activator sequences (UAS)** (discussed earlier in this chapter) can function upstream at variable distances and in either orientation as a regulatory element.

Like promoters, enhancers contain modules composed of short DNA sequences. The SV40 enhancer contains two 72 base pair sequences. If one or the other is deleted, there is no effect, but if both are missing, transcription is greatly reduced. Other mutations reduce the efficiency of the enhancer by 40 to 80 percent.

The most intriguing question about enhancers is how they are able to exert control over transcription at a great

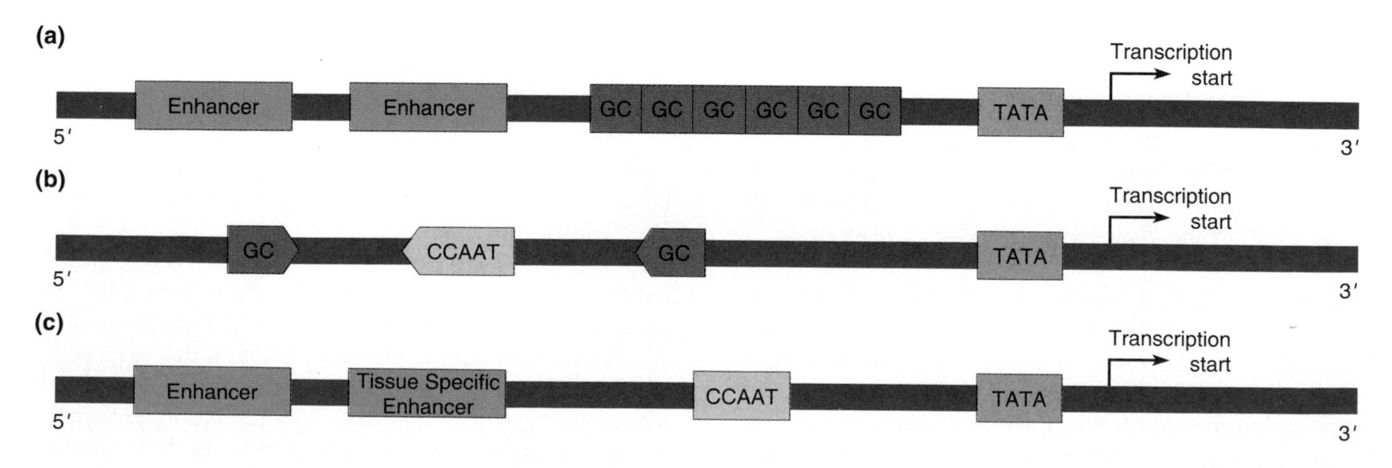

FIGURE 16.8
Upstream organization of several eukaryotic genes, illustrating the variable nature, number, and arrangement of controlling elements. (a) SV40 control region; (b) Thymidine kinase gene; (c) Insulin gene.

distance from either promoters or the transcriptional start site. While there is no clear answer yet, the consensus of opinion is that transcription factors bind to enhancers and alter the structure of chromatin and/or DNA to facilitate the binding of RNA polymerase. The binding of transcriptional factors to enhancers may alter the configuration of chromatin by bending or looping the DNA to bring distant enhancers and promoters into direct contact. A model depicting such a configuration is shown in Figure 16.9. In the new configuration, the basal level of transcription may be optimized, increasing the rate of RNA synthesis.

Transcription Factors

Proteins that are not part of the RNA polymerase molecule itself, but are needed for the initiation of transcription, are called **transcription factors.** These proteins have at least two functional domains: one that binds to DNA sequences present in promoters and enhancers, and another that activates transcription via protein–protein interaction by binding to RNA polymerase or to other transcription factors.

While there have been a variety of such factors that have been isolated and studied, no clear picture has yet emerged as to how these "activators" of transcription work. However, it is believed that the various binding interactions with DNA and with one another provide a functional link between the **transcriptional apparatus** (i.e., the promoter-RNA polymerase complex) and the **regulatory apparatus** (i.e., the various DNA units essential for stimulating transcription to occur).

Some insights into the transcriptional apparatus of type II genes are now available and involve a series of transcriptional factors (e.g., TFIIA, TFIIB, TFIIC, TFIID, and TFIIE). It appears that in order to initiate formation of the apparatus, TFIID must first bind to the TATA box of class II promoters. About 20 base pairs of DNA are involved. Benjamin Lewin has described this as the *"commitment"* stage. Then, other TFII factors and RNA polymerase II bind to the promoter. At this point, transcription of the DNA downstream may ensue at a minimal *"basal"* level. Stage 3 in Lewin's scheme involves the achievement of the *"induced"* state, where transcription is stimulated above the basal level. This state is not yet well defined, but involves other areas of the promoter region, enhancers, and numerous transcription factors. These conceptual stages are diagrammed in Figure 16.9.

The domains of eukaryotic transcription factors take on several forms. The **DNA binding domains** or **structural motifs** of eukaryotic transcription factors analyzed to date have been classified into three major types: **zinc fingers, helix-turn-helix motifs,** and **leucine zippers** (Figure 16.10). This classification is not exhaustive, and other new groups will undoubtedly be established as new factors are characterized.

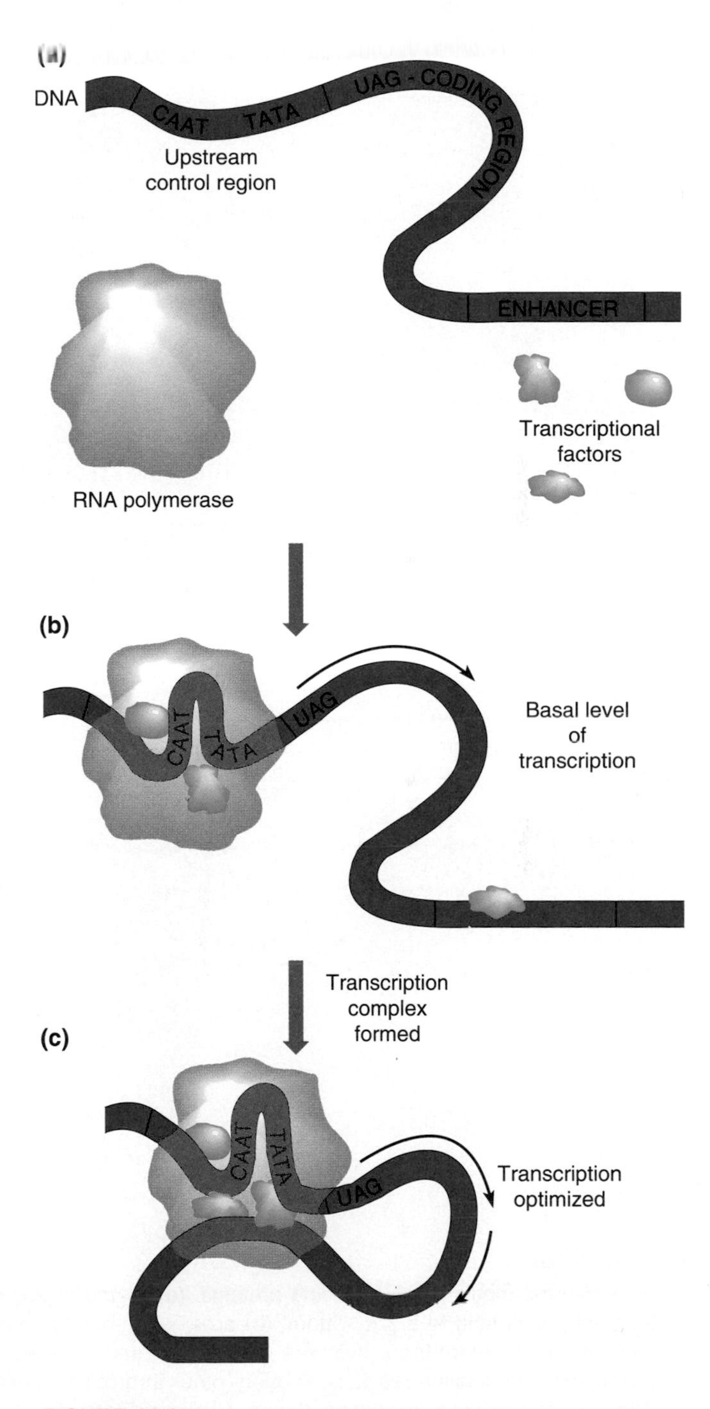

FIGURE 16.9
A model depicting the components (part [a]) and interactions (parts [b] and [c]) of transcriptional factors with different DNA control regions of a gene. A basal level of transcription is achieved initially as a result of interactions involving the CAAT and TATA boxes, producing a conformation that attracts RNA polymerase to the promoter region. Further interaction involving the enhancer region optimizes transcription.

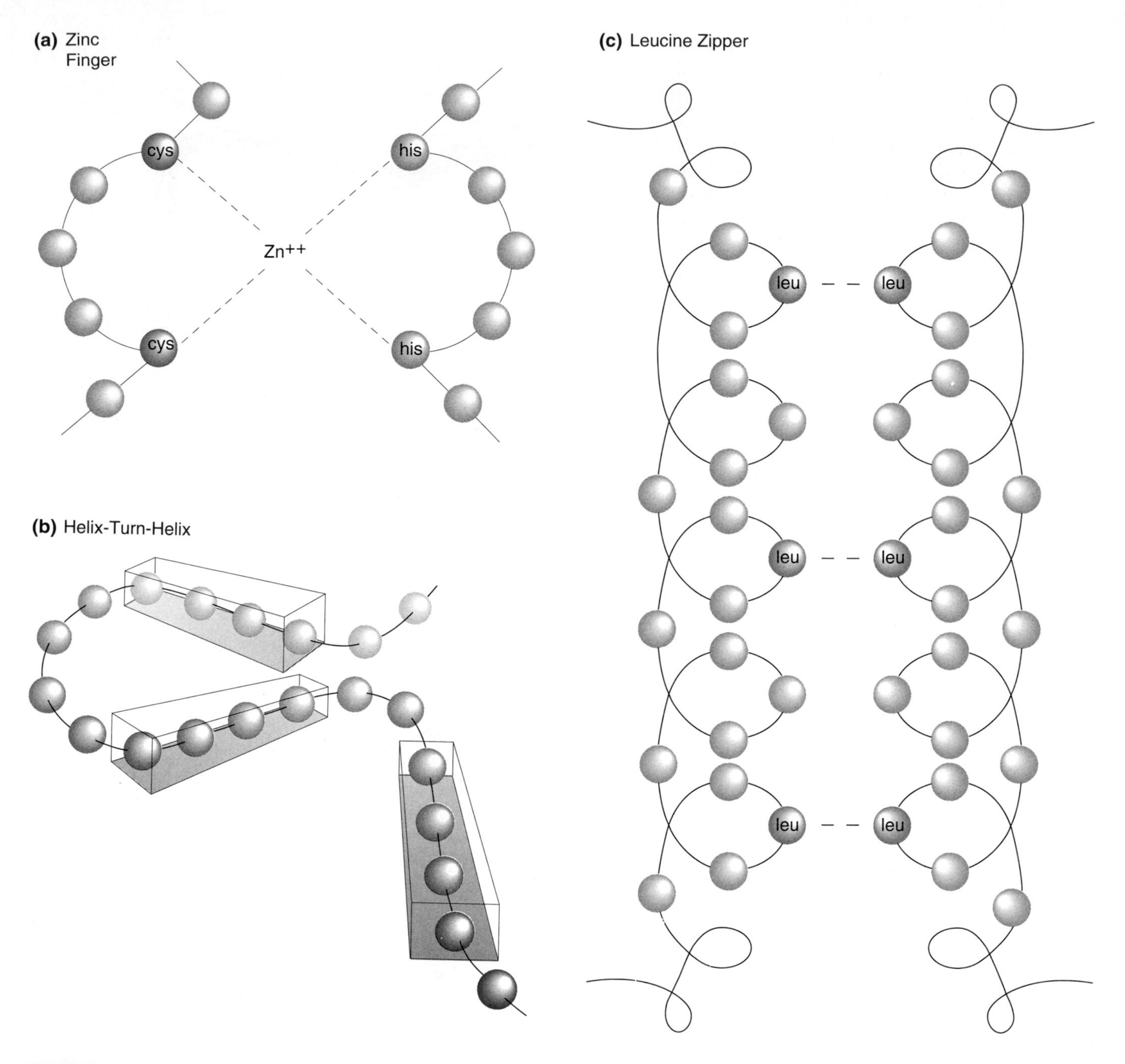

FIGURE 16.10
Three structural motifs of DNA-binding proteins. (a) A *zinc finger,* where cysteine and histidine residues bind to a Zn^{++} atom; (b) a *helix-turn-helix* or *homeodomain,* where three different planes of the α helix are established; and (c) a *leucine zipper,* where dimers result from leucine residues at every other turn of the α helix in facing stretches of one or two polypeptide chains. Each motif provides a unique way to bind to DNA.

Zinc fingers consist of amino acid sequences containing two cysteine and two histidine residues at repeating intervals. The cysteine and histidine residues are interspersed and covalently bind zinc atoms, folding the amino acids into loops known as zinc fingers (Figure 16.10). Each finger consists of approximately 23 amino acids, with a loop of 12 to 14 amino acids between the Cys and His residues, and a linker between loops consisting of 7 or 8 amino acids. The amino acids in the loop interact with and bind to specific DNA sequences. The number of fingers in a transcription factor varies form 2 to 13, as does the length of the DNA binding sequence.

The mammalian transcription factor Sp1, for example, contains three zinc-finger regions and binds selectively to GC-box modules in promoters. Similar proteins with zinc-finger motifs have been isolated from yeast, *Drosophila, Xenopus,* and humans.

The second DNA binding domain is called the helix-turn-helix motif and was first identified in the phage λ *cro* repressor molecule. This motif is characterized by its geometric conformation rather than a distinctive amino acid sequence. Two adjacent α helicies separated by a "turn" of several amino acids provide the binding potential of the protein to DNA (and after which the motif is named). In the *cro* repressor: DNA binding, one of the helicies binds within a major groove containing six bases, while the other helix stabilizes the interaction by binding to the DNA backbone.

The potential for forming helix-turn-helix geometry has been recognized in distinct regions of a large number of eukaryotic genes known to regulate developmental processes. Present almost universally in eukaryotic organisms and called the **homeobox,** a stretch of 180 bp specifies a 60-amino acid **homeodomain** sequence that can potentially form a helix-turn-helix structure. Of the 60 amino acids, many are basic (arginine and lysine), and a conserved sequence is found among these many genes.

We shall return to homeobox genes in Chapter 18 because of their significance to developmental processes. Important here is the recognition of a particular binding motif that appears to be characteristic of a large class of genes, called **homeotic genes,** the regulation of which plays a critical role during the differentiation of organisms.

The third type of domain is represented by the so-called leucine zipper. First seen as a stretch of 35 amino acids in a nuclear protein in rat liver, four leucine residues are spaced seven amino acids apart, and flanked by basic amino acids. The leucine-rich regions form a helix with leucine residues protruding at every other turn. When two such molecules dimerize (Figure 16.10), the leucine residues "zip" together, forming the DNA-binding domain. This is possible because if the alpha helix of the protein is viewed as a wheel, the leucine residues all protrude on the same side of the helix at every other turn. Leucine zippers almost exclusively involve protein dimer binding to DNA.

Overall, the picture of transcriptional regulation in eukaryotes is complex, but a number of generalizations can be drawn. First, regulation of transcription by protein factors is based largely on positive control mechanisms. The binding of one or more factors is a prerequisite to transcriptional activation of a locus. Promoter and enhancer sequences are recognized and bound by transcription factors, rather than by RNA polymerase. Because their recognition sequences are modular, different transcription factors may compete for binding to a given DNA sequence, or the same site in different cells may bind different factors in a cell-specific fashion. Because of the variability in the location and modular nature of promoters and enhancers, and the variety of transcription factors, it is likely that an initiation complex (consisting of transcription factors and RNA polymerase II) can be constructed in a number of ways, all involving protein–protein interaction. Because of this variability, however, it is difficult to provide a single transcriptional activation scheme that applies for all genes.

Gene Regulation by Steroid Hormones

Evidence that gene expression in eukaryotes is regulated by effector molecules originating outside the cell was elegantly provided by Ulrich Clever and Peter Karlson with their discovery that the steroid hormone **ecdysone** induces specific changes in the puffing pattern in polytene chromosomes. Ecdysone is a hormone that plays a crucial role in the progress of metamorphosis. In eukaryotes, steroid hormones are used to regulate growth and development and to maintain homeostasis. The major sex hormones are all steroids, as is vitamin D, and the homeostatic adrenal hormones (more than 30 different types) that regulate glucose metabolism and mineral utilization are also steroids.

Steroid hormones enter the cell by passing through the plasma membrane and binding to a specific receptor protein in the cytoplasm. The receptor-hormone complex is translocated to the nucleus and activates transcription of one or more specific genes. Hormone receptors and the DNA sequences to which they bind have been studied intensively over the last decade. The structural genes encoding virtually every known hormone receptor protein have been cloned, and enough is now known to provide a general picture of how these receptors work. All receptors analyzed to date have three functional domains: a variable N-terminus domain; a short, highly conserved central domain; and a fairly well-conserved C-terminus domain (Figure 16.11). The central domain contains two zinc fingers. The zinc fingers in hormone receptors bind to specific DNA sequences known as **hormone-responsive elements** or **HREs.**

In steroid receptors, the finger closer to the N-terminus binds in a base-specific way to the HRE and determines the specificity of binding to a particular target gene. The C-terminus finger contacts the sugar-phosphate residues adjacent to the HRE and may have other functions. In general, HREs share some characteristics with promoters and enhancers. They are composed of short consensus sequences that are related but not always identical (Table 16.2). HREs are often located several hundred bases upstream from the transcription start site and may be present in multiple copies. Often, HREs are present within promoter or enhancer sequences.

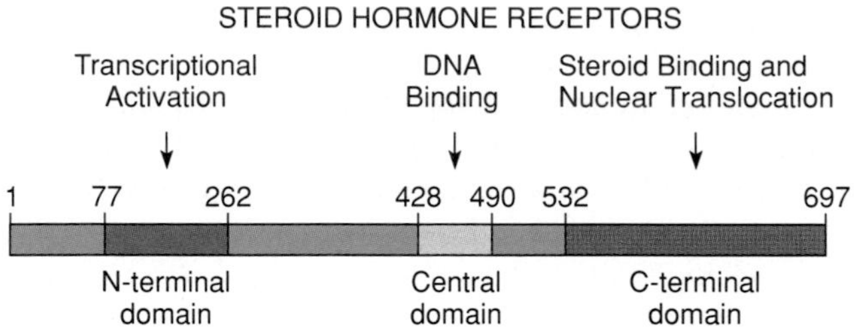

FIGURE 16.11
Steroid hormone receptors have three functional domains, as exemplified by the glucocorticoid receptor: the N-terminal domain (amino acids 77–262 in the glucocorticoid receptor), the central domain (amino acids 428–490), and the C-terminal domain (amino acids 532–697). The central domain binds DNA sequences known as the glucocorticoid response element (GRE) by means of two zinc fingers. The C-terminal domain binds to the steroid hormone and may be responsible for allowing the hormone-receptor complex to be translocated to the nucleus. This region is also responsible for regulating the activity of the receptor.

While binding of the receptor to the HRE is *necessary* for activation of a specific gene, it may not be sufficient on its own to cause activation. Binding of the receptor to the HRE may simply serve to facilitate the interaction of other transcription factors by altering the chromatin structure in the HRE and adjacent regions.

Gene activation by steroid hormones requires alteration of chromatin structure near the regulated gene. Genes in chromatin that are actively transcribed or are able to be transcribed are hypersensitive to digestion with the enzyme DNase I, while inactive or repressed genes are relatively resistant to DNase I. *In vitro* studies of steroid activation of genes in chromatin indicate that following hormone administration, DNase I sensitivity greatly increases in the regulated gene and its flanking regions, including the hormone receptor binding site. The interpretation of these results is consistent with the idea that the binding of one or more receptor-hormone complexes to the HRE causes a change or shift in the structure of chromatin at the binding site, perhaps involving nucleosomes. Such an alteration may make other binding sites, including the promoter, available to transcription factors and RNA polymerase II, resulting in the initiation of transcription.

In some ways, then, the biology of steroid hormone regulation of gene transcription in eukaryotes is similar to the *lac* operon system in prokaryotes in that an external effector binds to a cytoplasmic receptor, changes the configuration of the receptor, and moves the receptor to the DNA, where it acts as a transcription factor. In other ways, however, the regulation of gene expression is quite different in that the DNA binding sequence is often at a great distance from the regulated gene, and that alterations in chromatin structure mediate the action of the receptor.

POST-TRANSCRIPTIONAL REGULATION OF GENE EXPRESSION

As outlined above, there are many opportunities in the processes of transcription and translation where regulation of genetic expression can occur. Although transcriptional control is perhaps the most obvious and widely used mode of regulation in eukaryotes, post-transcriptional modes of regulation are also employed in many organisms. For example, nuclear RNA transcripts are modified prior to translation, noncoding **introns** are removed, and the remaining **exons** are precisely spliced together. This processing step offers several possibilities for regulation.

Processing Transcripts

In discussing gene regulation through processing of transcripts, we will first consider the case in which a single gene is transcribed, but following processing, two mRNAs, differing in their untranslated leader sequences, are produced. In mice, the enzyme alpha-amylase is pro-

TABLE 16.2
DNA receptor sequences for hormone binding proteins.

Hormone	Consensus Sequence	Element Name
Glucocorticoid	GGTACANNNTGTTCT	GRE
Estrogen	AGGTCANNNTGACCT	ERE
Thyroid	AGGTCA . . . TGACCT	TRE

duced in both salivary glands and liver. The alpha-amylase gene contains two promoters, one used in salivary glands, and the other in liver (Figure 16.12). The salivary gland promoter is about 2850 bp upstream from the liver promoter, and two different exons are adjacent to the promoters. The use of the two different promoters creates two different pre-mRNAs that are differentially spliced to yield tissue-specific mRNAs. It is important to emphasize that the protein product in both tissues is identical. The differences lie in the exons at the 5′ end of the mRNA, a region that constitutes part of the 5′ untranslated leader sequence.

Researchers are especially interested in cases where splicing variations might generate different mRNAs responsible for directing the synthesis of different polypeptides. Several instances have now been documented demonstrating that splicing may include or exclude an entire exon. This leads to distinctive mRNAs that specify different polypeptides. This is the case in several systems, including the expression of the myosin gene in *Drosophila,* the α-crystallin gene in rodents, and the immunoglobulin genes in vertebrates.

Figure 16.13 illustrates still another example of this phenomenon where the polypeptide products are clearly distinct from one another. Shown is a diagram of the initial bovine RNA transcript that is processed into one of two **preprotachykinin mRNAs (PPT mRNA).** This mRNA molecule potentially includes the genetic information specifying two neuropeptides called **P** and **K.** These two peptides are members of the family of sensory neurotransmitters referred to as **tachykinins** and are believed to play different physiological roles. While the P neuropeptide is restricted largely to tissues of the nervous system, the K neuropeptide is found more predominantly in the intestine and thyroid.

The RNA sequences for both neuropeptides are derived from the same gene. However, processing of the initial RNA transcript can occur in two different ways. In one case, exclusion of the K-exon during processing results in the α-PPT mRNA, which upon translation yields

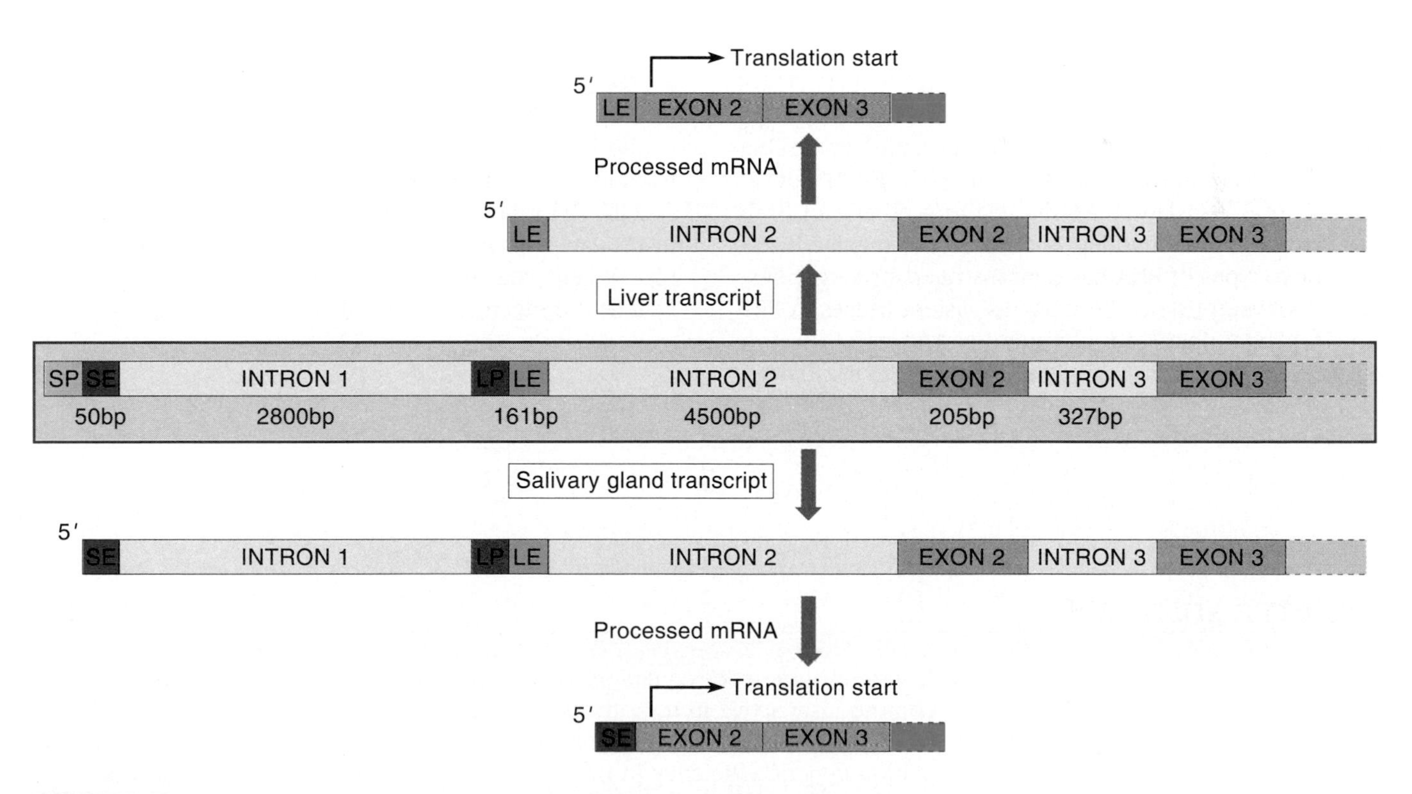

FIGURE 6.12
Organization of the mouse alpha-amylase gene. The salivary promoter (SP) and its adjacent exon (SE) are located approximately 3000 bp upstream from the liver promoter (LP) and the liver exon (LE). The use of alternate promoters results in the transcription of different pre-mRNAs in liver and salivary. glands. Processing of these transcripts results in mRNAs that start with different exons in the 5′ untranslated region.

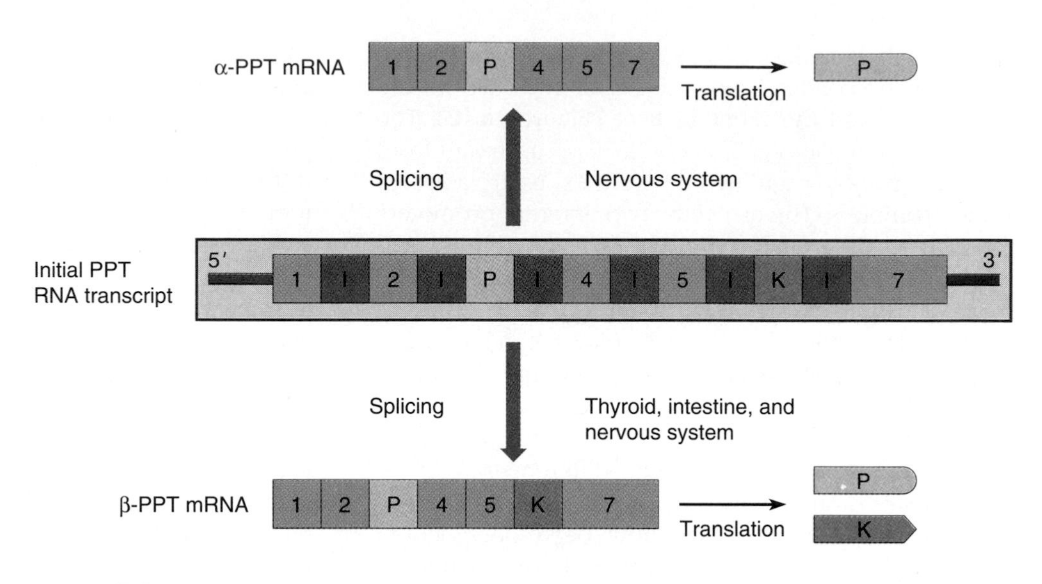

FIGURE 16.13
Diagrammatic representation of the alternate splicing of the initial RNA transcript of the preprotachykinin gene (PPT). Introns are labeled I. Exons are either numbered or designated by the letters P or K. Inclusion of the P and K exons leads to β-PPT mRNA, which, upon translation, yields both the P and K tachykinin neuropeptides. When the K exon is excluded, α-PPT mRNA is produced, and only the P neuropeptide is synthesized.

neuropeptide P but not K. On the other hand, processing that includes both the P and K exons yield β-PPT mRNA, which upon translation results in the synthesis of both the P and K neuropeptides. Analysis of the relative levels of the two types of RNA has demonstrated striking differences between tissues. In nervous system tissues, α-PPT mRNA predominates by as much as a threefold factor, while β-PPT mRNA is the predominant type in the thyroid and intestine.

Alternate splicing of transcripts from a single gene, leading to two distinct products, is governed by tissue-specific alteration of the initial RNA transcript. Further investigations will very likely uncover still other examples of gene regulation at the post-transcriptional level as well as the mechanism by which such regulation is controlled.

CHAPTER SUMMARY

1 A system of genetic regulation must exist if the complete genome is not to be continuously active in transcription throughout the life of every cell of all species. Highly refined mechanisms have evolved that regulate transcription, maintaining genetic efficiency in prokaryotes and the specialization associated with differentiation in multicellular eukaryotes.

2 Both inducible and repressible operons, illustrated by the *lac* and *trp* gene complexes, respectively, have been documented and studied in bacteria. They involve genes of a regulatory nature in addition to structural genes that code for the enzymes of the system. Both operons are under the negative control of a repressor molecule.

3 An additional regulatory step, referred to as attenuation, has been studied in the *trp* operon, whereby transcription begins, but is interrupted under appropriate cellular conditions.

4 The *ara* regulator protein exerts both positive and negative control over expression of the genes specifying the enzymes that metabolize the sugar arabinose.

5 Repression of transcription in phage lambda has been shown to be due to one of its own gene products, the repressor *cI*. Another gene product, *cro,* acts to inhibit the transcription of *cI* and promotes the lytic pathway when the activity of *cII* protein is reduced.

6 Mechanisms controlling gene regulation in eukaryotes are constrained by the properties of multicellularity, expanded genome size, and the spatial and temporal separation between transcription and translation.

7 Transcription in eukaryotes is controlled by regulatory DNA sequences known as promoters and enhancers. Regulatory sequences, including the TATA box and the CCAAT box, are elements of promoters found near the transcriptional starting point. Enhancer elements, which appear to control the degree of transcription, can be located before, after, or within the gene expressed.

8 Transcription factors are proteins that bind to DNA recognition sequences within the promoters and enhancers and activate transcription through protein–protein interaction. Such factors display various types of motifs that are related to their potential binding to DNA.

9 Several types of post-transcriptional control of gene expression are possible in eukaryotes. One such mechanism is alternate processing of a single class of pre-mRNAs to generate different mRNA species.

KEY TERMS

attenuation
B-galactosidase
CAAT box
constitutive mutations
cro
DNA binding domains
exons
enhancers
GC box
helix-turn-helix motif
homeobox
homeodomain
homeotic genes
hormone-responsive elements (HRE)
inducer
inducible enzymes
introns
lac operon model
leucine zipper
lysogenic phage
lytic cycle
negative control
operator region
operon model
polycistronic message
positive control
post-transcriptional regulation
promoter
repressible enzymes
repressor gene
repressor molecule
repressible system
steroid hormones
structural gene
TATA Box
transcription factors
transcriptional control
tryptophan synthetase
zinc finger
λ repressor protein

1 A theoretical operon (*theo*) in *E. coli* contains several structural genes encoding enzymes that are involved sequentially in the biosynthesis of an amino acid. Unlike the *lac* operon, where the repressor gene is separate from the operon, the gene encoding the regulator molecule is contained within the *theo* operon. When the end product (the amino acid) is present, it combines with the regulator molecule, and this complex binds to the operator, repressing the operon. In the absence of the amino acid, the regulatory molecule fails to bind to the operator, and transcription proceeds.

Characterize this operon; then consider the following mutations as well as the situation in which the wild-type gene is present along with the mutant gene in partially diploid cells (F′). In each case, will the operon be active or inactive in transcription, assuming that the mutation affects the regulation of the *theo* operon? Compare each response to the equivalent situation in the *lac* operon.

(a) Mutation in the operator gene
(b) Mutation in the promoter region
(c) Mutation in the regulator gene

SOLUTION: The operon is under negative control and is repressible. The regulatory molecule, when bound to the amino acid, binds to the operator region and inhibits gene expression.

(a) As in the *lac* operon, a mutation in the *theo* operator gene inhibits binding with the repressor complex, and transcription occurs constitutively. The presence of an F′ plasmid bearing the wild-type allele would have no effect.

(b) A mutation in the *theo* promoter region would no doubt inhibit binding to RNA polymerase and therefore inhibit transcription. This would also happen in the *lac* operon. A wild-type allele present in an F′ plasmid would have no effect.

(c) A mutation in the *theo* regulator gene, as in the *lac* system, may either inhibit its binding to the repressor or its binding to the operator gene. In both cases, transcription will be constitutive because the *theo* system is repressible. Both cases result in the failure of the regulator to bind to the operator, allowing transcription to proceed. In the *lac* system, failure to bind the corepressor lactose would permanently repress the system. The addition of a wild-type allele would restore repressibility, provided that this gene was transcribed constitutively.

PROBLEMS AND DISCUSSION QUESTIONS

1 Contrast the need for the enzymes involved in the metabolism of lactose and tryptophan in bacteria in the presence and absence of lactose and tryptophan, respectively.

2 Contrast positive and negative control systems.

3 Contrast the role of the repressor in an inducible system and in a repressible system.

4 For the following *lac* genotypes, predict whether the structural genes (z) are constitutive, permanently repressed, or inducible in the presence of lactose.

Genotype	Constitutive	Repressed	Inducible
$i^{+}o^{+}z^{+}$			×
$i^{-}o^{+}z^{+}$			
$i^{+}o^{c}z^{+}$			
$i^{-}o^{+}z^{+}/F'i^{+}$			
$i^{+}o^{c}z^{+}/F'o^{+}$			
$i^{s}o^{+}z^{+}$			
$i^{s}o^{+}z^{+}/F'i^{+}$			

5 For the following genotypes and condition (lactose present or absent), predict whether functional enzymes are made, nonfunctional enzymes are made, or no enzymes are made.

Genotype	Condition	Functional Enzyme Made	Nonfunctional Enzyme Made	No Enzyme Made
$i^{+}o^{+}z^{+}$	No lactose	—	—	×
$i^{+}o^{c}z^{-}$	Lactose			
$i^{-}o^{+}z^{-}$	No lactose			
$i^{-}o^{+}z^{-}$	Lactose			
$i^{-}o^{+}z^{+}/F'i^{+}$	No lactose			
$i^{+}o^{c}z^{+}/F'o^{+}$	Lactose			
$i^{+}o^{+}z^{-}/F'i^{+}o^{+}z^{+}$	Lactose			
$i^{-}o^{+}z^{-}/F'i^{+}o^{+}z^{+}$	No lactose			
$i^{s}o^{+}z^{+}/F'o^{+}$	No lactose			
$i^{+}o^{c}z^{-}/F'o^{+}z^{+}$	Lactose			

6 In a theoretical operon, genes *a, b, c,* and *d* represent the repressor gene, the promoter sequence, the operator gene, and the structural gene, but not necessarily in that order. This operon is concerned with the metabolism of a theoretic molecule (tm). From the data given below, first decide if the operon is inducible or repressible. Then assign *a, b, c,* and *d* to the four parts of the operon. (AE = active enzyme; IE = inactive enzyme; NE = no enzyme)

Genotype	tm Present	tm Absent
$a^+b^+c^+d^+$	AE	NE
$a^-b^+c^+d^+$	AE	AE
$a^+b^-c^+d^+$	NE	NE
$a^+b^+c^-d^+$	IE	NE
$a^+b^+c^+d^-$	AE	AE
$a^-b^+c^+d^+/F'a^+b^+c^+d^+$	AE	AE
$a^+b^-c^+d^+/F'a^+b^+c^+d^+$	AE	NE
$a^+b^+c^-d^+/F'a^+b^+c^+d^+$	AE + IE	NE
$a^+b^+c^+d^-/F'a^+b^+c^+d^+$	AE	NE

7 If the *cI* gene of phage λ contained a mutation with effects similar to the i^- mutation in *E. coli,* what would be the result?

8 Why is gene regulation assumed to be more complex in a multicellular eukaryote than in a prokaryote? Why is the study of this phenomenon in eukaryotes more difficult?

9 List and define the levels of gene regulation discussed in this chapter.

10 Distinguish between the regulatory elements referred to as promoters and enhancers.

11 A bacterial operon is responsible for the production of the biosynthetic enzymes needed to make a theoretical amino acid tisophane (tis). The operon is regulated by a separate gene, *r*. Deletion of the *r* gene causes the loss of synthesis of the enzymes. In the wild-type condition, when tis is present, no enzymes are made. In the absence of tis, the enzymes are made. Mutations in the operator gene (O^-) result in repression regardless of the present of tis.
Is the operon under positive or negative control? Propose a model for (1) repression of the genes in the presence of tis in wild-type cells; and (2) the O^- mutations.

SELECTED READINGS

BEATO, M. 1989. Gene regulation by steroid hormones. *Cell* 56:335–44.

BUSCH, S. J., and SASSONE-CORSI, P. 1990. Dimers, leucine zippers and DNA-binding domains. *Trends in Genet.* 6:36–40.

DYNAN, W. S. 1988. Modularity in promoters and enhancers. *Cell* 58:1–4.

GILBERT, W., and MULLER-HILL, B. 1966. Isolation of the *lac* repressor. *Proc. Rev. Acad. Sci.* 56:1891–98.

———. 1967. The *lac* operator in DNA. *Proc. Natl. Acad. Sci.* 58:2415–21.

GILBERT, W., and PTASHNE, M. 1970. Genetic repressors, *Scient. Amer.* (June) 222:36–44.

HERSHEY, A. D., ed. 1971. *The bacteriophage lambda.* Cold Spring Harbor, NY: Cold Spring Harbor Laboratory.

JACOB, F., and MONOD, J. 1961. Genetic regulatory mechanisms in the synthesis of proteins. *J. Mol. Biol.* 3:318–56.

JOHNSON, A. D., et al. 1981. λ repressor and *cro*-components of an efficient molecular switch. *Nature* 294:217–23.

KAKIDANI, H., and PTASHNE, M. 1988. GAL4 activates gene expression in mammalian cells. *Cell* 52:161–67.

LEWIN, B. 1974. Interaction of regulator proteins with recognition sequences of DNA. *Cell* 2:1–7.

MANIATIS, T., GOODBOURN, S., and FISCHER, J. A. 1987. Regulation of inducible and tissue-specific expression. *Science* 236:1237–45.

MANIATIS, T., and PTASHNE, M. 1976. A DNA operator-repressor system. *Scient. Amer.* (Jan.) 234:64–76.

MILLER, J. H., and REZNIKOFF, W. S. 1978. *The operon.* Cold Spring Harbor, NY: Cold Spring Harbor Laboratory.

MITCHELL, P. J., and TJIAN, R. 1989. Transcriptional regulation in mammalian cells by sequence-specific DNA binding proteins. *Science* 245:371–78.

PTASHNE, M., and GANN, A. A. F. 1990. Activators and targets. *Nature* 346:329–31.

PTASHNE, M., JOHNSON, A. D., and PABO, C. O. 1982. A genetic switch in a bacterial virus. *Scient. Amer.* (Nov.) 247:128–40.

STRINGER, K. F., INGLES, C. J., and GREENBLATT, J. 1990. Direct and selective binding of an acidic transcriptional activation domain to the TATA-box factor TFIID. *Nature* 345:783–86.

YANOFSKY, C. 1981. Attenuation in the control of expression of bacterial operons. *Nature* 289:751–58.

17 The Genetic Basis of Immunity

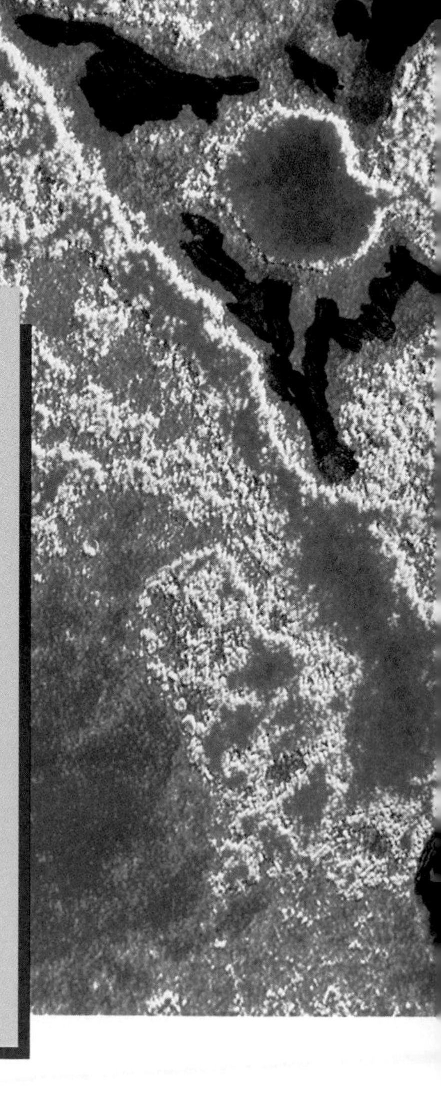

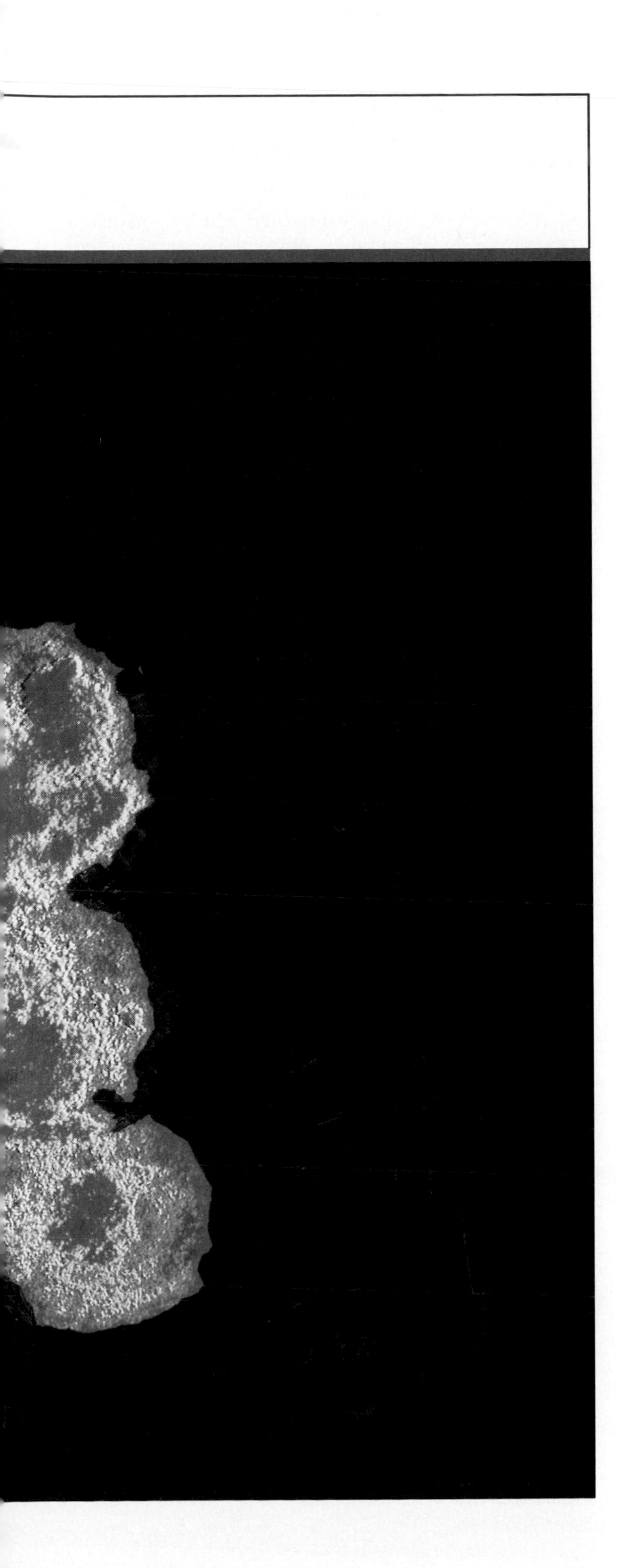

Chapter Concepts

The immune system of vertebrates is a genetically controlled mechanism that defends the body against disease-causing organisms. Extensive genetic recombination in somatic cells reshuffles a limited number of genes into new combinations that produce literally millions of different antibodies and cell surface receptors. Cell surface antigens play a role in determining blood types and in the success of transfusions and organ transplants. The proliferation, differentiation, and function of cells in the immune system is under genetic control, and mutations in these genes result in disorders of the immune system.

As advanced organisms such as vertebrates evolved, a unique and complex genetic mechanism developed that is critical to the survival of such organisms. Called the *immune system,* this mechanism provides a series of defensive responses to the entry of foreign substances or the invasion of viruses and other microorganisms into the body. Whether such a response is cell mediated or antibody mediated, it is highly specific and involves two phases: a *primary response* to the initial exposure, and a *secondary response* to subsequent exposures to the same agent.

From a genetic standpoint, the immune system has two fundamental characteristics. First, in every individual, it must recognize "self" so that an organism's cells and tissues are not destroyed by its immune system. This recognition involves a set of genes, which in humans, are located on chromosome 6. Second, the immune system must be able to produce specific molecules that will neutralize and subsequently destroy "nonself" agents. This response involves the production of *antibodies* against foreign substance or agents called *antigens.* By definition, an antigen elicits the response leading to the production of specific antibodies.

Antibody production against any and every potential antigen is one of the most remarkable genetic accomplishments of advanced organisms. From a collection of less than 300 genes, recombination events within developing antibody-producing cells create a vast immune potential. As a result of shuffling this limited number of genes, tens of millions of different antibodies can be produced.

In this chapter, we will examine the cells that make up the immune system and how these cells and their gene products are mobilized to mount an immune response. We will examine the role of the immune system in determining blood groups and mother–fetus incompatibility, and how cell-surface markers are used in transplants and how these markers can be used in a predictive way to determine risk factors for a wide range of diseases. Finally, we will consider a number of genetic disorders of the immune system, including mutations in single genes that inactivate the entire immune system. We will also describe how acquired immunodeficiency syndrome (AIDS) associated with infection by human immunodeficiency virus (HIV) acts to cripple the immune response of infected individuals. Taken together, work on the immune system constitutes one of the most exciting areas of genetics; one where new information and new breakthroughs occur with increasing frequency.

COMPONENTS OF THE IMMUNE SYSTEM

In a manner of speaking, the immune system provides an effective barrier against the successful invasion of potentially harmful foreign substances. In reality, the system represents a state of resistance whereby invading foreign substances and microorganisms are recognized as nonself and subsequently sequestered, inactivated, and destroyed. This **immune response** involves **antibodies** specific to foreign substances. Agents that elicit antibody production are termed **antigens.** All antibodies are proteins and are produced by specific cells of the immune system. Antigens are often proteins, although they are frequently combined with polysaccharide molecules. These invading antigens may be separate entities, or they may be part of the surface of a cell or microorganism. Whatever the case, organisms with an immune system have the capacity to make antibodies against any antigen they encounter.

There are two main branches of the immune system: 1) the antibody-mediated system, involving the production of antibodies by a form of white blood cell known as a *B cell,* and 2) a system in which another type of white blood cell, the *T cell,* directly attacks and inactivates antigens in what is called *cell-mediated immunity.* Other aspects of the immune response are also controlled by white blood cells, and a complete picture of the immune system requires knowledge of the origin, distribution, and function of these cell types.

Cells of the Immune System

One of the key cells in the immune system is the **phagocyte,** a cell that engulfs and destroys foreign molecules and microorganisms (Figure 17.1). Phagocytes play a critical role in both antibody-mediated and cell-mediated immunity by signalling other cells of the immune system that foreign antigens are present. In this signalling role, phagocytes are known as *antigen-presenting cells.* All phagocytes arise from stem cells, which are mitotically active cells found in bone marrow (Figure 17.2). Mature phagocytes have the ability to leave the circulatory system by squeezing through capillary walls to enter inflamed or damaged tissues to destroy antigens.

A second cell type, the T cell, is also produced by stem cells in bone marrow (Figure 17.2), and while still immature, T cells migrate to the thymus where they develop into several different subtypes, including *helper cells, suppressor cells,* and *killer cells.* During this period of maturation in the thymus, T cells become programmed to produce a unique type of cell receptor. Each T cell produces one and only one type of receptor that, in turn, will bind to only one type of antigen. There are literally tens or hundreds of millions of potential antigens, and there are a similar number of differently programmed T cells. Mature T cells can divide, and all the offspring of a single T

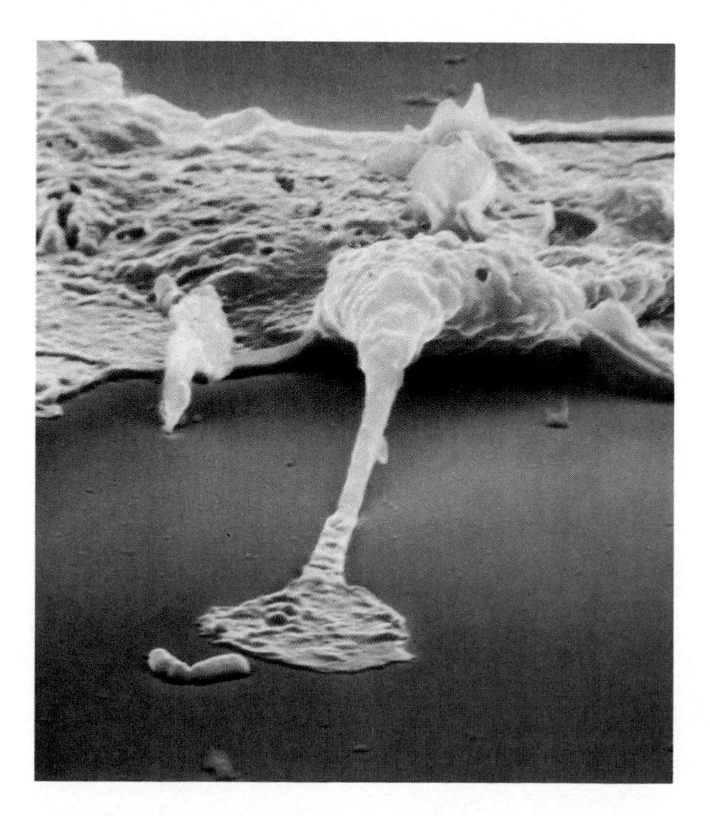

FIGURE 17.1
A phagocyte with a cell process extended to engulf a rod-shaped bacterium in the foreground.

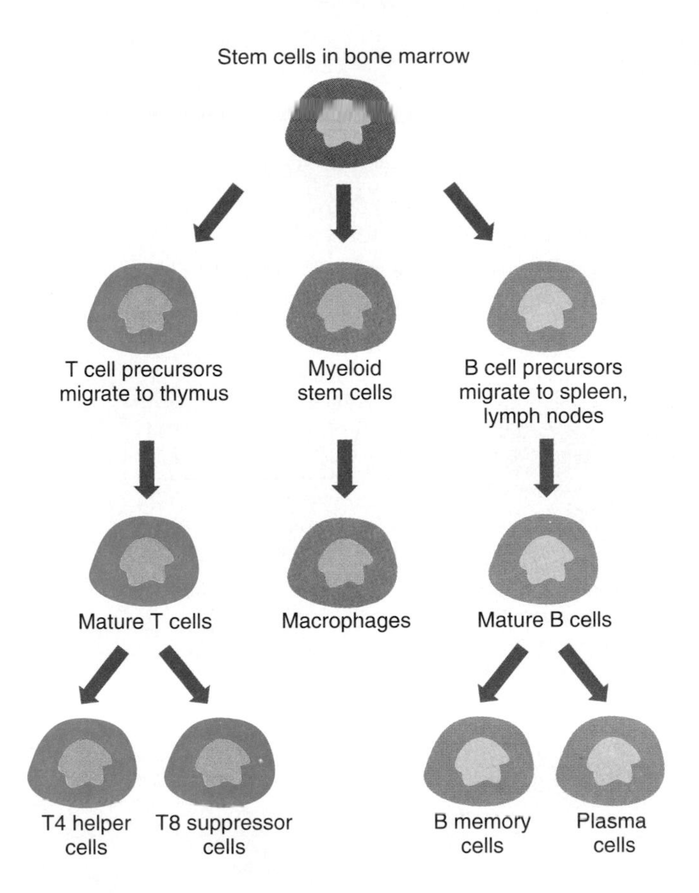

FIGURE 17.2
Stem cells in bone marrow give rise to the precursors of T cells, macrophages and B cells of the immune system.

TABLE 17.1
Cells of the immune system.

Cell Type	Function
T4 Cells	Also called helper T cells. Has role in both cell-mediated and antibody-mediated immune responses. Master switch of the immune system.
T8 Cells	Suppressor cells that participate in cell- and antibody-mediated immune response. Serves to slow down and/or turn off the immune response.
B Cells	Precursors to plasma cells, recognize antigens, and are activated by T4 cells.
Plasma cells	Produce large quantities of a single antibody. Are produced mitotically from an activated B cell.
Memory cells	Derived from either T cells or B cells by mitosis, these cells remain in circulation. Allow a rapid response to a second encounter with a specific antigen.
Killer cells	A specialized T cell that can detect and destroy vrius-infected cells.

cell form a family group or clone. From the thymus, mature T cells become distributed throughout the circulatory and lymph systems, and also become sequestered in the lymph nodes and spleen. There are two general classes of T cells: T4 helper/inducer cells and T8 cytotoxic/suppressor cells. The T4 helper cells are the master switch for the immune system; they turn on the immune response. The T8 suppressor cells are the "off" switch of the immune system; they stop or slow down the immune response. Other types of T cells are listed in Table 17.1.

A third component of the immune system, the B cells, originate in bone marrow (Figure 17.2) and migrate to the spleen and lymph nodes. These cells are genetically programmed to produce antibodies; each B cell produces only one type of antibody. Antibody production is triggered in B cells by the action of T4 helper cells.

THE IMMUNE RESPONSE

The immune response involves three stages: 1) the detection of foreign antigens, 2) activation of cells of the immune system, and 3) inactivation of the antigen. Each of these stages in the immune response is directed by a specific cell type. After exposure to an antigen and the generation of an immune response, subsequent challenges to the immune system by the same antigen are directed by a *secondary immune response,* also called immunological memory. We will first examine antibody- and cell-mediated immune responses and then consider how immunological memory protects the body against re-exposure to an antigen.

Antibody-Mediated Immunity

A type of phagocyte known as a **macrophage,** wanders through the circulatory system and inflamed tissues and is capable of responding to foreign molecules, viruses, or microorganisms. When an antigen is encountered, it is engulfed and ingested by the macrophage (Figure 17.3). After ingestion, fragments of the antigen consisting of approximately 20 amino acids are incorporated into the plasma membrane of the macrophage and displayed on its outer surface, along with the cell surface markers that identify the cell as a macrophage. As the cell moves through the circulatory system or intercellular spaces, it may encounter a helper T4 cell. Surface receptors on the T4 cell react with the antigen, causing activation of the helper T cell. In turn, the activated T cell identifies and activates B cells that manufacture an antibody against the

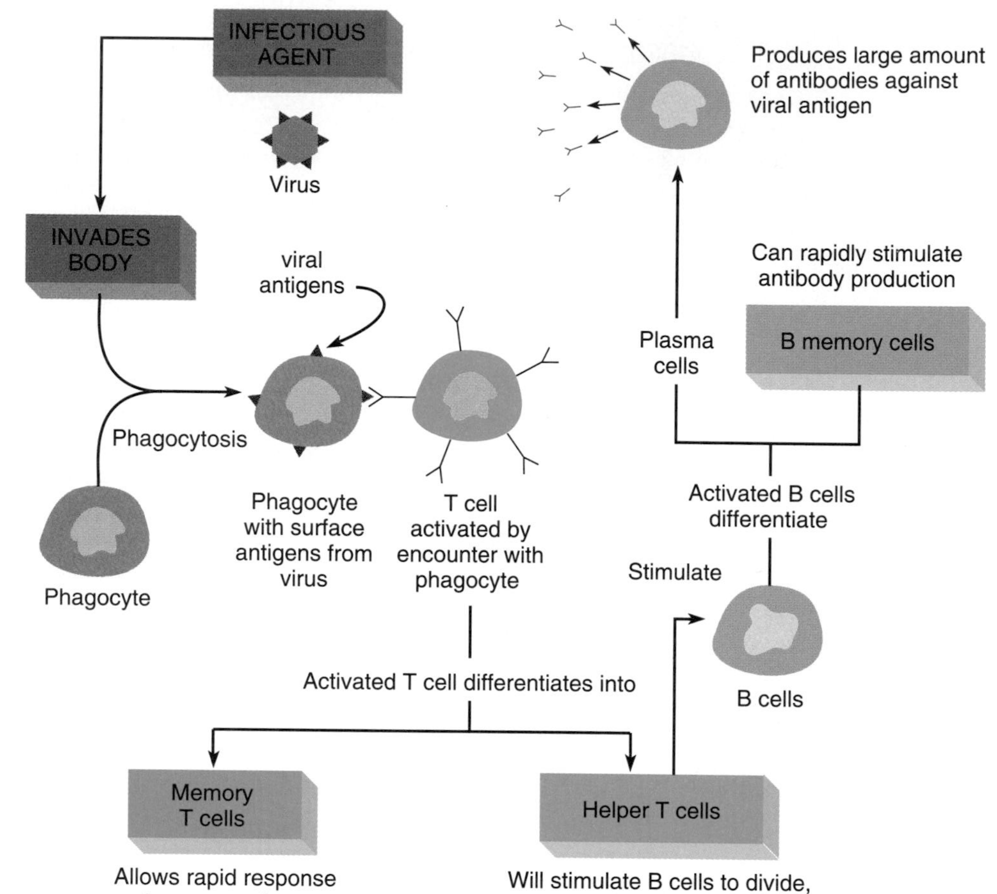

FIGURE 17.3
The pathway of responses in the humoral immune response. When an infectious agent such as a virus encounters a macrophage, it is ingested and destroyed. Some of the viral antigens are displayed on the surface of the macrophage and activate T4 cells. Activated T cells can differentiate into T memory cells or helper T cells. The helper T cells stimulate B cells in the lymph nodes that can produce an antibody against the viral antigen to become mitotically active. The progeny of the activated B cells can differentiate into B memory cells and plasma cells, which produce and secrete large amounts of a single antibody. The secreted antibody binds to the viral antigen, marking the viral particle for destruction by phagocytes.

antigen presented to the T cell. The stimulated B cells begin to divide, forming two types of daughter cells. One is the **memory cell,** and the other is the **plasma cell,** which synthesizes and releases 2000–20,000 antibody molecules per *second* into the bloodstream.

Antibodies (Figure 17.4) are effective against extracellular antigens such as bacterial cells, free virus particles, and protozoan parasites in the bloodstream, or virus-infected cells that display viral antigens on their surface. Antibodies work in several ways to inactivate antigens. They can interact with the antigen to form antigen–antibody complexes. Clumping of the antigen–antibody complex tags the antigens for destruction, and they are ingested and destroyed by phagocytes. Antibodies can also interact with antigens to destroy their ability to function. For example, some viruses have proteins on their surface that help attach the virus to the cell surface as a

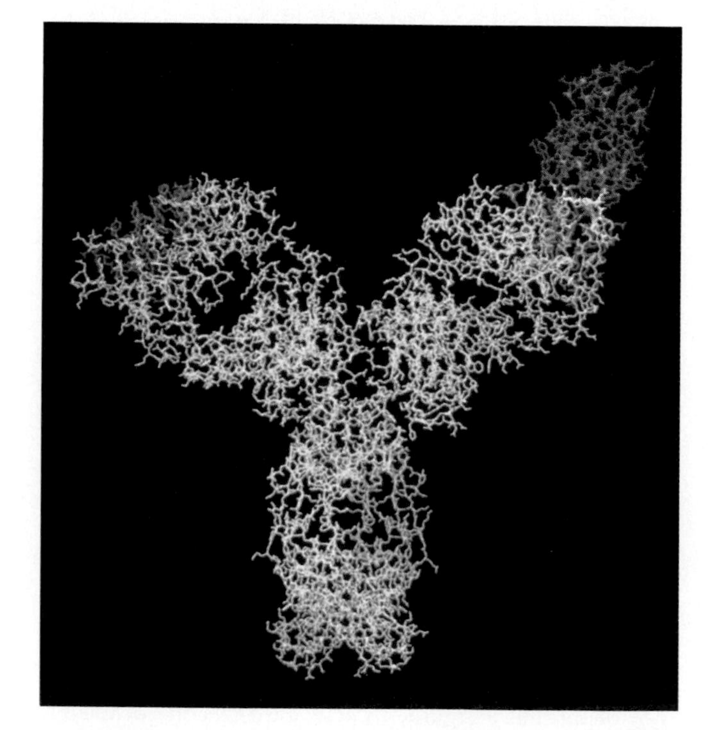

FIGURE 17.4
Computer graphic representation of the structure of an antibody. The backbone of the antibody molecule is represented in green, and the antigen-combining site is shown in blue.

prelude to viral infection. Antibodies can react with this attachment protein and inactivate it, leaving the virus exposed to ingestion and destruction by phagocytes.

The second type of cell produced by activated B cells is the memory cell. These cells also produce antibodies, but instead of a life span of several days, as is the case for plasma cells, memory cells have an extended life span of several months to years. These cells play an important role in the secondary immune response.

The entire immune response is monitored by T8 suppressor cells, and when detectable antigens have been inactivated or destroyed, the T8 cells stop proliferation of B cells and turn off antibody production. Thus, T8 cells act as the "off" switch for the immune system.

Cell-Mediated Immunity

In contrast to the action of antibodies, which tag antigens for destruction by phagocytes, direct cell–cell interactions called **cell-mediated immunity,** carried out by a class of T cells, can also result in antigen inactivation. The T cells that participate in this form of the immune reaction are known as *killer* T cells or *cytotoxic* T cells (Figure 17.5). Killer T cells originate in bone marrow and mature in the thymus gland. While the antibody-mediated immune response is effective against antigenic molecules and bacteria, cytotoxic T cells and cell-mediated immunity protect the body against viral infections. In cases where a cell is infected by a virus, some viral protein fragments become incorporated into the cell's plasma membrane and are displayed as a new set of antigens on the surface of the infected cell. This array of new antigens allows killer T cells to identify and kill virus-infected cells. Killer cells attach to their target cells (Figure 17.6) and secrete a protein called *perforin,* which inserts into plasma membrane of the virus-infected cell. Perforin molecules link to form cylinders that function as pores in the plasma membrane. The cytoplasmic contents of the target cell leak out through these pores, and the virus-infected cell dies. Once detached, the killer cell is capable of detecting and killing other infected cells. In addition to killing virus-infected cells, cytotoxic T cells and a second type of killer cell known as a natural killer cell attack and kill cancer cells, fungi, and some types of parasites. Cytotoxic T cells will also attack and kill cells introduced into the body during tissue or organ transplants if the transplanted cells are recognized as foreign. The role of the immune system in transplants will be discussed in a subsequent section.

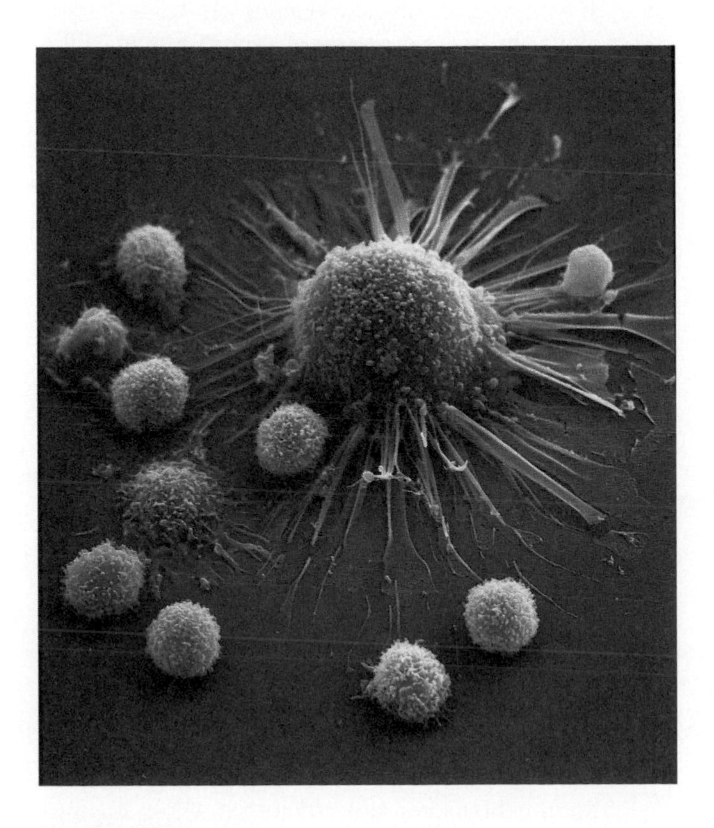

FIGURE 17.5
A large cancer cell, with cytoplasmic processes extended, has been detected and surrounded by an array of killer cells.

Immunological Memory and Immunization

Even in ancient times it was noted that exposure to certain diseases confers immunity to subsequent exposures to the same disease. For example, those who have had measles (an infectious disease caused by a virus) will not

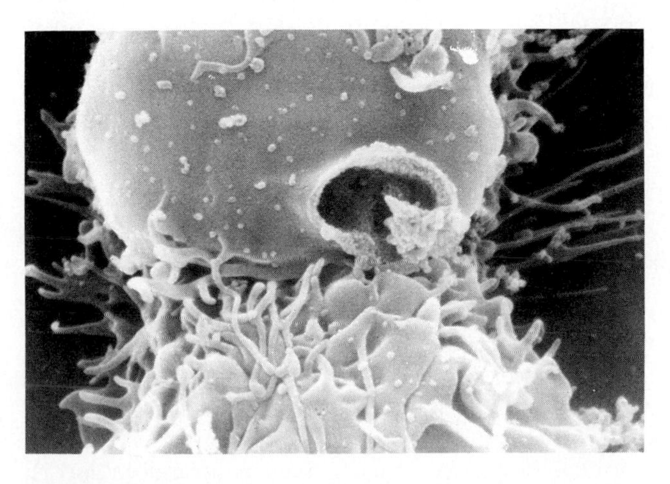

FIGURE 17.6
A killer cell (at bottom), has attacked a cancer cell and punctured its membrane. The contents of the tumor cell will leak out through this hole, and the cell will die. The remaining debris will be scavenged by phagocytes.

get measles again, even upon repeated exposure to infected individuals. This resistance to second infections is controlled by the *secondary immune response* and is the result of the production of B and T memory cells during the first exposure to the antigen. A second exposure to a molecular or microorganismic antigen results in an immediate, large-scale production of antibodies and killer T cells. Because of the presence of the memory cells, this reaction is more rapid, on a larger scale, and lasts longer than the primary immune response.

The existence of the secondary immune response is the basis for vaccination against a number of infectious diseases. A vaccine is a preparation designed to stimulate the production of memory cells against a pathogenic agent. In this procedure, the antigen is administered orally or by injection and provokes a primary immune response, including the production of memory cells. Often, a second or booster dose is administered to elicit a secondary response that raises or boosts the number of memory cells. Vaccines can be prepared from killed pathogens or weakened strains that are able to stimulate the immune system, but unable to produce symptoms of the disease. Modern recombinant DNA techniques have also been used in devising vaccines. For example, purified fragments of the protein coat of the hepatitis B virus have been prepared in this way for use as a vaccine.

ANTIBODY DIVERSITY

Mammals can produce millions of different antibodies, each responding to a different antigen. It is now clear that the basis for this molecular diversity lies in the various amino acid sequences composing the chains of the antibodies. In humans, antibodies or immunoglobulins (Ig) have been divided into five classes: **IgG, IgA, IgM, IgD,** and **IgE.** The first class, IgG, represents about 80 percent of the antibodies found in blood and is the most extensively characterized.

Each antibody molecule consists of two different types of polypeptide chains, each present in duplicate. The larger or **heavy chain (H)** consists of approximately 440 amino acids, and the smaller or **light chain (L)** contains approximately 220 amino acids. The first 110 amino acids at the N-terminus of each chain vary in sequence in different immunoglobulin molecules and are thus known as the **variable region (V).** The remaining C-terminal amino acid sequence of each chain is invariable and is called the **constant region (C).** The variable regions are responsible for antibody specificity and contain the **antibody combining sites** that bind to the antigens. This general scheme, illustrated in Figure 17.7, was first outlined by Rodney Porter and Gerald Edelman in the early 1960s.

If a foreign organism invades the body, a small part of a surface protein or glycoprotein will serve as an antigen. Once specific antibodies are produced, they combine with the antigenic determinants in a lock-and-key configuration. Since each antibody has two combining sites (Figure 17.7), large complexes are formed that can be engulfed by white blood cells and effectively removed from the body.

Within the variable region are areas that vary more than others. There are three such **hypervariable regions** in both light chains and heavy chains (Figure 17.7). They are most important to the three-dimensional configuration of the combining site.

The question of major genetic interest is how the vast array of molecular variability demonstrated by antibodies is encoded in the genome. Three hypotheses have been proposed and supported to one degree or another over the past several decades: the **germline theory,** the **somatic mutation theory,** and the **recombination theory.** The latter has received the greatest experimental support. Basically, this theory proposes that there are a limited number of genes that encode each of the various portions of the different types of H and L chains. As antibody-forming B lymphocytes develop, a mechanism involving DNA recombination rearranges these genes so that each lymphocyte comes to encode only one specific type of antibody. In the presence of a specific antigen, the corresponding lymphocyte is stimulated to differentiate and proliferate further. As a result, antibodies containing the appropriate combining sites are produced and interact with the antigen.

To comprehend how the diversity of antibodies is generated, we must delve into the structure of immunoglobulins and the genes encoding the regions of their polypeptide components. Recall that there are light and heavy polypeptide chains, each containing variable and constant regions (V_L, C_L, V_H, and C_H). Also, there are five classes of immunoglobulins. As shown in Table 17.2, each class contains one of two types of light chains, λ or κ. Furthermore, each class is characterized by its own specific heavy chain, designated α, γ, δ, ϵ, or μ. Thus, there are ten different general classes of immunoglobulins that can be formed.

Although there is some variation found in the C region of different chains, the major variation is found in the V region (Figure 17.8). The V_L region appears to be encoded by two sets of genes, designated V and J. **V genes** specify the N-terminal portion, including the first two hypervariable regions and part of the third. The **J genes** specify the remainder of the V region, including the latter part of the third hypervariable sequence. When an antibody-producing cell is formed, different combinations of V and J genes are fused together through recombination. If numerous V and J genes combine at random, an extremely large number of V_L sequences can be encoded.

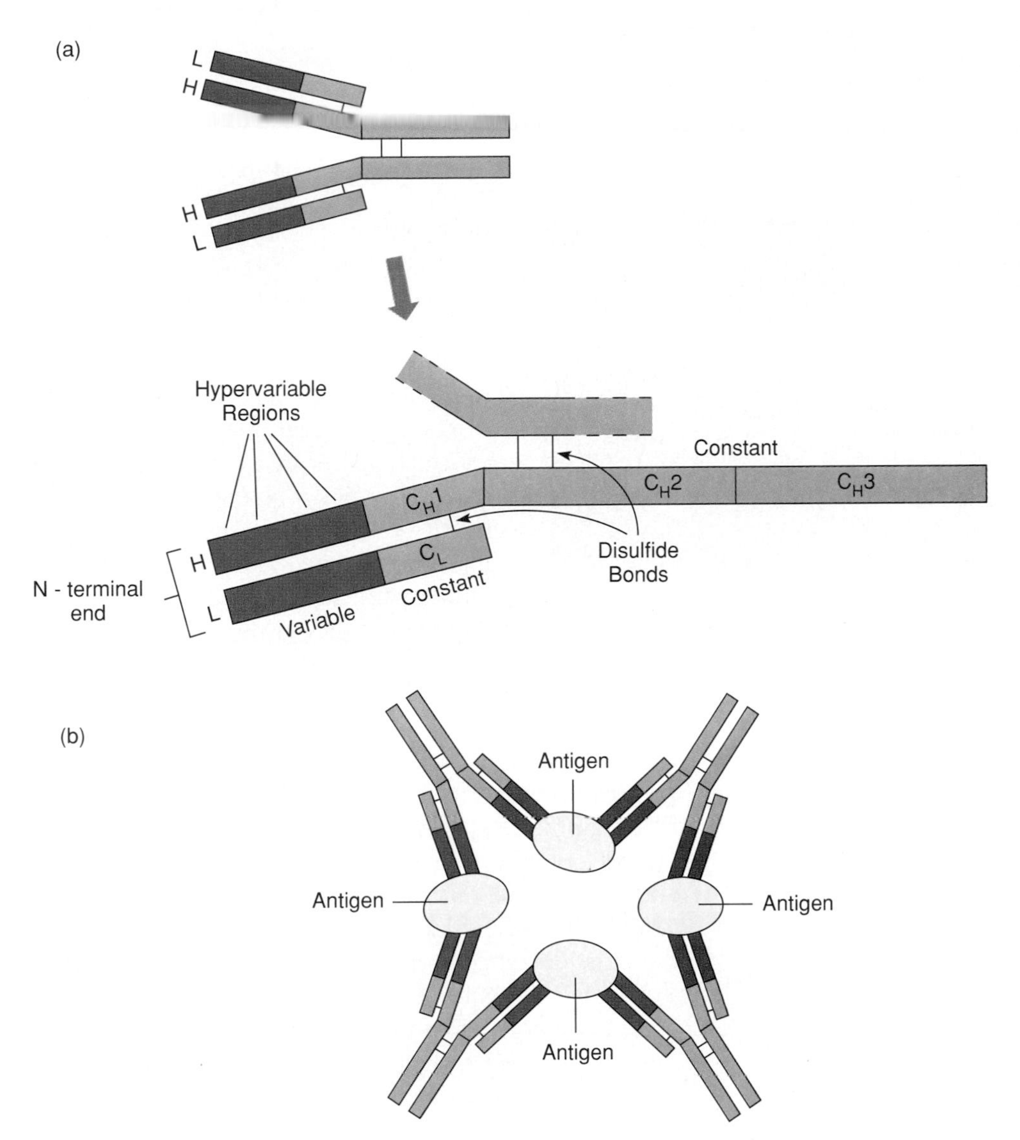

FIGURE 17.7
(a) General scheme of the IgG immunoglobulin showing the constant, variable, and hypervariable regions of the heavy (H) and light (L) chains. (b) Antibody-binding sites of the variable regions and complexes that may be formed.

Does this really happen? In 1976, Susumo Tonegawa applied the techniques of recombinant DNA cloning and nucleotide sequencing to demonstrate that it does. He compared a cloned fragment specifying a λ V_L chain derived from embryonic germline DNA with that derived from DNA of an antibody-producing cell. In the embryonic DNA, the region coding for the first 95 amino acids of the V_L region was separated by a 4.5-kb region from that coding for the remaining 13 amino acids (the J region) and the C region. However, in the DNA of the antibody-producing cell, the V and J segments were joined contiguously to form a single gene encoding a specific λ V_L chain. This observation strongly supports the hypothesis that immunoglobulin genes are formed through a process of recombination. Similar processes lead to the formation of DNA encoding the κ light chain, illustrated in Figure 17.9.

Based on what is now known, a general explanation of the origin of antibody diversity is emerging. In germ cells, certain clusters or gene families (e.g., the Ig heavy chain, kappa and lambda light chains) encode segments of heavy and light chains. During early stages of B lym-

TABLE 17.2
Categories and components of immunoglobins.

Ig Class	Light Chain	Heavy Chain	Tetramers	
IgA		α	$\kappa_2\alpha_2$	$\lambda_2\alpha_2$
IgD		δ	$\kappa_2\delta_2$	$\lambda_2\delta_2$
IgE	κ or λ	ϵ	$\kappa_2\epsilon_2$	$\lambda_2\epsilon_2$
IgG		γ	$\kappa_2\gamma_2$	$\lambda_2\gamma_2$
IgM		μ	$\kappa_2\mu_2$	$\lambda_2\mu_2$

MONOCLONAL ANTIBODIES: HARNESSING THE IMMUNE RESPONSE

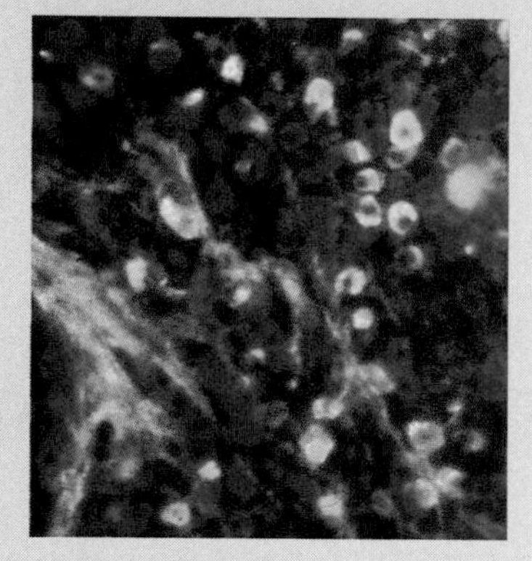

Antibody secretion is dependent upon the formation of specialized cells in the immune system called *plasma cells.* These cells become antibody factories, secreting several thousand antibody molecules per second over their brief lifetimes. One of the peculiar features of the immune system is that each plasma cell synthesizes and secretes only one type of antibody directed against a single antigen.

To harness the synthetic ability of the plasma cell, Cesar Milstein and Georges Kohler first injected a mouse with a specific antigen and then isolated a single plasma cell from the mouse's spleen. This plasma cell was then fused with a type of cancer cell called a **myeloma cell.** Myeloma cells, like many cultured cancer cells, divide continuously and are easy to grow. The cell resulting from this fusion, called a **hybridoma,** retains two properties of the parental cells: Like the plasma cell, it synthesizes large quantities of a single antibody; like the myeloma cell, it divides continuously. Antibodies synthesized and secreted from this single cell and its descendants are called **monoclonal antibodies (MAb).** Monclonal antibodies are now important tools in research, in diagnosis, and in treatment of human disease. For their work on MAbs, Milstein and Kohler received the 1984 Nobel Prize in medicine.

In research, MAbs have been used to help identify cells that take part in the immune response. For example, MAbs have been used to identify and categorize the types and functions of T cells. In addition, MAbs are used to precisely localize proteins in cells. For example, the gene product of the cystic fibrosis locus was thought to be involved in the transport of ions across the plasma membrane of the cell. The use of MAbs confirmed this location, verifying the function of this gene product. In diagnosis, there are now many diagnostic procedures, including home pregnancy tests, that use MAbs. In use, a woman applies a drop of her urine to a strip coated with MAb. If a hormone (hCG) found only during pregnancy is present, it binds to the MAb, causing a color change. In clinical settings, MAbs are used to identify infectious agents, such as the HIV virus associated with AIDS, and to identify the organisms associated with other sexually transmitted diseases such as chlamydia and gonorrhea. In addition, MAbs are used to screen for drugs and signal the presence of cancer long before a tumor becomes detectable. Such screening is rapid, sensitive, and highly specific, reducing costs, time, and false results. Since some forms of cancer are treated more easily than others, early, accurate diagnosis and immediate initiation of therapy are critical.

In treatment, the use of MAbs takes advantage of the fact that many cancer cells display tumor-specific antigens on their surface. MAbs against these antigens can be used to mark these cells for destruction by the immune system. When MAbs bind to the cancer cell, phagocytes are drawn to the site and engulf and destroy the cancer cell. MAbs can also be used to deliver therapeutic drugs or isotopes directly to the targeted cancer cells. In use, the MAb is coupled to a drug molecule or to a radioactive isotope. The complex is introduced into the body and the MAbs interact with the antigens on the tumor cell surface, binding the drug or isotope to the target cell, resulting in the destruction of the tumor cell, with little or no damage to normal cells.

In the future, researchers hope to be able to use MAbs to treat autoimmune diseases in which the immune system attacks normal cells and tissues. Diseases that may be treated in this way include multiple sclerosis, systemic lupus ertheymatosus, and myasthenia gravis. By selectively killing the immune cells that activate the autoimmune response, it may be possible to stop or even reverse the damage caused by these diseases. Other advances, especially the use of recombinant DNA technology to generate custom-made antibody genes and MAbs for cancer therapy appear promising. MAb technology, already a useful tool in research and clinical settings, will become an even more versatile and important means of diagnosing and treating human diseases in the future.

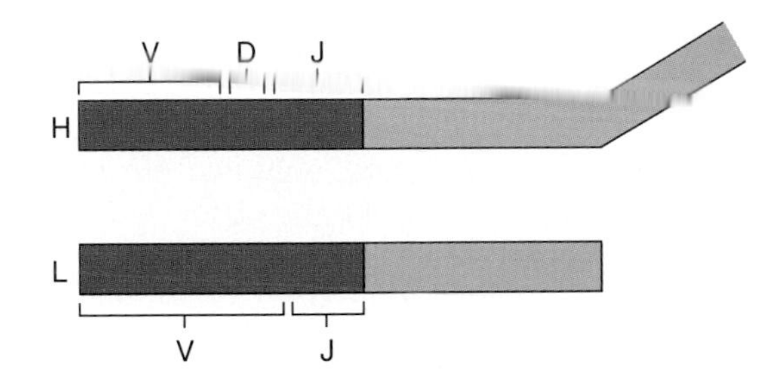

FIGURE 17.8
Schematic diagram of the V and J regions of light chains and the V, D, and J regions of heavy chains.

phocyte differentiation, recombination of the V, D, and J elements brings together various gene segments, creating a composite gene that encodes a specific chain.

This shuffling of DNA segments allows the production of literally millions of different antibodies from a restricted amount of genetic information (Figure 17.10). The larger question is, what controls the shuffling and reshuffling of genes in developing cells of the immune system to generate the multitude of antibodies? Analysis of unrearranged V, D, and J segments indicates that they are flanked by *recognition sequences* (RS) that presumably serve as targets for an enzyme that recombines gene segments. Further work indicates that this enzyme, termed **V(D)J recombinase** has a number of activities that include: recognition of the RS regions, endonucleolytic cleavage at the borders of RS and V(D)J gene segments, and polymerase and ligase activity to rejoin segments in new combinations.

Two closely linked genes in the human genome whose products act in a synergystic fashion to carry out V(D)J recombination have been identified. One gene, **RAG-1** (recombination activating gene) is expressed only in lymphocytes and only in those stages when V(D)J recombination is taking place. When transferred in an active

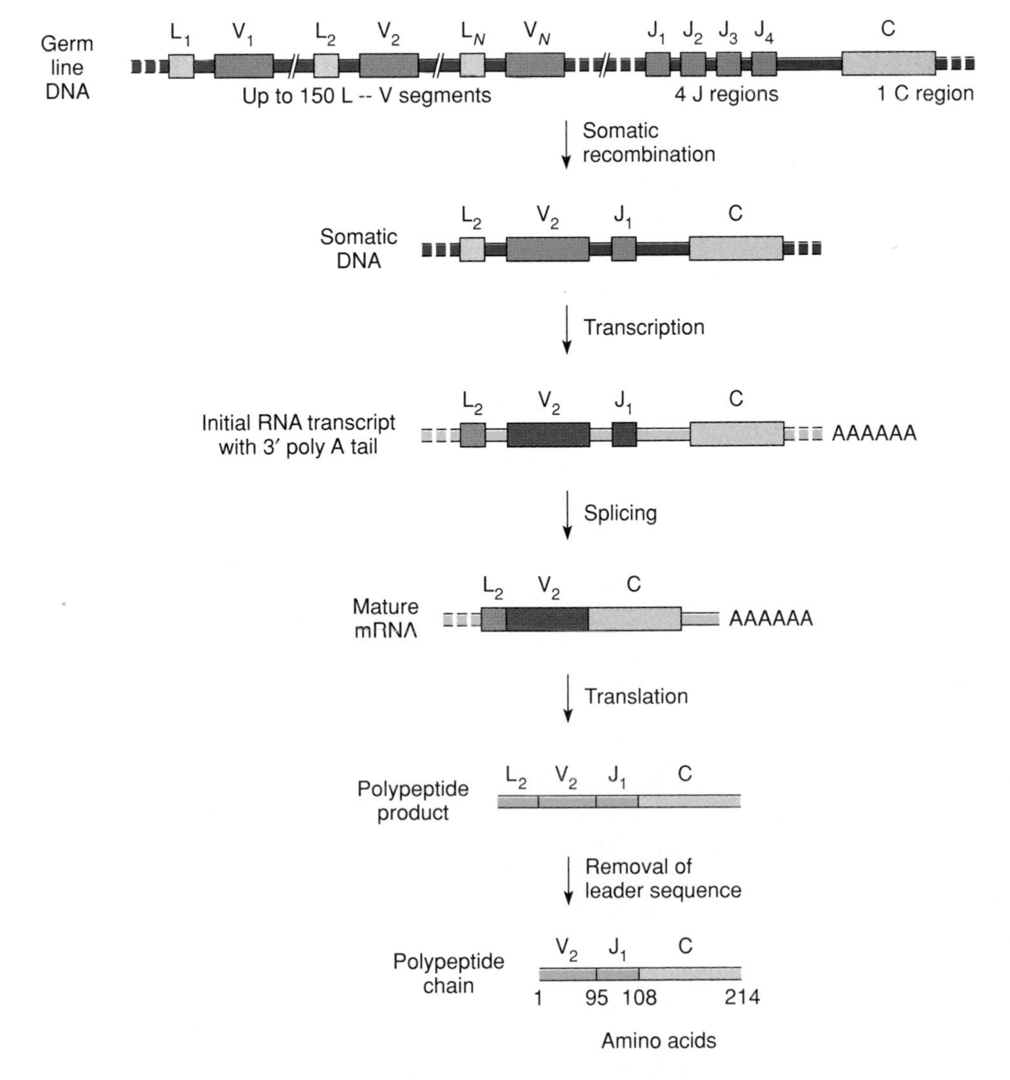

FIGURE 17.9
The formation of the DNA segments encoding a human κ light chain and the subsequent transcription, mRNA splicing, and translation leading to the final polypeptide chain. In germline DNA, up to 150 different L-V (Leader-Variable) segments are present. These are separated from the J regions by a long noncoding sequence. The J regions are separated from a single *C* gene by an intervening sequence (intron) that must be spliced out of the initial mRNA transcript. Following translation, the amino acid sequence derived from leader RNA is cleaved off as the mature polypeptide chain passes across the cell membrane.

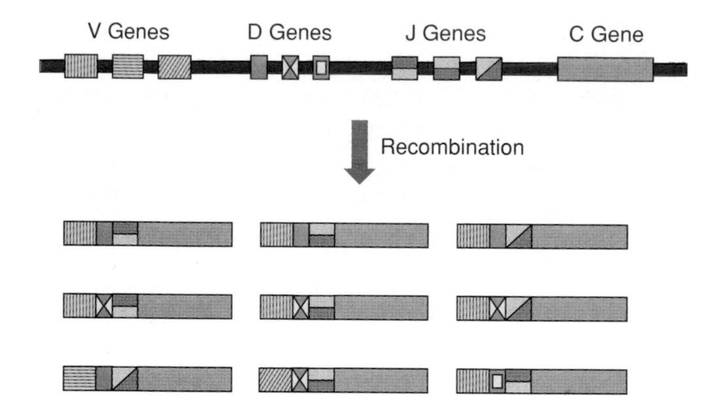

FIGURE 17.10
The theoretical formation of various heavy chain genes during the maturation of an antibody-forming cell.

form into cultured human cells, this gene confers the ability to bring about the assembly of variable (V), joining (J), and diversity (D) segments, a property not normally found in fibroblasts. RAG-1-deficient mice have no mature B or T lymphocytes, and maturation of these cells is arrested at the stage during which V(D)J recombination takes place. Similarly, mice deficient for the second gene **RAG-2** are unable to initiate recombination and have no mature B or T cells.

As yet, little is known about the precise function of the RAG-1 and RAG-2 gene products, but the lack of recombination in RAG-deficient mice suggests that these genes encode subunits or components of the recombinase. It is not clear whether these genes work independently or are part of a coordinate pattern of activity involving a number of regulatory genes. Further work may reveal whether or not there is yet another level of regulation that switches on the activity of the recombinases.

BLOOD GROUPS

Thus far, we have concentrated on the genetic basis of antibody production and cell-mediated immunity as components of an immunological defense system. We now turn our attention to the genetic basis of antigens. These molecules establish the chemical identity that underlies the immunological concept of self. As we shall see, this is an equally important, closely related topic. For example, knowledge of the genetic determination of molecules found on the surface of blood cells is needed to carry out blood transfusions. To date, more than thirty such genes have been identified, and each (along with its alleles) constitutes a blood group or blood type. To be safely transfused, the blood types of the donor and recipient must be matched. If the transfused red blood cells have a foreign antigen on their surface, the body of the recipient will produce antibodies against this antigen, causing the transfused red cells to clump together, blocking circulation in capillaries and other blood vessels, with severe and often fatal consequences. Although there are a large number of blood groups, only two are of major immunological significance: 1) the ABO system, and 2) the Rh-blood group.

ABO System

The genetics of the ABO system were described in Chapter 4 as an example of a gene with multiple alleles. To briefly review this sytem, recall that the gene *I* (for isoagglutinin) encodes a cell surface glycoprotein and has three alleles: I^A, I^B, and I^O. The I^A and I^B alleles are codominant, each producing a slightly different version of the gene product, while I^O is a recessive null allele, producing no effective antigen on the cell surface. The gene products produced by the A and B alleles behave as antigens and may stimulate the production of antibodies. For example, individuals with type A blood (genotypes AA and AO), carry the A antigen on red blood cells, so they will not make antibodies against this cell surface marker, but can make antibodies against the antigen encoded by the B allele (Table 17.3). Those with type B blood have the B antigen on their red cells and can make antibodies against the A antigen. Individuals with AB blood have both antigens and do not make either antibody, while those with type O blood have neither antigen and can make antibodies against both the A and B antigen. Thus, in transfusions, AB individuals have no serum antibodies against A or B and can receive blood of any type, while type O individuals have neither red cell antigen and thus are universal donors.

Rh Incompatibility

The Rh blood group (named for the rhesus monkey in which it was discovered) consists of at least three genes (C, D, and E), each with two alleles (C, c; D, d; and E, e), making a large number of genotypic combinations possible. Each allele (except for d, which produces no gene product) encodes the information for a distinct antigen present on the surface of red blood cells. However, because the antigens encoded by the C and E genes are weak, they invoke only a minimal immune response, and for practical purposes, the Rh system can be considered as a simple one gene, two allele system, involving gene D. Those of genotypes DD or Dd are said to be *Rh positive* (Rh+), and those of genotype dd are said to be *Rh negative* (Rh−).

Although the Rh blood group can play a role in transfusions, it is of major concern in immunological incompatibility between mother and fetus, a condition known as *hemolytic disease of the newborn (HDN)*. This condition occurs when the mother is Rh− (genotype dd) and the fetus is Rh+ (genotypes DD or Dd). If the mother is Rh−, and Rh+ blood from the fetus enters the maternal circu-

TABLE 17.3
Antigens, antibodies of ABO system.

Blood Type	Antigens on Red Blood Cells	Antibodies in Blood	Safe Transfusions To	From
A	A	B	A, AB	A, O
B	B	A	B, AB	B, O
AB	A, B	—	AB	A, B, AB
O	—	A, B	A, B, AB, O	O

lation, antibodies against the Rh antigen will be made (Figure 17.11). The most common way fetal blood enters the maternal circulation is during the process of birth, so that the first Rh+ positive child is not affected. However, the maternal circulation now contains antibodies against the antigen, and a subsequent Rh+ fetus evokes a secondary response from the maternal immune system, producing massive amounts of antibodies that cross the placenta in late stages of the pregnancy and destroy the red blood cells of the fetus. The anemia that results can kill the fetus before birth, or the hemoglobin released from the lysed red blood cells can be converted into a degradation product that builds up in the brain causing neonatal death or, in survivors, severe mental retardation.

To circumvent this problem, after the birth of the first and all subsequent Rh+ children, Rh− mothers are given

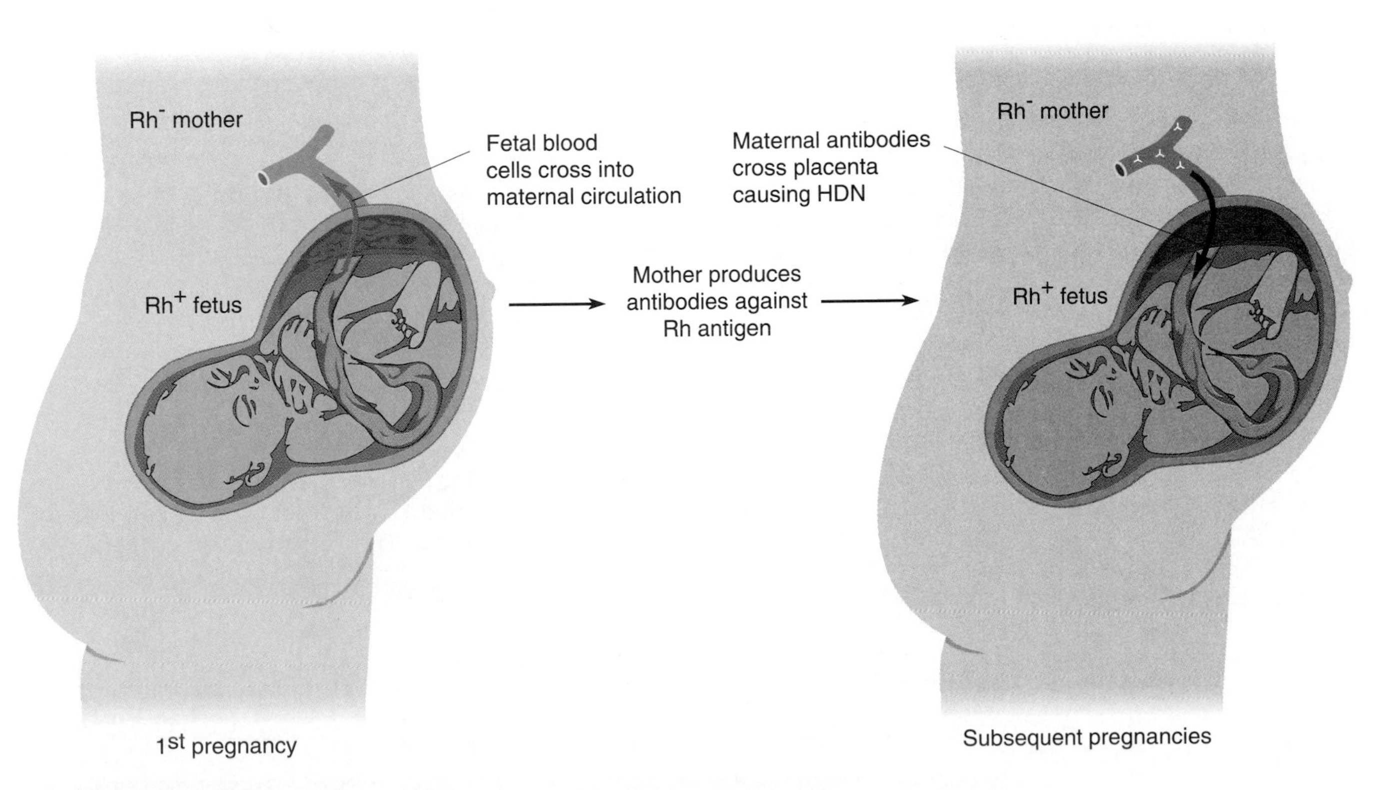

FIGURE 17.11
The development of hemolytic disease of the newborn (HDN). (a) When the mother is Rh− and the fetus is Rh+, any fetal blood that enters the maternal circulation will cause the production of antibodies against Rh+ red blood cells. Usually, this happens during birth, so the first Rh+ child escapes HDN. (b) In a subsequent pregnancy involving an Rh+ child, antibodies against Rh+ positive cells cross the placenta from the maternal circulation and destroy the red blood cells of the fetus. This causes HDN, which is characterized by anemia, jaundice, and, in extreme cases, can result in mental retardation and/or death.

an Rh-antibody preparation. These antibodies move through the maternal circulatory system and destroy any fetal cells before the maternal immune system has a chance to make its own antibodies against the Rh antigen. The antibody preparation is administered to Rh− mothers before the first birth because she can also be sensitized by a miscarriage, abortion, or blood transfusion with Rh+ blood.

THE HLA SYSTEM

Tissue transplants and skin grafts between unrelated individuals are usually rejected over a period of a few weeks. Second attempts at a transplant are rejected in a matter of days. The speed of the second rejection indicates that the immune system is involved in accepting or rejecting grafts and transplants. The interaction of genetically encoded cell-surface antigens of the donor with the immune system of the recipient determines whether grafts will be accepted or rejected. In laboratory mice, these antigens, known as *histocompatibility antigens,* are encoded by 20–40 genes, many of which have a large number of alleles, producing an enormous number of genotypic combinations.

Inbred strains of mice have been widely used to study the genetics of histocompatibility. Some strains are so highly inbred that all members of the strain can donate or receive grafts from any other member of the same strain. Crosses with other strains have allowed the identification and isolation of the histocompatibility genes. Results of such experiments indicate that one group of histocompatibility genes, known as the *major histocompatibility complex (MHC)* is the primary set of genes responsible for success or failure in transplants. The other minor genes are scattered at other loci in the genome.

In humans, a group of closely linked genes on chromosome 6 known as the HLA complex (equivalent to the mouse MHC), plays a critical role in histocompatible transplants.

HLA Genes

The HLA complex in humans consists of four closely linked, major genes: HLA-A, HLA-B, HLA-C, and HLA-D (Figure 17.12). These genes fall into two classes: Class I antigens found on the surface of all cells in the body (HLA-A, HLA-B, and HLA-C), and Class II antigens represented by HLA-D, which encodes antigens found only on cells of the immune system, such as B lymphocytes, activated T lymphocytes, and macrophages. Actually, HLA-D is subdivided into the HLA-DR, HLA-DQ, and HLA-DP genes. Class I antigens are glycoproteins; Class II antigens are composed of two polypeptide chains that enable cells of the immune system to identify each other.

A large number of alleles have been identified in each of the HLA genes; it is one of the most highly polymorphic gene systems in the human genome. There are 23 alleles in HLA-A, 47 in HLA-B, 8 in HLA-C, 14 alleles in HLA-DR, 3 in DQ, and 6 in DP, making literally millions of genotypic combinations possible. Each of these alleles encodes a distinct antigen identified by a letter and number. For example, A6 is allele number 6 at the A locus, B2 is allele 2 at the B locus, etc.

The genes of the HLA system are closely linked and are inherited in a codominant fashion. This close linkage means that recombination in this region is a rare event, and that the allelic combination on a single chromosome tends to be inherited as a unit. The array of HLA alleles on a given copy of chromosome 6 is known as a **haplotype.** Since we have two copies of chromosome 6, we each have two HLA haplotypes (Figure 17.13). The codominant pattern of inheritance means that each haplotype is fully and completely expressed.

The inheritance of HLA haplotypes is shown in Figure 17.13. Because of the large number of alleles that are possible, it is rare that anyone will be homozygous at any of these loci. In the example shown, the unrelated parents have completely different haplotypes, and each child receives one haplotype from each parent, carried by the copy of chromosome 6, which is incorporated into the parental sperm or egg. The result is four new haplotype

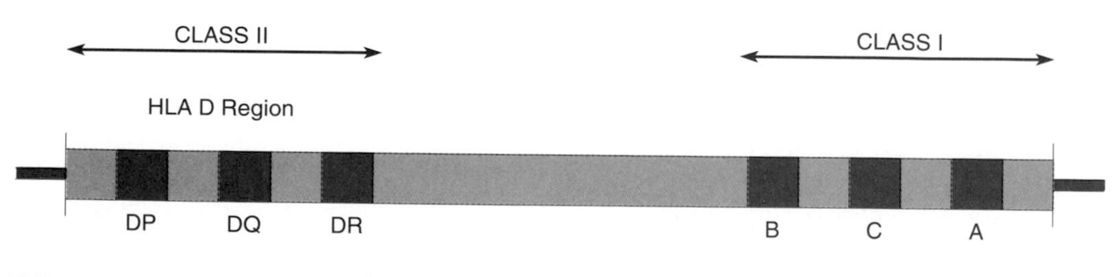

FIGURE 17.12
The HLA region on human chromosome 6, showing the organization of the class I and class II regions.

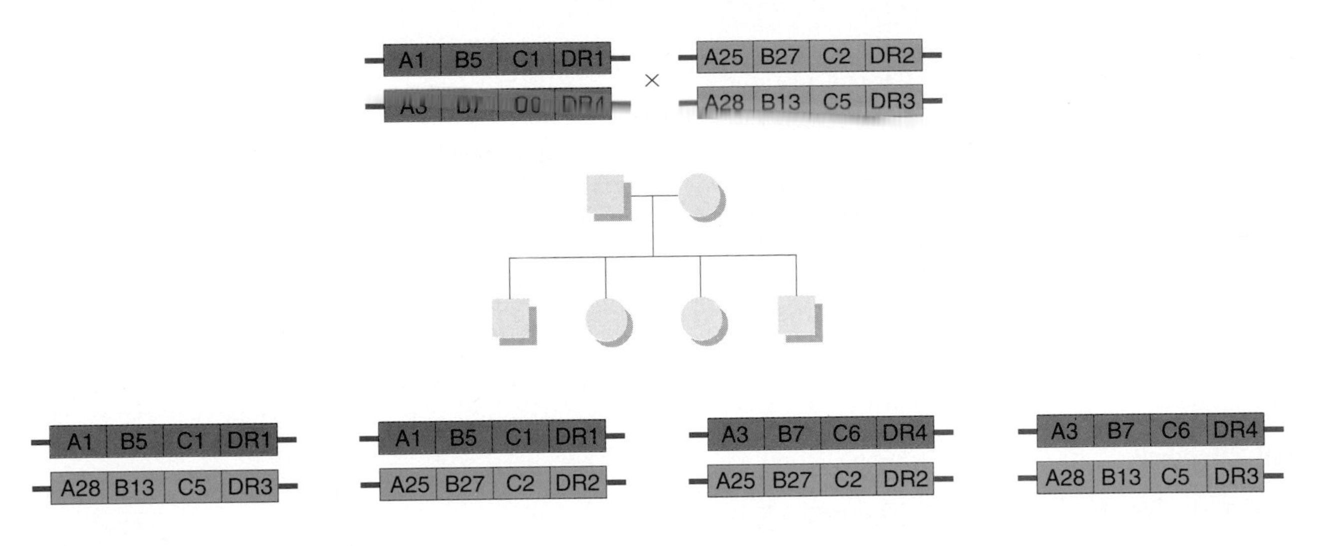

FIGURE 17.13
The transmission of HLA haplotypes. The cluster of HLA alleles on a single chromosome 6 is called a haplotype, each containing four major loci, encoding a different cell surface antigen. The arrangement of HLA haplotypes and their distribution to the offspring result in new combinations of haplotypes in each generation.

combinations represented in the offspring. As is apparent from this information, siblings have a one-in-four chance of sharing the same combination of haplotypes.

Organ and Tissue Transplantation

Successful transplantation of organs and tissues depends to a large extent on matching HLA haplotypes between donor and recipient. Because of the large number of HLA alleles, the best chances for a match is between siblings and other relatives. Identical twins always have a perfect match. Parents have only one haplotype in common with a child. Thus, the order of preference for organ and tissue donors among relatives is: identical twin > sibling > parent > unrelated donor. Among unrelated donors and recipients, the chances for a successful match are between 1 in 100,000 and 1 in 200,000. Because HLA allele frequency differs widely between racial and ethnic groups (for example, B27 is found in 8% of American whites, but only 4% of American blacks), matches between these groups is often difficult.

When HLA types are matched, the survival of transplanted organs is improved dramatically. The major causes of rejection are mismatching of HLA and/or ABO alleles. Other causes are more subtle and often difficult to define. For example, if the recipient has had blood transfusions, memory cells against HLA antigens might be present. If these antibodies act against HLA antigens on the grafted tissue, rejection is more likely. Figure 17.14 shows the 4-year survival rates for matched and unmatched kidney transplants. In HLA-matched transplants, over 90% of the transplanted kidneys survived the 4-year period, but in unmatched transplants, less than 50% of the kidneys were functional after 4 years.

More recently, drugs have been used to improve the survival of transplants even when HLA matching is some-

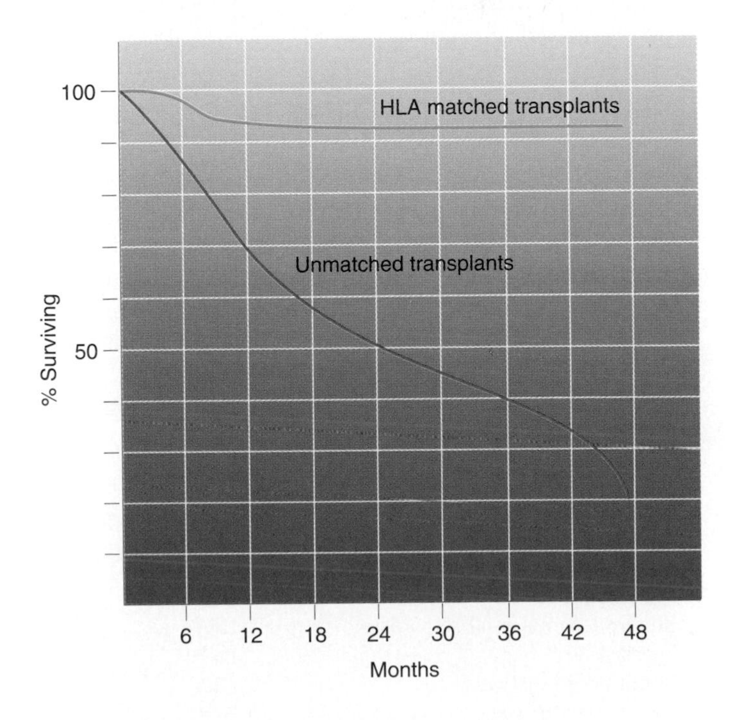

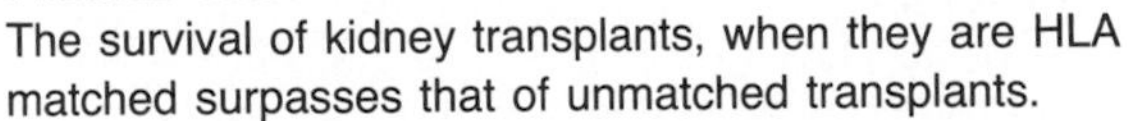

FIGURE 17.14
The survival of kidney transplants, when they are HLA matched surpasses that of unmatched transplants.

what imperfect. The most widely used drug is cyclosporin, first isolated from a soil fungus. Its mechanism of action is not yet known with certainty, but it selectively inactivates the T cell subpopulation (the killer T cells) that is most active in tissue rejection, while not impairing other cells of the immune system. Other drugs to suppress the immune rejection of transplants are now under development, and if successful, may reduce the need for precise HLA matching for transplants, making it possible for more individuals to receive needed organs without prolonged waits for a proper donor.

HLA and Disease

In a number of diseases, certain HLA haplotypes are found at a much higher frequency than would be expected by chance alone. For example, although allele B27 is found in 8% of the U.S. white population, more than 90% of those afflicted with a chronic inflammatory condition known as *ankylosing spondylitis* carry the B27 allele. Similarly, 70% of those with rheumatoid arthritis carry allele DR4, while only 28% of those in the general population carry this allele. A number of diseases and their HLA associations are shown in Table 17.4.

The nature of the relationship between HLA alleles and certain diseases is unknown, but a correlation exists and can often be used in diagnosing the condition. Several hypotheses have been put forward to explain this relationship, but none have proven entirely satisfactory.

DISORDERS OF THE IMMUNE SYSTEM

Genetically determined immunodeficiency disorders are caused by mutations that inactivate or destroy some component of the immune system. Most often, these disorders cause deficiencies in the production and functioning of one or more of the cell types in the immune system (Table 17.5). Much has been learned about how the immune system works by studying disorders that remove or disable single components. Unfortunately, not all the genetic disorders of the immune system have been described in enough detail to allow insights into the underlying molecular mechanisms. We will consider three disorders, one that affects B cells, one that affects T cells, and one that affects both B and T cells.

Not all disorders of the immune system are due to single gene mutations, however. Other factors including infections, cancer chemotherapy, and developmental errors can result in impaired or absent immune functions. Two of these, infection and developmental errors, are considered below.

The Genetics of Immunodeficiency

A genetic disorder in which the B cells are missing is **X-linked agammaglobulinemia** or Bruton disease. In classic fashion, affected individuals are almost always male, with an onset somewhere between 6 and 12 months of age. B cells and plasma cells are completely absent or immature B cells are present, but they never become functional. The result is a complete lack of circulating antibodies and an inability to make antibodies. Thus, affected individuals have no antibody-mediated immunity and are therefore highly susceptible to infections from microorganisms such as streptococcus and pneumonia. However, these same individuals have normal levels of T cells and retain an intact cell-mediated immune system that confers immunity to most viral infections. Currently, the most effective method of treatment is a bone marrow transplant to provide a stem cell population that can give rise to functional B cells. Once these are established, they confer antibody-mediated immunity, which permanently corrects the condition.

A genetically controlled form of **T cell immunodeficiency** has been described, although the number of reported cases is very small. Affected individuals have a deficiency of the enzyme nucleoside phosphorylase, part

TABLE 17.4
HLA alleles and disease.

Disease	HLA Allele	Risk Factor
Ankylosing spondylitis	B27	>100
Systemic lupus erythrematosus	DR3	3
Psoriasis	B17	6
Rheumatoid arthritis	DR4	6
Reiter's syndrome	B27	50
Multiple sclerosis	A3	3
Chronic active hepatitis	B8	6

TABLE 17.5
Cell types affected in immune disorders.

Immune Disorder	Affected Cell Type
X-linked agammaglobulinemia	B cells missing
Nucleoside phosphorylase deficiency	T cells reduced after birth
Severe combined immunodeficiency	T and/or B cells missing or non-functional
DiGeorge syndrome	T cells missing
Acquired immunodeficiency syndrome	T cells decline after infection

of a biochemical pathway that salvages nucleotides. The number of B and T cells in these individuals is normal at birth, but thereafter, there is a gradual decrease in the number of T cells and cell-mediated immunity. The number of B cells is not affected, and antibody-mediated immunity is maintained. Without cell-mediated immunity, there is an increased frequency of viral infections and a high risk of certain forms of cancer. It has been suggested that the decline in T cells is caused by the build-up of T-cell toxic purines as a result of the enzyme deficiency, but the mechanism of action remains somewhat obscure. The structural gene for nucleoside phosphorylase maps to the long arm of human chromosome 14, and the condition is inherited as an autosomal dominant trait.

In **severe combined immunodeficiency syndrome (SCID),** T- and B-cell populations are either absent or nonfunctional, and affected individuals have neither antibody-mediated nor cell-mediated immunity. As a result, they are susceptible to recurring and severe bacterial, viral, and fungal infections. There are autosomal and X-linked forms of SCID. About 50 percent of the cases of SCID in the United States are X linked, 35 percent are autosomal recessive with an unknown cause, and 15 percent are autosomal recessive cases associated with a deficiency of the enzyme adenosine deaminase. In the X-linked form (XSCID), there are persistent infections beginning about 6 months after birth, with no T cells present. There are normal or even elevated levels of B cells present, but there is no antibody production by these cells. Experimental results indicate that the unknown XSCID gene product is necessary in both T and B cells and is required for B cell maturation. The longest-surviving individual with this form of SCID was a boy named David, who was born in Texas and isolated from the outside world by being placed in germ-free isolation (Figure 17.15). David died at age 12 of complications following a bone marrow transplant.

In cases of autosomally inherited SCID associated with a deficiency of the enzyme *adenosine deaminase (ADA),* there are no T cells present, and B cells (if present) are not functional. Again, the result is a complete lack of cell-mediated and antibody-mediated immunity, and severe, recurring infections lead to death, usually by the age of 7 months. Some children affected with ADA-deficient SCID are currently receiving gene therapy to provide them with a normal copy of the gene. In this procedure, samples of their white blood cells are removed from the body and transfected with a retrovirus containing a copy of the normal gene for ADA. The cells are maintained in culture to ensure that the gene is active, and then the genetically modified white blood cells are replaced into their circulatory systems with the hope that ADA expression will stimulate the development of functional T and B cells and at least partially restore a functional immune system. Although these treatments are still in their early stages, results to date have been encouraging. This represents the first attempt at gene therapy in humans; a tech-

FIGURE 17.15
David, a boy born with severe combined immunodeficiency (SCID) survived by being isolated in a germ-free environment. He died at the age of 12 years following a bone marrow transplant undertaken to provide him with a functional immune system.

nique that will undoubtedly become widely used as a treatment for certain genetic disorders in the future. The recombinant DNA techniques used in gene therapy are reviewed in Chapter 15.

Acquired Immunodeficiencies: DiGeorge Syndrome and AIDS

A nongenetic form of T cell deficiency, called **DiGeorge syndrome** is the result of a mishap early in prenatal development. In this condition, the thymus and parathyroid glands fail to develop. The result is a complete lack of T cells that depend on the cellular environment of the thymus in order to mature. As predicted, no cell-mediated immunity results.

Acquired immunodeficiency syndrome (AIDS) is a collection of disorders that develop as a result of infection with a retrovirus known as the human immunodeficiency virus (HIV). The HIV virus consists of a protein coat, enclosing an RNA molecule that serves as the genetic material and an enzyme, reverse transcriptase. The entire viral particle is enclosed in a lipid coat derived from the plasma membrane of a T cell (Figure 17.16). The virus selectively infects the T4 helper cells of the immune system. Inside the cell, the RNA is transcribed into a DNA molecule by reverse transcriptase, and the viral DNA is inserted into a human chromosome, where it can remain for months or years.

At a later time, when the infected T cell is called upon to participate in an immune response, the cellular and viral DNAs are transcribed. The viral RNA transcript is translated into viral proteins, and new viral particles are formed (Figure 17.17). These bud off the surface of the T cell, rupturing and killing the cell and setting off a new round of T cell infection. Gradually, over the course of HIV infection, there is a decrease in the number of helper T4 cells. Recall that these cells act as the master "on" switch for the immune system. As the T4 cell population falls, there is a decrease in the ability to mount an immune response. The results are increased susceptibility to infection and increased risk of certain forms of cancer. Eventually, the outcome is premature death brought about by any of a number of diseases that overwhelm the body and its compromised immune system. The relationship between the loss of T4 cells and the progression of HIV infection is strong and can be used to monitor the status of infected individuals (Figure 17.18).

HIV is transmitted by the transfer of bodily fluids from infected individuals to noninfected individuals; these fluids include blood, semen, vaginal secretions, and breast milk. The virus is not viable for more than 1 or 2 hours outside the body and cannot be transmitted by food, water, or casual contact. At this time, treatment options are rather limited and include the use of drugs that limit the reproduction of the virus.

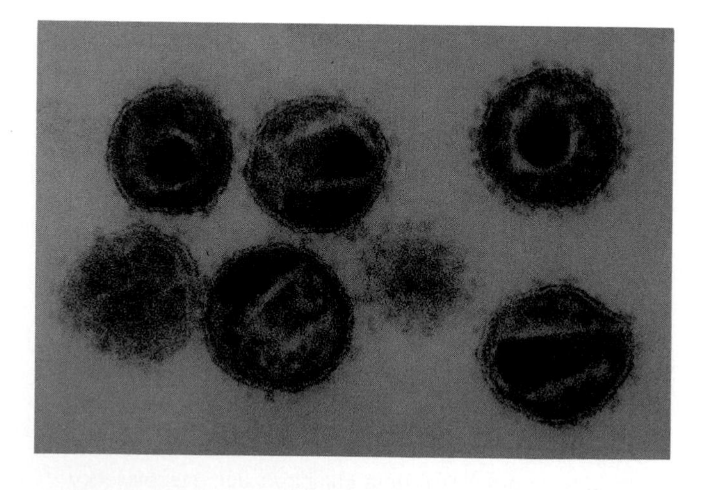

FIGURE 17.16
Transmission electron micrograph of the human immunodeficiency virus (HIV).

Autoimmunity

As the immune system matures, a state of **immune tolerance** develops between cells of the body and those of the immune system. This distinction between self and nonself prevents the immune system from attacking and destroying body tissues. Occasionally, immune tolerance breaks down, causing an autoimmune response in which the immune system turns against cells, tissues, and/or organs in the body. This breakdown of immune tolerance can take place in two ways: 1) a new antigen, previously unknown to the immune system can appear on cells, or 2) a new antibody produced against a foreign antigen ends up also attacking a pre-existing antigen on cells of the body. For example, normally the eye is an immunologically privileged site, meaning that many proteins in the eye are never exposed to the immune system when immunological tolerance develops, and thus are not recognized by the immune system as self. In a traumatic injury, however, antigens from the eye can be exposed to the immune system, and since these are not recognized as self, antibodies against eye proteins are made. These antibodies may attack and cause blindness even in the uninjured normal eye; a condition that is known as *sympathetic ophthalmia*. In other cases, such as infection with certain strains of streptococcus, the antibodies produced against the bacterial antigens are capable of attacking and destroying body cells that carry antigens similar to those of streptococcus. In rheumatic fever, an infection with streptococcus causes antibodies to be produced that kill

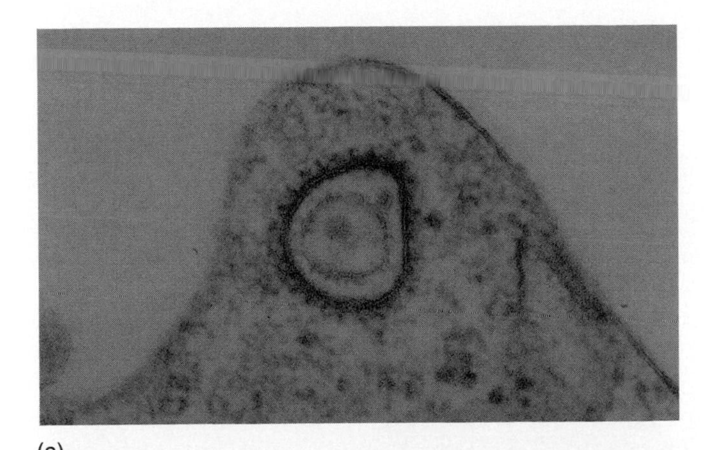

(a)

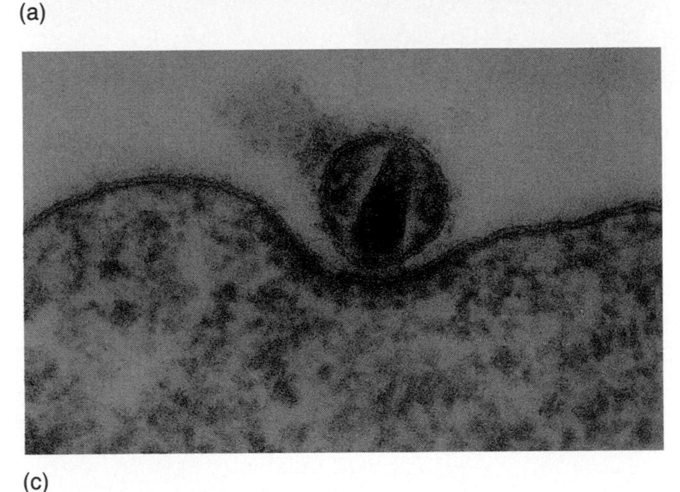

(c)

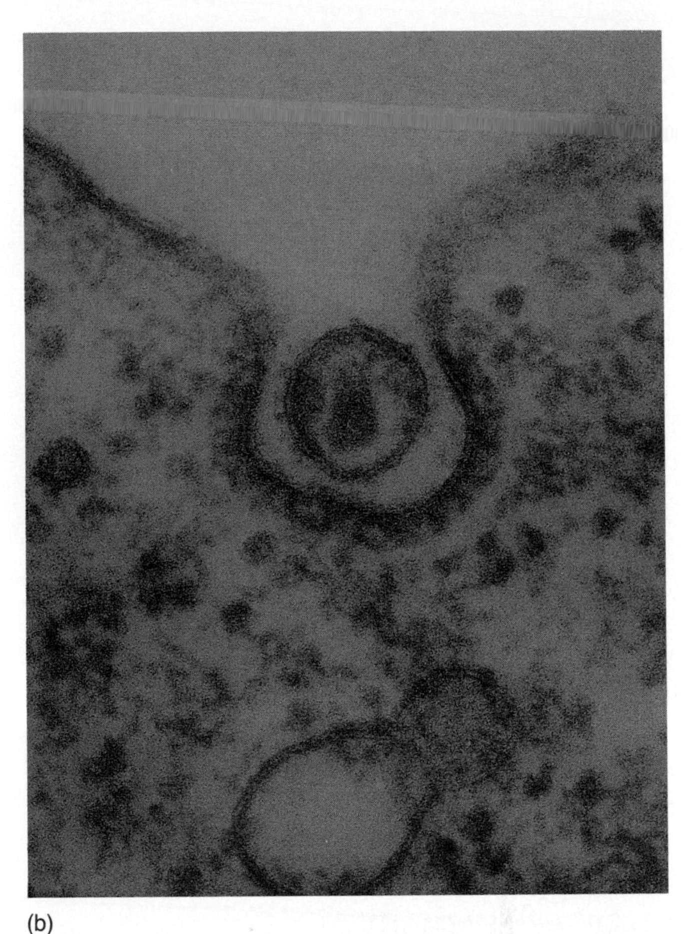

(b)

FIGURE 17.17
The release of new viral particles from an infected cell. (a) The virus approaches the cell surface. (b) The virus buds off from the cell surface. (c) The virus, free from the cell surface can now infect another T cell. Each round of replication releases several dozen new virus particles, and gradually decreases the level of T4 cells, reducing the body's ability to mount an immune response.

the invading bacteria, but these antibodies also attack cells in the valves of the heart, causing the heart problems typically associated with rheumatic fever. Some autoimmune disorders are systemic, affecting many organs, while others are specific to individual tissues and organs. There appears to be a large number of autoimmune disorders that affect connective tissues, and these may be the underlying bases for many forms of arthritis and other connective tissue diseases.

One autoimmune disease with a specific target is insulin-dependent diabetes (IDDM), or juvenile diabetes. As the name implies, the disease starts in childhood and requires daily injections of insulin as a therapy. Insulin is a hormone produced by clusters of cells, called *islets,* embedded in the pancreas. IDDM develops when the immune system attacks and destroys the islet cells that produce insulin. Without insulin, there is no control over sugar levels in the blood. Initial symptoms include thirst and high levels of blood sugar. Without treatment, the disease progresses to kidney failure, blindness, heart disease, and premature death. More than 50 percent of affected individuals die within 40 years of onset.

The HLA alleles DR-3 and DR-4 are associated with an increased risk for IDDM. Each allele is associated with a 15-fold greater risk, and the combination of DR-3 and DR-4 R is associated with a 30-fold greater risk. The mechanism by which the immune system attacks is not yet clear, and other factors are involved, since carrying a DR-3 or DR-4 allele alone is not sufficient to cause IDDM. At least one more gene, and possibly two, outside the HLA locus may play a role in stimulating the immune system, to destroy the islet cells.

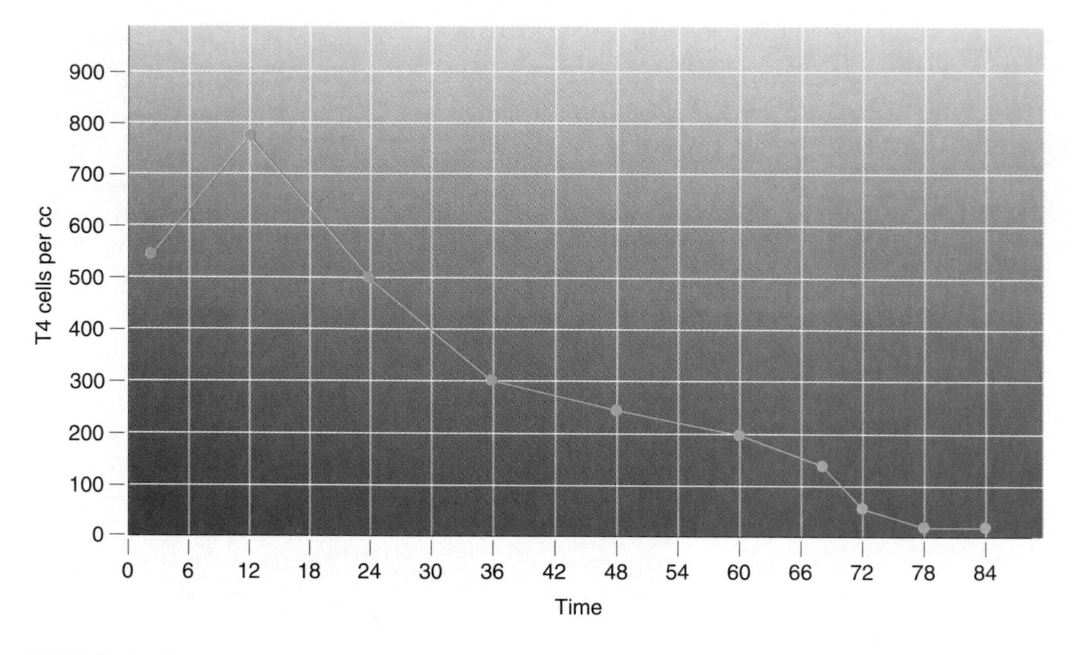

FIGURE 17.18
T4 cell levels during the course of an HIV infection. Following HIV infection, the number of T4 cells rise, then begin a slow decline. When levels fall below 200 cells per cc., the clinical symptoms of AIDS usually appear.

CHAPTER SUMMARY

1 The immune system is composed of two branches; one utilizes cell-mediated immunity and T cells, the other involves B cells and antibody-mediated immunity. The cells of the immune system originate from stem cells in bone marrow, and after maturation, move through the circulatory and lymph system as components of the immune system.

2 One of the most complex families of genes specifying proteins are those participating in the immune response. Human antibodies or immunoglobulins are characterized by amino acid sequence diversity. DNA shuffling during maturation of B cells produces the genetic variation required to match millions of different antigens with corresponding antibodies.

3 The immune response depends on the functional interrelationships of macrophages, which present foreign antigens to helper T cells, which, in turn, activate appropriate B cells to divide and produce large quantities of antibodies. The progress of the immune response is monitored by T8 suppressor cells that modulate the response and shut it down when it is no longer needed. T and B memory cells remain in circulation after the primary immune response and serve as the basis for a rapid and massive response if the immune system is challenged by the same antigen again. The formation of the memory cells is the basis for vaccination against pathogenic organisms.

4 The genetic basis of blood groups is the presence of cell surface antigens and the ability to make antibodies against foreign antigens. The success or failure of blood transfusions depends on matching blood types. Rh incompatibility between mother and fetus can result from antibodies produced in the maternal circulation against antigens on the blood cells of the fetus.

5 A class of cell surface antigens is produced by a complex of genes on human chromosome 6 called the HLA complex. With a large number of alleles, the combination carried by an individual as haplotypes constitutes a genetic signature for each individual. Successful tissue and organ transplantation depends on matching the HLA haplotypes of the donor and recipient. Certain HLA alleles are associated with specific diseases and can be used to diagnose and predict risk factors for such diseases.

6 Mutations that disrupt the immune system work by altering the development and/or function of the component cells that participate in the immune response. These mutations can affect one cell type, a subsystem, or the entire immune system. In addition to genetic forms of disruption, the immune system can be inactivated by external factors, most notably by infection with the human immunodeficiency virus (HIV). This virus selectively infects and destroys the T cells that act as the "on" switch for the immune system, producing a gradual loss in the ability of an infected individual to mount an immune response. As a result, infectious diseases that would ordinarily be suppressed by the immune system can pose a serious risk to the life of HIV-infected individuals.

KEY TERMS

antibodies
antibody combining sites
antigens
cell-mediated immunity
constant region
DiGeorge syndrome
germline theory
heavy chain H
hypervariable region
IgG, IgA, ImM, IgD, IgE
immune response
J genes
light chain L
macrophage
memory cell
monoclonal antibodies
phagocyte
plasma cell
RAG-1
recombination theory
somatic mutation theory
V genes
variable region
V (D) J recombinase

1 An antibody molecule contains two identical H chains, and two identical L chains. This results in antibody specificity, with the antibody binding to a specific antigen. Recall that there are five classes of H genes and two classes of L genes. Because of the variability in the genes encoding the H and L chains, it is possible that an antibody-producing cell contains two different alleles of the H gene and two different alleles of the L gene. Yet the antibodies produced by the plasma cell contains only a single type of H chain and a single type of L chain. How can you account for this, based on the number of classes of H and L genes and the possibility of heterozygosity?

ANSWER: While an antibody-producing cell contains different classes of H and L genes, and while it is entirely possible and even likely that a given antibody-producing cell will contain different alleles for the H gene and for the L gene, a phenomenon known as *alleleic exclusion* allows expression of only one H allele and one L allele in any given plasma cell at a given time. That is not to say that the same antibody is produced over the life span of the antibody-producing cell, however. At first, many plasma cells produce antibodies of the IgM class, and at later times, may produce antibodies of a different class (e.g., IgG). Even in switching between different H genes, allelic exclusion is maintained, with only one allele of an H gene (or L gene) being expressed at a given time.

2 In the mouse κ L gene, there are 250 V regions, 4 J regions, and 3 D regions. H chains have 250 V, 10 D, and 4 J regions. What are the number of combinations of these immunoglobulin components that are possible?

ANSWER: First, calculate the number of different L combinations, the number of different H combinations, then the number of H and L combinations. For the L genes: 250 V × 4 J × 3 D = 3000. For the H genes: 250 V × 10 D × 4 J = 10,000. The combination of 3000 L genes with 10,000 H genes gives a total of 3×10^6 combinations of immunoglobulins. This actually is a great underestimate, since it does not take into account several other sources of diversity, including: a phenomenon called junctional diversity that produces imprecise joining between V-J or V-D-J segments; the insertion of extra nucleotides into the V-D or V-J segments of H chains (N regions), lambda L genes, the 5 classes of H chains, or the high level of point mutations that arise in splice antibody genes (somatic hypermutation). When these are added in, there is probably a 10,000-fold increase in diversity for about 3×10^{11} different combinations of immunoglobulins.

PROBLEMS AND DISCUSSION QUESTIONS

1 What do the following symbols represent in immunoglobulin structure: V_L, C_H, IgG, J, and D? What is the most acceptable theory for the generation of antibody diversity?

2 Figure 17.10 shows 9 unique gene sequences formed by recombination. How many other unique sequences can be formed?

3 If germline DNA contains 10 V, 30 D, 50 J, and 3 C genes, how many unique DNA sequences can be formed by recombination?

4 If there are 5 V, 10 D, and 20 J genes available to form a heavy-chain gene, and 10 V and 100 J genes available to form a light chain, how many unique antibodies can be formed?

5 Distinguish between an antigen and an antibody.

6 What are the functions of helper T cells and suppressor T cells?

7 If a woman has a child with hemolytic disease of the newborn (HDN), what are the possible genotypes of the mother, father, and child? After discovering that her child has HDN, the mother requests treatment with antibodies in the belief that it will prevent HDN in subsequent children. Is she correct? Why?

8 Why is type AB blood regarded as a universal recipient, and type O blood regarded as the universal donor? Which is more important in blood transfusion matching: The antibodies of the donor and recipient, or the antigens of the donor or recipient? Why?

9 How many polypeptide chains are present in an IgG antibody molecule? How many different polypeptide chains are represented in this molecule? How many gene segments from a germline cell were combined to produce the polypeptide chains in this molecule?

10 What is an HLA haplotype? How many of these haplotypes do you carry? How are these haplotypes inherited?

11 Describe the immunological processes involved in the rejection of an organ transplant.

ent class (e.g., IgG). Even in switching between different H genes, allelic exclusion is maintained, with only one allele of an H gene (or L gene) being expressed at a given time.

2 In the mouse κ L gene, there are 250 V regions, 4 J regions, and 3 D regions. H chains have 250 V, 10 D, and 4 J regions. What are the number of combinations of these immunoglobulin components that are possible?

ANSWER: First, calculate the number of different L combinations, the number of different H combinations, then the number of H and L combinations. For the L genes: 250 V × 4 J × 3 D = 3000. For the H genes: 250 V × 10 D × 4 J = 10,000. The combination of 3000 L genes with 10,000 H genes gives a total of 3×10^6 combinations of immunoglobulins. This actually is a great underestimate, since it does not take into account several other sources of diversity, including: a phenomenon called junctional diversity that produces imprecise joining between V-J or V-D-J segments; the insertion of extra nucleotides into the V-D or V-J segments of H chains (N regions), lambda L genes, the 5 classes of H chains, or the high level of point mutations that arise in splice antibody genes (somatic hypermutation). When these are added in, there is probably a 10,000-fold increase in diversity for about 3×10^{11} different combinations of immunoglobulins.

PROBLEMS AND DISCUSSION QUESTIONS

1 What do the following symbols represent in immunoglobulin structure: V_L, C_H, IgG, J, and D? What is the most acceptable theory for the generation of antibody diversity?

2 Figure 17.10 shows 9 unique gene sequences formed by recombination. How many other unique sequences can be formed?

3 If germline DNA contains 10 V, 30 D, 50 J, and 3 C genes, how many unique DNA sequences can be formed by recombination?

4 If there are 5 V, 10 D, and 20 J genes available to form a heavy-chain gene, and 10 V and 100 J genes available to form a light chain, how many unique antibodies can be formed?

5 Distinguish between an antigen and an antibody.

6 What are the functions of helper T cells and suppressor T cells?

7 If a woman has a child with hemolytic disease of the newborn (HDN), what are the possible genotypes of the mother, father, and child? After discovering that her child has HDN, the mother requests treatment with antibodies in the belief that it will prevent HDN in subsequent children. Is she correct? Why?

8 Why is type AB blood regarded as a universal recipient, and type O blood regarded as the universal donor? Which is more important in blood transfusion matching: The antibodies of the donor and recipient, or the antigens of the donor or recipient? Why?

9 How many polypeptide chains are present in an IgG antibody molecule? How many different polypeptide chains are represented in this molecule? How many gene segments from a germline cell were combined to produce the polypeptide chains in this molecule?

10 What is an HLA haplotype? How many of these haplotypes do you carry? How are these haplotypes inherited?

11 Describe the immunological processes involved in the rejection of an organ transplant.

5 A class of cell surface antigens is produced by a complex of genes on human chromosome 6 called the HLA complex. With a large number of alleles, the combination carried by an individual as haplotypes constitutes a genetic signature for each individual. Successful tissue and organ transplantation depends on matching the HLA haplotypes of the donor and recipient. Certain HLA alleles are associated with specific diseases and can be used to diagnose and predict risk factors for such diseases.

6 Mutations that disrupt the immune system work by altering the development and/or function of the component cells that participate in the immune response. These mutations can affect one cell type, a subsystem, or the entire immune system. In addition to genetic forms of disruption, the immune system can be inactivated by external factors, most notably by infection with the human immunodeficiency virus (HIV). This virus selectively infects and destroys the T cells that act as the "on" switch for the immune system, producing a gradual loss in the ability of an infected individual to mount an immune response. As a result, infectious diseases that would ordinarily be suppressed by the immune system can pose a serious risk to the life of HIV-infected individuals.

KEY TERMS

antibodies
antibody combining sites
antigens
cell-mediated immunity
constant region
DiGeorge syndrome
germline theory
heavy chain H
hypervariable region
IgG, IgA, ImM, IgD, IgE
immune response
J genes
light chain L
macrophage
memory cell
monoclonal antibodies
phagocyte
plasma cell
RAG-1
recombination theory
somatic mutation theory
V genes
variable region
V (D) J recombinase

1 An antibody molecule contains two identical H chains, and two identical L chains. This results in antibody specificity, with the antibody binding to a specific antigen. Recall that there are five classes of H genes and two classes of L genes. Because of the variability in the genes encoding the H and L chains, it is possible that an antibody-producing cell contains two different alleles of the H gene and two different alleles of the L gene. Yet the antibodies produced by the plasma cell contains only a single type of H chain and a single type of L chain. How can you account for this, based on the number of classes of H and L genes and the possibility of heterozygosity?

ANSWER: While an antibody-producing cell contains different classes of H and L genes, and while it is entirely possible and even likely that a given antibody-producing cell will contain different alleles for the H gene and for the L gene, a phenomenon known as *alleleic exclusion* allows expression of only one H allele and one L allele in any given plasma cell at a given time. That is not to say that the same antibody is produced over the life span of the antibody-producing cell, however. At first, many plasma cells produce antibodies of the IgM class, and at later times, may produce antibodies of a differ-

18

Developmental Genetics and Neurogenetics

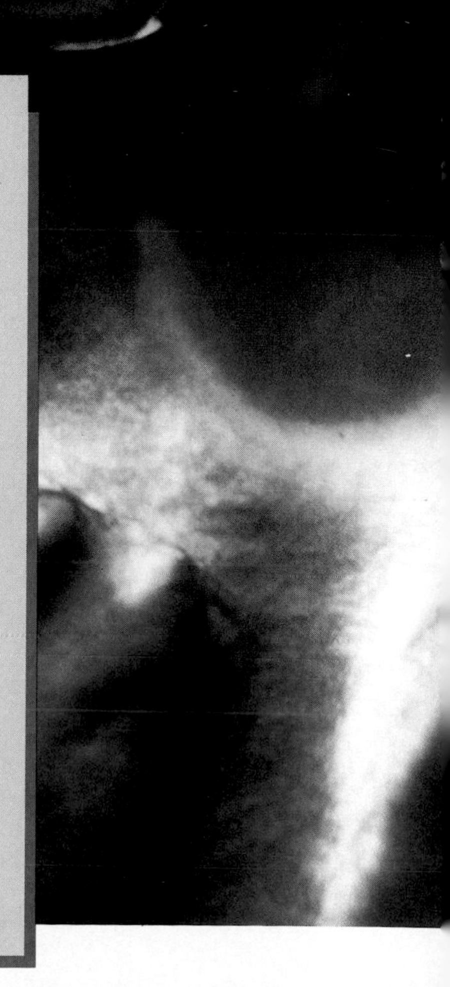

SELECTED READINGS

ARNETT, F. 1986. HLA genes and predisposition to rheumatic diseases. *Hosp. Pract.* 20:89–100.

FRENCH, D., LASKOV, R., and SCHARFF, M. 1989. The role of somatic hypermutation in the genration of antibody diversity. *Science* 244:1152–57.

GONDA, M. 1986. The natural history of AIDS. *Natural History.* 95:78–81.

GOLDE, D. 1991. The stem cell. *Scient. Amer.* 265:(December) 86–93.

HUNKAPILLER, T., and HOOD, L. 1986. The growing immunoglobulin gene superfamily. *Nature* 323:15–16.

LEDER, P. 1982. The genetics of antibody diversity. *Scient. Amer.* 246:(May) 102–15.

McDIVITT, H. 1986. The molecular basis of autoimmunity. *Clin. Res.* 34:163–75.

THOMSON, G. 1988. HLA disease associations: models for insulin-dependent diabetes mellitus and the study of complex human genetic disorders. *Ann. Rev. Genet.* 22:31–50.

TONEGAWA, S. 1985. The molecules of the immune system. *Scient. Amer.* 253:(October) 122–31.

VAN BOEHMER, H., and KISIELOW, P. 1991. How the immune system learns about itself. *Scient. Amer.* 265:(October) 74–86.

YAMAMOTO, F., CLAUSEN, H., WHITE, T., MARKEN, J., and HAKAMORI, S. 1990. Molecular genetic basis of the histo-blood group ABO system. *Nature* 345:229–33.

YANCOPOULOS, G., and ALT, F. 1988. Reconstruction of an immune system. *Science* 241:1581–83.

YOUNG, J., and COHN, Z. 1988. How killer cells kill. *Scient. Amer.* 258:38–44.

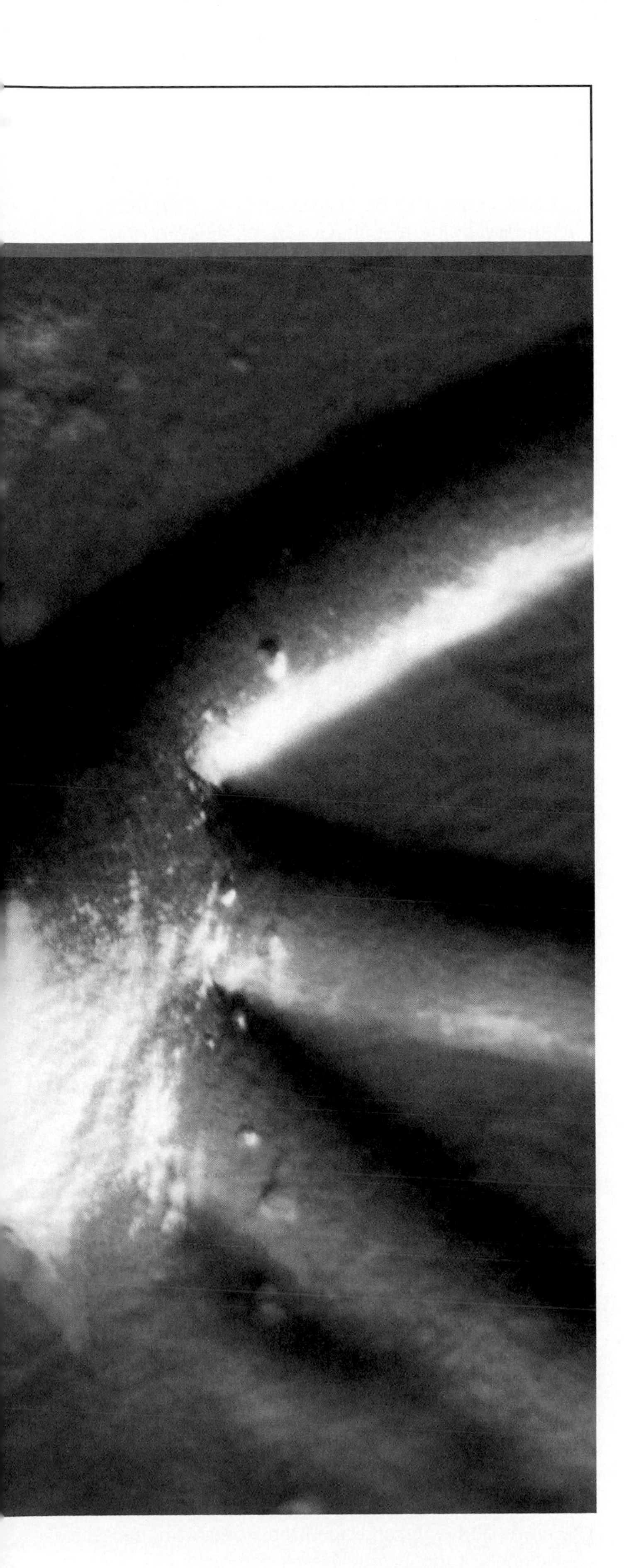

CHAPTER CONCEPTS

The genetic basis of development is differential gene action. Gene actions vary over time in the same tissue and between different tissues present at the same developmental stage and are required to bring about and maintain adult structures. Behavior in many organisms is controlled by both single genes and polygenic systems. At the biochemical level, these genes work by controlling cascades of metabolic reactions.

In multicellular plants and animals, a fertilized egg, without further stimulus, begins a cycle of developmental events that ultimately give rise to an adult member of the species from which the egg and sperm were derived. Thousands, millions, or even billions of cells are organized into a cohesive and coordinated unit that we perceive as a living organism. The heterogeneous series of events whereby organisms attain their final adult form is studied by developmental biologists. This area of study is perhaps the most intriguing in biology because comprehension of developmental processes requires knowledge of many biological disciplines.

In the past one hundred years, investigations in embryology, genetics, biochemistry, molecular biology, cell physiology, and biophysics have contributed to the study of development. Largely, the findings have pointed out the tremendous complexity of developmental processes. Unfortunately, this description of what actually happens does not answer the "why" and "how" of development. Nevertheless, further hypotheses and experimentation will be based on this knowledge and will lead us closer to a comprehensive understanding of development.

Geneticists and neurobiologists interested in the study of behavior have increasingly focused their attention on the analysis of gene action as a means of explaining how organisms react to stimuli and/or their environment. In humans, it is hoped that this approach will allow an objective evaluation of the genetic elements that contribute to intelligence and behavior. The use of genetics to study behavior minimizes the effect of environment on a behavioral response, and allows a specific pattern of action to be dissected into its components. In turn, many behavioral phenotypes are actually the result of developmental events that alter or restrict the function of the nervous system or its interactions with other body systems, emphasizing the relationship between cellular structure and function. In this chapter, we will first focus on the role of genetic information during development. After establishing how the genome controls the cellular phenotype, we will turn our attention to how genetics is being used to probe the relationship between the genotype and behavior.

DEVELOPMENTAL CONCEPTS

These processes govern the formation of embryonic and adult structures from the fertilized egg: *determination, differentiation,* and *cell-to-cell interaction.* Determination can be defined as one or more regulatory events that establish a specific pattern of gene activity and developmental fate for a given cell. In other words, through determination, an undifferentiated embryonic cell is programmed to develop into, for example, a muscle cell, a liver cell, or an epidermal cell. Differentiation is the process of genetic and morphological change by which cells attain their adult form and function. It is the mechanism by which the determined state becomes expressed. Cell-to-cell interactions represent a form of intercellular communication and can be involved in both processes of determination and differentiation.

The current thrust in developmental genetics is to provide a molecular explanation of these phenomena in order to establish a relationship between the presence or absence of molecules and the process of development. To explain the sequence of developmental events that take place in any organism, geneticists rely on certain heuristic concepts. These include the theory of variable gene activity, and the principles of differential transcription and the stability of the differentiated state. These concepts have been derived from the study of a wide range of organisms used as model systems, and these ideas form the theoretical framework that is used to explain how the developmental potential present in a single cell becomes elaborated into a recognizable yet individual form with a certain behavioral repertoire we identify as an adult organism.

THE VARIABLE GENE ACTIVITY THEORY

From a genetic perspective, development may be described as the attainment of a differentiated state. For example, an erythrocyte active in hemoglobin synthesis is differentiated, but a cell in a blastula-stage embryo is undifferentiated. In order to accomplish this specialization, certain genes are actively transcribed but many other genes are not.

The concept of differential transcription has led to the **variable gene activity hypothesis** of differentiation. The theory holds that the forms of differentiation assumed by any specific cell type are determined by those genes that are actively transcribed. Its underlying assumption is that each cell contains the entire diploid genome in the nucleus and that only certain gene products characteristic of the cell type are produced. The rest of the genome is actively shut down and not transcribed.

Given current knowledge of molecular biology, the variable gene activity theory is readily acceptable. Every cell is a biochemical entity composed of macromolecules that are either informational, structural, catalytic, or metabolic. These molecules constitute the cell. The presence or absence of any molecule in the cell is thus directly or indirectly influenced by gene activity or inactivity. Therefore, a cell is what its active genes direct it to be.

The variable gene activity theory is a very useful model system for experimental design. The first part of this chapter examines the validity of this theory and its premises and provides examples that offer experimental support. It is important to remember, however, that this type of approach initially tells us what happens. We still do not know why certain genes are active and others inactive, and why different gene activity occurs among different cells.

Genomic Equivalence

The variable gene activity hypothesis is based on the premise that in multicellular organisms, somatic cells all contain a complete diploid set of genetic information. Analysis has shown that the quantity of DNA is equal to the diploid content in cells of most species studied. But is it possible to show that differentiated cells are qualitatively equivalent? In other words, is a complete set of genes present in each cell?

One approach this question would be to show that a differentiated cell is **totipotent,** that is, capable of giving rise to a complete organism under the proper conditions. In the early part of this century, Hans Spemann demonstrated that a nucleus derived from one cell in the 16-cell stage of a newt embryo is capable of supporting total development. By constricting a newly fertilized egg prior to the first cell division, Spemann partitioned the zygote nucleus into one-half of the cell. The half containing the nucleus proceeded to undergo division and produce 16 cells. At that point, the constriction was loosened, and one of the 16 nuclei was allowed to pass back into the non-nucleated cytoplasm on the other side. Both halves subsequently produced complete but separate embryos. Therefore, at the 16-cell stage in this organism, genomic equivalence of the 16 nuclei was demonstrated.

Beginning in the 1950s, more sophisticated experiments were performed by Robert Briggs and Thomas King using the grass frog *Rana pipiens*, and by John Gurdon in the 1960s using an African frog *Xenopus laevis*. After inactivating or surgically removing the nucleus from an egg, these investigators were able to test the developmental capacity of somatic nuclei by transplanting nuclei isolated from cells at various stages of differentiation into enucleated oocytes. Nuclei derived from the blastula stage of development were capable of supporting the development of complete and normal adults. In *Rana* (Figure 18.1), transplanted nuclei derived from later stages such as the gastrula and neurula usually allowed only partial development. In *Xenopus*, however, Gurdon's experiments showed that epithelial gut nuclei of tadpoles are able to support the development of an adult frog when transplanted into enucleated oocytes.

To test whether the nucleus of a highly differentiated adult cell is irreversibly specialized or can support the development of a normal embryo, Gurdon used nuclei from adult frog skin cells in serial transplant experiments (Figure 18.2). In these experiments, a donor nucleus was transplanted into an enucleated egg, and the recipient was allowed to develop for a short time, for example, to the blastula stage. The blastula cells were then dissociated and a nucleus removed from one of them. This nucleus was transplanted into still another enucleated egg and development was allowed to occur. Such serial transfers were repeated a number of times. Subsequently, the blastula was not dissociated, but instead was allowed to continue development as far as it would go. After several such transfers, Gurdon found that 30 percent of the nuclear transplants resulted in the formation of advanced

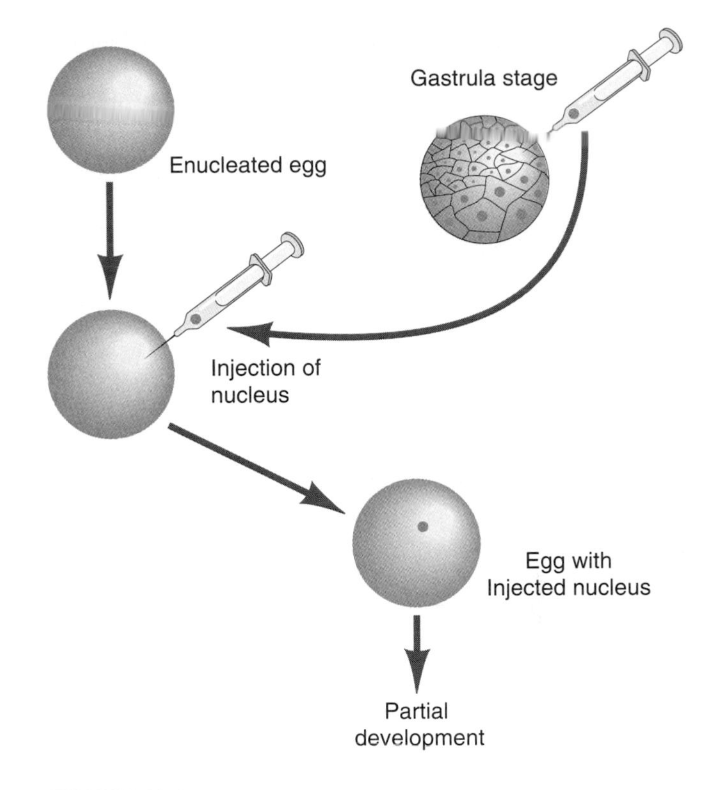

FIGURE 18.1
The process of nuclear transplantation in the frog.

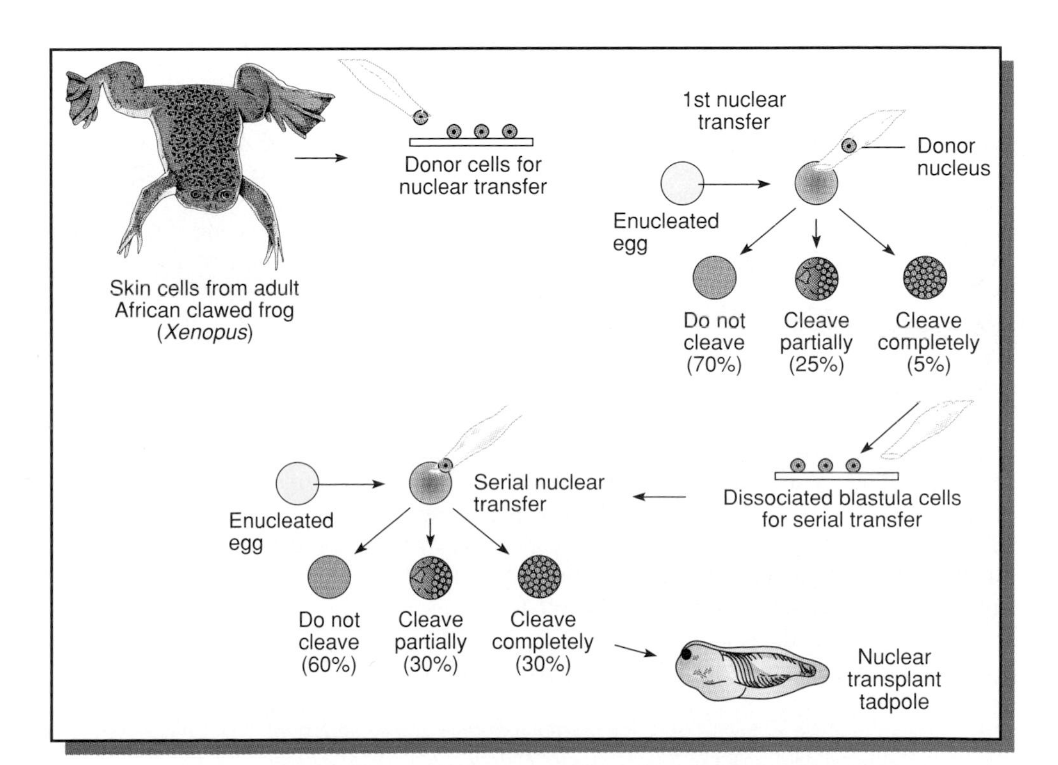

FIGURE 18.2
Serial transplantation experiments using adult frog skin cell nuclei.

tadpoles. Because nuclei from specialized epidermal cells can eventually direct the synthesis of gene products and promote the organization of cells and tissues into a tadpole, we can be confident that such cells contain a complete copy of the genome. Under the proper circumstances, it appears that genes not expressed in specialized cells can be reactivated.

In plants, the totipotency of differentiated cells has been demonstrated using the carrot. Fredrick Steward observed that when individual phloem cells are transferred to a liquid culture medium, each cell divides and eventually forms a mass called a callus. Under appropriate conditions, this cell mass will differentiate into a mature plant.

These studies argue convincingly that differentiated adult cells have not lost any of the genetic information present in the zygote, the original genome. Instead, some genes in any given cell type are shut off or repressed, but can be reprogrammed to direct normal development.

DIFFERENTIAL TRANSCRIPTION DURING DEVELOPMENT

We have now come to the central issue of developmental biology. We consider here the question of how cells eventually acquire one differentiated state or another. As alluded to earlier, there is no simple answer to this question. However, information derived from the study of many different organisms provides a starting point. We shall examine differential transcription in two organisms *Dictyostelium discoideum* and *Drosophila melanogaster* as examples of the direction and scope of our knowledge in this area.

Slime Mold Development

The eukaryotic cellular slime mold *Dictyostelium discoideum* is a unique organism in studies of development. As shown in Figure 18.3, the slime mold exists in one of two alternating forms during its life cycle. Germinating spores give rise to amoeboid cells called **myxamoebae.** These cells feed on soil bacteria and decaying vegetation, grow, and multiply in numbers. However, if nutrients become depleted, free-living cells aggregate together and form a multicellular body or slug consisting of about 10^5 cells. Two distinct cell types form in the slug, depending on their position in the aggregate. Cells in the posterior 75 to 80 percent of the slug form prespore cells, while the cells at the anterior tip become prestalk cells. The slug, which is initially migratory, differentiates into stalks and spores. When the spores mature and are released, they germinate into myxamoebae, and the cycle is repeated.

The slime mold life cycle is an ideal model for investigation of differentiation and cell-to-cell interactions because only two cell types are involved. Cellular structure and function are completely transformed during the conversion of a myxamoeba into a portion of the fruiting body. Proteins present in individual cells are broken down, and synthesis of different gene products occurs. For example, the posterior prespore cells express a characteristic set of proteins that are absent from the prestalk cells. Because the fate of any given amoeba is determined solely by its position in the aggregate, all cells must have the genetic potential for becoming any part of the fruiting body. In addition, the anterior tip of the slug has properties similar to the embryonic organizer found in vertebrates. If the tip (anterior 25 percent of the slug, containing the prestalk cells) is cut off, the posterior fragment mounds up, and a new tip is formed before morphogenesis is restarted. If a tip is grafted to another slug, it forms a new axis, causing the slug to separate into two slugs, each with an anterior tip.

Mutations that affect each step in the process of aggregation and differentiation are known. For example, over 100 different mutants exhibit an aggregation-minus phenotype. In this group, *frigid A* mutants show no chemotactic or signalling response and cannot be induced to differentiate. On the other hand, *synag* mutants can be induced to differentiate if supplied with cAMP. A general scheme for the timing of gene expression is shown in Figure 18.4. It is apparent that a programmed sequence of gene activation leading to the production of differentiation-inducing factor (DIF) is essential for the complete life cycle of *Dictyostelium.* Because this organism is less complex than higher eukaryotic forms, further study may help decipher the role of cell-to-cell interaction in development.

The Effects of the Cellular and Extracellular Environment

Eukaryotic organisms develop from a *zygote,* a cell formed by the fusion of sperm and oocyte. The oocyte cytoplasm is complex, heterogeneous, and nonuniform in distribution within the cell. Evidence suggests that these different forms of cytoplasm exert influences on the genetic material of different cells, causing differential transcription at some point during development. Recall that when a specific developmental fate for cells becomes fixed prior to the actual events of differentiation, the cells are said to have undergone determination. Although such cells show no immediate evidence of structural or functional specialization, their position in the developing embryo or organism seems to determine the ultimate differentiated form they will assume. It is as if their fate has been programmed prior to the actual events leading to specialization.

FIGURE 18.3
Stages in the life cycle of the slime mold *Dictyostelium discoideum.*

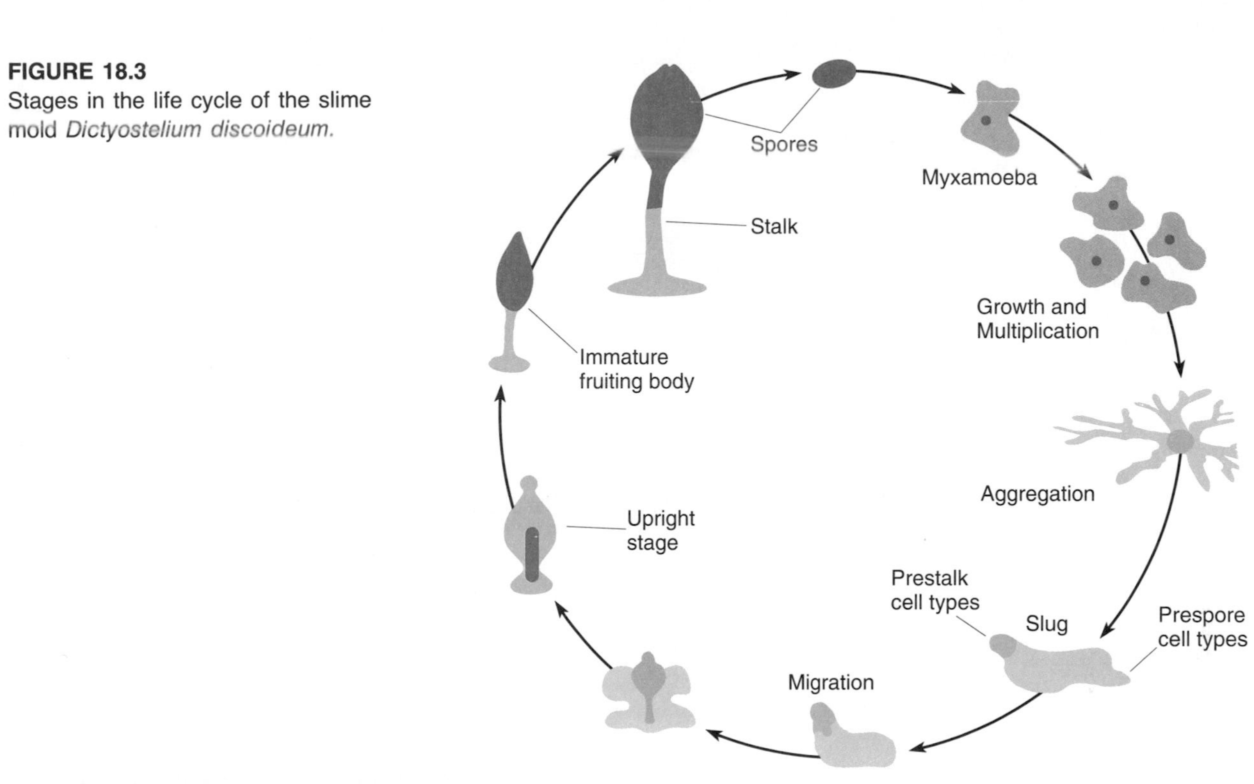

Aggregation stage

Slug stage

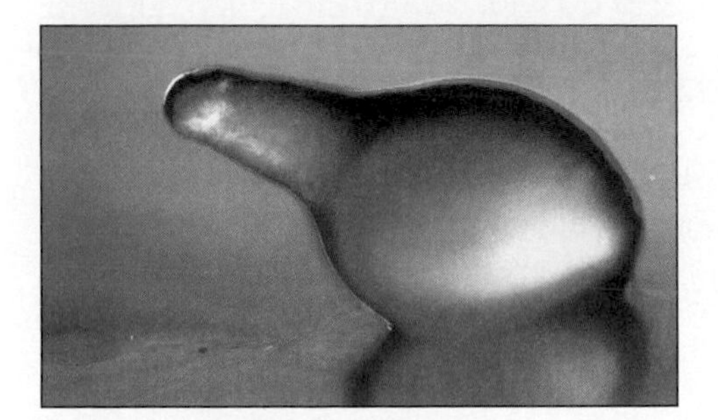
Upright stage following culmination

Immature fruiting body

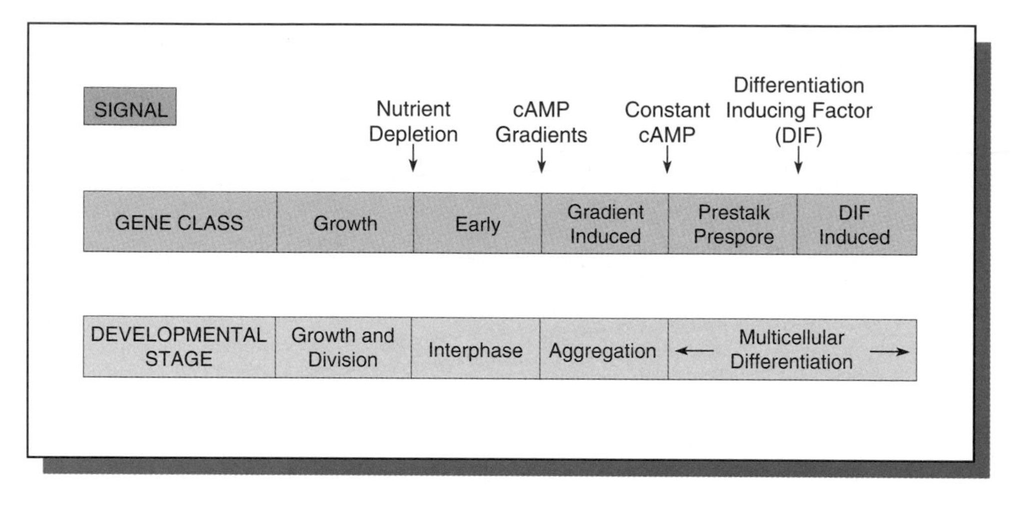

FIGURE 18.4
Cycle of gene expression during development in *Dictyostelium.* The signals that induce gene response are shown at the top. Aside from the first, which is environmental, the signals are produced by the cells themselves during development. Each signal causes a class of genes to become active during a developmental stage, and the products of each class are required for transition to the next stage.

Early gene products further alter the total cytoplasmic content of each cell, producing a still different cellular environment that may, in turn, lead to the activation of other genes and so on. In this way, cells embark on different pathways as development proceeds. As the number of cells increases, they influence one another. The total environment acting on the genetic material, therefore, now includes the cell's individual cytoplasm as well as the influence of other cells. As the developing organism becomes more complex, so does the total environment. In this way, different forms of determination and differentiation occur during development.

We have already discussed several examples that lend credence to the concept of cytoplasmic influence on nuclear activity. For example, Gurdon's nuclear transplantation work with *Xenopus* shows that nuclei from differentiated cell types are capable of reversing their role and serving as zygote nuclei. It may be concluded that the cytoplasmic environment of the enucleated egg exerts a profound influence on nuclear activity.

Similar conclusions may also be drawn from studies of somatic cell hybridization. When cultured somatic cells are experimentally fused, two nuclei may exist in a common cytoplasm, forming a heterokaryon. One of the two nuclei usually conforms to the cytoplasm native to the other nucleus so that both nuclei become uniform in their degree of genetic activity.

Overview of *Drosphila* Development

The formation of the body plan of the larva and adult of *Drosophila melanogaster* is an elegant example of the interaction between cytoplasm and nucleus. The body of *Drosophila* is composed of a number of head, thoracic, and abdominal segments. During embryogenesis, a series of nuclear divisions occurs immediately after fertilization. When nine divisions have occurred, the approximately 512 nuclei migrate to the egg's outer surface or cortex, where further divisions take place. The nuclei become enclosed in membranes, forming a single layer of cells (blastoderm) over the embryo surface. Each segment of the adult body is formed from the descendants of cells set aside at the blastoderm stage into discrete structures called **imaginal disks** (Figure 18.5). These disks, formed from hollow sacs of cells, are determined to form specific parts of the adult body. There are 12 bilaterally paired disks and one genital disk. For example, there are eye-antennal disks, leg disks, wing disks, and so forth. Other cells in the blastoderm will form the internal and external structures of the larva. During pupation, most of the body parts of the larva histolyze, or break down, and the imaginal disks differentiate into adult structures of the head, thorax, and abdomen.

Studies using a combination of physical and genetic techniques have revealed that the major features of the

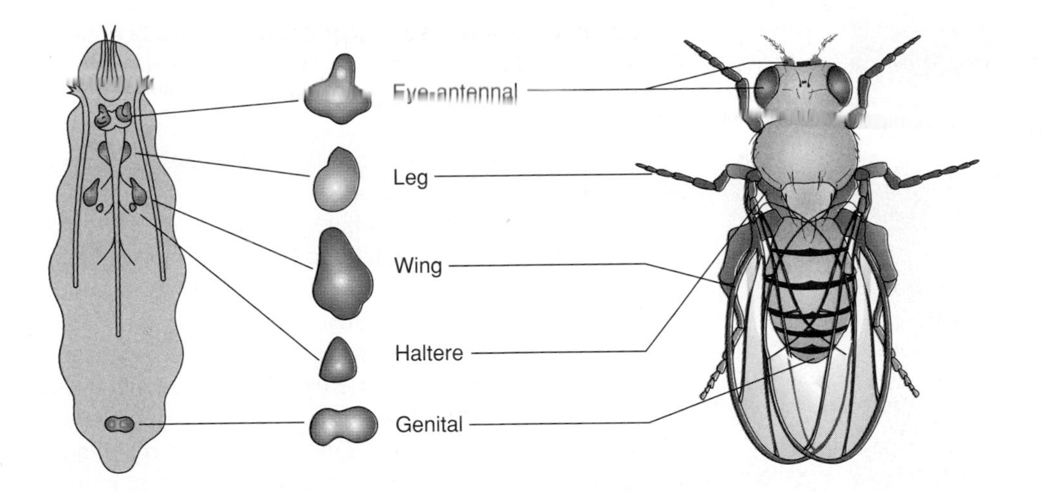

FIGURE 18.5
Imaginal disks of the *Drosophila* larva and the adult structures that are derived from them.

larval and adult body plans are present by the blastoderm stage of embryonic development in *Drosophila.* The two-dimensional projections of the external adult body parts on the surface of the blastoderm are known as **fate maps** (Figure 18.6).

One of the most striking features of fate maps is that the embryo is organized along an anterior-posterior axis that reflects the body axis of the larva, pupa, and adult (Figure 18.6). The existence of a fate map implies that developmental fate of a nucleus is governed by the location to which it migrates during blastoderm formation.

At a slightly later stage of development, the precursors of the adult body structures are present as distinct morphological landmarks in the embryo (Figure 18.7). Even at this stage of development, cells in each segment are determined to form either an anterior or a posterior half or **compartment** of each segment. Progressive restrictions continue to occur throughout development so that cells in each compartment are committed to forming a limited number of adult structures.

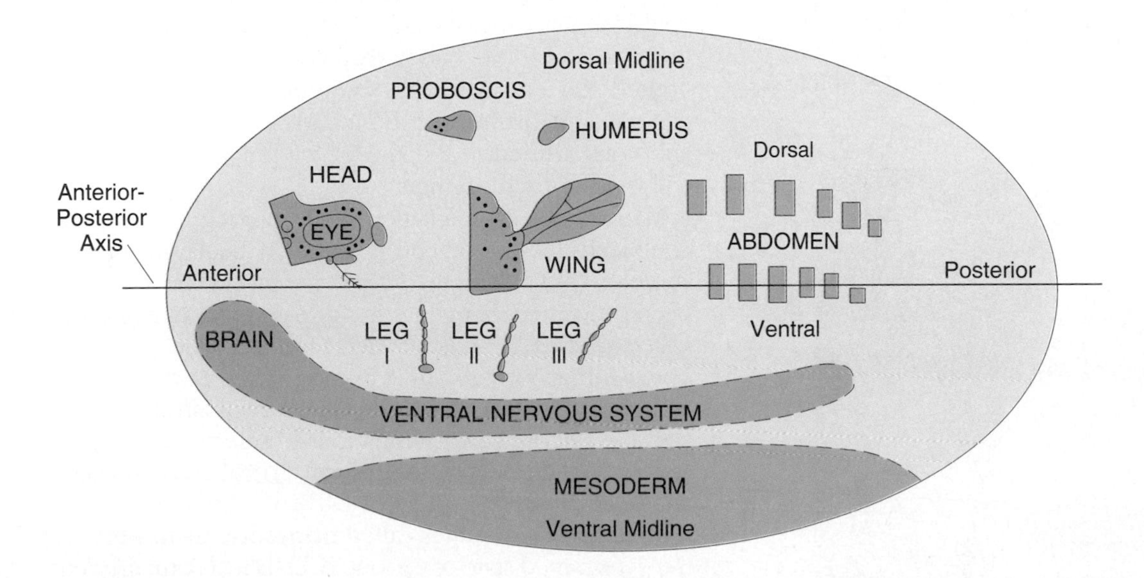

FIGURE 18.6
Fate maps showing the origin of some adult *Drosophila* structures from the embryonic blastoderm.

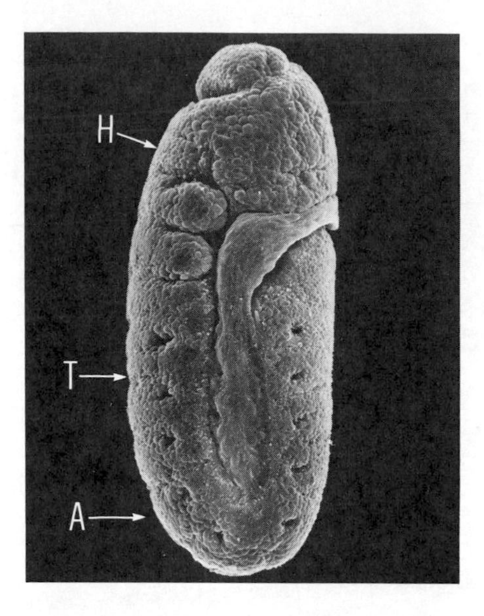

FIGURE 18.7
Scanning electron micrograph of the *Drosophila* embryo. Head (H), thorax (T), and abdominal (A) structures are shown.

Gene Expression During Early *Drosophila* Development

The shape of the *Drosophila* egg provides a set of morphological landmarks that define the anterior-posterior and dorsal-ventral axes of the egg (Figure 18.8). The anterior pole of the egg contains the micropyle, a specialized structure through which sperm enter the egg, while the posterior pole is rounded and marked by a series of aeropyles, which are openings where gas exchange occurs. The dorsal side of the egg is flattened and contains the chorionic appendages, while the ventral side is curved. These axes are associated with the distribution of molecular gradients within the egg that play a key role in directing the development of nuclei that migrate into a specific region of the embryo.

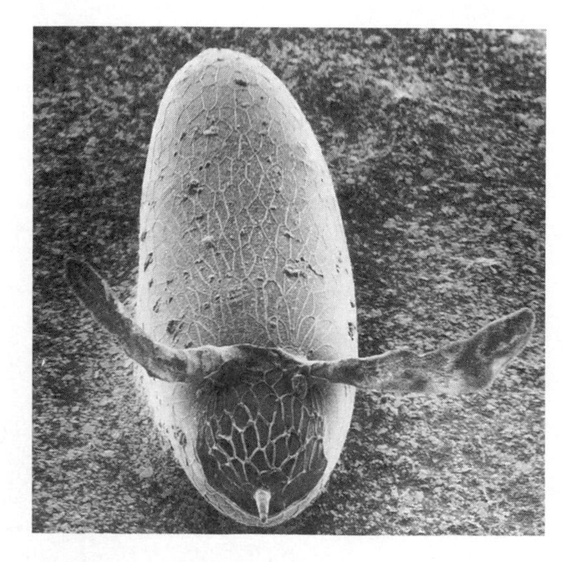

FIGURE 18.8
Scanning electron micrograph of the anterior pole of the newly oviposited *Drosophila* egg.

As nuclei migrate into different regions of the cytoplasm, gene products localized in these regions initiate a program of gene expression in the nuclei that results in the formation of the anterior-posterior axis of the embryo. Table 18.1 lists maternal genes required for formation of the anterior-posterior embryonic axis. One of the genes required for the formation of the anterior pattern of body segments is *bicoid.* Embryos of *bicoid* mutants lack head and thoracic structures and also have abnormalities in the first four segments of the abdomen. After fertilization, *bicoid* protein appears in a gradient in the anterior two-thirds of the embryo. The *bicoid* protein may be a transcription control factor. In fact, the *bicoid* protein is known to bind to the upstream region of at least one gene involved in the formation of body segments, indicating that this gene product acts as a positive transcription regulator on nuclei within the anterior portion of the embryo.

Together, the three groups of maternal genes listed in Table 18.1 establish a spatial pattern of segment-organizing activities during oogenesis. Following fertilization and the appearance of embryonic nuclei, these genes regulate the transcriptional activity of nuclei along the anterior-posterior axis of the egg.

The transcription of the dorsal-ventral segmentation genes begins at the blastoderm stage and is restricted to a number of narrow bands that extend circumferentially around the embryo. The temporal expression of this gene set establishes the boundaries of body segments and compartments, and the developmental fate of the cells within each segment.

Many of the segmentation genes belong to two gene complexes located on chromosome 3: the *Antennapedia* complex (*Ant-C*) and the *bithorax* complex (*BX-C*). It has been postulated that genes in these complexes represent *selector* genes. Such genes determine a particular developmental pathway for cells in a restricted portion of the embryo. In different segments and compartments, the activation of unique combinations of selector genes produces the differentiated segmental structures of the larva and adult bodies.

Mutant selector genes called **homeotic mutants** shift the determined state of a group of cells in a compartment or segment to form structures normally seen in another segment. For example, the mutant gene *tumorous head* transforms the head into the last abdominal segment and genitalia. The *Antennapedia* complex (*ANT-C*) controls segmentation of the head and anterior regions of the fly, while the **bithorax complex (BX-C)** acts to regulate the development of the middle and posterior regions of the

TABLE 18.1
Maternal effect genes involved in anterior–posterior axis formation in *Drosophila*.

Anterior Group	Posterior Group	Terminal Group
bicoid	*nanos*	*torso*
swallow	*pumilio*	*(l)polehole*
exuperantia	*vasa*	*torsolike*
	staufer	*trunk*
	cappuccino	*fs(l)nasrat*
	spire	*fs(l)polehole*
	valois	
	oskar	
	tudor	

body. The genes of the *bithorax* complex are arranged on the chromosome in an order that corresponds to the anterior-posterior arrangement of body segments in which they are active (Figure 18.9). In the mutant condition, each of these genes transforms a segment into structures normally formed by other segments, presumably by activating the wrong number of loci in the complex (Figure 18.10).

Organization and Regulation of Homeotic Genes

Because of their important role in controlling development, *ANT-C* and *BX-C* genes have been analyzed using recombinant DNA techniques and molecular cloning. Some 260 kilobases in the distal region of the *ANT-C* complex have been cloned, and the entire *BX-C* region, spanning almost 300 kb of DNA, has been cloned. The results to date have been somewhat surprising in several respects. The recombinant DNA studies have largely confirmed the order of genes in each complex, but they show that the complexes are very large—on the order of 200 to 300 kilobases of DNA, instead of the 8 to 10 kb predicted from genetic studies. An average gene with no introns is usually thought to be about 1 kb in length, so these regions have enormous coding capacities. Many of the genes in these complexes are also very large. The *Antennapedia* gene covers more than 100 kb of DNA and includes two promoters, eight exons, and two terminator processing regions. One of the introns in this gene covers almost 60 kb. Similarly, in *BX-C,* the *Ultrabithorax (Ubx)* gene covers some 70 kb.

Another unexpected and important discovery is that most of the mutations in this region are not point mutations involving a change in one or even a few bases, but instead are insertions or deletions covering several thousand base pairs of DNA. The ends of these insertions have

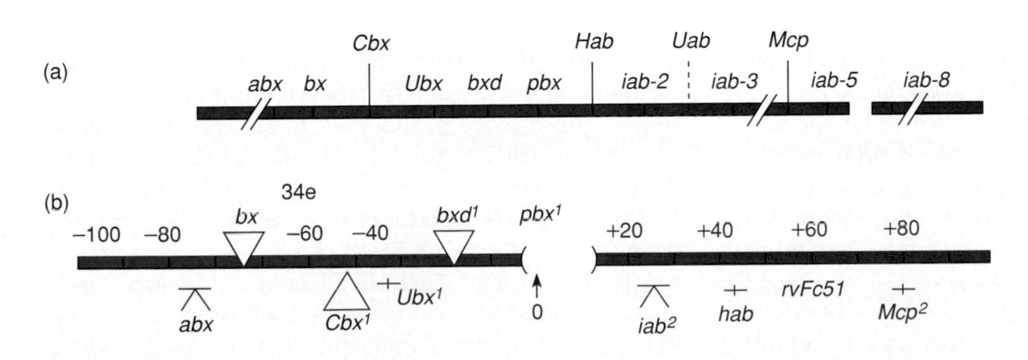

FIGURE 18.9
(a) Map of the *bithorax* locus in *Drosophila* derived from genetic studies. Mutations in the upper row (*Cbx, Hab, Uab, Mcp*) transform anterior segments into posterior ones. Mutations in the lower row cause posterior segments to become more like anterior ones. (b) Map derived from studies using recombinant DNA. The zero and the arrow indicate the location of the first clone recovered. Triangles and inverted Vs are insertions; lines are inversions; and *pbx* is a deletion.

FIGURE 18.10
Wild-type fly (left) compared to a fly with two pairs of wings (right) produced by a combination of *bithorax* mutants.

characteristics in common with the transposable elements found in the genomes of other higher organisms.

Not only is there evidence that some DNA regions in the complex may be transposable, but it appears that changing the position of a section of DNA alters its pattern of expression. The pbx^1 mutation is caused by the removal of a 17-kb DNA segment, and Cbx^1 is caused by the insertion of the same segment some 40 kb upstream from its original position. While pbx^1 transforms the rear of the haltere into the rear of the wing, Cbx^1 produces the opposite effect. The conclusion is that the 17-kb piece encodes the information for the rear half of the haltere (or, more generally, the rear half of the third thoracic segment), and the segment in which this information is expressed depends on the location of the DNA within the *bithorax* complex.

Molecular studies of the transcription and translation of genes in the *bithorax* cluster provide evidence for the mechanism of action of homeotic genes. Experiments show that homeotic genes within a complex are expressed in order and are maximally expressed in the segments which they specify. The results of such experiments have led to a model of homeotic gene action, which proposes that the action of a homeotic gene, which is active in a more anterior segment, is repressed in more posterior segments by the activity of genes located more posteriorly in the complex. This idea is supported by genetic experiments demonstrating that anterior genes are strongly expressed in more posterior segments when the next (more posterior) gene in the complex is deleted.

All of the steps in determining the segmental arrangement of the body as well as the activation of the correct combination of selector genes depend on gene products arranged into a positional array in the cortex of the egg. Some of these genes must act to control the action of the homeotic selector genes by providing positional cues in the early embryo. Such genes should be active in the maternal genome and, in the mutant condition, have some effect on the expression of homeotic genes. A number of genes fulfilling these criteria have been identified, including *extra sex combs (esc), Polycomb (Pc), super sex combs (sxc),* and *trithorax (trx).* In the second chromosomal mutant *extra sex combs (esc),* some of the head and all of the thoracic and abdominal segments develop as posterior abdominal segments (Figure 18.11), indicating that this gene normally controls the expression of *BX-C* genes in all body segments. These findings demonstrate that the esc^+ gene product that is synthesized and stored in the egg by the maternal genome may be required for the correct interpretation of the information gradient in the egg cortex.

Genes in the *bithorax* complex, and perhaps other genes involved in segmentation, may have arisen from a common ancestral gene by tandem duplication and subsequent divergence of function. In fact, recent evidence has shown that genes of the *BX-C* and *ANT-C* share a highly conserved DNA sequence of about 200 bp called a *homeo box* or *homeodomain.* Similar sequences have been found in the genomes of other eukaryotes with segmented body plans, including *Xenopus,* chicken, mice, and humans. Homeodomains from all organisms examined to date are very similar in DNA sequence and encode a protein associated with the transcriptional regulation of a specific gene set. This suggests that the metameric or segmented body plan may have evolved only once.

The examples cited represent the striking influence of the cellular and extracellular environment on the fate of cells during development. In many other species, deter-

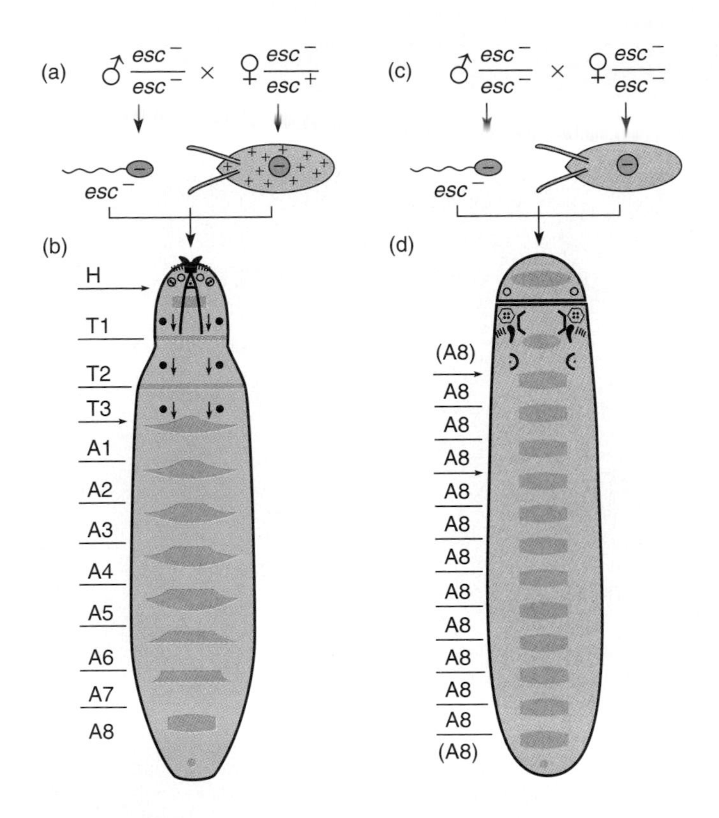

FIGURE 18.11
Action of the *extra sex combs* mutation in *Drosophila*.
(a) Heterozygous females form wild-type *esc* gene product and store it in the oocyte. (b) Fertilization of esc^- egg formed at meiosis by the heterozygous female by esc^- sperm still produces wild-type larva with normal segmentation pattern (maternal rescue). (c) Homozygous esc^- females produce defective eggs which, when fertilized by esc^- sperm, (d) produce a larva in which most of the segments of the head, thorax, and abdomen are transformed into the eighth abdominal segment. Maternal rescue demonstrates that the esc^- gene product is produced by the maternal genome and is stored in the oocyte for use in the embryo. (H, head; T1–3, thoracic segments; A1–8, abdominal segments). Borderlines between the head and thorax and between thorax and abdomen are marked with arrows.

mination is not demonstrated as early as it is in *Drosophila,* and it is not as easy to pinpoint and document. This does not imply, however, that determination does not occur. Rather, it seems just a matter of timing during development.

THE STABILITY OF THE DIFFERENTIATED STATE

Once a cell has taken on a specific structural and functional capacity, under normal circumstances in an adult organism the cell maintains that biological status. That is, kidney epithelial cells, red blood cells, muscle cells, and so on are not normally converted into other cell types. Such observations have led to the question of whether or not differentiation can be reversed under experimental conditions. Differentiation is a two-step process, beginning with determination and followed by the biochemical and morphological changes that lead to the differentiated state. Thus the question must be addressed in two steps: 1) can we reverse determination, and 2) can we reverse differentiation?

Recall that we have already established that nuclear differentiation in specialized cell types is not necessarily stable. This conclusion was drawn based on the nuclear transplantation experiments discussed earlier in this chapter. As we will see in the following examples, there are certain cases where differentiated cells, and not just nuclei, may alter their developmental status.

Transdetermination

Ernst Hadorn demonstrated that if imaginal disks are explanted from a larva and implanted into the abdomen of an adult, these primordia will continue to proliferate by cell division but will not differentiate. The disk subsequently may be recovered from the adult, cut in half, and reimplanted—part to another adult abdomen and part back into a larva (Figure 18.12). The half that was implanted into a second adult continues to proliferate and can be serially propagated in this way. The part placed into the larva will undergo metamorphosis and differentiate into an adult structure.

Hadorn showed that during the early transfers, a particular disk primordium always maintained its determined state (i.e., a leg disk always differentiated into leg structures) following experimental manipulation. After several passages through adult abdomens, atypical structures were occasionally discovered; that is, an antennal disk sometimes produced wing structures, a genital disk gave rise to leg structures, and so forth. This shift in determination is referred to as **transdetermination.**

The major forms of transdetermination are summarized in Figure 18.13. All disks can be transdetermined into one or another disk type, and certain sequences of transdetermination occur; a disk can give rise only to a limited number of other disks.

Hadorn's experiments showed that developmental programming or determination can be altered by serially "subculturing" disk cells in adult abdomens. This transdetermination involves complete cells, not just the nucleus (as in nuclear transplantation experiments). Thus, it may be concluded that irreversible changes do not occur during preadult development of imaginal disks. Homeotic mutants show many of the properties of transdetermination; in fact, most of the transdeterminations are also observed as homeotic mutations, emphasizing that these developmental programs are under genetic control.

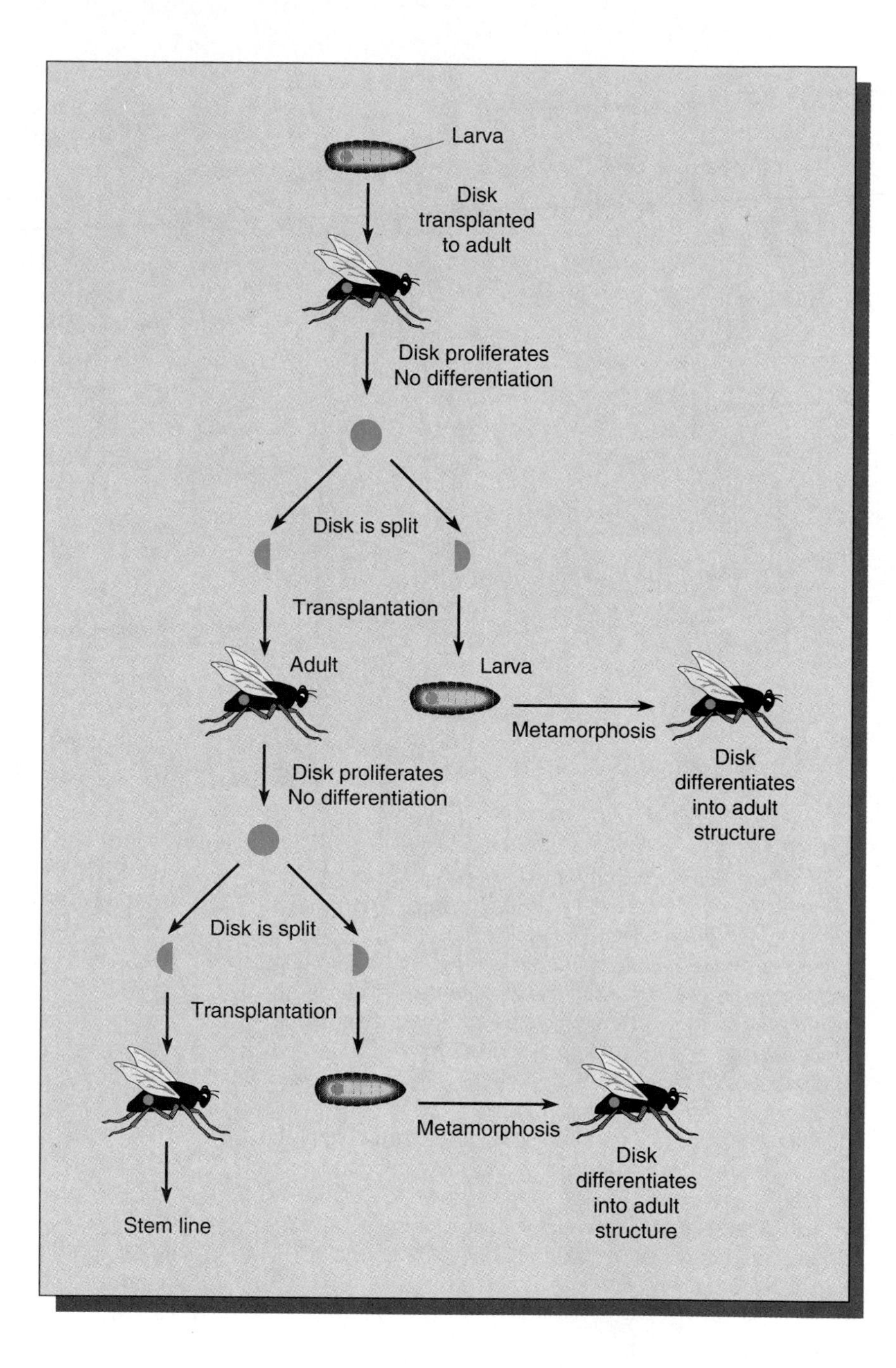

FIGURE 18.12
In vitro culturing of imaginal disk fragments. In each generation, a disk fragment is implanted into a larva to test its developmental potential and fate.

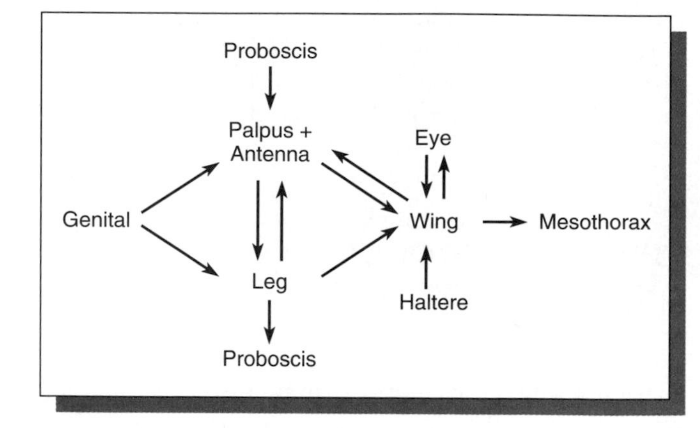

FIGURE 18.13
Observed sequence of transdeterminations from each imaginal disk.

Knowledge gained from the study of the morphological changes and the evidence for differential transcription in *Drosophila* development has had an impact on the study of genetic control of behavior in this organism. In addition, investigators on a nematode *Caenorhabditis elegans* have underscored the relationship between development and behavior. In the following section, we will explore the genetic analysis of behavior in selected organisms, often employing techniques and methodologies of developmental biology.

GENETIC ANALYSIS OF BEHAVIOR

Behavior is defined generally as reaction to stimuli or environment. In broad terms, every action, reaction, and response represents a type of behavior. Animals run, remain still, or counterattack in the presence of a predator; birds build complex and distinctive nests; fruit flies execute intricate courtship rituals; plants bend toward light; and humans reflexively avoid painful stimuli as well as "behave" in a variety of ways as guided by their intellect, emotions, and culture.

Even though clear-cut cases of genetic influence on behavior were known in the early 1900s, the study of behavior was of greater interest to psychologists who were concerned with learning and conditioning. Since about 1950, studies of the genetic component of behavioral patterns have intensified, and support for the importance of genetics in understanding behavior has increased. The prevailing view is that all behavior patterns are influenced both genetically and environmentally. The genotype provides the physical basis essential to execute the behavior and further determines the limitations of environmental influences. Behavior genetics has blossomed into a distinct specialty within the larger field of genetics as more and more behaviors have been found to be under genetic control.

DROSOPHILA BEHAVIOR GENETICS

The Genetic Dissection of Behavior in *Drosophila*

In 1967, Seymour Benzer and his colleagues initiated a comprehensive study of behavior genetics in *Drosophila*. Benzer's approach is an excellent example of the use of genetic techniques to "dissect" a complex biological phenomenon into its simpler components. Furthermore, the use of such techniques allows geneticists to study the underlying basis of the phenomenon—in this case, behavior. Benzer's goals in this research illustrate the "genetic dissection" approach:

1. To discern the genetic components of behavioral responses by the isolation of mutations that disrupt normal behavior.
2. To identify the mutant genes by chromosome localization and mapping.
3. To determine the actual site within the organism at which the gene expression influences the behavioral response.
4. To learn, if possible, how the particular gene expression influences behavior.

All four steps are illustrated in a discussion of **phototaxis,** one of the first behaviors studied by Benzer. Normal flies are positively phototactic; that is, they move toward a light source. Mutations are induced by feeding male flies sugar water containing **EMS** (ethylmethanesulfonate, a potent mutagen) and mating them to attached-X virgin females. As shown in Figure 18.14, the F_1 males receive their X chromosome from their fathers. Because they are hemizygous, any sex-linked recessive mutations are expressed.

The F_1 males are tested for their response to light, and those that are not attracted to it are isolated. Benzer found *runner* mutants, which move quickly to and from light; *negatively phototactic* mutants, which move away from light; and *nonphototactic* mutants, which show no preference for light or darkness. He established that the behavior changes were due to mutations by mating these F_1 males to attached-X virgin females. Male progeny of this cross also showed the abnormal phototactic responses, confirming them as products of sex-linked recessive mutations.

We shall now delve into further work concerning just one group, the *nonphototactic* mutants. Such flies behave in the light as normal flies do in the dark. They can walk normally, but respond to light as if they were blind. Benzer and Yoshiki Hotta tested the electrical activity at the surface of mutant eyes in response to a flash of light. The pattern of electrical activity was recorded as an **electroretinogram.** Various types of abnormal responses were detected, and in none of the mutants was a normal pattern observed. When these mutations were mapped, they were not all allelic; instead, they were shown to occupy several loci on the X chromosome. Thus, it can be concluded that several gene products contribute to the formation of the behavioral response to light.

Where, within the fly, must adequate gene expression occur to yield a normal pattern? In an ingenious approach aimed at answering this question, Benzer turned to the use of mosaics. In mosaic flies, some tissues are mutant and others are wild type. If it can be ascertained

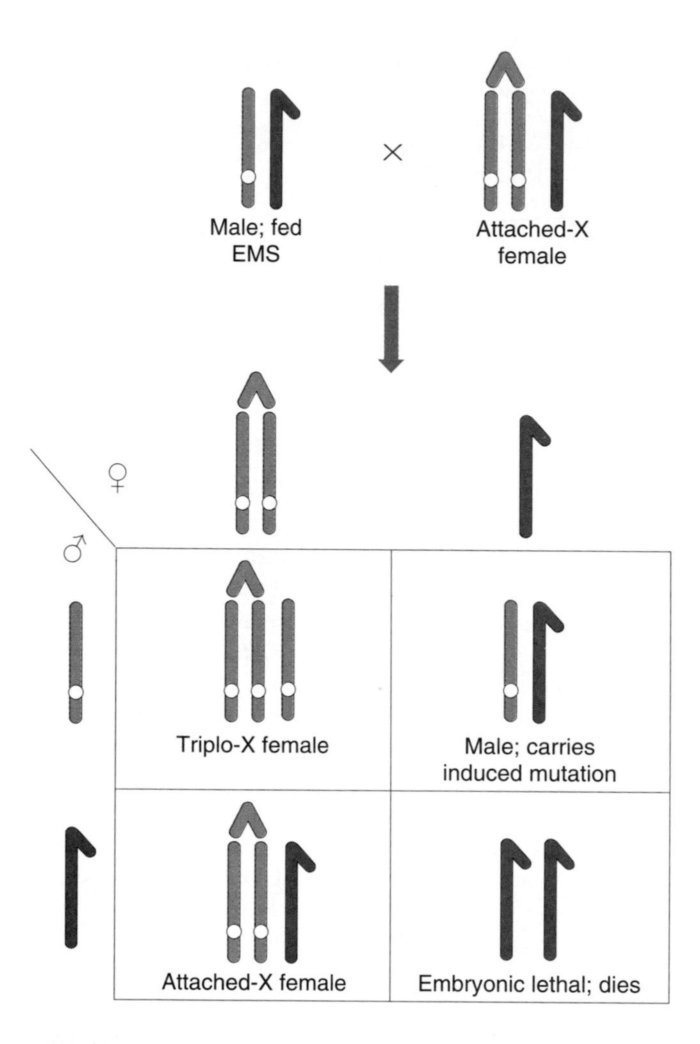

FIGURE 18.14
Genetic cross in *Drosophila* that facilitates the recovery of X-linked induced mutations. The female parent contains two X chromosomes that are attached, in addition to a Y chromosome. In a cross between this female and a normal male that has been fed the mutagen ethylmethanesulfonate, all surviving males receive their X chromosomes from their father and express all mutations induced on that chromosome.

which part must be mutant in order to yield the abnormal behavior, the **primary focus** of the genetic alteration can be determined.

To facilitate the production of mosaic flies, Benzer used a strain that has one of its X chromoxomes in an unstable ring shape. When present in a zygote undergoing cell division, the ring-X is frequently lost by nondisjunction. If the zygote is female and has two X chromosomes (one normal and one ring-X), loss of the ring-X at the first mitotic division will result in two cells—one with a single X (normal X) and one with two X chromosomes (one normal and one ring-X). The former cell goes on to produce male tissue (XO) and expresses all alleles on the remaining X, while the latter produces female tissue and does not express heterozygous recessive X-linked genes. Such an occurrence is illustrated in Figure 18.15. One can see that loss of the ring-X will produce mosaic flies with male and female parts that express or do not express sex-linked recessive genes, respectively.

In the embryo, when and where the ring-X is lost determines the pattern of mosaicism. The loss usually occurs early in development, before the cells migrate to the surface of the blastula to form the **blastoderm.** As shown in Figure 18.16, depending on the orientation of the spindle when the loss occurs, different types of mosaics will be created. If the stable X chromosome contains the behavior mutation and an obvious mutant gene (*yellow,* for example), the pattern of mosaicism will be readily apparent. For example, the fly may consist of a mutant head on a wild-type body, a normal head on a mutant body, one normal and one mutant eye on a normal or mutant body, and so on.

When such mosaics for *nonphototactic* mutants were studied, it was found that the **focus** of the genetic defect was in the eye itself. In mosaics where every part of the fly except the eye was normal, abnormal behavior was still detected. Even if only one eye was mutant, a modified abnormal behavior was observed. Instead of crawling straight up toward light as the normal fly does, the single mutant-eyed fly crawls upward to light in a spiral pattern. In the dark, such a fly will move in a straight line. Thus, mosaic studies have established that the focus of *nonphototactic* gene expression is in the eye itself. Additionally, the abnormal behavior is due to altered electrical conductance of the cells of the eye.

Benzer and other workers in the field have identified a large number of genes affecting behavior in *Drosophila.* As shown in Table 18.2, mutants have been isolated that affect locomotion, response to stress, circadian rhythm, sexual behavior, visual behavior, and even learning. The mutations have received very descriptive and often humorous names.

Many mutations have been analyzed with the mosaic technique in order to localize the focus of gene expression. While it was easy to predict that the focus of the *nonphototactic* mutant would be in the eye, other mutants are not so predictable. For example, the focus of mutants affecting circadian rhythms has been located in the head, presumably in the brain. The *wings up* mutant might have a defect in the wings, articulation with the thorax, the thorax musculature, or the nervous system. Mosaic studies have pinpointed the indirect flight mus-

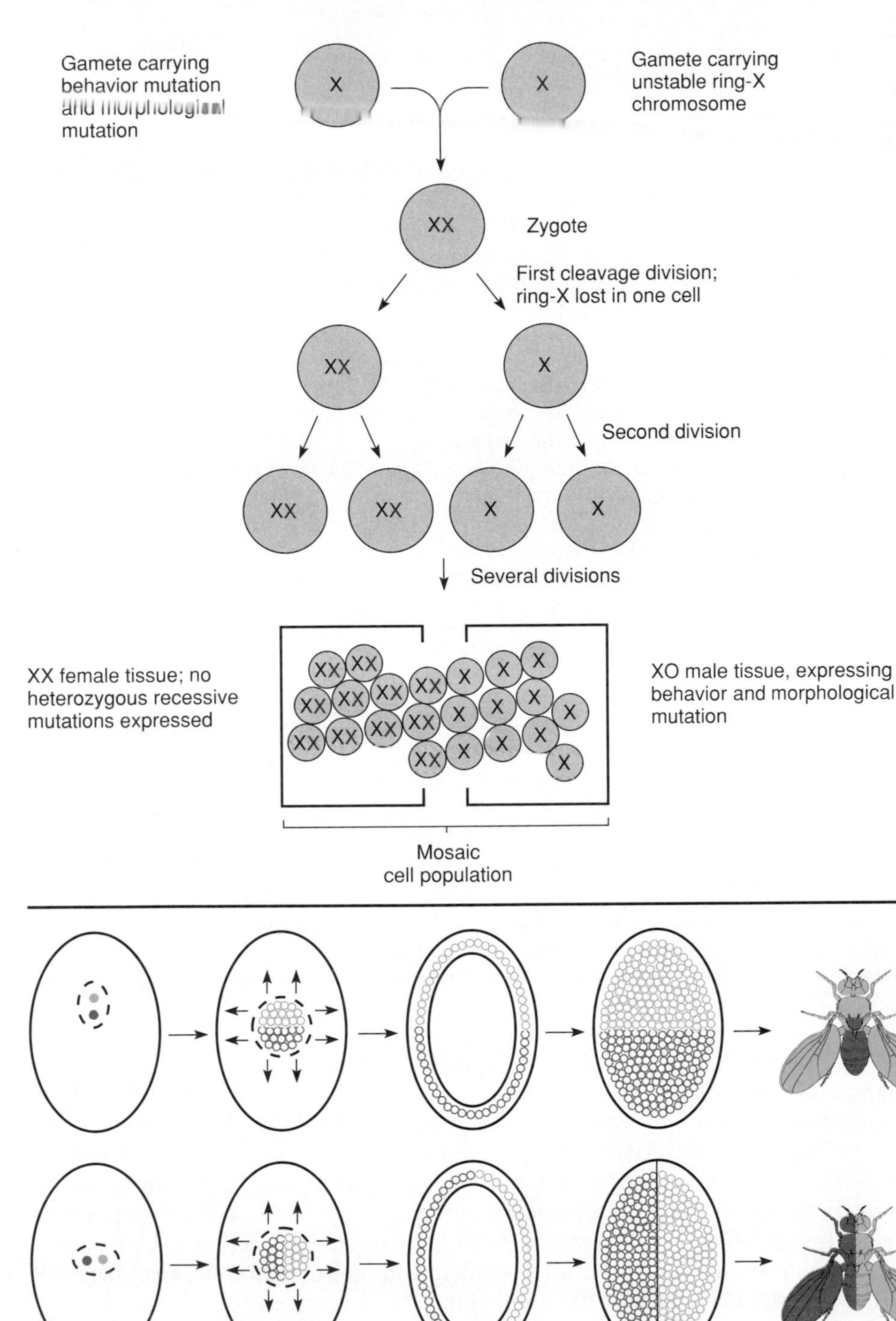

FIGURE 18.15
Production of a mosaic fruit fly as a result of fertilization by a gamete carrying an unstable ring-X chromosome (shown in color). If this chromosome is lost in one of the two cells following the first mitotic division, the body of the fly will consist of one part which is male (XO) and the other part, which is female (XX). The male side will express all mutations contained on the X chromosome.

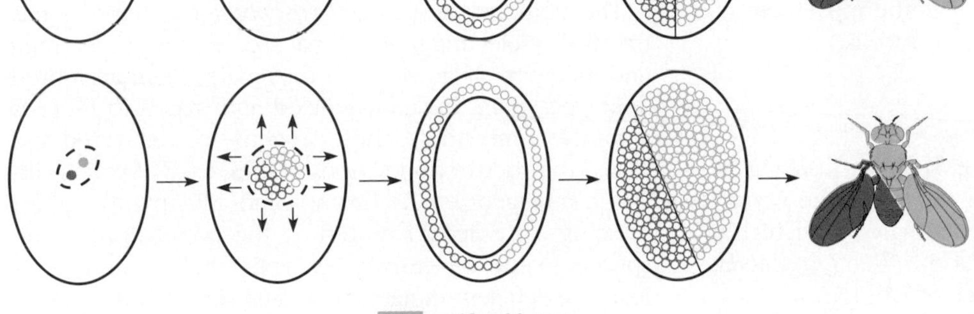

FIGURE 18.16
The effect of spindle orientation on the production of mosaic flies as shown in Figure 18.15.

TABLE 18.2
Some behavioral mutants of *Drosophila*.

Class	Name	Characteristics
Locomotor	*sluggish*	Moves slowly
	hyperkinetic	High consumption of O_2; shaking of legs, early death
	wings up	Wings perpendicular to body
	flightless	Does not fly well, although wings are well developed
	uncoordinated	Lacks coordinated movements
	nonclimbing	Fails to climb
Response to stress	*easily shocked*	Mechanical shock induces "coma"
	stoned	Stagger induced by mechanical shock
	shaker	Vibrates all legs while etherized
	freaked out	Grotesque, random gyrations under the influence of ether
	paralyzed	Collapses above a critical temperature
	parched	Dies quickly in low humidity conditions
	tko	Epilepticlike response
	comatose	Paralyzed by cool temperature
	out-cold	Similar to comatose
Circadian rhythm	$period^o$	Eclosion at any time; locomotor activity spread randomly over the day
	$period^s$	19-hour cycle rather than 24-hour cycle
	$period_f$	28-hour cycle rather than 24-hour cycle
Sexual	*savoir-faire*	Males unsuccessful in courtship
	fruity	Males pursue each other
	stuck	Male is often unable to withdraw after copulation
	coitus interruptus	Males disengage in about half the normal time
Visual	*nonphototactic*	Blind
	negatively phototactic	Moves away from light
Learning	*dunce*	Fails to learn conditioned response

cles of the thorax as the focus. Cytological studies have confirmed this finding, showing a complete lack of myofibrils in these muscles. Temperate-sensitive *paralytic* mutants are paralyzed at a relatively high temperature (29°C), but recover rapidly if the temperature is lowered. Mosaic studies have revealed that both the brain and thoracic ganglia represent the focus causing this abnormal behavior. More recent work has shown that mutant flies have defective sodium channels, and that the *paralytic* locus encodes a protein that controls the movement of sodium across the membrane of nerve cells.

Learning in *Drosophila*

To study the genetics of behavior such as learning, it would be advantageous to use an organism like *Drosophila,* in which methods of genetic analysis are highly advanced. However, the first question is: Can *Drosophila* learn? Recent work from a number of laboratories indicates that organisms such as *Drosophila* are, in fact, capable of learning. Using a simple apparatus, flies are presented with a pair of olfactory cues, one of which is associated with an electrical shock. Flies learn quickly to avoid the odor associated with the shock. That this response is learned is indicated by a number of factors: first, performance is associated with the pairing of a stimulus/response with a reinforcer; second, the response is reversible. Flies can be trained to select an odor that they previously avoided, and flies exhibit short-term memory for the training they have received.

The demonstration that *Drosophila* can learn opens the way to selecting mutants that are defective in learning and memory. To accomplish this, males from an inbred wild-type strain are mutagenized and mated to females from the same strain. Their progeny are recovered and mated to produce populations of flies, each of which carries a mutagenized X chromosome. Mutants that affect learning are selected by testing the population for response in the olfactory/shock apparatus. A number of learning-deficient mutants including *dance, turnip, rutabaga,* and *cabbage* have been recovered. In addition, a memory-deficient mutant *amnesiac* learns normally, but forgets four times faster than normal. Each of these mutations represents a single gene defect that affects a specific form of behavior. Because of the method used to recover them, all the mutants found so far are X-linked genes. Presumably similar genes controlling behavior are also located on autosomes.

Molecular Biology of Behavior

Since, in many cases, mutation results in the alteration or abolition of a single protein, biochemical study of the learning mutants described previously can provide a link between behavior and molecular biology. The *dunce* mutation is one of the first for which this link has been established. The locus for *dunce* has been shown to encode the structural gene for the enzyme cyclic AMP phosphodiesterase. It appears that *rutabaga* is a mutation in the structural gene for adenylate cyclase and that the *turnip* mutation is in the gene for a GTP-binding protein associated with adenylate cyclase. The unexpected clustering of these independently derived mutations in the biochemical pathway of the adenyl cyclase system suggests a role for cyclic nucleotides in learning. This conclusion is consistent with work in the sea hare *Aplysia,* indicating that short-term memory is associated with an increase in cyclic nucleotides within neurons. Further work will be necessary to clarify the role of such molecules in behavior.

Because behavior is controlled and modified by cells of the nervous system, recent studies in *Drosophila* have centered on the isolation and characterization of mutations that affect the generation and propagation of electrical signals in the nervous system. These mutations have been identified by screening behavioral mutants for electrophysiological abnormalities. Two general classes of mutations have been isolated: those affecting sodium channels, and those affecting potassium channels (Table 18.3). Using recombinant DNA techniques, the genes resonsible for these mutations are being cloned; the *Shaker* mutation, originally identified over 40 years ago as a behavioral mutant, has been cloned and found to encode a potassium channel structural protein. These cloned sequences can be used to answer fundamental questions about the function of the nervous system and the molecular basis of memory and learning. Certainly, work undertaken in this field in the next few years will be important and exciting.

Behavior Genetics of a Nematode

In 1968, in one of the boldest attempts to define the total genetic influence on the behavior of a single organism, Sidney Brenner began an investigation of the nematode *Caenorhabditis elegans.* He hoped that it would be possible to dissect genetically the nervous system of *C. elegans,* using techniques previously applied successfully to other organisms.

He chose this nematode because it was possible to determine the complete structure of the nervous system. Adult worms are about 1 mm long and are composed of only 959 cells, about 300 of which are neurons. As a result, it is possible to create a three-dimensional model of the entire organism and its nervous system. Brenner then hoped to induce large numbers of behavioral mutations and to correlate aberrant behavior with structural and biochemical alterations in the nervous system. Because of the vast scope of this research endeavor, work is still underway. Some progress has been made, particularly in isolating large numbers of mutations. Three classes of behavioral mutants have been characterized. The worms are positively chemotactic to a variety of stimuli (cyclic AMP and GMP; anions such as Cl^-, Br^-, and I^-; and cations such as Na^+, Li^+, K^+, and Mg^{++}). As shown in Figure 18.17, positive attraction can be tracked on agar

TABLE 18.3
Behavioral mutants of *Drosophila* affecting nerve impulse transmission.

Mutation	Map Location	Ion Channel Affected	Phenotype
*nap^ts^*	2–56.2	sodium	Adults, larvae paralyzed at 37.5°C. Reversible at 25°C.
*para^ts^*	1–53.9	sodium	Adults paralyzed at 29°C, larvae at 37°C. Reversible at 25°C.
tip-E	3–13.5	sodium	Adults, larvae paralyzed at 30–40°C. Reversible at 25°C.
*sei^ts^*	2–106	sodium	Adults paralyzed at 38°C, larvae unaffected. Adults recover at 25°C.
Sh	1–57.7	potassium	Aberrant leg shaking in adults exposed to ether.
eag	1–50.0	potassium	Aberrant leg shaking in adults exposed to ether.
Hk	1–30.0	potassium	Ether-induced leg shaking.
slo	3–85.0	potassium	At 22°C, adults are weak fliers; at 38°C, adults are weak, uncoordinated.

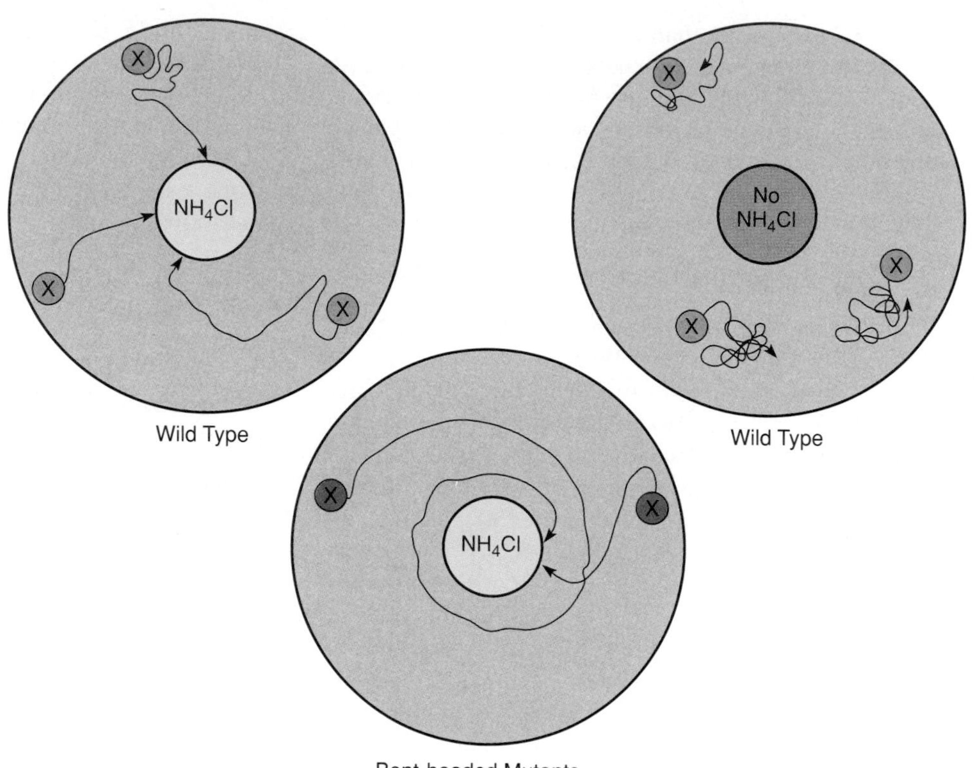

FIGURE 18.17
Chemotactic response to ammonium chloride (NH_4Cl) of wild-type and mutant *Caenorhabditis*.

plates. The study of mutants has shown that sensory receptors in the head alone mediate the orientation responses to attractants.

A second class of behavior studied involves *thermotaxis*. Cryophilic mutants move toward cooler temperatures, and thermophilic mutants move toward warmer temperatures. However, this behavior has not yet been correlated with the responsible component of the nervous system.

The third class of behavior involves generalized movement on the surface of an agar plate. Of 300 induced mutations, 77 affected the movement of the animal. While wild-type worms move with a smooth, sinuous pattern, mutants are either **uncoordinated** (*unc*) or **rollers** (*rol*). Those that are uncoordinated vary from the display of partial paralysis to small aberrations of movement, including twitching. Rollers move by rotating along their long axis, creating circular tracks on an agar surface. Of these mutants, some have been correlated with defects in the dorsal or ventral nerve cord or in the body musculature.

Linkage mapping has also begun. All mutants are distributed on six linkage groups, corresponding to the haploid number of chromosomes characteristic of this organism. Numerous *unc* mutants are found on each of the six chromosomes, indicating extensive genetic control of nervous system development.

Brenner's work is being extended in many laboratories throughout the world. In fact, an international congress now meets regularly to discuss work on *C. elegans*. Research now includes genetic analysis of development, and nongenetic problems are also being approached. This organism promises to rank with *Drosophila* in the amount of information acquired about it. It is hoped that study of *C. elegans* will unlock the mysteries of how genes control the structure of the nervous system.

HUMAN BEHAVIOR GENETICS

The genetic input to behavior in humans is more difficult to characterize than that of other organisms. Not only are humans unavailable as experimental subjects in genetic investigations, but the types of responses considered to be interesting behavior are extremely difficult to study. The most popular forms of studied behavior all include some aspects of **intelligence, language, personality,** or **emotion.** There are two problems in examining such traits. First, all are difficult to define objectively and to measure quantitatively. Second, they are the traits most affected by the environment. In each case, while there is undoubtedly a genetic basis, it is a complex one. Furthermore, the environment is extremely important in shaping, limiting, or facilitating the final phenotype for each trait.

The study of human behavior genetics has also been hampered by two other factors. Many studies of human behavior have been performed by psychologists without adequate input from the biologist or geneticist. Second, traits involving intelligence, personality, and emotion have the greatest social and political significance. As such, these traits are more likely to be the subject of sensationalism when reported to the public. Because their study comes closest to infringing upon individual liberties such as the right to privacy, these traits are the basis of the most controversial investigations.

In lamenting the gulf between psychology and genetics in explaining human behavior genetics, C. C. Darlington in 1963 wrote, "Human behavior has thus become a happy hunting ground for literary amateurs. And the reason is that psychology and genetics, whose business it is to explain behavior, have failed to face the task together." Since 1963, some progress has been made in bridging this gap, but the genetics of human behavior remains a controversial area.

Mental Disorders with a Clear-Cut Genetic Basis

Many genetic disorders in humans result in some behavioral abnormality. One of the most prominent examples is **Huntington disease (HD).** Inherited as an autosomal dominant disorder, it affects the nervous system, including the brain. Symptoms of HD usually appear in the fifth decade of life with a gradual loss of motor function and coordination. Degeneration of the nervous system is progressive, and personality changes occur. The affected individual soon is unable to care for himself. Most victims die within 10 to 15 years after onset of the disease. Since onset usually occurs after a family has been started, all children of an affected person must live with the knowledge that they face a 50 percent probability of developing the disorder. The gene for HD maps to the tip of the short arm of chromosome 4 and is associated with elevated brain levels of quinolinic acid, a naturally occurring neurotoxin. Although the gene has not yet been isolated, nearby molecular markers can be used in RFLP analysis (Chapter 15) to diagnose those carrying the dominant allele before symptoms appear.

The **Lesch-Nyhan syndrome** is inherited as a sex-linked recessive disorder. Onset is within the first year, and the disease is most often fatal early in childhood. The disorder is of a metabolic nature, involving purine biosynthesis. Affected individuals lack hypoxanthine-guanine phosphoribosyltransferase (HGPRT), and accumulate high levels of uric acid. Mental and physical retardation occur, and these individuals demonstrate uncontrolled self-mutilation.

Human Behavior Traits with Less-Defined Genetic Bases

Other aspects of human behavior, notably **schizophrenia** and **manic-depressive illness,** have been the subject of extensive study. Investigations have sought to relate the development of the mental disorder or the display of intelligence to the closeness of family relationships or twin studies. In all cases, it has been concluded that a genetic component influences the trait, but that environment also plays a substantial role.

The presence of schizophrenia in monozygotic and dizygotic twins has been the subject of many studies. In almost every investigation, concordance has been higher in monozygotic twins than in dizygotic twins reared together. Although these results suggest that a genetic component exists, they do not reveal the precise genetic basis of schizophrenia. Simple monohybrid and dihybrid inheritance as well as multiple gene control have been proposed for schizophrenia. However, it seems unlikely that only one or two loci are involved, nor is it likely that the control is strictly quantitative, as in polygenic inheritance. In both schizophrenia and manic-depressive illness, it is most sound to conclude that each individual is endowed with a genetic predisposition for normal or abnormal behavior, and that environmental factors can serve to alter the final phenotype.

There is a long-standing controversy about genetic differences in intelligence between races. While IQ testing has established intelligence differences in populations of different races, there is currently no strong evidence to support the conclusion that this is due to a genetic component. With regard to intelligence, genetic factors may provide an upper and lower potential range, but an individual's environment may modify the development of intelligence. The environment undoubtedly has a profound effect on the type of intelligence measured by the various forms of IQ tests. It seems likely that the genetic component of intelligence does not differ any more significantly between races than it does between individuals within the same race.

CHAPTER SUMMARY

1 The varible gene activity theory, which applies to higher organisms, assumes that all somatic cells in an organism contain equivalent genetic information, but that it is differently expressed. Studies conducted on genomic equivalence have demonstrated that somatic cells undergoing differentiation retain the entire gene set, which can be reprogrammed to direct the development of the entire organism.

2 During embryogenesis, specific gene activity appears to be affected by the total environment of the cell, which includes the cell's individual cytoplasm as well as the maternal cytoplasm. As development proceeds, the cell's environment becomes further altered by the presence of other cells and early gene products.

3 In *Drosophila,* genetic and molecular studies have confirmed that the egg contains information that specifies the body plan of the larva and adult, and that interaction of embryonic nuclei with the maternal cytoplasm initiates a transcriptional program characteristic of specific developmental pathway.

4 Behavioral genetics has emerged as an important specialty within the field of genetics because both genotype and environment have been found to have an impact in determining an organism's behavioral response.

5 By isolating mutations that cause deviations from normal behavior, the role of a corresponding wild-type allele in the respective response is established. Numerous examples have been examined, including general movement in nematodes and a variety of behaviors in *Drosophila.*

6 Any aspect of human behavior is difficult to study because the individual's environment makes an important contribution toward trait development. In humans, studies using twins have shown that while a family may have a predisposition to schizophrenia, the expression of this disorder may be modified by the environment. General intelligence is also a product of both genetic and environmental influences.

KEY TERMS

bithorax complex (BX-C)
blastoderm
compartment
electroretinogram
fate maps
focus
homeotic mutants
Huntington disease (HD)
imaginal disks
Lesch-Nyhan syndrome
manic-depressive illness
myxamoebae
phototaxis
primary focus
rollers (*rol*)
schizophrenia
totipotent
transdetermination
uncoordinated (*unc*)
variable gene activity hypothesis

The timing of differential gene action during development is a key factor in the normal developmental program. If a given gene has been cloned, the time of action and range of cell types in which the gene is active can be determined using a number of molecular techniques. However, when a cloned gene or its transcripts are not available, genetic methods can be used to establish a comparative order of gene action among two or more genes. By extending this analysis to include several genes, a relative order of gene action can be established. This order can serve as a starting point for investigations at the molecular level by providing a developmental time scale for gene action. The following example shows how such a relative time scale for the action of two genes can be constructed.

In *Drosophila,* the autosomal recessive gene *lozenge-clawless* (lz^{cl}) produces multiple abnormalities of the female genitals, eyes, and tarsal regions of the legs. Homozygous females have abnormal genitals, no sperm storage organs, no ovarian glands, and are sterile. Homozygous mutant males, on the other hand, have normal genitals and are fully fertile. A second autosomal recessive gene, *transformer (tra)* converts XX females into phenotypic males (recall that in *Drosophila* XX flies are female, XY flies are male). XX flies homozygous for both lz^{cl} and *tra* are phenotypically male with normal genitals.

What do these results say about the normal sequence of action of these two genes? Which one acts first during development?

ANSWER: The flies in question are genetically female, since they have two X chromosomes. Females homozygous for the lz^{cl} gene are expected to have abnormal genitalia. In this case, the action of the *tra* gene in changing the female phenotype into the male phenotype (with the development of male genitalia) must take place before the action of the lz^{cl} gene, since the XX flies have a normal male phenotype.

PROBLEMS AND DISCUSSION QUESTIONS

1 Carefully distinguish between the terms *differentiation* and *determination.* Which phenomenon occurs initially during development?

2 The *Drosophila* mutant *spineless aristapedia* (ss^a) results in the formation of a miniature tarsal structure (normally part of the leg) on the end of the antenna. Such a mutation is referred to as *homeotic.* From your knowledge of imaginal disks and Hadorn's transdetermination studies, what insight is provided by ss^a concerning the role of genes during determination?

3 In the sea urchin, early development up to gastrulation may occur even in the presence of actinomycin D, which inhibits RNA synthesis. However, if actinomycin D is removed at the end of blastula formation, gastrulation does not proceed. In fact, if actinomycin D is present only between the 6th and 11th hours of development, gastrulation (normally occurring at the 15th hour) is arrested. What conclusions can be drawn concerning the role of gene transcription between hours 6 and 15?

4 In the slime mold, the enzyme UDPG-pyrophosphorylase is essential to carbohydrate metabolism necessary for formation of the fruiting body. At a specific time during fruiting-body formation, mRNA specific to the enzyme is produced, translation of the enzyme ensues, and transcription ends. This is an example of specific gene activity during development. After transcription

of UDPG-pyrophosphorylase mRNA has stopped, the cells of the fruiting body may be dissociated and allowed to reaggregate. Again, at the same stage in the newly formed fruiting body, transcription and translation begin on schedule. What conclusions can you draw concerning the activation of the UDPG-pyrophosphorylase gene during development?

5 Nuclei from almost any species may be injected into frog oocytes. Studies have shown that these nuclei remain active in transcription and translation. How can such an experimental system be useful in developmental genetic studies?

6 The concept of epigenesis indicates that an organism develops by forming cells that acquire new structures and functions, which become greater in number and complexity as development proceeds. This theory is in contrast to the preformationist doctrine that miniature adult entities are contained in the egg that must merely unfold and grow to give rise to a mature organism. What sorts of isolated evidence presented in this chapter might have led to the preformation doctrine? Why is the epigenetic theory held as correct today?

7 Contrast the advantages of using *Drosophila* versus *Caenorhabditis* for studying behavior genetics.

8 Assume that you discovered a fruit fly that walked with a limp and was continually off balance as it moved. Describe how you would determine if this behavior was due to an injury (induced in the environment) or was an inherited trait. Assuming that it is inherited, what are the various possibilities for the focus of gene expression causing the imbalance? Describe how you would locate the focus experimentally if it were X linked.

9 In humans, the chemical phenylthiocarbamide (PTC) is either tasted or not. When the offspring of various combinations of taster and nontaster parents are examined, the following data are obtained:

PARENTS	Both tasters	Both tasters	Both tasters	One taster One nontaster	One taster One nontaster	Both nontasters
OFFSPRING	All tasters	½ tasters ½ nontasters	¾ tasters ¼ nontasters	All tasters	½ tasters ½ nontasters	All nontasters

Based on these data, how is PTC tasting behavior inherited?

10 Discuss why the study of human behavior genetics has lagged behind that of other organisms.

SELECTED READINGS

BASTOCK, M. 1956. A gene which changes a behavior pattern. *Evolution* 10:421–29.

BEACHY, P., HELFAND, S., and HOGNESS, D. 1985. Segmental distribution of bithorax complex proteins druing *Drosophila* development. *Nature* 313:545–551.

BEERMAN, W., and CLEVER, U. 1964. Chromosome puffs. *Scient. Amer.* (April) 210:50–58.

BENZER, S. 1973. Genetic dissection of behavior. *Scient. Amer.* (December) 229:24–37.

BERG, H. 1988. A physicist looks at bacterial chemotaxis. *Cold Spring Harbor Sympos. Quant. Biol.* 53:(Part 1) 1–9.

BRENNER, S. 1974. The genetics of *Caenorhabditis Elegans*. *Genetics* 77:71–94.

BRIGGS, R., and KING, T. 1952. Transplantation of living nuclei from blastula cells into enucleated frog eggs. *Proc. Nat. Acad. Sci.* 38:455–63.

DeROBERTIS, E., OLIVER, G., and WRIGHT, C. 1990. Homeobox genes and the vertebrate body plan. *Scient. Amer.* (July) 262:46–52.

DEVOR, E., and CLONINGER, C. 1989. The genetics of alcoholism. *Ann. Rev. Genet.* 23:19–36.

FOSTER, S., and JOHNSTONE, K. 1990. Pulling the trigger: the mechanism of bacterial spore germination. *Mol. Microbiol.* 4:137–41.

GODFREY, S., and SUSSMAN, M. 1982. The genetics of development in *Dictyostelium discoideum*. *Ann. Rev. Genet.* 16:385–404.

GURDON, J. 1968. Transplanted nuclei and cell differentiation. *Scient. Amer.* (December) 219:24–35.

HAKE, S. 1992. Unraveling the knots in plant development. *Trends Genet.* 8:109–14.

KAPPEN, C., SCHUGART, K., and RUDDLE, F. 1989. Organization and expression of homeobox genes in mouse and man. *Ann. N.Y. Acad. Sci.* 567:243–52.

LOSICK, R., YOUNGMAN, P., and PIGGOT, P. 1986. Genetics of endospore formation in *Bacillus subtilis*. *Ann. Rev. Genet.* 20:625–69.

MANSEAU, L., and SCHUPBACH, T. 1989. The egg came first, of course! Anterior-posterior pattern formation in *Drosophila* embryogenesis and oogenesis. *Trends Genet.* 5:400–05.

McGINNIS, W., and KRUMLAUF, R. 1992. Homeobox genes and axial patterning. *Cell* 68:283–302.

SLACK, J., and TANNAHILL, D. 1992. Mechanism of anteroposterior axis specification in vertebrates. Lessons from the amphibians. *Development* 114:285–302.

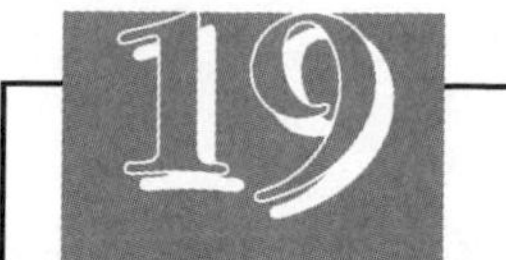

The Genetics of Cancer

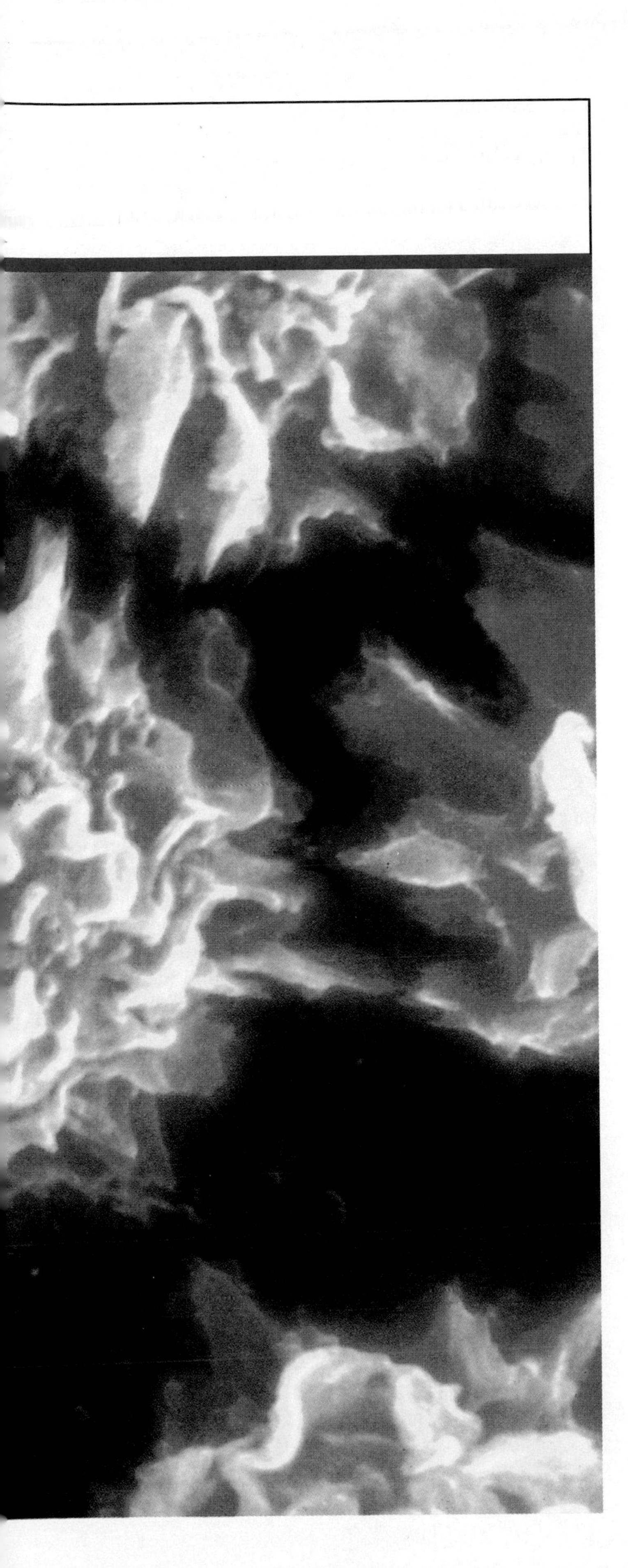

Chapter Concepts

Cancer is now recognized as a genetic disorder at the cellular level that involves the mutation of a small number of genes. Many of these genes normally act to suppress or stimulate progression through the cell cycle, and loss or inactivation of these genes causes uncontrolled cell division and tumor formation. Environmental factors and viruses also play a role in the genetic alterations that are necessary for the transition of normal cells into cancerous cells.

Although often viewed as a single disease, cancer is actually a group of complex diseases that affect a wide range of cells and tissues. A genetic link to cancer was first proposed early in this century, and this idea has served as one of the foundations of cancer research. Mutation that results in an alteration of the genome or an altered expression of gene products is now regarded as a common feature of all cancers. In some cases, such mutations are part of the germ line and are inherited. More often, mutations arise in somatic cells and are not passed on to future generations through the germ cells. Sometimes, the inherited mutation must be accompanied by a somatic mutation at the homologous locus, creating homozygosity. Whichever the case, *cancer is now considered a genetic disorder at the cellular level.*

Genomic alterations associated with cancer can involve changes as small as a single nucleotide substitution or large-scale changes such as chromosome rearrangement, chromosome gain or loss, or even the integration of viral genomes into chromosomal sites. Large-scale alterations are a common feature of cancer; indeed the majority of human tumors are characterized by visible chromosomal changes. Some of these chromosomal changes, particularly in leukemia, are so specific they can be used to diagnose and classify the disorder and to make accurate predictions about the severity and course of the disease.

Familial forms of cancer have been known for over two hundred years. In many of these cases, no well-defined pattern of inheritance can be established. In a small number of cases, however, a Mendelian pattern of dominant or recessive inheritance can be established, indicating the hereditary nature of the cancer. Perhaps the best-known example is **familial adenomatous polyposis**

(FAP), a dominant condition that, without treatment, develops into cancer of the colon.

If mutation is the underlying cause of cancer, there will always be a baseline rate of cancer, just as there is a background rate of spontaneous mutation. In addition, environmental agents that promote mutation should also play a role in the development of cancer. There is, in fact, clear evidence that environmental factors such as ionizing radiation, chemicals, and viruses are cancer-causing agents, and almost all of these act by generating mutations. Given that mutations play a role in cancer, it is reasonable to ask a series of questions about how mutations initiate the changes that convert normal cells into malignant tumors, which mutant genes are most likely to result in cancer, and how many mutations are required to cause cancer.

One approach to answering these questions is to consider what properties cancer cells possess that distinguish them from normal cells and to ask what genes may control these properties. All cancer cells have two properties in common: 1) uncontrolled growth and 2) the ability to spread or metastasize from their original site to other locations in the body. Cell proliferation is the result of cells traversing the cell cycle and is regulated in a cell-specific fashion. In cancer cells, control over the cell cycle is lost, and cells rapidly proliferate. Investigations into the genetic control of the cell cycle are beginning to provide clues concerning the origins of cancer.

The metastasis of cancer cells is apparently controlled by gene products that become localized on the cell surface, and the genetics of how metastasis is produced or suppressed is related to an understanding of how cells interact with the extracellular matrix and with other cells through cell surface molecules. Though this field is less well-developed than the study of the cell cycle, it is beginning to provide some insights into the secondary events in tumor progression.

In this chapter, we will consider the relationship between genes and cancer, with emphasis on the relationship between the cell cycle and genetic disorders associated with cancer. We will also examine the relationship between mutation and cancer, the identification of genes that in mutant form initiate the transformation of cells, and an estimate of the number of mutations involved in tumor formation. We will also discuss the relationship between chromosomal changes and cancer, and the role of environmental agents in the genesis of cancer.

THE CELL CYCLE AND CANCER

As described in Chapter 2, the cell cycle represents the sequence of events that occurs between mitotic divisions in a eukaryotic cell. In years past, work on control of the cell cycle has been conducted mainly by two groups: 1) geneticists working with yeast, especially *Saccharomyces cerevisiae* and *Schizosaccharomyces pombe*; and 2) developmental biologists, studying the newly fertilized eggs of organisms such as frogs, sea urchins, and newts. In the last few years, as these groups have succeeded in identifying and characterizing genes involved in the cell cycle, it has become apparent that their work is converging and overlapping with important areas of cancer biology, particularly studies on growth factors and the genes that suppress or promote tumor formation. This consolidation of fields has had a synergistic effect, resulting in new insights into the processes that control cell division and how regulation of the cell cycle is coupled to the transcription of selected genes. Because of these recent developments, it is necessary to spend some time describing what is now known about the events in the cell cycle and the genes that regulate progression through the cycle before undertaking a discussion of the genetics of cancer.

Ins and Outs of the Cell Cycle

Many cells of higher organisms pass continuously from one cell cycle to another, alternating between mitosis and interphase. The interphase portion of the cycle is a period of growth and synthesis, events that occur during stages termed G_1, S and G_2 (see Figure 2.5). The G_1 stage begins after mitosis; the synthesis of many cytoplasmic elements including ribosomes, enzymes, and membrane-derived organelles occurs at this time. In the S phase, DNA synthesis takes place, producing a duplicate copy of each chromosome. Then a second period of growth and synthesis known as G_2 takes place as a prelude to mitosis.

Because mitosis occurs rapidly, usually lasting less than an hour, cells spend most of the cell cycle in interphase. However, the duration of the cell cycle (the time between mitotic divisions) can vary widely in the life cycle of an organism and between different cell types in the same organism. For example, animal cells exhibit *in vivo* cycles ranging from a few minutes to several months. Most of this variation can be traced to the time spent in G_1, while the time necessary to complete S and G_2 remains relatively constant in different cell types.

While some cells such as meristematic cells in plants and dermal cells in human skin cycle continuously, other cell types, including many nerve cells, withdraw from the cycle and permanently enter a nondividing state known as G_0. Still other cell types such as white blood cells can be recruited to return from G_0 and re-enter the cell cycle.

Taken together, these observations suggest that the cell

cycle is tightly regulated and is dependent on the life history and differentiated state of a given cell. A great deal of information about the regulation of the cell cycle has become available in the last few years. A summary of what is currently known about the genetic regulation of the cell cycle will serve as a prelude to an overview of the genetics of cancer.

Control of the Cell Cycle

As work on the cell cycle in several organisms has converged, a universal model of the cell cycle and its regulation is beginning to emerge. While the details of all molecular events or even the exact number and sequence of steps are not yet known, all eukaryotic cells probably employ a common series of biochemical pathways to regulate events in the cell cycle. If the universal aspect of the current model of the cell cycle is upheld, results gathered from the study of yeasts or clam eggs can be used to understand and predict events in normal human cells and in mutated cells that have become cancerous. As a result, discoveries in cell cycle research will probably continue to be fast-moving and spectacular.

The emerging picture indicates that the cell cycle is regulated at two, and most probably three, main points: late in G_1 at a point called "Start," which is defined as the time when cells become committed to proceeding through the cycle and the subsequent steps of mitosis (Figure 19.1). A second regulatory point occurs early in S phase, and a third control point is present at the border between G_2 and mitosis.

At each of these points, the decision is made to proceed or halt progression through the cell cycle. This decision is controlled through the interaction of two types of proteins. One is a group of enzymes called **kinases;** they regulate other proteins by adding phosphate groups to them. The second group of regulatory proteins called **cyclins** bind to kinases, switching on or off kinase activity (Figure 19.1). For example, in the beginning of S phase, a cyclin combines with a kinase to stimulate the phosphorylation of a protein that is critical to the initiation of DNA synthesis. Cyclins are produced intermittently and degraded rapidly, resulting in a pulse of regulatory activity at each point in the cell cycle.

The kinase active at the 'start point' in G_1 and at the G_2/mitosis junction is called *cdc2* kinase, after the mutation in yeast that led to its discovery. The specificity of action for these kinases apparently resides in the cyclin that is produced at each of these stages of the cell cycle. In G_1, the *cdc2* kinase combines with cyclin to exert its effects. At the G_2/mitosis border, the *cdc2* kinase combines with another cyclin called cyclin B. In S phase, a cyclin called cyclin A combines with a kinase to regulate initiation of DNA synthesis. Altogether, almost a dozen

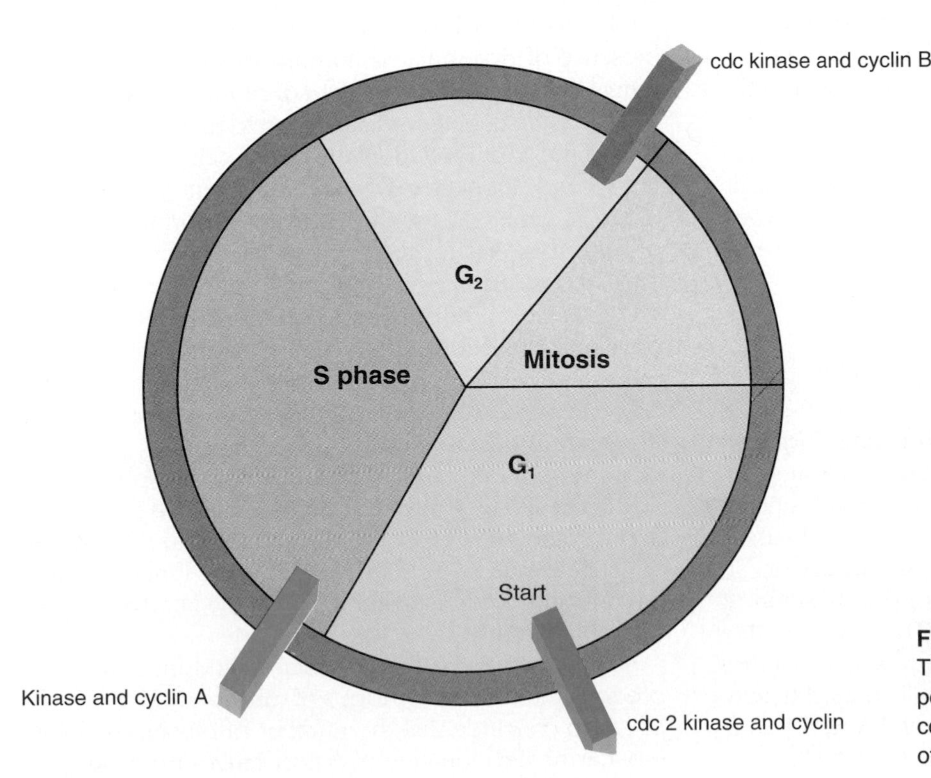

FIGURE 19.1
The cell cycle showing three major control points, and the kinase and cyclin components involved in regulating passage of cells through these points.

different cyclins have been discovered, and almost the same number of kinases have been identified, indicating that more control points remain to be characterized, or that kinases and cyclins may have multiple functions in the cell.

Obviously, mutations that disrupt any step in cell cycle regulation would be prime candidates for the study of cancer-causing genes. For example, mutations in genes that regulate the synthesis and turnover of the kinases and cyclins may be important in generating malignant transformations in cells.

In the next section, we will summarize what is known about the genetics of selected cancers, and wherever possible, relate this information to what has been discovered about the cell cycle and its regulation.

GENES AND CANCER

Genetic studies in several different cancers have identified a small number of genes that must be mutated in order to bring about the development of cancer or maintain the growth of malignant cells. It is clear that the two main properties of cancer, uncontrolled cell division and the ability to spread or metastasize are the result of genetic alterations. These alterations can involve large-scale mutational events such as chromosome loss, rearrangement or the insertion of foreign (often viral) DNA sequences into loci on human chromosomes. Smaller-scale alterations, such as changes in nucleotide sequence or more subtle modifications that alter only the amount of a gene product that is present or the time over which the gene product is active may also be involved. Two interrelated questions arise from considering cancer as a genetic disease at the cellular level: 1) are there mutant alleles that predispose an organism or specific cell types to cancer, and 2) if so, how many other mutational events are necessary to cause cancer?

Genes That Predispose Cells to Cancer

Single genes do indeed predispose cells to cancer, and it is possible to identify families in which certain forms of cancer are inherited. Many studies have documented families with high frequencies of certain forms of cancer, such as breast, colon, or kidney cancers. In most cases, however, it is difficult to identify a clear, simple pattern of inheritance; but one example of such a predisposition is the inheritance of **retinoblastoma (RB),** a cancer of the retinal cells of the eye. Retinoblastoma occurs with a frequency that ranges from 1 in 14,000 to 1 in 20,000, and most often appears between the ages of 1 and 3 years. Two forms of retinoblastoma are known. One is a familial form (about 40 percent of all cases) inherited as an autosomal dominant trait. Because the trait is dominant, family members have a 50 percent chance of receiving the mutant allele. Those that inherit the mutant RB allele are predisposed to develop eye tumors, and 90 percent of these individuals will develop retinal tumors, usually in both eyes. In addition, those family members that inherit the mutant allele are predisposed to developing other forms of cancer, even if they do not develop retinoblastoma.

A second form of retinoblastoma is also known, amounting to 60 percent of all cases. In this form, retinoblastoma is not familial, and these cases apparently develop spontaneously. In such cases, retinoblastoma usually develops only in one eye, and onset occurs at a much later age than in the familial form.

Other types of familial cancer are also known. These include **Wilms tumor (WT),** a cancer of the kidney inherited as an autosomal dominant condition, and **Li-Fraumeni syndrome,** a rare autosomal dominant condition that predisposes to a number of different cancers, including breast cancer.

How Many Mutations Are Needed?

The study of genes that predispose to cancer has provided insight into the number and sequence of mutational events that are necessary to cause different forms of cancer. By studying the two different forms of retinoblastoma, Alfred Knudson and his colleagues developed a model that suggests that retinoblastoma is caused by the presence of two mutated homologous loci in the same retinal cell. In the familial form of retinoblastoma, a mutant RB allele is inherited, and therefore, is carried by all cells of the body, including the retinal cells (Figure 19.2). If a second, spontaneous mutational event in the retinoblastoma gene occurs in the remaining normal RB allele in any retinal cell, formation of retinal tumors will result. Thus, those carrying an inherited single mutation of the RB gene are predisposed to develop retinoblastoma, since only one additional mutational event is required to cause retinoblastoma. This does not happen in all cases, however, since 10 percent of those that inherit a mutant RB gene do not develop retinoblastoma.

In the sporadic cases of retinoblastoma, two independently occurring somatic mutations of the RB gene must occur in the same retinal cell for a tumor to develop. As might be expected from this model, these random events are rare and occur at a later age than the familial form of retinoblastoma.

Similar studies on the distribution and inheritance of predispositions to other forms of cancer have led investigators to conclude that the number of mutations necessary for the development of cancer ranges from 2 to perhaps as many as 20 (Table 19.1).

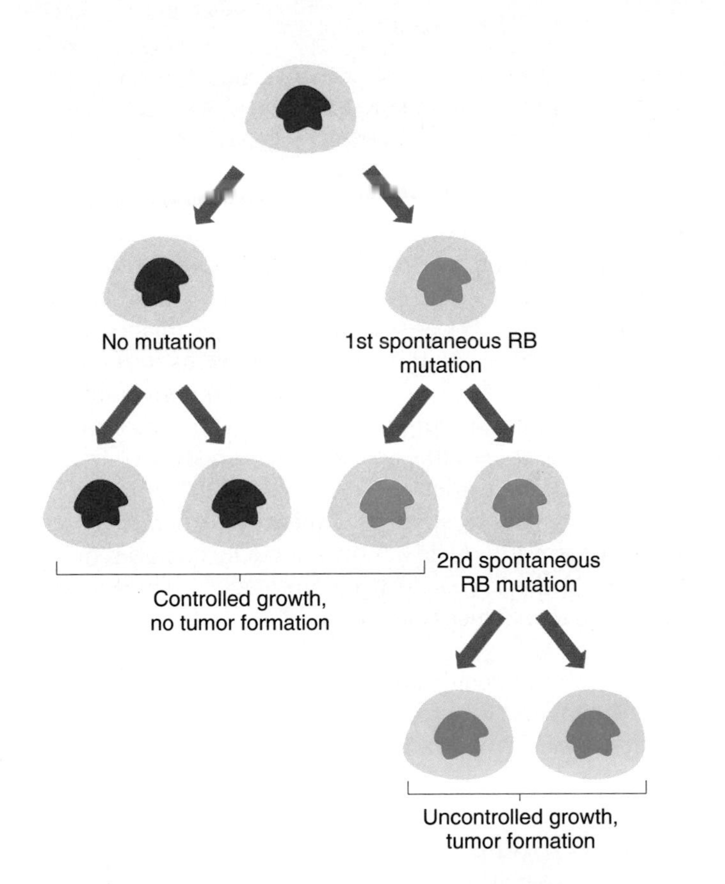

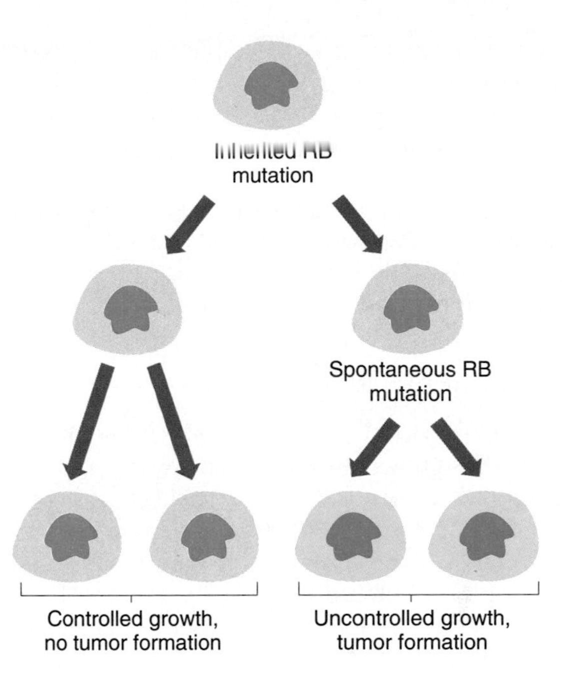

FIGURE 19.2
In spontaneous cases of retinoblastoma (left), two mutations in the retinoblastoma gene are acquired in a single cell, causing uncontrolled cell growth and division, resulting in tumor formation. In familial cases of retinoblastoma (at right), one mutation is inherited and present in all cells. A second mutation at the retinoblastoma locus in any retinal cell will result in uncontrolled cell growth and tumor formation.

TABLE 19.1
Minimum number of mutations for specific cancers.

Cancer	Chromosome Sites	Minimum Number of Mutations Required
Retinoblastoma	13q	2
Wilms Tumor	11p	2
Colon cancer	5p, 12p, 17p, 18q	4–5
Small cell lung cancer	3p, 11p, 13q, 17p	10–15

TUMOR SUPPRESSOR GENES

Mitosis and the rest of the cell cycle can be regulated in two ways: 1) by genes that normally function to suppress cell division, and 2) by genes that normally function to promote cell division. The first class, called **tumor suppressor genes,** normally function to inactivate or repress cell division. These genes and/or their gene products must be inactivated for cell division to take place. If these genes become permanently inactivated through mutation, control over cell division is lost, and the cell begins to proliferate in an uncontrollable fashion.

The second class of genes, called **proto-oncogenes,** normally function to promote cell division, and these genes and/or their gene products must be inactivated in order to halt cell division. If these genes become permanently switched on through mutation, then cell division occurs in an uncontrolled fashion, leading to tumor formation. The mutant forms of proto-oncogenes are known as **oncogenes.**

We shall consider some examples of how mutations in tumor suppressor genes can lead to a loss of control over cell division and the development of cancer. Following that discussion, we will consider proto-oncogenes and oncogenes, how the normal alleles act to regulate cell division, and how the mutant oncogenes work to promote tumor formation.

Retinoblastoma (RB)

As described previously, RB is a tumor of the retinal layer of the eye. The RB gene is located on chromosome 13. It encodes a protein that is 928 amino acids long and confined to the nucleus. The RB gene product is present in all cell and tissue types examined to date and is found in both resting (G_0) cells and cells active in the cell cycle. Moreover, the RB protein is present at all stages of the cell cycle. Because the gene is functionally active at all times, the gene is regulated by reversible modifications to the RB gene product by alternately adding or removing phosphate groups from the protein. The RB protein becomes phosphorylated in the S and the G_2-M stages of the cell cycle, but is without phosphate groups in the G_0 and G_1 stages of the cycle.

From these observations about alternating states of the RB protein, it has been suggested that when the RB protein is not phosphorylated, it acts to suppress cell division. Conversely, when phosphate groups are added to the RB protein, it becomes inactive and permits cells to enter the S phase and the subsequent events in the cell cycle. The inactivation of RB protein and the initiation of tumor formation in retinoblastoma provides indirect evidence for the role of this protein in suppressing cell proliferation. Direct evidence for the role of the RB protein in inhibiting cell division has come from experiments in which a normal RB gene has been introduced into retinoblastoma cells. Under such circumstances, synthesis of the RB gene product is associated with cessation of cell division.

Although the details are not yet clear, recent findings suggest how the RB protein acts to control cell division. At the G_1 stage of the cell cycle, the RB protein is not

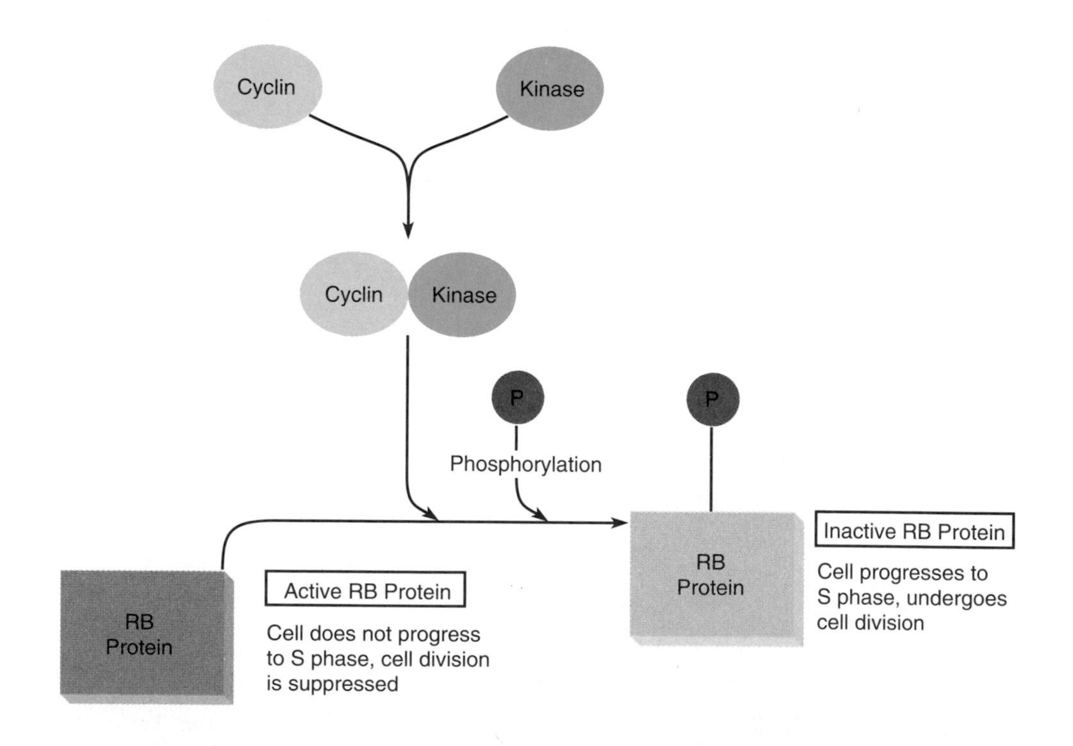

FIGURE 19.3
The proposed role of the retinoblastoma (RB) protein in regulation of the cell cycle. If the protein is phosphorylated by the interaction of a cyclin and a kinase, the protein is inactivated, and the cell progresses to S phase and undergoes cell division. The active form of the protein is not phosphorylated, and suppresses cell division. Therefore, any mutation that causes the inactivation or loss of the RB protein will cause cells to continuously cycle through S phase and division, resulting in uncontrolled growth, division and tumor formation.

phosphorylated. This form of the RB protein binds to a transcription factor known as EF2, causing EF2 to be inactivated. Recall from Chapter 16 that transcription factors are proteins that bind to specific DNA sequences and activate the expression of nearby genes.

According to the model now being proposed, phosphorylation of RB is controlled by one or more of the cell cycle kinases (Figure 19.3). When cyclins are produced at the G_1 "Start" point, they bind to and mobilize cell cycle kinase molecules that phosphorylate, and therefore, inactivate the RB protein. The inactivated RB protein releases the EF2 transcription factor, resulting in the expression of genes required for progression through the cell cycle and mitosis. The RB protein is therefore a link that connects the cell cycle to the regulation of gene transcription. When the RB protein is missing or inactivated by mutation, there is no regulation of the EF2 transcription factor, and thus, the genes for progression through the cell cycle are permanently expressed, resulting in tumor formation.

Wilms Tumor

Wilms tumor (WT) or nephroblastoma is a cancer of the kidney found primarily in children. It occurs with a frequency of about 1 in 10,000 births, and like retinoblastoma, is found in two forms, a noninherited spontaneous form and a familial form conferring a predisposition to WT. This form is inherited as an autosomal dominant trait. According to the model developed by Knudson and colleagues, familial cases inherit one mutant allele through the germ line and develop a second mutation in the remaining normal allele in a somatic cell. Sporadic cases, on the other hand, require two independent mutations of the WT gene within the same cell and more often develop tumors that involve only one kidney. Because mutation of the gene and/or the loss of a functional gene product are associated with the development of tumors, the normal gene is regarded as a tumor suppressor.

The WT gene has been mapped to the short arm of chromosome 11. The gene product encoded by the WT gene contains four contiguous zinc finger domains. Recall from Chapter 16 that these domains are regions with interspersed cysteine and histidine residues that covalently bind zinc atoms, folding the amino acid chains into loops known as zinc fingers. Such motifs are characteristic of DNA binding proteins that regulate transcription. In addition to this feature, the WT protein has a strong similarity to two growth-regulating proteins known as **EGR1** and **EGR2.** And the amino acid sequences of the WT protein upstream from the zinc fingers are similar to those of other proteins that are transcription factors. Unlike the RB gene, the WT gene has a very restricted pattern of expression; it is activated only in mesenchymal cells of the fetal kidney that give rise to the nephron (the basic filtration unit of the kidney), and in the tumorous nephroblastoma cells. Expression of this gene is barely detectable in cells of the adult kidney and is absent in all other adult cell and tissue types tested.

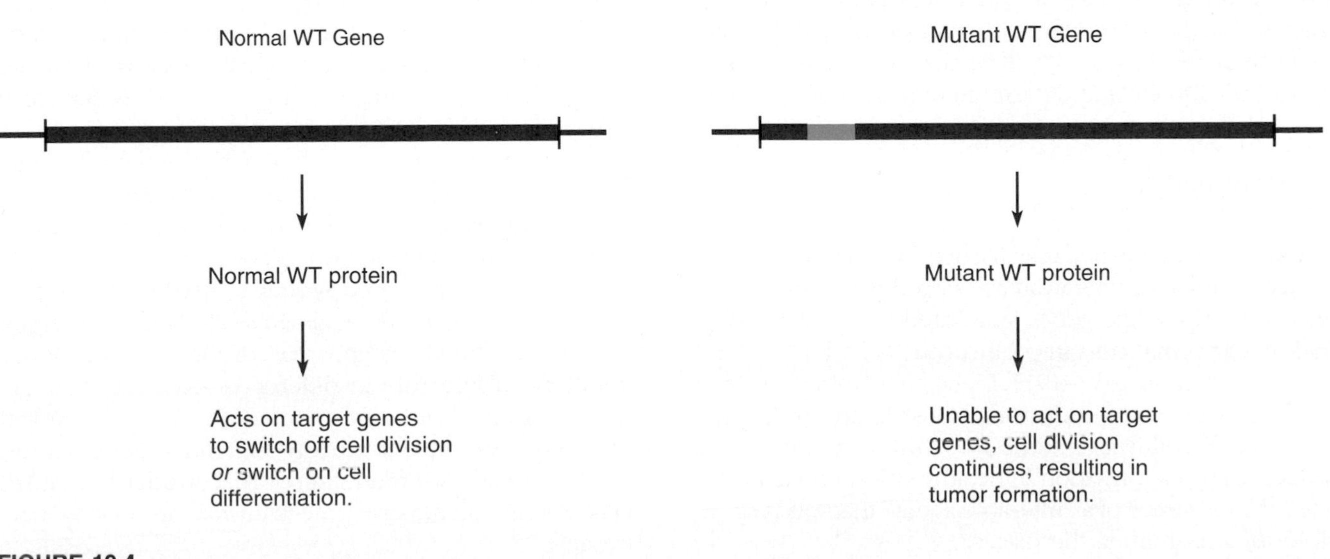

FIGURE 19.4
The proposed role of the WT protein in regulating cell division. In normal cells (at left), the WT protein is produced and acts on a set of target genes to switch off cell division or switch on cell differentiation. Any mutation that causes the loss or inactivation of the WT protein (at right) will result in a failure to regulate the target genes, allowing cell division and tumor formation to occur.

The structure of the WT gene product, its pattern of expression and mutant phenotype all suggest that the WT gene encodes a nuclear protein that functions to turn off genes that sustain cell proliferation. Alternately, the gene product may switch on genes that begin the process of differentiation of the mesenchymal cells into kidney structures. In either case, it appears that in the mutation causing Wilms tumor, the gene is switched on at the appropriate time in the appropriate cells, but the altered gene product is unable to regulate its target genes, resulting in continued proliferation of the mesenchymal cells, aberrant differentiation, and tumor formation (Figure 19.4). To completely resolve the chain of events in Wilms tumor, it will be necessary to identify the loci regulated by the WT gene and determine the nature of their gene products and modes of action.

Although both the RB and WT genes encode gene products restricted to the nucleus that are normally involved in the suppression of tumor formation, there are conspicuous differences in their properties and modes of action. The RB protein does not bind to DNA, but instead, regulates cell division by interacting with other nuclear proteins. The WT gene product has all the characteristics of a DNA-binding protein that regulates cell division by acting as a transcription factor. The RB protein is expressed in all dividing cells, while expression of WT is restricted to a small number of cell types during a restricted developmental period. Thus, RB appears to be a general regulator of cell division, while WT is probably a cell or tissue specific regulator of gene activity during fetal or neonatal development. These differences emphasize the unique aspects of tumor suppressor genes and suggest that it will be difficult to develop a general model for such genes. Instead, it will be necessary to track them down individually and determine their modes of action.

ONCOGENES

Oncogenes are genes that induce or maintain uncontrolled cellular proliferation associated with cancer. The existence of specific genes associated with the transformation of normal cells into cancerous cells was inferred by Peyton Rous in 1910–1911. Using cells from a tumor of chickens known as a **sarcoma,** Rous demonstrated that injection of cell-free extracts from these tumors could induce sarcoma formation in healthy chickens. He postulated the existence of a "filterable agent" that was responsible for transmitting the disease.

Decades later, his filterable agent was shown by other investigators to be a virus, now known as the Rous sarcoma virus (RSV). This virus belongs to a group known as **retroviruses** because the RNA genome must be transcribed by an enzyme, **reverse transcriptase,** into single-stranded DNA. The single-stranded DNA is converted to double-stranded DNA and integrated into the genome of the infected cell, where it can be transcribed to produce an infectious RNA. In the case of RSV, the tumor-forming ability results from the presence of a single gene, called the *src* gene in the viral genome. This single gene, responsible for the induction of sarcoma formation in chicken cells, is designated as an oncogene, and retroviruses that carry oncogenes are known as **acute transforming viruses.** Other retroviruses that can cause tumor formation do not carry oncogenes, but are able to induce the activity of cellular genes that brings about tumor formation. They are designated as **non-acute** or **nondefective viruses.**

The oncogenes carried by acute transforming viruses are acquired from the host's cellular genome during the process of infection, when a portion of the viral genome is exchanged for a cellular gene (Figure 19.5), as in bacterial transduction. The oncogenes (*onc*) carried by retroviruses are called *v-onc*, and the cellular version of the gene is called a *c-onc* gene or **proto-oncogene.** Retroviruses that carry a *v-onc* gene are able to infect and transform a specific type of host cell. For RSV, the oncogene captured from the chicken genome is called *v-src*, and it confers the ability to infect chicken cells and transform them into tumorous growths known as sarcomas. The cellular version of the same gene, found in the chicken genome, is called *c-src*. More than 20 oncogenes have been identified by their presence in retroviral genomes, and over 50 oncogenes have been identified overall. Some of these are listed in Table 19.2.

Two questions about oncogenes come to mind: 1) are *v-onc* and *c-onc* different, and 2) what mechanisms allow oncogenes to bring about cellular transformation and tumor formation? Comparison of the DNA sequence of *v-onc* genes with the corresponding *c-onc* sequence reveals several examples (such as *v-ras*, *v-mos*) in which there are only minimal differences, probably generated by point mutations. In other oncogenes, portions of the *c-onc* sequence have been replaced or lost. In *v-src*, 19 amino acids of the *c-onc* sequence have been replaced with 12 different amino acids. In *v-erb*, the *v-onc* version retains only the C-terminus half of the gene. Mutations, therefore, play a role in differentiating some *v-onc* sequences from their *c-onc* counterparts. However, not all oncogenes are mutant versions of cellular genes carried by retroviruses. We must, therefore, consider both qualitative and quantitative mechanisms in oncogenetic events.

Oncogenes and Gene Expression

At least three mechanisms can be invoked to explain the conversion of proto-oncogenes into oncogenes. These

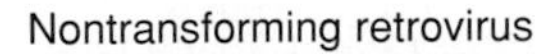

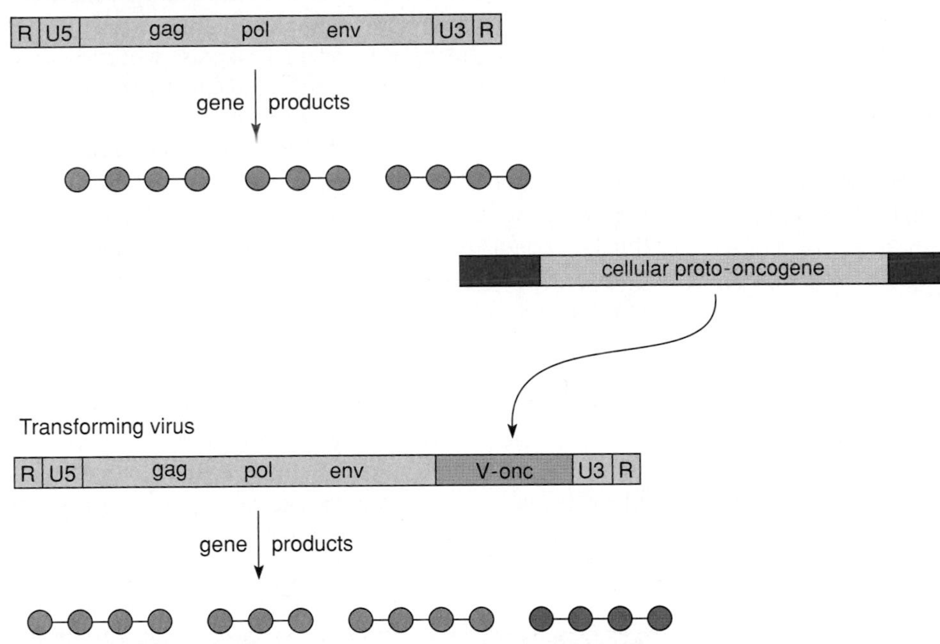

FIGURE 19.5
A transforming retrovirus has acquired a copy of a gene from the host genome, converting it from a c-onc or proto-oncogene into an oncogene that confers on the virus the ability to transform a specific type of host cell into a cancerous cell.

mechanisms, which are related to the range of functions carried out by the proto-oncogenes in the cellular genome, include **point mutations, translocations,** and **overexpression.** While some of these are mediated by viruses, others operate in the absence of retroviruses, and are generated solely by intracellular events.

The *ras* gene family, encoding a protein of 189 amino acids involved with the transduction of external signals across the cell membrane, illustrates how point mutations as small as single base changes underly the differences between the normal *c-onc* sequence and the oncogenic version of the gene. The normal *ras* protein

TABLE 19.2
Oncogenes.

c-onc	Origin	Species	Human Chromosome Location
A. Cellular Oncogenes with Viral Equivalents			
src	Rous sarcoma virus	Chicken	20
fos	FBJ osteosarcoma virus	Mouse	2
sis	Simian sarcoma virus	Monkey	22
fes	ST feline virus	Cat	15
abl	Abelson murine leukemia virus	Mouse	9
erb-B	Avian erythroblastosis virus	Chicken	7
Ha-ras-1	Harvey murine sarcoma virus	Mouse	11
myc	Avian MC29 myelocytomatosis virus	Chicken	8
B. Oncogenes without Viral Equivalents			
N-ras	Neuroblastoma, leukemia	Human	1
N-myc	Neuroblastoma	Human	12
neu	Neuroglioblastoma	Human, rat	17
man	Mammary carcinoma	Human, mouse	?

functions as a molecular switch, alternating between an "on" and "off" state (Figure 19.6). In some cases, the mutational event causes the *ras* protein to be stuck in the "on" position, stimulating cell growth. In some human tumors, this event was not mediated by a retrovirus, and is presumed to be the result of a somatic mutation. Comparison of the amino acid sequence of *ras* proteins from a number of different human carcinomas reveals that all *ras* mutations involve a single amino acid substitution at position 12 or 61 that is caused by a single base change (Figure 19.7).

One of the well-characterized examples of oncogene activation by translocation is that of *c-abl* associated with chronic myelogenous leukemia (CML). In this case, described in detail in a later section, the translocation results in altered gene activity that causes tumor formation.

At least three separate mechanisms of proto-oncogene activation are associated with overexpression. First, the *c-onc* may acquire a new promoter, causing an increase in the level of transcript production, or activating a silent locus. This is the case in avian leukosis, where strong viral promoters become integrated upstream from a proto-oncogene, causing an increase in transcript production and an increase in the amount of the gene product. The second is overexpression by acquisition of new upstream regulatory sequences including enhancers. The third mechanism involves the amplification of the proto-oncogene. In human tumors, members of the *myc* family of oncogenes are frequently amplified. The *c-myc* proto-oncogene is found amplified up to several hundred copies in some human tumors.

The protein products of proto-oncogenes are found in and associated with the plasma membrane, cytoplasm, and nucleus. While their specific functions vary widely, all products characterized to date alter gene expression in a direct or an indirect manner.

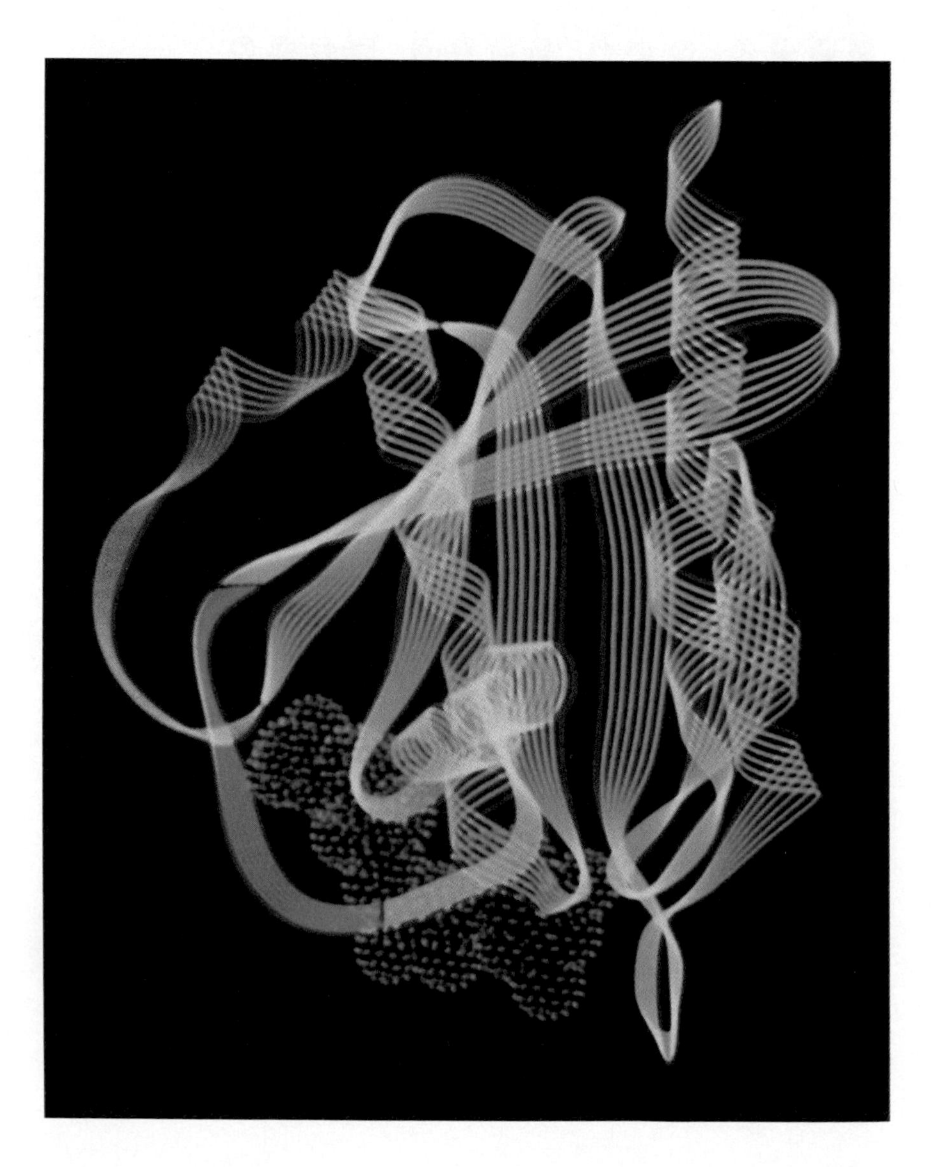

FIGURE 19.6
A three-dimensional computer-generated image of *ras* proteins in two different conformations. Normal *ras* proteins act as molecular switches controlling cell growth and differentiation. The switch is "on" when GTP binds to the protein, and "off" when the GTP is hydrolyzed to GDP. Switching the protein between states alters the conformation of the protein in the two regions shown in blue and yellow. Oncogenic mutations of *ras* are stuck in the "on" state, continuously signalling for cell growth.

GENETICS AND METASTASIS

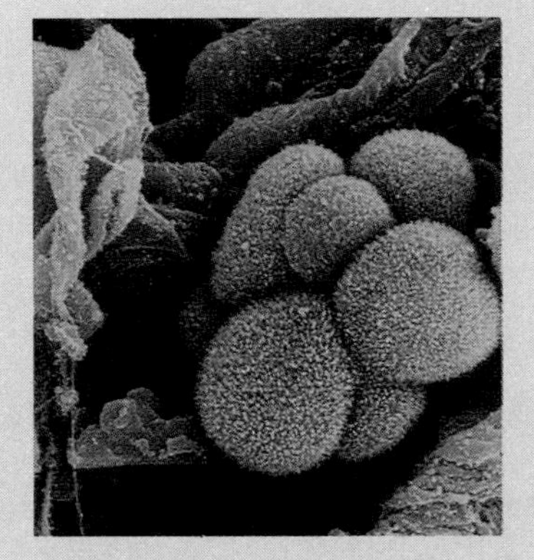

Cancer cells have lost the ability to regulate their own growth and division. As a result, they can develop into malignant tumors that invade neighboring tissue. Sometimes, cancer cells detach from the primary tumor and settle elsewhere in the body, where they again grow and divide, producing secondary tumors. This process, called *metastasis,* is often the cause of death of cancer patients.

A metastatic cancer cell can spread from a primary tumor by entering the blood or lymphatic circulatory system. These cells are then carried in the general circulation until they eventually become lodged in a capillary bed where more than 99 percent die. The cells that do survive must then invade nearby tissue and begin dividing in order to form a secondary tumor. To reach a new site, the cell needs to pass through the layer of epithelial cells that line the interior wall of the capillary (or lymph vessel) and then penetrate the surrounding extracellular matrices.

The extracellular matrix is a meshwork of proteins and carbohydrate molecules. It serves as a means to separate tissues and as a scaffold for tissue growth. Normally, it also prevents inappropriate migration of cells. To establish a secondary tumor, a metastatic cell must overcome inhibition of migration by the extracellular matrix. Initially, a tumor cell binds to the epithelial cells of the capillary wall, apparently causing them to retract and expose the matrix layer beneath, called the basement membrane. The basement membrane would normally constitute an impenetrable barrier, but metastatic cells secrete enzymes that can cut the protein molecules in the basement membrane and create holes through which they can move. The cell can "tunnel" through the matrix, enter new tissues, and establish a secondary tumor.

The invasive ability of the cancer cell is sometimes characteristic of normal cells. For example, the implantation of the embryo in the wall of the uterus during pregnancy requires that tissues coalesce, and for white blood cells to reach the site of infection, they must penetrate capillary walls. The mechanism of invasion is probably the same in normal cells as it is in cancer cells; but in normal cells, the invasive ability is controlled and tightly regulated.

In order to understand the process of invasion, scientists are looking for genetic factors that promote invasiveness, for example, genes that might increase the production of matrix-cutting enzymes. Recent studies have shown that protein-cutting enzymes called *metalloproteinases* are necessary for cell invasion. Tumor cell lines, which have high invasive ability, produced greater amounts of certain metalloproteinases than tumor cell lines with low invasive properties. However, the regulation of metalloproteinase activity may involve other genes. Several studies have shown that activated metalloproteinases can be inhibited by the presence of a tissue inhibitor of metalloproteinase (TIMP). Normal cells produce TIMP and suppress inappropriate activity of metalloproteinase. In tumor cells, however, the TIMP protein will only inhibit the metalloproteinase if the number of TIMP molecules is greater than the number of metalloproteinase molecules. Thus, TIMP could be considered a metastasis suppressor protein.

Another potential metastasis suppressor protein is the "nonmetastatic 23" protein (nm23). This protein is absent or produced at reduced levels in many metastatic tumor cell lines and is produced in high levels in nonmetastatic tumor lines. Clinical studies in samples of primary breast cancers give similar results. Low levels of nm23 protein were observed in metastatic cancers compared to nonmetastatic tumors and to normal tissue. Measurements of nm23 levels may be used to identify patients with undetected metastases.

We can see that definition of the biological and genetic basis of cancer not only leads to a more refined understanding of the symptoms of the disease, but also provides clues to improved diagnostic techniques and potentially more successful therapy.

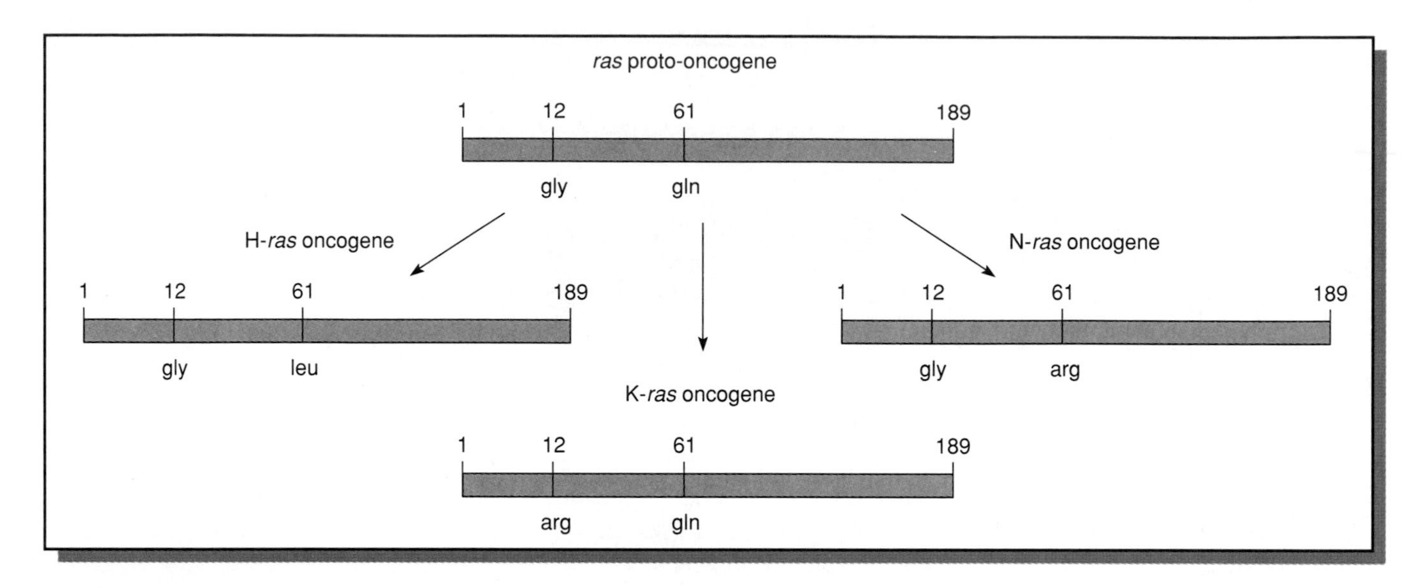

FIGURE 19.7
The *ras* proto-oncogene encodes a protein of 189 amino acids. In the normal protein, glycine is encoded at position 12, and glutamine at position 61. Analysis of *ras* oncogene proteins from several tumors shows a single amino acid substitution at one of these positions. A single amino acid substitution resulting from a single base change can convert a proto-oncogene into a tumor-promoting oncogene.

COLON CANCER: A GENETIC MODEL FOR CANCER

Even in the small number of cases where it has been studied in detail, it is clear that cancer is a multistep process resulting from a number of specific genetic alterations. Although the study of tumors such as RB and WT have been useful in establishing that a limited number of steps are required to transform a normal cell into a malignant one, these tumors are of limited value in establishing the molecular nature and order of the genetic events that lead to tumor formation and metastasis. For these questions, the study of colon cancer offers several advantages. First, malignant tumors of the colon and rectum develop from pre-existing benign tumors, and tumors at all stages of development are available for study. Moreover, two forms of colon cancer are known: one in which a predisposition is inherited in an autosomal dominant fashion (known as **familial adenomatous polyposis** or **FAP**), and one that is completely spontaneous, making it possible to study the interaction of genetic and environmental factors in the genesis of tumors.

Through an analysis of mutations in tumors at various stages, ranging from small benign growths known as **adenomas,** through intermediate stages, to **malignant tumors** and **metastatic tumors,** it has been possible to define the number and nature of the genetic and molecular steps involved in changing normal intestinal epithelial cells into tumor cells and to develop a genetic model for colon cancer. This model is shown in Figure 19.8. The first feature of this model to be noted is that multiple mutations are required. Mutations in four or five specific genes are needed to bring about malignant growth. If fewer changes are present, benign growths or intermediate stages of tumor formation result. Second, based on the analysis of many tumors, the order of mutations usually follow a preferred sequence (as indicated in the figure). Ultimately, however, it is the accumulation of specific mutations that is more important than the order in which they occur.

The first mutation in this process converts normal epithelium into a rapidly proliferating epithelial region and one or more benign tumors. In cases of FAP, this mutation is inherited and results in the development of dozens or hundreds of benign adenomas in the colon and rectum. In spontaneous cases, the evidence suggests that the initial mutational event takes place in a single cell and that the resulting adenoma consists of a clone of cells, all carrying the mutation. This first mutation takes place in a gene called **APC,** located on the long arm of chromosome 5. The loss of the corresponding allele on the homologous copy of chromosome 5 is *not* necessary for proliferation and adenoma formation, suggesting that the APC gene can be classified as a tumor suppressor gene. The relative order of subsequent mutations is shown in

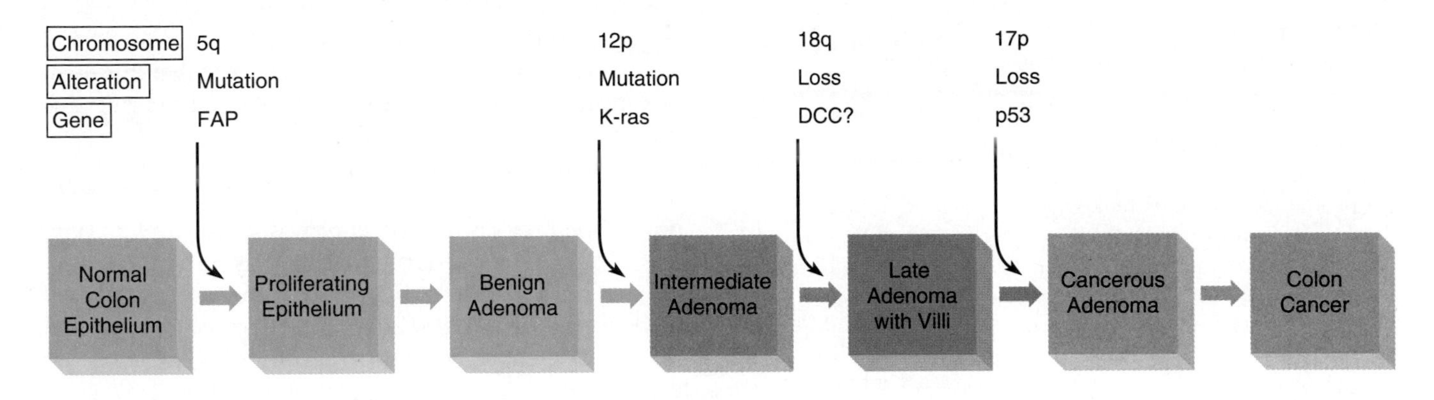

FIGURE 19.8
A model for the step-wise production of colon cancer. The first step is the loss or inactivation of both alleles of the FAP gene on chromosome 5. In familial cases, one mutation of the FAP gene is inherited. Loss of both alleles leads to the formation of benign adenomas. Subsequent mutations involving genes on chromosomes 12, 18 and 17 in cells of the benign adenomas can lead to a malignant transformation resulting in colon cancer. Although the mutations in chromosomes 12, 17 and 18 usually occur at a later stage than those involving chromosome 5, the sum of the changes is more important than the order in which they occur.

Figure 19.8. Mutations in the *ras* oncogene may precede or follow the loss of a segment of chromosome 18p. In either case, the accumulation of these two mutations in adenoma cells with a pre-existing mutation on chromosome 5 causes the adenoma to grow larger and develop a number of finger-like villous outgrowths. Finally, a mutation in 17p, involving the loss or inactivation of a gene called *p53* causes the transition to a cancerous cell. Mutations in the *p53* gene are pivotal to the development of a number of cancers, including lung, brain, and breast cancers. It is a tumor suppressor gene that normally functions to regulate the passage of cells from late G_1 to S phase, perhaps by regulating transcription of a set of genes required at this stage. The *p53* gene product has DNA binding properties, and mutations in *p53* may alter DNA binding to confer a new function on the gene product.

The process of metastasis occurs after the formation of cancerous cells and involves an unknown number of mutational steps, although work on other tumors suggests that one or two mutations may be sufficient to allow tumor cells to detach and spread to secondary sites.

In sum, the genetic model for colon cancer involves mutations in oncogenes, tumor suppressor genes, and disruption of the cell cycle at a specific transition point, although the nature and function of normal and mutant gene products of the *p53* gene have not yet been identified with certainty. In cases of predisposition to colon cancer, the first mutation is inherited, and the rest occur as a result of the action of environmental agents, pointing up the role of the environment in the development of cancer. This multistep model appears to have applications to other forms of cancer, but many questions remain unanswered, including the normal functions of the genes involved and the molecular mechanisms by which the mutated genes and/or their gene products bring about tumor formation.

CHROMOSOMES AND CANCER

Alterations in chromosome structure and/or number are associated with many forms of cancer. In most cases, the relationship between changes in chromosome number or arrangement and the development of cancer is not clear. For example, individuals with Down syndrome carry an extra copy of chromosome 21. This quantitative alteration in genetic content is associated with a 20-fold increased risk of leukemia as compared to the general population. In a limited number of cases where specific chromosome rearrangements are associated with cancer, more direct information is available to indicate how the rearrangement is associated with the development and/or maintenance of the cancerous state. We shall first address this topic.

Chromosome Rearrangements and Cancer

The relationship between chromosome aberrations and cancer is most clearly seen in leukemias, where the presence of specific chromosome changes is well defined and often diagnostic (Table 19.3). In solid tumors, the task of identifying the primary cytogenetic alteration is often

TABLE 19.3
Specific Chromosome Aberrations and Cancer.

Cancer	Chromosome Alteration
Chronic myelogenous leukemia	t(9; 22)
Acute promyelocytic leukemia	t(15; 17)
Acute lymphocytic leukemia	t(4; 11)
Acute myelogenous leukemia	t(8; 21)
Prostate cancer	del(10q)
Synovial sarcoma	t(x; 18)
Testicular cancer	i(12p)
Retinoblastoma	del(13q)
Wilms tumor	del(11p)

obscured by subsequent chromosomal changes, but some progress is being made in this field. One of the best-studied examples of the association between a chromosome rearrangement and the development of cancer is the translocation between chromosomes 9 and 22 that is associated with **chronic myelocytic leukemia (CML)** (Figure 19.9). Originally, this translocation was described as an abnormal chromosome 21 and called the **Philadelphia chromosome.** Later, it was shown that the Philadelphia chromosome results from the exchange of genetic material between chromosomes 9 and 22. This translocation is never seen in a normal cell and is found only in the white blood cells involved in CML. These observations indicate that the translocation is a primary and causal event in the generation of CML, and that this form of cancer may originate from a single cell bearing this translocated chromosome.

By examining a large number of cases involving this translocation, it was possible to establish the exact location of the break points on chromosomes 9 and 22 (Figure 19.9). Genetic mapping studies using recombinant DNA techniques established that a proto-oncogene *c-abl* maps to the breakpoint region on chromosome 9, and the gene *bcr* maps near the breakpoint on chromosome 22. In the translocation event, a portion of the *c-abl* gene is translocated to a region within the *bcr* gene. This generates a hybrid *bcr/c-abl* gene that is transcriptionally active and produces a hybrid 200 kd protein product (Figure 19.10) that has been implicated in the generation of CML.

A combination of cytogenetic and molecular approaches has been applied to the cytogenetic events in

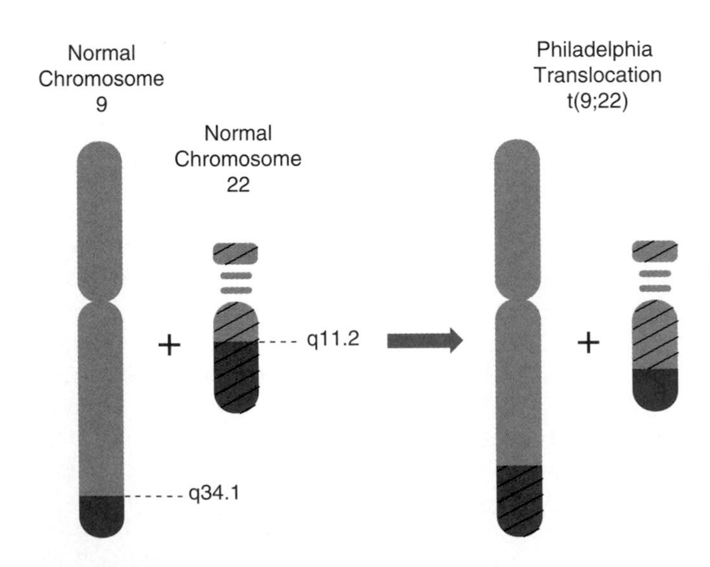

FIGURE 19.9
A reciprocal translocation involving the long arms of chromosomes 9 and 22 result in the production of a characteristic chromosome, the Philadelphia chromosome that is associated with cases of chronic myelocytic leukemia (CML).

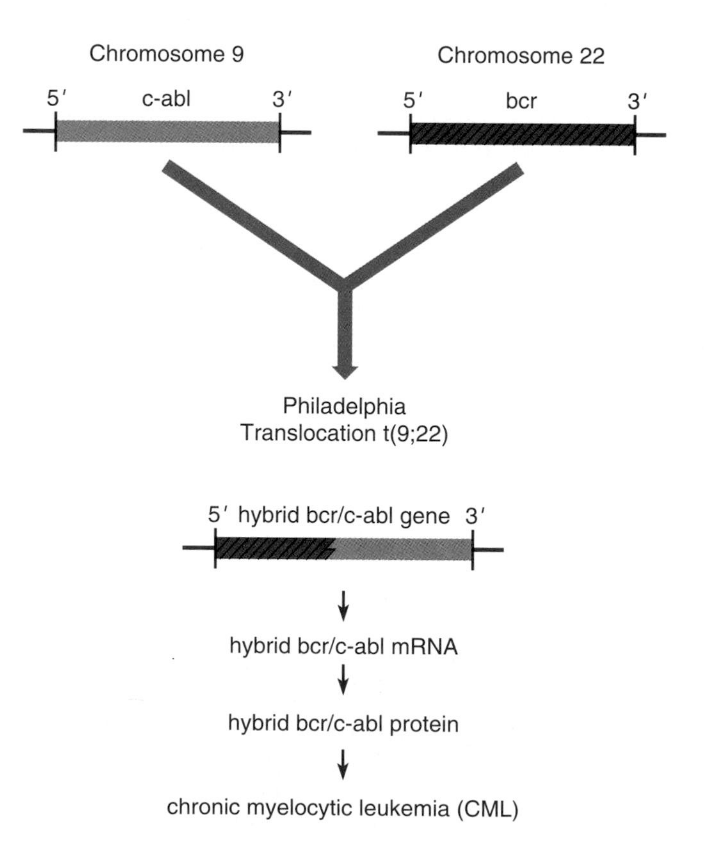

FIGURE 19.10
The t(9;22) translocation associated with chronic myelogenous leukemia (CML) results in the fusion of the *c-abl* oncogene on chromosome 9 with the *bcr* gene on chromosome 22. The normal *c-abl* protein is a kinase and the normal *bcr* protein activates a phosphorylation reaction. The fusion protein is a powerful hybrid that allows cells to escape control of the cell cycle, resulting in leukemia.

lymphomas involving translocation break points on chromosome 8 (these include t(8;14), t(8;22), and t(2;8)). The breakpoint on chromosome 8 in all these cases is the same, and the proto-oncogene *c-myc* has been mapped to this locus. The loci at the breakpoint on the other chromosomes involved in this series of translocations all have various immunoglobulin genes at the break points. The movement of the *c-myc* gene to a position near these immunoglobulin genes leads to an alteration in the expression of the immunoglobulin genes, resulting in the transformation of the lymphoid cells into leukemic cells.

Other genes at translocation breakpoints have also been isolated and characterized. In these cases and in CML, the translocation results in the formation of an abnormal gene product that causes the cell to undergo a malignant transformation even though a second and presumably normal copy of the gene is present and active in synthesizing a normal gene product. If the abnormal gene products at other translocation loci associated with leukemia can be identified, it is hoped that therapeutic strategies can be developed to administer high levels of the normal gene product, or that the abnormal gene product can be inactivated.

CANCER AND THE ENVIRONMENT

The relationship between environmental agents and the genesis of cancer is often elusive, and in the early stages of investigation, this relationship is based on indirect evidence. Such studies often begin with an epidemiological survey, comparing the cancer death rates among different geographic populations. When differences are found in the death rate for a given type of cancer, or a given age group, or a cluster of related occupations such as chemical workers, further work is necessary to identify one or more environmental factors that correlate with these cancer deaths. These correlations are not conclusive, but serve to identify factors that upon additional investigation may turn out to be directly related to the development of cancer. Finally, extensive laboratory investigations may establish the mechanism by which an environmental agent generates cancer. In other cases, such as certain viruses, the relationship between cancer and the environment is more straightforward.

Viruses and Cancer

Epidemiological surveys have shown that individuals who develop a form of liver cancer known as **hepatocellular carcinoma (HCC)** are infected with the **hepatitis B virus (HBV)** (Figure 19.11). In fact, for those carrying HBV, the risk of cancer is increased by a factor of 100.

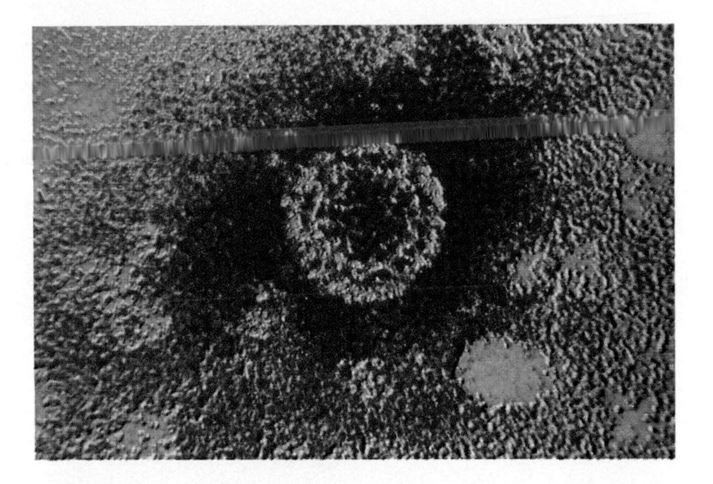

FIGURE 19.11
A false color transmission electron micrograph of a hepatitis B virus. Infection with this virus often results in cancer of the liver. World-wide, viral infections of all kinds are thought to be responsible for about 15% of the total number of cancers.

Aside from the risk of cancer, HBV infection is a public health risk that affects some 300 million persons worldwide. It is most prevalent in Asia and tropical regions of Africa. Infection produces a wide range of responses, from a chronic, self-limiting infection with few symptoms, to active hepatitis and cirrhosis, to fatal liver disorders and hepatocellular carcinoma. In the last few years, efforts have centered on understanding the mechanism by which HBV replicates and its role in causing liver cancer. The genome of HBV is a mostly double-stranded DNA molecule of 3200 nucleotides. After cellular infection, the DNA moves to the nucleus where it is copied into an RNA molecule and packaged into a viral capsid. The capsid containing the RNA pregenome moves to the cytoplasm where it is reverse-transcribed into a DNA strand, which in turn is made double stranded. The copied genome is repackaged into a new capsid for release from the cell or returns to the nucleus for another round of replication.

The key to the role of HBV in carcinogenesis apparently resides in its entry into the nucleus. While in the nucleus, the HBV genome can insert itself at many different sites into human chromosomes (Figure 19.12). This insertion process often results in cytogenetic alterations of the host genome that include translocations, deletions, or amplification of adjacent regions. As outlined earlier, most forms of cancer are associated with chromosomal rearrangements of these types, and it is possible that such rearrangements trigger the development of a cancerous transformation. Alternately, the HBV genome has been shown to integrate into the cyclin A gene, and abnormal expression of this gene can disrupt the cell cycle and

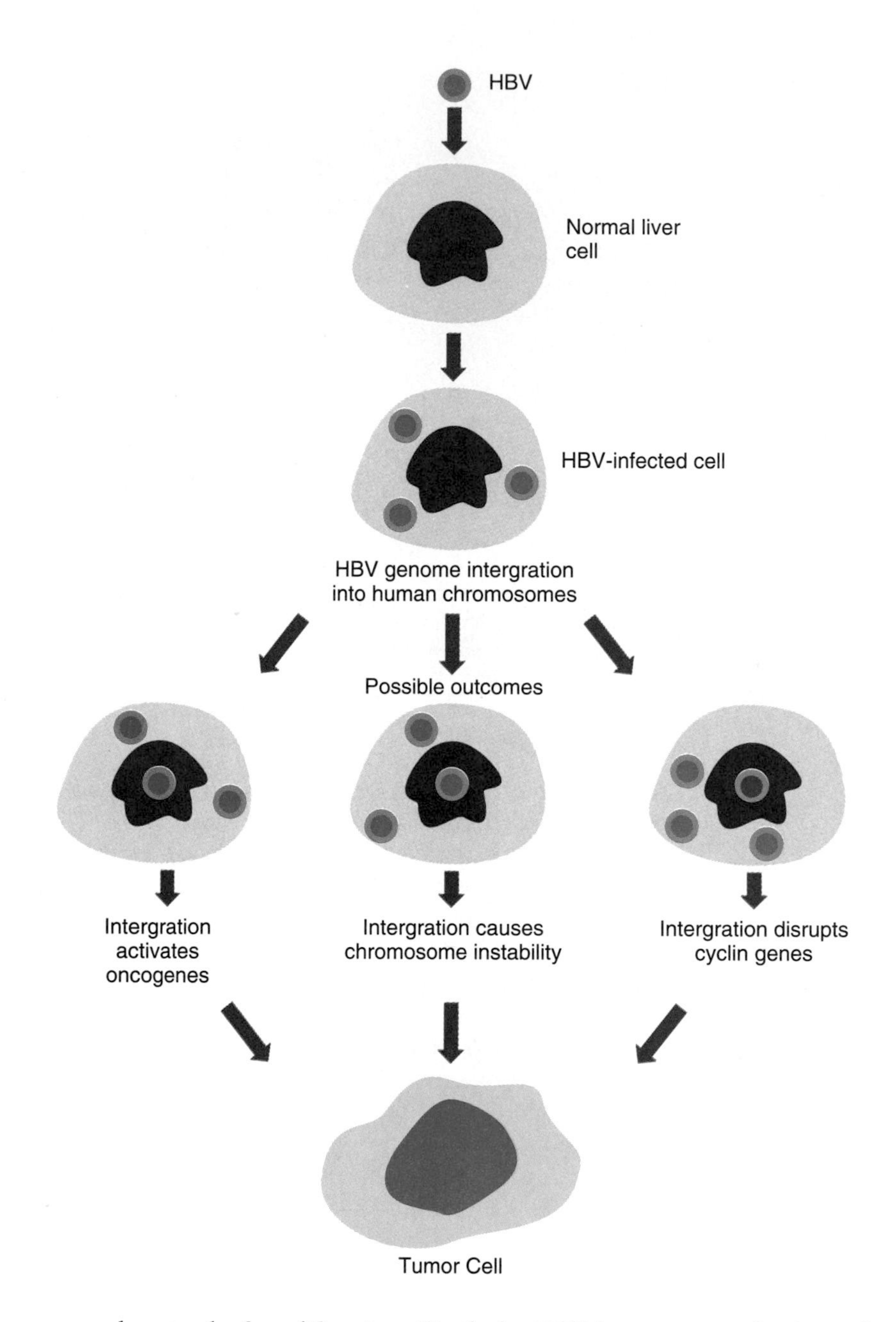

FIGURE 19.12
The possible outcomes following cellular infection with hepatitis B virus (HBV). Following integration of the viral chromosome into the human genome, oncogenes can be activated, resulting in tumor formation. Integration of the HBV genome can also cause chromosome instability, including the production of translocations and deletions that trigger a malignant cyclin A gene, causing abnormal regulation of the cell cycle and cellular proliferation resulting in cancer formation.

normal control of proliferation. Similarly, HBV integration has resulted in activation of proto-oncogene loci, and altered expression of such loci has been clearly implicated in the development of malignant cell growth. Thus, there is a strong link between the HBV virus as an environmental agent and the development of cancer in infected individuals.

Other Environmental Agents

Many surveys of cancer incidence have pointed to the role of the physical environment and personal behavior as factors that contribute to the development of cancer. The precise role of the environment in the genesis of cancer can be difficult to ascertain. Obviously, such induction of cancer involves an interaction between the genotype and environmental agents. The differential expression of a specific genotype in various environments is often neglected in making calculations about the role of the environment in cancer, but it has been estimated that at least 50% of all cancers are environmentally induced. The environmental agents responsible for cancer include background levels of radiation, occupational exposure to physical and chemical agents, exposure to sunlight, and personal behavior such as diet and the use of tobacco. To show how environmental factors are identified, let us consider a recent study on the risk of colon cancer.

Epidemiological surveys reveal that cancer of the colon occurs with a much higher frequency in North America and Western Europe than in Asia, Africa, or other parts of the developing world (Table 19.4). In addition, when people migrate from low-risk areas to high-risk areas, their rate of colon cancer increases to match that of the native residents. These and other observations suggest that dietary factors may play a role in the incidence of colon cancer. In a study of more than 88,000 registered nurses across the United States, dietary habits have been monitored by questionnaires since 1980. In that population, 150 cases of colon cancer were observed through 1986. Detailed analysis of dietary intake adjusted for diet composition indicated that the risk of colon cancer was positively associated with the intake of animal fat. The women who had daily meals of pork, beef, or lamb had a 2.5-fold higher risk of colon cancer than women who ate such meals less than once a month. This correlation strongly suggests that animal fat in the diet is an environmental risk factor for colon cancer.

Because there was a negative correlation between skinned chicken meat and colon cancer, what remains to be sorted out from this study is the role of red meat and/or fat alone as risk factors. Finally, if animal fat is identified as an important risk factor for colon cancer, laboratory studies to identify the mechanisms by which fat brings about a cancerous transformation of the intestinal epithelium will be undertaken. As described in an earlier section, these transformations are associated with mutations involving six different genes. However, even in the absence of information about the mechanism of action, it would seem prudent to reduce intake of animal fat in order to reduce the risk of colon cancer.

The accumulated results of research on the role of external factors as a cause of cancer indicate that diet, tobacco, and other agents, including medical and dental x-rays, viruses, drugs, and ultraviolet light, play significant roles and are responsible for up to 50% of all cases of cancer. This means that it is not the environment in general and pollution in particular that is responsible for a large fraction of cancer cases. Instead, personal choices (e.g., the composition of food in the diet, tobacco use, suntanning) are responsible for a large proportion of all cancers. Education and more judicious choices on the part of individuals could lead to the prevention of a high percentage of all human cancers.

TABLE 19.4
Epidemiology of cancer.

	Death Rate per 100,000	
Country	**Male**	**Female**
Australia	212	125
Canada	214	136
Dominican Republic	54	48
Egypt	39	18
England	248	156
Greece	188	103
Israel	174	145
Japan	190	109
Nicaragua	22	35
Portugal	180	108
Singapore	249	130
United States	216	137
Venezuela	135	128

CHAPTER SUMMARY

1 Cancer can be defined as a genetic disorder at the cellular level that can result from the mutation of a given subset of genes, or from alterations in the timing and amount of gene expression. Although some forms of cancer show familial patterns of inheritance, few show clear evidence for Mendelian inheritance.

2 Cancer results from the uncontrolled proliferation of cells and from the ability of such cells to metastasize or migrate to other sites and form secondary tumors. Mutant forms of genes involved in the regulation of the cell cycle are obvious candidates for cancer-causing genes. The cell cycle is apparently regulated at three sites, with two types of gene products primarily involved: cyclins and kinases. The action of these genes and the genes they control are important in regulating cell division, and links between these genes and the process of tumor formation are being discovered.

3 Although Mendelian patterns of cancer development are not clear-cut, there are genes that predispose to cancer. Study of these genes, such as retinoblastoma, has provided insight into the number and sequence of mutations that result in the development of tumors.

4 Tumor suppressor genes normally act to suppress cell division. When these genes are mutated or have altered expression, control over cell division is lost. Molecular analysis of two such genes retinoblastoma (RB) and Wilms tumor (WT) indicates that these genes work to control gene expression in different ways, either by affecting the activity of transcription factors or by acting as transcription factors.

5 Oncogenes are genes that normally function to maintain cell division, and these genes must be mutated or inactivated to halt cell division. If these genes escape control and become permanently switched on, cell division occurs in an uncontrolled fashion.

6 The cells of most tumors have visible chromosomal alterations, and the study of these aberrations has provided some insight into the steps involved in the development of cancer. This relationship between cancer and chromosome alterations has been best studied in leukemias, where the formation of hybrid genes or substitution of regulatory sequences is associated with the transformation of normal cells into malignant tumors.

7 Although cancer is the result of genetic alterations, it appears that the environment is an important factor in cancer induction. Environmental agents include occupational exposure to physical or chemical substances, viruses, diet, and other personal choices such as the use of tobacco.

KEY TERMS

acute transforming viruses
APC
chronic myelocytic leukemia (CML)
cyclins
familial adenomatous polyposis (FAP)
Hepatitis B virus (HBV)
hepatocellular carcinoma (HCC)
kinases
Li-Fraumeni syndrome
malignant tumor
metastatic tumor
nonacute; nondefective viruses
oncogenes
overexpression
Philadelphia chromosome
point mutations
proto-oncogene
retinoblastoma (RB)
retrovirus
sarcoma
reverse transcriptase
tumor suppressor
Wilms tumor (WT)

In disorders such as retinoblastoma, a mutation in one allele of the retinoblastoma (RB) gene can be inherited from the germ line, causing an autosomal dominant predisposition to the development of eye tumors. In order to develop tumors, a somatic mutation in the second copy of the RB gene is necessary. In sporadic cases, two independent mutational events, involving both RB alleles is necessary for tumor formation. Given that the first mutation can be inherited, what are the ways in which a second mutational event can occur?

ANSWER: In considering how this second mutation can arise, several levels of mutational events need to be considered, including changes in nucleotide sequence as well as events that involve whole chromosomes or chromosome parts. Retinoblastoma results when both copies of the RB locus have been lost or inactivated. With this in mind, perhaps the best way to proceed is to prepare a list of the phenomena that can result in a mutational loss or inactivation of a gene.

The first and most obvious way for the second RB mutation to occur is by a nucleotide alteration that converts the remaining normal RB allele to a mutation. This alteration may occur through a base substitution or by a frameshift mutation caused by the insertion or deletion of one or more nucleotides. A second method involves loss of the chromosome carrying the normal allele. Since the retinal cells are active in cell division, this event would take place during mitosis, resulting in chromosome 13 monosomy, with the mutant copy of the gene left as the only RB allele. This mechanism does not necessarily have to involve loss of the entire chromosome; deletion of the long arm (RB is on 13q) or an interstitial deletion involving the RB locus and some surrounding material would have the same result.

As an alternative, a chromosome aberration involving loss of the normal copy of the RB gene might be followed by a duplication of the chromosome carrying the mutant allele, restoring two copies of chromosome 13 to the cell, but no normal allele of the RB gene would be present. Lastly, a recombination event followed by chromosome segregation could produce a homozygous combination of mutant RB alleles.

Having made such a list, the next step is to analyze the cells from tumors using a combination of cytogenetic and molecular techniques (such as RFLP analysis and hybridizations to look for deletions) to see which of these mechanisms are actually found in tumors and to what extent they play a role in generating the second mutation. While such analysis is still in the preliminary stages, all mechanisms proposed have been found in tumors, indicating that a variety of spontaneous events can bring about the second mutation that triggers retinoblastoma.

PROBLEMS AND DISCUSSION QUESTIONS

1 As a genetic counselor, you are asked to assess the risk for a couple who plans to have children, but where there is a family history of retinoblastoma. In this case, both the husband and wife are phenotypically normal, but the husband has a sister with bilateral, familial retinoblastoma. What is the probability that this couple will have a child with retinoblastoma? Are there any tests you can think of that you could recommend to help in this assessment?

2 Review the stages of the cell cycle. What events occur in each of these stages? Which stage is most variable in length?

3 Where are the major regulatory points in the cell cycle?

4 Progression through the cell cycle depends on the interaction of two types of regulatory proteins: kinases and cyclins. List the functions of each and describe how they interact with each other to cause cells to move through the cell cycle.

5 What is the difference between saying that cancer is inherited and saying that predisposition to cancer is inherited?

6 Define tumor suppressor genes. Why are most tumor suppressor genes expected to be recessive?

7 Review the differences between the transcriptional activity, tissue distribution, and function of the tumor suppressor genes associated with retinoblastoma and Wilms tumor? What properties do they have in common? Which are most different?

8 Distinguish between oncogenes and proto-oncogenes. In what ways can proto-oncogenes be converted to oncogenes?

9 How do translocations such as the Philadelphia chromosome lead to oncogenesis?

10 Given that 50% of all cancers are environmentally induced, and most of these are caused by actions such as smoking, suntanning and diet choices, how much of the money spent on cancer research do you think should be devoted to research and education on the prevention of cancer rather than on finding a cure?

SELECTED READINGS

AMES, B., MAGRAW, R., and GOLD, L. 1990. Ranking possible cancer hazards. *Science* 236:71–80.

AMES, B., PROFET, M., and GOLD, L. 1990. Dietary pesticides (99.99% all natural). *Proc. Nat. Acad. Sci.* 87:7777–81.

BENEDICT, W., XU, H., HU, S., and TAKAHASHI, R. 1990. Role of the retinoblastoma gene in the initiation and progression of a human cancer. *J. Clin. Invest.* 85:988–93.

BIRRER, M. and MINNA, J. 1989. Genetic changes in the pathogenesis of lung cancer. *Ann. Rev. Med.* 40:305–17.

BISHOP, J. 1987. The molecular genetics of cancer. *Science* 235:306–11.

CAIRNS, J. 1985. The treatment of diseases and the war against cancer. *Scient. Amer.* (November) 253:51–59.

CROCE, C., and KLEIN, G. 1985. Chromosome translocations and human cancer. *Scient. Amer.* (March) 252:54–60.

FELDMAN, M., and EISENBACH, L. 1988. What makes a tumor cell metastatic? *Scient. Amer.* (November) 259:60–85.

HABER, D., and BUCKLER, A. 1992. WT1: a novel tumor suppressor gene inactivated in Wilms tumor. *New Biol.* 4:97–106.

LASKO, D., CAVANEE, W., and NORDENSKJOLD, M. 1991. Loss of constitutional heterozygosity in human cancer. *Ann. Rev. Genet.* 25:281–314.

MURRAY, A., and KIRSCHNER, M. 1991. What controls the cell cycle. *Scient. Amer.* (March) 264:56–63.

RUSTGI, A., and PODOLSKY, D. 1992. The molecular basis of colon cancer. *Ann. Rev. Med.* 43:61–68.

SAGER, R. 1989. Tumor suppressor genes: the puzzle and the promise. *Science* 246:1406–1412.

SOLOMON, E., VOSS, R., HALL, V., BODMER, W., JARS, J., JEFFREYS, A., LUCIBELLO, F., PATEL, I., and RIDER, S. 1987. Chromosome 5 allele loss in human colorectal carcinomas. *Nature* 328:616–29.

STANBRIDGE, E. 1992. Functional evidence for human tumor suppressor genes: chromosome and molecular genetic studies. *Cancer Surv.* 13:43–57.

SUGIMURA, T. 1990. Cancer prevention: underlying principles. *Basic Life Sci.* 52:225–32.

WEINBERG, R. 1985. The action of oncogenes in the cytoplasm and nucleus. *Science* 203:770–776.

WHITE, R. 1992. Inherited cancer genes. *Curr. Opin. Genet. Dev.* 2:53–57.

Quantitative Genetics

Chapter Concepts

Some traits are controlled by a number of genes and are classified as polygenic traits. Such traits are studied using biometric and statistical methods. Statistical methods are also used to determine how genetic variation interacts with the environment to produce the phenotype. This fraction of phenotypic variation that is due to genetic variation is its heritability.

In Chapter 4 we considered how the classical patterns of Mendelian inheritance (the 3:1 and 9:3:3:1 F_2 ratios) are modified because of gene interaction or the action of multiple genes on a single trait. At that time, we discussed examples of polygenic traits such as the inheritance of grain color in wheat, a trait that is controlled by three gene pairs. These traits were discussed to illustrate that even in complex modes of inheritance, the fundamental principles of segregation and independent assortment discovered by Mendel are operational.

Having considered the patterns of inheritance that are characteristic of polygenic traits in that earlier discussion, in this chapter we will describe the methods used by geneticists to study traits that are controlled by several genes. These methods are often statistical and involve analysis of traits using mathematical tools in addition to those of biochemistry or molecular biology. We will also discuss the effects of some extrinsic factors such as temperature, nutrition, and age on phenotypic expression. Finally, we will examine the concept of **heritability,** which is used by geneticists to estimate the degree to which phenotypic variation is a reflection of genetic differences, and to estimate indirectly the degree of environmental influence on the expression of complex traits.

Polygenic traits are at the heart of several disciplines in genetics, including plant breeding, livestock breeding, and wildlife management. Polygenic traits are also an important part of human genetics; traits such as intelligence, skin color, obesity, and predisposition to certain diseases are thought to be under polygenic control. In the following discussion, remember that in polygenic inheritance, as in monogenic inheritance, barring accidents, the genotype is fixed at the moment of fertilization, while the phenotype is more flexible and changes over the life span of the organism as the genotype interacts with the environment.

CONTINUOUS VARIATION AND POLYGENES

Continuous versus Discontinuous Variation

The traits studied by Gregor Mendel in the garden pea could each be separated into two classes: For height, the plants were either tall or dwarf; for color, either green or yellow, and so forth. Such traits are known as **discontinuous traits** because they fall into a limited number of distinct or discontinuous groups (Figure 20.1a). Other investigators working on the mechanisms of inheritance in the nineteenth century studied traits that showed a continuous gradation of phenotypes. Over a period of years, Sir Francis Galton used the sweet pea in a series of genetic crosses and measured the diameter of the pea as a phenotypic trait. When he crossed plants with large peas to plants with small peas, the F_1 peas were all intermediate in size to the parents. When the F_1 were crossed among themselves, the resulting F_2 generation contained peas of many different diameters. Some of the F_2 peas were as large or as small as the parents, but most were intermediate in size when compared to the parents (Figure 20.1b).

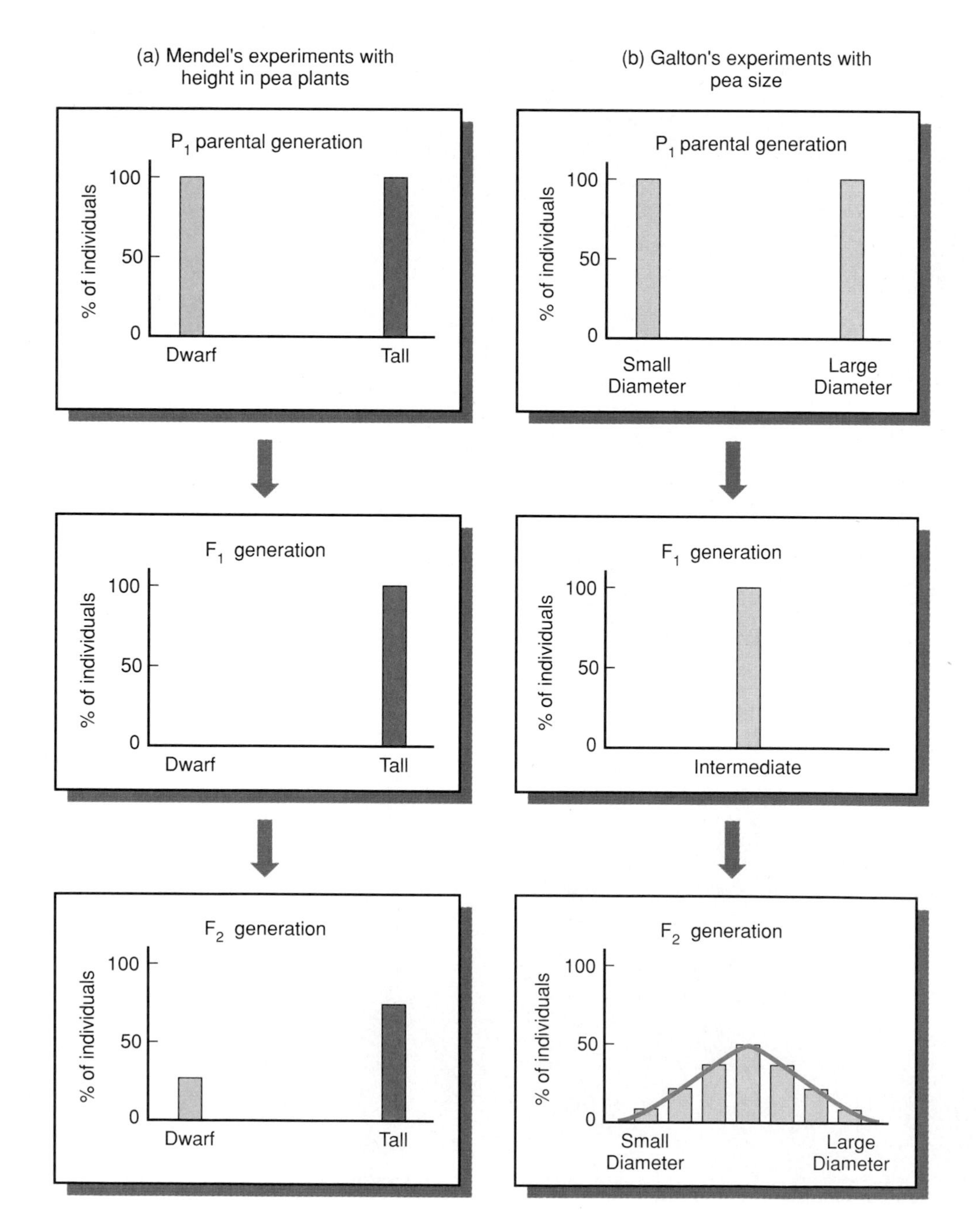

FIGURE 20.1
Histograms comparing phenotypic distributions in two crosses involving pea plants. (a) The distribution of phenotypes in the F_1 and F_2 for a trait exhibiting discontinuous variation. (b) The distributions of phenotypes in the F_1 and F_2 for a trait showing continuous variation.

In contrast to the two distinct phenotypic groups in the F_2 of Mendel's experiments, the F_2 phenotypes in Galton's experiments were continuously distributed over a wide range of sizes. Traits that are continuously distributed in the F_2 generation exhibit a pattern called **quantitative** or **polygenic** inheritance. Compared with classical Mendelian traits, polygenic phenotypes are usually measured in some way. This quantitation can be expressed as weight, stature, volume, or other units of measure. In addition, it is assumed that the phenotypes of such traits are controlled by more than one gene, that is, the trait is controlled by **multiple factors.**. For simplicity, it is further assumed that each gene pair contributing to the trait consists of one additive and one nonadditive allele, and that the total effect of each additive allele is small and roughly equal to all others. These alleles are also influenced by phenotypic and environmental variations. In sum, if phenotypes are quantitated and continuous in distribution, they are quantitative or polygenic traits. In transmission from generation to generation, such traits do not show segregation into clear-cut groups showing classical Mendelian ratios. Such traits are controlled by more than one gene.

Galton devised statistical methods to study these **continuous traits,** and the resulting field became known as **biometry.** While these mathematical techniques provided an important tool for the analysis of experimental data, they offered no explanation for the underlying mechanisms of inheritance. Not surprisingly, this led to many erroneous conclusions about the characteristics of continuous variation and to a subsequent debate as to whether Mendelian principles could explain this form of inheritance. As outlined in Chapter 4, this debate was settled in a series of experiments that demonstrated the role of Mendelian factors in continuous variation. The resulting **multiple-factor hypothesis** proposed that many genes can contribute to the phenotype in a cumulative fashion. Recall that in multiple-factor inheritance, the traits are quantified in some way, and that, for simplicity, it is usually assumed that each gene pair consists of one additive and one nonadditive allele. When two or more genes contribute to a phenotypic trait in an additive way, the trait is said to be **polygenic.**

Mapping Polygenic Traits

In discussing polygenic inheritance, it is useful to know how to approximate the number of genes that control a given trait. In a given polygenic cross, the phenotype of the F_2 generation will show greater phenotypic variability than the P_1 or F_1 generation. The multiple-factor hypothesis suggests that the number of allelic pairs of genes controlling a trait is inversely proportional to the number of F_2 individuals with the most extreme phenotype. That relationship is expressed as: $1/4^n$, where n equals the number of gene pairs that control the trait.

While the number of genes involved in a polygenic trait can be estimated from observation of phenotypes, such calculations provide no information about the physical location and relationships of these genes. Because polygenes act on a single trait, it is not unreasonable to ask whether such genes might be physically clustered on a contiguous portion of a single chromosome, or scattered throughout the genome.

In *Drosophila,* resistance to the insecticide DDT has been shown to be a polygenic trait. To find the loci responsible, strains selected for resistance to DDT and strains selected for sensitivity were crossed to a stock carrying dominant markers on each chromosome (Figure 20.2). Backcrosses and $F_1 \times F_1$ matings were used to create offspring carrying combinations of marker chromosomes and chromosomes from resistant strains. These combinations were then tested for resistance by exposure to DDT. The results indicate that each of the chromosomes in the *Drosophila* karyotype contains genes that contribute to resistance. In other words, the loci that control DDT resistance are not clustered together on a single chromosome, but instead are scattered throughout the genome. This does not mean that in all cases polygenes controlling a trait are scattered throughout the genome, but it does show that polygenes need not be clustered in order to affect a single trait.

It is possible to map genes responsible for DDT resistance in *Drosophila* because identifiable genetic markers are present on each chromosome. In many other organisms, especially those of agricultural importance, genetic markers are not available for each chromosome, so that systematic mapping of quantitative traits is not feasible. However, the development of restriction fragment length polymorphisms (RFLPs) as genetic markers (see Chapter 15 for a description of RFLPs) provides a new approach to the problem of enumerating and mapping the loci responsible for quantitative traits.

RFLPs were first used in mapping human genes (described in Chapter 15), but, in principle, can be used in almost any organism. For example, an RFLP map covering all 12 chromosomes of the tomato has been constructed, using some 300 RFLP markers. This provides a map of the tomato genome with genetic markers spaced at an average distance of 5 cM (recall that 1 cM, or centimorgan, is the distance that allows 1 percent recombination between markers). Using this map and some newly derived analytical methods, quantitative trait loci (QTLs) for fruit weight and soluble solids have been mapped in the tomato. Genes for fruit weight were found on six chromosomes (1, 4, 6, 7, 9, and 11), and four loci for soluble solids were identified (on chromosomes 3, 4, 6, and 7). Further work using RFLP markers on each of these chro-

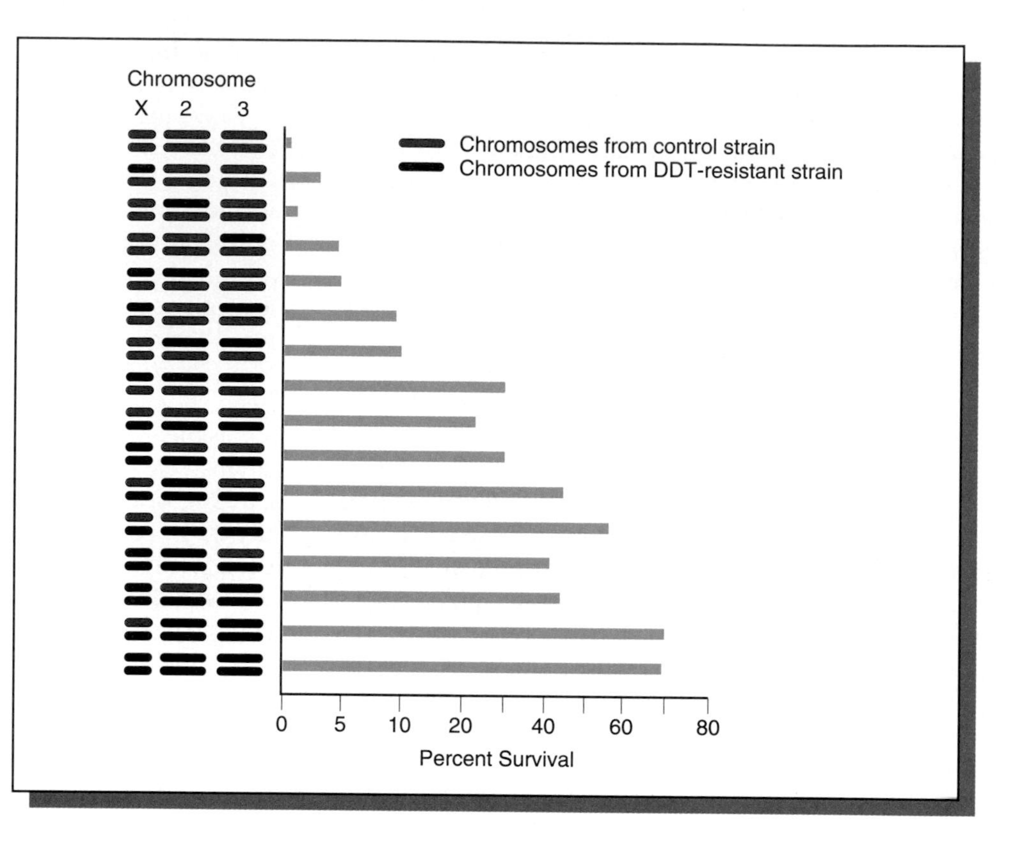

FIGURE 20.2
The differential survival of *Drosophila* carrying combinations of chromosomes from DDT-resistant and DDT-sensitive strains when exposed to DDT. The results indicate that DDT resistance is polygenic, with genes on each of the major chromosomes making a major contribution. Chromosome 4 carries few genes and was omitted from this analysis.

mosomes has allowed localization of several loci to relatively short chromosomal segments.

The use of chromosome-specific RFLP markers and identified loci for a desirable trait will make it possible to transfer traits such as disease and heat resistance or improved flavor from wild type or other domestic strains into agriculturally important strains of domesticated plants. Eventually such methods can be used to isolate and characterize the genes that control quantitative traits in the same way that RFLP analysis has led to the identification and isolation of genes associated with specific human genetic disorders such as cystic fibrosis or neurofibromatosis.

ANALYSIS OF POLYGENIC TRAITS

Because traits inherited in a quantitative fashion cannot be described in the same way as Mendelian traits, it is necessary to provide a method for evaluating the inheritance of quantitative characters. In most cases, the outcome of quantitative genetic experiments is a set of measurements that can be expressed as a frequency diagram (Figure 20.3). These measurements, made from one or a series of crosses, are taken as a sample of all possible crosses that could have been made in the experimental population. If we made an infinite number of such measurements, the frequency distribution would take on the shape of the curve in Figure 20.3, which is a normal, or bell-shaped, distribution curve. Obviously, in many cases, it is inconvenient or impossible to count all the individuals in an experimental population or to collect exhaustive amounts of data. But gathering a smaller data sample that is unbiased, and yet representative of the measured group, is often a problem. In addition, individual variation is a universal attribute of living systems, and this variability is reflected in the data gathered from observations of organisms. How can we distinguish between normal variation that is strictly due to chance and that due to an experimental variable?

The solution to these and other problems with data analysis has been to employ the techniques of **statistics.** Statistical analysis serves three purposes:

1 Data can be mathematically reduced to provide a **descriptive summary** of the sample.

2 Data from a small but random sample can be utilized to infer information about groups larger than those from which the original data were obtained **(statistical inference).** The sample values are called **statistics** (symbolized by Roman letters) and are used as an estimate of the value for the population. These population values are called **parameters** (symbolized by Greek letters).

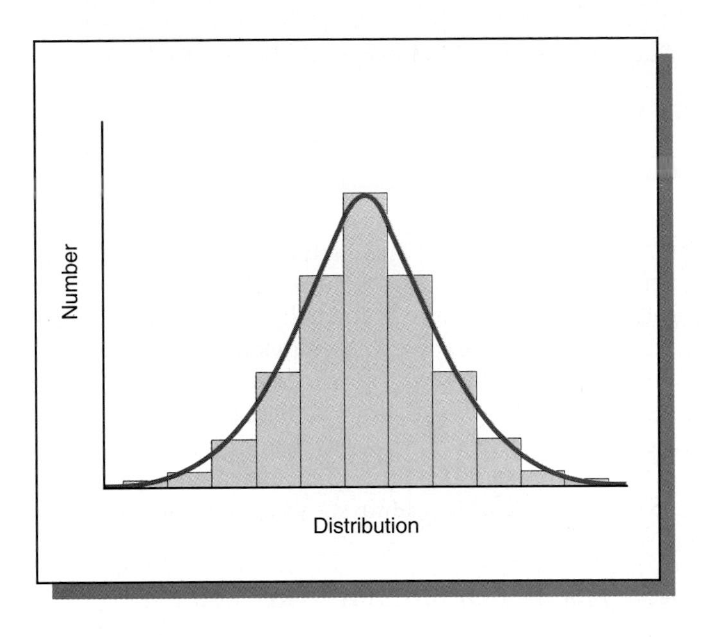

FIGURE 20.3
A normal frequency distribution.

3 Two or more sets of experimental data may be compared to determine whether they represent significantly different populations of measurements.

Several statistical methods are useful in the analysis of traits that exhibit a normal distribution, including the mean, variance, standard deviation, and standard error of the mean.

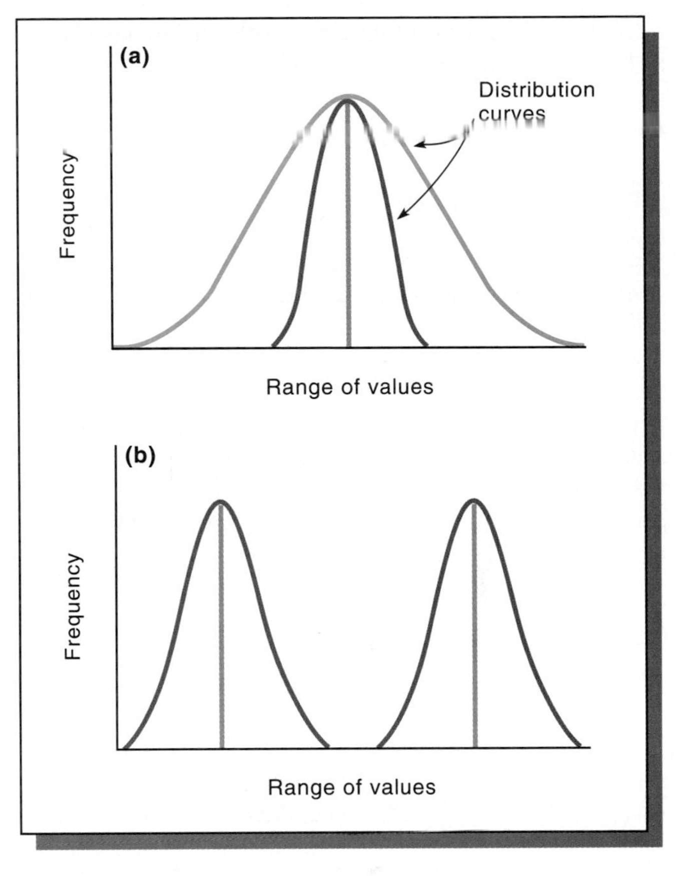

FIGURE 20.4
(a) Two normal frequency distributions with the same mean but different amounts of variation. (b) Two normal distributions with different means but the same amount of variation.

The Mean

The distribution of phenotypic values in Figure 20.4(a) tends to cluster around a central value. This clustering is called a **central tendency,** the most common measurement of which is the mean ($\overline{X}$). The mean is simply the arithmetic average of a set of measurements or data and is calculated as

$$\overline{X} = \frac{\Sigma X_i}{n} \qquad (20.1)$$

where $\overline{X}$ is the mean, ΣX_i represents the sum of all individual values in the sample, and n is the number of individual values.

Although the mean provides a descriptive summary of the sample, it is in itself of limited value. All values in the sample may be clustered near the mean, or they may be distributed widely around it. Figure 20.4 shows normal or symmetrical distributions with identical means, but a widely different range of values. These contrasting conditions represent a different type and amount of variation within the sample. Whether due to chance or to one or more experimental variables, such variation creates the need for methods to describe variation around the mean.

Variance

As seen in Figure 20.4(a), the range of values on either side of the mean will determine the width of the distribution curve. Measurement of this variation in a sample is called the sample variance (s^2 or V) and is used as an estimate of the variation present in an infinitely large population. The variance for a sample is calculated as:

$$s^2 = V = \frac{\Sigma(X_i - \overline{X})^2}{n - 1} \qquad (20.2)$$

where the sum (Σ) of the squared differences between each measured value (X_i) and the mean ($\overline{X}$) is divided by one less than the total sample size ($n - 1$). To avoid the numerous subtraction functions necessary in calculating s^2 for a large sample, we can convert the equation to its algebraic equivalent:

$$s^2 = V = \frac{\Sigma X_i^2 - n\overline{X}^2}{n - 1} \qquad (20.3)$$

The variance is a valuable measure of sample variability. Two distributions may have identical means ($\overline{X}$), yet vary considerably in their concentration around the mean. The variance represents the average squared deviation of the measurements from the mean. The larger the variance, the greater the range of values on either side of the mean. The estimation of variance has been particularly valuable in determining the degree of genetic control of traits where the immediate environment also influences the phenotype.

Standard Deviation

Since the variance is a squared value, its unit of measurement is also squared (cm^2, mg^2, etc.). To express variation around the mean in the original units of measurement, it is necessary to calculate the square root of the variance, a term called the **standard deviation (s):**

$$s = \sqrt{s^2} \qquad (20.4)$$

Table 20.1 shows what percentage of the individual values within a normal or bell-shaped distribution is included with different multiples of the standard deviation. The mean plus or minus one standard deviation ($\overline{X} \pm 1s$) includes 68 percent of all values in the sample. Over 95 percent of all values are found within two standard deviations ($\overline{X} = 2s$). As such, the standard deviation provides an important descriptive summary of a set of data.

Standard Error of the Mean

To estimate how much the means of other similar samples drawn from the same population might vary, we can calculate the **standard error of the mean ($S_{\overline{X}}$):**

$$S_{\overline{X}} = \frac{s}{\sqrt{n}} \qquad (20.5)$$

where s is the standard deviation, and $\sqrt{n}$ is the square root of the sample size. The standard error of the mean is a measure of the accuracy of the sample mean, that is, the variation of sample means in replications of the experiment. Because it can be expected that the standard deviation of mean values will reflect less variance than the standard deviation of a set of individual measurements, the standard error is always less than the standard deviation. Since the standard error of the mean is computed by dividing s by $\sqrt{n}$, it is always a smaller value.

TABLE 20.1
Sample inclusion for various s values.

Multiples of s	% of Sample included
$\overline{X} \pm 1s$	68.3
$\overline{X} \pm 1.96s$	95
$\overline{X} \pm 2s$	95.5
$\overline{X} \pm 3s$	99.7

Analysis of a Quantitative Character

Many characteristics of importance in livestock and crop plants are controlled by polygenic systems. In an earlier chapter, we discussed such phenotypes as ear length in corn and kernel color in wheat. To illustrate how biometric methods are used in the analysis of quantitative characters, we will consider a simplified example involving fruit weight in tomatoes. Let us assume that fruit weight is a quantitative character, and that one highly inbred strain produces tomatoes averaging 18 oz. in weight, and another highly inbred strain produces fruit averaging 6 oz. in weight. These two varieties are crossed and produce an F_1 generation with weights ranging from 10 oz. to 14 oz. (Table 20.2). The F_2 population contains individuals that produce fruit ranging from 6 oz. to 18 oz.

The mean value for the fruit weight in the F_1 generation can be calculated as:

$$\overline{X} = \frac{\Sigma X_i}{n} = \frac{626}{52} = 12.04$$

Similarly, the mean value for fruit weight in the F_2 generation is calculated as:

$$\overline{X} = \frac{\Sigma X_i}{n} = \frac{872}{72} = 12.11$$

Average fruit weight is 12.04 oz. in the F_1 generation and 12.11 oz. in the F_2 generation. While these mean values are similar, it is apparent that there is more variation present in the F_2 generation, because fruit weight ranges from 6 to 18 oz. in the F_2 generation, but only from 10 to 14 oz. in the F_1 generation.

To assist in quantitating the amount of variation present in each generation, we can construct a frequency

TABLE 20.2
Distribution of F_1 and F_2 progeny.

		Weight in oz.												
		6	7	8	9	10	11	12	13	14	15	16	17	18
Number of	F_1:					4	14	16	12	8				
Individuals	F_2:	1	1	2		9	13	17	14	7	4	3		1

table (Table 20.3). The first column (x) represents the midpoint of a class interval. For example, all tomatoes weighing between 5.5 and 6.5 oz. are classified as 6 oz. The second column (marked f) lists the number of tomatoes that fall into that category. The last column $f(x)$ computes the product of the first two columns ($x \times f$) and lists the total weight in oz. for each class. The sum of the f column (Σf) equals the total number of tomatoes counted, and the sum of the $f(x)$ column [$\Sigma f(x)$] equals the total weight of all the tomatoes in the sample. The mean value for fruit weight in the F_1 and F_2 generations taken from the frequency table can be calculated as:

$$F_1: \quad \frac{\Sigma f(x)}{\Sigma f} = \frac{626}{52} = 12.03$$

$$F_2: \quad \frac{\Sigma f(x)}{\Sigma f} = \frac{842}{72} = 12.11$$

As noted above, the sample variance can be calculated as the sum of the squared differences between each value and the mean, divided by one less than the total number of observations (Equation 20.2). However, in the case where a number of observations (f) have been grouped into representative classes (x), the variance can be calculated according to the formula:

$$s^2 = V = \frac{n\Sigma f(x^2) - (\Sigma fx)^2}{n(n-1)}$$

As shown in Table 20.4, the value for the F_1 generation is 1.29, and for the F_2 generation, it is 4.28. When converted to the standard deviation ($s = \sqrt{s^2}$), the values become 1.13 and 2.06, respectively. Thus, the distribution of tomato weight in the F_1 generation can be described as 12.04 ± 1.13, and in the F_2 as 12.11 ± 2.06. This analysis indicates that the mean fruit weight of the F_1 is identical to that of the F_2, but that the F_2 generation shows greater

TABLE 20.3
Frequency distribution in F_1 and F_2.

F_1:	x	f	$f(x)$	F_2:	x	f	$f(x)$
	6				6	1	6
	7				7	1	7
	8				8	2	16
	9				9		
	10	4	40		10	9	90
	11	14	154		11	13	143
	12	16	192		12	17	204
	13	12	156		13	14	182
	14	6	84		14	7	98
	15				15	4	60
	16				16	3	48
	17				17		
	18				18	1	18
		$\Sigma f = 52$	$\Sigma f(x) = 626$			$\Sigma f = 72$	$\Sigma f(x) = 872$

TABLE 20.4
Calculation of variance.

F_1: x	f	$f(x)$	$f(x^2)$	F_2: x	f	$f(x)$	$f(x^2)$
6				6	1	6	36
7				7	1	7	48
8				8	2	16	128
9				9			
10	4	40	400	10	9	90	900
11	14	154	1,694	11	13	143	1,573
12	16	192	2,304	12	17	204	2,448
13	12	156	2,028	13	14	182	2,366
14	6	84	1,176	14	7	98	1,372
15				15	4	60	900
16				16	3	48	768
17				17			
18				18	1	18	324
	$52 = n$	$626 = \Sigma f(x)$	$7{,}602 = \Sigma f(x^2)$		$72 = n$	$872 = \Sigma f(x)$	$10{,}864 = \Sigma f(x^2)$

$$s^2 = \frac{n\Sigma f(x^2) - (\Sigma fx)^2}{n(n-1)}$$

For F_1:

$$s^2 = \frac{52 \times 7{,}602 - (626)^2}{52(52-1)} = \frac{395{,}304 - 391{,}876}{2{,}652} = 1.29$$

For F_2:

$$s^2 = \frac{72 \times 10{,}864 - (872)^2}{72(72-1)} = \frac{782{,}208 - 760{,}304}{5{,}112} = 4.28$$

variability in the distribution of weights than the F_1.

The observations about the inheritance of fruit weight in crosses between these two strains of tomatoes meet the expectations for polygenic traits. For the sake of this example, if we assume that each parental strain is homozygous dominant or homozygous recessive for the genes that control fruit weight, we can calculate the number of gene pairs involved in controlling fruit weight in these two strains of tomatoes. Since 1/72 of the F_2 offspring have a phenotype that overlaps one of the parental strains (72 total F_2 offspring, one weighs 6 oz., one weighs 18 oz.; see Table 20.2), the formula $1/4^n = 1/72$ indicates that 3–4 genes control fruit weight in these tomato strains.

PHENOTYPIC EXPRESSION

Gene expression is often discussed as if the genes operate in a closed "black box" system in which the presence or absence of functional products directly determines the collective phenotype of an individual. The situation is actually much more complex. Most gene products function within the internal milieu of the cell, cells interact with one another in various ways, and the organism must survive under diverse environmental influences. Thus, gene expression and the resultant phenotype are often modified through the interaction between an individual's particular genotype and its internal and external environment.

The degree of environmental influence may vary from inconsequential to subtle to very strong. Subtle interactions are the most difficult to detect and document and have led to unresolvable "nature–nurture" conflicts in which scientists debate the relative importance of genes versus environment. How easily such conflicts are resolved depends on the characteristic being investigated, and even then it is sometimes impossible to provide definitive information. In this section, we will deal with some of the variables known to modify gene expression.

Penetrance and Expressivity

Some mutant genotypes are always expressed as a distinct phenotype, while others produce a proportion of individuals whose phenotypes cannot be distinguished from wild type. The variable expression of a particular trait may be quantitatively studied by determining the degree of penetrance and expressivity. The percentage of individuals that shows at least some degree of expression of a mutant genotype defines the **penetrance** of the mutation. For example, many mutant alleles in *Drosophila* are said to overlap with wild type. Flies homozygous for the recessive mutant gene *eyeless* yield phenotypes that range from the presence of normal eyes to the complete absence of one or both eyes (Figure 20.5). If 15 percent of mutant flies show the wild-type appearance, the mutant gene is said to have a penetrance of 85 percent. On the other hand, the range of expression of the genotype defines the **expressivity**. In the case of *eyeless*, although the average reduction of eye size is one-fourth to one-half, the range of expressivity is from complete loss of both eyes to completely normal eyes.

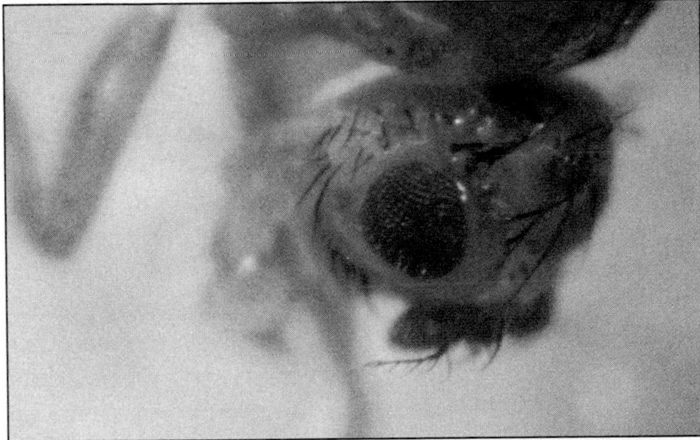

FIGURE 20.5
Gradations in phenotype, ranging from wild-type to eyeless, associated with the *eyeless* gene in *Drosophila.*

Examples such as the expression of the *eyeless* phenotype have provided the basis for experiments designed to determine the causes of phenotypic variation. If the laboratory environment is held constant and extensive variation is still observed, the genetic background may be investigated. It is possible that other genes are influencing or modifying the *eyeless* phenotype. On the other hand, if it is experimentally proven that the genetic background is not the cause for the phenotypic variation, environmental variables such as temperature, humidity, and nutrition may be examined. In the case of the *eyeless* phenotype, it has been determined experimentally that both genetic background and environmental factors influence its expression.

Genetic Background: Suppression and Position Effects

With only certain exceptions, it is difficult to assess the specific relationship between the **genetic background** and the expression of a gene responsible for determining the potential phenotype.

The influence of genetic background may occur in two ways. First, the expression of other genes throughout the genome may have an effect on the phenotype produced by the gene in question. The phenomenon of **genetic suppression** is an example. Mutant genes such as *suppressor of vermilion* (*su-v*), *suppressor of forked* (*su-f*), and *suppressor of Hairy-wing* (*su-Hw*) in *Drosophila* completely or partially restore the normal phenotype to an organism that is homozygous (or hemizygous) for these recessive genes. For example, flies homozygous for both *vermilion* (a bright red eye color mutation) and *su-v* have eyes with wild-type color.

The phenomenon of suppression occurs in a wide variety of organisms. In microorganisms, where molecular studies are easier to perform, some suppressor gene products are molecules that function in the genetic translation process by correcting certain errors produced by the mutation of other genes. Transfer RNA in mutant form has been shown to have such an effect. It has also been hypothesized that a suppressor gene product might provide an alternate metabolic route to bypass a block in a biosynthetic pathway caused by the primary mutation. Suppressor genes are excellent examples of the genetic background modifying primary gene effects.

Second, the physical location of a gene in relation to other genetic material may influence its expression. Such a situation is called a **position effect.** For example, if a gene is included in a **translocation** or **inversion** event (in which a region of a chromosome is relocated), the expression of the gene may be affected. This is particularly true if the gene is translocated to or near an area of the chromosome composed of **heterochromatin.** These regions are genetically different and often appear condensed in interphase.

An example of a position effect involves female *Drosophila* heterozygous for the sex-linked recessive eye-color mutant *white* (w). If the region of the X chromosome containing the wild-type w^+ allele is translocated so that it is close to heterochromatin, expression of the w^+ allele is modified (Figure 20.6). Before translocation, a w^+/w genotype results in a wild-type brick red eye color. After translocation, the dominant effect of the w^+ allele is reduced. Instead of having a red color, the eyes are variegated, or mottled with red and white patches. A similar position effect is produced if a heterochromatic region is relocated next to the *white* locus on the X chromosome. Apparently, heterochromatic regions inhibit the expression of adjacent genes. Loci in many other organisms also exhibit position effects, providing proof that alteration of the normal arrangement of genetic information can modify its expression.

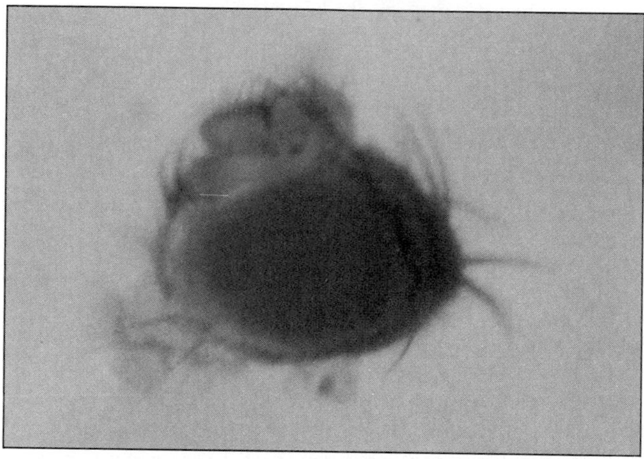

(a)

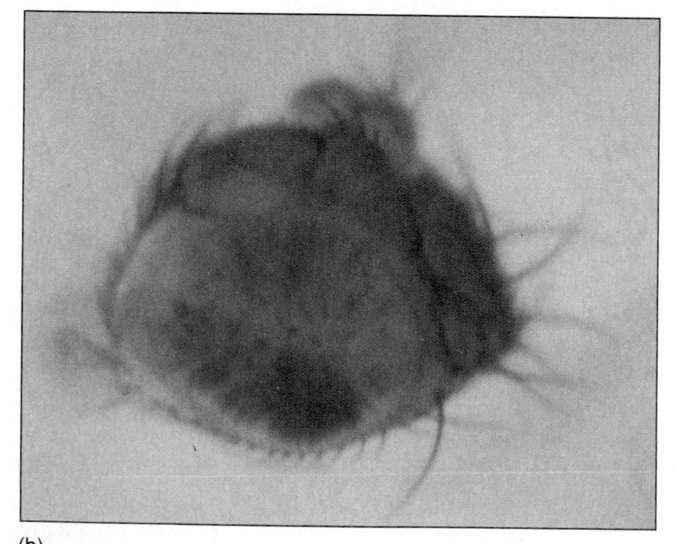

(b)

FIGURE 20.6
Eye phenotype in two female *Drosophila* heterozygous for the gene *white.* (a) Normal dominant phenotype showing brick red eye color. (b) Variegated color of an eye caused by transposition of the *white* gene to another location in the genome.

Temperature Effects

Because chemical activity depends on the kinetic energy of the reacting substances, which, in turn, depends on the surrounding temperature, we can expect temperature to influence phenotypes. For example, the evening primrose produces red flowers at 23°C and white flowers at 18°C. Siamese cats and Himalayan rabbits exhibit dark fur in regions of the nose, ears, and paws because the body temperatures of these extremities are slightly cooler (Figure 20.7). In these cases, it appears that an enzyme that functions in pigment production at the lower temperatures found in the extremities loses its catalytic function at the slightly higher temperatures throughout the rest of the body.

The most striking effect of temperature is the complete inhibition of mutant expression. Organisms carrying a certain mutant allele may show the mutant phenotype at one temperature, but if they are grown under a second temperature, they may exhibit a wild-type phenotype. Mutations of this type are called **temperature sensitive;** they have been discovered in numerous organisms, including viruses, fungi, and *Drosophila.*

The effects of temperature are used in studying mutations that interrupt essential processes during development and are thus normally lethal to the organism. If viruses carrying a **temperature-sensitive lethal mutation** are allowed to infect bacteria cultured at 42°C, infection progresses until the gene product essential to the virus is needed and then arrests, because this is the **restrictive temperature.** If cultured at the **permissive temperature** of 25°C, infection occurs normally, new viruses are produced, and the bacteria are lysed. The use of temperature-sensitive mutations, which may be induced and isolated, has added immensely to the study of viral genetics.

Similarly, many temperature-sensitive mutations have been discovered in *Drosophila*. Not only have dominant

FIGURE 20.7
Part A: A Himalayan rabbit. Part B: A Siamese cat. Both show dark fur color at the extremities of the ears, nose, and paws. These patches are due to expression of a temperature-sensitive allele for dark fur at the lower temperatures found in the extremities.

and recessive conditional lethal mutations been recovered in great numbers, but nonlethal temperature-sensitive mutations affecting development, morphology, and behavior have also been recovered. These nonlethal mutants have proven to be valuable in the dissection of genetic control mechanisms.

Nutritional Effects

Another example of conditional mutations involves nutrition. In microorganisms, mutations that prevent synthesis of nutrient molecules are quite common. These **nutritional mutants** arise when an enzyme essential to a biosynthetic pathway becomes inactive. If the end product of a biochemical pathway can no longer be synthesized, and if that molecule is essential to normal growth and development, the mutation is lethal. If, for example, the bread mold *Neurospora* can no longer synthesize the amino acid leucine, proteins cannot be synthesized unless leucine is added to the growth medium. If leucine is added, the lethal effect is overcome. Nutritional mutants have been crucial to molecular genetic studies and also served as the basis for George Beadle and Edward Tatum's proposal, in the early 1940s, that one gene functions to produce one enzyme.

In humans, a slightly different set of circumstances is known. The presence or absence of certain dietary substances, which normal individuals may consume without harm, may adversely affect individuals with abnormal genetic constitutions. Often, a mutation may prevent an individual from metabolizing some substance commonly found in normal diets. For example, those afflicted with the genetic disorder **phenylketonuria** cannot metabolize the amino acid phenylalanine. Those with **galactosemia** cannot metabolize the sugar galactose. Other individuals are intolerant of the milk sugar lactose. Those with **diabetes** or **hypoglycemia** cannot adequately metabolize glucose. In each case, excessive amounts of the molecule accumulate in the body and become toxic, and a characteristic phenotype results. However, if the dietary intake of the molecule is drastically reduced or eliminated, the associated phenotype may be reversed.

The case of **lactose intolerance** illustrates the general principles involved. Lactose is a disaccharide consisting of a molecule of glucose and a molecule of galactose. It is present as 7 percent of human milk and 4 percent of cow's milk. To metabolize lactose, humans require the enzyme **lactase,** which cleaves the disaccharide. Adequate amounts of lactase are produced during the first few years after birth. However, in many ethnic groups, the levels of this enzyme soon drop drastically, and adults become intolerant of milk. The major phenotypic effect involves severe intestinal diarrhea, including flatulence and cramps. This condition is particularly prevalent in Eskimos, Africans, Asiatics, and Americans with these heritages. In these cultures, milk is usually used for food in different forms, including cheese, butter, and yogurt. In these forms, the amount of lactose is reduced significantly and adverse effects are largely eliminated. Thus, by altering the diet, phenotypic expression of a genetic trait can be modified.

Onset of Genetic Expression

Not all genetic traits are expressed at the same time during an organism's life span. In most cases, the age at which a gene is expressed corresponds to the normal sequence of growth and development. In humans, the prenatal, infant, preadult, and adult phases require different kinds of genetic information. In a similar way, many genetic disorders can be expected to manifest themselves at different stages of life. Lethal genes account for many of the frequent spontaneous abortions and miscarriages occurring in the human population. These mutations are believed to alter genetic products essential to prenatal development. **Tay-Sachs disease** and the **Lesch-Nyhan syndrome** are severe disorders of lipid and nucleic acid metabolism, respectively, but do not cause lethality until early in childhood. **Huntington disease** demonstrates a wide range of age of onset, but not usually before age 20, and most often after age 40. Observations of the development of these conditions support the concepts that the need for certain gene products may be more essential at certain times and that the internal environment of an organism changes with age. In the broadest sense, aging must be thought of as a process that begins following conception and continues until death.

Heredity versus Environment

Many genetically determined traits are influenced by the environment and have considerable impact on the organism's phenotype. How do geneticists attempt to define the relative importance of heredity compared with that of the environment? Some traits demonstrate minimal variation under normal environmental circumstances, but other traits show distinct variability under the same environmental conditions. The latter case is particularly true for traits controlled by polygenic inheritance. Often, polygenic traits show such a large degree of variation in a population that geneticists can only speculate that a large but unknown number of genes provide the genetic potential, which is then modified by the environment.

Provided that a trait can be quantitatively measured, we can approach analytically the question of genetics versus environment. Experiments on plants and animals other than humans can test the causes of variation. Inbred strains containing individuals of a relatively homogeneous or constant genetic background can be generated, thus producing an **isogenic** population. Experiments are then designed to test the effects of the range of prevailing environmental conditions on phenotypic variability.

In studies with such isogenic lines, the **heritability index** or **ratio (H^2)** for a given trait can be calculated. Also called the **broad heritability** of the character being studied, H^2 measures the degree to which phenotypic variation (V_P) is due to genetic factors in a single population under the limits of environmental variability during the study. It is important to emphasize here that H^2 *does not* measure the proportion of the total phenotype attributed to genetic factors, but only the observed variation in the phenotype. Technically, the term H^2 is an expression of phenotypic variance, which is due to the sum of three components: environmental variance (V_E), genetic variance (V_G), and variance resulting from the interaction of genetics and environment (V_{GE}). Therefore, phenotypic variance (V_P) is theoretically expressed as

$$V_P = V_E + V_G + V_{GE} \qquad (20.6)$$

Since V_{GE} is nearly impossible to analyze, it is usually omitted. Therefore, the simpler equation is generally used:

$$V_P = V_E + V_G \qquad (20.7)$$

Broad heritability expresses that proportion of variance due to the genetic component:

$$H^2 = \frac{V_G}{V_P} \qquad (20.8)$$

A very high H^2 value indicates that the environmental conditions have had little impact on phenotypic variance in the population studied. A very low H^2 value indicates that the variation in the environment has been almost solely responsible for the observed phenotypic variation. Table 20.5 lists estimates of heritability for traits in a variety of organisms.

It is not possible to obtain an absolute H^2 value for any given character. If measured in a different population under a greater or lesser degree of environmental variability, H^2 might well change for that character.

Practically, heritability is most useful in animal and plant breeding as a measure of potential selection. Breeders are most interested in the improvement of economically important characters. From the standpoint of selection, V_G must be examined in a more complex way:

$$V_G = V_D + V_A + V_I \qquad (20.9)$$

TABLE 20.5
Estimates of heritability of several traits.

Trait	Percent Heritability (H^2)
Mouse	
Tail length	60
Litter size	15
Drosophila	
Abdominal bristle number	52
Wing length	45
Egg production	18
Humans	
Asthma	80
Diabetes (late onset)	70
Ulcers	37

where V_D is the genetic effect of dominant genes, V_A is due to the effect of additive genes, and V_I is the effect of interactive or epistatic genes. Of these components, V_A is most important in selection. This more limited estimate has been called **narrow heritability (h^2).** Therefore, heritability in breeding experiments becomes effectively

$$h^2 = \frac{V_A}{V_P} \tag{20.10}$$

Narrow heritability estimates, while characterizing a particular population, have been very useful in agricultural programs. Based on these estimates, selection techniques have led to vast improvements in the quality and quantity of plant and animal products.

In humans, isogenic strains are obviously not available for study, nor can the environment be controlled for individuals or populations. However, twins are very useful subjects for studying the heredity versus environment question in humans. **Monozygotic** or **identical twins,** derived from the division and splitting of a single egg following fertilization, are identical in their genetic composition. Although most identical twins are reared together and are exposed to very similar environments, some pairs are separated and raised in different settings. For any particular trait, average similarities or differences can be investigated. Such an analysis is particularly useful because characteristics that remain similar in different environments are believed to be inherited. These data can then be compared with a similar analysis of **dizygotic** or **fraternal twins,** who originate from two separate fertilization events. Dizygotic twins are thus no more genetically similar than any two siblings.

A form of quantitative analysis of characteristics of twins reared together may also be pursued. Twins are said to be **concordant** for a given trait if both express it or neither expresses it and **discordant** if one shows the trait and the other does not. Table 20.6 lists concordance values for various traits in both types of twins.

These data must be examined very carefully before any conclusions are drawn. If the concordance value approaches 90 to 100 percent with monozygotic twins, we might be inclined to interpret this value as indicating a large genetic contribution to the expression of the trait. In some cases—blood types and eye color, for example—we know this is indeed true. In the case of measles, however, a high concordance value merely indicates that the trait is almost always induced by a factor in the environment—in this case, a virus.

Therefore, it is more meaningful to compare the difference between the concordance values of monozygotic and dizygotic twins. If these values are significantly higher for monozygotic twins than dizygotic twins, we suspect that there is a genetic component involved in the determination of the trait. We reach this conclusion because monozygotic twins, with identical genotypes, would be expected to show a greater concordance than genetically related, but not genetically identical, dizygotic twins. In the case of measles, where concordance is high in both types of twins, the environment is assumed to contribute significantly.

Even though a particular trait may be determined to be influenced substantially by genetic factors, it is often difficult to formulate a precise mode of inheritance based on available data. In many cases, the trait is considered to be controlled by multiple-factor inheritance. However, when the environment is also exerting a partial influence, such a conclusion is particularly difficult to prove.

TABLE 20.6
A comparison of concordance for various traits between monozygotic (MZ) and dizygotic (DZ) twins.

Trait	Concordance	
	MZ	DZ
Blood types	100%	66%
Eye color	99	28
Mental retardation	97	37
Measles	95	87
Idiopathic epilepsy	72	15
Schizophrenia	69	10
Diabetes	65	18
Identical allergy	59	5
Tuberculosis	57	23
Cleft lip	42	5
Club foot	32	3
Mammary cancer	6	3

CHAPTER SUMMARY

1 Discontinuous variation is represented by those traits that fall into discrete phenotypic categories. Traits that demonstrate considerably more variation and are not easily categorized into distinct classes are examples of continuous variation. Polygenic inheritance, which controls characters that can be measured, involves genes with additive alleles that influence a character in a quantitative way.

2 Polygenic characteristics can be analyzed using statistical methods, which include the mean, the variance, the standard deviation, and the standard error of the mean. Such statistical analysis can be descriptive, can be used to make inferences about a population, or can be used to compare sets of data.

3 The penetrance of a mutant allele is measured by the percentage of a population that exhibits evidence of the mutant phenotype. Expressivity, on the other hand, measures the range of expression within a population.

4 Phenotypic expression may be modified by factors such as genetic background, temperature, nutrition, and other environmental factors. The phenomena of genetic suppression and position effects have been used to illustrate the existence of genetic background factors. Together, all such factors constitute the total environment in which genetic information is expressed.

5 The time of onset of gene expression varies as the need for certain gene products occurs at different periods during the processes of development, growth, and aging. This observation is particularly evident in the study of human disorders.

6 The degree of impact of genetic and environmental factors in the establishment of a given phenotype is difficult to ascertain. Heritability can be calculated for many characters but is especially useful in selective breeding of commercially valuable plants and animals.

7 Studies involving twins are aimed at resolving the question of heredity versus environment in human traits. The degree of concordance of a trait may be compared in monozygotic (identical) and dizygotic (fraternal) twins raised together or apart.

KEY TERMS

biometry
broad heritability
central tendency
concordant
continuous traits
descriptive summary
diabetes
discontinuous traits
discordant
dizygotic (fraternal) twins
expressivity
galactosemia
genetic background
genetic suppression
heritability index
heritability
heterochromatin
Huntington disease
hypoglycemia
inversion
isogenic
lactose
lactose intolerance
Lesch-Nyhan syndrome
monozygotic (identical) twins
multiple-factor hypothesis
narrow heritability
parameters
penetrance
permissive temperature
phenylketonuria
position effect
restrictive temperature

ratio
standard deviation
statistical inference
statistics
standard error of the mean
translocation
Tay-Sachs disease
temperature sensitive
temperature sensitive lethal mutation

The following results were recorded for ear length in corn:

	Length of ear in cm.																
	5	**6**	**7**	**8**	**9**	**10**	**11**	**12**	**13**	**14**	**15**	**16**	**17**	**18**	**19**	**20**	**21**
Parent A	4	21	24	8													
Parent B									3	11	12	15	26	15	10	7	2
F_1					1	12	12	14	17	9	4						
F_2			1	10	19	26	47	73	68	68	39	25	15	9	1		

1 For each of the parental strains, and the F_1 and F_2, calculate the mean values for ear length.

ANSWER: The mean values can be calculated from equation 20.1.

$$\bar{X} = \frac{\Sigma X_i}{n} \qquad P_A: \; \bar{X} = \frac{\Sigma X_i}{n} = \frac{378}{57} = 6.63$$

$$P_B: \; \bar{X} = \frac{\Sigma X_i}{n} = \frac{1697}{101} = 16.80$$

$$F_1: \; \bar{X} = \frac{\Sigma X_i}{n} = \frac{836}{69} = 12.11$$

$$F_2: \; \bar{X} = \frac{\Sigma X_i}{n} = \frac{\quad}{\quad} =$$

2 Compare the mean of the F_1 with that of each parental strain. What does this tell you about the type of gene action involved?

ANSWER: The F_1 mean (12.11) is almost midway between the parental means of 6.63 and 16.80. This indicates that the genes in question are probably additive in effect.

3 If the F_1 mean was skewed toward one or the other parental mean value, what genetic factor or factors might be at work?

ANSWER: If the mean values are skewed (especially upwards), it may indicate that the F_1 mean approximates the geometric ($X = \sqrt{X_A \times X_B}$), rather than the arithmetic mean of the parental strains. In this case, it would suggest that there is a multiplying rather than an additive effect of some genes on the trait in question.

PROBLEMS AND DISCUSSION QUESTIONS

1 Distinguish between epistasis and polygenic inheritance, and between discontinuous and continuous variation.

2 List as many human traits as you can that are likely to be under the control of a polygenic mode of inheritance.

3 Describe the difference between penetrance and expressivity.

4 Define and discuss the significance of the following terms: (a) position effect, (b) suppressor genes, (c) monozygotic and dizygotic twins, (d) concordance and discordance, and (e) heritability.

5 In the following table, average differences of height and weight between monozygotic twins (reared together and apart), dizygotic twins, and siblings are compared. Draw as many conclusions as you can concerning the effects of genetics and the environment in influencing these human traits.

Trait	MZ Reared Together	MZ Reared Apart	DZ Reared Together	Sibs Reared Together
Height (cm)	1.7	1.8	4.4	4.5
Weight (kg)	1.9	4.5	4.5	4.7

SOURCE: Newman, Freeman, and Holzinger, 1937.

6 In *Drosophila,* the sex-linked recessive mutation *vermilion* (v) causes bright red eyes, which is in contrast to brick red eyes of wild type. A separate autosomal recessive mutation *suppressor of vermilion* (*su-v*) causes flies homozygous or hemizygous for v to have wild-type eyes. In the absence of vermilion alleles, *su-v* has no effect on eye color. Determine the F_1 and F_2 phenotypic ratios from a cross between a female with wild-type alleles at the *vermilion* locus, but who is homozygous for *su-v*, with a *vermilion* male who has wild-type alleles at the *su-v* locus.

7 Corn plants from a test plot are measured, and the distribution of heights at 10-cm intervals is recorded below:

Height (cm)	100	110	120	130	140	150	160	170	180
Plants (no.)	20	60	90	130	180	120	70	50	40

Calculate (a) the mean height, (b) the variance, (c) the standard deviation, and (d) the standard error of the mean.

8 A dark red strain and a white strain of wheat are crossed to produce an intermediate or medium red F_1. The F_1 plants are self-crossed to produce an F_2 in a ratio of 1 dark red:4 medium-dark red:6 medium red:4 light red:1 white. Further crosses reveal that the dark red and white F_2 plants are true breeding.

(a) Based on the ratio of offspring in the F_2, how many loci are involved in the production of color? How many alleles at each locus?

(b) Are the allele effects additive?

(c) How many units of color are produced by each allele?
(d) How many genotypes are represented by each phenotypic class? What are they?

9 Height in humans depends on the additive action of genes. Assume that this trait is controlled by the four loci *R*, *S*, *T*, *U* and that environmental effects are negligible. Dominant alleles contribute two units and recessive alleles contribute one unit to height.
(a) Can two individuals of moderate height produce offspring that are much taller or shorter than either parent? How?
(b) If an individual with the minimum height specified by these genes marries an individual of intermediate or moderate height, will any of their children be taller than the tall parent? Why?

10 The mean length and variance of corolla length in two true-breeding strains of *Nicotiana* and their progeny are as shown below.

Strain	Mean length (mm)	Variance
P_1 short	40.47	3.124
P_2 long	93.75	3.876
$F_1(P_1 \times P_2)$	63.90	4.743
$F_2(F_1 \times F_1)$	68.72	47.708

Calculate heritability of flower length in three strains.

SELECTED READINGS

BRINK, R., ed., 1967. *Heritage from Mendel.* Madison: University of Wisconsin Press.
FARBER, S. 1980. *Identical Twins Reared Apart.* New York: Basic Books.
FALCONER, D. 1989. *Introduction to Quantitative Genetics.* 3rd ed. Harlow, England: Longman Scientific and Technical Publications.
HALEY, C. 1991. Use of DNA fingerprints for the detection of major genes for quantitative traits in domestic species. *Anim. Genet.* 22:259–77.
LANDER, E., and BOTSTEIN, D. 1989. Mapping mendelian factors underlying quantitative traits using RFLP linkage maps. *Genetics* 121:185–99.
LEWONTIN, R. 1974. The analysis of variance and the analysis of causes. *Amer. J. Hum. Genet.* 26:400–11.
PATERSON, A., DEVERNA, J., LANINI, B., and TANKSLEY, S. 1990. Fine mapping of quantitative traits loci using selected overlapping recombinant chromosomes in an interspecific cross of tomato. *Genetics* 124:735–42.
PLOMIN, R., MCCLEARN, G., GORA-MASLAK, G., and NEIDERHISER, J. 1991. Use of recombinant inbred strains to detect quantitative trait loci associated with behavior. *Behav. Genet.* 21:99–116.
WILHAM, R., and WILSON, D. 1991. Genetic predictions of racing performance in quarter horses. *J. Anim. Sci.* 69:3891–94.

21 Population Genetics & Evolution

Chapter Concepts

A group of individuals can carry a larger number of different alleles than an individual, giving rise to a reservoir of genetic diversity that can be measured by the Hardy-Weinberg equation. Factors such as mutation, drift, migration, and selection can alter the amount of genetic variation in populations.

Evolution is concerned with the effects of gene frequency changes within the gene pool of a population. Speciation divides an originally homogeneous gene pool into two or more reproductively separate gene pools, and this division may be accompanied by changes in morphology, physiology, and adaptation to the environment.

When Charles Darwin published *The Origin of Species* in 1859 following the work of Alfred Russel Wallace on natural selection, the foundation for the modern interpretation of evolution was established. Although organisms are capable of reproducing in a geometric fashion, Wallace and Darwin observed that this growth potential of species is not realized. Instead, population numbers remain relatively constant in nature. They deduced that some form of competitive struggle for survival must therefore occur and that there is a natural variation between individuals within a species. To quote Darwin: ". . . any being, if it vary however slightly in any manner profitable to itself . . . will have a better chance of surviving." Darwin included in his concept of survival ". . . not only the life of the individual, but success in leaving progeny."

While Mendel was familiar with Darwin's work, Darwin was unaware of any underlying mechanism to account for the morphological variation he had observed. However, with the development of the concept of genes and alleles, the genetic basis of inherited variation was established.

As others pursued the study of evolution, it became apparent that the population rather than the individual is the functional unit in this process. In order to study the role of genetics in the process of evolution, therefore, it is necessary to consider gene frequencies in populations rather than offspring from individual matings. Thus arose the discipline of **population genetics.**

Early in the twentieth century various workers, including Gudny Yule, William Castle, Godfrey Hardy, and Wilhelm Weinberg, formulated ideas that became the basic principles of this field. Over the past forty years, experimentalists and field workers have studied gene frequencies and forces such as mutation, migration, selection, and random genetic drift, that alter such frequencies. In this chapter, we shall consider some general aspects of population genetics and focus on three topics central to evolution: (1) the extent of genetic diversity and its potential contribution to speciation; (2) evolution at the molecular level, and (3) lastly, the role of mutation in evolution.

POPULATIONS, GENE POOLS, AND GENE FREQUENCIES

Members of a species are often distributed over a wide geographic range. A **population,** however, is a local group belonging to a single species, within which mating is actually or potentially occurring. The set of genetic information carried by all interbreeding members of the population is called the **gene pool.** In population genetics, the focus is on the measurement of gene (allelic) and genotype frequencies from generation to generation rather than on the distribution of genotypes resulting from a single mating. The term **allelic frequency** will be used throughout the chapter and will represent the frequency of alleles in contrast to genotype frequencies.

One approach used to study a population's genetic structure is to measure the frequency of a given gene controlling a known trait. This is possible once the mode of inheritance and the number of different alleles present in the population have been established. Gene frequencies cannot always be determined directly because, in many cases, only phenotypes, and not genotypes, can be observed. However, if alleles expressed in a codominant fashion are considered, phenotypes are equivalent to genotypes. Such is the case with the autosomally inherited **MN blood group** in humans.

In this case, the gene *L* on chromosome 2 has two alleles, L^M and L^N (often referred to as *M* and *N*, respectively).* Each controls the production of a distinct antigen on the surface of red blood cells. Thus, an individual may be type M (L^ML^M), N (L^NL^N), or MN (L^ML^N).

Because they are codominant, the frequency of *M* and *N* in a population can be determined simply by counting the number of alleles for each phenotype. For example, in a population of 100 individuals, where 36 are *MM*, 48 are *MN*, and 16 are *NN*, the 36 *MM* individuals represent 72 *M* alleles, and the 48 *MN* heterozygotes represent 48 additional *M* alleles. Therefore, the frequency of the *M* allele in this population is 0.6 (120/200 = 60% = 0.6). Table 21.1 illustrates two methods for computing the frequency of *M* and *N* alleles in a hypothetical population of 100 individuals.

*L stands for Karl Landsteiner, the geneticist for whom the locus is named.

THE HARDY–WEINBERG LAW

In the example of the MN blood group, *M* and *N* are codominant. If, on the other hand, one allele had been recessive, the heterozygotes would have been phenotypically identical to the homozygous dominant individuals, and the frequency of the alleles could not have been directly determined. However, a mathematical model developed independently by the British mathematician Godfrey H. Hardy and a German physician, Wilhelm Weinberg, can be used to calculate allele frequencies in this case. The **Hardy–Weinberg law (HWL)** is one of the fundamental concepts in population genetics. In the Hardy–Weinberg law, the following conditions are presumed:

1. The population is infinitely large, or large enough that sampling error is negligible.
2. Mating within the population occurs at random.
3. There is no selective advantage for any genotype; that is, all genotypes are equally viable and fertile.
4. There is an absence of other factors such as mutation, migration, and random genetic drift.

In an ideal population, suppose that a locus has two alleles, *A* and *a*. The frequency of the dominant allele *A* in both eggs and sperm is represented by p, and the frequency of the recessive allele in gametes is represented by q. Because the sum of p and q represents 100 percent of the alleles for that gene in the population, then $p + q = 1$. A Punnett square can be used to represent the random combination of gametes containing these alleles and the resulting phenotypes:

Eggs \ Sperm	$A(p)$	$a(q)$
$A(p)$	*AA* (p^2)	*Aa* (pq)
$a(q)$	*Aa* (pq)	*aa* (q^2)

TABLE 21.1
Methods of determining allele frequencies for codominant alleles.
A. Counting Alleles

Genotype/Phenotype	*MM*	*MN*	*NN*	Total
No. of individuals	36	48	16	100
No. of *M* alleles	72	48	0	120
No. of *N* alleles	0	48	32	80
Total no. of alleles	72	96	32	200

Frequency of *M* in population: $\frac{120}{200} = 0.6 = 60\%$

Frequency of *N* in population: $\frac{80}{200} = 0.4 = 40\%$

B. From Genotypes

Genotype/Phenotype	*MM*	*MN*	*NN*	Total
No. of individuals	36	48	16	100
Genotype frequency	36/100 = 0.36	48/100 = 0.48	16/100 = 0.16	1.00

Frequency of *M* in population: 36 + (1/2)48 = 0.36 + 0.24 = 0.60 = 60%
Frequency of *N* in population: 16 + (1/2)48 = 0.16 + 0.24 = 0.40 = 40%

In the random combination of gametes in the population, the probability that sperm and egg both contain the *A* allele is $p \times p = p^2$. Similarly, the chance that gametes will carry unlike alleles is $(p \times q) + (p \times q) = 2pq$, and the chance that a homozygous recessive individual will result is $q \times q = q^2$. Note that these terms also describe an important aspect of the HWL: allelic frequencies determine genotype frequencies. In other words, while the value p^2 is the probability that both gametes in a fertilization event will carry the *A* allele, it is also a measure of the frequency of *AA* homozygotes in the ensuing generation. In a similar way, $2pq$ describes the frequency of *Aa* het erozygotes, and q^2 is a measure of the frequency of homozygous recessive (*aa*) zygotes. Thus, the distribution of genotypes in the next generation can be expressed as

$$p^2 + 2pq + q^2 = 1 \quad (21.1)$$

Let us consider a population in which 70 percent of the alleles for a given gene are *A*, and 30 percent are *a*. Thus, $p = 0.7$ and $q = 0.3$, and $p + q = 0.7 + 0.3 = 1$.

The distribution and frequency of genotypes produced by random mating are shown in the following Punnett square.

		Sperm *A*(p = 0.7)	Sperm *a*(q = 0.3)
Eggs	*A* (p = 0.7)	*AA* p^2 (0.49)	*Aa* pq (0.21)
Eggs	*a* (q = 0.3)	*Aa* pq (0.21)	*aa* q^2 (0.09)

In this new generation, 49 percent (p^2) of the individuals will be homozygous dominant, 42 percent ($2pq$) will be heterozygous, and 9 percent (q^2) will be homozygous recessive. By inspection of the Punnett square, the frequency of the *A* allele in the new generation can be calculated as

$$p^2 + \tfrac{1}{2}\,2pq \quad (21.2)$$
$$0.49 + \tfrac{1}{2}\,(0.42)$$
$$0.49 + 0.21 = 0.70$$

For a, the frequency is

$$q^2 + \frac{1}{2}\, 2pq \qquad (21.3)$$
$$0.09 + \frac{1}{2}\,(0.42)$$
$$0.09 + 0.21 = 0.30$$

Since $p + q = 1$, the value for a could have been calculated as

$$q = 1 - p \qquad (21.4)$$
$$q = 1 - 0.70 = 0.30$$

The values of A and a in the new generation are the same as in the previous generation.

A population in which the frequency of a given gene remains constant from generation to generation is said to be in state of **genetic equilibrium** for that gene. In this case, since the frequencies of A and a remain constant, we can conclude that the conditions presumed for the Hardy-Weinberg law hold true in this population.

Several important points are relevant to the preceding example. First, while the hypothetical genes we have considered were at equilibrium, not all genes in a population are. This is particularly true when the assumptions made in the Hardy-Weinberg law do not hold. Second, the examples illustrate why dominant traits do not tend to increase in frequency as new generations are produced. Finally, the examples demonstrate that genetic equilibrium maintains a state of **genetic variety** in a population. Once fixed in a population, allelic frequencies remain unchanged during equilibrium—a factor important to the evolutionary process.

Sex-Linked Genes

In considering genotype and gene frequencies for autosomal recessive traits, we assumed that the frequency of A is the same in both sperm and eggs. What about sex-linked genes? In species that have two X chromosomes, such as *Drosophila,* females carry two copies of all genes on the X chromosome, while males carry only one copy of all X-linked genes. Genes on the X chromosome are therefore distributed unequally in the population, with females carrying two-thirds and males one-third of the total number.

It is easy to determine frequencies for X-linked genes in males, because the phenotype of both dominant and recessive alleles is expressed. Therefore, the frequency of an X-linked allele of a gene is the same as the phenotypic frequency. As an example, in Western Europe, a form of sex-linked color blindness occurs with a frequency of 8 percent in males ($q = 0.08$). Since color blindness in males has a frequency of 0.08, the expected frequency in females is q^2, or 0.0064. This means that 800 out of 10,000 males surveyed would be expected to be color-blind, but only 64 of 10,000 females would show this trait. Expected values of the frequency of X-linked traits in males and females are compared in Table 21.2.

TABLE 21.2
Expected relative frequency of X-linked traits in males and females.

Frequency of Males with Trait	Expected Frequency in Females
90/100	81/100
50/100	25/100
10/100	1/100
1/100	1/10,000
1/1000	1/1,000,000
1/10,000	1/100,000,000

Multiple Alleles

In addition to autosomal recessive and sex-linked genes, it is common to find several alleles of a single locus in a population. The ABO blood group in humans is such an example. The single locus I (isoagglutinin) has three alleles (I^A, I^B, and I^O), yielding six possible genotypic combinations (I^AI^A, I^BI^B, I^OI^O, I^AI^B, I^AI^O, I^BI^O). Recall that in this case A and B are codominant alleles, and both of these are dominant to O. The result is that homozygous AA and heterozygous AO individuals are phenotypically identical, as are BB and BO individuals, so we can distinguish only four phenotypic combinations.

By adding another term to the Hardy–Weinberg equation, we can calculate both genotype and gene frequencies for the situation involving three alleles. In an equilibrium population, the frequency of the three genes can be described by

$$p(A) + q(B) + r(O) = 1 \qquad (21.5)$$

and the distribution of genotypes will be given by

$$(p + q + r)^2 \qquad (21.6)$$

In our hypothetical population, the genotypes AA, AB, AO, BB, BO, and OO will be found in the ratio

$$p^2(AA) + 2pq(AB) + 2pr(AO) + q^2(BB) + 2qr(BO) + r^2(OO) = 1 \qquad (21.7)$$

Having measured the frequencies of A, B, and O blood types in a population, we can then calculate both the genotypic and phenotypic frequencies for these three alleles. For example, in an Armenian population, the frequency of A (p) is 0.38; B (q) is 0.11; and O (r) is 0.51. By using

TABLE 21.3
Calculating genotypic and phenotypic frequencies for multiple alleles where the frequency of $A(p)$ is 0.38, $B(q)$ is 0.11, and $O(r)$ is 0.51.

Genotype	Genotypic Frequency	Phenotype	Phenotypic Frequency
AA	$p^2 = (0.38)^2 = 0.14$	*A*	0.53
AO	$2pr = 2(0.38 \times 0.51) = 0.39$		
BB	$q^2 = (0.11)^2 = 0.01$	*B*	0.12
BO	$2qr = 2(0.11 \times 0.51) = 0.11$		
AB	$2pq = 2(0.38 \times 0.11) = 0.082$	*AB*	0.08
OO	$r^2 = (0.51)^2 = 0.26$	*O*	0.26

the formula $(p + q + r)^2 = 1$, we can calculate the genotypic and phenotypic frequencies for this population as shown in Table 21.3, combining phenotypically identical genotypes to obtain the phenotypic frequencies.

Heterozygote Frequency

One of the practical applications of the Hardy-Weinberg law is the calculation of heterozygote frequency in a population. The frequency of a deleterious recessive phenotype usually can be determined by counting such individuals in a sample of the population. For example, albinism, an autosomal recessive trait, has an incidence of about 1/10,000 (0.0001) in some populations. Albinos are easily distinguished from the population at large by a lack of pigment in skin, hair, and iris. Since this is a recessive trait, albino individuals must be homozygous. Their frequency in a population is represented by q^2, provided that mating has been at random, and all Hardy-Weinberg conditions have been met in the previous generation.

The frequency of the recessive allele therefore is

$$\sqrt{q^2} = \sqrt{0.0001} \qquad (21.8)$$
$$q = 0.01, \text{ or } 1/100$$

Since $p + q = 1$, then the frequency of p is

$$p = 1 - q \qquad (21.9)$$
$$= 1 - 0.01$$
$$= 0.99, \text{ or } 99/100$$

In the Hardy-Weinberg equation, the frequency of heterozygotes is given as $2pq$. Therefore, we have

$$\text{Heterozygote frequency} = 2pq \qquad (21.10)$$
$$= 2[(0.99)\ (0.01)]$$
$$= 0.02, \text{ or } 2\%, \text{ or } 1/50$$

Thus, heterozygotes for albinism are rather common in the population (2%), even though the incidence of homozygous recessives is only 1/10,000.

In general, the frequencies of all three genotypes can be calculated once the frequency of either allele is known. The relationship between genotype frequency and gene frequency is shown in Figure 21.1. It is important to note how fast heterozygotes increase in a population as the values of p and q move away from zero. This observation confirms our conclusion that when a recessive trait like albinism is rare, the majority of those carrying the allele are heterozygotes. In some cases, where the frequencies of p and q are between 0.33 and 0.67, heterozygotes actually constitute the major class in the population.

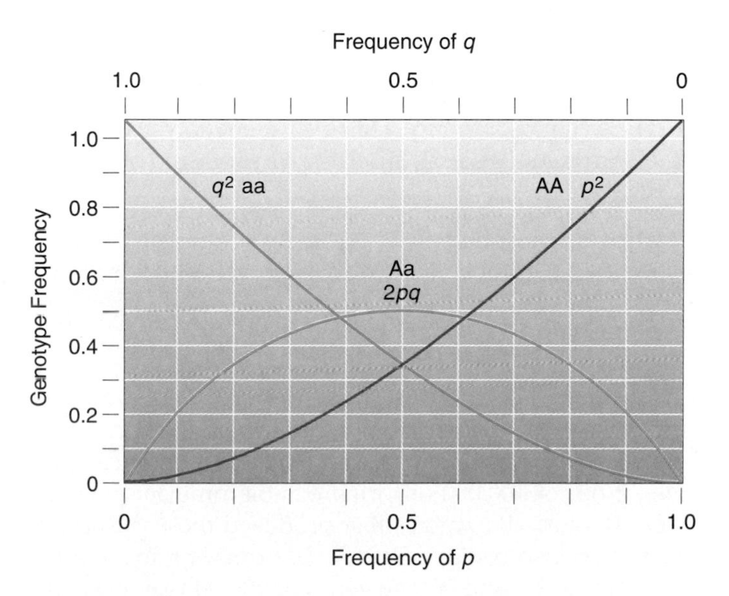

FIGURE 21.1
Relationship between genotype frequency and gene frequencies derived from the Hardy-Weinberg equation.

FACTORS THAT ALTER GENE FREQUENCIES

The Hardy-Weinberg law allows us to compare gene frequencies and genotype frequencies in populations in which our initial assumptions about random mating, absence of selection and mutation, and equal viability and fecundity hold true. It is difficult—if not impossible—to find natural populations in which all these conditions are met. Populations in nature are dynamic, and changes in size and structure are part of their life cycles. This does not mean that the Hardy-Weinberg relationship is invalid; rather, it provides an opportunity to study those forces that introduce and maintain genetic diversity in populations. In this section, we will discuss factors that prevent populations from reaching equilibrium and the relative contribution of these factors to evolutionary change.

Mutation

Within a population, the gene pool is reshuffled by assortment and recombination in each generation to produce new combinations in the genotypes of the offspring. Assortment and recombination do not produce any new alleles. **Mutation** alone acts to create new alleles and is therefore a force in increasing genetic variation. For our purposes, we need to understand that mutational events occur at random—that is, without regard for any possible benefit or disadvantage to the organism. To know whether mutation is a significant force in changing gene frequencies, we must measure the rate at which mutations are produced. Most mutations are recessive, so it is difficult to observe mutation rates directly in diploid organisms. But some mutations are dominant, and for certain dominant mutations, a direct method of measurement can be used if several conditions are met:

1 The trait must produce a distinctive phenotype that can be distinguished from similar traits produced by recessive alleles.

2 The trait must be fully expressed or completely penetrant so that mutant individuals can be identified.

3 An identical phenotype must never be produced by nongenetic agents such as drugs or chemicals.

Mutation rates can be expressed as the number of new mutant genes per given number of gametes. Suppose for a given gene that undergoes mutation to a dominant allele, 2 out of 100,000 offspring exhibit a mutant phenotype. Because the zygotes that produced these offspring each carry two copies of the gene, we have actually surveyed 200,000 copies of the gene (or 200,000 gametes). If we assume that the affected individuals are each heterozygous, we have uncovered 2 mutant alleles out of 200,000. Thus the mutation rate is 2/200,000 or 1/100,000,

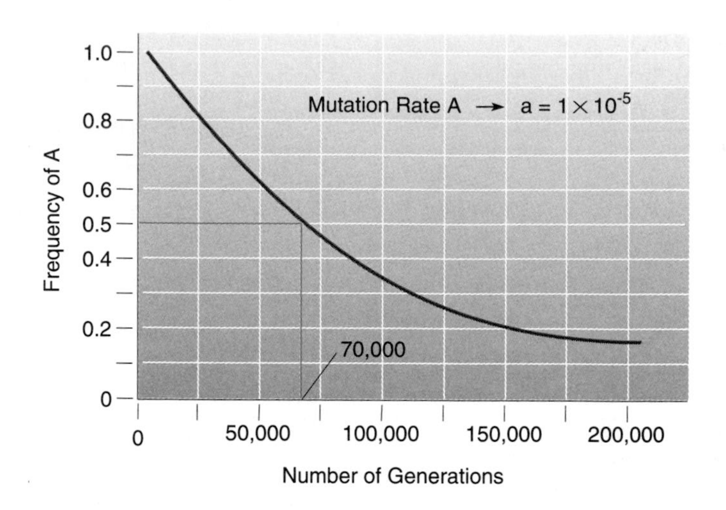

FIGURE 21.2
Rate of replacement of an allele by mutation alone, assuming an average mutation rate of 1.0×10^{-5}.

which in scientific notation would be written as 1×10^{-5}.

As an example, let us assume that the mutation rate for genes in the human genome is 1.0×10^{-5}. With such a low mutation rate, changes in gene frequency brought about by mutation alone are very small. Figure 21.2 shows that if *A* is the only allele of the *a* locus in the population ($p = 1$), with a mutation rate of 1.0×10^{-5} from *A* to *a*, it would require about 70,000 generations to reduce the frequency of *A* to 0.5. This example emphasizes that mutation is a major force in generating genetic variability, but by itself has an insignificant role in changing gene frequencies.

Migration

Frequently, a species of plant or animal becomes divided into subpopulations that to some extent are geographically isolated. Differences in mutation rate and selective pressures can establish different allele frequencies in the subpopulations. **Migration** occurs when individuals move between these populations. Consider an individual carrying a single pair of alleles, *A* and *a*. The change in the frequency of *A* in one generation can be expressed as

$$\Delta p = m(p_m - p) \quad (21.11)$$

where

p = frequency of *A* in existing population
p_m = frequency of *A* in immigrants
Δp = change in one generation
m = coefficient of migration (proportion of migrant genes entering the population per generation)

For example, assume that $p = 0.4$ and $p_m = 0.6$, and that ten percent of the parents giving rise to the next generation are immigrants ($m = 0.1$). Then, the change in the frequency of A in one generation is

$$\begin{aligned}\Delta p &= m(p_m - p)\\ &= 0.1\ (0.6 - 0.4)\\ &= 0.1\ (0.2)\\ &= 0.02\end{aligned}$$

In the next generation, the frequency of A (p_1) will increase as follows:

$$\begin{aligned}p_1 &= p + \Delta p \qquad (21.12)\\ &= 0.40 + 0.02\\ &= 0.42\end{aligned}$$

If either m is large and/or if p is much larger or smaller than p_m, then a rather large change in the frequency of A will occur in a single generation. All other factors being equal (including migration), an equilibrium will be attained when $p = p_m$. These calculations reveal that the change in gene frequency attributable to migration is proportional to the differences in frequency between the donor and recipient populations. Since the coefficient of migration can have a wide range of values, the effect of migration can substantially alter gene frequencies in populations. Although migration can be somewhat difficult to quantify, it can often be estimated.

Migration can also be regarded as the flow of genes between two populations that were once but are no longer geographically isolated. For example, West African populations were the source of most of the blacks brought to the United States as slaves. In Africa, the frequency of Duffy blood group allele Fy^0 is almost 100 percent. In Europe, the source of most U.S. whites, the frequency is close to zero percent, and almost all individuals are Fy^a or Fy^b. By measuring the frequency of Fy^a or Fy^b among U.S. blacks, we can estimate the amount of gene flow into the black population. Figure 21.3 shows the frequency of the Fy^a allele among blacks in three African nations and in several regions of the United States. Using the average frequency and assuming an equal rate of gene flow in each generation, we calculate the migration of the Fy^a allele at 5 percent per generation.

Selection

Mutations and migration introduce new alleles into populations. **Natural selection,** on the other hand, is the principal force that shifts gene frequencies within large populations and is one of the most important factors in evolutionary change.

Darwin's main contribution to the study of evolution was his recognition that natural selection is the mechanism that leads to the divergence of populations into distinct species. In any population at a given time, there are individuals with different genotypes. Because of these inherent genetic differences, some of the individuals in this population will be better adapted to the environment than others, leading to the differential survival and reproduction of some genotypes over others. Natural selection

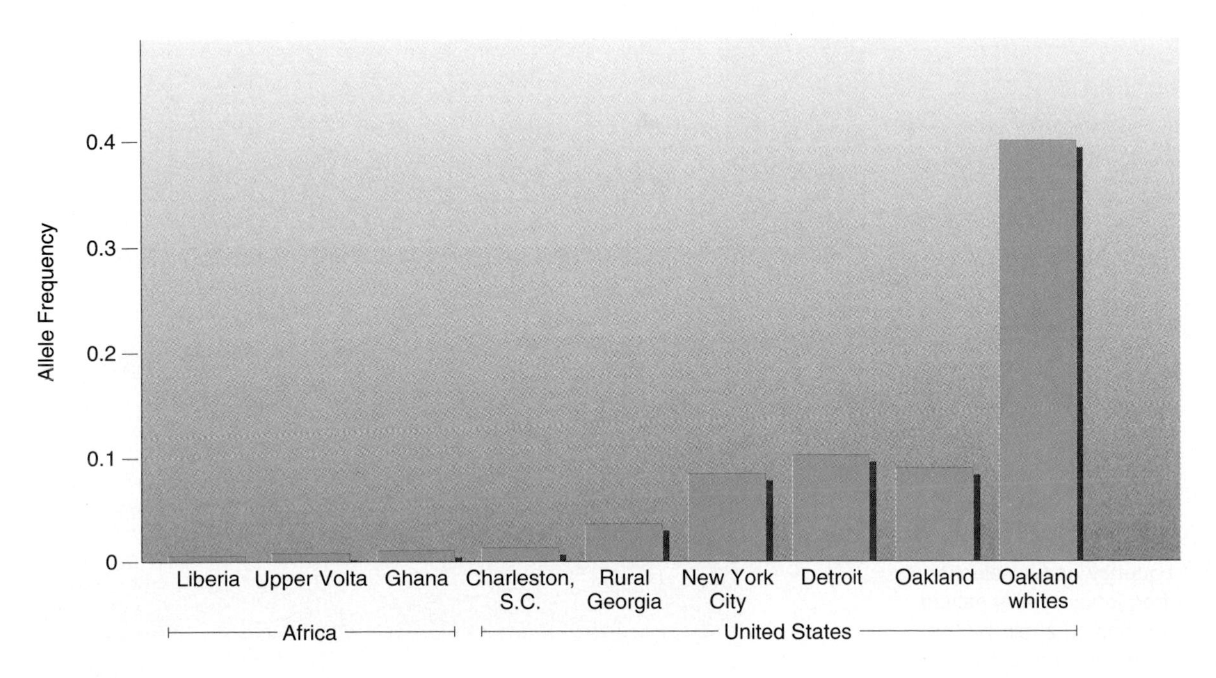

FIGURE 21.3
Frequencies of the Fy^a allele in African and U.S. black populations.

is the differential reproduction of genotypes. Therefore, allelic frequencies will change over time. Natural selection, then, is a major force in changing gene frequencies and represents a departure from the Hardy-Weinberg assumption that all genotypes have equal viability and fertility.

When a particular genotype/phenotype combination confers an advantage to organisms in competition with others harboring an alternate combination, selection occurs. The relative strength of selection varies with the amount of advantage provided. The probability that a particular phenotype will survive and leave offspring is a measure of its **fitness.** Fitness refers to a total reproductive potential or efficiency. The concept of fitness is usually expressed in relative terms by comparing a particular genotypic/phenotypic combination with one regarded as optimal. Fitness is a relative concept because as environmental conditions change, so does the advantage conferred by a particular genotype.

Mathematically, the difference between the fitness of a given genotype and another genotype regarded as optimal is called the **selection coefficient (s).** For a phenotype conferred by the genotype *aa*, when only 99 of every 100 organisms successfully reproduce, $s = 0.01$. If the *aa* genotype is a homozygous lethal, then $s = 1.0$, and the *a* allele is propagated only in the heterozygous carrier state. Starting with any original frequencies of A (p_0) and a (q_0) in a population, when $s = 1.0$, the effect of selection on any successive generation can be calculated using the formula

$$q_n = \frac{q_0}{1 + nq_0} \tag{21.13}$$

Generation	p	q	p^2	$2pq$	q^2
0	0.50	0.50	0.25	0.50	0.25
1	0.67	0.33	0.44	0.44	0.12
2	0.75	0.25	0.56	0.38	0.06
3	0.80	0.20	0.64	0.32	0.04
4	0.83	0.17	0.69	0.28	0.03
5	0.86	0.14	0.73	0.25	0.02
6	0.88	0.12	0.77	0.21	0.01
10	0.91	0.09	0.84	0.15	0.01
20	0.95	0.05	0.91	0.09	<0.01
40	0.98	0.02	0.95	0.05	<0.01
70	0.99	0.01	0.98	0.02	<0.01
100	0.99	0.01	0.98	0.02	<0.01

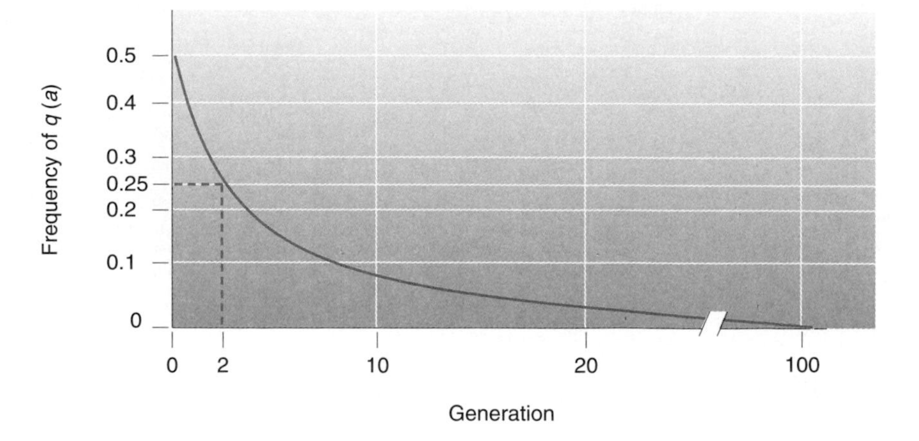

FIGURE 21.4
Changes in allele frequency under selection where $s = 1.0$. The frequency of *a* is halved in two generations and halved again by the sixth generation. Subsequent reductions occur slowly because the majority of *a* alleles are carried by heterozygotes.

where n is the number of generations elapsing since p_0 and q_0.

Figure 21.4 demonstrates the change in allele frequencies when A (p_0) = a (q_0) = 0.5. Initially, because of the high percentage of aa genotypes, the frequency of the a allele is reduced rapidly. The frequency of a is halved (0.25) in only two generations. By the sixth generation, the frequency of a is again reduced twofold (0.12). By now, however, the majority of a alleles are carried by heterozygotes. Since selection operates on phenotypes, this means that heterozygotes are not selected against. Subsequent reductions in frequency occur very slowly in successive generations and depend on the rate at which heterozygotes are removed from the population. For this reason, it is difficult to eliminate a recessive gene from the population by selection as long as heterozygotes continue to mate.

There has been much study of the action of selection on both laboratory and natural populations. A classic example of selection in natural populations is that of the peppered moth *Biston betularia* in England. Before 1850, 99 percent of the moth population was light colored, allowing the moths to fit well into their surroundings. As toxic gases produced by industry killed the lichens and mosses growing on trees and buildings, and soot deposits darkened the landscape, the light-colored moths became easy targets for their predators. The rare dark-colored moths suddenly gained a great advantage because of their natural camouflage. A rapid shift in frequency of this phenotype occurred, probably in 50 generations or less.

Laboratory experiments have demonstrated that the dark form is due to a single dominant allele, *C*. Figure 21.5 demonstrates the protective advantage from predators provided by the dark phenotype. Following the adoption of laws to restrict environmental pollution in 1964, the frequency of nonmelanic forms of the moth has started to increase around the Manchester area, showing the close relationship between the environment and selection.

Because selection acts on the organism's genotypic/phenotypic combination, it also acts on the phenotypes of polygenic or quantitative traits—those that are controlled by a number of genes and may be susceptible to environmental influences. Such quantitative traits, including adult body height and weight in humans, often demonstrate a continuously varying distribution resembling a bell-shaped curve. Selection for such traits can be classified as (1) directional, (2) stabilizing, or (3) disruptive.

Directional selection, used by plant and animal breeders, often shifts the population mean towards an extreme phenotype. If the trait is polygenic, the most extreme phenotypes will appear in the population only after prolonged selection. An example is the selection for oil content in domestic corn kernels. C. M. Woodworth and others were able to raise the oil content of corn more than threefold, from about 4 percent to just over 15 percent in 50 generations, with no sign of a plateau being reached (Figure 21.6). Similarly, they used selection to reduce oil content to about 1 percent. Directional selection favors an extreme phenotype and tends to produce genetic uniformity in the population. In nature, directional selection can occur when one of the phenotypic extremes becomes selected for or against, usually as a result of changes in the environment.

Stabilizing selection, on the other hand, tends to favor intermediate types, with both extreme phenotypes being selected against. One of the clearest demonstrations of stabilizing selection is the data of Mary Karn and Sheldon Penrose on human birth weight and survival.

FIGURE 21.5
The speckled and melanic forms of the peppered moth, *Biston,* seen on a normal, lichen-covered tree trunk (left) and on a darkened, soot-covered trunk (right).

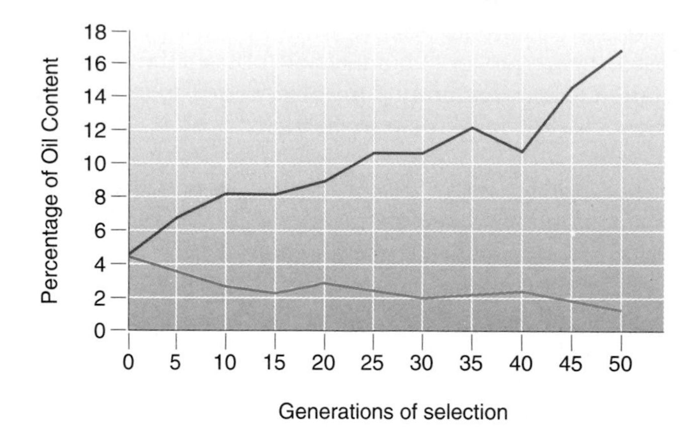

FIGURE 21.6
Long-term selection to alter the oil content of corn kernels.

Figure 21.7 shows the distribution of birth weight for 13,730 children born over an 11-year period and the percentage of mortality at four weeks of age. Infant mortality increases on either side of the optimal birth weight of 7.5 pounds. At the genetic level, stabilizing selection represents a situation where a population is adapted to its environment.

Disruptive selection is selection against intermediates and for both phenotypic extremes. It may be viewed as the opposite of stabilizing selection. John Thoday applied disruptive selection to progeny of *Drosophila* strains with high and low bristle number. Matings were carried out using females from one strain and males from the other. Progeny were selected for high and low bristle number, and matings were carried out for a number of generations. Despite the fact that opportunities for gene flow between the two lines through random mating were experimentally provided at each generation, the two lines diverged rapidly. In natural populations, such a situation may exist for a population in a heterogeneous environment. Gene flow may occur between organisms in two niches in which selection is favoring opposite extremes.

The types and effects of selection are summarized in Figure 21.8.

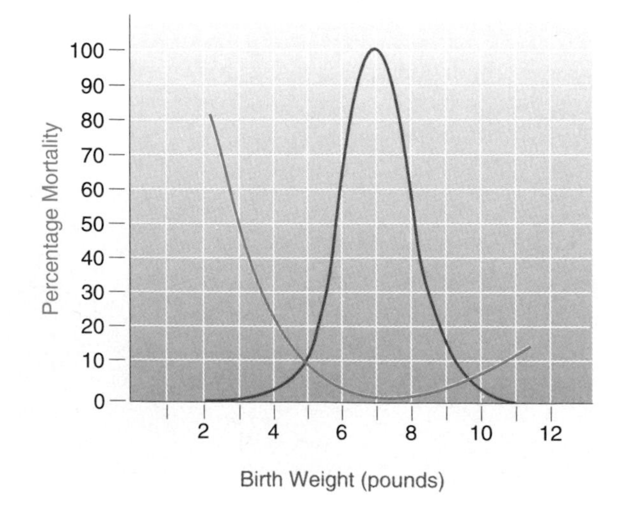

FIGURE 21.7
Relationship between birth weight and mortality in humans.

Genetic Drift

In laboratory crosses, one condition essential to realizing theoretical genetic ratios (i.e., 3:1, 1:2:1, 9:3:3:1) is a

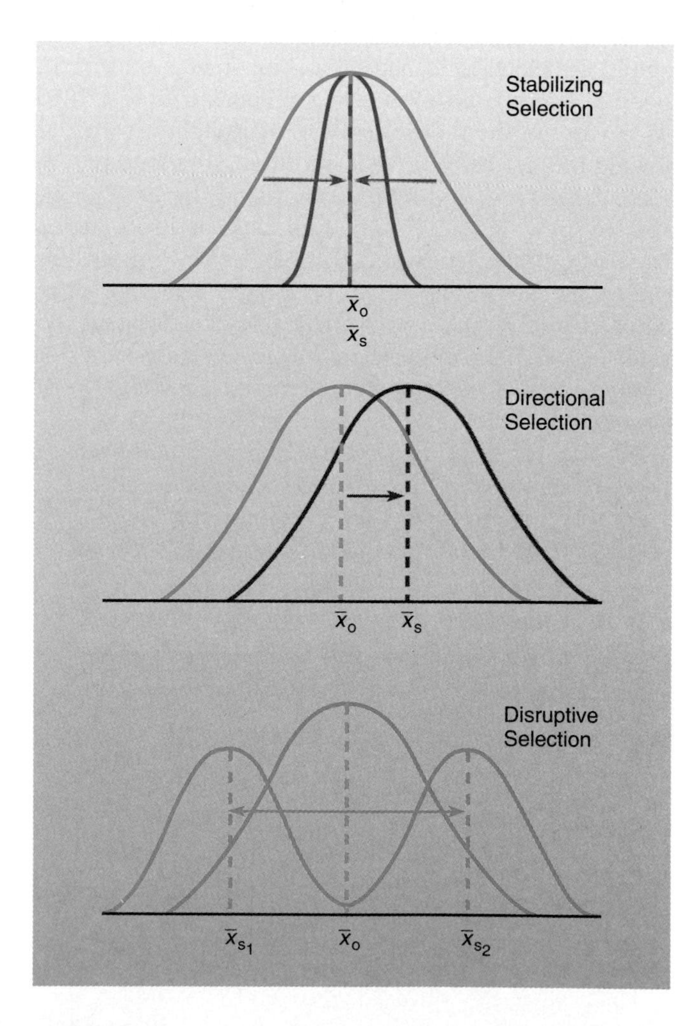

FIGURE 21.8
Comparison of the impact of stabilizing, directional, and disruptive selection. In each case, $\overline{x}_o$ is the mean of an original population, and $\overline{x}_s$ is the mean of the population following selection.

fairly large sample size. This condition is equally important to the study of population genetics as gene and genotype frequencies are examined or predicted.

However, if a population is formed from only one set of heterozygous parents and they produce only two offspring, gene frequency can change drastically. In this case, the probability of the occurrence of genotypes and gene frequencies in the subsequent generation can be predicted as summarized in Table 21.4. As shown, in 10 of 16 times when such a cross is made, the new allele frequencies would be altered dramatically. In 2 of 16 times, either the *A* or *a* allele would be eliminated in a single generation.

The conditions used to generate these calculations are extreme, but they illustrate why large interbreeding populations are essential to the Hardy-Weinberg equilibrium. In small populations, significant random fluctuations in allelic frequencies are possible strictly on the basis of chance deviation. The degree of fluctuation will increase as the population size decreases. Such changes illustrate the concept of **genetic drift.** In the extreme case, genetic drift may lead to the chance fixation of one allele to the exclusion of another allele.

How are small populations created in nature? In one instance, a large group of organisms may be split by some natural event, creating a small, isolated subpopulation. A natural disaster such as an epidemic might also occur, leaving a small number of survivors to constitute the breeding population. Third, a small group might emigrate from the larger population, as founders, to a new environment, such as a volcanically created island.

Allelic frequencies in certain human populations on isolated islands best support the role of drift as an evolutionary force in natural populations. The Pingelap atoll in the western Pacific Ocean has in the past been devastated by typhoons and famine, and in 1775, there were only 30 survivors. Today there are fewer than 2000 inhabitants, and their ancestry can be traced to the typhoon survivors.

TABLE 21.4
All possible pairs of offspring produced by two heterozygous parents (*Aa* × *Aa*).

Possible Genotypes of Two Offspring	Probability	Allele Frequency	
		A	*a*
AA and *AA*	(1/4)(1/4) = 1/16	1.00	0.00
aa and *aa*	(1/4)(1/4) = 1/16	0.00	1.00
AA and *aa*	2(1/4)(1/4) = 2/16	0.50	0.50
Aa and *Aa*	(2/4)(2/4) = 4/16	0.50	0.50
AA and *Aa*	2(2/4)(1/4) = 4/16	0.75	0.25
aa and *Aa*	2(1/4)(2/4) = 4/16	0.25	0.75

Approximately 5 percent of the current population is affected by the recessively determined disorder **achromatopsia,** which causes ocular disturbances and a form of color blindness. This disorder is extremely rare in the human population as a whole. However, the responsible allele is present at a relatively high frequency in the Pingelap population. From genealogical reconstruction, it was found that one of the original survivors was a chief who was heterozygous for the condition. If we assume that he was the only carrier in the founding population, the initial gene frequency was 1/60, or 0.016. Since some 5 percent of the current population is affected, they represent homozygous recessives, and the frequency of the gene in the population is calculated as 0.23.

Inbreeding and Heterosis

One of the assumptions of the Hardy-Weinberg equilibrium is random mating within the population. This is an ideal condition and as such does not often occur in nature. Within a widespread population, individuals tend to mate with those in physical proximity rather than with those separated by some distance. The result is the restriction of gene flow within that population.

Nonrandom mating occurs in many species, including humans. One form of nonrandom mating is called **assortative.** In humans, pair bonds may be established by religious practices, physical characteristics, professional interests, and so forth on a nonrandom basis. In nature, phenotypic similarity plays the same role.

Another form of nonrandom mating is **inbreeding,** where mating occurs between relatives. Inbreeding results in increased homozygosity. To illustrate this concept, we shall consider the most extreme form of inbreeding, **self-fertilization.**

Figure 21.9 shows the consequences of four generations of self-fertilization starting with a single individual heterozygous for one pair of alleles. By the fourth generation, only about 6 percent of the individuals are still heterozygous. Note that the frequencies of *A* and *a* still remain at 50 percent.

In humans, inbreeding (called **consanguineous marriages**) is related to population size, mobility, and social customs governing marriages among relatives. To determine the probability that two alleles at the same locus in an individual are derived from a common ancestor, Sewall Wright devised the **coefficient of inbreeding.** Expressed as F, the coefficient can be defined as the probability that two alleles of a given gene in an individual are derived from a common allele in an ancestor. If $F = 1$, all genotypes are homozygous and derived from a common ancestor. If $F = 0$, no alleles present are derived from a common ancestor. Shown in Figure 21.10 are pedigrees of first- and second-cousin marriages. A

P_1:	Self-Fertilization *Aa*		
	AA	*Aa*	*aa*
F_1:	0.250	0.500	0.250
F_2:	0.375	0.250	0.375
F_3:	0.437	0.125	0.437
F_4:	0.468	0.063	0.468
F_n:	$\frac{1-\frac{1}{2}n}{2}$	$\frac{1}{2^n}$	$\frac{1-\frac{1}{2}n}{2}$

FIGURE 21.9
Reduction in heterozygote frequency brought about by self-fertilization. After *n* generations, the frequencies of the genotypes can be calculated according to the formulas in the bottom row.

lethal allele *a* is assumed to be present in one heterozygous parent. For each relative, the probability of the presence of this lethal allele is shown. The pedigrees then culminate in the final probability *F* that the lethal allele will become homozygous in the offspring of first or second cousins. Since every population carries as its genetic burden a number of heterozygous recessive alleles that are lethal or deleterious in homozygotes, mortality will rise in matings between relatives.

Thus, from an evolutionary standpoint, if a population is split into smaller subpopulations, the effects of inbreeding will become evident. **Inbreeding depression,** an expression of reduced fitness in a population caused by inbreeding, may occur; and because an increase in homozygous individuals results, genetic variability will decrease. Both favorable and unfavorable alleles will become more frequently homozygous. From the standpoint of Darwinian fitness, the adaptive nature of the population will decrease as well.

This knowledge has been applied in the breeding of domesticated plant and animal species. First, an inbreeding program is initiated. As homozygosity increases, some populations will become fixed for favorable alleles and others for unfavorable alleles. By selecting the more viable and vigorous plants or animals, the proportion of individuals carrying desirable traits can be increased.

If members of two favorable inbred lines are mated, hybrid offspring are often more vigorous in desirable traits than is either of the parental lines. This phenomenon has been called **hybrid vigor** or **heterosis.** Such an approach was used in the breeding programs established for corn, and crop yields increased tremendously. Unfortunately, the hybrid vigor extends only through the first generation. Many hybrid lines are sterile, and those that are fertile show subsequent declines in yield. Conse-

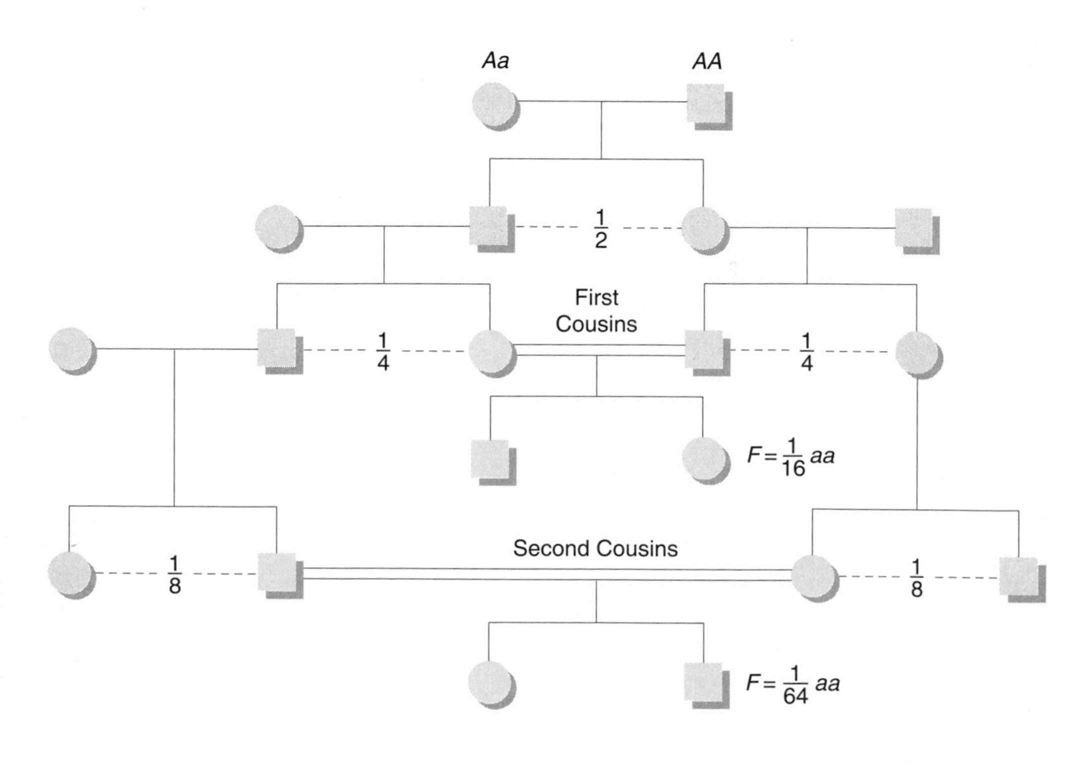

FIGURE 21.10
Changes in the coefficient of inbreeding (*F*) brought about by consanguineous matings. Numbers indicate the probability of carrying the recessive allele *a*.

quently, the hybrids must be regenerated each time by crossing the inbred parental lines.

Heterosis has been explained in two ways. The first hypothesis, the **dominance hypothesis,** incorporates the reversal of inbreeding depression, which inevitably must occur in out-crossing. Consider a cross between two strains of corn with the following genotypes:

Strain A Strain B
aaBBCCddee × *AAbbccDDEE*
↓
AaBbCcDdEe

The F_1 hybrids are heterozygotes at all loci shown. Deleterious recessive alleles present in the homozygous form in the parents will be masked by the more favorable dominant alleles in the hybrids. Such masking is thought to cause hybrid vigor or heterosis.

The second theory **overdominance** holds that in many cases the heterozygote is superior to either homozygote. This may be related to the fact that in the heterozygote, there may be two forms of a protein present, providing a form of biochemical diversity. Thus, the cumulative effect of heterozygosity at many loci accounts for the hybrid vigor. Most likely, such vigor results from a combination of phenomena explained by both hypotheses.

GENETIC DIVERSITY

We might assume that members of a population are fit because the most favorable allele at each locus has become fixed homozygously. Certainly, an examination of most populations of plants and animals reveals them to be phenotypically similar, if not identical. However, current evidence supports the opinion that a high percentage of heterozygosity is maintained within the genome of diploid individuals in a population. This built-in diversity is concealed, so to speak, because it is not necessarily apparent in the phenotype. Such diversity may better adapt a population to the inevitable changes in the environment and may promote the survival of the species.

The detection of this concealed genetic variation is not a simple task. Nevertheless, such investigation has been successful using several techniques.

Protein Polymorphisms

The use of Mendelian genetics to estimate genetic diversity is both laborious and slow. Gel electrophoresis on the other hand, is a simpler technique that can be used to separate protein molecules on the basis of differences in electrical charge (Figure 21.11). If variation in a structural gene results in the substitution of a charged amino acid such as glutamic acid for an uncharged amino acid such as glycine, the net charge on the protein will be altered. This change in charge can be detected by electrophoresis.

When two alleles from one locus produce slightly different electrophoretic forms, yet perform identical functions, the resultant proteins are designated as **allozymes.** Electrophoretic studies have shown that a surprisingly large percentage of loci from diverse species exhibit allozymes. The presence of allozymes can be used to estimate the degree of heterozygosity in a population. Table 21.5 lists the heterozygosity detected as allozymes in populations of four species. An approximate average of 10 percent heterozygosity per diploid genome was revealed.

These values apply only to genetic variation detectable by altered protein migration in an electric field. Electrophorsis probably detects only about 30 percent of the actual variation that is due to amino acid substitutions because many substitutions do not change the net electric charge on the molecule. Richard Lewontin has thus estimated that about two-thirds of all loci are polymorphic in a population, and that in any individual within that group about one-third of the loci exhibit genetic variation in the form of heterozygosity.

There is some controversy over the significance of genetic variation as detected by electrophoresis. Some argue that alleles producing allozymes are effectively neutral and therefore do not play any role in evolution. We shall address this argument later in this chapter.

Chromosomal Polymorphisms

While the chromosome number is usually invariant for a given species, the arrangement of chromosomal material is often polymorphic as a result of chromosomal inversions and translocations. These chromosomal aberrations usually have little direct effect on the phenotype because gene content is rearranged but not altered.

Inversions are produced by two breaks in a chromosome followed by rejoining of the broken fragment rotated 180° (see Chapter 6). Because of pairing difficulties in meiosis, inversion heterozygotes rarely produce viable gametes that separate the genes contained within an inversion. Therefore, inversions tend to preserve specific allele arrangements along the inverted stretch of the chromosome involved. The net effect of inversions in populations, therefore, is the prevention of an increase in genetic variability. In addition, inversions reduce heterozygote fitness because of the production of inviable crossover gametes. Therefore, if an inversion is present in a species, it implies the presence of selective factors that favor inversion heterozygotes. In other words, per-

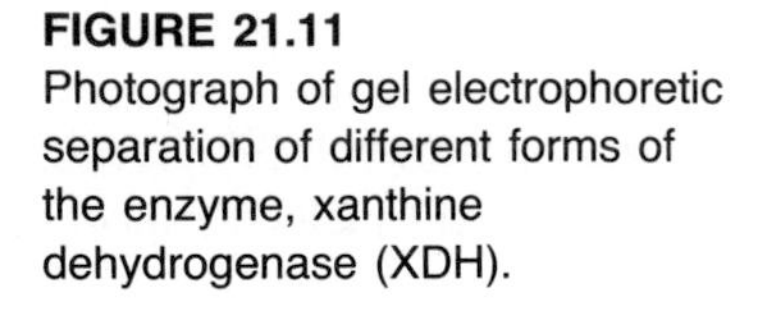

FIGURE 21.11
Photograph of gel electrophoretic separation of different forms of the enzyme, xanthine dehydrogenase (XDH).

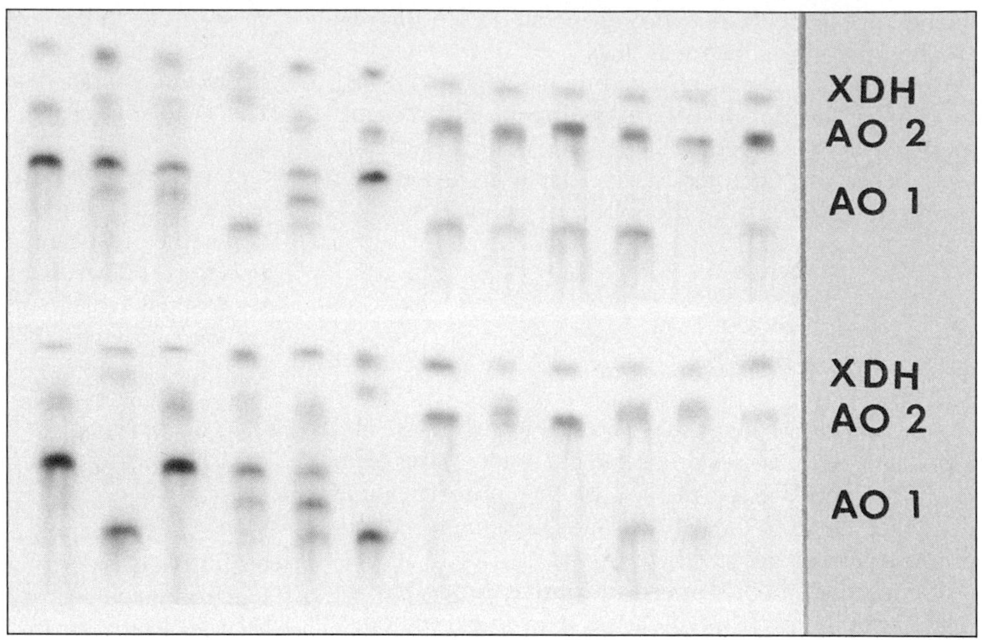

haps the inversion has preserved a favorable gene arrangement, which at least partially accounts for the fitness of the population in a specific habitat. The role of inversions in natural populations has been intensively studied in *Drosophila* and will be discussed later in this section.

Translocations, in which chromosome segments move to other nonhomologous chromosomes are also found in populations. Translocations also reduce genetic variability in populations because translocation heterozygotes are usually subfertile, producing 50 percent chromosomally abnormal gametes. Although somewhat rare in animal populations, translocation heterozygotes are found in many plants.

DNA Sequence Polymorphisms

The most direct way to estimate genetic variation is by examination of the actual nucleotide sequence diversity between individuals of a population. With the development of techniques for cloning and sequencing DNA, nucleotide sequence variations have been cataloged for an increasing number of gene systems. Using restriction endonucleases to detect polymorphisms (see Chapter 15), Alec Jeffrey surveyed 60 unrelated individuals to estimate the total number of DNA sequence variants in humans. His results show that within the genes of the beta-globin cluster, 1 in 100 base pairs shows polymorphic

TABLE 21.5
Heterozygosity at the molecular level.

Species	Number of Populations Studied	Number of Loci Examined	Loci per Population	Heterozygosity per Locus
Homo sapiens (humans)	1	71	28	6.7
Mus musculus (house mouse)	4	41	29	9.1
Drosophila pseudoobscura (fruit fly)	10	24	43	12.8
Limulus polyphemus (horseshoe crab)	4	25	25	6.1

SOURCE. From Lewontin, 1974 p. 117.

THE CHEETAH
A GENETICALLY ENDANGERED SPECIES?

Beginning in the mid 1960's scientists began investigating genetic variation in populations using gel electrophoresis. This technique identifies protein variants based on their differential mobility in an electric field. (Proteins are, of course, gene products and some alleles generate proteins with altered electrical charge and different mobility.) When a given population has more than one common allele of a given gene (i.e. contains individuals with different forms of the corresponding protein), the gene is said to be "polymorphic".

Polymorphism is quite common. Protein electrophoresis of many species has demonstrated that most populations contain a large amount of genetic variation. Between 10% to 60% of the genes within most populations are polymorphic and between 1% to 36% of the genetic loci in any one individual are heterozygous. Genetic variation is essential if populations are to evolve in response to changing environmental conditions. However, some populations appear to lack genetic variability. This is commonly observed in populations with a history of inbreeding. Inbreeding is the result of matings between close relatives and occurs by necessity in small populations. Inbreeding generates genetic uniformity by chance, much as the chance of spinning five "heads" in a row is much greater than spinning five hundred. A related phenomenon called "inbreeding depression" is often found in inbred populations, which may show decreased fertility, increased juvenile mortality and increased susceptibility to diseases.

A notable example of genetic uniformity is the African cheetah. This species numbers about 20,000 and is divided into two subspecies; the subspecies, *Acinonyx jubatus jubatus,* from southern Africa and *Acinonyx jubatus raineyi* from east-central Africa. In an initial study carried out by Stephen O'Brien and his associates on the southern African cheetah, analysis of 52 proteins (the products of 52 genes) showed no variation among the 55 animals tested; thus all animals were genetically identical and were homozygous for each of the 52 genes.

Further tests were done to see if any variation existed at the major histocompatibility complex which is the most extensively polymorphic locus known in mammals. In humans, the chance that two individuals are identical at this locus is less than 1 in 10,000. This complex is primarily responsible for acceptance or rejection of organ transplants between non-matching (non-identical) humans and provides an extremely sensitive test of genetic similarity. To analyze genetic variation in the cheetah at this locus, skin grafts were exchanged between unrelated animals. All grafts were accepted, indicating an unprecedented level of genetic uniformity in an out-bred mammalian population. As a control, it was determined that the cheetahs are competent to reject grafts, because skin grafts from domestic cats do not take.

Studies of the eastern African cheetah gave very similar results. The eastern cheetah shows diminished genetic variation, although two polymorphic loci were detected. To account for the apparent lack of genetic variability in the cheetah, O'Brien suggests that, perhaps as long as 10,000 years ago, the cheetah went through one or more population "bottlenecks" in which the population numbers were severely reduced. The consequent inbreeding led to the nearly complete loss of genetic variation.

Does this homozygosity have any consequences for the cheetah populations? It must be assumed that the most deleterious mutant alleles in the founder populations were removed by selection early in the process leading to homozygosity. But both populations do show evidence of the reproductive impairment characteristic of inbred animals: cheetah semen contains relatively low concentrations of sperm, of which more than 70% are structurally abnormal. In domestic cattle, a semen sample with 10–20% abnormal sperm indicates a bull that is subfertile or sterile.

Whether the lack of genetic diversity will affect the long-term evolutionary fate of the cheetah is not yet known. While some populations are stable, we have to wonder whether the cheetah will lack the genetic ability to cope with the current human disruption of its environment. Since many polymorphic species have failed to do so, should the cheetah, which lacks such variability, be classified as a "genetically endangered" species?

variation. If this region is representative of the genome, this indicates that at least 3×10^7 nucleotide variants per genome are possible. Data from other organisms such as *Drosophila*, the rat, and the mouse have produced similar estimates of nucleotide diversity, indicating that there is an enormous reservoir of genetic variability within a population, and that at the level of DNA, most and perhaps all genes exhibit diversity from individual to individual.

SPECIATION

In the classical sense, the process of splitting a genetically homogeneous population into two or more populations that undergo genetic differentiation and eventual reproductive isolation is called **speciation.** According to Ernst Mayr, species originate in two predominant ways. In the first mode, often called **phyletic evolution,** species A over a long period of time becomes, through genetic change, transformed into species B. In the second method, one species gives rise to one or more derived species, bringing about multiplication of species. This second process can occur over a long period of time or abruptly in a generation or two. Table 21.6 summarizes the principal methods of speciation.

The most developed classical model of speciation is **geographic** or **allopatric speciation.** Physical isolation of populations by geographic features such as lakes, rivers, or mountains that act as barriers to gene flow is the first step toward species formation. In a second step, these isolated populations undergo independent evolution and may continue to diverge to produce two distinct species.

TABLE 21.6
Modes of speciation.

I. Transformation of Species (phyletic evolution)
  1. Autogenous speciation
II. Reduction in Number of Species (fusion of two species)
III. Multiplication of Species (true species)
  A. Instant speciation (through individuals)
    1. Genetic
      (a) Single mutation in asexual species
      (b) Macrogenesis
    2. Cytological
      (a) Chromosomal mutation (translocations, etc.)
      (b) Autopolyploidy
      (c) Amphidiploidy
  B. Gradual speciation (through populations)
    1. Sympatric speciation
    2. Semigeographic speciation
    3. Geographic speciation (allopatric)

SOURCE: From Mayr, 1963, Table 15.1.

In this model, the first step requires that populations become separated; that is, gene flow must be interrupted. The absence or interruption of gene flow is a prerequisite for the development of genetic differences by adaptation to local conditions. Genetic diversity brought about by natural selection, or by random genetic drift, will be reflected in the presence of new alleles, changes in allele frequency, or the presence of new chromosomal arrangements. Eventually, a point will be reached when the populations have enough differences that they can be identified as distinct races. This process of divergence may continue uninterrupted until two or more species are present.

If at any time during the process of genetic divergence, the conditions that prevent gene flow between the populations are removed, two outcomes are possible: (1) the two populations may fuse into a single gene pool, because hybridization does not reduce fertility or viability; or (2) the gene pools of the populations may have diverged to the point where isolating mechanisms may have arisen.

The various biological and behavioral properties of organisms that act to prevent or reduce interbreeding are called **reproductive isolating mechanisms.** These mechanism are classified in Table 21.7. For example, genetic divergence may have reached the stage where the viability or fertility of hybrids is reduced. Hybrid zygotes may be formed, but all or most may be inviable. Alternately, the hybrids may be viable, but have reduced fertility or be sterile. In another possibility, the hybrids themselves may be fertile, but their progeny may have lowered viability or fertility. These mechanisms, called **postzygotic,** are a byproduct of genetic divergence. Such isolating mechanisms waste gametes and zygotes and lower the fitness of hybrid survivors. Selection will therefore favor the spread of alleles that will reduce the formation of hybrids, leading to the development of **prezygotic isolating mechanisms.**

Speciation and the development of isolating mechanisms may gradually occur in the absence of selection, as when populations remain permanently isolated on two or more islands. Selection accelerates speciation in those cases where some reproductive isolation has resulted from genetic diversity, but not all isolating mechanisms are represented in each speciation event. Usually at least two isolating mechanisms are developed by natural selection drawing on the genetic variability present in the evolving populations.

Formation of Races and Species

To illustrate the first step in speciation, the formation of races, we shall consider the case of *Drosophila pseudoob-*

TABLE 21.7
Reproductive isolating mechanisms

Prezygotic Mechanisms (prevent fertilization and zygote formation)

1. Geographic or ecological: The populations live in the same regions but occupy different habitats.
2. Seasonal or temporal: The populations live in the same regions but are sexually mature at different times.
3. Behavioral (only in animals): The populations are isolated by different and incompatible behavior before mating.
4. Mechanical: Cross-fertilization is prevented or restricted by differences in reproductive structures (genitalia in animals, flowers in plants).
5. Physiological: Gametes fail to survive in alien reproductive tracts.

Postzygotic Mechanisms (fertilization takes place and hybrid zygotes are formed, but these are nonviable or give rise to weak or sterile hybrids)

1. Hybrid nonviability or weakness.
2. Developmental hybrid sterility: Hybrids are sterile because gonads develop abnormally or meiosis breaks down before completion.
3. Segregational hybrid sterility: Hybrids are sterile because of abnormal segregation to the gametes of whole chromosomes, chromosome segments, or combinations of genes.
4. F_2 breakdown: F_1 hybrids are normal, vigorous, and fertile, but F_2 contains many weak or sterile individuals.

SOURCE: From G. Ledyard Stebbins. *Processes of Organic Evolution*. 3rd edition. © 1977, p. 143. Reprinted by permission of Prentice-Hall, Inc. Englewood Cliffs, N.J.

scura as studied by Theodosius Dobzhansky and his colleagues. This species is found over a wide range of habitats in the western and southwestern United States. Although the flies throughout this range are morphologically similar, Dobzhansky discovered that populations from different locations varied substantially in the arrangement of genes on chromosome 3. He found a variety of different inversions in this chromosome that could be seen by loop formations in the salivary chromosomes. Each particular inversion sequence was named after the locale in which it was first detected (e.g., AR = Arrowhead, British Colombia; CH = Chiricahua Mountains) and was compared with one standard sequence, designated ST.

Figure 21.12 shows a comparison of three arrangements detected in populations found at three different elevations in the Sierra Nevada in California's Yosemite region. The ST arrangement is most common at low elevations, but declines in frequency as elevation increases. At 8000 feet, AR is most common and ST least common. In these same populations, the frequency of the CH arrangement gradually increases with elevation. The gradual change in frequencies of arrangements is probably the result of natural selection and thus parallels the gradual environmental changes occurring at ascending elevations.

Dobzhansky also found that if populations are collected at a single site throughout the year, inversion frequencies also change. During the different seasons cyclic variation occur, as shown in Figure 21.13. Such variation was observed to be repeated over a period of several years. ST was always observed to decline during the spring, with a concomitant increase in CH during the same period.

To test the hypothesis that this cyclic change is a response to natural selection, Dobzhansky devised the following laboratory experiment. He constructed a large population cage in which samples could be periodically removed and studied. He began with a population of a known inversion frequency, 88 percent CH and 12 percent ST. He reared it at 25°C and sampled it over a one-year period. As shown in Figure 21.14, the frequency of ST increased gradually until it was present at a level of 70 percent. At this point, an equilibrium between ST and CH was reached. When the same experiment was performed at 16°C, no change in inversion frequency occurred. It can be concluded that the equilibrium reached at 25°C was in response to the elevated temperature, the only variable in the experiment.

The results of Dobzhansky's experiment are strong evidence that a balanced condition of the two inversions and their respective gene arrangements is superior to either inversion by itself. The equilibrium attained presumably represents the greatest degree of fitness in the population under varying laboratory conditions. This interpretation of the experiment suggests that natural selection is the driving force toward equilibrium. If so, then we may conclude that the existence of various inversions is representative of genetic variation.

In a more extensive study, Dobzhansky went on to sample populations over a much broader geographic range. In Figure 21.15, the frequencies of five inversions are shown according to geographic location. Some general trends are apparent. ST increases in frequency from east to west, while just the opposite is true of PP. AR is least common in the east-west extremes, yet predominant in Arizona and New Mexico.

Because these locations represent varied environments, and the inversions preserve different gene combi-

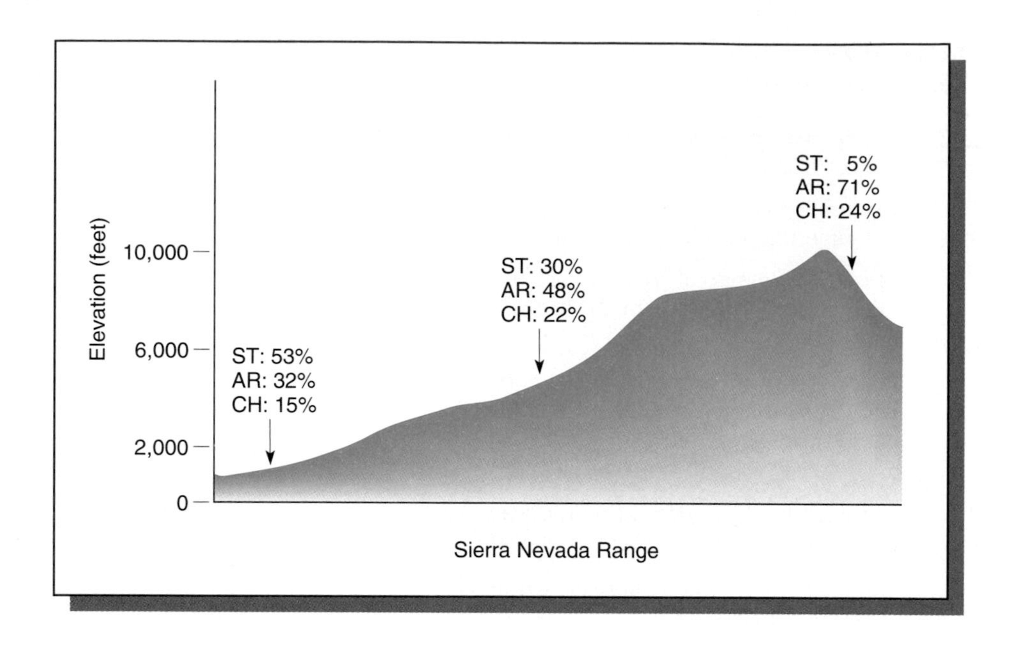

FIGURE 21.12
Inversions in chromosome 3 of *D. pseudoobscura* found at different elevations in the Sierra Nevada near Yosemite National Park.

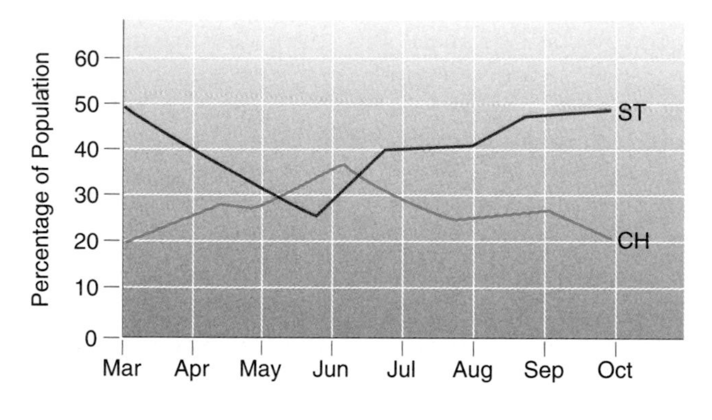

FIGURE 21.13
Changes in the ST and CH arrangements in *D. pseudoobscura* throughout the year.

nations, we can conclude that numerous races of *D. pseudoobscura* have been formed. Not only do the studies of *D. pseudoobscura* illustrate the principles of the initial step in speciation, but they also strongly support the concept that the adaptive norm consists of balanced genotypes within a population.

For speciation to occur, the development of races must be followed by a second step, reproductive isolation. We might then ask whether the evolution of *D. pseudoobscura* has gone beyond the formation of races. Dobzhansky investigated this question by examining the chromosome structure of other related **sibling species.** Sibling species have become reproductively isolated, but remain very similar morphologically.

One sibling species, *Drosophila persimilis*, has provided very interesting information. *D. persimilis*, like *pseudoobscura*, contains five pairs of chromosomes and carries inversions within chromosome 3. One, ST, is also found in *D. pseudoobscura*. Based on the common arrangement, it was possible to construct a phylogenetic sequence for all arrangements found in both species. As shown in Figure 21.16, ST is shared by both species. It appears that an ancestral population with the ST arrangement gave rise to many different inversions. Some were incorporated into races as members of the *pseudoobscura* species, and others gave rise to the *persimilis* species.

Today, even when the geographic distributions of these sibling species overlap, several isolating mechanisms keep them from interbreeding. They are isolated by prezygotic mechanisms such as habitat selection and courtship behavior.

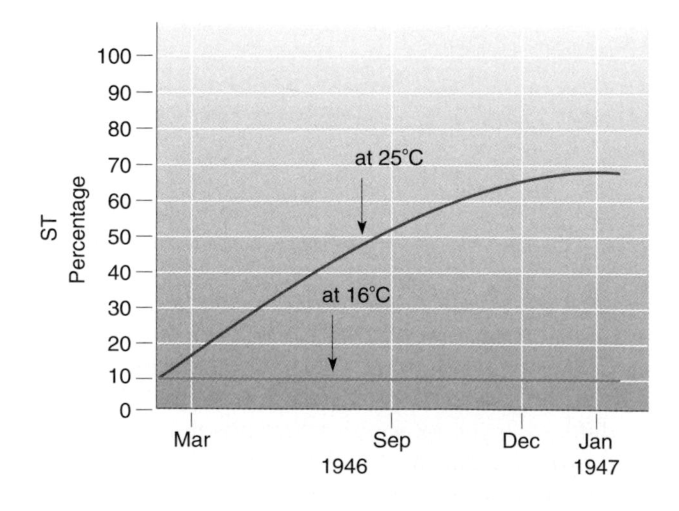

FIGURE 21.14
Increase in the ST arrangement of *D. pseudoobscura* in population cages under laboratory conditions.

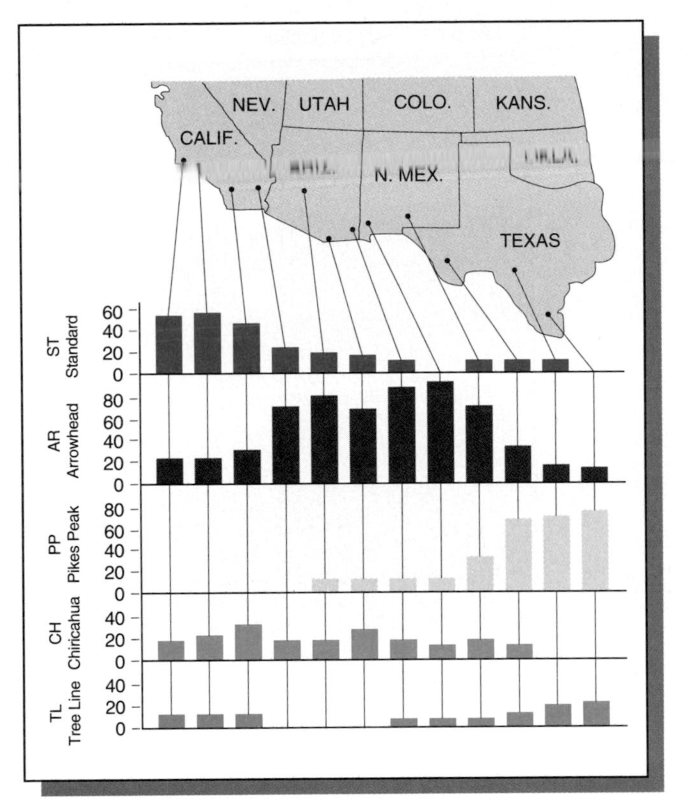

FIGURE 21.15
Frequencies of five chromosomal inversions in *D. pseudoobscura* in different geographic regions.

Quantum Speciation

At the heart of the classical theory of speciation is **gradualism.** According to this concept, speciation is a microevolutionary event resulting from the accumulation of many minute gene differences over a long period of time under the influence of natural selection. Recently, however, more **stochastic** or **catastrophic models of speciation** have been proposed. These models emphasize the role of evolutionary events, which occur suddenly and intermittently, and are therefore referred to as **quantum speciation.**

Drawing on evidence from the fossil record, Niles Eldredge and Stephen Jay Gould have proposed a mechanism of species formation known as **punctuated equilibria.** According to this model, evolutionary changes are not gradual and continuous, but occur intermittently and

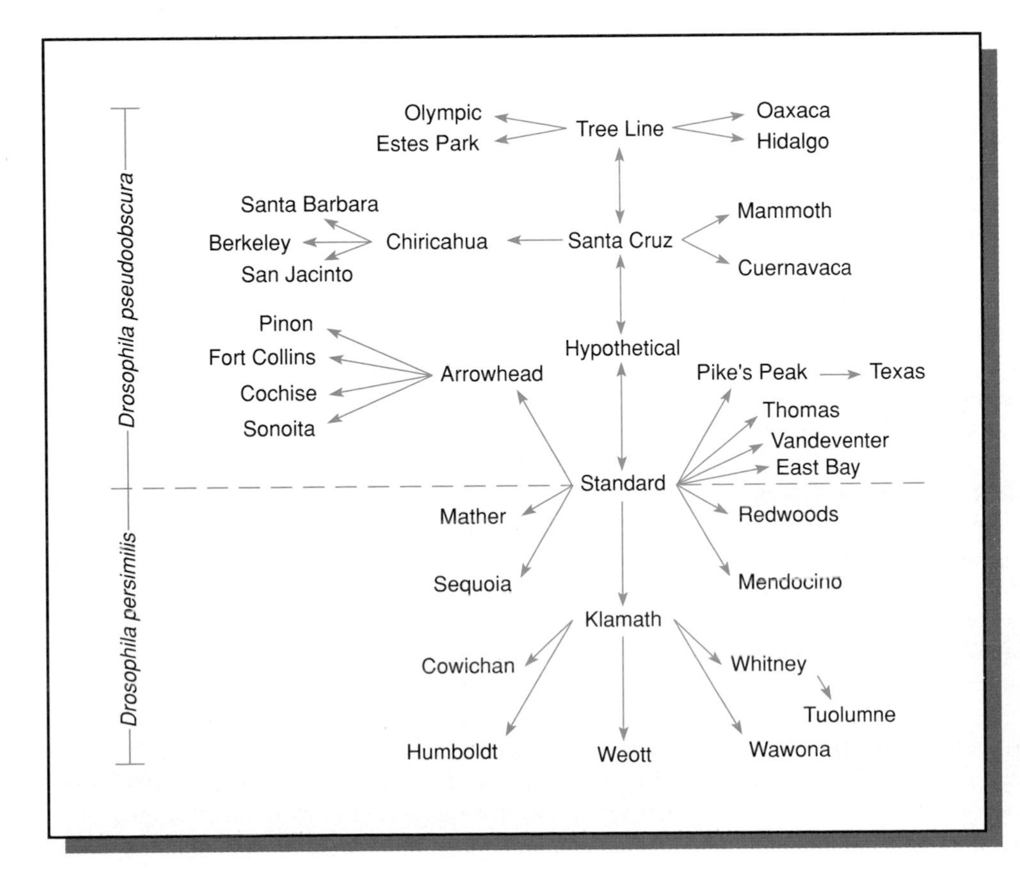

FIGURE 21.16
Inversion phylogeny for arrangement of chromosome 3 in *D. pseudoobscura* and *D. persimilis*. The ST (standard) arrangement is shared by both species.

rapidly as events that punctuate or interrupt long periods of evolutionary equilibrium during which little or no change occurs. In paleontological terms, *rapidly* means within thousands or even hundreds of thousands of years. The results of such evolutionary alterations are thought to cause abrupt changes in the fossil record, since relatively few fossils will be formed during such changes, compared to the large numbers of fossils formed during long periods (extending into millions of years) of evolutionary equilibrium.

The evolutionary spurts that punctuate periods of stasis are thought to be events that produce mass extinctions, or events that take place in small populations located at the fringes of the species range. In these locales, environmental stresses disrupt evolutionary stability, eventually causing the emergence of new species. Proponents of this hypothesis point out that such changes leading to speciation take place through natural selection of individual variations, and do not invoke any unusual genetic mechanisms. On the other hand, some population geneticists argue that the mechanism of punctuated equilibria unnecessarily separates the process of species formation from other evolutionary events that generate diversity between members of a species.

In his study of the evolution of Hawaiian *Drosophila*, Hampton Carson has proposed a **founder-flush speciation theory.** According to this theory, populations and even species can be started by a single individual. In Carson's proposal, gradualism is not a necessary component of speciation, and more importantly, reproductive isolation *precedes* adaptation rather than being a consequence of genetic diversification. In other words, Carson has changed the order of steps in the classical model of speciation. According to the founder-flush theory, a single fertilized female can colonize an isolated territory previously unoccupied by members of this species. If conditions are favorable, the population founded by this individual will undergo a **flush,** or rapid expansion. After several generations, it is likely that the population growth will outstrip the environment's carrying capacity, causing a **population crash.** The crash, a completely random event, causes the death or dispersal of almost the entire population. A single survivor, or at most a few survivors, may rebuild the population, which eventually undergoes several flush-crash cycles before coming to equilibrium with the environment. This cycle is diagrammed in Figure 21.17. Through the genetic revolution it has undergone, the colony will, in all probability, have acquired adaptations making it unable to interbreed with the parental population.

According to Carson's theory, genetic changes are brought about in two ways. First, the original founding of the population by one or a very few individuals can establish allele frequencies different from those in the ancestral population. Second and most important, selection is relaxed during a population flush.

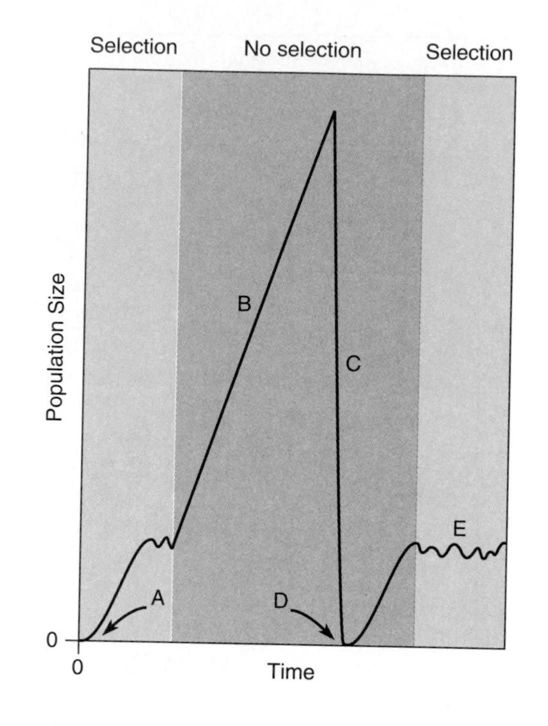

FIGURE 21.17
The flush–crash cycle. Small populations descended from a single founder (A) undergo a population flush (B), followed by a crash (C). A single survivor in the form of a fertilized female (D) builds a new population (E).

The evolution of *Drosophila* species in the Hawaiian Islands appears to have followed such a founder-flush cycle. The relationships between species of *Drosophila* can be traced by mapping the location and frequency of inversions in the banded polytene chromosomes of larval salivary glands. One group, the *planitibia* complex, has three species with the same basic set of chromosome inversions: *D. planitibia, D. heteroneura,* and *D. silvestris. D. planitibia* is found on the older island of Maui, and the other two are found on the younger island of Hawaii. From this evidence, it is postulated that an immigrant fertilized female belonging to an ancestral stock on Maui, chromosomally related to the presentday *D. planitibia*, crossed the Alenuihaha Channel to Hawaii. Subsequent flush-crash cycles led to the reconstruction of the colony's genotype, giving rise to the two species found on Hawaii—*D. heteroneura* and *D. silvestris* (Figure 21.18). Similar evidence indicates that two other groups on Maui have given rise to a total of five species on Hawaii.

In laboratory tests of this theory, Jeffrey Powell has found that 15 generations after several flush-crash cycles, laboratory strains of *Drosophila* species show significant prezygotic (behavioral) isolation from other strains. Subsequent testing several months later show that these dif-

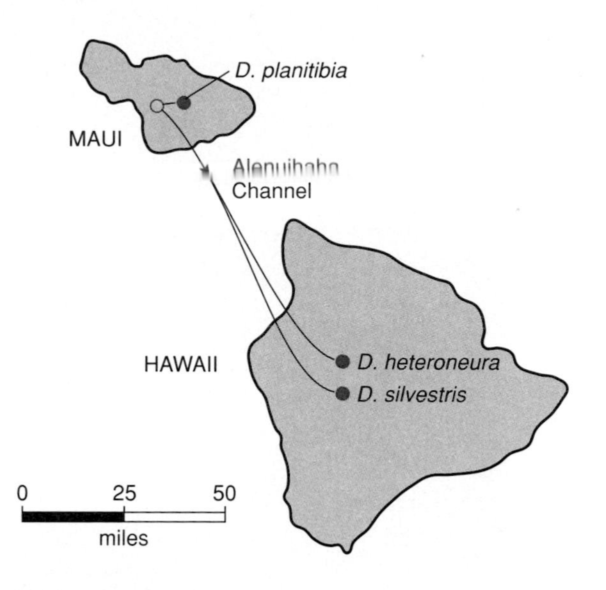

FIGURE 21.18
Proposed pathway of colonization of Hawaii by members of the *D. planitibia* species complex. Open circle represents a population ancestral to the three present-day species.

ferences are not transitory. These experimental results are important in establishing that the first stages of speciation can occur rapidly under certain circumstances.

A final example illustrating quantum speciation involves polyploidy in plants. The formation of species by polyploidy in animals is rare, but has been an important factor in the evolution of plants. It is estimated that one-half of all flowering plants have evolved by polyploidy. One such form of polyploidy is **allopolyploidy** (see Chapter 6), produced by doubling the chromosome number in an interspecific hybrid. If two species of related plants have the genetic constitution *SS* and *TT*, where *S* and *T* represent the haploid set of chromosome in each species, then the F_1 hybrid would have the chromosome constitution *ST*. Normally, such a plant would be sterile because there are few or no homologous chromosome pairs, and aberrations would arise during meiosis. If the hybrid undergoes a spontaneous doubling of chromosome number, however, a tetraploid *SSTT* would be produced. This might occur in somatic tissue, giving rise to a partially tetraploid plant that would produce some tetraploid flowers. Alternatively, aberrant meiotic events may produce *ST* gametes, which when fertilized, would yield *SSTT* zygotes. The *SSTT* plants would be fertile because they would possess homologous chromosomes producing viable *ST* gametes. This new, true-breeding tetraploid would have a combination of characters derived from the parental species, and would be reproductively isolated from them because F_1 hybrids between diploids and tetraploids would be triploids and consequently sterile. The tobacco plant *Nicotiana tabacum* ($2n = 48$) is the result of the doubling of the chromosome number in the hybrid between *N. otophora* ($2n = 24$) and *N. silvestris* ($2n = 24$).

MOLECULAR EVOLUTION

The diversity of life forms inhabiting the earth is overwhelming. Nearly two million species of plants and animals have been described, and surely there are many yet to be classified. Nevertheless, all organisms share the same molecular features: They are composed principally of carbon, hydrogen, nitrogen, and oxygen atoms; they use nucleic acids to store and transfer chemical information; and proteins are, for them, the indispensable products of the stored genetic information.

The recent development of techniques to analyze and sequence proteins and nucleic acids has allowed biologists to infer relatedness of organisms and to construct molecular phylogenetic trees. In this section we shall review some of the findings in this area of evolutionary study and examine the genetic variability demonstrated at the molecular level within populations.

Amino Acid Sequence Phylogeny

Evolutionary relatedness or divergence can be measured by comparing the amino acid sequences of proteins common to various organisms. The first protein to be sequenced was **insulin,** composed of only 51 amino acids. In the early 1950s, the amino acid sequence of insulin was examined in a variety of mammals, including cattle, pigs, horses, sperm whales, and sheep. With the exception of a stretch of three amino acids, the protein was shown to contain an identical sequence in each of these mammals.

Cytochrome *c* is another commonly investigated protein. It is a respiratory pigment found in the mitochondria of eukaryotes. The molecule consists of 104 amino acids in many vertebrates and a slightly higher number in most other organisms. Cytochrome *c* has changed very slowly during evolution. For example, the amino acid sequence in humans and chimpanzees is identical; between humans and rhesus monkeys only one amino acid is different. This is remarkable considering that lines leading to humans and monkeys diverged from a common ancestral form approximately 20 million years ago.

Table 21.8a shows the number of amino acid differences between a variety of organisms, using human cytochrome *c* as a standard. Even as distantly removed from humans as yeasts are, only 38 amino acids are different.

In a comparison of a large number of organisms, more than 15 percent of the sequences remain unchanged.

In addition to the number of amino acid differences, it is possible to assess the minimum number of nucleotide changes that must have occurred during the evolution of a protein. This assessment requires knowledge of the genetic code and is a more refined analysis. For example, more than one nucleotide change may have been necessary to establish any given amino acid change found between two organisms. Thus, over evolutionary time, two or more independent mutations may have been essential in order to produce the observed change. When all nucleotide changes necessary for all amino acid differences are totaled, the **minimal mutational distance** between any two species is established. Table 21.8b shows such an analysis of the genes coding for cytochrome *c* in ten organisms. One can see that, as expected, these values are larger than the corresponding number of amino acids separating humans from the other nine organisms.

The phylogenetic tree shown in Figure 21.19 is based on the sequences used to derive the data of Table 21.8. The tree is plotted so that the ordinate represents proportional amounts of distance. If a constant mutation rate is assumed, the ordinate represents a relative estimate of geologic time. When phylogenetic trees are constructed in this way, they agree remarkably well with trees constructed using more conventional approaches such as morphological and paleontological evidence. Perhaps their greatest value is that these studies are performed directly at the level of genetic variation, which is the underlying resource supporting evolutionary change.

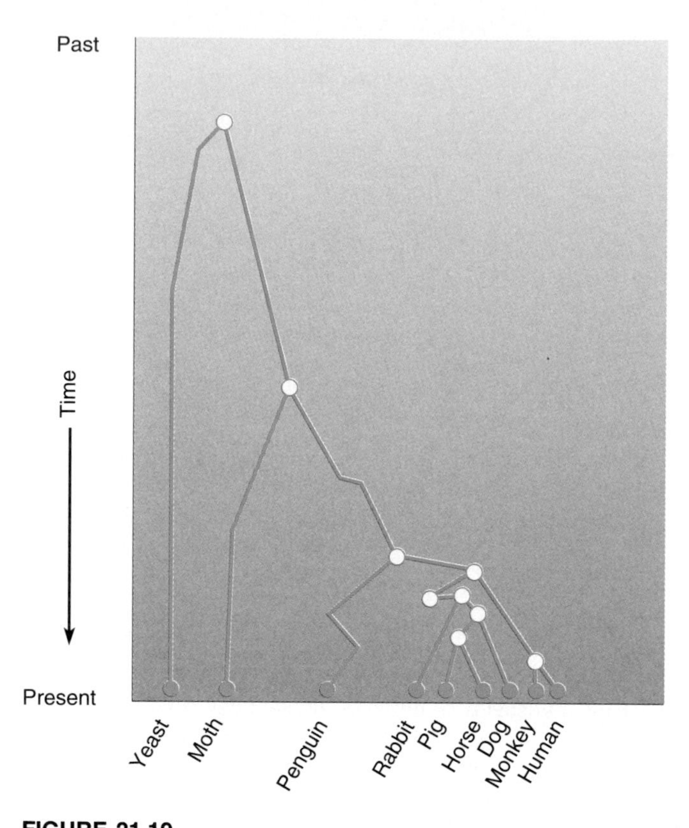

FIGURE 21.19
Phylogenetic sequence constructed by comparison of homologies in cytochrome c amino acid sequences.

TABLE 21.8
A comparison of (a) the number of amino acid differences and (b) the minimal mutational distance in cytochrome *c*.

Organism	a. Number of Amino Acid Differences	b. Minimal Mutational Distance
Human	0	0
Chimpanzee	0	0
Rhesus monkey	1	1
Rabbit	9	12
Pig	10	13
Dog	10	13
Horse	12	17
Penguin	11	18
Moth	24	36
Yeast	38	56

SOURCE: From W. M. Fitch and E. Margoliash. "Construction of Phylogenetic Trees." *Science* 155:279–84, 20 January 1967.

Nucleotide Sequence Phylogeny

The molecular hybridization of DNA from different sources has been a valuable technique in evolutionary research. In studies of sequence diversity between species, radioactive DNA from one species is prepared and those sequences present once in the haploid genome (single copy fraction) are isolated. This single copy tracer is dissociated into individual strands and can be reassociated with melted single copy DNA from the same species (homologous reaction) or from other species (heterologous reaction). The difference in thermal stability (ΔT_m) between homologous and heterologous duplex molecules is a measure of the nucleotide sequence divergence between the two species. For example, a ΔT_m of 1°C roughly corresponds to 1 percent mismatching in nucleotide sequences.

Where the nucleotide sequence divergence can be compared with other indicators of genetic variability, such as protein polymorphisms and chromosomal rearrangements, the evidence indicates that nucleotide sequence diversity is a more sensitive indicator of evolutionary divergence than amino acid replacements. *Drosophila heteroneura* and *D. silvestris*, found only on the island of Hawaii, are estimated to have diverged only

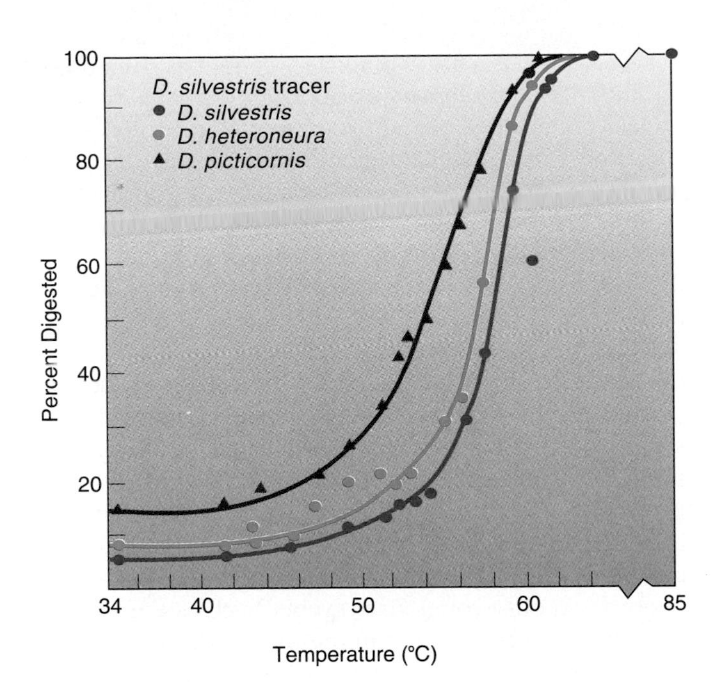

FIGURE 21.20
Nucleotide sequence diversity in the *planitibia* species complex. The degree of shift to the left by the heterologous hybrids is an indication of the degree of nucleotide diversity.

about 300,000 years ago, but it is difficult to demonstrate significant differences between these species in chromosomal inversion patterns or protein polymorphisms. Figure 21.20 shows the results obtained when labeled single copy DNA from *D. silvestris* is hybridized with itself and with DNA from *D. heteroneura* and *D. pictornis*. (*D. picticornis* is another member of the *planitibia* group and is found only on Kauai, a much older island.) The sequence diversity between the two species from Hawaii is about 0.55 percent, but *D. silvestris* and *D. picticornis* show a 2.1 percent difference in nucleotide sequences. Thus, nucleotide sequence diversity may precede the development of protein or chromosomal polymorphisms.

It may seem paradoxical that *D. heteroneura* and *D. sylvestris* share identical chromosome arrangements, have almost no detectable protein differences, are 99 percent homologous in DNA sequence, and yet are classified as separate species. However, the two species are clearly separated from each other by different and incompatible courtship and mating behaviors (a prezygotic isolating mechanism), by morphology, and by pigmentation of the body and wings (Figure 21.21). The available evidence suggests that these differences are controlled by a relatively small number of genes. In a genetic analysis of the differences in head shape and pigmentation, F. C. Val has estimated that only 15 to 19 loci are responsible for the morphological differences between these species, demonstrating that the process of speciation need only involve a small number of genes.

RFLP Analysis of Nucleotide Phylogenies

The question of where the human species originated has spawned several theories. Africa, Asia, Europe, and the Americas have been suggested as the site of human origin, and another theory proposes that modern humans arose simultaneously in several locations. The available fossil record indicates that *Homo sapiens* was present in

(a) (b)

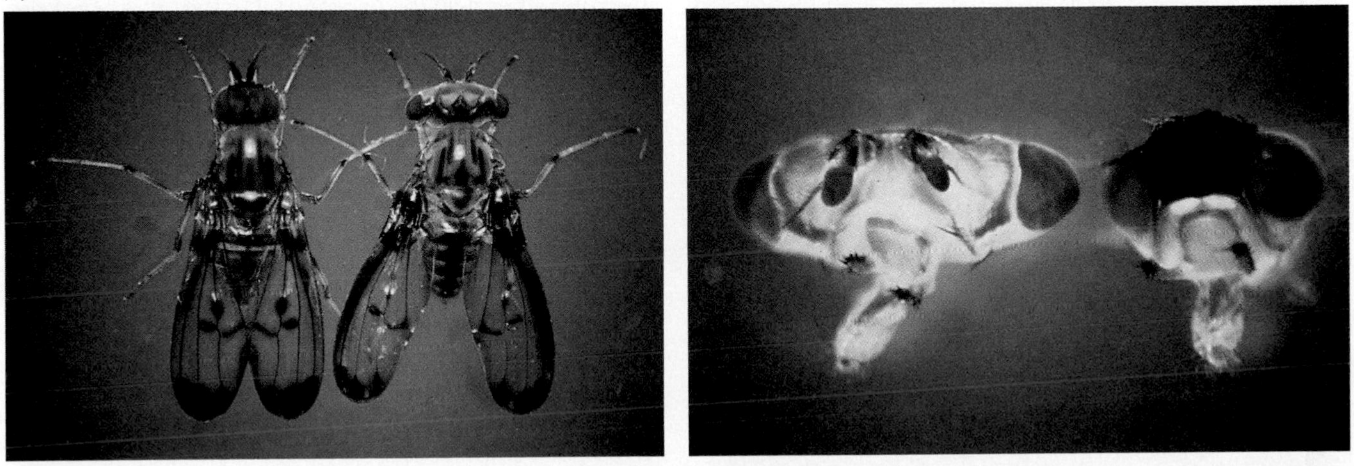

FIGURE 21.21
(a) Differences in pigmentation patterns in *D. sylvestris* (left) and *D. heteroneura* (right). (b) Head morphology in *D. sylvestris* (left) and *D. heteroneura* (right).

Southern Africa more than 100,000 years ago, and was found in Asia at least 50,000 years ago, and it appears that *Homo sapiens* replaced Neanderthals in Europe about 30,000 to 40,000 years ago. The bulk of evidence from fossils supports the origin of humans in Africa, probably by phyletic evolution from *Homo erectus*, with a rapid spread to non-African regions. Over the past decade, molecular techniques have been used to provide additional evidence for the origin and spread of the human species across the globe.

In Chapter 15, we discussed the use of restriction fragment length polymorphisms (RFLPs) to map genes to specific chromosomes and/or chromosomal regions. RFLPs can be generated by single nucleotide base changes in a DNA sequence that is recognized by a restriction enzyme. Since restriction enzymes make double-stranded cuts in DNA, alterations in recognition sequences result in changes in the pattern of cuts made in the DNA. These detectable changes or polymorphisms are inherited in a codominant Mendelian fashion. RFLP analysis has been used to construct evolutionary branching patterns known as dendrograms and to reconstruct evolutionary events.

A study of restriction polymorphisms in mitochondrial DNA (which is maternally inherited) by Cavelli-Sforza and his colleagues first suggested that African populations are ancestral to others. More recent work on mitochondrial RFLPs has suggested that Caucasoid populations may have diversified from the ancestral African population and served as the source of all other groups.

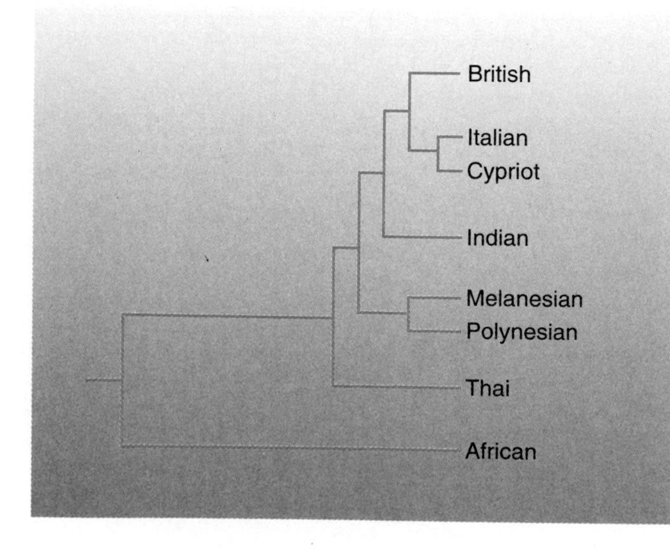

FIGURE 21.22
Evolutionary dendrogram of human populations.

Recently, RFLP analysis of nuclear genes has been used to study the evolutionary relationships among various human populations. Using five different restriction sites clustered within the beta globin complex, as markers, J. S. Wainscoat and colleagues studied the distribution of these sites in eight population groups. In all, 32 combinations of these five restriction sites are possible, and 14 of these were actually observed in the surveyed populations.

Using the fossil evidence and the RFLP data, it is possible to construct a dendrogram for the eight populations studied (Figure 21.22). The model is consistent with the suggestion that *Homo sapiens* originated in Africa, and that a small founder population migrated from Africa and later gave rise to all non-African populations. Further RFLP screening of other nuclear genes will help to resolve the issues raised in this study and to answer other questions about evolutionary events, such as the size of the founding population and the time scale over which these evolutionary events may have transpired.

THE ROLE OF MUTATION IN EVOLUTION

Selectionists versus Neutralists

When we discussed the concept of genetic variation in an earlier section, it was pointed out that there is a controversy over the importance of extensive genetic variation within a population. This controversy centers particularly on the variation detected at the molecular level. On the one hand, the **classical hypothesis** proposes that natural selection favors the fixation of the most favorable alleles at each locus. Homozygosity, therefore, should be the rule rather than the exception. On the other hand, the **balance hypothesis** favors the maintenance through natural selection of a high degree of heterozygosity in populations.

The classical hypothesis is supported by the observation that induced mutations almost invariably lower fitness. Therefore, if most mutations are detrimental, the accumulation of genetic variability will result in a substantial genetic burden to the fitness of a population. According to the classical theory, this detrimental variation will be purged by natural selection.

In contrast, Dobzhansky and his colleagues provided the initial support for the balance hypothesis. Having observed a high degree of heterozygosity in adapted *Drosophila* populations, they postulated that populations displaying genetic polymorphism will have added fitness. The observation of protein polymorphisms in *Drosophila* and in humans, and the evidence for numerous amino acid substitutions in single proteins throughout evolution, attest to a high degree of heterozygosity, particularly at the molecular level. These findings lend support to the balance hypothesis and argue against the classical theory.

However, the existence of a high degree of genetic variation does not prove that it is the basis of evolutionary fitness. A third theory, the **neutralist hypothesis** proposed by James Crow and Motoo Kimura, argues that mutations leading to amino acid replacement are most often neutral or genetically equivalent to the allele that is replaced. Those that are favorable or detrimental are either preserved or removed from the population, respectively, by natural selection. However, neutral genetic changes will not be affected by selection and will instead become randomly fixed in the population. Their frequency will be determined by the rate of mutation and the principles of random genetic drift.

Those who oppose the neutralist theory have been called **selectionists.** They point out examples where enzyme or protein polymorphism is associated with adaptation to certain environmental conditions. The well-known advantage of sickle-cell anemia carriers in malarial regions is such an example.

Selectionists also stress that enzyme polymorphism may often appear to offer no advantage, but exists in such a frequency that it is impossible to explain as a random occurrence. Thus, even though no currently available analytical technique can detect any physiological difference, it cannot be proved that some slight advantage associated with any given amino acid substitution does not exist.

It is difficult to argue against the notion that some genetic variation must be neutral. The difference between the two theories is in the degree of neutrality that exists. While current data are insufficient to resolve the problem one way or the other, one important point has emerged from the arguments. In considering natural selection, there are at least two levels that must be examined: the phenotypic level, including all morphological and physiological characteristics imparted by the genotype; and the molecular level, represented by the precise nucleotide and amino acid sequences of DNA and proteins. There is no question about selection acting at the phenotypic level. The extent to which selection occurs at the molecular level, however, is in question.

CHAPTER SUMMARY

1 The idea that the population rather than the individual is the effective unit in the process of evolution has led to the study of gene frequencies and to the emergence of the discipline of population genetics.

2 The Hardy-Weinberg law established the conditions for genetic equilibrium within a population. These conditions include a large, randomly breeding population and the absence of selection, mutation, and migration. If any of these conditions are not met, gene or genotype frequencies will change from generation to generation.

3 The process of evolution requires conditions leading to change, such as changing gene frequencies that provide genetic variation. The mathematical derivation of the Hardy-Weinberg equilibrium has served to assess the forces of evolution: selection, mutation, migration, genetic drift, and inbreeding.

4 Mutation and migration introduce new alleles into a population, but the retention or loss of these alleles depends on the degree of fitness conferred and the action of selection.

5 Speciation, which depends on genetic variation and forces controlling its distribution (e.g., natural selection), is initiated when a population becomes separated into smaller, reproductively semi-isolated breeding groups.

6 The partitioning of a population's gene pool results in a better fitness for each group in its new environment. Races form as each group acquires its own unique set of genetic variations. Races may become reproductively isolated from one another and form new species as evolution proceeds. The formation of races and speciation are illustrated by several studies of *Drosophila*.

7 Two approaches have been especially fruitful in evolutionary study at the molecular level: (1) comparison of complementary sequences present in the DNA of different organisms, and (2) comparison of amino acid substitutions in proteins common to a variety of organisms. Such comparisons provide not only a measure of evolutionary relatedness, but further allow the assessment of genetic variation during evolution.

8 The central question concerning the importance of variation at the molecular level is whether or not all nucleotide and amino acid changes preserved through evolution in DNA and proteins are adaptive to the population. Some believe that the majority of these changes are neutral or genetically equivalent, while others adhere to the theory of adaptive evolution through natural selection.

KEY TERMS

achromatopsia
allelic frequency
allopatric speciation
allopolyploidy
assortative
balance hypothesis
classical hypothesis
coefficient of inbreeding
consanguineous marriages
cytochrome *c*
directional selection
disruptive selection
dominance hypothesis
fitness
flush
founder-flush speciation theory
genetic equilibrium
genetic drift
genetic variety
gradualism
Hardy-Weinberg law
hybrid vigor (heterosis)
inbreeding
inbreeding depression
insulin
inversions
migration
minimal mutational distance
mutation
MN blood group
natural selection
neutralist hypothesis
overdominance
phyletic evolution
population
population crash
population genetics
postzygotic
prezygotic isolating mechanisms
punctuated equilibria
quantum speciation
reproductive isolating mechanisms
selection coefficient
self-fertilization
sibling species
stabilizing selection
stochastic or catastrophic models of speciation
speciation
translocations

Eugenics is the term employed for the selective breeding of humans to bring about improvements in the species. As a eugenic measure, it has been suggested (sometimes by force of law) that individuals suffering from serious genetic disorders should be prevented from reproducing (by sterilization, if necessary) in order to reduce the frequency of the disorder in future generations. Suppose that such a trait were present in the population at a frequency of 1 in 40,000, and that affected individuals did not reproduce. In 100 generations, or about 2500 years, what would be the frequency of the condition? What are your assumptions about this condition? Are the eugenic measures effective in this case?

ANSWER: Let us assume that the trait is recessive, and that homozygous recessive individuals are prevented from reproducing. In this case, selection against the homozygous recessive phenotype is equal to 1. Using formula 21.13, we can calculate the frequency after 100 generations as follows:

$$q_n = \frac{q_0}{1 + nq_0}$$

$$q_n = \frac{0.000025}{1 + 100\ (0.000025)}$$

$$= \frac{0.000025}{1.0025}$$

where

$$q_n = 1/41{,}666$$
$$s = 1$$
$$n = 100$$
$$q_0 = 0.000025$$

The frequency of the condition has been reduced from 1 in 40,000 to 1 in 41,666 in 100 generations, indicating that this eugenic measure is ineffective.

PROBLEMS AND DISCUSSION QUESTIONS

1 The ability to taste the compound PTC is controlled by a dominant allele T, while individuals homozygous for the recessive allele t are unable to taste this compound. In a genetics class of 125 students, 88 were able to taste PTC, 37 could not. Calculate the frequency of the T and t alleles in this population and the frequency of the genotypes.

2 If 4 percent of the population in equilibrium expresses a recessive trait, what is the probability that the offspring of two individuals who do not express the trait will express it?

3 If initial gene frequencies are $p = 0.5$ and $q = 0.5$, and $s = 1.0$, what will be the gene frequencies after 1, 5, 10, 25, 100, and 1000 generations?

4 In a breeding program to improve crop plants, which of the following mating systems should be employed to produce a homozygous line in the shortest possible time?
(a) self-fertilization
(b) brother–sister matings

(c) first cousin matings
(d) random matings
Illustrate your choice with pedigree diagrams.

5 One of the first Mendelian traits identified in man was a dominant condition known as *brachydactyly*. This gene causes an abnormal shortening of the fingers and/or toes. At the time, it was thought by some that this dominant trait would spread until 75 percent of the population would be affected (since the phenotypic ratio of dominant to recessive is 3:1). Show how this line of reasoning is incorrect.

6 Discuss the rationale behind the statement that inversions in chromosome 3 of *Drosophila pseudoobscura* represent genetic variation.

7 Determine the minimal mutational distances between the following amino acid sequences of cytochrome *c* from various organisms. Compare the distance between humans and each organism.

Human:	Lys	Glu	Glu	Arg	Ala	Asp
Horse:	Lys	Thr	Glu	Arg	Glu	Asp
Pig:	Lys	Gly	Glu	Arg	Glu	Asp
Dog:	Thr	Gly	Glu	Arg	Glu	Asp
Chicken:	Lys	Ser	Glu	Arg	Val	Asp
Bullfrog:	Lys	Gly	Glu	Arg	Glu	Asp
Fungus:	Ala	Lys	Asp	Arg	Asn	Asp

8 The genetic difference between *D. heteroneura* and *D. sylvestris* as measured by nucleotide diversity is about 1.8 percent. The difference between chimpanzees (*Pan troglodytes*) and humans (*Homo sapiens*) is about the same, yet the latter species are classified in different genera. In your opinion, is this valid? If so, why; if not, why not?

9 As an extension of the previous question, consider the following: In sorting out the complex taxonomic relationships among birds, species with $\Delta T_{50}H$ values of 4.0 are placed in the same genus, even by traditional taxonomy based on morphology. Using the data in Figure 26.17, construct a phylogeny that obeys this rule, using any of the appropriate genus names (*Pongo, Pan, Homo*), or constructing new ones.

10 Of what value to our understanding of genetic variation and evolution is the debate concerning the neutral mutation theory?

SELECTED READINGS

BARTON, N. H., and HEWITT, G. M. 1989. Adaptation, speciation and hybrid zones. *Nature* 341:497–503.
BARTON, N. H., and TURELLI, M. 1989. Evolutionary quantitative genetics: How little do we know? *Ann. Rev. Genet.* 23:337–70.
DAVID, J. R., and CAPY, P. 1988. Genetic variation of *Drosophila melanogaster* natural populations. *Trends Genet.* 4:106–11.
DAYHOFF, M. O. 1969. Computer analysis of protein evolution. *Scient. Amer.* (July) 221:86–95.
DOVER, G. A., and FLAVELL, R. B., eds. 1982. *Genome evolution*. Orlando: Academic Press.
FREIRE-MAIA, N. 1990. Five landmarks in inbreeding studies. *Am. J. Med. Genet.* 35:118–20.

KIMURA, M. 1979. The neutral theory of molecular evolution. *Scient. Amer.* (Nov.) 241:98–126.
———. 1989. The neutral theory of molecular evolution and the world view of the neutralists. *Genome* 31:24–31.
LANDE, R. 1989. Fisherian and Wrightian theories of speciation. *Genome* 31:221–27
MAYR, E. 1963. *Animal species and evolution.* Cambridge, MA: Harvard University Press.
METTLER, L. E., GREGG, T., and SCHAFFER, H. E. 1988. *Population genetics and evolution.* 2nd ed. Englewood Cliffs, NJ: Prentice-Hall.
SPIESS, E. B. 1989. *Genes in populations.* 2nd ed. New York: Wiley.

APPENDIX A

ANSWERS TO SELECTED PROBLEMS

CHAPTER 1

4 Darwin's theory of natural selection proposed that more offspring are produced than can survive, and that in the competition for survival, those with favorable variations survive. Over many generations, this will produce a change in the genetic make-up of populations if the favorable variations are inherited. Darwin did not understand the nature of heredity and variation which led him to lean toward theories of pangenesis and inheritance of acquired inheritance.

9 In the last 40 years human transmission, cytological, and molecular genetics have provided an understanding of many human diseases. There is promise that a certain amount of human suffering will be minimized by the application of genetics to medicine. Major areas of activity include genetic counseling, gene mapping and identification, disease diagnosis, and genetic engineering.

CHAPTER 2

3 Overall length and centromere position are but two factors required for homology. Most importantly, genetic content in non-homologous chromosomes is expected to be quite different. Other factors including banding pattern and time of replication during S phase would also be expected to vary among non-homologous chromosomes.

5 Since each chromosome in prophase is doubled (having gone through an S phase) and is visible at the end of prophase, there should be 32 chromatids. There should be 16 chromosomes moving to each pole.

11 **(a)** 8 tetrads
(b) 8 dyads
(c) 8 monads

13 **(a)** C^1, C^2, M^1, M^2, and S^1, S^2
(b) any one of 8 configurations will occur as homologous chromosomes synapse to produce daughter cells where each cell has one representative of each homologous chromosome (C^1, M^2, and S^1, for example).
(c) There will be eight different meiotic (haploid) products: C^1, M^2, and S^1; C^2, M^1, and S^2, *etc.*

CHAPTER 3

2 **(a)** The parents must both be heterozygous (*Aa*).
(b) The normal male *could* have either the *AA* or *Aa* genotype. The female must be *aa*.

4 *P* = checkered; *p* = plain. Checkered is tentatively assigned the dominant function because in a casual examination of the data, especially cross (b), we see that checkered types are more likely to be produced than plain types.
Cross (a): *PP* × *PP* or *PP* × *Pp*
Cross (b): *PP* × *pp*
Cross (c): The genotype of both parents is *pp*

F_1 Progeny of cross (b): *Pp*

6 Symbolism:
w = wrinkled seeds *g* = green cotyledons
W = round seeds *G* = yellow cotyledons
(a) *WwGG* × *WwGG* or *WwGG* × *WwGg*
(b) *wwGg* × *WwGg*

12 *E* = grey body color *V* = long wings
e = ebony body color *v* = vestigial wings
(a) P_1: *EEVV* × *eevv*
F_1: *EeVv* (grey, long)

F_2: *Phenotypes*	*Ratio*	*Genotypes*	*Ratio*
grey, long	9/16	*EEVV*	1/16
		EEVv	2/16
		EeVV	2/16
		EeVv	4/16
grey, vestigial	3/16	*EEvv*	1/16
		Eevv	2/16
ebony, long	3/16	*eeVV*	1/16
		eeVv	2/16
ebony, vestigial	1/16	*eevv*	1/16

(b) P_1: *EEvv* × *eeVV*
F_1: Results from this cross will be exactly the same as those in part (a) above.
F_2: same as (a) also.

(c) P_1: *EEVV* × *EEvv*
F_1: *EEVv* (grey, long)

F_2: *Phenotypes*	*Ratio*	*Genotypes*	*Ratio*
grey, long	3/4	*EEVV*	1/4
		EEVv	2/4
grey, vestigial	1/4	*EEvv*	1/4

15 F_1: *WwGg* × *wwgg*
1/4 *WwGg* (round, yellow)
1/4 *Wwgg* (round, green)
1/4 *wwGg* (round, yellow)
1/4 *wwgg* (round, green)

17 **(a)**

Expected ratio	*Observed (o)*	*Expected (e)*
9/16	315	312.75
3/16	108	104.25
3/16	101	104.25
1/16	32	34.75

$\chi^2 = 0.5$ with a probability greater than 0.90 for 3 degrees of freedom. The observed and expected values do not deviate significantly.

(b) and **(c)** it is easier to see the observed values for the monohybrid ratios if the phenotypes are listed:

smooth, yellow 315
smooth, green 108
wrinkled, yellow 101
wrinkled, green 32

For the smooth: wrinkled *monohybrid component,* the smooth types total 423 (315 + 108), while the wrinkled types total 133 (101 + 32).

Expected ratio	*Observed (o)*	*Expected (e)*
3/4	423	417
1/4	133	139

The χ^2 value is 0.35 and for 1 degree of freedom, the p value is greater than 0.50 and less than 0.90. We fail to reject the null hypothesis.

(c)

Expected ratio	*Observed (o)*	*Expected (e)*
3/4	416	417
1/4	140	139

The χ^2 value is 0.01 and for 1 degree of freedom, the p value is greater than 0.90. We fail to reject the null hypothesis.

21 For the test of a 3:1 ratio, the χ^2 value is 33.3 with an associated p value of less than 0.01 for 1 degree of freedom. For the test of a 1:1 ratio, the χ^2 value is 25.0 again with an associated p value of less than 0.01 for 1 degree of freedom. Based on these probability values, both null hypotheses are rejected.

CHAPTER 4

1 Incomplete dominance is the likely mode of inheritance.
Symbolism: *AA* = red, *aa* = white, *Aa* = roan
Results of crosses:

$AA \times AA \longrightarrow AA$
$aa \times aa \longrightarrow aa$
$AA \times aa \longrightarrow Aa$
$Aa \times Aa \longrightarrow$ 1/4 *AA*; 2/4 *Aa*; 1/4 *aa*

3 Individuals with blood type B can have the genotype I^BI^B or I^BI^o and those with blood type A, genotypes I^AI^A or I^AI^o.
Male Parent: must be I^BI^o.
Female Parent: must be I^AI^o.
Offspring: I^AI^B(AB), I^AI^o(A), I^BI^o(B), I^oI^o(O). The ratio would be 1:1:1:1.

5 **(a)** *RRPPDD* × *rrppdd* ⟶ *RrPpDd* (pink, personate, tall)
(b) 2/4 pink × 3/4 personate × 3/4 tall = 18/64

7 Notice that the distribution of observed offspring fits a 9:3:4 ratio quite well. This suggests that two independently assorting gene pairs with epistasis are involved.
A = pigment; *a* = pigmentless (colorless);
B = purple; *b* = red

AaBb × *AaBb* gives:
A–B– = purple
A–bb = red
aaB– = colorless
aabb = colorless

precursor ⟶ cyanidin ⟶ purple pigment
(colorless) *aa* (red) *bb*

8 **(a)** *AaBbCc* ⟶ gray (*C* allows pigment)
(b) *A–B–C–* ⟶ gray (*C* allows pigment)
(c) 16/32 albino; 9/32 gray; 3/32 yellow; 3/32 black; 1/32 cream

10 In this problem, each upper-case letter contributes 5 cm above 20 cm.
(a) Because there are four upper-case letters, the plant should be 40 cm high.
(b) Since the F_1 plant is AaBb (two upper-case letters), the plant should be 30 cm high. There would be 2n + 1 (n = number of heterozygous gene pairs) classes of F_2 plants with 0, 1, 2, 3, and 4 upper-case letters. The ratio is 1:4:6:4:1.
(c) The possibilities for a 25 cm plant are $A^1a^2b^1b^2$, $a^1A^2b^1b^2$, $a^1a^2B^1b^2$, and $a^1a^2b^1B^2$ and for a 35 cm plant, $a^1A^2B^1B^2$, $A^1a^2B^1B^2$, $A^1A^2b^1B^2$, and $A^1A^2B^1b^2$.

12 **(a)** 1/4 **(b)** 1/2 **(c)** 1/4 **(d)** zero

13 $1/2 \times 1 \times 1/4 = 1/8$

15 B = black coat color; b = yellow coat color

Female tortoise-shell = X^BX^b (mosaic)
Male black = X^BY

$X^BX^b \times X^BY \longrightarrow$

X^BX^B = female black
X^BX^b = female tortoise-shell
X^BY = male black
X^bY = male yellow

Normally there is no way for a tortoise-shell male to be produced, however with nondisjunction of the X chromosome in the female parent, thus producing a gamete containing the two X chromosomes, the coat color genes could be in the heterozygous state and mosaicism may result. The nondisjunctional event must have occurred in meiosis I. If such a female gamete is fertilized by a Y-bearing sperm, then an XXY tortoise-shell male could result.

19 **(a)** One can conclude that some phenotypic blending is occurring which is probably the result of several gene pairs acting in an additive fashion. Because the extreme phenotypes (6cm and 30cm) each represent 1/64 of the total, it is likely that there are three gene pairs in this cross. The genotypes of the parents would be combinations of alleles which would produce a 6cm (*aabbcc*) tail and a 30cm (*AABBCC*) tail while the 18cm offspring would have a genotype of *AaBbCc*.

(b) A mating of an 18 cm *AaBbCc* pig with the 6 cm *aabbcc* pig would result in the following offspring:

AaBbCc (18 cm)
AaBbcc (14 cm)
AabbCc (14 cm)
Aabbcc (10 cm)
aaBbCc (14 cm)
aaBbcc (10 cm)
aabbCc (10 cm)
aabbcc (6 cm)

In this case, a 1:3:3:1 ratio is the result.

CHAPTER 5

4 The genetic map would be as *follows:*

dp – *cl*———————*ap*
3 mu. 39 mu.

5 *RY/ry* × *ry/ry* with the two dominant genes on one chromosome and the two recessives on the homologue. Seeing that there are 20 crossover progeny among the 200, or 20/200, the map distance would be 10 map units (20/200 × 100 to convert to percentages) between the *R* and *Y* loci.

7 **(a)** P_1: *sc s v /sc s v* × + + +/Y
F_1: + + +/*sc s v* × *sc s v*/Y

(b) <u>*sc v s*</u>
+ + +

sc – *v* = 33% (map units)
v – *s* = 10% (map units)

(c) The coefficient of coincidence is 0.727, which indicates that there were fewer double [illegible] than expected, therefore positive chromosomal interference is present.

8 **(a)** P_1: *D* + +/+ + + × + *e p*/+ *e p*
F_1: *D* + +/+ *e p* × + *e p*/+ *e p*
F_2: *D* + +/+ *e p* Dichete
+ *e p*/+ *e p* ebony, pink
D e +/+ *e p* Dichete, ebony
+ + *p*/+ *e p* pink
D + *p*/+ *e p* Dichete, pink
+ *e* +/+ *e p* ebony
D e p/+ *e p* Dichete, ebony, pink
+ + +/+ *e p* wild type

(b) F_1: *D* + +/+ *p e* × + *p e*/+ *p e*
D – *p* = 3.0 map units
p – *e* = 18.5 map units

9 In the data given, the percentage of second division segregation is 20/100 or 20%. Dividing by 2 (because only two of the four chromatids are involved in any single crossover event) gives 10 map units.

12 Since *Stubble* is a dominant mutation (and homozygous lethal) one can determine whether it is heterozygous (*Sb*/+) or homozygous type (+/+). One would use the typical test cross arrangement with the *curled* gene so the arrangement would be: + *cu*/ + *cu*

13 **(a)** *y w* +/+ + *ct* × *y w* +/Y

(b) *y* – *w* = 1.5 map units and *w* – *ct* = 18.5 map units

y———*w*———————*ct*
0.0 1.5 20.0

(c) There were $.185 \times .015 \times 1000 = 2.775$ double crossovers expected.

(d) Because the cross to the F_1 males included the normal (wild type) gene for *cut* wings it would not be possible to unequivocally determine the genotypes from the F_2 phenotypes for all classes.

CHAPTER 6

4 A significant maternal age effect associated with Down syndrome indicates that nondisjunction in older females contributes disproportionately to the number of Down syndrome individuals. In addition, certain genetic and cytogenetic marker data indicate the influence of female nondisjunction.

7 Temperature shock applied during meiosis or colchicine applied during mitosis may lead to chromosome doubling. Colchicine interferes with spindle fiber formation.

9 If the diploid chromosome number is 18, 2n = 18, then in the somatic nuclei of
haploid individuals n = 9,
triploid (3n) = 27,
tetraploid (4n) = 36.

12 The minimum distance between loci *d* and *e* can be estimated as 10 map units. However, this is actually the distance from the *e* locus to the breakpoint which includes the inversion.

14 If the father had hemophilia it is likely that the Turner syndrome individual inherited the X chromosome from the father and no sex chromosomes from the mother. If nondisjunction occurred in the mother, either during meiosis I or meiosis II, an egg with no X chromosome can be the result.

16 In the trisomic, segregation will be "2 × 1" as illustrated below:

P_1: *b/b/b* × *B/B*

F_1: *B/b/b* (normal bristles) *and B/b* (normal bristles)

CHAPTER 7

5 The early evidence would be considered circumstantial in that at no time was there an experiment, like transformation in bacteria, in which genetic information in one organism was transferred to another using DNA. Rather, by comparing DNA content in various cell types (sperm and somatic cells) and observing that the action and absorption spectra of ultraviolet light were correlated, DNA was considered to be the genetic material. This suggestion was supported by the fact that DNA was shown to be the genetic material in bacteria and some phage. Direct evidence for DNA being the genetic material comes from a variety of observations including gene transfer which has been facilitated by recombinant DNA techniques.

6 Some viruses contain a genetic material composed of RNA. The tobacco mosaic virus is composed of an RNA core and a protein coat. Retroviruses contain RNA as the genetic material and use an enzyme known as reverse transcriptase to produce DNA which can be integrated into the host chromosome.

13 The reassociation of separate complementary strands of a nucleic acid, either DNA or RNA, is based on hydrogen bonds forming between A-T (or U) and G-C.

CHAPTER 8

2 After one round of replication in the ^{14}N medium the conservative scheme can be ruled out. After one round of replication in ^{14}N under a dispersive model, the DNA is of intermediate density, just as it is in the semiconservative model. However, in the next round of replication in ^{14}N medium, the density of the DNA is between the intermediate and "light" densities.

6 ϕX174 is a well-studied single-stranded virus (phage) which can be easily isolated. It has a relatively small DNA genome (5500 nucleotides) which, if mutated, usually alters its reproductive cycle.

7 The *polAI* mutation was instrumental in demonstrating that DNA polymerase I activity was not necessary for the *in vivo* replication of the *E. coli* chromosome. Such an observation opened the door for the discovery of other enzymes involved in DNA replication.

14 Gene conversion is likely to be a consequence of genetic recombination in which nonreciprocal recombination yields products in which it appears that one allele "converted" to another. Gene conversion is now considered a result of heteroduplex formation which is accompanied by mismatched bases. When these mismatches are corrected, the "conversion" occurs.

CHAPTER 9

2 **(a)** To determine the likelihood that each base will occur in each position of the codon (first, second, third), multiply the individual probabilities (fractions) for a final probability (fraction).

GGG = 3/4 × 3/4 × 3/4 = 27/64
GGC = 3/4 × 3/4 × 1/4 = 9/64
GCG = 3/4 × 1/4 × 3/4 = 9/64
CGG = 1/4 × 3/4 × 3/4 = 9/64
CCG = 1/4 × 1/4 × 3/4 = 3/64
CGC = 1/4 × 3/4 × 1/4 = 3/64
GCC = 3/4 × 1/4 × 1/4 = 3/64
CCC = 1/4 × 1/4 × 1/4 = 1/64

(b) Glycine: GGG and one G_2C (adds up to 36/64)
Alanine: one G_2C and one C_2G (adds up to 12/64)
Arginine: one G_2C and one C_2G (adds up to 12/64)
Proline: one C_2G and CCC (adds up to 4/64)

(c) With the wobble hypothesis, variation can occur in the third position of each codon.

Glycine: GGG, GGC
Alanine: CGG, GCC, CGC, GCG
Arginine: GCG, GCC, CGC, CGG
Proline: CCC, CCG

4 Given that AGG = arg, then information from the AG copolymer indicates that AGA also codes for arg and GAG must therefore code for glu. Coupling this information with that of the AAG copolymer, GAA must also code for glu, and AAG must code for lys.

15 **(b)** TCCGCGGCTGAGATGA (use complementary bases, substituting T for U)

(c) GCU

(d) Assuming that the AGG . . . is the 5′ end of the mRNA, then the sequence would be *arg-arg-arg-leu-tyr*

CHAPTER 10

2 **(a)** If one assumes that homozygosity for either or both of the two loci gives white, then let strain *A* = *aaBB* and strain *B* = *AAbb*. The F_1 is *AaBb* and pigmented (purple). The typical F_2 ratio would be as follows: 9/16 *A–B– Purple*; 3/16 *aaB–* white; 3/16 *A–bb* white; 1/16 *aabb* white

(b)

X —*aa*→ Y —*bb*→ Purple pigment
(white) (pink)

9/16 *A–B–* Purple
3/16 *aaB–* white
3/16 *A–bb* pink
1/16 *aabb* white

3 AA⟶IGP⟶I⟶TRY

4

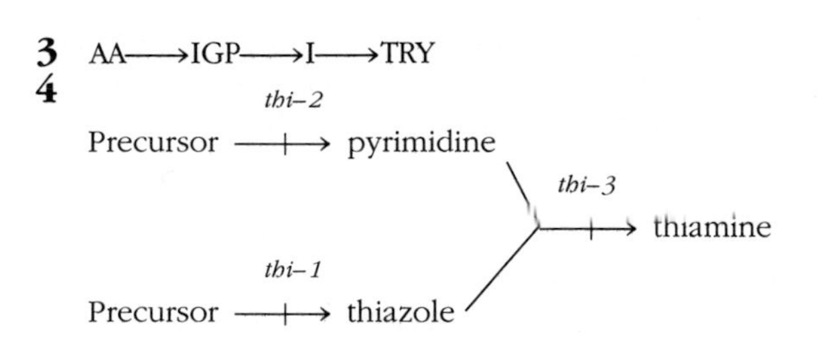

7 Colinearity refers to the sequential arrangement of subunits, amino acids and nitogenous bases in proteins and DNA, respectively. Sequencing of genes and products in MS2 phage and studies on mutations in the *A* subunit of the tryptophan synthetase gene indicate a colinear relationship.

CHAPTER 11

1 The term chromosomal mutation refers to changes in chromosome number or structure, such as duplications, deletions, inversions, and translocations. A gene mutation is a change in the nucleotide sequence of a single gene.

6 Frameshift mutations are likely to change more than one amino acid in a protein product because as the reading frame is shifted, different codons are generated. In addition, there is the possibility that a nonsense triplet could be introduced, thus causing premature chain termination. If a single pyrimidine or purine has been substituted, then only one amino acid is influenced.

9 *(XP1, XP2, XP3) (XP4) (XP5, XP6, XP7)*
The groupings (complementation groups) indicate that there are at least three "genes" which form products necessary for unscheduled DNA synthesis. All of the cell lines which are in the same complementation group are defective in the same product.

10 $2(5 \times 10^{-5})(1 \times 10^{5}) = 10$ Assuming 4.3×10^{9} individuals, there would be 4.3×10^{10} new mutations in the current populace.

CHAPTER 12

5 **(a)** 5×10^{6} nucleosomes
(b) 1×10^{6} solenoids.
(c) $9(5 \times 10^{6})$ histone molecules: 4.5×10^{7}.
(d) Since there are 10^{9} base pairs present and each base pair is 3.4 Å the overall length of the DNA is 3.4×10^{9} Å. Dividing this value by the packing ratio (50) gives 6.8×10^{7}.

6 The first part of this problem is to convert all of the given values to cubic Å remembering that 1 μm = 1000 Å. Using the formula πr^2 for the area of a circle and 4/3 πr^3 for the volume of a sphere, the following calculations apply:
Volume of DNA: $3.14 \times 10\,Å \times 10\,Å \times (50 \times 10^{4}\,Å) = 1.57 \times 10^{8}\,Å^{3}$
Volume of capsid: 4/3 $(3.14 \times 400\,Å \times 400\,Å \times 400\,Å) = 2.67 \times 10^{8}\,Å^{3}$
Because the capsid head has a greater volume than the volume of DNA, the DNA will fit into the capsid.

10 The C value paradox recognizes that with evolutionary divergence there has been a dramatic divergence in the amount of DNA among different taxa. It is likely that with the phylogenetic divergence there is not a commensurate requirement for *that many* different, additional genes. Indeed, some organisms which are quite closely related phylogenetically have vastly different DNA contents. So what is this "extra" DNA doing in various genomes? Recent advances in molecular biology have shown that much of this "extra" DNA is found in heterochromatic regions, introns, regions flanking genes, and intergenic spacer regions.

CHAPTER 13

2 Since both of the parents are *Dd*, the parent contributing the eggs must be *Dd*. Therefore, all of the offspring must have the phenotype of the mother's genotype, which is dextral.

4 **a)** green
b) white
c) variegated (patches of white and green)
d) green

5 As with any description of dominance, one looks to the phenotype of the diploid heterozygote. In this problem, the heterozygote is of normal phenotype, therefore the *petite* gene is recessive.

8 **(a)** There are many similarities among mitochondrial, chloroplast, and prokaryotic molecular systems. It is likely that mitochondria and chloroplasts evolved from bacteria in a symbiotic relationship; therefore it is not surprising that certain antibiotics which influence bacteria will also influence all mitochondria and chloroplasts. **(b)** Clearly, the mt^{+} strain is the donor of the cpDNA since the inheritance of resistance or sensitivity is dependent on the status of the mt^{+} gene.

9 Parents: *Dd* × *dd*
Offspring (F_1): 1/2 *Dd*, 1/2 *dd* (all dextral because of the maternal genotype)
Progeny (F_2): All those from *Dd* parents will be dextral while all of those from *dd* parents will be sinistral.

CHAPTER 14

3 **(a)** In an $F^{+} \times F^{-}$ cross, the transfer of the F factor produces a recipient bacterium which is F^{+}. Any gene may be transferred, and the frequency of transfer is relatively low. Crosses which are Hfr × F^{-} produce recombinants at a higher frequency than the $F^{+} \times F^{-}$ cross. The transfer is oriented (non-random) and the recipient cell remains F^{-}. **(b)** Bacteria which are F^{+} possess the F factor, while those that are F^{-} lack the F factor. In Hfr cells the F factor is integrated into the bacterial chromosome and in F′ bacteria, the F factor is free of the bacterial chromosome yet possesses a piece of the bacterial chromosome.

4 For each F type, the point of insertion and the direction of transfer are fixed, therefore breaking the conjugation tube at different times produces partial diploids with corresponding portions of the donor chromosome being transferred. The length of the chromosome being transferred is contingent on the duration of conjugation, thus mapping of genes is based on time.

5 *Hfr Strain Order*

1 TCHRO>>
2 HROM B>>
3 <<CHROM
4 M BAKT>>
5 <<BAKTC

Notice that all of the genes can be linked together to give a consistent map and that the ends overlap, indicating that the map is circular. The order is reversed in two of the crosses indicating the orientation of transfer is reversed.

10 Starting with a single bacteriophage, one lytic cycle produces 200 progeny phage, three more lytic cycles would produce $(200)^4$ or 1,600,000,000 phage.

11 **(a)** Culture #1 = 240 cells/ml × 10^9 (recall that only 0.1ml of the various dilutions was plated) which had been irradiated. Both *leu*$^+$ and *leu*$^-$ cells will grow. Culture #2 = 120 cells/ml × 10^2 which had been irradiated. Only *leu*$^+$ cells will grow.
Culture #3 = 120 cells/ml × 10^9 which had not been irradiated. Both *leu*$^+$ and *leu*$^-$ cells will grow.
Culture #4 = 30 cells/ml × 10^1 which had not been irradiated. Only *leu*$^+$ cells will grow. One would expect the values to be similar in cultures #1 and #3 because, with leucine added to the medium, one can not differentiate between *leu*$^+$ and *leu*$^-$ cells. However, it is likely that new non-leucine related nutritional mutations might be induced by the irradiation. If anything therefore, we might expect to see fewer colonies from culture #1 when compared to #3. The difference of 240 cells/ml × 10^9 and 120 cells/ml × 10^9 may be within the limits of experimental error.

(b) $(30 \times 10^1)/(120 \times 10^9) = 0.25 \times 10^{-8}$
The induced mutation rate would be calculated as follows:
$(120 \times 10^2)/(240 \times 10^9) = 0.5 \times 10^{-7}$

CHAPTER 15

2 Reverse transcriptase is often used to make cDNA (complementary DNA) from a mRNA molecule. Eukaryotic mRNAs typically have a 3′ poly A tail. The poly dT segment provides a double-stranded section which serves to prime the production of the complementary strand.

3 **(a)** Because the *Drosophila* DNA has been cloned into the *Eco R1* site in the ampicillin resistance gene of the plasmid, the gene will be mutated and any bacterium with the plasmid will be ampicillin sensitive. The tetracycline resistance gene remains active however. Bacteria which have been transformed with the recombinant plasmid will be resistant to tetracycline and therefore tetracycline should be added to the medium.

(b) Colonies which grow on a tetracycline medium should be tested for growth on an ampicillin medium. Those bacteria which do not grow on the ampicillin medium probably contain the *Drosophila* DNA insert.

(c) First, if cleavage with the *Eco R1* was incomplete, then no change in biological properties of the uncut plasmids would be expected. Also, it is possible that the cut ends of the plasmid were ligated together in the original form with no insert.

4 Apply the formula where *P* is the probability of recovering a given sequence and *f* represents the fraction of the genome present in each clone.

$$\begin{aligned} N &= \ln(1 - P)/\ln(1 - f) \\ &= \ln(1 - 0.99)/\ln(1 - [5 \times 10^3/1.5 \times 10^8]) \\ &= \ln(.01)/\ln(.9999667) \\ &= -4.605/-0.0000333 \\ &= 1.38 \times 10^5 \end{aligned}$$

5 When the insulin gene is "manufactured" it is made as a complementary copy of the mRNA, which is void of introns. Therefore the insulin mRNA which is made from the cDNA does not have introns.

CHAPTER 16

2 Under negative control, the regulatory molecule interferes with transcription while in positive control, the regulatory molecule stimulates transcription.

4 $i^+ o^+ z^+$ = Inducible
$i^- o^+ z^+$ = Constitutive
$i^+ o^c z^+$ = Constitutive
$i^- o^+ z^+ / F' i^+$ = Inducible
$i^+ o^c z^+ / F' o^+$ = Constitutive
$i^s o^+ z^+$ = Repressed
$i^s o^+ z^+ / F' i^+$ = Repressed

5 $i^+ o^+ z^+$ = No Enzyme Made.
$i^+ o^c z^-$ = Nonfunctional Enzyme Made.
$i^- o^+ z^-$ = Nonfunctional Enzyme Made.
$i^- o^+ z^-$ = Nonfunctional Enzyme Made.
$i^- o^+ z^+ / F' i^+$ = No Enzyme Made.
$i^+ o^c z^+ / F' o^+$ = Functional Enzyme Made.
$i^+ o^+ z^- / F' i^+ o^+ z^+$ = Functional and Nonfunctional Enzyme Made.
$i^- o^+ z^- / F' i^+ o^+ z^+$ = No Enzyme Made.
$i^s o^+ z^+ / F' o^+$ = No Enzyme Made.
$i^+ o^c z^- / F' o^+ z^+$ = Nonfunctional Enzyme Made and Functional Enzyme Made.

6 Gene *c* codes for the structural gene and *b* is the promoter. The *a* locus is the operator and the *d* locus is the repressor gene.

CHAPTER 17

3 10 V × 30 D × 50 J × 3 C = 45,000.

4 For the heavy chain 5 V × 10 D × 20 J = 1000, and for the light chain 10 V × 100 J = 1000. The final total would be 1000 × 1000 = 10^6.

7 The mother is Rh^- (*dd*). The father must be either *DD* or *Dd*, and each HDN child must be *Dd*.

CHAPTER 18

1 Determination refers to early developmental and regulatory events which set eventual patterns of gene activity. Differentiation follows determination and is the manifestation, in terms of genetic, physiological, and morphological changes, or the determined state.

9 A dominant gene (*TT* or *Tt*) providing taste of PTC.

CHAPTER 19

1 Assuming complete penetrance of the RB gene, the chance that a child born to this couple having RB is no higher than the frequency of sporadic occurrence. However, because the gene is 90% penetrant, there is a chance that the husband has the gene but does not express it. Therefore one would get a different answer when computing in the penetrance factor: 0.0225 or just over 2%. It is possible in some forms of RB to identify (by Southern blot) a defective or missing DNA segment. Otherwise, one might attempt to assay the RB product in cells to see if it is present and functional at normal levels.

3 The major regulatory points of the cell cycle include the following:

1. "Start" in late G1
2. A point in early S phase
3. The border between G2 and mitosis

8 Oncogenes are genes that induce or maintain uncontrolled cellular proliferation associated with cancer. They are mutant forms of proto-oncogenes which normally function to regulate cell division.

CHAPTER 20

5 These data indicate that genetics plays a major role in determining height. However, for weight, notice that MZ twins reared together have a much smaller (1.9 kg) difference than MZ twins reared apart indicating that the environment also has a considerable impact on weight.

6 F_1: v^+/v; $su\text{–}v^+/su\text{–}v$ (females, wild type)
v^+/Y; $su\text{–}v^+/su\text{–}v$ (males, wild type)
F_2: Overall, 1/2 of the offspring are wild type females, 5/16 are wild type males, and 3/16 are vermilion males.

7 **(a)** 140 cm
(b) 374.18
(c) 19.34.
(d) about 0.70.

8 **(a)** There are two alleles at each locus for a total of four alleles.
(b) yes
(c) Each gene provides an equal unit amount to the phenotype and the colors differ from each other in multiples of that unit amount.
(d) 1/16 = dark red = *AABB*
4/16 = medium-dark red = 2*AABb*
2*AaBB*
6/16 = medium red = *AAbb*
4*AaBb*
aaBB
4/16 = light red = 2*aaBb*
2*Aabb*
1/16 = white = *aabb*

9 (a) It is possible that two parents of moderate height can produce offspring that are much taller or shorter than either parent because segregation can produce a variety of gametes; therefore offspring as illustrated below:
rrSsTtuu × *RrSsTtUu*
(moderate) (moderate)
Offspring from this cross can range from very tall *RrSSTTUu* (14 units) to very short *rrssttuu* (8 units).
(b) If the individual with a minimum height, *rrssttuu*, is married to an individual of intermediate height *RrSsTtUu*, the offspring can be no taller than the height of the tallest parent.

10 The formula for estimating heritability is
$H^2 = V_G/V_P$ where V_G and V_P
are the genetic and phenotypic components of variation, respectively. The main issue in this question is obtaining some estimate of two components of phenotypic variation: genetic and environmental. V_P is the combination of genetic and environmental variance. Because the two parental strains are true-breeding, they are assumed to be homozygous and the variance of 3.124 and 3.876 considered to be the result of environmental influences. The average of these two values is 3.5. The F_1 is also genetically homogeneous and gives us an additional estimation of the environmental factors. By averaging with the parents
[(3.500 + 4.743)/2 = 4.123]
we obtain a relatively good idea of environmental impact on the phenotype. The phenotypic variance in the F_2 is the sum of the genetic (V_G) and environmental (V_E) components. We have estimated the environmental input as 4.123, so 47.708 minus 4.123, gives us an estimate of (V_G) which is 43.585. Heritability then becomes 43.585/47.708 or 0.914. This value, when viewed in percentage form indicates that about 91% of the variation in corolla length is due to genetic influences.

CHAPTER 21

1 $q = .544$
$p = .456$
Frequency of *Aa*=.208 or 20.8%
Frequency of *Aa*=.496 or 49.6%
Frequency of *aa*=.296 or 29.6%

2 Given that $q^2 = .04$, then $q = .2$, $2pq = .32$, and $p^2 = .64$ which is the frequency of heterozygotes in the population. Of those not expressing the trait, only a mating between heterozygotes can produce an offspring which expresses the trait, and then only at a frequency of 1/4. The different types of matings possible (those without the trait) in the population, with their frequencies, are given below:

AA × *AA*=.64 × .64 = .4096
AA × *Aa*=.64 × .32 = .2048
Aa × *AA*=.64 × .32 = .2048
Aa × *Aa*=.32 × .32 = .1024
Aa × *Aa*=.32 × .32 = .1024

Notice that of the matings of the individuals who do not express the trait, only the last two (about 20%) are capable of producing offspring with the trait. Therefore one would arrive at a final likelihood of 1/4 × 20% or 5% of the offspring with the trait.

3 The general equation for responding to this question is
$q_n = q_o/(1 + nq_o)$
where n = the number of generations, q_o = the initial gene frequency, and q_n = the new gene frequency.
(a) $q_n = .33$
(b) $q_n = .143$
(c) $q_n = .083$
(d) $q_n = .037$
(e) $q_n = .0098$
(f) $q_n = .00099$

7 horse, 3; pig, 2; dog, 3; chicken, 3; bullfrog, 2; fungus, 6.

GLOSSARY

A-DNA An alternate form of the right-handed double-helical structure of DNA in which the helix is more tightly coiled, with 11 base pairs per full turn of the helix. In this form, the bases in the helix are displaced laterally and tilted in relation to the longitudinal axis. It is not yet clear whether this form has biological significance.

abortive transduction An event in which transducing DNA fails to be incorporated into the recipient chromosome (See *transduction.*)

acentric chromosome Chromosome or chromosome fragment with no centromere.

acquired immunodeficiency syndrome (AIDS) An infectious disease caused by a retrovirus designated as human immunodeficiency virus (HIV). The disease is characterized by a gradual depletion of T lymphocytes, recurring fever, weight loss, multiple opportunistic infections, and rare forms of pneumonia and cancer associated with collapse of the immune system.

acrocentric chromosome Chromosome with the centromere located very close to one end. Human chromosomes 13, 14, 15, 21, and 22 are acrocentric.

active immunity Immunity gained by direct exposure to antigens followed by antibody production.

active site That portion of a protein, usually an enzyme, whose structural integrity is required for function (e.g., the substrate binding site of an enzyme).

adaptation A heritable component of the phenotype which confers an advantage in survival and reproductive success. The process by which organisms adapt to the current environmental conditions.

additive genes See *polygenic inheritance.*

albinism A condition caused by the lack of melanin production in the iris, hair, and skin. In humans, most often inherited as an autosomal recessive.

aleurone layer In seeds, the outer layer of the endosperm.

alkaptonuria An autosomal recessive condition in humans caused by the lack of an enzyme, homogentisic acid oxidase. Urine of homozygous individuals turns dark upon standing due to oxidation of excreted homogentisic acid. The cartilage of homozygous adults blackens from deposition of a pigment derived from homogentisic acid. Such individuals often develop arthritic conditions.

allele One of the possible mutational states of a gene, distinguished from other alleles by phenotypic effects.

allele frequency Measurement of the proportion of individuals in a population carrying a particular allele.

allelic exclusion In plasma cell heterozygous for an immunoglobulin gene, the selective action of only one allele.

allopatric speciation Process of speciation associated with geographic isolation.

allopolyploid Polyploid condition formed by the union of two or more distinct chromosome sets with a subsequent doubling of chromosome number.

allosteric effect Conformational change in the active site of a protein brought about by interaction with an effector molecule.

allotetraploid Diploid for two genomes derived from different species.

allozyme An allelic form of a protein that can be distinguished from other forms by electrophoresis.

alpha fetoprotein (AFP) A 70-kd glycoprotein synthesized in embryonic development by the yolk sac. High levels of this protein in the amniotic fluid are associated with neural tube defects such as spina bifida. Lower than normal levels may be associated with Down syndrome.

***Alu* sequence** An interspersed DNA sequence of approximately 300 bp found in the genome of primates that is cleaved by the restriction enzyme *Alu* I. *Alu* sequences are composed of a head to tail dimer, with the first monomer approximately 140 bp and the second approximately 170 bp. In humans, they are dispersed throughout the genome and are present in 300,000 to 600,000 copies, constituting some 3 to 6 percent of the genome. See *SINES.*

amber codon The codon UAG, which does not code for an amino acid but for chain termination.

Ames test An assay developed by Bruce Ames to detect mutagenic and carcinogenic compounds using reversion to histidine independence in the bacterium *Salmonella typhimurium.*

amino acid Any of the subunit building blocks that are covalently linked to form proteins.

aminoacyl tRNA Covalently linked combination of an amino acid and a tRNA molecule.

amniocentesis A procedure used to test for fetal defects in which fluid and fetal cells are withdrawn from the amniotic layer surrounding the fetus.

amphidiploid See *allotetraploid.*

anabolism The metabolic synthesis of complex molecules from less complex precursors.

analogue A chemical compound structurally similar to another, but differing by a single functional group (e.g., 5-bromodeoxyuridine is an analogue of thymidine).

anaphase Stage of cell division in which chromosomes begin moving to opposite poles of the cell.

aneuploidy A condition in which the chromosome number is not an exact multiple of the haploid set.

angstrom Unit of length equal to 10^{-10} meter. Abbreviated Å.

antibody Protein (immunoglobulin) produced in response to an antigenic stimulus with the capacity to bind specifically to the antigen.

anticodon The nucleotide triplet in a tRNA molecule which is complementary to, and binds to, the codon triplet in a mRNA molecule.

antigen A molecule, often a cell surface protein, that is capable of eliciting the formation of antibodies.

antiparallel Describing molecules in parallel alignment, but running in opposite directions. Most commonly used to describe the opposite orientations of the two strands of a DNA molecule.

apoenzyme The protein portion of an enzyme that requires a cofactor or prosthetic group to be functional.

ascospore A meiotic spore produced in certain fungi.

ascus In fungi, the sac enclosing the four or eight ascospores.

asexual reproduction Production of offspring in the absence of any sexual process.

assortative mating Nonrandom mating between males and females of a species. Selection of mates with the same genotype is positive; selection of mates with opposite genotypes is negative.

ATP Adenosine triphosphate.

attached-X chromosome Two conjoined X chromosomes that share a single centromere.

attenuator A nucleotide sequence between the promoter and the structural gene of some operons that can act to regulate the transit of RNA polymerase and thus control transcription of the structural gene.

autogamy A process of self-fertilization resulting in homozygosis.

autoimmune disease The production of antibodies that results from an immune response to one's own molecules, cells, or tissues. Such a response results from the inability of the immune system to distinguish self from nonself. Diseases such as arthritis, scleroderma, systemic lupus erythematosus, and perhaps diabetes are considered to be autoimmune diseases.

autopolyploidy Polyploid condition resulting from the replication of one diploid set of chromosomes.

autoradiography Production of a photographic image by radioactive decay. Used to localize radioactively labeled compounds within cells and tissues.

autosomes Chromosomes other than the sex chromosomes. In humans, there are 22 pairs of autosomes.

auxotroph A mutant microorganism or cell line which requires a substance for growth that can be synthesized by wild-type strains.

B-DNA See *double helix.*

back-cross A cross involving an F_1 heterozygote and one of the P_1 parents (or an organism with a genotype identical to one of the parents).

bacteriophage A virus that infects bacteria (synonym is *phage*).

bacteriostatic A compound that inhibits the growth of bacteria, but does not kill them.

balanced lethals Recessive, nonallelic lethal genes, each carried on different homologous chromosomes. When organisms carrying balanced lethal genes are interbred, only organisms with genotypes identical to the parents (heterozygotes) survive.

balanced polymorphism Genetic polymorphism maintained in a population by natural selection.

Barr body Densely staining nuclear mass seen in the somatic nuclei of mammalian females. Discovered by Murray Barr, this body is thought to represent an inactivated X chromosome.

base analogue See *analogue.*

base substitution A single base change in a DNA molecule that produces a mutation. There are two types of substitutions: *transitions,* in which a purine is substituted for a purine or a pyrimidine for a pyrimidine; and *transversions,* in which a purine is substituted for a pyrimidine, or vice versa.

biotechnology Commercial and/or industrial processes that utilize biological organisms or products.

bivalents Synapsed homologous chromosomes in the first prophase of meiosis.

BrdU (5-bromodeoxyuridine) A mutagenically active analogue of thymidine in which the methyl group at the 5′ position in thymine is replaced by bromine.

buoyant density A property of particles (and molecules) that depends upon their actual density, as determined by partial specific volume and degree of hydration. Provides the basis for density gradient separation of molecules or particles.

CAAT box A highly conserved DNA sequence found about 75 base pairs 5′ (upstream) to the initiation site of transcription in eukaryotic genes.

canonical sequence See *consensus sequence.*

cAMP Cyclic adenosine monophosphate. An important regulatory molecule in both prokaryotic and eukaryotic organisms.

CAP Catabolite activator protein. A protein that binds cAMP and regulates the activation of inducible operons.

carcinogen A physical or chemical agent that causes cancer.

carrier An individual heterozygous for a recessive trait.

cassette model First proposed to explain mating type interconversion in yeast, this model proposes that both genes for mating types, *a* and *alpha* are present as silent or unexpressed genes in transposable DNA segments (cassettes) that are activated (played) by transposition to the mating type locus.

catabolism A metabolic reaction in which complex molecules are broken down into simpler forms, often accompanied by the release of energy.

catabolite activator protein See *CAP.*

catabolite repression The selective inactivation of an operon by a metabolic product of the enzymes encoded by the operon.

***cdc* mutation** A class of mutations in yeast that affect the timing and progression through the cell cycle.

cDNA DNA synthesized from an RNA template by the enzyme reverse transcriptase.

cell cycle Sum of the phases of growth of an individual cell type; divided into G_1 (gap 1), S (DNA synthesis), G_2 (gap 2), and M (mitosis).

cell-free extract A preparation of the soluble fraction of cells, made by lysing cells and removing the particulate matter, such as nuclei, membranes, and organelles. Often used to carry out the synthesis of proteins by the addition of specific, exogenous mRNA molecules.

CEN In yeast, fragments of chromosomal DNA, about 120 bp in length, that when inserted into plasmids confer the ability to segregate during mitosis. These segments contain at least three types of sequence elements associated with centromere function.

centimeter A unit of length equal to 10^{-2} meter. Abbreviated cm.

centimorgan A unit of distance between genes on chromosomes. One centimorgan represents a value of 1 percent crossing over between two genes.

central dogma The concept that information flow progresses from DNA to RNA to proteins. Although exceptions are known, this idea is central to an understanding of gene function.

centric fusion See *Robertsonian translocation.*

centriole A cytoplasmic organelle composed of nine groups of microtubules, generally arranged in triplets. Centrioles function in the generation of cilia and flagella and serve as foci for the spindles in cell division.

centromere Specialized region of a chromosome to which the spindle fibers attach during cell division. Location of the centromere determines the shape of the chromosome during the anaphase portion of cell division. Also known as the primary constriction.

centrosome Region of the cytoplasm containing the centriole.

character An observable phenotypic attribute of an organism.

charon phage A group of genetically modified lambda phage designed to be used as vectors for cloning foreign DNA. Named after the ferryman in Greek mythology who carried the souls of the dead across the River Styx.

chemotaxis Negative or positive response to a chemical gradient.

chiasma (pl., chiasmata) The crossed strands of nonsister chromatids seen in diplotene of the first meiotic division. Regarded as the cytological evidence for exchange of chromosomal material, or crossing over.

chi-square (χ^2) analysis Statistical test to determine if an observed set of data fits a theoretical expectation.

chloroplast A cytoplasmic self-replicating organelle containing chlorophyll. The site of photosynthesis.

chorionic villus sampling (CVS) A technique of prenatal diagnosis that intravaginally retrieves fetal cells from the chorion and uses them to detect cytogenetic and biochemical defects in the embryo.

chromatid One of the longitudinal subunits of a replicated chromosome, joined to its sister chromatid at the centromere.

chromatin Term used to describe the complex of DNA, RNA, histones, and nonhistone proteins that make up chromosomes.

chromatography Technique for the separation of a mixture of solubilized molecules by their differential migration over a substrate.

chromocenter An aggregation of centromeres and heterochromatic elements of polytene chromosomes.

chromomere A coiled, beadlike region of a chromosome most easily visible during cell division. The aligned chromomeres of polytene chromosomes are responsible for their distinctive banding pattern.

chromosomal aberration Any change resulting in the duplication, deletion, or rearrangement of chromosomal material.

chromosomal mutation See *chromosomal aberration.*

chromosomal polymorphism Alternate structures or arrangements of a chromosome that are carried by members of a population.

chromosome In prokaryotes, an intact DNA molecule containing the genome; in eukaryotes, a DNA molecule complexed with RNA and proteins to form a threadlike structure containing genetic information arranged in a linear sequence.

chromosome banding Technique for the differential staining of mitotic or meiotic chromosomes to produce a characteristic banding pattern or selective staining of certain chromosomal regions such as centromeres, the nucleolus organizer regions, and GC- or AT-rich regions. Not to be confused with the banding pattern present in unstained polytene chromosomes, which is produced by the alignment of chromomeres.

chromosome map A diagram showing the location of genes on chromosomes.

chromosome puff A localized uncoiling and swelling in a polytene chromosome, usually regarded as a sign of active transcription.

***cis* configuration** The arrangement of two mutant sites within a gene on the same homologue, such as

$$\frac{a^1\, a^2}{+\ +}$$

Contrasts with a *trans* arrangement, where the mutant alleles are located on opposite homologues.

***cis* dominance** The ability of a gene to affect the expression of other genes adjacent to it on the chromosome.

***cis-trans* test** A genetic test to determine whether two mutations are located within the same cistron.

cistron That portion of a DNA molecule that codes for a single polypeptide chain; defined by a genetic test as a region within which two mutations cannot complement each other.

cline A gradient of genotype or phenotype distributed over a geographic range.

clonal selection Theory of the immune system that proposes that antibody diversity precedes exposure to the antigen, and that the antigen functions to select the cells containing its specific antibody to undergo proliferation.

clone Genetically identical cells or organisms all derived from a single ancestor by asexual or parasexual methods. For example, a DNA segment that has been enzymatically inserted into a plasmid or chromosome of a phage or a bacterium and replicated to form many copies.

cloned library A collection of cloned DNA molecules representing all or part of an individual's genome.

code See *genetic code.*

codominance Condition in which the phenotypic effects of a gene's alleles are fully and simultaneously expressed in the heterozygote.

codon A triplet of bases in a DNA or RNA molecule which specifies or encodes the information for a single amino acid.

coefficient of coincidence A ratio of the observed number of double-crossovers divided by the expected number of such crossovers.

coefficient of selection A measurement of the reproductive disadvantage of a given genotype in a population. If for genotype *aa,* only 99 of 100 individuals reproduce, then the selection coefficient (s) is 0.1.

colchicine An alkaloid compound that inhibits spindle formation during cell division. Used in the preparation of karyotypes to collect a large population of cells inhibited at the metaphase stage of mitosis.

colicin A bacteriocidal protein produced by certain strains of *E. coli* and other closely related bacterial species.

colinearity The linear relationship between the nucleotide sequence in a gene (or the RNA transcribed from it) and the order of amino acids in the polypeptide chain specified by the gene.

competence In bacteria, the transient state or condition during which the cell can bind and internalize exogenous DNA molecules, making transformation possible.

complementarity Chemical affinity between nitrogenous bases as a result of hydrogen bonding. Responsible for the base pairing between the strands of the DNA double helix.

complementation test A genetic test to determine whether two mutations occur within the same gene. If two mutations are introduced into a cell simultaneously and produce a wild-type phenotype (i.e., they complement each other), they are often nonallelic. If a mutant phenotype is produced, the mutations are noncomplementing and are often allelic.

complete linkage A condition in which two genes are located so close to each other that no recombination occurs between them.

complexity The total number of nucleotides or nucleotide pairs in a population of nucleic acid molecules as determined by reassociation kinetics.

complex locus A gene within which a set of functionally related pseudoalleles can be identified by recombinational analysis (e.g., the *bithorax* locus in *Drosophila*).

concatemer A chain or linear series of subunits linked together. The process of forming a concatemer is called concatenation (e.g., multiple units of a phage genome produced during replication).

concordance Pairs or groups of individuals identical in their phenotype. In twin studies, a condition in which both twins exhibit or fail to exhibit a trait under investigation.

conditional mutation A mutation that expresses a wild-type phenotype under certain (permissive) conditions and a mutant phenotype under other (restrictive) conditions.

conjugation Temporary fusion of two single-celled organisms for the sexual exchange of genetic material.

consanguine Related by a common ancestor within the previous few generations.

consensus sequence A nucleotide sequence most often found in a defined segment of DNA.

cosmid A vector designed to allow cloning of large segments of foreign DNA. Cosmids are hybrids composed of the cos sites of lambda inserted into a plasmid. In cloning, the recombinant DNA molecules are packaged into phage protein coats, and after infection of bacterial cells, the recombinant molecule replicates and can be maintained as a plasmid.

coupling conformation See *cis configuration.*

covalent bond A nonionic chemical bond formed by the sharing of electrons.

cri-du-chat syndrome A clinical syndrome in humans produced by a deletion of a portion of the short arm of chromosome 5. Afflicted infants have a distinctive cry which sounds like that of a cat.

crossing over The exchange of chromosomal material (parts of chromosomal arms) between homologous chromosomes by breakage and reunion. The exchange of material between nonsister chromatid during meiosis is the basis of genetic recombination.

cross-reacting material (CRM) Nonfunctional form of an enzyme, produced by a mutant gene, which is recognized by antibodies made against the normal enzyme.

C-terminal amino acid The terminal amino acid in a peptide chain which carries a free carboxyl group.

cytogenetics A branch of biology in which the techniques of both cytology and genetics are used to study heredity.

cytokinesis The diversion or separation of the cytoplasm during mitosis or meiosis.

cytological map A diagram showing the location of genes at particular chromosomal sites.

cytoplasmic inheritance Non-Mendelian form of inheritance involving genetic information transmitted by self-replicating cytoplasmic organelles such as mitochondria, chloroplasts, etc.

dalton A unit of mass equal to that of the hydrogen atom, which is 1.67×10^{-24} gram. A unit used in designating molecular weights.

Darwinian fitness See *fitness.*

deficiency (deletion) A chromosomal mutation involving the loss or deletion of chromosomal material.

degenerate code Term used to describe the genetic code, in which a given amino acid may be represented by more than one codon.

deletion See *deficiency.*

deme A local interbreeding population.

denatured DNA DNA molecules that have been separated into single strands.

de novo Newly arising; synthesized from less complex precursors rather than having been produced by modification of an existing molecule.

density gradient centrifugation A method of separating macromolecular mixtures by the use of centrifugal force and solvents of varying density. In sedimentation velocity centrifugation, macromolecules are separated by the velocity of sedimentation through a preformed gradient such as sucrose. In density gradient equilibrium centrifugation, macromolecules in a cesium salt solution are centrifuged until the cesium solution establishes a gradient under the influence of the centrifugal field, and the macromolecules sediment until the density of the solvent equals their own.

deoxyribonuclease A class of enzymes that breaks down DNA into oligonucleotide fragments by introducing single-stranded breaks into the double helix.

dermatoglyphics The study of the surface ridges of the skin, especially of the hands and feet.

deoxyribonucleic acid (DNA) A macromolecule usually consisting of antiparallel polynucleotide chains held together by hydrogen bonds, in which the sugar residues are deoxyribose. The primary carrier of genetic information.

determination A regulatory event that establishes a specific pattern of gene activity and developmental fate for a given cell.

diakinesis The final stage of meiotic prophase I in which the chromosomes become tightly coiled and compacted and move toward the periphery of the nucleus.

dicentric chromosome A chromosome having two centromeres.

differentiation The process of complex changes by which cells and tissues attain their adult structure and functional capacity.

dihybrid cross A genetic cross involving two characters in which the parents possess different forms of each character (e.g., tall, round × short, wrinkled peas).

diploid A condition in which each chromosome exists in pairs; having two of each chromosome.

diplotene A stage of meiotic prophase I immediately after pachytene. In diplotene, one pair of sister chromatids begins separating from the other, and chiasmata become visible. These overlaps move laterally toward the ends of the chromatids (terminalization).

directional selection A selective force that changes the frequency of an allele in a given direction, either toward fixation or toward elimination.

discontinuous variation Phenotypic data that fall into two or more distinct classes that do not overlap.

discordance In twin studies, a situation where one twin shows a trait but the other does not.

disjunction The separation of chromosomes at the anaphase stage of cell division.

disruptive selection Simultaneous selection for phenotypic extremes in a population, usually resulting in the production of two discontinuous strains.

dizygotic twins Twins produced from separate fertilization events; two ova fertilized independently. Also known as fraternal twins.

DNA See *deoxyribonucleic acid.*

DNA footprinting See *footprinting.*

DNA gyrase One of the DNA topoisomerases that functions during DNA replication to reduce molecular tension caused by supercoiling. DNA gyrase produces, then seals, double-stranded breaks.

DNA ligase An enzyme that forms a covalent bond between the 5′ end of one polynucleotide chain and the 3′ end of another polynucleotide chain. Also called polynucleotide-joining enzyme.

DNA polymerase An enzyme that catalyzes the synthesis of DNA from deoxyribonucleotides and a template DNA molecule.

DNase Deoxyribonuclease, a class of enzymes that degrades or breaks down DNA into fragments or constitutive nucleotides.

dominance The expression of a trait in the heterozygous condition.

dosage compensation A genetic mechanism that regulates the levels of gene products at certain autosomal loci; this results in homozygous dominants and heterozygotes having the same amount of a gene product. In mammals, random inactivation of one X chromosome in females leads to equal levels of X chromosome-coded gene products in males and females.

double-crossover Two separate events of chromosome breakage and exchange occurring within the same tetrad.
double helix The model for DNA structure proposed by James Watson and Francis Crick, involving two antiparallel, hydrogen-bonded polynucleotide chains wound into a right-handed helical configuration, with 10 base pairs per full turn of the double helix. Often called B-DNA.
duplication A chromosomal aberration in which a segment of the chromosome is repeated.
dyad The products of tetrad separation or disjunction at the first meiotic prophase. Consists of two sister chromatids joined at the centromere.

effector molecule Small, biologically active molecule that acts to regulate the activity of a protein by binding to a specific receptor site on the protein.
electrophoresis A technique used to separate a mixture of molecules by their differential migration through a stationary phase in an electrical field.
endogenote The segment of the chromosome in a partially diploid bacterial cell (merozygote) which is homologous to the chromosome transmitted by the donor cell.
endomitosis Chromosomal replication that is not accompanied by either nuclear or cytoplasmic division.
endonuclease An enzyme that hydrolyzes internal phosphodiester bonds in a polynucleotide chain or nucleic acid molecule.
endoplasmic reticulum A membranous organelle system in the cytoplasm of eukaryotic cells. The outer surface of the membranes may be ribosome-studded (rough ER) or not (smooth ER).
endopolyploidy The increase in chromosome sets that results from endomitotic replication within somatic nuclei.
endosymbiont theory The proposal that self-replicating cellular organelles such as mitochondria and chloroplasts were originally free-living organisms that entered into a symbiotic relationship with nucleated cells.
enhancer Originally, a 72-bp sequence in the genome of the virus, SV40, that increases the transcriptional activity of nearby structural genes. Similar sequences that enhance transcription have been identified in the genomes of eukaryotic cells. Enhancers can act over a distance of thousands of base pairs and can be located 5′ or 3′ to the gene they affect, and thus are different from promoters.
enhanson The DNA sequence that represents the core sequence of an enhancer.
environment The complex of geographic, climatic, and biotic factors within which an organism lives.
enzyme A protein or complex of proteins that catalyzes a specific biochemical reaction.
epigenesis The idea that an organism develops by the appearance and growth of new structures. Opposed to preformationism, which holds that development is the growth of structures already present in the egg.
episome A circular genetic element in bacterial cells that can replicate independently of the bacterial chromosome or integrate and replicate as part of the chromosome.
epitope That portion of a macromolecule or cell that acts to elicit an antibody response; an antigenic determinant. A complex molecule or cell can contain several such sites.
epistasis Nonreciprocal interaction between genes such that one gene interferes with or prevents the expression of another gene. For example, in *Drosophila,* the recessive gene *eyeless,* when homozygous, prevents the expression of eye color genes.
equatorial plate See *metaphase plate.*
euchromatin Chromatin or chromosomal regions that are lightly staining and are relatively uncoiled during the interphase portion of the cell cycle. The region of the chromosomes thought to contain most of the structural genes.
eukaryotes Those organisms having true nuclei and membranous organelles and whose cells demonstrate mitosis and meiosis.
euploid Polyploid with a chromosome number that is an exact multiple of a basic chromosome set.
evolution The origin of plants and animals from preexisting types. Descent with modifications.
excision repair Repair of DNA lesions by removal of a polynucleotide segment and its replacement with a newly synthesized, corrected segment.
exogenote In merozygotes, the segment of the bacterial chromosome contributed by the donor cell.
exon The DNA segment(s) of a gene that are transcribed and translated into protein.
exonuclease An enzyme that breaks down nucleic acid molecules by breaking the phosphodiester bonds at the 3′ or 5′ terminal nucleotides.,
expression vector Plasmids or phage carrying promoter regions designed to cause expression of cloned DNA sequences.
expressivity The degree or range in which a phenotype for a given trait is expressed.
extranuclear inheritance Transmission of traits by genetic information contained in cytoplasmic organelles such as mitochondria and chloroplasts.

F^+ cell A bacterial cell having a fertility (F) factor. Acts as a donor in bacterial conjugation.
F^- cell A bacterial cell that does not contain a fertility (F) factor. Acts as a recipient in bacterial conjugation.
F factor An episome in bacterial cells that confers the ability to act as a donor in conjugation.
F′ factor A fertility (F) factor that contains a portion of the bacterial chromosome.
F_1 generation First filial generation; the progeny resulting from the first cross in a series.
F_2 generation Second filial generation; the progeny resulting from a cross of the F_1 generation.
F pilus See *pilus.*
facultative heterochromatin Chromatin that may alternate in form between euchromatic and heterochromatic. The Y chromosome of many species contains facultative heterochromatin.
familial trait A trait transmitted through and expressed by members of a family.

fate map A diagram or "map" of an embryo showing the location of cells whose developmental fate is known.

fertility (F) factor See *F factor.*

filial generations See *F_1, F_2 generations.*

fingerprint The pattern of ridges and whorls on the tip of a finger. The pattern obtained by enzymatically cleaving a protein or nucleic acid and subjecting the digest to two-dimensional chromatography or electrophoresis.

fitness A measure of the relative survival and reproductive success of a given individual or genotype.

fixation In population genetics, a condition in which all members of a population are homozygous for a given allele.

fixity of species The idea that members of a species can give rise only to other members of the species, thus implying that all species are independently created.

fluctuation test A statistical test developed by Salvadore Luria and Max Delbruck to determine whether bacterial mutations arise spontaneously or are produced in response to selective agents.

fMet See *formylmethionine.*

footprinting A technique for identifying a DNA sequence that binds to a particular protein, based on the idea that the phosphodiester bonds in the region covered by the protein are protected from digestion by deoxyribonucleases.

formylmethionine (fMet) A molecule derived from the amino acid methionine by attachment of a formyl group to its terminal amino group. This is the first amino acid inserted in all bacterial polypeptides. Also known as N-formyl methionine.

founder effect A form of genetic drift. The establishment of a population by a small number of individuals whose genotypes carry only a fraction of the different kinds of alleles in the parental population.

fragile site A heritable gap or nonstaining region of a chromosome that can be induced to generate chromosome breaks.

frameshift mutation A mutational event leading to the insertion of one or more base pairs in a gene, shifting the codon reading frame in all codons following the mutational site.

fraternal twins See *dizygotic twins.*

gamete A specialized reproductive cell with a haploid number of chromosomes.

gene The fundamental physical unit of heredity whose existence can be confirmed by allelic variants and which occupies a specific chromosomal locus. A DNA sequence coding for a single polypeptide.

gene amplification The process by which gene sequences are selected for differential replication either extrachromosomally or intrachromosomally.

gene conversion A directional process that changes one allele into another specific allele. In fungi, this process involves meiosis and a recombinational step, but in other organisms, such as trypanosomes, meiosis may not be required.

gene duplication An event in replication leading to the production of a tandem repeat of a gene sequence.

gene flow The gradual exchange of genes between two populations, brought about by the dispersal of gametes or the migration of individuals.

gene frequency The percentage of alleles of a given type in a population.

gene interaction Production of novel phenotypes by the interaction of alleles of different genes.

gene mutation See *point mutation.*

gene pool The total of all genes possessed by reproductive members of a population.

generalized transduction The transduction of any gene in the bacterial genome by a phage.

genetic burden Average number of recessive lethal genes carried in the heterozygous condition by an individual in a population. Also called genetic load.

genetic code The nucleotide triplets that code for the 20 amino acids or for chain initiation or termination.

genetic counseling Analysis of risk for genetic defects in a family and the presentation of options available to avoid or ameliorate possible risks.

genetic drift Random variation in gene frequency from generation to generation. Most often observed in small populations.

genetic engineering The technique of altering the genetic constitution of cells or individuals by the selective removal, insertion, or modification of individual genes or gene sets.

genetic equilibrium Maintenance of allele frequencies at the same value in successive generations. A condition in which allele frequencies are neither increasing nor decreasing.

genetic fine structure Intragenic recombinational analysis that provides mapping information at the level of individual nucleotides.

genetic load See *genetic burden.*

genetic polymorphism The stable coexistence of two or more discontinuous genotypes in a population. When the frequencies of two alleles are carried to an equilibrium, the condition is called balanced polymorphism.

genetics The branch of biology that deals with heredity and the expression of inherited traits.

genome The array of genes carried by an individual.

genotype The specific allelic or genetic constitution of an organism; often, the allelic composition of one or a limited number of genes under investigation.

Goldberg-Hogness box A short nucleotide sequence 20 to 30 bp 5′ to the initiation site of eukaryotic genes to which RNA polymerase II binds. The consensus sequence is TATAAAA. Also known as TATA box.

graft versus host disease (GVHD) In transplants, reaction by immunologically competent cells of the donor against the antigens present on the cells of the host. In human bone marrow transplants, often a fatal condition.

gynandromorph An individual composed of cells with both male and female genotypes.

gyrase One of a class of enzymes known as topoisomerases. Gyrase converts closed circular DNA to a negatively supercoiled form prior to replication, transcription, or recombination.

H substance The carbohydrate group present on the surface of red blood cells. When unmodified, it results in blood type O; when modified by the addition of monosaccharides, it results in type A, B, and AB.

haploid A cell or organism having a single set of unpaired chromosomes. The gametic chromosome number.

haplotype The set of alleles from closely linked loci carried by an individual and usually inherited as a unit.

Hardy-Weinberg law The principle that both gene and genotype frequencies will remain in equilibrium in an infinitely large population in the absence of mutation, migration, selection, and nonrandom mating.

heat shock A transient response following exposure of cells or organisms to elevated temperatures. The response involves activation of a small number of loci, inactivation of previously active loci, and selective translation of heat shock mRNA. Appears to be a nearly universal phenomenon observed in organisms ranging from bacteria to humans.

helicase An enzyme that participates in DNA replication by unwinding the double helix near the replication fork.

hemizygous Conditions where a gene is present in a single dose. Usually applied to genes on the X chromosome in heterogametic males.

hemoglobin (Hb) An iron-containing, conjugated respiratory protein occurring chiefly in the red blood cells of vertebrates.

hemophilia A sex-linked trait in humans associated with defective blood-clotting mechanisms.

heredity Transmission of traits from one generation to another.

heritability A measure of the degree to which observed phenotypic differences for a trait are genetic.

heterochromatin The heavily staining, late replicating regions of chromosomes that are prematurely condensed in interphase.

heteroduplex A double-stranded nucleic acid molecule in which each polynucleotide chain has a different origin. These structures may be produced as intermediates in a recombinational event, or by the *in vitro* reannealing of single-stranded, complementary molecules.

heterogametic sex The sex that produces gametes containing unlike sex chromosomes.

heterogenote A bacterial merozygote in which the donor (exogenote) chromosome segment carries different alleles than does the chromosome of the recipient (endogenote). A heterozygous merozygote.

heterogeneous nuclear RNA (hnRNA) The collection of RNA transcripts in the nucleus, representing precursors and processing intermediates to rRNA, mRNA, and tRNA. Also represents RNA transcripts that will not be transported to the cytoplasm, such as snRNA (small nuclear RNAs).

heterokaryon A somatic cell containing nuclei from two different sources.

heterosis The superiority of a heterozygote over either homozygote for a given trait.

heterozygote An individual with different alleles at one or more loci. Such individuals will produce unlike gametes and therefore will not breed true.

Hfr A strain of bacteria exhibiting a high frequency of recombination. These stains have a chromosomally integrated F factor that is able to mobilize and transfer all or part of the chromosome to a recipient F^- cell.

histocompatibility antigens See *HLA.*

histones Proteins complexed with DNA in the nucleus. They are rich in the basic amino acids arginine and lysine and function in the coiling of DNA to form nucleosomes.

HLA Cell surface proteins, produced by histocompatibility loci, which are involved in the acceptance or rejection of tissue and organ grafts and transplants.

hnRNA See *heterogeneous nuclear RNA.*

holandric A trait transmitted from males to males. In humans, genes on the Y chromosome are holandric.

homeotic mutation A mutation that causes a tissue normally determined to form a specific organ or body part to alter its differentiation and form another structure. Alternately spelled: homoeotic.

homogametic sex The sex that produces gametes that do not differ with respect to sex chromosome content; in mammals, the female is homogametic.

homogeneously staining regions (hsr) Segments of mammalian chromosomes that stain lightly with Giemsa following exposure of cells to a selective agent. These regions arise in conjunction with gene amplification and are regarded as the structural locus for the amplified gene.

homogenote A bacterial merozygote in which the donor (exogenote) chromosome carries the same alleles as the chromosome of the recipient (endogenote). A homozygous merozygote.

homologous chromosomes Chromosomes that synapse or pair during meiosis. Chromosomes that are identical with respect to their genetic loci and centromere placement.

homozygote An individual with identical alleles at one or more loci. Such individuals will produce identical gametes and will therefore breed true.

homunculus The miniature individual imagined by preformationists to be contained within the sperm or egg.

human immunodeficiency virus (HIV) A human retrovirus associated with the onset and progression of acquired immunodeficiency syndrome (AIDS).

hybrid An individual produced by crossing two parents of different genotypes.

hybridoma A somatic cell hybrid produced by the fusion of an antibody-producing cell and a cancer cell, specifically, a myeloma. The cancer cell contributes the ability to divide indefinitely, and the antibody cell confers the ability to synthesize large amounts of a single antibody.

hybrid vigor See *heterosis.*

hydrogen bond An electrostatic attraction between a hydrogen atom bonded to a strongly electronegative atom such as oxygen or nitrogen and another atom that is electronegative or contains an unshared electron pair.

identical twins See *monozygotic twins.*

Ig See *immunoglobulin.*

imaginal disk Discrete groups of cells set aside during embryogenesis in holometabolous insects which are determined to form the external body parts of the adult.

immunoglobulin The class of serum proteins having the properties of antibodies.

inborn error of metabolism A biochemical disorder that is genetically controlled; usually an enzyme defect that produces a clinical syndrome.

inbreeding Mating between closely related organisms.

inbreeding depression A loss or reduction in fitness that usually accompanies inbreeding of organisms.

incomplete dominance Expression of heterozygous phenotype which is distinct from, and often intermediate to, that of either parent.

incomplete linkage Occasional separation of two genes on the same chromosome by a recombinational event.

independent assortment The independent behavior of each pair of homologous chromosomes during their segregation in meiosis I. The random distribution of genes on different chromosomes into gametes.

inducer An effector molecule that activates transcription.

inducible enzyme system An enzyme system under the control of a regulatory molecule, or inducer, which acts to block a repressor and allow transcription.

initiation codon The triplet of nucleotides (AUG) in an mRNA molecule that codes for the insertion of the amino acid methionine as the first amino acid in a polypeptide chain.

insertion sequence See *IS element.*

***in situ* hybridization** A technique for the cytological localization of DNA sequences complementary to a given nucleic acid or polynucleotide.

intercalating agent A compound that inserts between bases in a DNA molecule, disrupting the alignment and pairing of bases in the complementary strands (e.g., acridine dyes).

interference A measure of the degree to which one crossover affects the incidence of another crossover in an adjacent region of the same chromatid. Positive interference increases the chances of another crossover; negative interference reduces the probability of a second crossover event.

interferon One of a family of proteins that act to inhibit viral replication in higher organisms. Some interferons may have anticancer properties.

interphase That portion of the cell cycle between divisions.

intervening sequence See *intron.*

intron A portion of DNA between coding regions in a gene which is transcribed, but which does not appear in the mRNA product.

inversion A chromosomal aberration in which the order of a chromosomal segment has been reversed.

inversion loop The chromosomal configuration resulting from the synapsis of homologous chromosomes, one of which carries an inversion.

in vitro Literally, in glass; outside the living organism; occurring in an artificial environment.

in vivo Literally, in the living; occurring within the living body of an organism.

IS element A mobile DNA segment that is transposable to any of a number of sites in the genome.

isoagglutinogen An antigenic factor or substance present on the surface of cells that is capable of inducing the formation of an antibody.

isochromosome An aberrant chromosome with two identical arms and homologous loci.

isolating mechanism Any barrier to the exchange of genes between different populations of a group of organisms. In general, isolation can be classified as spatial, environmental, or reproductive.

isotopes Forms of a chemical element that have the same number of protons and electrons, but differ in the number of neurons contained in the atomic nucleus. Unstable isotopes undergo a transition to a more stable form with the release of radioactivity.

isozyme Any of two or more distinct forms of an enzyme that have identical or nearly identical chemical properties, but differ in some property such as net electrical charge, pH optima, number and type of subunits, or substrate concentration.

kappa particles DNA-containing cytoplasmic particles found in certain strains of *Paramecium aurelia.* When these self-reproducing particles are transferred into the growth medium, they release a toxin, paramecin, which kills other sensitive strains. A nuclear gene, *K*, is responsible for the maintenance of kappa particles in the cytoplasm.

karyokinesis The process of nuclear division.

karyotype The chromosome complement of a cell or an individual. Often used to refer to the arrangement of metaphase chromosomes in a sequence according to length and position of the centromere.

kilobase A unit of length consisting of 1000 nucleotides. Abbreviated kb.

kinetochore A fibrous structure with a size of about 400 mn, located within the centromere. It appears to be the location to which microtubules attach.

Klenow fragment A part of bacterial DNA polymerase that lacks exonuclease activity, but retains polymerase activity. It is produced by enzymatic digestion of the intact enzyme.

Klinefelter syndrome A genetic disease in human males caused by the presence of an extra X chromosome. Klinefelter males are XXY instead of XY. This syndrome is associated with enlarged breasts, small testes, sterility, and, occasionally, mental retardation.

lagging strand In DNA replication, the strand synthesized in a discontinuous fashion, 5′ to 3′ away from the replication fork. Each short piece of DNA synthesized in this fashion is called an Okazaki fragment.

lampbrush chromosomes Meiotic chromosomes characterized by extended lateral loops, which reach maximum extension during diplotene. Although most intensively studied in amphibians, these structures occur in meiotic cells of organisms ranging from insects through humans.

leader sequence That portion of an mRNA molecule from the 5′ end to the beginning codon; may contain regulatory or ribosome binding sites.

leading strand During DNA replication, the strand synthesized continuously 5′ to 3′ toward the replication fork.

lectins Carbohydrate-binding proteins found in the seeds of leguminous plants such as soybeans. Although their physiological role in plants is unclear, they are useful as probes for cell surface carbohydrates and glycoproteins in animal cells. Proteins with similar properties have also been isolated from organisms such as snails and eels.

leptotene The initial stage of meiotic prophase I, during which the chromosomes become visible and are often arranged in a bouquet configuration, with one or both ends of the chromosomes gathered at one spot on the inner nuclear membrane.

lethal gene A gene whose expression results in death.

leucine zipper A structural motif in a DNA binding protein that is characterized by a stretch of leucine residues spaced at every seventh amino acid residue, with adjacent regions of positively charged amino acids. Leucine zippers on two polypeptides may interact to form a dimer that binds to DNA.

LINES Long interspersed repetitive sequences found in the genomes of higher organisms, such as the 6-kb KpnI sequences found in primate genomes.

linkage Condition in which two or more nonallelic genes tend to be inherited together. Linked genes have their loci along the same chromosome, do not assort independently, but can be separated by crossing over.

linkage group A group of genes that have their loci on the same chromosome.

linking number The number of times that two strands of a closed, circular DNA duplex cross over each other.

locus The site or place on a chromosome where a particular gene is located.

long period interspersion Pattern of genome organization in which long stretches of single copy DNA are interspersed with long segments of repetitive DNA. This pattern of genome organization is found in *Drosophila* and the honeybee.

long terminal repeat (LTR) Sequence of several hundred base pairs found at the ends of retroviral DNAs.

Lutheran blood group One of a number of blood group systems inherited independently of the ABO, MN, and Rh systems. Alleles of this group determine the presence or absence of antigens on the surface of red blood cells. Gene is on human chromosome 19.

Lyon hypothesis The idea proposed by Mary Lyon that random inactivation of one X chromosome in the somatic cells of mammalian females is responsible for dosage compensation and mosaicism.

lysis The disintegration of a cell brought about by the rupture of its membrane.

lysogenic bacterium A bacterial cell carrying a temperate bacteriophage integrated into its chromosome.

lysogeny The process by which the DNA of an infecting phage becomes repressed and integrated into the chromosome of the bacterial cell it infects.

lytic phase The condition in which a temperate bacteriophage loses its integrated status in the host chromosome (becomes induced), replicates, and lyses the bacterial cell.

major histocompatibility loci See *MHC.*

map unit A measure of the genetic distance between two genes, corresponding to a recombination frequency of 1 percent. See *centimorgan.*

maternal effect Phenotypic effects on the offspring produced by the maternal genome. Factors transmitted through the egg cytoplasm which produce a phenotypic effect in the progeny.

maternal influence See *maternal effect.*

maternal inheritance The transmission of traits via cytoplasmic genetic factors such as mitochondria or chloroplasts.

mean The arithmetic average.

median The value in a group of numbers below and above which there is an equal number of data points or measurements.

meiosis The process in gametogenesis or sporogenesis during which one replication of the chromosomes is followed by two nuclear divisions to produce four haploid cells.

melting profile See T_m.

merozygote A partially diploid bacterial cell containing, in addition to its own chromosome, a chromosome fragment introduced into the cell by transformation, transduction, or conjugation.

messenger RNA See *mRNA.*

metabolism The sum of chemical changes in living organisms by which energy is generated and used.

metacentric chromosome A chromosome with a centrally located centromere, producing chromosome arms of equal lengths.

metafemale In *Drosophila,* a poorly developed female of low viability in which the ratio of X chromosomes to sets of autosomes exceeds 1.0. Previously called a superfemale.

metamale In *Drosophila,* a poorly developed male of low viability in which the ratio of X chromosomes to sets of autosomes is less than 0.5. Previously called a supermale.

metaphase The stage of cell division in which the condensed chromosomes lie in a central plane between the two poles of the cell, and in which the chromosomes become attached to the spindle fibers.

metaphase plate The arrangement of mitotic or meiotic chromosomes at the equator of the cell during metaphase.

MHC Major histocompatibility loci. In humans, the HLA complex; and in mice, the H2 complex.

micrometer A unit of length equal to 1×10^{-6} meter. Previously called a micron. Abbreviated μm.

micron See *micrometer.*

migration coefficient An expression of the proportion of migrant genes entering the population per generation.

millimeter A unit of length equal to 1×10^{-3} meter. Abbreviated mm.

minimal medium A medium containing only those nutrients that will support the growth and reproduction of wild-type strains of an organism.

missense mutation A mutation that alters a codon to that of another amino acid, causing an altered translation product to be made.

mitochondrion Found in the cells of eukaryotes, a cytoplasmic, self-reproducing organelle that is the site of ATP synthesis.

mitogen A substance that stimulates mitosis in nondividing cells; e.g., phytohemagglutinin.

mitosis A form of cell division resulting in the production of two cells, each with the same chromosome and genetic complement as the parent cell.

mode In a set of data, the value occurring in the greatest frequency.

monohybrid cross A genetic cross between two individuals involving only one character (e.g., $AA \times aa$).

monosomic An aneuploid condition in which one member of a chromosome pair is missing; having a chromosome number of $2n - 1$.

monozygotic twins Twins produced from a single fertilization event; the first division of the zygote produces two cells, each of which develops into an embryo. Also known as identical twins.

mRNA An RNA molecule transcribed from DNA and translated into the amino acid sequence of a polypeptide.

mtDNA Mitochondrial DNA.

multiple alleles Three or more alleles of the same gene.

multiple-factor inheritance See *polygenic inheritance.*

multiple infection Simultaneous infection of a bacterial cell by more than one bacteriophage, often of different genotypes.

mu phage A phage group in which the genetic material behaves like an insertion sequence, capable of insertion, excision, transposition, inactivation of host genes, and induction of chromosomal rearrangements.

mutagen Any agent that causes an increase in the rate of mutation.

mutant A cell or organism carrying an altered or mutant gene.

mutation The process which produces an alteration in DNA or chromosome structure; the source of most alleles.

mutation rate The frequency with which mutations take place at a given locus or in a population.

muton the smallest unit of mutation in a gene, corresponding to a single base change.

nanometer A unit of length equal to 1×10^{-9} meter. Abbreviated nm.

natural selection Differential reproduction of some members of a species resulting from variable fitness conferred by genotypic differences.

nearest-neighbor analysis A molecular technique used to determine the frequency with which nucleotides are adjacent to each other in polynucleotide chains.

neutral mutation A mutation with no immediate adaptive significance or phenotypic effect.

noncrossover gamete A gamete which contains no chromosomes that have undergone genetic recombination.

nondisjunction An accident of cell division in which the homologous chromosomes (in meiosis) or the sister chromatids (in mitosis) fail to separate and migrate to opposite poles; responsible for defects such as monosomy and trisomy.

nonsense codon The nucleotide triplet in an mRNA molecule that signals the termination of translation. Three such codons are known: UGA, UAG, and UAA.

nonsense mutation A mutation that alters a codon to one which encodes no amino acid; i.e., UAG (amber codon), UAA (ochre codon), or UGA (opal codon). Leads to premature termination during the translation of mRNA.

NOR See *nucleolar organizer region.*

normal distribution A probability function that approximates the distribution of random variables. The normal curve, also known as a Gaussian or bell-shaped curve, is the graphic display of the normal distribution.

np See *nucleotide pair.*

N-terminal amino acid The terminal acid in a peptide chain that carries a free amino group.

nu body See *nucleosome.*

nuclease An enzyme that breaks bonds in nucleic acid molecules.

nucleoid The DNA-containing region within the cytoplasm in prokaryotic cells.

nucleolar organizer region (NOR) A chromosomal region containing the genes for rRNA; most often found in physical association with the nucleolus.

nucleolus A nuclear organelle that is the site of ribosome biosynthesis; usually associated with or formed in association with the NOR.

nucleoside A purine or pyrimidine base covalently linked to a ribose or deoxyribose sugar molecule.

nucleosome A complex of four histone molecules, each present in duplicate, wrapped by two turns of a DNA molecule. One of the basic units of eukaryotic chromosome structure. Also known as a nu body.

nucleotide A nucleoside covalently linked to a phosphate group. Nucleotides are the basic building blocks of nucleic acids. The nucleotides commonly found in DNA are deoxyadenylic acid, deoxycytidylic acid, deoxyguanylic acid, and deoxythymidylic acid. The nucleotides in RNA are adenylic acid, cytidylic acid, guanylic acid, and uridylic acid.

nucleotide pair The pair of nucleotides (A and T, or G and C) in opposite strands of the DNA molecule that are hydrogen-bonded to each other.

nucleus The membrane-bounded cytoplasmic organelle of eukaryotic cells that contains the chromosomes and nucleolus.

nullisomic Describes an individual with a chromosomal aberration in which both members of a chromosome pair are missing.

ochre codon A codon that does not code for the insertion of an amino acid into a polypeptide chain, but signals chain termination. The ochre codon is UAA.

Okazaki fragment The small, discontinuous strands of DNA produced during DNA synthesis.

oncogene A gene whose activity promotes uncontrolled proliferation in eukaryotic cells.

operator region A region of a DNA molecule that interacts with a specific repressor protein to control the expression of an adjacent gene or gene set.

operon A genetic unit that consists of one or more structural genes (that code for polypeptides) and an adjacent operator gene that controls the transcriptional activity of the structural gene or genes.

orphon Single copies of tandemly repeated genes found dispersed in the genome. For example, histone genes are members of a multigene family present in several hundred copies clustered in a tandem array. A single copy of a histone gene found elsewhere in the genome is said to have lost its family and is regarded as an orphon.

overlapping code A genetic code first proposed by George Gamow in which any given nucleotide is shared by two adjacent codons.

pachytene The stage in meiotic prophase I when the synapsed homologous chromosomes split longitudinally (except at the centromere), producing a group of four chromatids called a tetrad.

palindrome A word, number, verse, or sentence that reads the same backward or forward (e.g., *able was I ere I saw elba*). In nucleic acids, a sequence in which the base pairs read the same on complementary strands ($5' \rightarrow 3'$). For example: 5′GAATTC3′, 3′CTTAAG5′. These often occur as sites for restriction endonuclease recognition and cutting.

pangenesis A discarded theory of development that postulated the existence of pangenes, small particles from all parts of the body that concentrate in the gametes, passing traits from generation to generation, blending the traits of the parents in the offspring.

paracentric inversion A chromosomal inversion that does not include the centromere.

parasexual Condition describing recombination of genes from different individuals which does not involve meiosis, gamete formation, or zygote production. The formation of somatic cell hybrids is an example.

parental gamete See *noncrossover gamete.*

parthenogenesis Development of an egg without fertilization.

partial diploids See *merozygote.*

partial dominance See *incomplete dominance.*

passive immunity A form of immunity produced by receipt of antibodies synthesized by another individual.

patroclinous inheritance A form of genetic transmission in which the offspring have the phenotype of the father.

pedigree In human genetics, a diagram showing the ancestral relationships and transmission of genetic traits over several generations in a family.

penetrance The frequency (expressed as a percentage) with which individuals of a given genotype manifest at least some degree of a specific mutant phenotype associated with a trait.

peptide bond The covalent bond between the amino group of one amino acid and the carboxyl group of another amino acid.

pericentric inversion A chromosomal inversion that involves both arms of the chromosome and thus involves the centromere.

permissive condition Environmental conditions under which a conditional mutation (such as a temperature-sensitive mutant) expresses the wild-type phenotype.

phage See *bacteriophage.*

phenocopy An environmentally induced phenotype (nonheritable) which closely resembles the phenotype produced by a known gene.

phenotype The observable properties of an organism that are genetically controlled.

phenylketonuria (PKU) A hereditary condition in humans associated with the inability to metabolize the amino acid phenylalanine. The most common form is caused by the lack of the liver enzyme phenylalanine hydroxylase.

phosphodiester bond In nucleic acids, the covalent bond between a phosphate group and adjacent nucleotides, extending from the 5′ carbon of one pentose (ribose or deoxyribose) to the 3′ carbon of the pentose in the neighboring nucleotide. Phosphodiester bonds form the backbone of nucleic acid molecules.

photoreactivation repair Light-induced repair of damage caused by exposure to ultraviolet light. Associated with an intracellular enzyme system.

phyletic evolution The gradual transformation of one species into another over time; vertical evolution.

pilus A filamentlike projection from the surface of a bacterial cell. Often associated with cells possessing F factors.

plaque A clear area on an otherwise opaque bacterial lawn caused by the growth and reproduction of phages.

plasmid An extrachromosomal, circular DNA molecule (often carrying genetic information) that replicates independently of the host chromosome.

platysome Term originally used in electron and X-ray diffraction studies to describe the flattened appearance of the DNA in the nucleosome core.

pleiotropy Condition in which a single mutation simultaneously affects several characters.

ploidy Term referring to the basic chromosome set or to multiples of that set.

point mutation A mutation that can be mapped to a single locus. At the molecular level, a mutation that results in the substitution of one nucleotide for another.

polar body A cell produced at either the first or second meiotic division in females which contains almost no cytoplasm as a result of an unequal cytokinesis.

polycistronic mRNA A messenger RNA molecule that encodes the amino acid sequence of two or more polypeptide chains in adjacent structural genes.

polygenic inheritance The transmission of a phenotypic trait whose expression depends on the additive effect of a number of genes.

polymerase chain reaction (PCR) A method for amplifying DNA segments that uses cycles of denaturation, annealing to primers, and DNA polymerase-directed DNA synthesis.

polymerases The enzymes that catalyze the formation of DNA and RNA from deoxynucleotides and ribonucleotides, respectively.

polymorphism The existence of two or more discontinuous, segregating phenotypes in a population.

polypeptide A molecule made up of amino acids joined by covalent peptide bonds. This term is used to denote the amino acid chain before it assumes its functional three-dimensional configuration.

polyploid A cell or individual having more than two sets of chromosomes.

polyribosome See *polysome.*

polysome A structure composed of two or more ribosomes associated with mRNA, engaged in translation. Formerly called polyribosome.

polytene chromosome A chromosome that has undergone several rounds of DNA replication without separation of the replicated chromosomes, forming a giant, thick chromosome with aligned chromomeres producing a characteristic banding pattern.

population A local group of individuals belonging to the same species, which are actually or potentially interbreeding.

position effect Change in expression of a gene associated with a change in the gene's location within the genome.

postzygotic isolation mechanism Factors that prevent or reduce inbreeding by acting after fertilization to produce nonviable, sterile hybrids or hybrids of lowered fitness.

preadaptive mutation A mutational event which later becomes of adaptive significance.

prezygotic isolation mechanism Factors that reduce inbreeding by preventing courtship, mating, or fertilization.

Pribnow box A 6-bp sequence 5′ to the beginning of transcription in prokaryotic genes, to which the sigma subunit of RNA polymerase binds. The consensus sequence for this box is TATAAT.

primary protein structure Refers to the sequence of amino acids in a polypeptide chain.

primary sex ratio Ratio of males to females at fertilization.

primer In nucleic acids, a short length of RNA or single-stranded DNA which is necessary for the functioning of polymerases.

prion An infectious pathogenic agent devoid of nucleic acid and composed mainly of a protein, PrP, with a molecular weight of 27,000 to 30,000 daltons. Prions are known to cause scrapie, a degenerative neurological disease in sheep, and are thought to cause similar diseases in humans, such as kuru and Creutzfeldt-Jakob disease.

probability Ratio of the frequency of a given event to the frequency of all possible events.

proband See *propositus.*

product law The law that holds that the probability of two independent events occurring simultaneously is the product of their independent probabilities.

progeny The offspring produced from a mating.

prokaryotes Organisms lacking nuclear membranes, meiosis, and mitosis. Bacteria and blue-green algae are examples of prokaryotic organisms.

promoter Region having a regulatory function and to which RNA polymerase binds prior to the initiation of transcription.

prophage A phage genome integrated into a bacterial chromosome. Bacterial cells carrying prophage are said to be lysogenic.

propositus (female, **proposita)** An individual in whom a genetically determined trait of interest is first detected. Also known as a proband.

protein A molecule composed of one or more polypeptides, each composed of amino acids covalently linked together.

proto-oncogene A cellular gene that normally functions to control cellular proliferation. Proto-oncogenes can be converted to oncogenes by alterations in structure or expression.

protoplast A bacterial or plant cell with the cell wall removed. Sometimes called a spheroplast.

prototroph A strain (usually microorganisms) that is capable of growth on a defined, minimal medium. Wild-type strains are usually regarded as prototrophs.

pseudoalleles Genes that behave as alleles to one another by complementation, but that can be separated from one another by recombination.

pseudodominance The expression of a recessive allele on one homologue caused by the deletion of the dominant allele on the other homologue.

pseudogene A nonfunctional gene with sequence homology to a known structural gene present elsewhere in the genome. They differ from their functional relatives by insertions or deletions and by the presence of flanking direct repeat sequences of 10 to 20 nucleotides.

puff See *chromosome puff.*

quantitative inheritance See *polygenic inheritance.*

quantum speciation Formation of a new species within a single or a few generations by a combination of selection and drift.

quaternary protein structure Types and modes of interaction between two or more polypeptide chains within a protein molecule.

R point The point (also known as the restriction point) during the G_1 stage of the cell cycle when either a commitment is made to DNA synthesis and another cell cycle, or the cell withdraws from the cycle and becomes quiescent.

race A phenotypically or geographically distinct subgroup within a species.

rad A unit of absorbed dose of radiation with an energy equal to 100 ergs per gram of irradiated tissue.

radioactive isotope One of the forms of an element, differing in atomic weight and possessing an unstable nucleus that emits ionizing radiation.

random mating Mating between individuals without regard to genotype.

reading frame Linear sequence of codons (groups of three nucleotides) in a nucleic acid.

reannealing Formation of double-stranded DNA molecules from dissociated single strands.

recessive Term describing an allele that is not expressed in the heterozygous condition.

reciprocal cross A paired cross in which the genotype of the female in the first cross is present as the genotype of the male in the second cross, and vice versa.

reciprocal translocation A chromosomal aberration in which nonhomologous chromosomes exchange parts.

recombinant DNA A DNA molecule formed by the joining of two heterologous molecules. Usually applied to DNA molecules produced by *in vitro* ligation of DNA from two different organisms.

recombinant gamete A gamete containing a new combination of genes produced by crossing over during meiosis.

recombination The process that leads to the formation of new gene combinations on chromosomes.

recon A term coined by Seymour Benzer to denote the smallest genetic units between which recombination can occur.

redundant genes Gene sequences present in more than one copy per haploid genome (e.g., ribosomal genes).

regulatory site A DNA sequence that is involved in the control of expression of other genes, usually involving an interaction with another molecule.

rem Radiation equivalent in man; the dosage of radiation that will cause the same biological effect as one roentgen of X-rays.

renaturation The process by which a denatured protein or nucleic acid returns to its normal three-dimensional structure.

repetitive DNA sequences DNA sequences present in many copies in the haploid genome.

replicating form (RF) Double-stranded nucleic acid molecules present as an intermediate during the reproduction of certain viruses.

replication The process of DNA synthesis.

replication fork The Y-shaped region of a chromosome associated with the site of replication.

replicon A chromosomal region or free genetic element containing the DNA sequences necessary for the initiation of DNA replication.

replisome The term used to describe the complex of proteins, including DNA polymerase, that assembles at the bacterial replication form to synthesize DNA.

repressible enzyme system An enzyme or group of enzymes whose synthesis is regulated by the intracellular concentration of certain metabolites.

repressor A protein that binds to a regulatory sequence adjacent to a gene and blocks transcription of the gene.

reproductive isolation Absence of interbreeding between populations, subspecies, or species. Reproductive isolation can be brought about by extrinsic factors, such as behavior, and intrinsic barriers, such as hybrid inviability.

resistance transfer factor (RTF) A component of R plasmids that confers the ability for cell-to-cell transfer of the R plasmid by conjugation.

resolution In an optical system, the shortest distance between two points or lines at which they can be perceived to be two points or lines.

restriction endonuclease Nuclease that recognizes specific nucleotide sequences in a DNA molecule, and cleaves or nicks the DNA at that site. Derived from a variety of microorganisms, those enzymes that cleave both strands of the DNA are used in the construction of recombinant DNA molecules.

restrictive condition Environmental conditions under which a conditional mutation (such as a temperature-sensitive mutant) expresses the mutant phenotype.

restrictive transduction See *specialized transduction.*

retrovirus Viruses with RNA as genetic material that utilize the enzyme reverse transcriptase during their life cycle.

reverse transcriptase A polymerase that uses RNA as a template to transcribe a single-stranded DNA molecule as a product.

reversion A mutation that restores the wild-type phenotype.

R factor (R plasmid) Bacterial plasmids that carry antibiotic resistance genes. Most R plasmids have two components: an r-determinant, which carries the antibiotic resistance genes, and the resistance transfer factor (RTF).

Rh factor An antigenic system first described in the rhesus monkey. Recessive *r/r* individuals produce no Rh antigens and are Rh negative, while *R/R* and *R/r* individuals have Rh antigens on the surface of their red blood cells and are classified as Rh positive.

ribonucleic acid A nucleic acid characterized by the sugar ribose and the pyrimidine uracil, usually a single-stranded polynucleotide. Several forms are recognized, including ribosomal RNA, messenger RNA, transfer RNA, and heterogeneous nuclear RNA.

ribosomal RNA See *rRNA.*

ribosome A ribonucleoprotein organelle consisting of two subunits, each containing RNA and protein. Ribosomes are the site of translation of mRNA codons into the amino acid sequence of a polypeptide chain.

RNA See *ribonucleic acid.*

RNA polymerase An enzyme that catalyzes the formation of an RNA polynucleotide strand using the base sequence of a DNA molecule as a template.

RNase A class of enzymes that hydrolyze RNA molecules.

Robertsonian translocation A form of chromosomal aberration that involves the fusion of long arms of acrocentric chromosomes at the centromere.

roentgen A unit of measure of the amount of radiation corresponding to the generation of 2.083×10^9 ion pairs in one cubic centimeter of air at 0°C at an atmospheric pressure of 760 mm of mercury. Abbreviated R.

rolling circle model A model of DNA replication in which the growing point of replication fork rolls around a circular template strand; in each pass around the circle, the newly synthesized strand displaces the strand from the previous replication, producing a series of contiguous copies of the template strand.

rRNA The RNA molecules that are the structural components of the ribosomal subunits. In prokaryotes, these are the 16*S*, 23*S*, and 5*S* molecules; and in eukaryotes, they are the 18*S*, 28*S*, and 5*S* molecules.

RTF See *resistance transfer factor.*

S_1 nuclease A deoxyribonuclease that cuts and degrades single-stranded molecules of DNA.

satellite DNA DNA that forms a minor band when genomic DNA is centrifuged in a cesium salt gradient. This DNA usually consists of short sequences repeated many times in the genome.

SCE See *sister chromatid exchange.*

secondary protein structure The alpha helical or pleated-sheet form of a protein molecule brought about by the formation of hydrogen bonds between amino acids.

secondary sex ratio The ratio of males to females at birth.

secretor An individual having soluble forms of the blood group antigens A and/or B present in saliva and other body fluids. This condition is caused by a dominant, autosomal gene unlinked to the *ABO* locus (*I* locus).

sedimentation coefficient See *Svedberg coefficient unit.*

segregation The separation of homologous chromosomes into different gametes during meiosis.

selection The force that brings about changes in the frequency of alleles and genotypes in populations through differential reproduction.

selection coefficient (*s*) A quantitative measure of the relative fitness of one genotype compared with another.

selfing In plant genetics, the fertilization of ovules of a plant by pollen produced by the same plant. Reproduction by self-fertilization.

semiconservative replication A model of DNA replication in which a double-stranded molecule replicates in such a way that the daughter molecules are composed of one parental (old) and one newly synthesized strand.

semisterility A condition in which a proportion of all zygotes are inviable.

sex chromosome A chromosome, such as the X or Y in humans, which is involved in sex determination.

sexduction Transmission of chromosomal genes from a donor bacterium to a recipient cell by the F factor.

sex-influenced inheritance Phenotypic expression that is conditioned by the sex of the individual. A heterozygote may express one phenotype in one sex and the alternate phenotype in the other sex.

sex-limited inheritance A trait that is expressed in only one sex even though the trait may not be X-linked.

sex linkage The pattern of inheritance resulting from genes located on the X chromosome.

sex ratio See *primary* and *secondary sex ratio.*

sexual reproduction Reproduction through the fusion of gametes, which are the haploid products of meiosis.

Shine-Dalgarno sequence The nucleotides AGGAGG present in the leader sequence of prokaryotic genes that serves as a ribosome binding site. The 16*S* RNA of the small ribosomal subunit contains a complementary sequence to which the mRNA binds.

short period interspersion Pattern of genome organization in which stretches of single copy DNA (about 1000 bp) are interspersed with short segments of repetitive DNA (300 bp). This pattern is found in *Xenopus,* humans, and the majority of organisms examined to date.

shotgun experiment The cloning of random fragments of genomic DNA into a vehicle such as a plasmid or phage, usually to produce a bank or library of clones from which clones of specific interest will be selected.

sickle-cell anemia A genetic disease in humans caused by an autosomal recessive gene, usually fatal in the homozygous condition. Caused by an alteration in the amino acid sequence of the beta chain of globin.

sickle-cell trait The phenotype exhibited by individuals heterozygous for the sickle-cell gene.

sigma factor A polypeptide subunit of the RNA polymerase which recognizes the binding site for the initiation of transcription.

SINES Short interspersed repetitive sequences found in the genomes of higher organisms, such as the 300-bp *Alu* sequence.

sister chromatid exchange (SCE) A crossing over event which can occur in meiotic and mitotic cells; involves the reciprocal exchange of chromosomal material between sister chromatids (joined by a common centromere). Such exchanges can be detected cytologically after BrdU incorporation into the replicating chromosomes.

site-directed mutagenesis A process that uses a synthetic oligonucleotide containing a mutant base or sequence as a primer for inducing a mutation at a specific site in a cloned gene.

small nuclear RNA (snRNA) Species of RNA molecules ranging in size from 90 to 400 nucleotides. The abundant snRNAs are present in 1×10^4 to 1×10^6 copies per cell. snRNAs are associated with proteins and form RNP particles known as snRNPs or snurps. Six uridine-rich snRNAs known as U1–U6 are located in the nucleoplasm, and the complete nucleotide sequence of these is known. snRNAs have been implicated in the processing of pre-mRNA and may have a range of cleavage and ligation functions.

snurps See *small nuclear RNA (snRNA).*

solenoid structure A level of eukaryotic chromosome structure generated by the supercoiling of nucleosomes.

somatic cell genetics The use of cultured somatic cells to investigate genetic phenomena by parasexual techniques.

somatic cells All cells other than the germ cells or gametes in an organism.

somatic mutation A mutational event occurring in a somatic cell. In other words, such mutations are not heritable.

somatic pairing The pairing of homologous chromosomes in somatic cells.

SOS response The induction of enzymes to repair damaged DNA in *E. coli.* The response involves activation of an enzyme that cleaves a repressor, activating a series of genes involved in DNA repair.

spacer DNA DNA sequences found between genes, usually repetitive DNA segments.

special creation An idea that each species originated through a special act of creation by a divine force.

specialized transduction Genetic transfer of only specific host genes by transducing phages.

speciation The process by which new species of plants and animals arise.

species A group of actually or potentially interbreeding individuals that is reproductively isolated from other such groups.

spheroplast See *protoplast.*

spindle fibers Cytoplasmic fibrils formed during cell division which are involved with the separation of chromatids at anaphase and their movement toward opposite poles in the cell.

spontaneous generation The origin of living systems from nonliving matter.

spontaneous mutation A mutation that is not induced by a mutagenic agent.

spore A unicellular body or cell encased in a protective coat that is produced by some bacteria, plants, and invertebrates; is capable of survival in unfavorable environmental conditions; and can give rise to a new individual upon germination. In plants, spores are the haploid products of meiosis.

stabilizing selection Preferential reproduction of those individuals having genotypes close to the mean for the population. A selective elimination of genotypes at both extremes.

standard deviation A quantitative measure of the amount of variation in a sample of measurements from a population.

standard error A quantitative measure of the amount of variation in a sample of measurements from a population.

sterility The condition of being unable to reproduce; free from contaminating microorganisms.

strain A group with common ancestry which has physiological or morphological characteristics of interest for genetic study or domestication.

structural gene A gene that encodes the amino acid sequence of a polypeptide chain.

sublethal gene A mutation causing lowered viability, with death before maturity in less than 50 percent of the individuals carrying the gene.

submetacentric chromosome A chromosome with the centromere placed so that one arm of the chromosome is slightly longer than the other.

subspecies A morphologically or geographically distinct interbreeding population of a species.

sum law The law that holds that the probability of one or the other of two mutually exclusive events occurring is the sum of their individual probabilities.

supercoiled DNA A form of DNA structure in which the helix is coiled upon itself. Such structures can exist in stable forms only when the ends of the DNA are not free, as in a covalently closed circular DNA molecule.

superfemale See *metafemale.*

supermale See *metamale.*

suppressor mutation A mutation that acts to restore (completely or partially) the function lost by a previous mutation at another site.

Svedberg coefficient unit A unit of measure for the rate at which particles (molecules) sediment in a centrifugal field. This unit is a function of several physico-chemical properties, including size and shape. A sedimentation value of 1×10^{-13} sec is defined as one Svedberg coefficient (*S*) unit.

symbiont An organism coexisting in a mutually beneficial relationship with another organism.

sympatric speciation Process of speciation involving populations that inhabit, at least in part, the same geographic range.

synapsis The pairing of homologous chromosomes at meiosis.

synaptonemal complex (SC) An organelle consisting of a tripartite nucleoprotein ribbon that forms between the paired homologous chromosomes in the pachytene stage of the first meiotic division.

syndrome A group of signs or symptoms that occur together and characterize a disease or abnormality.

synkaryon The nucleus of a zygote that results from the fusion of two gametic nuclei. Also used in somatic cell genetics to describe the product of nuclear fusion.

syntenic test In somatic cell genetics, a method for determining whether or not two genes are on the same chromosome.

T_m The temperature at which a population of double-stranded nucleic acid molecules is half-dissociated into single strands. This is taken to be the melting temperature for that species of nucleic acid.

target theory In radiation biology, a theory which states that damage and death from radiation is caused by the inactivation of specific targets within the organism.

TATA box See *Goldberg-Hogness box.*

tautomeric shift A reversible isomerization in a molecule brought about by a shift in the localization of a hydrogen atom. In nucleic acids, tautomeric shifts in the bases of nucleotides can cause changes in other bases at replication and are a source of mutations.

telocentric chromosome A chromosome in which the centromere is located at the end of the chromosome.

telomere The terminal chromomere of a chromosome.

telophase The stage of cell division in which the daughter chromosomes reach the opposite poles of the cell and re-form nuclei. Telophase ends with the completion of cytokinesis.

temperate phage A bacteriophage that can become a prophage and confer lysogeny upon the host bacterial cell.

temperature-sensitive mutation A conditional mutation that produces a mutant phenotype at one temperature range and a wild-type phenotype at another temperature range.

template The single-stranded DNA or RNA molecule that specifies the nucleotide sequence of a strand synthesized by a polymerase molecule.

teratocarcinoma Embryonal tumors that arise in the yolk sac or gonads and are able to undergo differentiation into a wide variety of cell types. These tumors are used to investigate the regulatory mechanisms underlying development.

terminalization The movement of chiasmata toward the ends of chromosomes during the diplotene stage of the first meiotic division.

tertiary protein structure The three-dimensional structure of a polypeptide chain brought about by folding upon itself.

test cross A cross between an individual whose genotype at one or more loci may be unknown and an individual who is homozygous recessive for the genes in question.

tetrad The four chromatids that make up paired homologues in the prophase of the first meiotic division. The four haploid cells produced by a single meiotic division.

tetrad analysis Method for the analysis of gene linkage and recombination using the four haploid cells produced in a single meiotic division.

tetranucleotide theory An early theory of DNA structure which proposed that the molecule was composed of repeating units, each consisting of the four nucleotides adenosine, thymidine, cytosine, and guanine.

tetraparental mouse A mouse produced from an embryo that was derived by the fusion of two separate blastulas.

thymine dimer A pair of adjacent thymine bases in a single polynucleotide strand between which chemical bonds have formed. This lesion, usually the result of damage caused by exposure to ultraviolet light, inhibits DNA replication unless repaired by the appropriate enzyme system.

topoisomerase A class of enzymes that convert DNA from one topological form to another. During DNA replication, these enzymes facilitate the unwinding of the double-helical structure of DNA.

totipotent The ability of a cell or embryo part to give rise to all adult structures. This capacity is usually progressively restricted during development.

trailer sequence A transcribed but nontranslated region of a gene or its mRNA that follows the termination signal.

trait Any detectable phenotypic variation of a particular inherited character.

***trans* configuration** The arrangement of two mutant sites on opposite homologues, such as

$$\frac{a^1 \; +}{+ \; a^2}$$

Contrasts with a *cis* arrangement, where they are located on the same homologue.

transcription Transfer of genetic information from DNA by the synthesis of an RNA molecule copied from a DNA template.

transdetermination Change in developmental fate of a cell or group of cells.

transduction Virally mediated gene transfer from one bacterium to another, or the transfer of eukaryotic genes mediated by retroviruses.

transfer RNA See *tRNA*.

transformation Heritable change in a cell or an organism brought about by exogenous DNA.

transgenic organism An organism whose genome has been modified by the introduction of external DNA sequences into the germ line.

transition A mutational event in which one purine is replaced by another, or one pyrimidine is replaced by another.

translation The derivation of the amino acid sequence of a polypeptide from the base sequence of an mRNA molecule in association with a ribosome.

translocation A chromosomal mutation associated with the transfer of a chromosomal segment from one chromosome to another. Also used to denote the movement of mRNA through the ribosome during translation.

transposable element A defined length of DNA that translocates to other sites in the genome, essentially independent of sequence homology. Usually such elements are flanked by short, inverted repeats of 20 to 40 base pairs at each end. Insertion into a structural gene can produce a mutant phenotype. Insertion and excision of transposable elements depends on two enzymes, transposase and resolvase. Such elements have been identified in both prokaryotes and eukaryotes.

transversion A mutational event in which a purine is replaced by a pyrimidine, or a pyrimidine is replaced by a purine.

triploidy The condition in which a cell or organism possesses three haploid sets of chromosomes.

trisomy The condition in which a cell or organism possesses two copies of each chromosome, except for one, which is present in three copies. The general form for trisomy is therefore $2n + 1$.

tritium (^{3}H) A radioactive isotope of hydrogen, with a half-life of 12.46 years.

tRNA Transfer RNA; a small ribonucleic acid molecule that contains a three-base segment (anticodon) that recognizes a codon in mRNA, a binding site for a specific amino acid, and recognition sites for interaction with the ribosome and the enzyme that links it to its specific amino acid.

Turner syndrome A genetic condition in human females caused by a 45,X genotype (XO). Such individuals are phenotypically female but are sterile because of undeveloped ovaries.

unequal crossing over A crossover between two improperly aligned homologues, producing one homologue with three copies of a region and the other with one copy of that region.

unique DNA DNA sequences that are present only once per genome. Single copy DNA.

unwinding proteins Nuclear proteins that act during DNA replication to destabilize and unwind the DNA helix ahead of the replicating fork.

up promoter A promoter sequence, often mutant, that increases the rate of transcription initiation. Also known as strong promoter.

variable region Portion of an immunoglobulin molecule that exhibits many amino acid sequence differences between antibodies of differing specificities.

variance A statistical measure of the variation of values from a central value, calculated as the square of the standard deviation.

variegation Patches of differing phenotypes, such as color, in a tissue.

vector In recombinant DNA, an agent such as a phage or plasmid into which a foreign DNA segment will be inserted.

viability The measure of the number of individuals in a given phenotypic class that survive, relative to another class (usually wild type).

virulent phage A bacteriophage that infects and lyses the host bacterial cell.

W, Z chromosomes Sex chromosomes in species where the female is the heterogametic sex (WZ).

wild type The most commonly observed phenotype or genotype, designated as the norm or standard.

wobble hypothesis An idea proposed by Francis Crick which states that the third base in an anticodon can align in several ways to allow it to recognize more than one base in the codons of mRNA.

writhing number The number of times that the axis of a DNA duplex crosses itself by supercoiling.

X inactivation In mammalian females, the random cessation of transcriptional activity of one X chromosome. This event, which occurs early in development, is a mechanism of dosage compensation. Molecular basis of inactivation is unknown, but loci on the tip of the short arm of the X can escape inactivation. (See *Barr body; Lyon hypothesis.)*

X linkage See *sex linkage.*

X-ray crystallography A technique to determine the three-dimensional structure of molecules through diffraction patterns produced by X-ray scattering by crystals of the molecule under study.

Y chromosome Sex chromosome in species where the male is heterogametic (XY).

Y linkage Mode of inheritance shown by genes located on the Y chromosome.

Z-DNA An alternate structure of DNA in which the two antiparallel polynucleotide chains form a left-handed double helix. Z-DNA has been shown to be present along with B-DNA in chromosomes and may have a role in regulation of gene expression.

zein Principal storage protein of corn endosperm, consisting of two major proteins, with molecular weights of 19,000 and 21,000 daltons.

zinc finger A DNA binding domain of a protein that has a characteristic pattern of cysteine and histidine residues that complex with zinc ions, throwing intermediate amino acid residues into a series of loops or fingers.

zygote The diploid cell produced by the fusion of haploid gametic nuclei.

zygotene A stage of meiotic prophase I in which the homologous chromosomes synapse and pair along their entire length, forming bivalents. The synaptonemal complex forms at this stage.

CREDITS

Chapter One: An Introduction to Genetics

CO.1 Biophoto Associates/Science Source/ Photo Researchers, Inc.
F1.1 The Metropolitan Museum of Art, gift of John D. Rockefeller, Jr., 1932. (32.143.3)
F1.2 (c) ARCHIV/Photo Researchers, Inc.
F1.3 K. G. Murti/Visuals Unlimited
F1.4 Dan McCoy/Rainbow
F1.5 D. Wilder/Tom Stack Associates
F1.6 Runk/Scheonberger/Grant Heilman Photography, Inc.
F1.7 Charles Seaborn/Odyssey/Frerck/Chicago
F1.8 Grant Heilman Photography

Chapter Two: The Cytological Basis of Inheritance

CO.2 Lionel I. Rebhun/University of Virginia
F2.2 CNRI/SPL/Science Source/Photo Researchers, Inc.
F2.4 CNRI/SPL/Science Source/Photo Researchers, Inc.
F2.7B Andrew S. Bajer/Department of Biology, University of Oregon
F2.8 C. A. Hasenkampf/Biological Photo Service
F2.10A Courtesy of David F. Comings and Tadashi A. Okada
F2.10B Courtesy of Walter F. Engler
F2.10C Biophoto Associates/Science Source/Photo Researchers, Inc.
F2.11 Andrew Syred/SPL/Photo Researchers, Inc.
F2.12 Science Source/Photo Researchers, Inc.
F2.13A Omikron/Science Source/Photo Researchers, Inc.
F2.13B Courtesy of Nicole Angelier
F2.13C Courtesy of Nicole Angelier

Chapter Three: Mendelian Inheritance

CO.3 R. W. Van Norman/Visuals Unlimited
F3.1 Larry Lefever/Grant Heilman Photography
F3.7A R. W. Van Norman/Visuals Unlimited
F3.7B R. W. Van Norman/Visuals Unlimited

Chapter Four: Neo-Mendelian Inheritance

CO.4 John Kaprelian/ Photo Researchers, Inc.
F4.1 John Kaprelian/ Photo Researchers, Inc.
F4.3 Stanton Short/The Jackson Library
F4.7 Wolfgang Kaehler Photography
F4.8 Norm Thomas/ Photo Researchers, Inc.
F4.11 Carolina Biological Supply Co.
F4.14 Hans Reinhard/Bruce Coleman, Inc.
F4.15 Debra P. Hershkowitz/ Bruce Coleman Inc.
page 89 Dr. Ralph G. Somes, Jr., University of Connecticut Department of Nutritional Sciences

Chapter Five: Linkage and Chromosomes

CO.5 Dr. Sheldon Wolff and Judy Bodycote, Laboratory of Radiobiology and Environmental Health, University of California, San Francisco.
F5.7 Biological Photo Service
F5.16 Dr. Sheldon Wolff and Judy Bodycote, Laboratory of Radiobiology and Environmental Health, University of California, San Francisco.

Chapter Six: Variations in Chromosome Number and Arrangement

CO.6 CNRI/SPL/Science Source/Photo Researchers, Inc.
F6.1A Courtesy of Catherine G. Palmer, Ph.D., Department of Medical Genetics, Indiana University.
F6.1B Courtesy of Catherine G. Palmer, Ph.D., Department of Medical Genetics, Indiana University.
F6.3 Gerald and Teresa Audesirk
F6.6A Courtesy of Irene Uchida, Director of Cytogenetics, Oshawa General Hospital, Ontario
F6.6B Courtesy of Irene Uchida, Director of Cytogenetics, Oshawa General Hospital, Ontario
F6.7 Jane Grushow/Grant Heilman Photography
F6.8A Courtesy of Catherine G. Palmer, Ph.D., Department of Medical Genetics, Indiana University.
F6.8B Brian Parker/Tom Stack & Associates
F6.11 Dr. John D. Cunningham/Visuals Unlimited
F6.12 Pfizer, Inc./Phototake
F6.19 From Yunis and Chandler 1979

Chapter Seven: DNA-The Physical Basis of Life

CO.7 Jean Claude Revy/Phototake
F7.2A Bruce Iverson
F7.2B Bruce Iverson
F7.4 Lee Simon/Stammers/SPL/Science Source Photo Researchers, Inc.
F7.11 From Franklin and Gosling, 1953. Reprinted by permission from Nature, Vol. 171, p 738. ©1953 Macmillan Journals Limited. Courtesy of M.H.F. Wilkens, Biophysics Department, King's College, London.
F7.13A Courtesy of Loren Williams, Department of Biology Massachusetts Institute of Technology.
F7.13B Courtesy of Loren Williams, Department of Biology Massachusetts Institute of Technology.
F7.13C Courtesy of Loren Williams, Department of Biology Massachusetts Institute of Technology.
F7.15 Courtesy of ONCOR, Inc.

Chapter Eight: Replication and Synthesis of DNA

CO.8 Lawrence Livermore Laboratory/ Science Photo Library/Photo Researchers, Inc.
F8.4 Top W.H. Hodge/Peter Arnold
F8.4A "Molecular Genetics," Pt.1, pp 74–75, J. H. Taylor (ed). Academic Press, NY.
F8.4B "Molecular Genetics," Pt.1, pp 74–75, J. H. Taylor (ed). Academic Press, NY.
F8.12 Dr. Ross Inman, University of Wisconsin, Biophysics Laboratory

Chapter Nine: Storage and Expression of Genetic Information

CO.9 Courtesy of E.V. Kiseleva
F9.13A From Rich et al, 1963 Reproduction by permission of Cold Spring Harbor Laboratory.
F9.13B Courtesy of E.V. Kiselva
F9.15 Courtesy of Bert W. O'Malley

Chapter Ten: Proteins-The End Product of Genes

CO.10 J. Gross/Biozentrum/SPL Photo Researchers, Inc.
F10.4A Dennis Kunkel/CNRI/Phototake
F10.4B Francis Leroy/Biocosmos/SPL/Photo Researchers, Inc.
F10.12 J. Gross/Biozentrum/SPL/Photo Researchers, Inc.

Chapter Eleven: Mutation and Mutagenesis

CO.11 Dr. Nina Fedoroff, Department of Embryology, Carnegie Institute of Washington
F11.3 Mary Evans Picture Library/Photo Researchers, Inc.
F11.13R W. Clark Lambert, M.D., Ph.D. University of Medicine and Dentistry of New Jersey Medical School.
F11.13L W. Clark Lambert, M.D., Ph.D. University of Medicine and Dentistry of New Jersey Medical School.
F11.16 Dr. Nina Fedoroff, Department of Embryology, Carnegie Institute of Washington

Chapter Twelve: Organization of Genes and Chromosomes

CO.12 Courtesy of ONCOR, Inc.
F12.1A Dr. M. Wurtz/Biozentrum, University of Basel/ SPL/ Science Source/Photo Researchers, Inc.
F12.1B Hans Ris from Westmoreland, 1969. Copyright (c)1969 by the American Association for the Advancement of Science.
F12.2 Science Source/Photo Researchers, Inc.
F12.3 Science Photo Library/ Science Source/Photo Researchers, Inc.
F12.4A Dr. Don Fawcett/Kahri David/Photo Researchers, Inc.

F12.4B Courtesy of Richard Kolodnar.
F12.6A From Olins and Olins, 1978, Figs. 1 and 4.
F12.6B From Olins and Olins, 1978, Figs. 1 and 4.
F12.8 From Chen and Ruddle, 1971, p.54
F12.9 Dept. of Clinical Cytogenetics,Adden-Brookes Hospital,
Cambridge/Science Photo Library/Photo Researchers, Inc.
F12.10 From Yunis, 1981.
F12.16 Rich et. al., 1963. Reproduced by permission of Cold Spring Harbor Laboratory.

Chapter Thirteen: Cytoplasmic Inheritance

CO.13 Courtesy of Dr. Ed Coe, Curtis Hall, University of Missouri, Columbia, MO
F13.3 Courtesy of Dr. Ed Coe, Curtis Hall, University of Missouri, Columbia, MO
F13.4 Raymond B. Otero/Visuals Unlimited
F13.5 Courtesy of Eduardo Bonilla M.D., College of Physicians
and Surgeons of Columbia University, NY.
F13.6 A. M. Siegelman/Visuals Unlimited

Chapter Fourteen: Genetics of Bacteria and Viruses

CO.14 Dr. L. Caro/SPL/Science Source/Photo Researchers, Inc.
F14.2 Visuals Unlimited
F14.5 Dr. L. Caro/SPL/Science Source/Photo Researchers, Inc.
F14.16 Professor Stanley Cohen/SPL/Science Source/ Photo Researchers, Inc.
F14.18 Dr. M. Wurtz/Biozentrum, University of Basel/ SPL/Science Source/Photo Researchers, Inc.
F14.19 Bruce Iverson
F14.21 From Hershey and Chase, 1951.
F14.22 Lennart Nilsson/Photo courtesy of Boehringer Ingelheim International GmbH

Chapter Fifteen: Recombinant DNA - Technology and Applications

CO.15 K. G. Murti/Visuals Unlimited
F15.2 K. G. Murti/Visuals Unlimited
F15.4 Dr. M. Wurtz/Biozentrum, University of Basel/ SPL/Science Source/Photo Researchers, Inc.
F15.7 Biophoto Associates/Photo Researchers, Inc.
F15.9 Courtesy of BIO-RAD
F15.11 Michael Gabridge/Visuals Unlimited
F15.13 NIH/Custom Medical Stock Photo
F15.16 Matt Meadows/Peter Arnold, Inc.
F15.20 David Parker/SPL/Photo Researchers, Inc.

Chapter Sixteen: Genetic Regulation

CO.16 Michael Gabridge/Visuals Unlimited

Chapter Seventeen: Genetic Basis of Immunity

CO.17 Prof. Luc Montagnier, Institut Pasteur/CNRI/SPL/ Photo Researchers, Inc.
F17.1 Lennert Nilsson/Boehringer Ingelheim International GmbH
F17.4 Institut Pasteur/Phototake
F17.5 Lennart Nilsson/Boehringer Ingelheim International GmbH
F17.6A Courtesy of Gilla Kaplan/ Rockefeller University
F17.15 Baylor Collection of Medical Photos/Peter Arnold, Inc.
F17.16 Hans Gelderblom/Visuals Unlimited
F17.17A-C Hans Gelderblom/Visuals Unlimited

Chapter Eighteen: Developmental Genetics and Neurogenetics

CO.18 Cabisco/Visuals Unlimited
F18.3A-D Cabisco/Visuals Unlimited
F18.7 F. R. Turner, Ph. D., Department of Biology, Indiana University.
F18.8 Dr. William S. Klug
F18.10A Runk Schoenberger/Grant Heilman Photography, Inc.
F18.10B E. B. Lewis, Division of Biology, California Institute of Technology, Pasadena, CA.

Chapter Nineteen: Genetics and Cancer

CO.19 Electra/CNRI/Phototake
F19.11 Science Photo Library/Photo Researchers, Inc.

Chapter Twenty: Quantitative Genetics

CO.20 Wolfgang Kaehler
F20.5 Joel C. Eissenberg, Ph. D., St. Louis University Medical Center.
F20.6 Joel C. Eissenberg, Ph. D., St. Louis University Medical Center.
F20.7A Jane Burton/Bruce Coleman, Inc.
F20.7B Runk/Schoenberger/Grant Heilman Photography, Inc.

Chapter Twenty-One: Population Genetics

CO.21 George Holton/Photo Researchers, Inc.
F21.5A-B Breck P. Kent
F21.21A-B Courtesy of Kenneth Y. Kaneshiro, Dept. of Entomology,
University of Hawaii.

INDEX

ISBN 0-02-364797-3
90000>
EAN
9 780023 647970